# MANUAL OF PTERIDOLOGY

Typical New Zealand lowland rainforest (Westland) with *Cyathea medullaris, C. dealbata, Hemithelia Smithii* etc., on left: *Freycinethia Banksii; Podocarpus dacrydioides* in background. — Photo I. L. COLLINS.

# MANUAL
## OF
# PTERIDOLOGY

*EDITED BY*

Fr. VERDOORN

*IN COLLABORATION WITH*

A. H. G. Alston, I. Andersson-Kottö, L. R. Atkinson, H. Burgeff, H. G. du Buy, C. Christensen, W. Döpp, W. M. Docters van Leeuwen, H. Gams, M. J. F. Gregor, M. Hirmer, R. E. Holttum, R. Kräusel, E. L. Nuern-bergk, J. C. Schoute, J. Walton, K. Wetzel, S. Williams, H. Winkler and W. Zimmermann

Foreword by F. O. BOWER

WITH 121 ILLUSTRATIONS

Springer-Science+Business Media, B.V.

1938

*ISBN 978-94-017-5743-0       ISBN 978-94-017-6111-6 (eBook)*
*DOI 10.1007/978-94-017-6111-6*

# CONTENTS

## CHAPTER I. Morphology

## CHAPTER II. Anatomy

CHAPTER III. **Experimental Morphology**

CHAPTER IV. **Associations with Fungi and other lower plants**

CHAPTER V. **Mycorhiza**

CHAPTER VI. **Zoocecidia**

CHAPTER VII. **Cytology**

CHAPTER VIII. **Karyologie**

# CHAPTER XVI. **Psilophytinae**

# CHAPTER XVII. **Lycopodiinae**

# CHAPTER XVIII. **Psilotinae**

# CHAPTER XIX. **Articulatae**

# CHAPTER XX. **Filicinae**

## CHAPTER XXI. Fossile Filicinae

## CHAPTER XXII. Pteridophyta Incertae Sedis

## CHAPTER XXIII. Phylogenie

## INTRODUCTION

*The old-fashioned taxonomists were interested to a limited degree in the morphology, anatomy, distribution and economic uses of their plants. Nowadays, however, it is impossible to carry out successful taxonomic research without a thorough knowledge and real understanding of the experimental morphology, cytology, ecology and general biology of the group with which one is working. This book is primarily, but by no means exclusively, designed for the taxonomist who is anxious to improve his methods and broaden his outlook. At the same time it offers to the general botanist working on the Pteridophytes the necessary fundamental facts about the group and a survey of the chief results of lines of investigation related to his own.*

*Besides this, certain branches of the subject are dealt with more fully, because no good surveys of them have been published. Though the existing literature has always been considered, no attempt has been made to review the whole of it. This manual is not, and never could be, an exhaustive monograph, it is rather a collection of essays. The editor has always urged his contributors to give more space to new ideas than to an academic summary of established knowledge. Comparatively little space has been allowed to subjects which have been treated at length in recent books, e.g. ontogeny, or the classification of the fossil Pteridophytes. In this way more room has been found for topics in greater need of treatment, and duplication has been avoided.*

*In aims as well as in method of treatment it follows closely the same editor's "Manual of Bryology" (The Hague, 1932).*

*It is hoped that these two Manuals will provide those working on the Archegoniatae with a new breadth of view and help them to put their problems and results in a truer perspective.*

## FOREWORD

by

F. O. BOWER, F.R.S.

An opportunity has been offered me, and gladly accepted, to write a brief Foreword on the Archegoniatae. It may serve as a bridge between the Manual of Bryology already produced by Dr. VERDOORN, and the Manual of Pteridology now presented by him. Collectively these volumes give a general conspectus of those primitive plants of the land which take a pivot-place in morphological comparison, and present in its most pronounced form the alternating life-cycle first fully disclosed by HOF-MEISTER. A special feature of these volumes is that they bring together studies of Mosses and Ferns, not only from the laboratory and experimental garden, but also from the herbarium and the field, while Palaeobotany takes an important place in them. Too often these branches of botanical science have been cultivated apart, and so are liable to be estranged: though all sources of the knowledge of Plants should be coordinated in leading to a general understanding of them, and to their natural seriation.

A scientific Morphology of Plants may be said to have dated from 1851, when HOFMEISTER's Comparative Studies were published. The reasoning of the earlier morphologists had been for the most part of the nature of deductive ideology, based primarily on study of the Higher Plants. Their method was corrected in 1841 by SCHLEIDEN who, together with NAEGELI, strove to convert Botany into an inductive science, on the same footing as Physics or Chemistry. This aspiration was realised in the brilliant research-es of HOFMEISTER on the Archegoniatae and Gymnosperms. His synthesis from below marked the birth of a Morphology founded on a comparative study of relatively primitive types. There followed a period of search into life-histories, not only in the Archegoniatae but also in the Thallophytes. Meanwhile cytological analysis having advanced, the quest was extended from somatic form to nuclear detail. In 1894 came STRASBURGER's announcement of Periodic Reduction as a constant feature in the normal nuclear cycle for organisms possessed of sex. But for Plants at large the somatic phases, or "generations", proved inconstant in their relation to it. There are in fact three forms which that relation may take. Examples of them are known respectively as I. *Haplobionts*, such as the Algae *Scinaia*, *Oedogonium*, or *Chara*, where there is no diploid soma: II. *Haplo-diplo-*

*bionts*, as in many Thallophytes, though variable in them: but typical of the normal Archegoniatae and of primitive Seed-Plants, where both gametophyte and sporophyte are present: and III. *Diplobionts*, where the soma is diploid, a state characteristic of Animals. In the Highest Plants practically a like condition has been attained by progressive reduction of the haploid phase. Thus for organisms possessing sex somatic development varies in relation to a more stable cytological cycle. It is an old principle in Morphology to accord weight to characters according to their stability. Hence the normal nuclear cycle will take precedence over such somatic developments as have been imposed upon it.

The development of a soma, or "generation", whether haploid or diploid, punctuating the nuclear cycle, is in fact an optional not an obligatory event in the life-history of sexual organisms at large, however constant such phases may be within limited circles of plant-affinity. OLTMANNS has concluded from wide comparison of the Algae that just as sexuality may have arisen repeatedly and independently in various groups of the lower organisms, so may the various higher families have carried out independently the establishment of two generations. In fact that homoplastic post-sexual phases have been developed in widely different plants, such as *Dictyota* and *Polysiphonia* among the Algae. But this does not imply their homogeny one with another, nor with that of any one of the Archegoniatae. This leads on to the statement of GOEBEL, that the doctrine of alternation as founded originally for the higher plants, from the Bryophyta upwards, cannot be extended to all plants. Further we may conclude that, since it is not possible to bring any one family of living Thallophytes into genetic relation with the Archegoniatae, the alternation of generations seen normally in them should be discussed on its own merits.

Many years earlier than 1894 irregularities had been recorded in the normal Hofmeisterian cycle. They are known as Apogamy and Apospory. At first sight such happenings seemed to be wholly subversive of regularised alternation, and a period of confused thought followed. But so long as the eye rests upon the normal cycle, with its constantly recurring features of syngamy and reduction, the Scylla of apogamy and the Charybdis of apospory may safely be passed, as relatively recent irregularities in the developmental history of a Land Flora. For we are not entitled to assume that the normal cytological cycle was arbitrarily regularised for all time, nor that the irregularities seen today took any part in the development of the remote past. For us they demonstrate present potentialities rather than past history.

Thus far this discussion has been purely morphological. But it may well be assumed that events, relatively so constant as those of the normal archegoniate cycle, have had some steady biological foundation to stabilise them. As early as 1890 I had suggested that amphibial life presents conditions that would favour the development of a diploid sporophyte. It would thus appear as a natural outcome of migration from water to land, though no consecutive series of living amphibians now exists illustrating that transition. It has always been held that the Archegoniatae probably sprang from some green algal source. Such types as *Ulothrix, Oedogonium, Coleochaete*, and *Chara* have commonly been quoted. All of these stand cytologically on the haplo-biontic plane, reduction being involved in the first division of the zygote. Not one of them seems to have hit off the innovation of postponing reduction, and interpolating a stable diploid phase. Their stolid conservatism has resulted in evolutionary inertia. Here lies the biological gap between green aquatic and green amphibial life. But if any race of littoral plants were to have initiated such a diploid phase as the more primitive sub-aerial plants possess, the capture of the land would have lain open before it, provided the diploid phase were fitted to endure sub-aerial life. It would have secured at one stroke three biological ends of supreme importance to any land-living plant: (I.) a multiplication of possible combinations of hereditary characters, as SVEDELIUS has shown, (II.) an opportunity for a wide spread on land by dissemination of spores; and (III.) relief from dependence on repeated syngamy for numerical increase. Probably this last may have been the most important. The superiority thus gained would favour a rapid advance of the sporophyte. The haploid ancestors would be left hopelessly behind, and a gap would widen between these and their successful rivals. The result of this is actually what is seen today.

A précis, biological rather than phylogenetic, may be given of the general course which appears to have been taken in the establishment of a Land Flora, with the sporophyte as its dominant factor. The normal Hofmeisterian cycle underlies the whole progression. There is no clear evidence how the protective archegonium originated, but its constancy shows its importance for nascent life on land. The zygote, as we see it in the Bryophytes, first assumes polarity, apex and base being defined in relation to the basal cleavage. Here, as comparison shows it to be in the archegoniate embryo at large, the apex lies centrally in the epibasal hemisphere, and the base in a corresponding position in the hypobasal. A primitive spindle is thus defined, having an oval or cylindrical form in the simplest types. In the Bryophytes this is elaborated structurally along well-known lines of specialisation, but without branching. It remains dependent on the gametophyte. An important step in its elaboration, following NAEGELI's fundamental law of organic development (Abstammungslehre, p. 352), has been the distinction of a sterile base from a distal fertile capsule. In this it prefigures the structural unit of vascular plants designated by ZIMMERMANN as the Telome.

Notwithstanding that the gametophyte reaches its highest state of organisation on land in the Bryophytes, these plants present a dwarfed habit. There was something wanting in their make-up. It appears to have been a lack of coordination of the factors of advance. The Moss-plant, on which the sporogonium depends, bears leaves and is usually profusely branched: but it is without internal ventilation. On the other hand, the sporogonium of a Moss possesses internal ventilation through stomata, but lacks appendages or branching. Thus neither phase of the cycle has been developed structurally to a full photosynthetic equipment, by combining elaboration of form with internal ventilation, as the Higher Plants have done. Moreover the sporogonium cannot draw its supplies directly from the soil. What is required for the advance of such a diploid phase as this on land is physiological independence, together with elaboration of form. Without the latter no great size can be reached. The simple sporogonium may be contrasted with the diffuse form of ordinary land-vegetation. The difference shows the importance of maintaining a due surface-volume-ratio as the size increases. Notwithstanding these disabilities the Bryophytes have achieved a limited success, and have made the best of the simple Spindle, — or Telome-unit.

Up to the turn of the Century the widest gap in the sequence of plants was held to lie between the Mosses and Vascular Plants. But its opening years were marked by the establishment of the ancient Class of the Psilophytales. In them we see fossil plants of very primitive type and early horizon, which have solved for themselves the biological problems that held back the Bryophyta. They present an independent and branched sporophyte. Probably these innovations were closely related in origin. The Psilophytales possessed cylindrical, photosynthetic, and forked shoots, with more or less marked tendency to dichopodial or even monopodial branching. Their distal and often large sporangia were solitary or grouped, while the whole plant was fixed in the soil by tuberous or elongated rhizomes. These plants suggest types of branched sporogonia, fixed independently in the soil. Of living plants the nearest related to them are the Psilotales. In the various known types of ancient fossils we may find suggestions of further lines of advance towards types still living, such as the Lycopods and Equiseta; or to others known only as fossils, such as the Sphenophylls. But above all, comparisons with the Ferns are the most interesting. Such comparisons raise questions of foliar origin and of the grouping of sporangia, on which considerable diversity of opinion has developed, especially in relation to microphyllous types. These problems may be resolved later by help of new discoveries among the fossils. Certainly the last word, whether on the origin of leaves or of their relation to sporangia, has not yet been spoken.

The Ferns present the best object-lesson in continuous evolution of any Class of Land Plants. From the earliest Coenopterids to the present day Ferns have been distinctively Ferns. We may trace many of those types now living in continuous sequence from pre-existent Ferns. They now appear not as vestigial relics, but as vital factors in vegetation: being

represented by many thousands of species. Up to 1890 it had been assumed
that the simplest Leptosporangiates were the most primitive of them all.
But in that year, on comparative grounds, CAMPBELL claimed priority in
evolution for the Eusporangiates. This claim is now amply supported by
fossil evidence. The Coenopterids, with their massive sporangia, sometimes
solitary and distal, sometimes grouped in simple sori resembling telome-
trusses, are essentially palaeozoic types. The Ophioglossaceae and Marattiaceae represent them today. Their cladode sporophylls give evidence by
venation of a dichopodial origin, to which even the axis has been ascribed.
It is still an open question whether or not there was in the ancient fossil
*Stauropteris* any axis at all; it is significant that its frond resembles
anatomically the stalk of *Asteroxylon*. Collectively the Coenopterids suggest
the origin of early *Filicales* in relation to a source such as the Psilophytales
might supply. From those Palaeozoic types to the modern Leptosporangiate Ferns the sequence is one of steadily advancing accommodation to
shaded sub-aerial life. Even at the critical period of transition at the end of
the Palaeozoic Age there was continuity of type. It is witnessed by the
Osmundaceae, Schizaeaceae, and Gleicheniaceae, types that still survive.
The story of the evolution of the modern Leptosporangiate Ferns has been
followed in detail in "The Ferns", Vols: I–III, 1923–1928, and stated in
brief in "Primitive Land Plants", 1935.

The data therein contained show how greatly the organisation of an
ancient phylum, such as that of the Ferns, may be transformed from that
of its original source. How much greater may we expect the transformation
to have been in the evolution of modern Seed-Plants. If this argument be
given its due weight, the attitude in further enquiry upwards will be, to
use the experience gained from simpler types, as suggesting valid methods
of analysis and comparison, in leading towards those plants that take the
highest place. In either case we may expect to find the traces of early
evolutionary steps disguised and overgrown by secondary adaptive
change. This appears to affect the vegetative more than the propagative
system. Spore-production is a constant feature in each normally completed
cycle, and the spore-producing parts retain their character. In particular,
the micro-sporangia of Flowering Plants show a degree of conservatism
that links the organisation of the whole series of Vascular Plants historical-
ly together, more effectively than any other feature of land-vegetation.
Such sporangia may be held as the correlatives of those of the homosporous
Pteridophytes, and of the distal capsules of the Psilophytales: or even of
some type resembling those of the living Bryophytes. The normal scheme
of life which we see in the amphibial Archegoniatae, though variously
balanced, is essentially the same for the whole series. It consists of an
underlying cytological cycle with somatic growth, developed or vestigial,
threaded between its alternating events.

CHAPTER I

MORPHOLOGY

by

J. C. SCHOUTE (Groningen)

**§ 1. Historical introduction.** — Since the task of any morphology is
the study of forms, it should always be based on an inventory of the occurring forms and a description of them by means of a suitable terminology.

The terminology for the Vascular Plant sporophyte, being recast and
improved by LINNÉ, in that shape offered a satisfactory basis for the
taxonomical description of the Pteridophyte species.

A judgment on the mutual relations of the different Pteridophyte orders
and on their relation to Bryophytes and Spermophytes, was still much
impeded by lack of knowledge about the gametophyte and the alternation
of generations; so LINNÉ himself still included *Lycopodium* in the *Musci*
and the Cycads in the *Filicinae* (Genera Plantarum).

In the first half of the nineteenth century botanists, stimulated by
HEDWIG's success in discovering the sexual reproduction of Bryophytes
(1782), diligently searched for sexual reproduction in Pteridophytes.
However as they were convinced that Pteridophytes were only imperfect
Spermophytes, their line of enquiry followed curious paths. The sporangia
usually being looked upon as fruits, fertilization was imagined to be
brought about in the most different ways, by the involucre, by the annulus,
or by certain hairs; the spores of *Equisetum* on the other hand were mostly
considered to be flowers, consisting of an ovary with four surrounding
stamens.

A. P. DE CANDOLLE gave an interesting survey of these opinions in
1827 [1]). The name of "spore or gongyle" is considered by him as a very
appropriate non-committal name for those organs of which it is still
unknown whether they are fecundated or not; if they are, of course they
really are seeds, if not, they are bulbils. The only Pteridophytes offering
less difficulty according to the opinion of these days were the heterospores,

---

[1]) A. P. DE CANDOLLE, Organographie végétale, Paris 1827, II, p. 119.

the megasporangia for instance of *Pilularia* and of *Marsilia* obviously being monospermous pistils and the microsporangia being sessile anthers containing a yellowish globular pollen. It was, however, inexplicable why *Selaginella* had male and female organs while *Lycopodium* and *Psilotum* only had one, presumable male, kind.

The elucidation of the real state of things only came gradually. The first observations were those on germinating spores, revealing the existence of a prothallus. This prothallus was generally taken to be a cotyledon, or in the case of a cordate prothallus, to represent two cotyledons.

BISCHOFF rightly saw a difference between spores and seeds in 1828 [1]), in the fact that spores did not contain any part of the future plant.

Shortly afterwards von MOHL in an excellent paper [2]), by comparison came to the conviction that the spores in no respect resembled seeds or ovules, but that morphologically they were related in a striking way to pollen grains. As however their life cycle was wholly different he was obliged to admit that they yet were different from pollen grains.

Real progress could only become possible after the discovery of the antheridia and at the same time of the spermatozoids by NÄGELI in 1844[3]); the author however did not conceal his doubt as to the importance of his discovery, the position of the antheridia on the cotyledon being so strange that it "fast nicht denkbar ist welche Beziehung sie hier zur Befruchtung der Sporenzellen haben könnten" [4]).

Four years afterwards Count LESZCZYC-SUMIŃSKI discovered the archegonia [5]); though his observations in many respects were still defective, he even saw the entrance of spermatozoids into the neck canal. His conclusion was that the Ferns no longer could be included in the Cryptogams, and that it would be better to transfer them to the Monocotyledons (on account of their single cotyledon, the prothallus).

When after the study of these papers we consult the famous paper by HOFMEISTER [6]) of 1851, we are struck with admiration for HOFMEISTER's genius. Indeed this modest volume of 142 pages text has been something like a flash of lightning. Without a single word of introduction it treats a

---

[1]) G. W. BISCHOFF, Über die Entwicklung der Equiseten, insbesondere des *Equisetum palustre*, aus den Sporen, Nova Acta 14, 1828, p. 779.

[2]) H. VON MOHL, Einige Bemerkungen über die Entwicklung und den Bau der Sporen der cryptogamischen Gewächse, Flora 16, 1833, p. 33.

[3]) C. NÄGELI, Bewegliche Spiralfaden (Saamenfaden?) an Farren. Zeitschr. f. wiss. Bot. 1. Heft 1844, p. 168.

[4]) l.c. p. 184.

[5]) J. LESZCZYC-SUMIŃSKI, Zur Entwickelungsgeschichte der Farnkräuter, Berlin 1848.

[6]) W. HOFMEISTER, Vergleichende Untersuchungen der Keimung, Entfaltung und Fruchtbildung höherer Kryptogamen und der Samenbildung der Coniferen, Leipzig 1851.

number of Mosses, Pteridophytes and Conifers from several aspects, especially dealing with the cell divisions giving rise to stem and leaves, and also with their reproduction. At the end a three page "Rückblick" follows, giving the full theory of alternation of generations in two pages, the third being reserved for the resignation of all hope of discovering differences between stem and leaf in the cell division order.

Not only by this work was the life history of Pteridophytes established, not only was their full cycle of morphological features made known, but moreover a high-light was thrown on their affinities to both Bryophytes and Spermophytes, clearing the way for a natural system of the great plant groups.

Theoretically the way was cleared too for the idea that Pteridophytes are not to be understood from the knowledge of flowering plants, but that on the contrary the flowering plants up to then had been wrongly supposed to be understood, and that they had to be taken as a highly specialized and hardly recognizable branch of the Pteridophytes. In practice it took more than half a century before this view became commonly held, as the flowering plants being dominant, were much more studied, and the feeble modern remnants of Pteridophytes were less accessible.

A great improvement in this respect was reached by the discovery of numerous new fossil types, especially of the *Psilophytinae*. By these discoveries the range of the Pteridophyte canon of morphology was much enlarged and a much better survey of the underlying principles became available. As we shall see in the following pages this survey shows a plant group in which a number of different solutions for one and the same problem have been elaborated, a single one of which has been preserved in the Spermophytes.

Some general references to literature: the only special Pteridophyte morphology as far as I know up to the present, has been written by VE-LENOVSKÝ [1]). For further literature we may refer to ENGLER & PRANTL [2]), GOEBEL [3]), HIRMER [4]) and specially to BOWER [5]), where the most recent papers may be found.

§ 2. **The common principles of Pteridophyte morphology.** — Because of the great range of variation in the Pteridophyte morphology the principles common to all members of the class are relatively few. They are:

[1]) J. VELENOVSKÝ, Vergl. Morph. d. Pflanzen, I, Prag 1905, p. 152–277.

[2]) ENGLER & PRANTL, Die natürlichen Pflanzenfamilien, I, 4, Leipzig 1902.

[3]) K. VON GOEBEL, Organographie der Pflanzen, 3rd ed. II, Jena 1930, p. 1039–1362.

[4]) M. HIRMER, Handbuch der Paläobotanik I, München und Berlin 1927, p. 147–692.

[5]) F. O. BOWER, Primitive land plants also known as the Archegoniatae, London 1935, see p. 111–646.

1. the occurrence of an antithetic alternation of generations, the two generations being a small and lowly organized haploid gametophyte, usually called the prothallus, still more or less adapted to aquatic life or to life in a moist medium, and a large and highly organized diploid sporophyte, adapted to terrestrial life and provided as such with vascular tissue, with a cuticle and with stomata;

2. the occurrence in the gametophyte of antheridia and archegonia of a special type of organization;

3. the occurrence in the sporophyte of a stem with apical growth;

4. the occurrence in the sporophyte of sporangia of a special type of organization.

Moreover a number of organs may be present in a larger or smaller proportion of the Pteridophytes, which by adding essential parts to the morphology of these groups deserve special treatment here. Firstly the embryo may be provided with special organs which evidently are adaptations to the conditions of embryonic life, the suspensor and the foot. In the second place the adult sporophyte in the large majority of cases produces roots, leaves and sporophylls, and often strobili.

Further a number of Pteridophytes bear organs about which the opinions of botanists are widely divergent: the *Psilotum* sporophyll, the *Selaginella* rhizophore, the stigmarian axis, the stigmarian rootlet, the *Isoetes* stock base, the *Pleuromeia* stem base, and the *Nathorstiana* stem base.

The following paragraphs of this chapter will accordingly be devoted successively to the prothallus (§ 3), the gametangia (§ 4), the stem (§ 5), the sporangium (§ 6), the root (§ 7), the leaf (sterile leaf and sporophyll) (§ 8), the strobilus (§ 9), special parts of the embryo (§ 10), organs which have given rise to controversies about their morphological nature (§ 11) and appendages of the surface, hairiness (§ 12).

**§ 3. The prothallus.** — Unless reduced the prothallus is a green autotrophic pluricellular body, of comparatively low organization.

Its shape varies to such a degree that no general rules can be given: it may be like a moss protonema, consisting of a branched system of filiform cell-rows (*Schizaea*); it may be like a foliaceous liverwort, flat, dorsiventral and prostrate (*Aspidium*), it may be cone-shaped or cylindrical (*Lycopodium* spp).

When it is attached to the substratum, this is brought about by means of rhizoids.

In some cases the prothallus shows a certain differentiation of its parts; in the different orders this differentiation does not follow the same lines and evidently it is of polyphyletic origin [1]). Thus there may be lobes

_____

[1]) R. ORTH criticizes (Morphologische und physiologische Untersuchungen an Farnprothallien, Planta 25, 1936, p. 104) the "artificial" taxonomical systems of

showing a certain resemblance to leaves (*Equisetum, Lycopodium cernuum*), or there may be separate parts with a peculiar shape in which the gametangia are formed (*Lycopodium clavatum*, apical part of top-shaped prothallus, *Trichomanes*, archegoniophores). In many cases brood-tubers may be formed, enabling the gametophyte to resist injurious conditions.

Many prothalli undergo a certain amount of reduction. A not uncommon feature' is the loss of chlorophyll, in which case free-living prothalli get their organic matter by symbiosis (*Psilotum*, many *Lycopodium* spp.).

Another kind of reduction, often combined with loss of chlorophyll, is the reduction in size and in number of cells in the prothalli of heterospores. The megagametophytes as well as the microgametophytes mostly remain inside the spore wall or slightly protrude from it; the number of cells may become so small that a microgametophyte only consists of one antheridium and a few, or even only one, prothallus cell.

In all these cases the development of the prothallus as well as that of the gametes is executed mainly or wholly by means of the nutritive substances deposited in the spore, from the sporophyte. In the megagametophyte even the development of the embryo and the young sporophyte to a certain extent is provided for from the same source.

§ 4. The gametangia. — The Pteridophyte gametangia on the whole are of a rather uniform construction, following mainly the same lines as in the Bryophyta; they always consist of a sterile cellular wall covering the gametes.

The antheridia, usually being more or less globular, may be more or less embedded in the prothallus tissue. At maturity they either open by sejunction of the wall cells at the apex or by disintegration of one large apical cell. They always produce two or more usually numerous spermatozoids, which are more or less spirally coiled cells with two (*Lycopodium, Selaginella*) or numerous cilia (*Equisetum, Psilotum, Isoetes, Filicinae*).

The archegonia, being always embedded with their basal part in the prothallus tissue and only the neck protruding more or less, do not quite show the typical flask shape of the Bryophyte archegonia. Every archegonium contains one large basal egg cell and a row of a varying number of canal cells. At maturity the neck opens at the top by sejunction of the apical cells while by the disorganization of the canal cells the way to the egg cell is made free for the spermatozoids. In some cases (*Psilotum*) the opening of the neck ensues by the throwing off of its upper part.

---

Ferns, based exclusively on the differences in the annuli, the construction of the sporangia and the indusia and on sorus distribution; he wants these systems to be replaced by a "natural" system, based on the developmental history and the morphology of the gametophyte (see especially p. 149).

Probably most taxonomists will not yet be inclined to go that length.

**§ 5. The stem. — A. General Properties.** In our § 2 the stem has already been characterized by its apical growth. Indeed the universal possession of a vegetative cone, adding again and again new young parts to the mature ones, is as essential for the stem morphologically as it is from a physiological point of view.

We cannot reverse the thesis and say that every part of the plant with an apical cone is a stem, as the root always and the leaf in many cases, at least temporarily, are formed in the same way. No more can we say that the apical cone in the Pteridophytes is a peculiar feature, a new acquisition, as the stems of many Algae like *Fucus* already have it in the same way.

The striking properties of vegetative cones of course have often captivated the attention of botanists; what we find in literature on the topic is however practically limited to observations on the presence or absence of an apical cell, on the order of cell division, the direction of partition walls, and further, on the earliest differentiations of new organs in the cones.

Yet it is not the cell network which gives the peculiar power to the cone, nor is the production of new organs its chief property. In the unicellular apex of *Caulerpa* or of the *Sphacelariales* the same forces are present without any cell network; cones with and without apical cells are alike in their achievements, and many cones, for instance most root cones, never produce new organs.

The vegetative cone is to be taken as another consequence of the general faculty of living beings to form their different parts according to different laws, by means of the activation of different morphogenetic forces. This faculty applies equally well to the different parts of a cell as to the parts of a tissue. It is the cause of the different plastics of different parts; it may also give rise to different physiological properties.

In the case of the vegetative cone its chief property, the power of adding new parts to the mature ones, is restricted to a very limited area. As in most similar cases, that of the cambium for instance, it is accompanied by a marked difference in plastics between the active cells and other cells of the plant, the former being meristematic. The apical meristem is further characterized by a most remarkable but quite unexplained regeneration and concentration power; the parts adjoining the mature tissues differentiating in their turn, the special forces in the centre of the meristem are always concentrating and never disappear.

It is exactly the mode of concentration which furnishes the great difference from intercalary meristems, cambia and marginal meristems: in the intercalary meristems and in the cambia the concentration is directed towards a plane, in the marginal meristems it is directed towards a line and in the apical meristems towards a point, the so-called vegetative point. As a consequence of this mode of growth we may take the paraboloid shape of the vegetative cone.

The backward concentration in a vegetative cone usually works with an astonishing perfection. Yet it seems to be a general property of all vegetative cones that in some abnormal specimens, more or less subject to exceptionally vigorous growth, the mode of concentration changes. The central area in such cases loses its circular outline, it becomes elliptical or even almost ribbon-shaped, the concentration now being obviously directed towards a line. In such a way the stem apex becomes cuneiform, and a fasciation of the stem is the result.

Continued growth of a fasciated stem usually tends to exaggeratation of the fasciated condition, the "vegetative line" growing longer and longer, and the stem becoming broader and broader. Probably this is due to the fact that the surrounding meristematic tissues which in the normal cone increase in all three dimensions, continue to grow in about the same way, but that their tangential growth for which in the normal paraboloid cone room is provided, here has the effect of stretching out the thin apical line.

Usually after some time the vegetative line is broken up and consequently the apex itself, instead of remaining cuneiform, is divided into two or more pieces which may themselves be elliptical or circular: in such a way the fasciation is accompanied by a dichotomous ramification [1]).

Fasciation has been observed in Pteridophytes as well as in Gymnosperms and Angiosperms, in stems as well as in roots. Though the real nature of fasciation may be quite unexplained as yet, we may be sure that its causes must be connected with the essential properties of the vegetative cone itself, and perhaps the study of fasciation may incidentally afford the means of getting some insight into the mysteries of the vegetative cone.

It need hardly be mentioned that the growth by an apical meristem confers a polar character on the stem, to be exact, a unipolar or heteropolar one.

**B. Form of stem.** The paraboloid shape of the vegetative cone usually entails a cylindrical form for adult stems; only in dorsiventral stems may the transverse section get an elliptical or other form owing to the different morphogenetic forces acting at opposed sides of the vegetative cone.

It is only in very rare cases that the stem form may become more complex; such a case is found in the stem of *Matteuccia Struthiopteris* and some other ferns, where "epidermal pockets" are developed [2]) over the leaf insertions. These cavities may extend even into the pith.

They have a superficial likeness to leaf-gaps, but as MEKEL has shown convincingly they are due only to growth differences, not to any intrusion. GWYNNE-VAUGHAN once speculated about the possibility of a fusion of

---

[1]) This paper being already in type, the point of view taken by the present author has been worked out in: Fasciation and dichotomy, Rec. trav. bot. néerl. 33, 1936, p. 649, where instances are given and the literature is reviewed.

[2]) J. C. MEKEL, Die Entwicklung des Stammes von *Matteuccia Struthiopteris* insbesondere die der Höhlungen, Rec. trav. bot. néerl. 30, 1933, p. 627.

these pockets [1]) which would give rise to a stem having the form of a lattice-work tube, in the same way as a fusion of leaf-gaps gives rise to a pith. The difference however is that while leaf-gaps are due to the centripetal extension of stimuli which may fuse, the endodermal pockets never can surpass the limit of the original circumference of the stem at the time of their formation.

**C. Origin of stem.** The main stem of a Pteridophyte always or nearly always is formed directly from the embryo. It is only in some *Lycopodium* species and in *Phylloglossum* that a protocorm, a tuberous body, is formed first which in its turn produces one or more stems; about this protocorm see § 11.

Usually every plant forms a number of stems the greater part of which arises by ramification of the main axis, as will be dealt with at length sub D. In a minority of cases however stems may be produced from roots or from leaves.

Cases in which roots form buds are rather rare: *Ophioglossum* furnishes a well-known instance [2]).

Buds on leaves are rare in microphyllous plants (*Isoetes* [3])), but frequent in ferns. Usually they are formed accurately in definite places, which may vary in the different species. According to Kupper [4]) these positions may be: at the lamina base only; on the main rachis in the pinna axils or at the pinna bases; on the lamina itself; on the terminal pinna only, at its base or near its apex; on a specially elongated main rachis with small pinnae; at the apex of the main rachis; on leaf-runners without any pinnae.

In all cases the stem seems to arise exogenously. In the literature known to me only one instance has been reported of an endogenous formation, for the sporeling of *Equisetum*. Barratt found that whereas all branches in *Equisetum* always arise exogenously, only the first branch of the main axis is formed in the interior of the stem, about at the level of the endodermis [5]).

The transverse section figured by Barratt (fig. 5) seems quite convincing as a proof. Yet in view of the remarkably exceptional character of the case caution seems advisable. The bud in question is formed either at the first node or under it, sometimes even a considerable distance under the first leaf sheath. Now it might be possible that the space in which the bud develops was connected by a canal or pocket with the axillary space

---

[1]) D. T. Gwynne-Vaughan, On the possible existence of a fern stem having the form of a lattice-work tube, New Phytol. 4, 1905, p. 211.

[2]) Goebel, op.c. p. 1221: Engler & Prantl, op.c. p. 461; M. W. Beyerinck, Beobachtungen und Betrachtungen über Wurzelknospen und Nebenwurzeln, Verh. Kon. Akad. v. Wetensch. Amsterdam, Afd. Nat. 25, 1886, see p. 15.

[3]) Goebel, op.c. p. 1220.

[4]) W. Kupper, Ueber Knospenbildung an Farnblättern, Flora 96, 1906, p. 337.

[5]) K. Barratt, A contribution to our knowledge of the vascular system of the genus *Equisetum*, Ann. of Bot. 34, 1920, p. 201, see p. 208.

of the first leaf sheath over it; in that case the conditions would be a parallel to the pseudo-endogenous buds of *Angiopteris* [1]) and of *Dracaena* [2]).

**D. Ramification.** A not strictly essential, yet most important character of the stem is its ramification.

Certainly in any plant at least part of the stems will remain unbranched throughout life. Yet the faculty of branching, of producing new vegetative cones giving rise to daughter branches, is of general occurrence and is quite indispensable to the life-cycle of nearly all Pteridophytes.

**I. Kinds of ramification.** As to the mode of production of new vegetative cones it is customary to distinguish between dichotomy, lateral and adventitious branching.

Dichotomy is the splitting up of the vegetative cone into two parts, recalling the simplest case of splitting in fasciation; lateral branching is the development of a new vegetative cone in a lateral part of the stem apex and adventitious branching is the production of a new cone in adult tissue.

This distinction, though seemingly absolutely definite and clear-cut, on closer examination loses much of its pregnancy. In the first place because of the fact that all graduations between dichotomy and lateral branching occur [3]). Dichotomy approaches lateral branching as soon as the two daughter branches are not of equal size. In such cases the stronger branch, as occupying the greater half of the podium, grows out in a direction less diverging from the original podium direction than the weaker branch which is crowded somewhat aside. This gradually leads to a condition in which the stronger branch is the direct continuation of the podium, the weaker branch being quite lateral in position.

In the second place we have transitional stages equally between lateral and adventitious buds; moreover in practice there is usually great uncertainty about the time of initiation of any given bud [4]).

Yet, notwithstanding these difficulties, there remain clear cases like the dichotomy of *Lycopodium Hippuris*, the lateral branching of *Equisetum* and the adventitious branches of *Rhynia Gwynne-Vaughani*, and on the whole the distinction is not only serviceable in most cases, but it even has some taxonomic value in so far as dichotomy, with transitions to lateral branching, is the prevailing mode of ramification in the *Psilophytinae*, the *Psilotinae* and the *Lycopodiinae*; lateral branching with transitions to

[1]) W. DOCTERS VAN LEEUWEN, Ueber die vegetative Vermehrung von *Angiopteris evecta* Hoffm. Ann. de Buitenzorg, 2nd ser, 10, 1912, p. 202.

[2]) J. C. SCHOUTE, Ueber die Verästelung bei monokotylen Bäumen III, Die Verästelung einiger baumartigen Liliaceen, Rec. trav. bot. néerl. 15, 1918, p. 263.

[3]) AD. BRONGNIART, Histoire des végétaux fossiles, II, Paris 1837, p. 3. See moreover J. C. SCHOUTE, Beiträge zur Blattstellungslehre II, Über verästelte Baumfarne und die Verästelung der Pteropsida im allgemeinen, Rec. trav. bot. néerl., 11, 1914, p. 94, especially in the object from Sendoro.

[4]) SCHOUTE 1914, l.c. p. 165.

dichotomy prevails in the *Filicinae*, whereas the *Articulatae* only have lateral branching, the *Protoarticulatae* with their more or less dichotomous branching only excepted.

**II. Spatial relations of branches.** The branches of a branch system usually are not distributed at random without any order. In most cases certain laws may be found governing their arrangement, laws to which up to the present botanists have not paid sufficient attention.

These spatial relations may be reciprocal relations between the branches themselves, relations to the nodes of the stem, relations to the leaves, or finally combinations of these.

*a.* Spatial relations between the branches of a branch system.

1. In plants with equally dichotomizing stems. In these cases the two daughter branches, equally dividing the available space at the top of the podium, are always opposed; there is no opportunity for additional rules. Successive dichotomies however may show a definite reciprocal orientation of their division planes.

So in *Psilotum triquetrum* the successive planes are at right angles to each other, ensuring in this way a satisfactory spatial allocation of the erect branches, and no doubt the same relation which has been recently termed cruciate dichotomy by TROLL [1]), occurs in many other plants too.

Another condition is that the successive dichotomies all have their planes parallel, so that all branches fall into a single plane. This condition, called flabellate dichotomy by TROLL, is observed in some dorsiventral branch systems as in *Lycopodium complanatum*, *Phegopteris Dryopteris* and *Lygodium scandens* [2]).

According to TROLL these two modes are due to the general symmetry, other conditions not being possible. For the flabellate dichotomy the dorsiventral nature of the stems may indeed be responsible. For the cruciate dichotomy however I cannot admit this, radiate symmetry not being itself a principle in the plant, but only a lack of differentiation of the morphogenetic forces at the different sides of the stem.

Indeed, though other stable conditions have not come to my notice, they might very well be possible, and in any case there are instances of radially symmetric stems in which the successive dichotomies are independent of each other in direction. In *Lycopodium carinatum* I found the perfectly equal dichotomies following each other at any angle; in one and the same shoot system rectangular, oblique and parallel positions of the planes being equally represented.

Similar conditions have been reported in literature for other species.

---

[1]) W. TROLL, Grundsätzliches zum Stigmarienproblem, Flora 129, 1934, see p. 99.

[2]) For illustrations of the two last mentioned cases see VELENOVSKÝ, op.c. p. 248 and 249.

CRAMER [1]) writes about *L. Selago*: "Die successiven Verzweigungsebenen bilden alle möglichen Winkel mit einander, die Verzweigungsrichtung unterliegt also keinem Gesetz. Ich habe mich davon auch bei der Untersuchung junger Zustände überzeugt", and illustrates this by three diagrams of three plant individuals (Pl. 32, fig. 13–15). And HEGELMAIER found in *L. alpinum* [2]) that the dichotomies usually follow under an oblique angle, sometimes rectangularly; in *L. Selago* he could not find a fixed relation at all.

Obviously the cruciate dichotomy, where it occurs, is due to an effect of the lower dichotomy on the next higher one, an effect not to be found in *L. carinatum* and *L. Selago*.

2. I n   u n e q u a l l y   d i c h o t o m i z i n g   s t e m s. Here the conditions are somewhat more complicated, as the successive larger branches form a more or less perfect sympodial axis, whereas the smaller branches may show reciprocal spatial relations.

Of these a single one is very well known, namely that of the flabellate dichotomy, where alternately the right and the left branch is the stronger. In the dorsiventral shoot systems of *Lycopodium volubile* we have fine instances; the relative strength of all branches being very accurately regulated in these cases, the shoot system in outline resembles a fern frond.

Of radial shoots with unequal dichotomy no reliable data were available, though the subject would certainly repay an investigation. There is some probability that at least in some cases the smaller branches will be arranged in a regular cladotactical system, though in other cases they may be arranged without any apparent regularity; cf. § 8 B IV.

3. I n   l a t e r a l l y   b r a n c h i n g   s t e m s. Here we have striking examples of regular cladotactical systems.

In *Ulodendron, Lepidodendron Veltheimianum* and other related forms the large scars which undoubtedly bore some kind of branches are placed in a regular distichous arrangement, quite independently of the complicated phyllotactical system. The main axis here sometimes follows a zigzag course from one branch to the other; this has been illustrated by SCHENCK [3]). It is clear that such a cladotaxis can only be due to determination of the place of higher branches by the lower ones, just as the place of higher leaves is determined by lower ones: these two parallel processes must have been present in the same stem, independently.

Perhaps still more striking is the case of the *Halonia* branches of *Lepidophloios* where the lateral branches are arranged in a more complicat-

---

[1]) C. CRAMER, Über *Lycopodium Selago*, Pflanzenphysiologische Untersuchungen von CARL NÄGELI and CARL CRAMER, 3. Heft, Zürich 1855, p. 10.

[2]) F. HEGELMAIER, Zur Morphologie der Gattung *Lycopodium*, Bot. Ztg 30, 1872, col. 773, see col. 826.

[3]) K. A. ZITTEL, Handbuch der Palaeontologie, München & Leipzig, II Palaeophytologie 1890, see p. 192.

ed spiral: a specimen pictured by GOLDENBERG [1]) of *Lepidophloios lari-cinus* with four branch parastichies seems to have had an arrangement after a system 4 + 5.

*b*. S p a t i a l   r e l a t i o n s   t o   n o d e s   o f   s t e m. Relations of this kind evidently are only possible in lateral branching. They occur only in the class of the *Articulatae*, where they seem to have been present in all members, probably only the *Protoarticulatae* excepted. The relation is that the production of branches is only possible at the nodes of the stem, so that a certain relation to the leaf production ensues.

In the distribution of the buds on the nodes great differences may be observed. We may have stems which are quite unbranched in the aerial part (*Calamites* spp. of the group *Stylocalamites*), where the ramification is limited to the subterranean stems, and to the production of lateral fructifications.

In other cases the stems may branch at every node (*Calamites* spp. group *Eucalamites*; *Equisetum arvense*, sterile shoots), the number of branches at a single node varying from two (or one?) to many.

In still other cases the production of branches is limited to some of the nodes, in a regular rhythmic order (*Asterocalamites Lohesti* [2]), *Calamites* spp. group *Calamitina*), every single fertile node following after one (*Asterocalamites*) or after a number (*Calamitina*) of sterile nodes, a rhythm accompanied in the latter case by an apically increasing length of the successive sterile internodes. The fertile node in these cases bears a number of branches.

*c*. S p a t i a l   r e l a t i o n s   t o   n o d e s   a n d   t o   o t h e r branches   a t   t h e   s a m e   t i m e. Branches borne at nodes at the same time may show mutual spatial relations, and indeed these occur in many instances. In the first place we may cite cases in which every node bears a fixed number of equidistant branches; in *Calamites carinatus* the number usually is two, in *C. multiramis* and in other members of the *Carinatus* group, as well as in *Asterocalamites Lohesti*, the number is higher.

In other cases however the branches, though being numerous in every branch whorl, seem to have been placed at unequal mutual distances (*Calamites* spp. group *Calamitina*, see for a good example *Calamophyllites verticillatus* as figured by ZEILLER [3]), and without any doubt these distances were strikingly different in *Asterocalamites radiatus* (see our fig. 1).

In the second place the branches of different nodes may show spatial relations. Often the branches of successive branch whorls alternate regu-

[1]) F. GOLDENBERG, Flora Saraepontana fossilis, Saarbrücken 1855–1862, Pl. 16. fig. 6.

[2]) A. RENIER, *Asterocalamites Lohesti* n.sp., du houiller sans houille (H 1*a*) du bassin d'Antiée, Ann. d. l. Soc. géol. de Belgique II 1910, p. 13.

[3]) R. ZEILLER, Bassin houiller de Valenciennes, Description de la flore fossile, Paris 1886, in Études des gîtes mineraux de la France; see Atlas Pl. 57, fig. 2.

larly (*Asterocalamites Lohesti*; *Calamites multiramis*, large branches of main axis). In other cases the successive whorls are superposed (*Calamites multiramis*, on the second order branches; *Asterophyllites equisetiformis*). In still other cases other relations occur; in *Calamites carinatus* according to HIRMER [1]) the planes of the successive pairs may meet at an angle of 60°, so that we get a cladotactical system 2 + 4 with six orthostichies.

On the contrary, in the group *Calamitina* the successive whorls, which are moreover often heteromerous, do not seem to have had any spatial relations.

Quite an isolated form of these cladotactical patterns is afforded by *Asterocalamites radiatus*. A study of two natural objects, both pictured from two sides in KIDSTON and JONGMANS [2]), reveals the following conditions. Fig. 1 renders one of these objects in a diagrammatic drawing; the

other object [3]) offered quite analogous conditions. From our figure it is obvious that the place of insertion of every branch lies over the largest gap between two lower ones, and further that when these two lower ones are inserted at different nodes, the higher branch lies towards the side of the lowest branch. The 16 internodes of the object show successively, as far as can be made out from the photographs, 3, 1, 3, 2, 4, 2, 4, 1, 4, 2, 6, 1, 5, 2, 5 and 3 scars. Perhaps one or two scars have been hidden at the

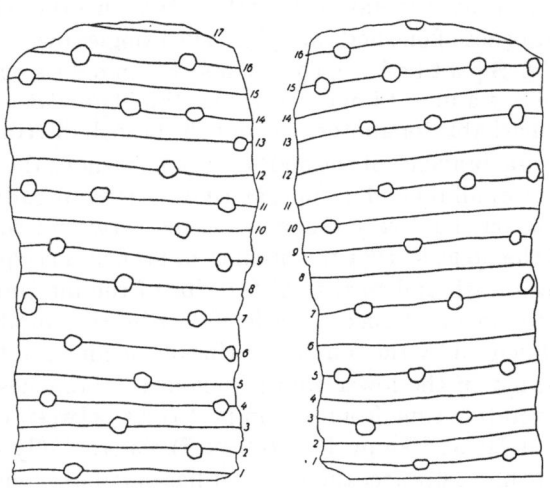

FIG. 1. *Asterocalamites radiatus*. After KIDSTON and JONGMANS, op.c. (f.n. below), Pl. 150, fig. 3 and Pl. 151, fig. 1 (same object from two sides). Only outline of specimen, internode lines and branch scars rendered. Internodes numbered. Reduced.

side or have become indistinct, but in any case the number of scars at the nodes varies very irregularly. The cladotactical system surely is not a whorled one, but is most like a spiral system. Its regularity however is not such as to make possible a counting of the parastichies.

The most remarkable property of this arrangement is that it clearly demonstrates that branch formation, though only possible at the nodes,

---

[1]) HIRMER, op. c. p. 445, fig. 542.

[2]) R. KIDSTON and W. J. JONGMANS, A monograph of the *Calamites* of Western Europe, the Hague 1915-1917.

[3]) KIDSTON and JONGMANS, op.c. Pl. 152, fig. 1 and 2.

is not a consequence of a certain disposition of a particular node, but that any part of any node is able to form a bud, the actual distribution of the buds being determined by the lower branches, just as is the case with leaves in a phyllotaxis.

In *Asterocalamites Lohesti* all this may be exactly the same, but as, at least in the specimen described by RENIER, really every second node is fertile, the intermediate ones being sterile, this can not be made out.

In still other forms we may observe another curious phenomenon, viz the occurrence of two kinds of branches. These branches are placed at different nodes in *Calamites discifer* and in *C. crassicaulis*, but at one and the same node in *C. multiramis*. Here we may observe that the two kinds of branches are independent to such a degree that the row of numerous small strobiliferous branches passes underneath the insertion of the row of larger, less numerous, vegetative branches, though otherwise in all *Articulatae* the branches of one node are placed in a single row.

*d*. Spatial relations to nodes and to leaves at the same time. Amongst the *Articulatae* the genus *Equisetum* is remarkable for another rule for the branch distribution. Though bearing all lateral branches at the nodes in a single whorl, the arrangement differs from that of all related groups, as BRONGNIART already described in 1828 [1]), by the fact that every branch is placed between two leaves of the leaf-sheath of the node, so that the number of branches is equal to that of the teeth of the sheath and to that of the ribs of the internode below. In the case of heteromery of successive leaf sheaths the number of branches is never influenced by the number of ribs of the higher internode but always keeps to that of the lower whorl, though the branches take their origin in the vegetative cone from a stem part right between the two nodes .

As far as I know JANCZEWSKI [2]) was the only author who ever pointed out this easily accessible fact; it has only been figured by MEYER [3]).

In other *Articulatae* there is no relation between the individual leaves and the branches of the node.

*e*. Spatial relations to leaves.

1. In plants with equally dichotomizing stems. All *Filicinae* with equally dichotomizing stems have a fixed relation, as far as observed, between the stem branching and a single one of the leaves, the so-called angular leaf of VELENOVSKÝ [4]). This relation amounts to the

---

[1]) AD. BRONGNIART, Histoire des végétaux fossiles, I, Paris 1828, p. 102.

[2]) E. DE JANCZEWSKI, Recherches sur le développement des bourgeons dans les Prêles, Mém. Soc. nationale des Sc. natur. de Cherbourg, 20, 1876, p. 69, see on p. 84.

[3]) F. J. MEYER, Das Leitungssystem von *Equisetum arvense*, Jahrb. f. wiss. Bot. 59, 1920, p. 263, see fig. 6.

[4]) VELENOVSKÝ, op. c.p. 249. See moreover SCHOUTE, l.c.p. 1914 (f.n. p. 9), p. 163.

fact that this angular leaf occurs on the podium, at one side of the dichotomy, just under and between the two daughter branches.

In the equally dichotomizing microphyllous Pteridophytes no such angular leaves are to be observed.

For *Lycopodium* VELENOVSKÝ mentions that no angular leaf is present in stems with a spiral phyllotaxis, but that in tetrastichous stems of *Lycopodium complanatum* an angular leaf is to be found on both sides of the dichotomy.

As in *L. complanatum* the leaf pairs are alternately placed laterally and medianly, of course every dichotomy has on both sides a leaf in the position of the angular leaf, but the fact that as soon as the phyllotaxis changes, even in the same plant, no angular leaves remain already demonstrates that these angular leaves are only angular by incidental position. The same conclusion is reached by an observation of the natural object. Our fig. 2 represents two cases being drawn from dichotomies of two sister branches, taken at random. In A and B we have a case corresponding to VELE-NOVSKÝ's decription, A as seen from the light side, B from the shadow side. In C and D we have quite another distribution owing to the fact that the last whorl of the podium and the first whorl of both shanks are trimerous. Consequently the light side C has no median leaf at all, the shadow side D ou the contrary has, but in quite another position from that of B. In the same way even those cases

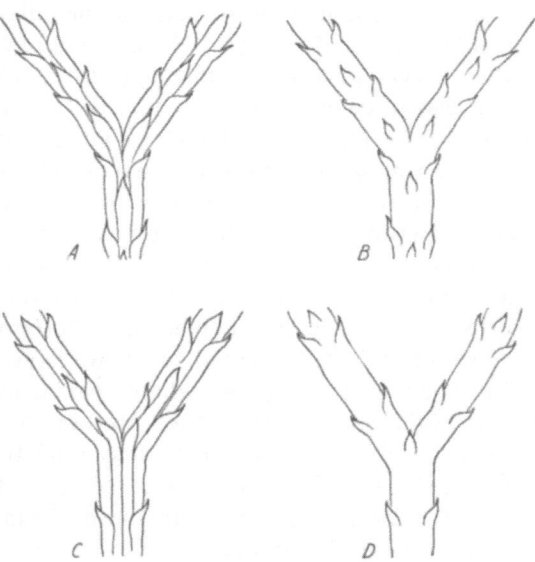

FIG. 2. *Lycopodium complanatum*. Herb. Univ. Groningen, specimen from Hängsberg bei Herrnhut, coll. Breutel. Two dichotomies from light side (*A,C*) and same from shadow side (*B,D*). Enlarged.

with a median leaf on both sides may vary; at the level of the dichotomy we may have a median or a lateral leaf pair or any intermediate condition [1]).

In *Psilotum* angular leaves are equally absent; this may already be concluded from the fact that often the successive dichotomies follow so

---

[1]) After having written this paragraph I have become aware of the fact that HEGELMAIER (l.c. col. 826) reports exactly the same conditions for *Lycopodium complanatum*, even with the incidental trimerous whorl under the forking.

closely, that many shanks have no leaves at all. Yet they dichotomize just
as well as others.

So the angular leaves may be considered as restricted to the *Filicinae*.
For this class their importance has been denied of late by Bower [1]) on
account of the varying mode of junction of the angular leaf trace with the
stele. This argument in my opinion is of no value, as the downward course
of all traces varies according to local circumstances and is no reliable basis
for the judgment of external organization.

For any botanist doubting the validity of this opinion I might cite
as an instructive instance the fine observations made by Stenzel on
*Dryopteris Filix-mas* [2]), in which he found that the bud trace may join the
stelar system of the leaf base in very different ways, namely after having
been simplified in its downward course to a single solid bundle (l.c. Pl. 4,
fig. 6, 8, 9), so that the branch medulla was "formed anew" as he expressed
it, or as a tubular bundle, so that the medulla of the branch was formed
from that of the leaf base (fig. 10), or finally in the form of three separate
bundles, decurrent from a solid ring (fig. 12, 13). In an analogous way the
angular leaf trace obviously may join the stelar system in different ways.

2. In unequally dichotomizing stems. Here the conditions are almost the same as in equal dichotomies.

In all *Filicinae* we observe an angular leaf between two unequal branches,
though sometimes the angular leaf may be shifted on to the weak branch
(*Pteridium aquilinum* [3])). And in *Lycopodium* the angular leaves are
wanting in the same way as in equally dichotomizing cases.

For *Selaginella*, where the unequal dichotomy prevails and which consequently has not been treated sub 1, we have to consider separately the
numerous dorsiventral species where angular leaves are to be found, and
the few radial species where they are absent. The former will be dealt
with below, sub *g*; for the latter I may refer to our fig. 3, representing a
branching shoot of *S. selaginoides* as seen from two sides. The phyllotaxis
of podium and shanks being a spiral one of the main series, it is particularly
clear that no angular leaf occurs.

This has already been recognized by Al. Braun [4]) who wrote for *Lycopodium* and for *Selaginella*: "Bei vielzeiliger Blattstellung lässt sich ein
constantes Verhältniss des Zweiges zu einem Blatte, das man als Tragblatt
bezeichnen könnte, nicht nachweisen, indem der Zweig bald genau über

---

[1]) F. O. Bower, The ferns, I, Cambridge 1923, see p. 74, and op. c. 1935 p. 300.

[2]) K. G. Stenzel, Untersuchungen über Bau und Wachsthum der Farne. II
Über Verjüngungserscheinungen bei den Farnen, Nova Acta 28, 1861, p. 3.

[3]) Velenovský op. c. p. 250 and especially M. Büsgen, Einige Eigentümlichkeiten des Adlerfarns, Zeitschr. Forst- und Jagdwesen 47, 1915, p. 235.

[4]) Al. Braun, Über die Blattstellung und Verzweigung der Lycopodiaceen,
insbesondere der Gattung *Selaginella*, Verh. bot. Ver. Prov. Brandenburg, 16, 1874,
p. 60.

ein unter ihm stehendes Blatt, bald in die Lücke zwischen 2 vorausgehende Blätter fällt".

3. In laterally branching stems. Here a striking relation between the branch and a leaf may occur, but again this is only to be found in the *Filicinae*, not in any microphyllous plant. The relation, as I tried to demonstrate at length on a former occasion [1]), is identical with that between a dichotomy and the angular leaf, though because of the great difference between the small lateral branch and the main stem it amounts to a leaf and a lateral bud next to it.

FIG. 3. *Selaginella selaginoides*. Herb. Scandinavicum (Häglund & Källström, Falun). Dichotomic branching from two sides; specimen soaked in water and more or less transparent. Leaves numbered according to spiral phyllotaxis. Enlarged.

As a rule the lateral branch is situated at the right or left side of the leaf insertion. In many cases however the position changes: the lateral branch may become quite axillary, it may be situated basally to the leaf, just under it, or it may shift on to the leaf-base, forming in this way a transition to the many stable forms of branch formation in the fern leaf mentioned above sub C. For particulars see my quoted paper and especially METTENIUS [2]). There may even be two branches near one single leaf insertion, one to the left, one to the right, or even four, two at each side.

For those starting from the well known axillary branching of the *Angiosperms* and trying to explain the *Filicinae* from that point of view, these conditions may seem very strange; we should however reverse the case and recognize that the general tendency of macrophyllous plants is the production of branches in relation to one particular leaf; as special forms of this tendency we have on one side the equal dichotomy with its angular leaf, on the other side the axillary branching, which in the *Filicinae* is represented in the *Hymenophyllaceae*.

*f*. Spatial relations uncertain or not fixed. Of course the presence of stable spatial relations is more easily stated than their absence. Perhaps many cases of branch systems without spatial relations are present in nature, but beyond the case of *Lycopodium carinatum*

[1]) SCHOUTE l.c., 1914, (f.n. p. 9) p. 161.

[2]) G. METTENIUS, Über Seitenknospen bei Farnen, Abh. d. math. phys. Kl. d. K. Sächs. Ges. d. Wiss. 5, 1861, p. 611.

treated above p. 10, no cases can be mentioned here for the above reason.

The only topic to be dealt with here is the position of lateral branches in the *Lycopodiales* of which the spatial relations have been described very differently. We shall deal successively with *Lycopodium Selago, Lepidodendron* and *Sigillaria*.

1. L y c o p o d i u m   S e l a g o. In *L. Selago* and in some allied species so-called bulbils are of common occurrence, short detachable branches serving as a means of vegetative propagation. About their arrangement as well as about their morphological origin [1]) the statements of the different authors are widely divergent.

METTENIUS wrote in 1860 [2]) that the bulbils are placed at the side of the leaf-bases. HEGELMAIER on the contrary found them to occur in the place of a leaf [3]). VELENOVSKÝ again reports them to arise "zwischen den Blättern ohne alle Orientierung zu den letzteren" in *L. inundatum* [4]), and in *L. Selago* he mentions them as placed "unregelmässig zwischen den Blättern an der Achse, indem sie die Ordnung der nächsten Parastichen stören". Lastly CZURDA like HEGELMAIER finds them in the place of the leaves when they are few in number; as soon as they become numerous the whole phyllotaxis becomes irregular [5]). In his diagrammatic drawing Fig. 1 we see that most bulbils take the place of a single leaf, but some of them of two superposed leaves.

My own observations on some *L. Selago* plants from Alp Palü in Switzerland belonged to the simpler cases of CZURDA, all bulbils taking the place of a single leaf without causing more disturbance.

Summarizing these data we may say that the bulbils in any case disturb the phyllotaxis in so far as by their presence one or sometimes two leaves are wholly omitted; when they are numerous the whole phyllotaxis loses its regularity. A spatial relation of any bulbil to any particular leaf is not recognizable; the bearing of these facts on the phyllotaxis will be discussed later in § 8 sub A IV.

---

[1]) R. WILSON SMITH (Bulbils of *Lycopodium lucidulum*, Bot. Gaz. 69, 1920, p. 426) tried to solve the question of their morphological origin on the basis of modern technique and serial sectioning; he concluded from the nature of the bulbil trace, that the bulbils should be metamorphosed leaves: a new argument for the inefficiency of anatomical research in morphological questions.

The same result was reached on an experimental basis by S. WILLIAMS (A contribution to the experimental morphology of *Lycopodium Selago*, with special reference to the development of adventitious shoots, Trans. Roy. Soc. Edinb. 57, 1933, p. 711); the descriptions of the regeneration products and the drawings do not however warrant any conclusion.

[2]) METTENIUS, l.c. p. 628.

[3]) F. HEGELMAIER, l.c. (f.n. p. 11), col. 841.

[4]) VELENOVSKÝ, op.c. (f.n. p. 3), p. 257.

[5]) V. CZURDA, Zur Kenntnis der Brutzwiebeln von *Lycopodium Selago* und *L.* *lucidulum*, Flora 116, 1923, p. 457.

2. L e p i d o d e n d r o n. The distichous ulodendroid scars of *L. Veltheimianum* have been quoted p. 11 as examples of a cladotactical system. It remains however to be investigated, especially after taking cognizance of the facts related for the *Lycopodium* bulbils, as to how far the large corresponding branches disturbed the phyllotaxis.

Many drawings or photographs of the large scars amidst the small

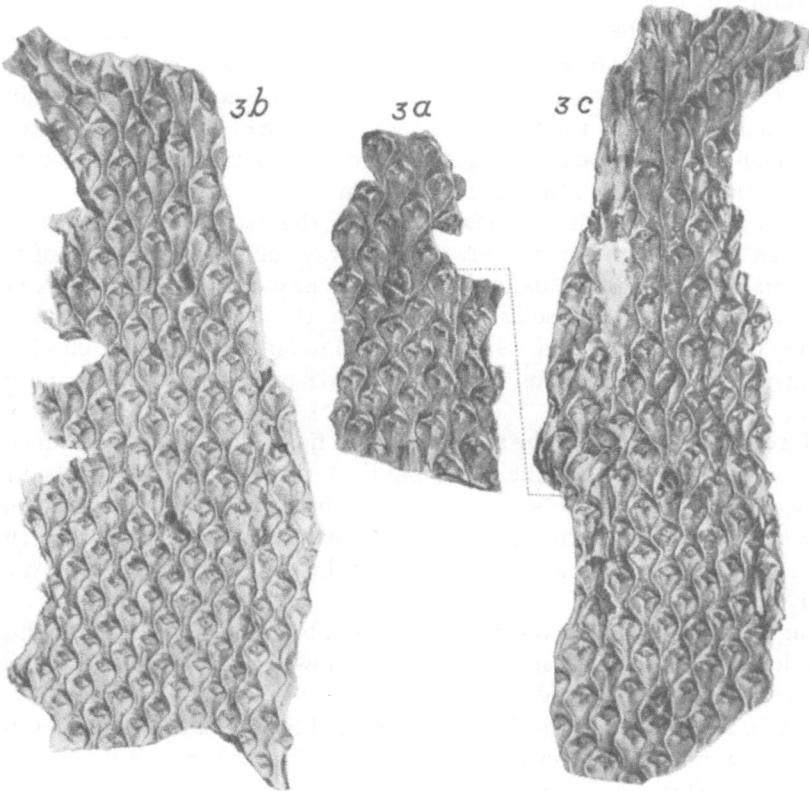

FIG. 4. *Lepidodendron Volkmannianum*, After STUR, op.c. (f.n. below), Pl. 23 (40), fig. 3*a*, (actual stem) 3*b*, 3*c* [impressions].

leaf scars seem to demonstrate that though the phyllotactical system above the branch scar resumes its original pattern, at the actual place of insertion a great number of leaf scars has been suppressed. The fine drawings by FAHRENBAUER in STUR [1]) however, clearly demonstrate that a large part of the apparent branch scar was clothed with epidermis and bore numerous small leaf scars, only the centre (the "Nabel") being the insertion of the branch. About the completeness of the phyllotactical pattern these drawings give no information.

[1]) D. STUR, Die Culm Flora, Abh. k.k. Geol. Reichsanst. 8, Wien 1875–77, see Pl. 21 (38), 22 (39).

As STUR himself already compared these large scars to the *Lycopodium* bulbils which he believed to be axillary productions, he tried to prove the axillary position of the *Lepidodendron* branch scars too. He therefore looked for objects with smaller "bulbils" amongst the *Lepidodendra*, and once succeeded in finding them in a specimen of *L. Volkmannianum*. The object consisted of the flattened stem itself and moreover of the two different impressions in the rock (our fig. 4); four "bulbil"-scars in total being present. Two of them are to be seen in the right impression, the cross higher up only indicating the place where a further scar might have been expected. Two others are present in the left impression, in such positions as to make clear that the four scars together formed a distichous system. The small preserved part of the actual stem only contains a single scar, the same as the lower one in the right impression.

From these figures it becomes clear that the branch scars are placed between the leaf scars in the only possible way: obliquely over one of the asymmetric leaf scars. A definite relation to any one of the leaves is not to be assumed, the more so as we see that in the left impression the insertion of the two branch scars is different in so far as the higher one is in contact with a lower scar on its left, the lower one with a leaf scar on its right.

Moreover — and this is the reason why the figures have been reproduced here — it is obvious that the whole phyllotactical pattern is disturbed in the vicinity of the branch scars, as only the higher branch scar of the left impression is surrounded by unchanging leaf parastichies, whereas the three other branch scars all have changes in the parastichy numbers just at their place of insertion.

Our conclusion may therefore be that these branch scars, whatever they may have been, in their spatial relations showed analogy to the bulbils of *Lycopodium* as described above.

3. S i g i l l a r i a. In the different species of *Sigillaria* the strobiliferous branches were borne in various ways, as may be gathered from the fine drawings in ZEILLER [1]), some of which have been reproduced in our fig. 5.

In *S. elegans* (5a) they were borne more or less exactly in whorls, their number in a whorl being far less than the number of leaf orthostichies. In the branch whorl the individual branches were placed at varying distances and indiscrimately in a leaf orthostichy, between two orthostichies or in an intermediate position.

In other species, like *S. tesselata*, the branches were locally more numerous and were placed in vertical rows between the leaf orthostichies, either without causing any disturbance of the latter (5b) or disturbing the phyllotaxis more or less (5c). The number of the branch scars here is somewhat greater than that of the leaves in the adjacent orthostichies:

---

[1]) R. ZEILLER, op.c. (f.n. p. 12).

special spatial relations between branches and individual leaves are lacking.

In *S. mamillaris* the branch scars are not confined to the spaces between the leaf orthostichies, but are often placed exactly in the leaf orthostichy itself; they exert a rather strong disturbing influence on the phyllotaxis.

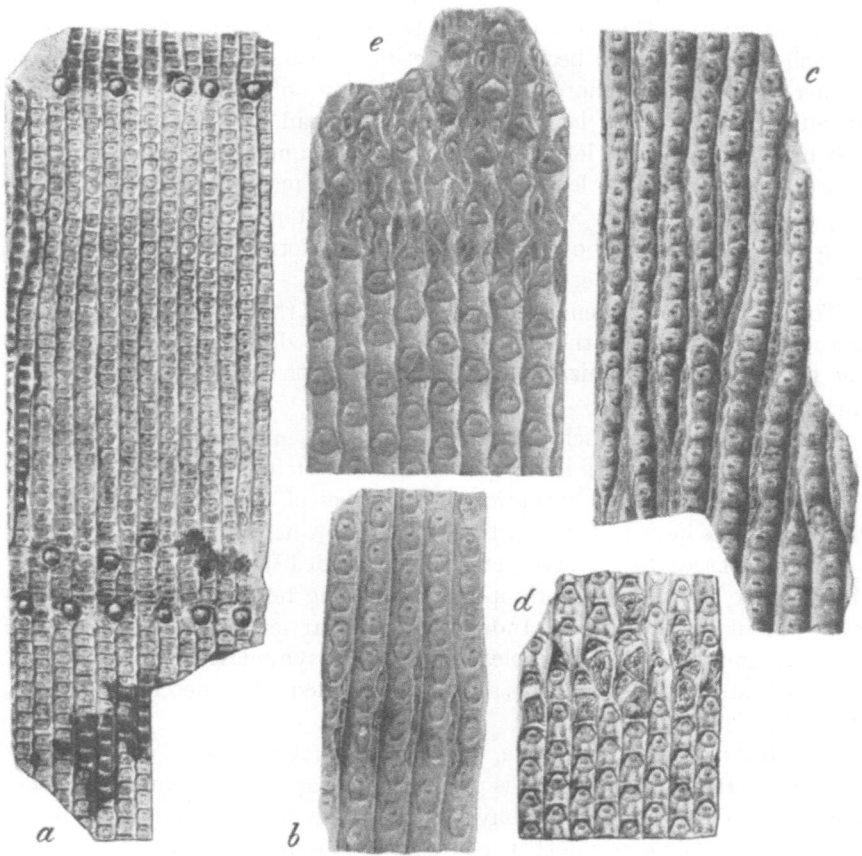

FIG. 5. *Sigillaria* spp. After ZEILLER, op.c. (f.n. p. 12). *a* = Pl. 87, fig. 1, *S. elegans*; *b,c* = Pl. 85, fig. 8 and 1, *S. tesselata*; *d* = Pl. 87, fig. 5, *S. mamillaris*; *e* = Pl. 82, fig. 9, *S. scutellata*.

In *S. scutellata* finally the phyllotaxis in the branch bearing zone is so strongly disturbed that it becomes wholly irregular.

4. C o m p a r i s o n   o f   s p a t i a l   r e l a t i o n s   o f   l a t e r a l b r a n c h e s   i n   L y c o p o d i i n a e. The phenomena in the three groups described above from the observations instituted quite independently by so many authors, all show a strong similarity. In all cases the branches, when numerous, disturb the phyllotactical pattern; when they are less

numerous they may be formed quite independently of any leaf insertion, or they may be formed at those places which have been as yet left free from leaf formation, i.e. either between the leaf orthostichies or in a single leaf orthostichy. In so doing the branches may either only crowd the leaves a little aside without influencing their number, or they may prevent the formation of one or more leaves, the place of which they have taken.

Our impression therefore is that in the *Lycopodiinae* leaf formation and branch formation may become competing processes; when branching is earlier, as in equal dichotomy, the leaves may only cover the space that is left for them; when leaf formation gets ahead, the branches have to put up with the space left between the leaves, and finally when the two processes are more or less simultaneous they may disturb each other.

g. T a x o n o m i c  v a l u e  o f  d i f f e r e n t  m o d e s o f b r a n c h-
i n g. As pointed out above p. 9, the different forms of branching have a certain taxonomic value.

To the arguments given there, we may now add that the relation between branches and nodes in the *Articulatae* which no doubt is connected with the peculiar node organization in that class, is of outstanding taxonomic importance.

The relation of a branch to a particular leaf seems equally to be of first importance, as it occurs in all *Filicinae* and not in microphyllous Pteri-dophytes, with the only exception of a number of *Selaginella* species. For that reason we have to consider these cases anew here.

When VELENOVSKÝ in 1905 enounced the idea [1]) of the importance of the angular leaf in the Pteridophyte branching he based his views to a large extent on *Selaginella*. Indeed the angular leaf in the dorsiventral stem systems is very remarkable; it is the only symmetrical leaf occurring in these systems. As such it has already been adequately described in 1874 by BRAUN [2]).

Of course it would be possible, that the same relation between a particu-lar leaf and ramification which is present in megaphyllous plants, had been developed in a group of microphyllous plants as well, as a parallel phe-nomenon. The fact however that only dorsiventral species display it, while it is lacking in radial species of the same genus, makes this rather improbable, and it may therefore be worth while to consider the facts in some more detail.

As far as I can see there is only one other possible supposition, namely that the dorsiventrality factors which influence both ramification and phyl-lotaxis, in doing so establish a certain relation between the two. To check the merits of such a view it will be necessary to review the facts for both ramification and phyllotaxis.

---

[1]) VELENOVSKÝ, op.c. p. 249; probably already earlier in a Czech paper of 1890.
[2]) AL. BRAUN l.c. (f.n. p. 16), p. 62.

The ramification in a dorsiventral *Selaginella* is of a very characteristic kind. In the aerial branching systems a sympodial main axis is formed out of the stout shanks of the successive unequal dichotomies, the smaller shanks being placed in a strictly alternating order at the right and left sides of the main axis. At those places where the *Selaginella* "frond" is quite flat, as usually is the case at least at the last ramifications, all branches are really placed exactly in the same plane. Often however the lateral branches, especially near the base of the "frond" are turned upwards, towards the sky, and in such places the insertions of these branches are also to be found on the upper side, so that the successive dichotomies are no longer flabellate, but cruciate.

In view of this fact the idea becomes tempting that perhaps all branches really are due to cruciate dichotomy, but that by the dorsiventrality factor all shanks are reduced more or less to the same plane. The stronger branches in that case would all be the under shanks, the smaller branches the upper shanks.

The lateral branches usually follow at about equal distances; sometimes however a very unequal, we might say an iambic, rhythm may be observed when alternately the successive parts of the main axis bear e.g. 3 or 1 pair of leaves, and some number of, say 13. I have never seen two successive lateral branches at the same side of the main axis.

The phyllotaxis is essentially diagonally decussate, the under leaves being shifted towards the sides and all leaves being often unequal-sided or falcate. In the last thin branches of a "frond" with their densely packed leaves the phyllotaxis is usually quite regular, in the main branches with their more diverging leaves it often becomes rather irregular by many longitudinal shifts. The last leaf pair of every shank consists of a peculiar, symmetrical under leaf, the angular leaf, which is situated in the fork and a normal upper leaf at the insertion of the stout shank or often shifted up on to that shank, even several mm. The under leaf of the next lower pair is then often also shifted on to that same shank.

Because of these shifts it is not always easy to determine the number of leaf pairs of a given shank, which may give some trouble in checking the correctness of the view enounced by BRAUN, that every shank always bears an odd number of leaf pairs. Yet I believe this statement to be correct, and as BRAUN remarks, the alternation of the lateral branches and the normal position of the angular leaf pair can only be reconciled in that way, as our schematic fig. 6A may illustrate. Between the two dichotomies two normal leaf pairs have been drawn and it is clear that only an even number of pairs between the two angular leaf pairs will be able to ensure their reverse positions; the shank therefore should contain an odd number, the angular leaf pair being included.

In my incidental observations I was not able to find exceptions to this curious rule. Yet BRAUN himself writes: "Von dieser Regel kommen jedoch

merkwürdige Ausnahmen vor, indem der Zweig auch nach einer geraden
Zahl von Blattpaaren (Unterblättern) eintreten kann, in welchem Falle er
mit dem vorausgehenden Zweige auf dieselbe Seite fällt. Dem Bestreben
nach Abwechslung wird alsdann (nicht immer, aber meist) dadurch Genüge
geleistet, dass der Zweig sich aufrichtet und die Stelle der Hauptachse
einnimmt, die Hauptachse dagegen, zur Seite gedrängt, die Rolle des
Zweiges spielt. Bei vielen Arten kommt ein solches Verhalten nur als
seltene Abweichung vor, bei einigen aber (*S. Wallichii*) wird es zur Regel".

In order to examine these curious but not very lucid facts I studied some specimens of *Selaginella Wallichii* from the University Herbarium at Utrecht. They indeed showed the even number of leaf pairs for every part of the sympodial axis of
the "fronds", in casu invariably 4 pairs. The distribution of these pairs was however not that of fig. 6C (derived from 6B), as might have been inferred from
BRAUN's description. In 6C the position of the angular leaf pair would be alternately

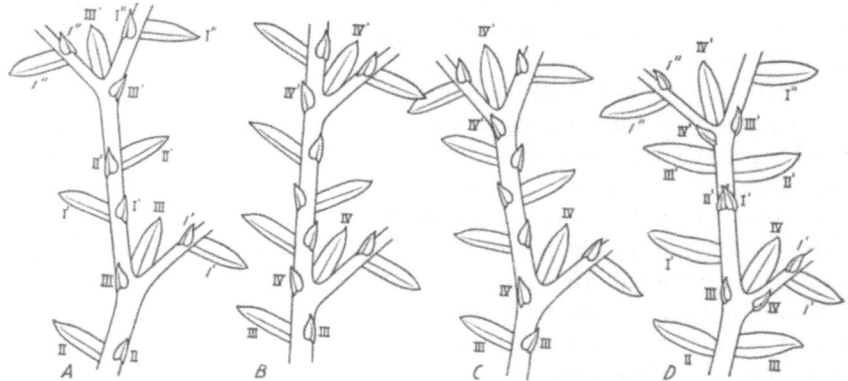

FIG. 6. *Selaginella. A =* diagrammatic representation of relation between phyllotaxis and branching in most dorsiventral species. *B =* same from description by
BRAUN for *S. Wallichii,* without change of size of main axis and lateral branches.
*C =* as above, with interchange of size. *D =* actual condition in *S. Wallichii.* Leaf
pairs of sympodial axis indicated by Roman numbers, of small shanks by Arabic
numbers.

different for any two successive dichotomies. It was however, with some variation
in the longitudinal shifts, the arrangement of 6D.

At first the idea might present itself that here we had some trimerous whorls
(III, IV, IV and III', IV', IV'), but as all leaves here, as in other dorsiventral
*Selaginella* spp. are inserted on the usual four orthostichies, this is not very probable. Moreover the leaves may very well be grouped in pairs, with some longitudinal
shifts; this has been done in 6D by designating the leaves in the same way as in the
diagrammatic figures *A–C.* The longitudinal metatopies of 6D are not stronger
there than in other cases.

From the above observations I am inclined to the following view on the relation
between branching and phyllotaxis in the dorsiventral *Selaginella* species. The
branching is an independent phenomenon; the smaller shanks of the unequal
dichotomy or, what amounts to the same, the lateral branches, are placed alternately
at the right and the left.

The phyllotaxis is a diagonally orientated decussation; one of the under leaves
always shifts into the angle of the dichotomy, and because of that peculiar position
the dorsiventrality factors do not get any hold on it, so that it develops symmetrically.

In the majority of species this angular leaf is one out of the orthostichy at the small shank side (6A). Therefore the upper leaf of that pair falls in front of the sympodial axis, and moreover any part of the sympodial axis has an odd pair of leaves.

The two dichotomy branches begin with symmetrically placed pairs of leaves, their upper leaves being turned towards the dichotomy angle: the stout shank therefore continues the phyllotaxis of the podium, the first leaf pair of the small shank being superposed over the last podium pair.

In *Selaginella Wallichii* the angular leaf is an under leaf out of the orthostichy at the stout shank side. The upper leaf of the same pair therefore falls in front of the small shank and the number of leaf pairs of the sympodial axis is even.

The two dichotomy branches begin their phyllotaxis exactly in the same way as in all other species; owing to the different position of the angular leaf pair it is now the small shank which continues the phyllotaxis of the podium.

After this view the hypothesis of the main axis and the lateral branch interchanging their rôles may be spared. At the same time, and this may be the excuse for this long digression, the way is cleared for the assumption that the angular leaf of the dorsiventral *Selaginellae* may be something quite different from the angular leaf of the *Filicinae*, playing no rôle in the branching itself.

**E. Kinds of stem plastics.** The term plastics (Plastik) has been framed by AL. BRAUN [1]) to indicate those form properties of the flower which do not depend on the mutual arrangement of the floral parts nor its division into floral belts: it therefore comprises all the special characters of form, colour, consistency, pubescence etc. The use of the term at present is almost limited to taxonomic papers, but it may be of great advantage in general morphology to denote the range of variation of differentiation.

For a survey of stem plastics in the Pteridophytes the reader may be referred to VELENOVSKÝ [2]). Here will be treated only the question of different kinds of stem plastics in one and the same plant. In many Pteridophytes such a difference does not occur, all stems being essentially alike (*Rhynia*, *Isoetes*, *Pteris*). In many other Pteridophytes however the fully developed plants exhibit two, three or even more different kinds.

Thus we may have sterile and fertile stems (*Equisetum arvense*), aerial stems and subterranean rhizomes (same instance), normal rhizomes and stem tubers (do.), normal leaf bearing stems and stolons (*Alsophila aculeata* [3])), normal rhizomes and special runners with root-hairs and root-function (*Trichomanes* sect. *Hemiphlebium*), normal stems and strobilus axes (*Lycopodium clavatum*).

**F. Nodes and internodes.** In leafless Pteridophytes nodes and internodes are wanting, every stem being a single unit. As soon as there are leaves however they may be distinguished, at least theoretically, and when

---

[1]) AL. BRAUN, Über den Blüthenbau der Gattung *Delphinium*, Jahrb. f. wiss. Bot. l. 1858, p. 307.

[2]) VELENOVSKÝ, op.c. p. 230–242.

[3]) STENZEL, l.c. (f.n. p. 16), Pl. 1.

the leaves form leaf traces, the structure of nodes and internodes even differs anatomically.

In a number of Pteridophytes the plastics of nodes and internodes is practically the same: in such cases the distinction properly has no value (*Lycopodium, Lepidodendron*). In other cases nodes and internodes may differ more or less in plastics; this occurs to a great extent in the *Articulatae*. The intercalary growth of the *Equisetum* stem, for instance, is wholly based on this differentiation. In such cases the distinction assumes real morphological importance.

In literature there has been much speculating about holocyclic and mericyclic nodes, including either a whole transverse disc of the stem or only a sector (ČELAKOVSKÝ [1])) or about the question whether an internode between two whorls might be taken as homologous to an internode between two scattered leaves. All these speculations are useless, and the simple opinion of AL. BRAUN is by far to be preferred, namely that a node is nothing but a part of the stem, differentiated more or less in a special way under the influence of the leaf insertion [2]).

**G. Telome.** The term telome as introduced by ZIMMERMANN [3]) designates the last ramification of a stem or shoot, as far as it is one single unbranched body, externally as well as in its stelar structure.

As soon as a stem develops a lateral branch, its basal part by that fact is no longer a telome or part of it, but it is a part of a syntelome; its unbranched apex on the contrary remains a telome.

Though the term has been accepted of late by highly competent authorities like HALLE and BOWER, yet in my opinion the use of it is not to be recommended. Our morphological conceptions should be read from nature as far as possible, our morphological terms should correspond to parts owing their origin to natural processes.

A stem, being formed by a given vegetative cone, responds to that requirement. But that a stem should change into two different parts by the development of a lateral branch to me seems an unnatural, artificial conception.

**§ 6. The sporangium.** — The sporangium of the Pteridophytes is a more or less globular, reniform or sack-shaped organ, usually freely projecting on the surface of the sporophyte, seldom imbedded, in which following reduction division haploid air-dispersed spores are formed. The sporangium being already known in the oldest land-plants as far as our record goes, its phylogeny is quite uncertain; perhaps it is derived from a group of tetrasporangia.

Its place on the sporophyte is either terminal on the stem (*Psilophytinae*)

---

[1]) L. J. ČELAKOVSKÝ, Die Gliederung der Kaulome, Bot. Ztg 59, 1901, p. 79.

[2]) AL. BRAUN, Tannenzapfen, Nova Acta 15, 1, 1831, see p. 347.

[3]) W. ZIMMERMANN, Die Phylogenie der Pflanzen, Jena 1930; see p. 65.

or in any position on the leaves (all recent Pteridophytes); the last condition seems to have been present from the beginning of land vegetation, as of late it has been reported by LANG and COOKSON [1]) for *Baragwanathia longifolia*.

Sporangia borne on microphylls are usually inserted on the adaxial leaf side; see moreover § 8 sub A III; those on megaphylls are usually placed on the abaxial side or on the margins, see § 8 sub B III.

Every sporangium consists of a pluricellular sterile wall within which an archesporium develops the spores in a smaller or greater number. In only a few cases is a central sterile columella present (*Hornea, Sporogonites*) at the basal part of the sporangium, being overarched by the dome-shaped spore mass. The organization of the different parts of the sporangial wall shows a great range of variation in the different groups; the taxonomy has to a certain extent been based on these differences, especially of the outer layers.

These outer layers often show complicated differentiations of various kinds, the function of which is the dehiscence of the sporangium. Amongst these the annulus, a complex of cells with partly thickened walls, may be cited. The inner layers have a nutritive function for the developing spores and are called the tapetum.

The sporangia of Pteridophytes are often produced in great numbers, and may be arranged in groups, called sori (see § 8 sub B III). The sporangia of a sorus may fuse into a unit of higher order, the plurilocular synangium; this seems to have already occurred in the Silurian age (*Yarravia* [2])).

Another differentiation of the sporangium amounts to the development of two kinds of sporangia in one and the same plant species; kinds differing in that from the spores of the one only male gametophytes arise and from the other, females. These two kinds are called microsporangia and megasporangia, and their spores microspores and megaspores.

The micro- and megasporangia may be of the same organization (*Selaginella*) or they may be different (*Azolla*). Their spores differ in all known cases notably in size, as is indicated by the names; it has however been pointed out of late (see e.g. PINCHER [3])) that such a difference is not essential for the phenomenon of heterospory.

**§ 7. The root.** — As remarked in § 2, the root does not occur in all Pteridophytes: it is lacking in the *Psilophytinae*, in the *Psilotinae* and in *Salvinia*. In the latter case its absence has always been recognized as

---

[1]) W. H. LANG and I. S. COOKSON, On a flora including vascular land plants, associated with *Monograptus*, in rocks of Silurian age, from Victoria, Australia, Phil. Trans. Roy. Soc. London, B, 224, 1935, p. 421.

[2]) LANG and COOKSON, l.c. p. 437.

[3]) H. C. PINCHER, A genetical interpretation of the origin of heterospory and related conditions, New Phyt. 34, 1935, p. 409.

being of a secondary nature, connected with the occurrence of the dissected submerged leaves with root function.

In the *Psilophytinae* on the other hand it is generally considered as being the primitive condition. In the *Psilotinae* it was formerly often suspected of being secondary, especially as the embryo long remained unknown and of course the embryo might have passed through a rooted stage. Since the discovery of the quite rootless embryos both of *Tmesipteris* [1]) and of *Psilotum* [2]) and the simultaneous acquisition of knowledge of the *Psilophytinae*, the general view is that roots probably never have been present in the ancestors of the *Psilotinae*.

The roots, where present, are always of a little varying organization. They grow by a vegetative cone much like the stem, a cone which however is covered by a calyptra or root-cap. They are cylindrical, never develop leaves and as a rule develop root-hairs, a feature which is rare in stems (*Rhynia*, *Hornea*, *Psilotum*, rhizomes, *Trichomanes*, runners and sometimes other stems too, *Psaronius*, stem after leaf fall) and which is still more rare in leaves (petioles and leaf under-surface of *Trichomanes Hildebrandtii* [3])).

Roots arising endogenously on the stem are formed in contact with the vascular tissue strands. Their distribution therefore in some cases may show a certain correspondence to that of the leaf traces [4]).

From the point of origin they make their way out in various directions, running sometimes through the stem tissues over considerable distances (*Lycopodium*, basipetal course, *Angiopteris*, geotropical course [5])).

Roots may be formed in some cases on leaves too.

The root may branch dichotomously (*Isoetes*, *Selaginella*, usually in *Lycopodium*, sometimes in *Ophioglossum* [6])) or laterally (*Equisetum*, *Filicinae*).

---

[1]) A. ANSTRUTHER LAWSON, The prothallus of *Tmesipteris tannensis*, Trans. Roy. Soc. Edinb. 51, 1917, p. 785, and: The gametophyte generation of the *Psilotaceae*, ibid. p. 93.

[2]) G. P. DARNELL SMITH, The gametophyte of *Psilotum*, Trans. Roy. Soc. Edinb. 52, 1918, p. 79.

[3]) ENGLER & PRANTL, op.c. p. 98.

[4]) J. P. LACHMANN, Contributions à l'histoire naturelle de la racine des Fougères, Ann. Soc. Botan. de Lyon, 16, Notes et Mémoires, 1889, p. 1, see p. 23; J. C. SCHOUTE, On the foliar origin of the internal stelar structure of the *Marattiaceae*, Rec. trav. bot. néerl. 23, 1926, p. 269, see p. 287, 288.

[5]) R. F. SHOVE, On the structure of the stem of *Angiopteris evecta*, Ann. of Bot. 14, 1900, p. 497, see p. 506.

[6]) HIRMER makes mention of dichotomous branching of the roots of *Asterocalamites scrobiculatus* (op.c. p. 377), without quoting his source. Perhaps this source has been D. STUR, who describes his *Archaeocalamites radiatus* as "radices irregulariter dichotomas emittens" (op.c. f.n. p. 19, p. 2). The figures (Pl. 1, fig. 3–5) as well as the further description of the text p. 5 are far from convincing: in the text it is expressly stated that the branching is not so regular nor so symmetrical as that of the leaves. So the question must remain unsettled.

The successive dichotomies may follow rectangularly (*Selaginella* [1]), usually in *Lycopodium* [2])) or occasionally in the same plane (*Lycopodium* [2])).

Transitions between dichotomy and lateral branching may occur; in *Lycopodium* branches arising at the apex but being smaller may be decidedly lateral; their arrangement may approach a regular one when the branches follow more or less in a $\frac{1}{2}$ or $\frac{1}{4}$ spiral, or in opposed pairs which themselves may be alternate or superposed [2]).

In *Equisetum* and in the *Filicinae* lateral roots arise endogenously; they are arranged in orthostichies, as their origin is confined to the presence of the vascular strands in the mother root.

The first root of the embryo usually is formed exogenously, in various positions, not always opposite the first leaf. Later roots nearly always arise endogenously, but in *Phylloglossum* also these are exogenous [3]). This difference of origin might be taken as a reason for distinguishing between a main root and adventitious roots, a practice which is sometimes followed, e.g. by VELENOVSKÝ [4]). Most authors however do not make much of the difference; perhaps on account of the repeated assertion of HOFMEISTER that Pteridophytes have no main root [5]), or perhaps from GOEBEL's consideration [6]), that a true main root should show secondary growth. In any case exogenous and endogenous roots are of exactly the same morphological description so that the distinction seems of little or no importance.

A phylogenetic origin of the root from a specialized kind of stem, differentiated for absorption and attachment purposes, has always been more or less probable [7]). Of such a differentiation we know many instances: the rhizomes of so many Pteridophytes, being only slightly modified subterranean stems, the rhizophore (see § 11 sub 2), the stigmarian axis (§ 11 sub 3) and finally the peculiar leafless runners of *Trichomanes*. These last dichotomously branching organs attach the plant to the substratum and have an absorptive function; they develop root hairs, but lack a calyptra. They fully replace the roots which are either entirely wanting (most spp. of *T.* sect. *Hemiphlebium*) or are rarely formed (*T. muscoides* [8])).

---

[1]) C. NÄGELI und H. LEITGEB, Entstehung und Wachsthum der Wurzeln, Beiträge zur wiss. Botan. von C. NÄGELI, 4. Heft, 1868, p. 73, see p. 128.

[2]) NÄGELI und LEITGEB, lc. 1868, p. 117.

[3]) F. O. BOWER, On the development and morphology of *Phylloglossum Drummondii*, Phil. Trans. 1885, no. 238 p. 665.

[4]) VELENOVSKÝ, op.c. p. 266.

[5]) HOFMEISTER, op.c. (f.n. p. 2), passim.

[6]) GOEBEL, op.c. (f.n. p. 3), p. 1145.

[7]) POTONIÉ, struck by the leaf-like characters of the stigmarian rootlets, suggested (H. POTONIÉ, Grundlinien der Pflanzen-morphologie im Lichte der Palaeontologie, Jena 1912, on p. 233), that all roots might be metamorph leaves. At present such a view is not likely to find supporters.

[8]) ENGLER & PRANTL l.c. p. 98.

It is not surprising therefore that LIGNIER in 1908 clearly expressed the view [1]) that roots are the oldest kind of rhizomes; it was however only several years later that this view was supported by direct fossil evidence, by the discovery of the *Psilophytinae*. In 1921 KIDSTON and LANG derived the roots from rhizome systems in *Rhynia* and in *Asteroxylon* [2]) and this derivation again was much strengthened by the discovery of *Asteroxylon elberfeldense* in which KRÄUSEL and WEYLAND observed root-like rhizome branches, differing from the other rhizomes of the plant [3]).

## § 8. The leaf (sterile leaf and sporophyll). — In the different groups

of *Algae* a differentiation between a special photosynthetic and usually flat part and a supporting and usually cylindrical part has evidently occurred several times. In the Pteridophytic sporophyte a successful differentiation of such a kind is nowadays generally assumed to have at least arisen twice, so that the leaves in this phylum are taken to be of a diphyletic origin. VELENOVSKÝ even distinguished between four leaf categories [4]), those of the *Articulatae, Lycopodiinae, Isoetinae* and *Filicinae*, but this view has not been confirmed since.

The two kinds of leaves are usually designated as microphylls (or Thursophyton leaves of BOWER [5])) and megaphylls (or cladode leaves of BOWER [6])). The latter are the leaves of the *Filicinae* and perhaps part of the leaves of the *Cladoxylinae*, while the leaves of all other Pteridophyte classes are usually taken together as microphylls.

In the present chapter the distinction will be accepted in this form, notwithstanding the fact that Lady ISABEL BROWNE in a recent paper [7]) questions the microphylly of the *Articulatae* and *Psilotinae*, mainly on account of the fact that some old forms, more or less related to these classes, had relatively large and forking leaves.

Though granting that our knowledge of the actual phylogeny is in-

---

[1]) O. LIGNIER, Essai sur l'évolution morphologique du règne végétal, C. R. du Congrès de Clermont Ferrand de l'Assoc. franç. 1908, reprinted in 1911 in: Bull. Soc. Linn. Normandie, 6th ser., 3, p. 35 and especially p. 41.

[2]) R. KIDSTON and W. H. LANG, On old red sandstone plants showing structure, from the Rhynie chert bed, Aberdeenshire, Part 4, Restorations of the vascular cryptogams, and dicussion of their bearing on the general morphology of the Pteridophyta and the origin of the organisation of land-plants, Trans. Roy. Soc. Edinb. 52, 1921, p. 831.

[3]) R. KRÄUSEL und H. WEYLAND, Beiträge zur Kenntnis der Devonflora II, Abh. d. Senckenberg. Naturf. Ges. 40, 1926, p. 115.

[4]) VELENOVSKÝ, op.c. p. 184.

[5]) BOWER, op.c. (f.n. p. 3), p. 552.

[6]) BOWER, op.c. p. 550.

[7]) Lady I. M. P. BROWNE, Some views on the morphology and phylogeny of the leafy vascular sporophyte, Botanical Review 1, 1935, p. 383, 427; see p. 440.

sufficient to settle the question directly, I believe that many facts point to the correctness of the usual view. The following general treatment of the leaves of *Psilophytinae, Lycopodiinae, Psilotinae* and *Articulatae* may show this, and moreover I might point to § 5 in which we saw that an important difference is found between the leaf categories so defined, in their relation to the stem branching.

## A. The microphyll.

**I. Phylogenetic origin.** According to our present knowledge the microphylls are the oldest. At least in *Baragwanathia* from the Silurian the microphyll is represented as clearly as in any later form, being a long, simple, lax leaf already ocurring in two forms, sterile and as a sporophyll with a reniform sporangium on its adaxial side.

According to a generally accepted and indeed very plausible supposition the microphyll has been developed as an emergence, or as BOWER expresses it, an enation. Whether this enation from the beginning was dorsiventral, with a different organization of the abaxial and the adaxial side, is unknown. In any case all known microphylls are always dorsiventral.

When they are formed as parts of a dorsiventral shoot, their right and left sides moreover may grow out differently, so that falcate and other asymmetrical forms ensue (*Selaginella* spp.).

In the oldest forms, the microphyll is quite simple without any incisions or any special parts. In many of the microphyllous groups it has remained so up to the present day, e.g. in *Lycopodium*. In other groups complications in shape have arisen; they will be dealt with separately for the sterile leaf and the sporophyll.

**II. The sterile microphyll.** Usually the microphyll is entire, without any lobing or dissection. In rare cases the margins may be provided with small teeth (*Selaginella*) or even may become in this way pinnatifid (*Selaginella Lyallii*, abnormal leaves, see BRUCHMANN [1])).

The microphyll however may become complicated by the occurrence of a dichotomous division. One forking occurs in *Protolepidodendron* and in many *Sphenophyllum* leaves, while other *Sph.* spp. have two or three forkings in their leaves; *Asterocalamites* has three or more. In *Pseudobornia* the leaf on the same basis has become flabellate, every long lobe being finely pinnately dissected.

In a number of related forms, known as the *Lycopodiinae ligulatae* and the *Isoetinae*, the leaf adaxially produces near its insertion a special laminar outgrowth, the ligule, usually inserted by a comparatively stout foot, the glossopodium, in a definite socket, the ligular pit. Other differentiations are rare. Stipules never occur and a sheath is seldom formed (*Isoetes*).

**III. The fertile microphyll.** In the sporophyll we meet

---

[1]) H. BRUCHMANN, Von den Vegetationsorganen der *Selaginella Lyallii* Spring, Flora 99, 1909, p. 436.

with the same differentiations as in the sterile leaves and of course also with the sporangia and eventually with their stalks.

The sporangia always or nearly always are borne on the adaxial side of the sporophyll, near its insertion on the stem: a single sporangium moreover being nearly always sessile (*Baragwanathia, Lycopodium, Selaginella*), seldom stalked (*Sphenophyllum angustifolium*). When a ligule is present in a sporophyll, the sporangium invariably is situated between axis and ligule.

When there are more sporangia on one sporophyll they are seldom sessile (*Sphenophyllum majus*), usually stalked, in which case the stalks may be branched (*Spenophyllum fertile*) and every branch may either bear one (*Sphenophyllum Dawsoni*) or two (*Sph. fertile*) or more sporangia, often four (*Cheirostrobus*). In the case of two or more sporangia to one stalk branch the latter tends to be terminally dilated, the sporangia pendent in a symmetrical distribution from the basal side of the shield-like dilation.

Cases in which the sporangia or their stalks are not inserted on the adaxial sporophyll side have been reported, but never satisfactorily proved to exist.

The best known case is that of *Sphenophyllum fertile* where the sporophyll, after SCOTT, consisted of two fertile storeys of two sporangium-bearing stalks each. Quite recently LECLERCQ [1]) however, was able to show that the sporophyll of *Sp. fertile* as in all other species possessed an abaxial sterile part and a number of fertile sporangium-bearing stalks.

A second case is that of *Pleuromeia* where an abaxial position of the sporangium was supposed to occur by SOLMS-LAUBACH [2]). However in the chapter by WALTON and ALSTON on *Lycopodiinae* in this Manual the statement has been called in question on good arguments [3]).

The union of one or more sporangia to a microphyll has often given rise to comment in literature. The botanists of the first half of the nineteenth century contented themselves with a comparison with the *Ophioglossum* sporophyll, as may be read in full in SOLMS-LAUBACH's paper on *Psilotum* [4]). By reduction this *Ophioglossum* sporophyll might have led to the

---

[1]) SUZ. LECLERCQ, Sur la structure réelle de *Sphenophyllostachys fertile* Scott, Proc. Int. Bot. Congr. I, 1936 p. 234.

[2]) H. GRAF ZU SOLMS-LAUBACH, Über das Genus *Pleuromeia*, Bot. Ztg, 57, 1899, p. 227, see p. 239.

[3]) HIRMER has tried to explain this supposed abaxial sporangium position (M. HIRMER, Rekonstruktion von *Pleuromeia Sternbergi* Corda, nebst Bemerkungen zur Morphologie der *Lycopodiales*, Palaeontographica 78 B, 1933, p. 47) by fancying an original form with a belt of sporangia all around the sporophyll; by reduction in the belt the normal case of *Lycopodium* as well as the case of *Pleuromeia* might have arisen.

[4]) H. Graf ZU SOLMS-LAUBACH, Der Aufbau des Stockes von *Psilotum triquetrum* und dessen Entwicklung aus der Brutknospe, Ann. d. Buitenzorg 4, 1884, p. 139.

condition of *Psilotum*, and by further reduction to the condition of *Lycopodium*. Even if such a "backward" view were still considered to be satisfactory, we should have to object that it implies a comparison between megaphylls and microphylls.

The question became especially interesting when the *Psilophytinae* seemed to prove that sporangia originally had been only borne on stems, terminally. With the hope of solving this problem, BOWER enounced his collocation theory [1]), claiming that for every sporophyll a union must have taken place between a Thursophyton-leaf and a Hostimella-branch, i.e. a stem with a terminal sporangium.

This conception hardly needs a detailed discussion. The improbability that, in plants where axillary branching is unknown, one single branch should arise exactly above every single leaf, without ever even betraying itself by teratological cases, is such that we must discard the idea altogether.

Moreover the discovery of *Baragwanathia* and *Yarravia* makes all such explanations superfluous. The original land flora did not only consist of leafless plants with terminal sporangia; there is no objection to supposing that the Pteridophyte sporangia originally were produced all over the plant, as in the *Laminariaceae*, where the sporangia are formed on the stalk and holdfast as well as on the frond.

The terminal group of sporangia in *Yarravia* and the single terminal sporangia of *Rhynia* may both represent original arrangements, but the leaf-borne sporangia of *Baragwanathia* have the same claim to originality.

The same view, that every sporophyll should owe its origin to the union of a leaf and a branch, has been upheld with much more stress and on the basis of quite different fossil evidence in the case of the *Articulatae*. Especially in this plant group do we meet with the stalked sporangia; and these stalks, often termed sporangiophores, have been considered as of a cauline nature by several botanists.

Not by all however. The foliar nature of the *Equisetum* sporophyll has been strongly emphasized by one of the best investigators of the *Equiseta*, MILDE [2]). From his extensive and excellent observations on the transitional forms between the sheath leaves and the annulus teeth, and between these teeth and the sporophylls, MILDE arrived at the firm conviction that sheath leaf, annulus tooth and sporophyll are all homologous organs. And as the *Equisetum* sporophyll consists of one single sporangiophore, the latter must be of a foliar nature too.

Starting independently from the same facts GOEBEL [3]) also came to the conclusion that the homology of leaf and sporangiophore in *Equisetum* is

---

[1]) BOWER, op.c. (f.n. p. 3), p. 598.

[2]) J. MILDE, Monographia Equisetorum, Nova Acta 32, 2, 1867; especially p. 164

[3]) GOEBEL, op.c. (f.n. p. 3), p. 1108.

so obvious that the assumption of an axial nature for the latter is untenable. As he rightly remarks, what can be stated with certainty in living plants is not to be doubted on the basis of much less known fossils. This view of GOEBEL is the more remarkable as he belonged to the zealous defenders of the axial nature of the *Psilotum* sporophyll.

The supporters of the axial nature of the sporangiophores are especially BOWER [1]) and BROWNE [2]); the arguments of both authors are entirely taken from the domain of phyllotaxis. We shall therefore postpone their discussion until we come to the microphyll phyllotaxis, sub IV; here I will only remark that our conclusion will be that the facts of phyllotaxis are not such as to require an axial sporangiophore nature. In the present paper all these complex sporophyll structures of the *Articulatae* will therefore be taken as real sporophylls; for *Psilotum* see § 11 sub 1.

The sporophyll may become forked or by repeated dichotomy more-lobed in the same way as the sterile microphyll; this occurs in *Psilotum, Sphenophyllum* and in *Cheirostrobus*; in the latter case the sporophyll is remarkably divided into three equal collateral parts, a curious deviation from the normal two, four or eight of dichotomous branching.

The collateral dedoublement of the sporophyll may equally well pertain to the sporangiophore; we may therefore have two or more sporangiophores as an upper storey over two or more sterile lobes (*Cheirostrobus*). In literature this division into two storeys is usually designated as a serial division, following in this respect the example of ZEILLER [3]). See for instance SCOTT [4]) and especially HIRMER [5]). Another explanation is that the under sterile and the upper fertile lobes should represent different phyllomes; a view especially advocated by TROLL [6]).

As far as I can see either explanation is equally unnecessary, as we may simply consider it as an elaboration of the sporophyll (the under lobe) with its sporangium (the upper lobe). This view is not at all new, as it has been discussed by many authors, see e.g. SEWARD [7]); but it has mostly been discarded on account of the complex organization and structure of the sporangiophore, and of its likeness to the *Equisetum* sporophyll.

For *Equisetum* the above view does not present any difficulty, as we may

---

[1]) F. O. BOWER, The origin of a land-flora, London 1908, see p. 383.

[2]) Lady I. M. P. BROWNE, A new theory of the morphology of the Calamarian cone, Annals of Bot. 41, 1927, p. 301.

[3]) R. ZEILLER, Étude sur la constitution de l'appareil fructificateur des *Sphénophyllum*, Mém. Soc. géol. de France, Paléont. 1893, p. 3.

[4]) D. H. SCOTT, Studies in fossil botany, 3rd ed. I, 1920, see p. 99.

[5]) M. HIRMER, Bemerkungen zur Theorie der serialen Spaltung der Blätter. Eine Erwiderung an Herrn W. TROLL, Berichte D. bot. Ges. 51, 1933, p. 127.

[6]) W. TROLL, Zur Deutung des Blütenbaues fossiler *Articulatales*, Ibid. p. 21.

[7]) A. C. SEWARD, Fossil plants, I, Cambridge 1898, see p. 405.

easily conceive how the sporophyll in these plants may have been reduced to the single sporangiophore [1]).

Finally the sporophyll of microphyllous plants may develop some special parts besides the sporangium and sporangiophore, in connection with the formation of a strobilus or even with the formation of seeds.

In some forms (*Selaginella* [2]), *Lepidostrobus, Spencerites, Cheirostrobus*) at the transition between the horizontal and the vertical tegumentary part of the sterile lobes a downward extending rim or knob may be present, the function of which is obviously the ensurance of an accurate fitting and closing of the strobilus parts in the bud condition.

In other forms where the seed habit was attained, a membrane was formed from the adaxial side of the sporophyll, more or less enveloping the megasporangium, and even sometimes (*Lepidocarpon*) in the microsporophyll, enveloping the microsporangium likewise. These membranes usually are designated as indusia.

**IV. Phyllotaxis.** Probably no fact has contributed more to the inclusion of microphyls and megaphylls in one single category of leaves than the regular phyllotaxis both often display. This regular arrangement must have been present very early during the phylogeny, as in *Baragwanathia* the parastichies, though not quite regular, are yet very conspicuous [3]). It is however quite possible that in other early or in still earlier forms the microphylls have been distributed irregularly, like many hairs.

The study of recent microphyllous Pteridophytes clearly reveals the fact that their leaves display exactly the same phyllotactical phenomena as the megaphylls in the *Filicinae* and in Spermophytes; evidently the so-called law of HOFMEISTER governs the phenomena in both. Yet there is this difference, that the phyllotactical patterns of the so-called main series whose parastichies occur in numbers belonging to the series 1, 1, 2, 3, 5 etc., are not nearly so frequent in microphyllous plants as e.g. in Conifers. This is due to the action of at least two disturbing factors, namely the dichotomy of the stem and the false whorl formation.

Where these factors are absent, systems of the main series may be found in perfect regularity, for instance in *Isoetes* [4]).

---

[1]) The strobili of *Calamostachys* and its allies are sometimes believed to display analogous phenomena of serial splitting; for literature see BROWNE l.c. 1927, p. 310 and especially HIRMER 1933 l.c.

In the discussion on phyllotaxis I hope to make clear why this view is not to be accepted, as the superposed parts in these strobili belong to different phyllomes.

[2]) M. G. SYKES and W. STILES, The cones of the genus *Selaginella*, Ann. of Bot. 24, 1910, p. 523.

[3]) LANG and COOKSON, l.c. (f.n. p. 27), fig. 6, 12, 14.

[4]) JOHANNA LIEBIG, Ergänzungen zur Entwicklungsgeschichte von *Isoetes lacustre* L., Flora 125, 1931, p. 321, see p. 335.

In a dichotomizing stem, however, only low systems of the main series, with low parastichy numbers, may occur regularly, as the following reasoning may show. In a vegetative cone where the young leaf primordia (or, according to my theory of phyllotaxis [1]), the dispersion circles) at the time of the determination of the next higher ones have a relative bulk (expressed in the stem circumference as unit) of between 0.50 and 0.58, no other phyllotactical systems are possible but a system 1 + 2, a low member of the main series.

From such a system higher members of the main series may be elaborated by a gradually diminishing bulk relation [2]), and this is what occurs in *Isoetes*. However as soon as a stem dichotomizes, the new systems of the two shanks are only partly the continuation of that of the podium; they are completed at the sides facing each other on a new and more or less horizontal base. When in the shanks the bulk relation happens to fall between 0.50 and 0.58, the new systems must be of the main series. When the bulk relation is smaller, the new systems can only belong to the main series when the supplementory part of the system is erected on a basis of the required obliquity; this not being the case it is not difficult to prove that the new system will be such a one in which the two sets of determining parastichies have a tendency to differ less in number than they do in the main series.

This proof may be given as follows. Let *a* and *b* in fig. 7*A* be two leaf centres,

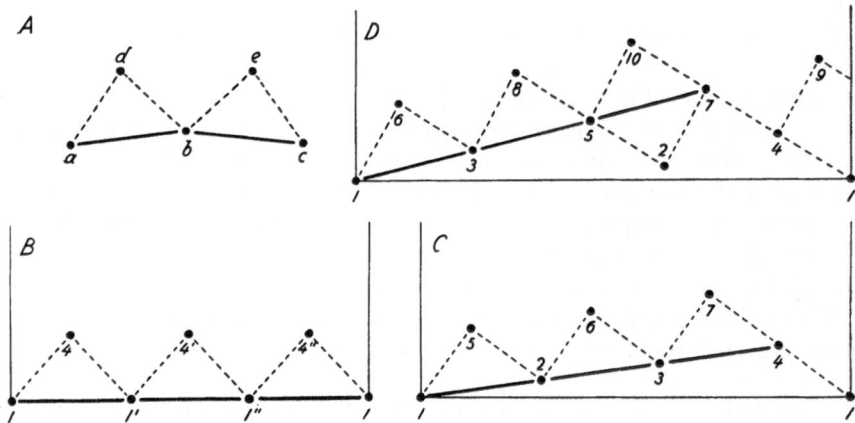

FIG. 7. *A* = part of a phyllotactical system. "Basal line" fully drawn, lines from determining to determined leaves dotted. *B,C,D* = regular systems 3 + 3, 3 + 4 and 3 + 5.

determining together the place of a third leaf *d*. We know that *b* at the other side cooperates with a third leaf *c* in determining the place of of a fifth leaf *e*, etc. Now

[1]) J. C. SCHOUTE, Beiträge zur Blattstellungstheorie. I, Die Theorie, Rec. trav. bot. néerl. 10, 1913, p. 153.

[2]) SCHOUTE 1913, l.c. p. 247.

we may call the line passing through the centres $a$, $b$ and $c$ a basal line; it is clear that we may extend such a line around the stem. For a realization of a whorled phyllotaxis it is required that the basal line on being continued from $c$ to the right, again abuts on $a$. (7B) When it leads to $d$, the system is of the form n + (n + 1) as in 7C, etc. In fig. 7B–D three regular systems have been drawn, 3 + 3, 3 + 4 and 3 + 5, each with its basal line.

Supposing now that in each of these three systems a certain part is taken away and replaced by a new part, built up on a horizontal base, it is obvious that all three will be likely to change into another system, but it is at the same time clear that in the new systems the basal lines will be inclined to be more or less horizontal, or in other terms will have determining parastichies whose numbers are not very different.

This theoretical conclusion is borne out by the facts to a large extent. Dichotomizing stems occur in:

*Asteroxylon.* Little of its phyllotaxis is known, but the only instance known to me in which it is possible to determine it from a photograph [1]), is that of a system 7 + 8, a very rare system for non-dichotomizing plants.

*Psilotum triquetrum.* The main axis of the aerial shoots has an irregular phyllotaxis, the thin terminal branches have systems 1 + 2.

*Lycopodium Selago.* According to AL. BRAUN [2]) very young plants have $^3/_8$. In six plantlets grown from bulbils I observed low systems of the main series. Adult specimens on the contrary are stated by BRAUN [3]) to have the systems 3 + 4, 4 + 4, 4 + 5, 5 + 5 and 5 + 6, all either systems of the form $^2/_7$, $^2/_9$, $^2/_{11}$ or whorled systems.

*Lycopodium* spp. In adult specimens the conditions usually are as in *L. Selago*: for *L. clavatum* BRAUN [3]) mentions all similar systems from 4 + 4 up to 8 + 9.

My own observations in some species showed the following results. No other systems were observed but n + n, n + (n + 1) and n + (n + 2), the last always being very rare. Of the former two the whorled systems often prevailed, especially in the thinner branches. Often in a shank the system was subject to one or more changes; in these cases there was no predilection to be found for changes towards a whorled condition [4]). A discussion whether the whorled or the spiral condition is

---

[1]) On Pl. 5, fig. 25, in KIDSTON and LANG, l.c. 1921 (f.n. p. 30).

[2]) AL. BRAUN 1831, l.c. (f.n. p. 26), p. 277.

[3]) AL. BRAUN 1831, l.c. p. 338.

[4]) My own observations pertained to *L. carinatum, dichotomum, Hippuris* and *Phlegmaria*. From one or more shoots the phyllotaxis was determined in all branches, all changes in system of the shanks being recorded; the results are plotted in the two following tables:

| Species | Number of observed shanks | Phyllotactical systems | | | Lowest system | Highest system |
|---------|------|-------|-----------|-----------|--------|--------|
|         |      | n+n | n+ (n+1) | n+ (n+2) |        |        |
| carinatum . . | 89 | 51 | 30 | — | 2+2 | 4+4 |
| dichotomum . | 31 | 40 | 33 | 2 | 5+5 | 13+13 |
| Hippuris . . . | 27 | 28 | 16 | 1 | 4+5 | 7+8 |
| Phlegmaria . . | 51 | 40 | 11 | — | 2+2 | 4+5 |

primitive in *Lycopodium* as may be found in BRONGNIART [1]) and in SPRING [2]), in view of such facts seems to be without any foundation; both the spiral and the whorled systems are incidental consequences of space relations, without any governing principles of their own. Both alike are original patterns and the whorled systems, in striking contrast with nearly all other whorls, are true whorls.

*Selaginella.* In the common dorsiventral species the phyllotaxis, though subject to horizontal shifts in connection with the dorsiventrality and often to longitudinal shifts too, is essentially decussate. In some radial species conditions are quite different. BRAUN writes about them [3]): "Nur wenige *Selaginellen* zeigen complicirtere Blattstellungsverhältnisse, .... *Sel. rupestris* zeigt an den vegetativen Sprossen $3/8$ oder $5/13$, in den Aehren dagegen das gewöhnliche vierzeilige Verhalten; *Sel. spinulosa* dagegen zeigt in allen Theilen complicirte Spiralstellungen oder complicirte Abwechslungsverhältnisse von Quirlen und bewegt sich dabei in einem nicht minder weiten Spielraume als manche *Lycopodien*".

*Lepidodendron.* Our knowledge of the phyllotaxis of the *Lepidodendron* stems is only scanty; a determination of the parastichies round the stem is usually impossible as loose parts of the stem surface are the common material.

In view of this difficulty STUR devised a method [4]) of determining a phyllotactical pattern from such loose surface parts. Though he thought he had obtained good results, his numerous observations are all worthless, as their basis, the determination of the orthostichies, was necessarily unsatisfactory. The only reason why he did not become aware of the unreliability was that he started from the preconceived idea that only patterns of the main series or its jugate forms were possible.

The only satisfactory determinations I know of have been made by DICKSON on 15 specimens which he was fortunate enough to obtain with the parastichies round the stem [5]). Three of these objects showed patterns of the main series: 8 + 13 + 21; 13 + 21 + 34; 55 + 89; five cases were conjugated systems of the same, namely one bijugate 16 + 26, two trijugate 15 + 24, 15 + 24; one quinquejugate 15 + 25, one septemjugate 35 + 56 + 91; one was of the first accessory series 18 + 29 + 47, two of the second accessory series, both 14 + 23, and four of three further anomalous series 6 + 11 + 17; 12 + 19 + 31; 13 + 23; 36 + 59.

As nothing is known about the lower part of the observed stem pieces these may have been main stems which had not yet forked. In that case we might expect patterns of the main series and perhaps, as most of the cases seem to be, irregular degenerations of that series. There was however one amongst the 15 specimens

Changes in phyllotactical system from

| Species | spiral to next whorl | whorl to next spiral | whorl to next whorl | spiral to next spiral | spiral to $n+(n+2)$ | $n+(n+2)$ to next spiral | $n+(n+2)$ to next whorl |
|---|---|---|---|---|---|---|---|
| carinatum | 7 | 1 | — | — | — | — | — |
| dichotomum | 18 | 21 | 2 | 1 | 1 | — | 1 |
| Hippuris | 9 | 6 | 4 | — | — | 1 | — |
| Phlegmaria | 2 | 3 | — | — | — | — | — |
| Total | 36 | 31 | 6 | 1 | 1 | 1 | 1 |

[1]) AD. BRONGNIART, op.c. II, (f.n. p. 9), p. 10.

[2]) A. SPRING, Monographie de la famille des Lycopodiacées, I, Mém. de l'acad. roy. de Belgique 15, 1841; do do II, ibid. 24, 1848; see II, p. 303.

[3]) AL. BRAUN 1874, l.c. (f.n. p. 16), p. 61.

[4]) STUR, l.c. (f.n. p. 19), p. 342 (236)–368 (262).

[5]) AL. DICKSON, On the phyllotaxis of *Lepidodendron*, Transactions of the Botanical Society 11, Edinburgh, 1873, p. 145.

which bore a lateral branch; this branch had a pattern 8 + 13. The latter case evidently is against our theoretical view.

*Sigillaria.* Here again we have numerous but worthless observations by STUR [1]) and moreover numerous determinations by NAUMANN [2]) and by GOLDENBERG [3]) who both tried a variation of STUR's method by the so-called "quincunx" method; a method which is as bad as the other. This judgment has already been enounced, in a very polite but nevertheless determined way, by AL. BRAUN [4]) to whom Prof. BRONN had sent NAUMANN's paper to report upon. For us BRAUN's paper is note-worthly as he added a positive contribution to his criticism, an observation on a stem piece of *S. Cortei* or *S. mamillaris*, in which the pattern successively changed from 42 + 43 into 43 + 43 and further into 43 + 44; being high systems of the kind mentioned above.

We might now ask reversely whether there are any plants with the same peculiar systems n + n and n + (n + 1) without dichotomous branching. As far as I know; such plants are not on record. The only case perhaps might have been given by *Asterochlaena laxa* where BERTRAND [5]) determined the phyllotaxis in three stem pieces as 10 + 11, 10 + 10 and 11 + 11. The mode of branching is unknown but as the longest stem piece on record is not yet twice its diameter in length, the supposition that the stem branched dichotomously as in the related *Coenopteridineae* is not very risky.

So we see, that the facts, as far as available, on the whole point towards the correctness of the theoretical view given above about the relation be-tween dichotomic branching and phyllotaxis. The facts related in § 5, p. 17 sub *f* tend to support the same view. As all lateral bulbils treated there may be compared to shanks in very unequal dichotomies, we may expect that the phyllotaxis of the podium will be continued on the large shank with only a small interruption at the side of the small shank; over this inter-ruption the system will soon close up again, and may do so in the same pattern as was present in the podium. In such a way it becomes clear why at the place of insertion of the small shank only one or two leaves of the system are lacking and, moreover, why a number of bulbils tends to disturb the phyllotaxis entirely.

As the other disturbing factor for phyllotaxis I mentioned the false whorl formation. Like the leaves in so many megaphyllous plants, the microphylls too may become secondarily connected so as to form whorls. This process has to be taken as an ontogenetic one, as it is brought about

---

[1]) STUR l.c. p. 400 (294).

[2]) C. F. NAUMANN, Über den Quincunx, als Gesetz der Blattstellung bei *Sigillaria* und *Lepidodendrum*, Neues Jahrbuch für Miner., Geogn., Geol. und Petrefakten-kunde, herausg. von VON LEONHARD und BRONN, 1842, p. 410; C. F. NAUMANN, Ueber den Quincunx als Grundgesetz der Blattstellung vieler Pflanzen, Dresden und Leipzig 1845; a third paper in Poggendorff' Annalen is alluded to by the author but not quoted.

[3]) F. GOLDENBERG, l.c. (f.n. p. 12), Heft 2 p. 2.

[4]) AL. BRAUN, Über die Blattstellung der Gewächse, mit Beziehung auf die fos-silen Formen und die vorangehende Abhandlung, same place as the first paper by NAUMANN, p. 418.

[5]) P. BERTRAND, Structure des stipes d'*Asterochlaena laxa* Stenzel, Mém. Soc. Géol. du Nord, 7, 1911, p. 5.

by metatopies, i.e. by shifts and one-sided extensions of the developing primordia or even of the still invisible preliminary stages.

In the *Articulatae* the phenomenon occurs on a large scale, and is of taxonomic interest. It is attended by a strong lateral binding of the neighbouring leaves in the whorl, as becomes obvious from the strong torsion to be observed in the rare cases of "Zwangsdrehungen" [1]). Two successive whorls have a certain tendency to isomery, though, especially in polymerous shoots, the numbers are hardly ever identical.

Successive whorls usually show a certain spatial relation of the members of the one to those of the other. In *Equisetum* and in many other *Articulatae* this relation amounts to an alternation. In strictly isomerous whorls the alternation is perfectly realized; in heteromerous whorls the rule is adhered to as far as possible. When for instance a whorl of 10 members follows on a whorl of 11, nine of the higher members alternate with the lower whorl, the tenth lying exactly over a member of the lower whorl with an unusually large space at both sides. Even in the torsion cases of *Equisetum* we may observe the same imperfect mode of alternation of successive windings of the spiral.

In *Asterocalamites* the relation is quite different; here superposition is the rule. Yet we have an absolutely analogous solution of the difficulty offered by the numerous deviations from strict isomery, especially in the whorls with very high numbers of members; in such cases nearly all leaves are placed exactly over leaves of the lower whorl, but for a number of places, where one or a few members occupy an alternate or an intermediate position.  ·

It is exactly from these deviations from the rule that the false whorl character may be concluded, as no true whorls could ever show such phenomena; every departure from the isomerous condition in true whorls must change the whole character of the system.

Alternation and superposition are not the only spatial relations of microphyll whorls; in the case of junction of whorls consisting of different organ categories, as may be observed in some strobili, more complicated relations are sometimes realized. We find this in *Calamostachys*, where two kinds of phyllomes, sporophylls and sterile scales [2]), both arranged in whorls, follow in turn. The numbers in the successive whorls have no tendency to isomery here, but the sporophyll whorls are nearly isomerous among each other. The bract whorls, lying between the sporophyll whorls, have about twice the number of scales each, or in other cases their normal

---

[1]) Already figured for *Equisetum fluviatile* by J. P. VAUCHER, in: Monographie des Prêles, Genève 1822.

[2]) Usually called bracts, a very inappropriate name, as the term bract has a strictly defined meaning in botanical morphology, namely of a subtending phyllome, especially in inflorescences.

number is one and a half that of the sporophyll whorls (*C. magnae crucis*). These relations may be paralleled by the junction of a stamen whorl of ten to a corolla of five, or by the junction of a dimerous gynoecium to a pentamerous androecium.

Usually in these cases the sporophyll whorls in themselves are superposed upon each other, whereas the scale whorls alternate amongst each other, in such a way that when half of the members of one scale whorl alternate with those of the lower sporophyll whorl, its other members being superposed to the same sporophyll whorl, the next scale whorl takes an intermediate position, no scale being alternate to nor superposed to the sporophylls.

All whorls in these remarkable cases only keep true to their normal number in a general way, not in detail; deviations occur and bring about unusual spatial relations locally. It is exactly from this lack of consistency that we conclude that we must abandon the old view of SCOTT, HICKLING and BROWNE, in later papers rejected by both SCOTT and BROWNE [1]), but defended again emphatically by HIRMER [2]), that "bracts" and "sporangiophores" should represent dorsal and ventral lobes of a phyllome, or as HIRMER calls it a "Sporophyll-Einheit". HIRMER tries to argue away the importance of the abnormalities by distinguishing between the "prinzipielle Organstellung" and the observed arrangement, affected by disturbances; a set of natural suppositions about the real processes determining the organogeny, might, however, prove difficult to correlate with such a view.

In *Sphenophyllum* and in *Cheirostrobus*, which are often compared with *Calamostachys* in this respect, conditions are quite different. Here a sporophyll really may consist of a dorsal and a ventral lobe, as dealt with above p. 34, and in such places where the successive whorls are heteromerous these two storeys always correspond.

Finally we may perhaps cite as a third factor disturbing phyllotaxis, ridge formation, the arrangement of leaves in false orthostichies.

This phenomenon is strikingly developed in the *Sigillariaceae*, but owing to the absence of sufficient data we can only point out its importance.

In phyllotactical respects it reminds one of the ridge formation in *Cactaceae*. Whenever the pattern changes, existing orthostichies may be discontinued or new ones may be intercalated. This is not only to be observed in *Sigillaria* in the strobiliferous zones dealt with above in § 5 sub *f*, p. 20, but also in the basal parts of the stout stems, see for instance the figure of *S. laevigata* as given by ZEILLER [3]).

---

[1]) BROWNE, l.c. 1927 (f.n. p. 34), p. 304.
[2]) HIRMER 1927, op. c. p. 471, 472.
[3]) R. ZEILLER, op.c. 1886 (f.n. p. 12). see Pl. 78, fig. 3.

**V. Kinds of microphyll plastics.** Of course the sterile and the fertile microphylls are already of different plastics; all Pteridophytes, only part of the *Psilophytinae* excepted, show this difference. Yet the organization of the two kinds of leaves may be the same, but for the formation of sporangia in the sporophylls (*Lycopodium Selago*). In other plants the sterile and fertile leaves may be different in most respects (*Equisetum*).

Some Pteridophytes however have two kinds of sterile microphyll plastics; some outstanding cases will be enumerated here.

*Lycopodium Selago.* In addition to the normal assimilating leaves there are developed some very thick and fleshy leaves in the bulbils, containing reserve materials.

*L. volubile* has a marked difference between the large falcate lateral leaves, and the small linear upper and under leaves.

*Selaginella* subgen. *Heterophyllum.* The small upper leaves and the large under leaves are different in shape[1]); the former turning their morphological under side towards the sky, the latter their upper side, their anatomical organization too is often different (*S. Martensii*).

*Lepidodendron* and *Sigillaria.* If the view on the stigmarian rhizome, to be defended in § 11, is to be confirmed, the stigmarian "rootlets" are nothing but metamorphosed leaves, widely diverging from the photosynthetic normal leaves by their root-like apical growth, their root-like anatomy and their root-function.

*Sphenophyllum verticillatum* and some other *Sph.* spp. have finely dissected leaves in the lower parts of the plant, and simple cuneate leaves in the higher parts.

*Equisetum.* Beyond the normal sheath leaves and the sporophylls, *Equisetum* has a third form of leaves, the teeth of the annulus, being transitional forms between the other two. Usually they are sterile, but in the *Equiseta cryptopora* they mostly bear a sporangium at the inner side of each tooth [2]).

Amongst the fertile microphylls dimorphism does not seem to have been developed, though of course, especially in monosporangeous sporophylls, mega- and microsporophylls may be distinguished (*Selaginella, Isoetes*).

---

[1]) SPRING (l.c. f.n. p. 38, II, p. 300) develops a remarkable theory on the polarity of the *Selaginella* leaves, trying to demonstrate that every form property of an upper leaf is the polar contrast of the same property of an under leaf. Without any doubt his views are too sketchy; moreover the exposition of his views is far from clear.

The fact is that the right and left sides of any leaf, developing in different positions with regard to the zygomorphy factors in the stem, differ in many respects; so most leaves are either unequal-sided or falcate or both at the same time. The position of the broad halves and the curvature of their midribs may be opposed in the two kinds of leaves or they may be similar; fixed rules such as SPRING claimed not being present.

[2]) MILDE, lc. (f.n. p. 33), p. 167.

## B. The megaphyll.

**I. Phylogenetic origin.** As the first author claiming the development of the megaphyll from a specialized branch system LIGNIER is usually cited [1]). The same view however has already been given expressively by HOFMEISTER in 1851 [2]) namely that the fern frond was a branch system, its paleae being its leaves.

In 1916 HALLE was inclined to adopt LIGNIER's view on account of his observation of Lower Devonian forms [3]), especially of *Psilophyton Goldschmidtii*, in which the lateral branch systems being already dorsiventral seemed an approach to megaphylls, while the main stem and the basal parts of the branches bore numerous small spines, the microphylls.

The subsequent discovery of *Asteroxylon Mackiei* with its specialized lateral branch systems and especially that of *A. elberfeldense* where these systems were very clear, and, moreover, appeared to have circinate tips in the young condition, strongly emphasized the correctness of the general view, and the fact that in *Asteroxylon* the incipient megaphyll did not yet display dorsiventrality only tends to corroborate it.

For it is clear that the original branch system may have been radially symmetrical, and that the acquisition of dorsiventrality factors is to be considered as an adaptation to the photosynthetic function.

Subsequent authors mostly accepted the same view: SCOTT [4]), KIDSTON and LANG [5]), HIRMER [6]); only GOEBEL and ZIMMERMANN did not. GOEBEL because he assumed the homology of all Pteridophytic leaves on account of the results of comparative morphology of recent forms [7]), and ZIMMERMANN because he derives both forms of leaves in a hypothetical way from a phylloid-complex, i.e. a sterile branch system; moreover he ascribes microphylls to the Conifers and megaphylls to *Sphenophyllum* [8]).

The elaboration of the fern leaf from the *Asteroxylon* branch system according to all authors has been accomplished by the following processes: the acquisition of dorsiventrality factors; the gradually perfected arrangement of all branches in a single plane, in connection with the dorsiventrality; the development of a lamina by a flattening out of the branches and by "webbing"; finally by the gradual change of the original dichotomous mode of branching into a monopodial branching habit.

---

[1]) LIGNIER, l.c. (f.n. p. 30).

[2]) HOFMEISTER, op.c. (f.n. p. 2), p. 87.

[3]) T. G. HALLE, Lower devonian plants from Röragen in Norway, K. Svenska Vet. Akad. Handl. 57, 1916, no. 1.

[4]) SCOTT, op.c. (f.n. p. 34), p. 414.

[5]) KIDSTON and LANG, l.c. (f.n. p. 30), p. 673.

[6]) HIRMER 1927, op.c. p. 688.

[7]) K. VON GOEBEL, Organographie der Pflanzen, 2nd ed. II, Jena, 1915–'18, see on p. 914; and op.c. 1930, p. 1045.

[8]) ZIMMERMANN op.c. (f.n. p. 26), p. 67.

The last mentioned process is even lacking in many recent fern leaves (*Rhipidopteris, Schizaea*) with their dichotomously branching veins; leaves without a lamina are to be met with in many *Coenopteridineae*[1]); leaves which though dorsiventral do not yet bear their different branches in one single plane occur in the same group, *Stauropteris* being the best example; a full discussion of these phenomena may be found in HIRMER[1]).

HIRMER moreover expresses the view that serial splitting may also play a rôle in the megaphyll, namely in the *Ophioglossum* sporophyll. There is however no reason for such an assumption. The adaxial position of the fertile part or parts in the *Ophioglossaceae* is usually explained by the assumption of fusion of the lowest pair of pinnae, as has already been enounced by ROEPER and defended by METTENIUS long ago; for quotations see TROLL[2]). It was CHRYSLER[3]) especially who brought forward good arguments in favour of this view.

TROLL rightly remarks that even a fusion of the two lowest pinnae is not to be assumed in all cases, as the two rows of lateral pinnae of the common leaf may be united by a basal curve at the adaxial side of the leaf, and a single pinna in the middle of this curve exactly assumes the required position[4]).

The megaphylls of the *Filicinae*, even when formed in dorsiventral shoots, do not seem to have ever acquired a notable asymmetry; at least no examples have come to my knowledge.

**II. The sterile megaphyll.** Though the sterile leaf may consist of a lamina only, a more or less clearly defined petiole is usually present.

The lamina may sometimes be simple (*Cyathea sinuata, Salvinia*), but in most cases it is compound, often being very finely dissected. It is further customary to call the leaflets pinnae, and the terete parts rachis. Amongst the compound leaves the pinnate up to multipinnate forms strongly prevail; of other types the following may be instanced: the palmate leaf (*Hemionitis palmata*), the pedate leaf (*Adiantum pedatum*), the dichotomously dissected leaf (*Schizaea*).

For details of the organization of the lamina and its venation the reader

---

[1]) HIRMER 1927, op.c. p. 688.

[2]) W. TROLL, Über die Blattbildung der Ophioglossaceen, insbesondere von *Ophioglossum*, Planta 19, 1933, p. 547.

[3]) M. A. CHRYSLER, The nature of the fertile spike in the *Ophioglossaceae*, Annals of Bot. 24, 1910, p. 1.

[4]) In his masterly doctor's thesis EICHLER describes (A. W. EICHLER, Zur Entwickelungsgeschichte des Blattes mit besonderer Berücksichtigung der Nebenblatt-Bildungen, Inaug. Diss. Marburg 1861) the same condition of two lateral rows of leaf-members, uniting basally by an adaxial "Transversalzone", such as he observed in plants with peltate leaves and also in many others like *Oxalis, Lupinus* and *Geranium* (op.c. p. 13).

may be referred to the works of GOEBEL [1]) and BOWER [2]); only a few points may be mentioned here.

In the first place the remarkable apical growth of the lamina and its pinnae may be mentioned, and the correlated circinate vernation of these parts, characters which do not occur in microphylls and which may be due to the phylogenetic origin of the megaphyll. This apical growth may even be unlimited (*Lygodium*), or it may act periodically, in which case the apical meristems may be protected by special leaflets, so that a kind of bud is formed in the leaf organization (*Gleichenia*).

In the second place a remark on the venation. In many cases every vein may be considered to represent a branch of the original branching system. This is only possible however for the "open" venation without commissural veins; in the "closed" venation and its elaboration, the reticulate venation, the veins obviously have lost this homology with branches, or in other terms, the morphogenetic forces inducing the veins are no longer modified branch forming factors.

Besides a blade and a petiole, in some forms stipules (*Marattiaceae*) or a sheath (*Ophioglossum*) may be present in the leaf; a ligule is never formed. The last parts to be mentioned are the aphlebiae, leaflets of quite different plastics from the other, normal leaflets. They often have a protective function and formerly were usually designated as adventitious pinnae. In accordance with the theory of the diphyletic origin of the Pteridophyte leaf they have later been regarded as microphylls, borne on the megaphyll (SCOTT [3]), HIRMER [4])). No doubt however BOWER is right in contending [5]) that though some of these aphlebiae may indeed be microphylls, most of these formations are only specially differentiated true leaflets of the megaphyll.

**III. The fertile megaphyll.** The demarcation line between the sterile and the fertile megaphyll is not very sharp. Not only in many cases are the plastics of both absolutely the same but for the production of sporangia in the sporophyll, but also many megaphylls are partly sterile, partly fertile, with all possible size relations between the two parts.

From the theoretical conception of the phylogenetic origin of the megaphyll it is clear that a marginal position of the sporangia must be held as primitive. The usual position of the sporangia however is on the leaf underside; this position is connected with the marginal position by so many transitional stages, that a phylogenetic shift of the position is easily assumed [6]); moreover marginal sporangia and sori even occur in recent

---

[1]) GOEBEL 1930, op.c. p. 1171–1209.
[2]) BOWER 1935, op.c. p. 302–316.
[3]) SCOTT, op.c. p. 414.
[4]) HIRMER, 1927, op.c. p. 690.
[5]) BOWER 1935, op.c. 628.
[6]) Compare BOWER 1935, op.c. p. 371.

ferns and in rare cases even the upper leaf surface may bear sori (*Poly-stichum anomalum* [1])).

The sporangia in relatively rare cases are produced singly; in most cases groups of sporangia are formed, called sori, or if the sporangia become concrescent, synangia. The sori are often protected by an indusium, an enation. For the manifold modes of organization and of position the works of GOEBEL and BOWER and, also, the taxonomic literature may be consulted.

In the case of the development of mega- and microsporangia the sori may also be differentiated into mega- and microsori (*Salviniaceae*) or they may be bisexual, mixed (*Marsiliaceae*); mega- and microsporophylls, so common in the seed plants, are unknown in the *Filicinae*.

The sporophyll of heterosporic *Filicinae* may develop fruit-like bodies, the sporocarps (*Marsiliaceae*), each consisting of a leaflet which by folding towards the adaxial side and concrescence of the free margins is converted into a hollow sac, like the ovary of a follicle, in the interior of which a number of indusia are produced and which by its peculiar differentiation strikingly reminds one of seed coats or fruit walls [2]).

**IV. Phyllotaxis.** Fern leaves evidently are arranged according to the same phyllotactical laws as those of Angiosperms. The systems of the main series strongly predominate; in dorsiventral rhizomes systems 1 + 1 are very frequent, in stout erect stems high spiral patterns prevail. Whorl formation rarely occurs: regularly in *Salvinia* (trimerous alternating whorls), in many *Psaronii* [3]) and side by side with spiral patterns in some tree ferns [4]).

The fact that the phyllotactical laws of megaphylls and of microphylls are the same, notwithstanding the diphyletic origin of the two kinds of leaves, may represent a parallel adaptation. It is however also possible, that the phyllotaxis of megaphylls is derived from the cladotaxis of the original smaller dichotomy shanks, and that in some ancient form with both microphylls and lateral branch systems both have been arranged independently according to the same laws.

**V. Kinds of megaphyll plastics.** All *Filicinae* possessing sterile and fertile leaves in a certain sense are heterophyllous. Yet the difference between the two kinds may only lie in the presence or absence of sporangia or sori (*Pteridium aquilinum*) or there may be other, often very great, differences (*Matteucia Struthiopteris*). Often a leaf may be

---

[1]) F. O. BOWER, The ferns, III, Cambridge 1928, see p. 259.

[2]) The sorus of the *Salviniaceae* is often wrongly called sporocarp too, as by SADEBECK in ENGLER & PRANTL op.c. 1902, p. 391 and even in the last edition (1936) of the well-known Bonn textbook.

[3]) K. G. STENZEL, Die Psaronien, Beobachtungen und Betrachtungen, Beiträge zur Paläontologie und Geologie Österreich-Ungarns und des Orients 19, 1906, p. 85, see p. 114.

[4]) See SCHOUTE 1914 l.c. (f.n. p. 9), p. 151.

partially sterile, partially fertile (*Osmunda regalis*), with great difference between the parts.

In some cases the sterile leaves are dimorphic too: the nest leaves and the foliage leaves of *Platycerium* being one instance, the floating leaves and the submerged leaves of *Salvinia* another. Moreover great differences may be found between juvenile and older leaves (*Asplenium epiphyticum* [1]).

**§ 9. The strobilus.** — A strobilus is a more or less condensed stem of limited growth, bearing a number of sporophylls.

In the bud condition the sporophylls are usually densely packed, forming a kind of investment protecting the sporangia. The investment may consist of the shield-like dilations of the sporangiophores (*Equisetum*) or of the upturned laminar part of the sporophylls (*Lepidostrobus, Sigillariostrobus*), or of separate sterile lobes of the sporophylls (*Sphenophyllum Dawsoni*), or finally of a number of separate sterile phyllomes, intercalated between the sporophylls (*Palaeostachya, Calamostachys*).

In the older literature the strobili were usually called spikes or cones, terms which have not yet entirely disappeared from modern literature. They are however to be avoided on account of their different morphological meaning in the Spermophytes. In some cases the strobili are called flowers, a designation to which less objection is to be made, as flowers are specialized strobili. It seems however more appropriate to retain the term strobilus for the less specialized reproductive shoots of the Pteridophytes.

Strobili are present in many *Lycopodiinae* and in most *Articulatae*; instances of forms without strobili belonging to these classes are *Lycopodium Selago* and *Sphenophyllum tenuissimum*. In the *Psilophytinae*, the *Psilotinae, Isoetinae* and *Filicinae* they are altogether lacking.

The strobili may be borne terminally (*Lycopodium, Lepidodendron obovatum, Selaginella, Equisetum*) or laterally (*Lepidophloios, Bothrodendron*).

In the case of heterosporous strobili there is usually a fixed rule for the distribution of mega- and microsporangia. In most cases the rule amounts to the production of megasporangia, casu quo of megasporophylls, at the basal part of the strobilus, the microsporangia (microsporophylls) occurring at the apical part (*Lepidostrobus Veltheimianus, Calamostachys Casheana, Selaginella* spp. with erect radial strobili [2])).

In other cases the rule may be different; in *Selaginella* spp. with horizontal strobili, the megasporophylls form the two ventral, the microsporophylls the two dorsal rows; in pendulous *Sel.* strobili the mega-

---

[1]) See GOEBEL 1930, op.c. p. 1188.
[2]) See ENGLER & PRANTL, op. c. p. 659.

sporophylls may occupy the morphological apical, the microsporophylls the basal part (ibid.).

**§ 10. Special parts of the embryo.** — In the second half of the last century great expectations were entertained about the results of comparative embryology. Developmental studies being in high esteem, the great success of animal embryology invited analogous work in botany.

So the succession of cell cleavage in the embryo was accurately studied and the cell divisions leading to organ formation were recorded. On the whole we may say however, that the results have fallen short of the expectations [1]).

Even such a fundamental distinction as that between exoscopic embryos, where the apex of the young sporophyte is directed towards the archegonium neck, and endoscopic embryos, where it is directed towards the base of the archegonium, did not prove to have the taxonomic value one might expect, the embryos of *Psilotum, Equisetum, Isoetes, Ophioglossum, Botrychium Lunaria* and of *Leptosporangiatae* being exoscopic, those of *Lycopodium, Selaginella, Marattiaceae, Botrychium obliquum* and *Helminthostachys* being endoscopic [2])).

Special organs are formed in all or nearly all embryos, as adaptations to the condition of development in the archegonium at the expense of the prothallus. Of these organs the foot and the suspensor will be dealt with here.

1. The foot. Contrary to its name, the foot serves as a haustorium, receiving nutritive substances from the prothallus. Accordingly in *Tmesipteris* it even has an irregular surface in contact with the prothallus. Sometimes as in *Lycopodium Selago* it is not conspicuous, but it seems to be present in all Pteridophytes. Its position in the embryonic body may vary, or as BOWER expresses it [3]): "the haustorial organs .... included under the term "foot" .... are, in fact, opportunist growths, formed in positions convenient for the plant".

2. The suspensor ("Embryoträger") is an organ of quite another function as it is destined to push the embryo farther into the prothallus. Accordingly it develops in the embryo at the side of the archegonium neck.

It occurs in *Lycopodium*, in *Selaginella* and in some *Filicinae* [2]), amongst the *Ophioglossaceae* we find both embryos with and without a suspensor; in *Angiopteris* we may even have specimens of both forms in the same species [4]).

---

[1]) Detailed surveys of the facts with full quotation of the extensive literature may be found in ENGLER & PRANTL, op. c. (1902), GOEBEL, op. c. (1930) and BOWER (1935).

[2]) All from BOWER 1935, op.c.

[3]) BOWER 1935, op.c. p. 542.

[4]) See BOWER 1935, op.c.

These facts all point to the unstable and morphologically not very important character of these adaptations. BRUCHMANN describes the interesting fact that the suspensor in *Selaginella Kraussiana* and *S. Poulteri* is much reduced and is replaced by another organ of analogous position, the embryo tube ("Embryo Schlauch" [1])) which however does not belong to the embryo but develops from the basal archegonium cell surrounding the egg cell.

**§ 11. Organs which have given rise to controversies about their morphological nature** (cf. p. 123). — In this paragraph the following organs will be dealt with: The *Psilotum* sporophyll, the rhizophore of *Selaginella*, the subterranean basal parts of some more or less related Pteridophytes, namely the stigmarian axis with its rootlets and the basal parts of *Pleuromeia, Isoetes* and *Nathorstiana*.

The protocorm, the tuberous stage of the *Lycopodium* embryo will not be treated, as it has become clear that these tubers are nothing but an occasional adaptation without phylogenetic importance; the reader may be referred for it to BOWER [2]) who after having cited the literature concludes that it will be best to drop the term protocorm altogether.

The aphlebiae are not treated either, as BOWER's conclusion has already been cited in § 8 on p. 45, that under this name two different categories have been included, namely microphylls and specialized megaphyll pinnae.

Before dealing with the indicated topics one general remark may be advanced on the so-called organs sui generis. It is a well-known fact that several authors have little objection to declaring an organ to be sui generis, when the morphological nature is difficult to determine.

A sui generis organ however should properly be a novelty, not having arisen by metamorphosis of existing organs, and such novelties are rare in living organisms. Of course they must have occurred sometimes, and perhaps the microphylls were such a novelty once, or the ligule of the *Ligulatae*.

But in most cases the so-called sui generis organs are evidently specialized forms of other organs, and often the assumption of the sui generis character is only an easy means of getting rid of the difficulty, an apparent solution which may retard the attempts to find a real solution.

The origin of a new specialized form of an organ must be supposed to take place by a change in the morphogenetic forces for the special organ. When, for instance, in a plant group the stem develops two forms with different plastics, the morphogenetic stem factors must have been specialized in two directions, or besides the old unchanged set of factors a somewhat different modification must have been segregated from it; it is clear that the two different organs arising in this way are homologous.

[1]) H. BRUCHMANN, Zur Reduktion des Embryoträgers bei Selaginellen, Flora 105, 1913, p. 337.

[2]) BOWER 1935, op. c. p. 272.

For the discrimination of homologous organs considerable weight has always been attached to the occurrence of transitional stages. Evidently this is due to the fact that the nearer the relation between the two sets of morphogenetic factors, the easier the occurrence of transitions. Especially in such cases where a number of consecutive organs are formed in which a change of character follows, as for instance in sporophylls following on foliage leaves, the change may often be executed gradually and the study of the transitional stages formed in this way may give excellent information on the homology of the different organs and their parts.

Most authors in declaring some organ to be sui generis, add their opinion that the whole question of the inclusion in one or another category is only of an academic order, or utterly unimportant, or some qualification like that. This only shows the more that they do not take it as a question of real biological processes, but only as a concern of idealistic morphology, a branch of science which for some unknown reason always wants to pigeon-hole all topics in a pragmatical way.

**1. The Psilotum sporophyll.** In § 8 on p. 34 the complex sporophyll of some *Articulatae* has been declared to be a single phyllome, and for *Psilotum* the same conclusion has been tacitly assumed. A number of authors however have claimed an axial nature for the fructificative complex; the first of them being JURÁNYI [1]), who derived his arguments from ontogenetic facts, from the presence of an apical cell for the synangium etc.

Many botanists later followed JURÁNYI, not only SACHS, STRASBURGER and GOEBEL, but even EICHLER; on the other hand LÜERSSEN, PRANTL and ČELAKOVSKÝ always adhered to the foliar nature of the complex. In an excellent paper SOLMS-LAUBACH clearly proved that JURÁNYI's facts were all ill-observed and that on the contrary the development was more in favour of the foliar nature than of the axial nature [2]). In the same paper, to which reference may be made for the older literature, SOLMS pointed out the transitions occurring between the sterile leaves and the sporophylls and concluded that if the latter should really be taken for lateral branches, then the sterile leaves would be monophyllous lateral branches too.

Accordingly after SOLMS's investigation the general opinion changed: VELENOVSKÝ [3]) and SEWARD [4]) again accepted the foliar view. The subsequent discoveries of the Rhynie fossils with their stem-borne sporangia, however, so strongly impressed all botanists, that several of them again

---

[1]) L. JURÁNYI, Über den Bau und die Entwickelung des Sporangiums von *Psilotum triquetrum* Sw., Bot. Ztg. 29, 1871, col. 177.

[2]) SOLMS-LAUBACH, 1884, l.c. (f.n. p. 32).

[3]) VELENOVSKÝ, op.c. p. 215.

[4]) A. C. SEWARD, Fossil Plants, II, Cambridge 1910, see p. 23.

took up JURÁNYI's view: as for instance SAHNI [1]) and KRÄUSEL and WEY-
LAND [2]). The last mentioned authors even put forward the old argument
that the fertile leaves of *Psilotum* possess a stele which is lacking in the
sterile leaves, although SOLMS had already shown that *Psilotum flaccidum*
has a leaf-trace for every leaf and that even in *Ps. triquetrum* part of the
leaves have weak traces.

As far as I can see there can be no doubt as to the true foliar nature of
the sporangium-bearing complex on account of the following arguments.

In the first place, because of the transitional stages already pointed out
by SOLMS, which clearly prove the homology of the two categories of
organs.

In the second place because of the analogy with *Sphenophyllum*, to
which SCOTT [3]) and SEWARD [4]) have pointed; the forked sporophyll with a
small number of adaxially borne sporangia occurring in both.

In the third place on account of the phyllotactical relations. In *Psilotum
triquetrum* the sterile leaves at the main branches are placed irregularly. In
the thin branches they are however placed in a tristichous system 1 + 2,
together with the transitional stages and the sporophylls; this has been
remarked already by GOEBEL [5]) and it would be very hard to explain if the
two categories of organs were not homologous.

The only other solution, that the sterile leaves are cauline too, is more or
less explicitly defended by ZIMMERMANN, who derives both microphylls and
megaphylls from overtopped dichotomy branches [6]), a view to which,
however, few botanists will have recourse at the present stage of our
knowledge.

2. **T h e   S e l a g i n e l l a   r h i z o p h o r e.**   The well-known
rhizophores of *Selaginella* share some properties with roots (positive ge-
otropy, lack of leaves), others with stems (exogenous origin, lack of root-
hairs and calyptra). Their anatomical constitution varies much in the
different species and is what might have been expected from leafless stems,
modified in the direction of roots; if they were true roots the structure in
some species would be quite uncommon, even unique in the plant kingdom
(see Ch. II § 2, A, III, e). They are easily converted into leaf-bearing shoots.

In view of these facts all three possible opinions have found their defenders, that rhizophores are transformed stems (e.g. PFEFFER, TREUB,
BRUCHMANN, VELENOVSKÝ, WORSDELL, TROLL); that they are roots (VAN

---

[1]) B. SAHNI, On *Tmesipteris Vieillardi* Dangeard, an erect terrestrial species
from New Caledonia. Phil. Trans. R. S. London, 1925, B 213, p. 143.

[2]) R. KRÄUSEL und H. WEYLAND, Beiträge zur Kenntnis der Devonflora, Senckenbergiana 5, 1923, p. 154, see on p. 169.

[3]) D. H. SCOTT, Studies in fossil botany, London 1900.

[4]) SEWARD, 1910, op.c. p. 17.

[5]) GOEBEL, 1930, op.c. p. 1252.

[6]) ZIMMERMANN, op.c. p. 125.

TIEGHEM, HARVEY-GIBSON, UPHOFF), or that they are organs sui generis (GOEBEL, BOWER). For references to the literature the reader may consult some of the later authors [1]).

About the improbability of the last opinion little remark need be made after what has been said about the sui generis organs in the introduction to this paragraph; the many features the rhizophores share with stems and roots completely exclude the idea of development in an independent way.

If rhizophores were transformed roots there would be no explanation for their exogenous origin and their lack of root-hairs and a calyptra; if, on the other hand, they are stems with a root-bearing function it is quite natural that they are transformed in the root-direction. There seems therefore no room for doubt that rhizophores are specialized stems.

**3. The stigmarian axis.** The large main root-like organs of *Lepidodendron* and *Sigillaria*, known as *Stigmaria* and in some cases as *Stigmariopsis*, have even given rise to more comment than the rhizophores. They are either thick and long, dichotomously branching horizontal organs very slowly tapering towards their apices (*Stigmaria*), or shorter and more downward sloping organs with stout positively geotropic lateral branches, the so-called tap-roots (*Stigmariopsis*). They arise from the stem base, and are usually four in number, a fact which is attributed to a once repeated dichotomy.

The stigmarian axes always bore a very large number of appendages, the so-called stigmarian rootlets (see sub 4), arranged in a high phyllotactical pattern. These rootlets were produced close to the dome-shaped apex of the stigmarian axis, and SOLMS-LAUBACH even wrote: "Indem sie sich vorwärts krümmen, neigen sie sich knospenartig um den Scheitel zusammen" [2]). This remark, based on personal observation, surely deserves consideration; it should however be confirmed by new and detailed investigation.

The appendages were formerly believed to arise exogenously, but LANG found in *Stigmaria bacupensis* that the two tissue zones of the rootlet outer cortex, which are continuous with two similar zones of the stigmarian axis, are covered in the latter by a third zone of three or four delicate and ill-defined cell-layers.

The rootlets are not only arranged in a phyllotactical pattern as leaves are, but they also cause distinct gaps in the axis stele, showing the same

---

[1]) J. C. TH. UPHOF, Contributions towards a knowledge of the anatomy of the genus *Selaginella*. The root. Ann. of Bot. 34, 1920, p. 493; S. WILLIAMS, An analysis of the vegetative organs of *Selaginella grandis* Moore, together with some observations on abnormalities and experimental results, Trans. Roy, Soc. Edinb., 57, 1931, p. 1.

[2]) H. Graf zu SOLMS-LAUBACH, Einleitung in die Paläophytologie, Leipzig 1887, see p. 276.

arrangement as the rootlets themselves; for lateral roots this would be an unparalleled phenomenon.

Some authors have concluded that the stigmarian axis was a root. WILLIAMSON strongly insisted upon this view [1]); his only argument however was the function, morphological facts not being taken into consideration. ZEILLER in his earlier years was inclined to the same view, though less decidedly [2]), and LINDINGER arrived at the same conclusion on the very slender basis of a comparison with the large roots of *Dracaena* [3]).

These views no doubt are erroneous, as a root never would bear organs in a phyllotactical pattern, causing gaps in its stele, particularly while its root function remained unaltered.

Of course the sui generis view has been held too: as its defenders we may quote GOEBEL [4]), POTONIÉ [5]), ZIMMERMANN [6]) and BOWER [7]). As remarked above, this view would imply that the stigmarian axis should have been formed anew by the plant, without borrowing its numerous cauline characters from the stem.

A third category of authors compares the stigmarian axis with the *Selaginella* rhizophore: SCOTT [8]) and especially TROLL [9]) may be instanced. As the rhizophore represents a modified stem, this opinion deserves some more attention that the above ones. Yet it seems unacceptable on the following ground.

The rhizophore is more advanced than the stigmarian axis in having lost all traces of lateral organs borne in a phyllotaxis. On the other hand it is less advanced in still bearing true roots; as we presently shall see the stigmarian axis was devoid of roots. It is therefore not probable that either of the two should be derived from the other.

---

[1]) W. C. WILLIAMSON, A monograph on the morphology and histology of *Stigmaria ficoides*, London 1887, in: The palaeontographical society, volume for 1886. See foot-note on p. 5.

[2]) R. ZEILLER, Végétaux fossiles du terrain houiller de la France, Paris 1880, as vol. 4 of "L'explication de la carte géologique de France".

[3]) L. LINDINGER, Die sekundären Adventivwurzeln von *Dracaena* und der morphologische Wert der Stigmarien, Jahrbuch der Hamb. wiss. Anst., 26, 1908, 3. Beiheft, p. 59, Hamburg 1909.

[4]) K. GOEBEL, Morphologische und biologische Bemerkungen, 16. Die Knollen der Dioscoreen und die Wurzelträger der Selaginellen, Organe welche zwischen Wurzeln und Sprosse stehen, Flora 95, Erg. Bd. 1905, p. 167.

[5]) H. POTONIÉ, Lehrbuch der Pflanzenpalaeontologie, Berlin 1899, see on p. 214; moreover POTONIÉ op.c. 1912, (f.n. p. 29), p. 236.

[6]) ZIMMERMANN, op.c. p. 148.

[7]) BOWER 1935, op.c. p. 236.

[8]) SCOTT 1920, op.c. p. 239.

[9]) TROLL 1934, l.c. (f.n. p. 10).

In the older literature many other improbable views have been given; for these Solms [1]) refers to a survey given by Goeppert.

So only the view which has been adopted by so many authors remains, namely that the stigmarian axis is a rhizome. This, in my opinion undoubtedly correct, view has been held by: W. Ph. Schimper, B. Renault, Solms-Laubach, Grand'Eury, Zeiller [2]), Hirmer, Weiss [3]), and Kubart [4]).

The objection raised by Troll [5]), that the stigmarian axis cannot be a rhizome, because of the lack of innovation buds producing aerial stems, does not seem well-founded. Even if it may turn out that the stigmarian axes never formed such innovation buds, the designation as a rhizome might be used for these specially adapted subterranean stems.

Moreover the problem as to how the aerial stem was produced remains quite unsolved up to the present day.

Goeppert, Renault and Grand'Eury came to the conclusion that the stigmarian axis in some way produced large tuberous thickenings giving rise to an aerial stem and also to new stigmarian axes; Goeppert compared it with the way in which moss stems are borne on a protonema.

Williamson, the best connoisseur of the English forms, was strongly opposed to this view. Later authors however admitted the possibility of such vegetative propagation. Solms in his excellent "Einleitung" discussed the question at length [6]) and concluded by saying that the conception, as given by these authors, is very plausible and explains everything completely. It is only a pity that the actual basis of observed facts is not worth much, the indications of a parent rhizome at the tree base especially being entirely lacking.

So by negative evidence the existance of vegetative propagation became less and less probable; yet Seward wrote in 1910 [7]) that he still believed the main contentions of Renault and Grand'Eury to have been correct.

It is only quite lately, by the discoveries of Walton [8]) that positive evidence has been adduced for the direct elaboration of the tree from the sporeling, for the case of *Lepidophloios Wünschianus*. Walton succeeded in

---

[1]) Solms-Laubach 1887, op.c. p. 288.

[2]) R. Zeiller, Éléments de paléobotanique, Paris 1900, see p. 202.

[3]) F. E. Weiss, A re-examination of the Stigmarian problem, Proc. Linn. Soc. London, session 144, 1932, p. 151. For further quotations see Hirmer 1927 op.c. and especially Solms-Laubach op.c. 1887, p. 270.

[4]) B. Kubart, Stigmaria Bgt., Mitth. naturw. Ver. Steiermark 71, 1934, p. 33.

[5]) Troll 1934, l.c. p. 94.

[6]) Solms-Laubach 1887, op.c. p. 299.

[7]) Seward 1910, op.c. (f.n. p. 50), p. 238.

[8]) J. Walton, Scottish lower carboniferous plants: the fossil hollow trees of Arran and their branches (*Lepidophloios Wünschianus* Carruthers). Trans. R. Soc. Edinb. 58, 1935, p. 313.

observing the well preserved primary xylem of the stem base in several specimens and he found it to consist of a solid core of very small dimensions, probably less than 2 mm in diameter at the base, and soon enlarging upwards in the stump to a medullated primary xylem cylinder of about 20 mm.

The slender primary xylem cylinder moreover showed two shallow grooves at its periphery, not present in other parts of the plant, which the author, in all likelihood correctly, connects with the stigmarian base.

These facts really are strong indications of a direct development of the stem from the sporeling. Of course many unsolved questions remain: thus we should like to know how the primary cortex of say 30 cm in diameter may have been elaborated, and especially we should like to have some knowledge about the origin of the stigmarian base.

As far as the facts go I should therefore be inclined to the supposition, that in many Lepidophytes the tree was directly elaborated from the sporeling, whereas in some others there may have been vegetative propagation with a gradual increase in the size of the successively produced stems.

There are actually some indications for the latter possibility. As such I may call attention to the numerous young specimens of *Pleuromeia*, to be described beneath sub 6, which are similar in all respects to the large specimens. But especially I might allege the fact that in the thick basal stem part of *Sigillaria* the number of leaf orthostichies may be much larger than in the normal part of the stem some feet higher, through the discontinuation of a considerable number of the orthostichies; see the drawing by BRONGNIART [1]). Such a condition would be natural if the stem arose from a large innovation bud, but if the stem was directly elaborated from the embryo the conditions must have been rather the reverse, as indeed has been observed by GOLDENBERG for *Sigillaria cactiformis* [2]) where between 16 orthostichies 4 new ones were intercalated and by KIDSTON [3]) in a *Sigillaria* of the Rhytidolepis-groups, where 29 orthostichies increased in number up to 45.

In any case new facts like those alleged by WALTON will be most useful; mere suppositions as given by LIGNIER [4]) or HIRMER [5]) will never be able to fill up the gaps in our knowledge.

---

[1]) BRONGNIART, l.c. 1828, (f.n. p. 14), Pl. 160.

[2]) F. GOLDENBERG, l.c. (f.n. p. 12), 1855–1862, 1. Heft p. 38. Pl. 4 fig. 1.

[3]) R. KIDSTON, Fossil Flora of the Yorkshire Coalfield II, Trans. R. Soc. Edinb. 39, 1897.

[4]) O. LIGNIER, Interprétation de la souche des Stigmaria, Bull. Soc. bot. de France, 60, 1913, p. 2. The author supposes that stigmarian rhizomes by dichotomous branching in a perpendicular plane, gave rise to the aerial stem at the upper side and to a new rhizome at the under side; the latter by two further dichotomies formed the usual stigmarian base. He further supposes that by apposition of secondary

**4. The stigmarian rootlet.** These well-known, more or less cylindrical or awl-shaped and sometimes dichotomizing organs, also called appendages (appendices), are usually regarded as true roots, notwithstanding their arrangement in a phyllotactical pattern, a unique feature for roots.

Their organization and function however are similar to that of roots; anatomically we have only the difference of the development of a transfusion tissue in the outer cortex, connected by small strands, through the trabecular tissue, with the stelar xylem [1]).

Apart from this small difference they have almost the same anatomy as the *Isoetes* roots; and as they are famous for their intrusive power, they must have possessed a strictly localized apical growth.

Accordingly WILLIAMSON, HIRMER, WEISS and many other authors declare them to be roots. Yet there are some characters deviating from what is to be found in other roots, beyond their arrangement: they lack root-hairs and a root-cap, in contrast with *Isoetes*. The absence of the root-cap is especially remarkable, in view of their normal root-function.

It is therefore no wonder that several other authors have seen metamorphosed leaves in these rootlets, especially from their arrangement; these are W. PH. SCHIMPER, RENAULT [2]) and GRAND'EURY. On the same arguments POTONIÉ [3]) and SCOTT [4]) recognize the homology of rootlets and leaves, but at the same time take them to be homologous with the *Isoetes* roots too. I shall come to the latter point below sub 5.

Of course the idea that the rootlets are modified leaves deserves full

---

tissues all these branching places became covered and withdrawn from the eye; a view for which it might be difficult to obtain a satisfactory anatomical foundation.

[5]) HIRMER (1927 op.c. p. 289; 1933 l.c. f.n. p. 32, p. 52 and 1934 in a paper „Grundsätzliches zur Rekonstruktion des Lepidophyten-embryos. Eine Erwiderung an Herrn Wilhelm Troll, Palaeontographica 79, B, p. 143) suggests, especially in the second paper, that in the embryo of *Lepidodendron* both epibasal and hypobasal halves may have been active, not only the epibasal one as in *Isoetes*, or the hypobasal one as in *Lycopodium* and *Selaginella*; in such a way two poles might have been produced, one for the aerial stem, the other for the stigmarian base. This view which does not seem very acceptable in itself and for which the facts mentioned above are not favourable, has been discussed by TROLL (1934, l.c. f.n. p. 10).

[1]) F. E. WEISS, The vascular branches of Stigmarian rootlets, Ann. of Bot. 16, 1902, p. 559 and same title 18, 1904, p. 180.

[2]) RENAULT believed that apart from the root-like leaves true roots also occurred on the older stigmarian axes, roots of the same form and properties as the leaves. His arguments in favour of this incredible view have been discussed and fully refuted by SOLMS-LAUBACH (1887, op. c. f.n. p. 52, p. 285, 297).

[3]) POTONIÉ, op.c. 1912 (f.n. p. 29), p. 238.

[4]) SCOTT, op.c. 1920, p. 238; this view has been adopted by KUBART (f.n. p. 54).

attention, as most facts mentioned above would be readily explained in this way: the arrangement, the presence of stelar gaps, the absence of root-hairs and root-cap, and their early formation at the stigmarian apex. It would only imply the assumption of a far-reaching adaptation to the root-function, including a striking approach to the root-anatomy.

There are however some objections. The occurrence of dichotomous branching would not be a serious one, as dichotomizing leaves are not lacking in the *Lycopodiinae*, nor even in the closely allied *Lepidophyta* (*Protolepidodendron*).

As a real objection we have in the first place the endogenous insertion as described by LANG for *Stigmaria bacupensis* [1]) and other species. Leaves are always exogenous, and for the realization of a phyllotaxis this is probably even an essential feature, the determination of leaf places being a surface phenomenon.

We may say however that an endogenous insertion does not necessitate an endogenous *origin*. What has been proved is that a few external cell layers of the mature axis did not extend over the surface of the mature rootlet which in any case must have been formed very near to the axis surface. If for instance there was a superficial tissue of the rootlet during the ontogeny, which disappeared during the stretching of the rootlet, then rootlets would have been exogenous in origin as has always been assumed. As long as nothing is known about the rootlet ontogeny, this point must remain unsettled, and the arrangement of the rootlets in parastichies in itself is an argument for an exogenous origin.

A second objection might be inferred from the communication of PO-TONIÉ [2]), that in rare cases on the *Sigillaria* stem between the leaf scars other scars of rootlets were to be observed; indeed if the rootlets occurred irregularly between the leaf scars whose phyllotactical pattern was unchanged, this would amount to a direct proof that the rootlets were not homologous with the foliage leaves.

POTONIÉ's observation however, as may be read in SOLMS [3]), was far from new, and its explanation has been very different, GERMAR already having declared these marks to be scars of spines. Unless proof is given that these scars really are rootlet scars, this objection may be left out of consideration.

In consideration of the known facts I am therefore inclined, though recognizing the need for further research, to assume that the rootlets are modified leaves. In any case we may say that this view is much more probable than any other one, and the comparison of the stigmarian axis

---

[1]) W. H. LANG, On the apparently endogenous insertion of the roots of *Stigmaria*, Mem. and Proc. Manch. Lit. and Phil. Soc. 67, 1923, p. 101.

[2]) POTONIÉ 1899, op.c. (p. 53), p. 212.

[3]) SOLMS-LAUBACH 1887, op.c. (f.n. p. 52), p. 252.

with parts bearing true roots, like the rhizophore or the *Isoetes* stem base, thus loses significance.

Finally there are still two divergent opinions to be mentioned. One of them of course is the unavoidable sui generis hypothesis, this time held by ZIMMERMANN [1]); it will be passed over in silence.

The other is a view of WEISS, giving an elegant solution for the difficulty of roots arranged in parastichies. WEISS supposes [2]) that the stelar gaps in the stigmarian axis are caused by wholly aborted leaves, whereas under every leaf a single root is developed. This solution, though attractive at first view, presents the greatest difficulties on closer examination. It implies for instance that for every wholly aborted leaf exactly one single root was formed, a root with the ordinary root function, but without root-hairs or root-cap, and that moreover the true root-trace would form the continuation, without any visible disturbance, of the well-developed trace of the aborted leaf: altogether an untenable set of suppositions.

**5. The Isoetes stock base.** The first to relate *Isoetes* with *Stigmaria* was BRONGNIART [3]). Since then it has become customary to see in *Isoetes* a last remnant of the Lepidophytes, the Mesozoic forms *Pleuromeia* and *Nathorstiana* representing intermediate links. This series has even been claimed by MÄGDEFRAU [4]) to be one of the best phylogenetic series known in the vegetable kingdom.

Other botanists on the contrary deny a close connection between *Isoetes* and the Lepidophytes, for instance WEST and TAKEDA [5]); in all papers by these authors the organization of the lower stem part plays a primary role.

A discussion of the morphological conditions of the *Isoetes* stock base is therefore necessary here; as we shall see it will reveal to us the existence of a large gap between the Lepidophytes and *Isoetes*.

The two- or three-lobed stem of *Isoetes* (sometimes five-lobed, rarely four-lobed) at first is more or less globular. The development of the lobes is not due to the action of supplementary vegetative cones at the basal side, but only to the local formation of secondary tissues by the cambium. This cambium surrounds the stem stele and even closes around its basal end, just as is the case in *Nolina recurvata* [6]). The lobes therefore are not comparable to stigmarian axes as all botanists formerly believed [7]), as

---

[1]) ZIMMERMANN, op.c. (f.n. p. 26), p. 149.

[2]) WEISS 1932, l.c. (f.n. p. 54), p. 156.

[3]) AD. BRONGNIART, Prodrome d'une histoire des végétaux fossiles, Paris 1828, see p. 82.

[4]) K. MÄGDEFRAU, Über *Nathorstiana*, eine Isoetacee aus dem Neokom von Quedlinburg a. Harz, B.B.C. 49, 2, 1932, p. 706; see p. 715.

[5]) C. WEST and H. TAKEDA, On *Isoetes japonica*, A. Br., Trans. Linn. Soc. London, 2nd ser. 8, Botany, 1915, p. 333.

[6]) J. C. SCHOUTE, Die Stammesbildung der Monokotylen, Flora 92, 1903, p. 33.

[7]) Thus for instance still in POTONIÉ 1912, op.c. (f.n. p. 29), p. 234.

these grew by a vegetative cone; for the same reason they can neither be stems or roots.

The place where the lobes arise is determined by the phyllotaxis of the young plant; in distichous plantlets the furrows alternate with the leaf orthostichies, under which the lobes are formed [1]). Once the lobes are constituted they retain their position, even on change of phyllotaxis.

Between the lobes the tissue is torn open by a longitudinal slit, forming in this way the well-known furrows; evidently these furrows are due to the growth of the lobes, by the dilatation of the tissue. The furrows meet at the stock base and unite there.

In the embryo the first root arises laterally and exogenously [2]); like all subsequent roots this first root is provided with root-hairs and with a root-cap. The position of later roots is always related to the place of the furrows, these roots according to all authors being formed in series alongside the furrows, and therefore more or less in the position of orthostichies.

In every orthostichy the development is acropetal, the youngest root being placed nearest to the furrow end, i.e. to the stem apex [3]).

The orthostichies however are soon discontinued, and a new orthostichy arises at either side of the furrow, between the last orthostichy and the furrow which is always becoming deeper.

This new orthostichy forms its members without any spatial relation to the roots of the previous row; there is no alternation or juxtaposition to be observed [4]); such a spatial relation moreover would hardly be expected, in view of the different age of the adjacent roots of different orthostichies.

---

[1]) First observed by AL. BRAUN, Weitere Bemerkungen über *Isoetes*, Flora 30, 1847, p. 33.

[2]) J. BRETLAND FARMER, On *Isoetes lacustris*, Ann. of Bot. 5, 1890/'91, p. 37.

[3]) This acropetal development has especially been demonstrated by D. H. SCOTT and T. G. HILL (The structure of *Isoetes Hystrix*, Ann. of Bot. 14, 1900, p. 413) who observed that the roots of any series are of different ages, the oldest root. traces having their vascular tissues more or less obliterated while the youngest are still wholly meristematic (l.c. p. 428).

WEST and TAKEDA however contend (l.c. p. 354) that all roots of a series are of the same age. No observations are given in proof of this aberrant statement; the fact that the roots in a series appear in a regular succession at the stem surface with considerable difference in time is duly recognized, but explained by the remark that the roots at the furrow end have to travel over a longer distance through the cortex.

This last statement is difficult to reconcile with the drawings of longitudinal sections through the stem in the furrow plane, as given by VON MOHL (Über den Bau des Stammes von *Isoetes lacustris*, Linnaea 14, 1840, p. 181, pl. 3, fig. 4, 9) and by W. H. LANG (Studies in the morphology of *Isoetes* I, The general morphology of the stock of *Isoetes lacustris*, Mem. & Proc. Manch. Lit. & Phil. Soc. 59, 1915, Memoir no. 3. See fig. 2).

[4]) SCOTT and HILL, l.c. fig. 11, 12; LANG, l.c. fig. 2.

For the rest the distribution and position of the root series cannot easily be grasped from literature. In particular the later position of the root series which is quite different from the original position alongside the furrow, has never been explained as far as I know [1]).

All these roots being formed endogenously, their traces are inserted on the base of the stem stele. Owing to their great number on the stem stele two, or in the three-lobed stems three, arms of stelar tissue are formed out

[1]) Though not having instituted any investigations myself I might suggest the following incomplete, but perhaps not improbable, explanation of the said difficulty, in the hope that it may stimulate some future student of *Isoetes* to complete it, or to amend or reject it.

Our fig. 8 represents two transverse sections through an *Isoetes* stock base, just under the stele, *A* being a two-lobed, *B* a three-lobed specimen. The position of the stele *s* as present in higher sections has been indicated by dotted lines; the lobes *l* and the cortex *ic* inside the furrows are represented.

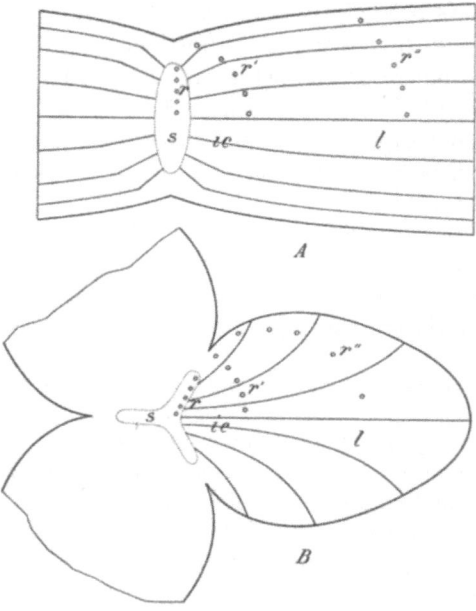

Radiating lines in *ic* and *l* are intended to indicate the presumed transverse growth directions; in *l* where the tangential tension is removed, these must be different from those in *ic*.

Now the root series arise at the under side of the upturned stelar arms; in our schematic figures this upward curve not being taken into account, *r* may represent a root series at the place of origin. By the formation of new cortex parts at the outside of the cambium, such a series will successively assume other positions, of which *r'* and *r''* may give an idea.

The position of the different root series in the drawings by WEST and TAKEDA of the three-lobed *I. japonica* (l.c. text-fig. 12 and 13) exactly corresponds to that in our fig. 8 *B*; the drawings by LANG of the two-lobed *I. lacustre* (l.c. fig. 2) contain, it is true, some series in the position of those in our fig. 8 *A*, but moreover some series in or near to the median line of the lobe; for this undoubtedly correct observation (from a photographed section series) the above explanation gives no clue.

FIG. 8. Schematic transverse sections just under stele through two-lobed (*A*) and three-lobed (*B*) stem of *Isoetes*; *s* = position of stele at slightly higher level; *ic* = cortex inside furrows; *l* = lobes of stem; *r*, *r'* and *r''* = successive positions of root series; radiating lines = presumed growth direction of parenchyma.

of the confluent root-traces, arms which FITTING has taken to be homologous with stigmarian axes [1]). The thorough description by SCOTT and HILL however leaves no doubt that the interpretation by these authors is correct [2]).

So the peculiar stem-formation of *Isoetes*, notwithstanding its unique feature of secondary growth in a recent Pteridophyte, does not offer any analogy to the Lepidophyte tree base; the fact that its roots are clearly true roots therefore is no longer a problem.

We cannot leave the topic without mentioning a widely accepted opinion from literature about the basal stock part. BRAUN believed that the lower part of the *Isoetes* stock was nothing but a telescoped main root; he based that view on the basipetal development of the new roots. Though this argument was wholly cancelled by the accurate observations on the root developmental order by SCOTT and HILL, similar views remained. So we read that LANG after intensive anatomical examination concluded that "the recognition of the lower region of the stock of *Isoetes* as a rhizophore in some way correlated with the upwardly growing shoot appears to be justified" [3]). And LIEBIG even declares [4]), that all her median longitudinal sections through the stock are proofs of BRAUN's thesis!

When we consider the arguments of these assertions we find that they are chiefly based on the fact that the stele in every monarch root is turned towards the furrow, a relation which surely is no basis for such far-reaching conclusions and which may be quite well correlated with the orientation of all roots towards the furrows.

So our conclusion is that the stock base of *Isoetes* is only peculiar in its constantly tearing furrows, a feature connected with the formation of local humps of secondary tissue, the lobes. Roots are formed, as always in all plants, in any suitable place; the presence of the furrows and the lack of

---

[1]) H. FITTING, Sporen im Buntsandstein — die Makrosporen von *Pleuromeia?* Ber. d. D. bot. Ges., 25, 1907, p. 434, see p. 441.

[2]) WEST and TAKEDA in this respect hold another view; they describe an apical meristem by which the stelar arms grow (l.c. p. 346). This meristem is compared to that of the stem, and on this basis the authors try to keep to the fore the conception of the downward growing "rhizophore" as an organ sui generis.

They admit however that this apical meristem differs from any other in being distributed over a large area, since it extends along the whole length of the curved lower edge of each of the stelar arms. They further admit that this primary meristem is not situated at the actual periphery of the stock, but that it is separated from the exterior by several regularly arranged layers of parenchyma cells, and finally they admit that it may also be regarded as a part of the cambium.

The natural conclusion therefore may be that this "primary meristem" is nothing but a specialized part of the cambium in which the adventitious roots are formed in great number.

[3]) LANG 1915, l.c. p. 21.

[4]) 'JOHA. LIEBIG, l.c. (f.n. p. 35), 329.

new higher stem parts cause these suitable places to be found in an unusual arrangement.

**6. The Pleuromeia stem base.** No comparison being allowed between the stigmarian axes and the *Isoetes* stock base, the question remains about the intermediate links.

In *Pleuromeia* the stem at its base usually developed four conical upturned lobes; sometimes only two have been present. In the normal case of four they were not placed in a regular cross, a single connecting part being present at the under surface between two opposed pairs [1]).

In rare cases some of the four lobes showed a beginning of further dichotomy[2]). Therefore there is no doubt that the lobes were due to a process of dichotomous branching; in this respect they entirely fall in line with the stigmarian axes and are in striking contrast with *Isoetes*.

The lobes were covered on all sides by scars left by appendages. About the distribution of these appendages we have but little information. Judging from the published figures it is not impossible that they were arranged in irregular parastichies. In any case we may say that their occurrence at all sides of the lobes agreed with the stigmarian rootlet distribution, and was in contrast to the arrangement of the *Isoetes* roots.

The lobes further were cone-shaped; when a vegetative cone has been present its place is absolutely clear. As, however, the scars cover the lobes up to the apex, it may have been that the lobes were of a determined growth and that their vegetative cones soon became exhausted.

Concerning the appendages which are sometimes found in situ our information is rather meagre. They had a small excentric stele like the stigmarian rootlets and the *Isoetes* roots; about root-cap and root-hairs nothing is known. There is however one important fact, described by SOLMS [3]), namely that the appendage traces caused gaps in the lobar stele; this correspondence to the stigmarian rootlets is an important argument for the homology of the lobes with the stigmarian axes and of the appendages with the stigmarian rootlets.

Apart from the adult stems of about 2 metres numerous small specimens have been observed. The most curious fact about them is that they were similar in all respects to the big ones, only much smaller. They had a stem of for instance 12—16 mm diameter instead of 6—9 cm, but their lobes were perfectly analogous, the only difference being [4]) that the lobe apices were not yet acute but more globular.

---

[1]) K. MÄGDEFRAU, Zur Morphologie und phylogenetischen Bedeutung der fossilen Pflanzengattung *Pleuromeia*, B.B.C., 48, 2, 1931, p. 119; already earlier observed by BISCHOFF 1855.

[2]) H. POTONIÉ, Abbildungen und Beschreibungen fossiler Pflanzenreste, Lief. 2, Berlin 1904.

[3]) SOLMS-LAUBACH, l.c. 1899 (f.n. p. 32), see p. 235.

[4]) TH. SPIEKER, *Pleuromeia*, eine neue fossile Pflanzengattung und ihre Arten,

These small specimens probably were nearly or quite as full-grown as the big ones; SOLMS in discussing them points out [1]) that it is extremely improbable that the small specimens could grow up by an evenly distributed growth, but that the only plausible conception, here as in the case of the Lepidophytes, was the development of the aerial stems from a rhizome.

Our conclusion can only be that the *Pleuromeia* stem base is really homologous to a Lepidophytic stem base, but on a reduced scale; and further that its appendages, in the same way as in the stigmarian rootlets, were modified leaves.

**7. The Nathorstiana stem base.** The stem of this recently described fossil [2]) had a basal part provided with some 12 to 20 longitudinal ridges, separated by shallow furrows. On the ridges appendages were borne, in the upper part of the ridge in a single, in the lower part in from two, up to four, rows.

The appendages which have been observed in situ left scars with a small excentric protuberance; evidently they had a small excentric stele like the appendages and roots in the former cases.

MÄGDEFRAU makes mention of indications of a partition of the stem base in old specimens into two or four lobes. His photographs however are very disappointing in that respect, showing only a somewhat irregular shape.

As far as the incomplete data allow I should be inclined to see in *Nathorstiana* a plant which has nothing whatever to do with the Lepidophytes, but which in its stem base is developed more or less along the same lines as *Isoetes*.

**§ 12. Appendages of the outer surface (Hairiness).** It is a remarkable fact that the stem and leaves of most microphyllous Pteridophytes are absolutely glabrous. Only the root and sometimes the rhizome with their root-hairs show other conditions.

Exceptions to this rule are rare; in the *Psilophytinae* they seem to be entirely lacking.

Amongst the *Articulatae* the only case I found mentioned was that of a "tomentum" developed in some *Equisetum* spp, consisting of simple and filiform hygrometric processes of the epidermis cells, up to 5 mm long. According to DUVAL-JOUVE this tomentum is present on the rhizomes of *E. maximum, sylvaticum* and *arvense*; in *E. littorale, limosum, palustre* and *pratense* it only occurs on the rhizome leaf sheaths [3]).

gebildet aus der *Sigillaria Sternbergi* Münst. des bunten Sandsteins zu Bernburg. Zeitschr. f. d. ges. Naturwiss., Halle, 3, 1854, p. 177 see p. 185.

[1]) SOLMS-LAUBACH 1899, l.c. p. 239.

[2]) K. MÄGDEFRAU 1932, l.c. (f.n. p. 58).

[3]) J. DUVAL-JOUVE, Histoire naturelle des *Equisetum* de France, Paris 1864, see on p. 18.

In *Lycopodium* the leaf margin may sometimes bear small teeth or cilia [1]), and in *Selaginella* the epidermis cells of the stem may form cylindrical outbulging processes giving rise to hairs [2]); only in some cases do we find independently formed hairs, borne on a podium of small cells.

In the *Filicinae* on the other hand hairs are abundant everywhere, mostly in the special form of paleae, being flat multicellulair hairs, though stellate and peltate forms occur too. These paleae have been rightly compared by several authors to microphylls, lately by BOWER [3]).

Indeed the likeness is often very striking. VELENOVSKÝ, giving a survey of these dermal appendages [4]), mentions the existence of a midrib consisting of special thick-walled cells in the paleae of the genus *Platycerium* and in *Asplenium Trichomanes*; moreover he mentions the fact that *Polypodium lycopodioides* owes its name to the scaliness of the rhizome, giving it a likeness to a *Lycopodium* shoot.

Besides the paleae there may be a second form of hairs, smaller and cylindrical, without any transitional forms between the two hair categories [5]); perhaps these hairs are other enations, independent of the line that once gave rise to the microphylls.

Groningen, May 1936.

Botanical Laboratory of
the Government University.

---

[1]) ENGLER & PRANTL, op.c. p. 583.

[2]) Ibidem p. 638 and moreover R. J. HARVEY-GIBSON, Contributions towards a knowledge of the anatomy of the genus *Selaginella* Spr., III, The leaf, Ann. of Bot., 11, 1897, p. 123, see pl. 9 fig. 1, 4, 14, 15.

[3]) BOWER 1935, op.c. p. 558.

[4]) VELENOVSKÝ op.c.p. 194, ibidem Part 4, 1913, p. 30.

[5]) VELENOVSKÝ 1905, op.c. p. 195.

# CHAPTER II

ANATOMY

by

J. C. SCHOUTE (Groningen)

**§ 1. Introduction. — A. Historical.** Anatomy, the study of internal construction, as compared to external morphology has the great disadvantage that its object usually is only open for observation after special preparation; often the studied spatial forms have to be synthesized by combination of the observations made on different sections.

When we further take into consideration that in vascular plants the internal structures as a rule are much more complex than the outer forms, it is no wonder that the knowledge of the Pteridophyte anatomy long remained behind the knowledge of the external forms of these plants.

Indeed we may say that the study of Pteridophyte anatomy only dates from the first half of the nineteenth century. The general works on plant anatomy from the beginning of that century [1]) only contain incidental and insignificant facts about Pteridophytes. Soon afterwards, however, we find a good anatomical description of the stem of *Equisetum fluviatile* by BRONGNIART [2]) and some years afterwards VON MOHL wrote his splendid paper on the stem structure of tree ferns [3]) which was to be followed by his equally excellent paper on *Isoetes* [4]).

The impression these papers make upon a reader of our times is that they

---

[1]) J. J. P. MOLDENHAWER, Beyträge zur Anatomie der Pflanzen, Kiel 1812; K. SPRENGEL, Von dem Bau und die Natur der Gewächse, Halle 1812; C. F. BRISSEAU-MIRBEL, Éléments de physiologie végétale et de botanique, Paris 1815.

[2]) AD. BRONGNIART, Histoire des végétaux fossiles, I, Paris 1828, see p. 100.

[3]) H. VON MOHL, De structura caudicis filicum arborearum, 1833, in: C. F. PH. DE MARTIUS, Icones plantarum cryptogamicarum quas in itinere annis 1817–20 per Brasiliam instituto collegit et descripsit, Monachii 1828–1834; shortened and translated into German in: H. VON MOHL, Vermischte Schriften botanischen Inhalts, Tübingen 1845, see p. 108.

[4]) H. VAN MOHL, Über den Bau des Stammes von *Isoetes lacustris*, Linnaea 1840, p. 181; Vermischte Schriften p. 122.

are quite modern in ground-plan and execution. This is due to the fact that these authors tried, without bias, to record the prevailing rules and laws, governing the construction of the tissues. Before 1825 most anatomists (or phytotomists as they were often called) were either led by mere curiosity and by the desire to make a trial of their lenses, or they at once tried to understand the internal structure as a base for the fundamental physiological phenomena in the plant.

As the former aims did not yield sufficient impulses for the continued efforts required to master the multifarious facts, and as the latter aim even at present for the greater part is still out of our reach, the results on the whole were disappointing. The study of the existing regularities however, being a safe and efficient guide in the labyrinth of phenomena, yielded much better results, and this is still the experience of our times.

Once the facts are established by a thorough description, we may try to make use of our knowledge for physiological, systematical or other purposes, and the better our description and the greater the detail the better the results will be. As MOLL expounds in his "Phytography" [1]), experience has shown that by making descriptions, as detailed as possible, without any prejudice, results may often be obtained as unsought for by-products, which would have been out of the reach of the author when trying purposely to solve the same problems without making such a description.

After VON MOHL a number of authors followed, of which VON NÄGELI, METTENIUS, DUVAL-JOUVE and RUSSOW may be instanced. Up to the year 1874 their work has been summarized and quoted in the masterly work by DE BARY [2]), an unrivalled monument in the anatomical field. From 1874 up to the present day the stream of anatomical work on Pteridophytes, both recent and fossil, has gradually risen in accordance with the increasing interest paid to Pteridophytes in general.

**B. Scope of the chapter.** In view of the extension of the field and the limited space available for the present chapter a severe restriction of the subject-matter is peremptorily required.

This will be attained in the first place by omitting all histological facts and dealing only with the gross anatomy, the topography. For information on histology the reader may be referred primarily to BARY who gives an astonishing number of facts, and for more recent work to ENGLER and PRANTL [3]) and to ZIMMERMANN [4]); for fossils to HIRMER [5]) and to HOF-

---

[1]) J. W. MOLL, Phytography as a fine art, Leiden 1934, see p. 465.

[2]) A. DE BARY, Vergl. Anatomie der Vegetationsorgane der Phanerogamen und Farne, Leipzig 1877.

[3]) ENGLER und PRANTL, Die natürlichen Pflanzenfamilien, I, 4, Leipzig 1902.

[4]) W. ZIMMERMANN, Die Phylogenie der Pflanzen, Jena 1930.

[5]) M. HIRMER, Handbuch der Paläobotanik, München und Berlin 1927.

MANN [1]). With the aid of these works and their literature lists the scattered original sources may be found.

In the second place even the topography will not be dealt with exhaustively, but only in so far as it has a more general importance either for phylogeny, or for the ontogenetical elaboration of the plant structure. It goes without saying that the choice of subjects has been rather arbitrary and has been determined by the private views, knowledge and experience of the present author. For more detailed information the same works quoted above may be recommended, with the exception of HOFMANN who only deals with histology, and with the important addition of BOWER [2]).

Developmental anatomy, in the sense of a study of cell division order, which was such a prominent feature in the researches of the nineteenth century will be left entirely out of consideration, as it has been recognized long since that it is in no way decisive for further differentiation; for striking new instances see BOWER [3]).

On the contrary due attention will be paid to the presumable order in which the plant structures are first induced. The process of induction not being open for direct observation itself, this has to be gathered from the results in the later stages in which the induced structures have become visible; in view of the easy observation of large complexes the adult condition often gives the best information.

### § 2. Anatomy of the stem. — A. Primary organization.

**I. E p i d e r m i s.** In such cases where the stem surface has not been wholly taken up by leaf insertions, but where a free stem surface is to be observed, the outer cell layer is called the epidermis.

Unlike the condition in Angiosperms, the boundary between epidermis and cortex is usually an irregular surface; this is due to the fact that the epidermis does not arise from an early defined dermatogen, but often is separated by late partition walls from the inner cells. Accordingly HEGELMAIER speaks of a most external, "die Stelle einer Epidermis vertretenden" cell layer [4]). The distinction between a one-layered and a many-layered epidermis, often to be made in Angiosperms, as a rule therefore is impossible in Pteridophytes.

When the plant forms different kinds of stem, the epidermis in these stems may be differentiated along divergent lines too, for instance in *Psilotum Bernhardi* in the rhizome and the aerial stem [5]). Two different

[1]) ELISE HOFMANN, Paläohistologie der Pflanze, Wien 1934.

[2]) F. O. BOWER, Primitive land plants also known as the Archegoniatae, London 1935.

[3]) BOWER, op.c. p. 328.

[4]) F. HEGELMAIER, Zur Morphologie der Gattung *Lycopodium*, Bot. Ztg 30, 1872, col. 773, see col. 797.

[5]) W. L. BEEKMAN, Über die Torsion des Stengels von *Psilotum Bernhardi*, Rec. trav. botan. néerl. 21, 1924, p. 1.

forms of epidermis may occur moreover in one and the same stem next to each other, when the underlying tissues differ locally, as in *Equisetum*.

Special kinds of organs may be present in the epidermis: stomata and hairs. Stomata are always present in such stems as are the chief photosynthetic part of the plant: *Rhynia, Psilotum, Equisetum*; when the leaves are well developed, stomata may be present (*Asteroxylon*) or they may be lacking (*Selaginella*, probably in most *Filicinae*).

**II. Cortex and ground-tissue.** The tissue between the epidermis and a monostele is called cortex; in the case of polystely the internal limit of the cortex being vague, the term ground tissue which includes the non-vascular part of the stele may be used instead.

The cortex may be homogeneous (*Ophioglossum* [1]), *Lycopodium Selago* [2])) or it may be homogeneous for the greater part, with locally differentiated other tissue parts (*Marattia*, parenchyma with mucilage ducts), but usually the cortex is divided into an inner and outer, or even into three concentric zones, the tissues of which are different in organization. Some instances are: *Psilotum*, aerial stem: outer cortex chlorenchyma, middle cortex sclerenchyma, inner cortex collenchyma; *Asteroxylon*, aerial leaf stem: outer cortex of tangentially extended cells, inner cortex divided into three layers: two zones of compact tissue and a zone of trabecular tissue with large intercellular spaces in between.

In other cases the differentiation is not in concentric layers, but in tangentially alternating tissue complexes (*Equisetum*, chlorenchyma and so-called vallecular cavities).

The cortex moreover in all Pteridophytes, the leafless *Rhyniaceae* only excepted, is traversed by the leaf-traces, sloping downward from the leaf insertion towards the stele, usually rather steeply. In the Lepidophytes these leaf-traces are accompanied at their abaxial side by a gutter-like half sheath of a loose tissue, called the parichnos.

The innermost layer of the cortex [3]) usually is of a particular organization and is called the endodermis; in *Rhynia* an endodermis is believed to have been absent. Otherwise it often occurs in the form of a protective sheath with Casparian strips.

In such cases where the vascular strands in the stele are separated by large parenchyma complexes, the endodermis instead of surrounding the whole stele may surround the individual strands separately (*Equisetum Heleocharis*). In this case, which rather inappropriately has been termed polystely, the cortex communicates with the pith without a definite

---

[1]) H. G. HOLLE, Über Bau und Entwickelung der Vegetationsorgane der Ophioglosseen, Bot. Ztg 33, 1875, col. 241, see col. 246.

[2]) HEGELMAIER l.c. (f.n. p. 67), col. 796.

[3]) For the reason why this layer may be taken as the innermost cortex layer see J. C. SCHOUTE, Die Stelär-Theorie, Groningen 1902, Groningen und Jena 1903, on p. 162.

boundary. The normal case in contrast with polystely is called monostely.

In *Equisetum hiemale* and *E. ramosissimum* a most curious complication has been observed by PFITZER in the rhizomes, consisting of the development of narrow closed accessory endodermis sheaths, surrounding only a limited amount of common parenchyma, and running in a longitudinal direction between the normal bundles; usually one singly, rarely two or three of them between two bundles.

These "Zwischenkernscheiden" as PFITZER calls them, may be either free at both ends or they may unite at the upper or lower end to the normal endodermis [1]).

**III. S t e l e.** The stele is the most complex and the most variably constructed part of the stem.

Its outermost cell layers — or in the case of polystely the outer layers of its meristeles — almost invariably are parenchymatous and are called the pericambium or pericycle.

Further it contains one or more vascular strands, composed of phloem and xylem strands, and often smaller or larger parenchyma strands. In the phloem and xylem strands a difference of organization may be present between the first differentiated parts, the protophloem and protoxylem, and the later differentiated metaphloem and metaxylem. The parenchyma may form strands between the peripheral vascular strands, called medullary commissures, and also a smaller or larger central massive strand, the pith or medulla.

In the case of medullated monostely the external pith layer may be differentiated as an internal endodermis (*Equisetum hiemale*).

The different arrangements of the several tissues in the stele have given rise to an intricate (and for the greater part superfluous) stelar nomenclature [2]). Of these terms only a few will be used here, namely:

protostele = a monostele with a peripheral phloem and a central massive xylem;

siphonostele = a medullated protostele wtih internal endodermis;

amphiphloic siphonostele = a siphonostele with a second phloem ring inside the xylem;

solenostele = a siphonostele, perforated by scattered leaf gaps;

dictyostele = a solenostele with more leaf- or other gaps, so as to show more than one interruption on any transverse section;

meristele = the vascular parts of a dictyostele between two neighbouring gaps, appearing in transverse section as separate strands.

Before entering into a consideration of the diverse forms of stelar

---

[1]) E. PFITZER, Ueber die Schutzscheide der deutschen Equisetaceen, Jahr. f. wiss. Bot. 6, 1867, p. 297, see p. 318.

[2]) See SCHOUTE, op.c. p. 144 and moreover F. J. MEYER, Die Stelärtheorie und die neuere Nomenklatur zur Beschreibung der Wasserleitungsbahnen der Pflanzen, B.B.C. 33, 1, 1916, p. 129.

structure in the different Pteridophyte groups, we may remark that as far as is known all Pteridophytes in the sporeling start with a protostelic stem. This fact has always been recognized as a proof of a phylogenetic derivation of all Pteridophytes from protostelic ancestors.

Notwithstanding this common origin, the resulting stelar structures in the adult plants are very divergent in the different groups. We shall therefore begin by considering a number of representative forms. To shorten the exposition only the distribution of the xylem will be dealt with, being the best known and giving a sufficient idea of the whole stelar structure. A comparison of the forms will follow afterwards.

*a.* R h y n i a c e a e. The stele in the *Rhyniaceae* is invariably a protostele, in all stages of the life cycle. This fact, only recognized later, of course has greatly corroborated the view, just mentioned, on the phylogenetic importance of the protostele. The organization of the protostele of the *Rhyniaceae* is very simple, as it is quite cylindrical and as the xylem only consists of tracheids without any parenchyma.

A difference between large tracheids at the outside and a central core of smaller tracheids is to be noticed in *Rhynia major* and in *Hornea*, not in *Rhynia Gwynne-Vaughani*: notwithstanding the similar organization of the larger and the smaller tracheids they are to be considered as proto- and metaxylem, especially as in *Hornea* the inner tracheids sometimes are found broken down inside the ring of intact larger tracheids.

In contrast in the early Devonian *Gosslingia* the protoxylem surrounds the metaxylem with a continuous cylindrical zone [1]); the conditions of endarchy (*Rhynia major*) and exarchy (*Gosslingia*) therefore have both been present in vascular plants as far as our record goes.

*b.* A s t e r o x y l o n. In the rhizome and in the leafless lateral branches, bearing the sporangia, the stelar structure is much like that of *Rhynia*, a central homogeneous xylem being present. In the leafy shoot however the xylem becomes stellate; as the spaces between the longitudinal xylem flanges are filled up with phloem tissue, the stele itself remains cylindrical. At the peripheral edges these flanges are somewhat thickened and in the centre of these thickenings an immersed protoxylem is present, the tracheids of which are of the same construction as those of the metaxylem, but smaller in size. In the young condition the protoxylem on which the mesarch leaftraces are inserted has clearly been demonstrated to be differentiated before the rest of the xylem.

*c.* P s i l o t a c e a e. In the young plant the stem is protostelic in *Psilotum* as well as in *Tmesipteris*; in the leafless rhizome of *Psilotum* it seems to lack a protoxylem. In the adult condition the stelar organization is

---

[1]) A. HEARD, On old red sandstone plants showing structure, from Brecon (South Wales), The Quarterly Journal of the Geological Society of London, 83, 1928 (volume for 1927), p. 195; see p. 199 and Pl. 13, fig. 7, Pl. 14, fig. 3, 7.

rather different in the two genera; in *Psilotum* the xylem is exarch, with a number of protoxylem strands, projecting from a tubular metaxylem mass with a central medulla. In *Tmesipteris* there is a ring zone of separate mesarch xylem strands, and sometimes one or a few similar strands in the medulla.

Leaf-traces connect the sporophylls of *Psilotum* with the stelar proto-xylem strands; the sterile leaves may have feeble traces or they may lack them. Yet the stelar protoxylem strands are related to the leaf arrangement. Some distance under the leaf insertion (3–8 mm) the stelar xylem has a prominent point, tapering slowly downwards [1]), and the stelar protoxylem strands more or less agree in number with that of the irregular leaf ortho-stichies [1]).

*d.* L y c o p o d i u m. The sporeling begins either with a protocorm without any stele, or it begins directly with a shoot; in the latter case a protostele with 2—4 peripheral protoxylem strands is formed [2]). In the adult stem the peripheral protoxylem strands usually are more numerous; the metaxylem may either form an irregularly fluted cylinder, the anastomosing ridges being situated inside the protoxylem strands, or it may form a small number of flat bands, extending vertically, parallel to one another and lined at their edges by protoxylem strands; finally the metaxylem may occupy the whole of the space inside the protoxylem strands but for a number of irregular meshes [3]).

In literature we find several indications of the relation between the stem protoxylem strands and phyllotaxis [4]). By the observations of these authors it has been established that the leaf-traces do not run straight downwards in the stele, but that every trace, before inserting on a stem strand, may shift laterally in order to meet that strand; accordingly their insertion on the strand may be either in the median plane or laterally.

The stem protoxylem strands on the whole follow a longitudinal course, with some tangential undulations; they may freely unite and split during their course. The number of these strands only corresponds in a very rough way with the phyllotaxis. In *Lycopodium Selago* stems with pentamer-

[1]) C. NÄGELI, Das Wachsthum des Stammes und der Wurzel bei den Gefäss-pflanzen und die Anordnung der Gefässstränge im Stengel, in: NÄGELI, Beitr. z. wiss. Botanik, 1, 1858, p. 1, see p. 52.

[2]) F. J. MEYER, Die diaplektischen Leitbündel der Lycopodien im Lichte der vergleichenden Anatomie und der Paläobotanik nebst einem Ausblick auf die übrigen Pteridophyten, *Engler's* Botan. Jahrb., 60, 1926, p. 317, see p. 324.

[3]) F. J. MEYER 1926, l.c. p. 319.

[4]) C. NÄGELI, Über das Wachsthum des Gefässstammes, in: SCHLEIDEN und NÄGELI, Zeitschr. f. wiss. Botan., 3 und 4, 1846, p. 129, see p. 133; C. CRAMER, Ueber *Lycopodium Selago*, in: NÄGELI und CRAMER, Pflanzenphysiologische Un-tersuchungen 3, 1855, p. 10, see p. 14, C. NÄGELI l.c. 1858, p. 53; HEGELMAIER l.c. (f.n. p. 67), col. 792.

ous whorls often have five protoxylem strands, the variation going from 4—6 or even up to 8; stems with tetramerous whorls usually have 4, sometimes 5 or 6. So here for every two leaf orthostichies about one protoxylem strand is present.

In *Lycopodium alpinum* on the other hand slender decussate stems usually have four strands, sometimes 3, and in *Lycopodium clavatum* stems with 4 + 5, 5 + 6, 6 + 7 systems have 10—17 strands, i.e. more than the number of leaf orthostichies [1]).

*e.* S e l a g i n e l l a. In the leafy stem many *Selaginella* spp. have a flattened stele, the edges being turned towards the sides of the dorsiventral stem. The stele is usually a protostele with two lateral protoxylem strands to which the leaf-traces run, one strand receiving the left row of under and upper traces, the other strand the right rows.

In other species the conditions may be different [2]). In the first place the xylem may become cylindrical and in many species at the same time polyarch, with a number of protoxylems all around; moreover the central metaxylem may be reduced so as to form a pith (*S. spinosa*, erect stem).

In the second place the xylem may become endarch instead of exarch (*S. spinosa*, trailing stem).

In the third place the stem may become polystelic, with 2 or 3, sometimes up to 12 meristeles (*S. laevigata* var. *Lyallii*, erect stem); at the branching places of the stem all meristeles are connected.

Lastly the stele may become siphonostelic by the development of an inner pericambium and endodermis around the pith, and if so, it may develop an inner ridge at the dorsal side which may be separated from the tube to form a new small "stele" inside the pith of the normal one (*S. laevigata* var. *Lyallii*, rhizome).

The rhizophores always have one single protostele, but otherwise their stelar structure may show remarkable differences in the different species [3]).

In *Selaginella Martensii* and some other spp. the xylem is exarch and monarch, the protoxylem strand being turned towards the abaxial side of the rhizophore at its origin, but turned towards the dichotomy centre in

---

[1]) See foot-note 4 p. 71.

[2]) A. DE BARY op.c. (f.n. p. 66) p. 293; R. J. HARVEY-GIBSON, Contributions towards a knowledge of the anatomy of the genus *Selaginella*, Spr. [The stem], Ann. of Bot. 8, 1894, p. 133.

[3]) See: C. NÄGELI und H. LEITGEB, Entstehung und Wachsthum der Wurzeln, in: C. NÄGELI, Beiträge zur wiss. Botan, 4, 1868, p. 73, esp. p. 124; PH. VAN TIEGHEM, Recherches sur la symétrie de structure des plantes vasculaires, Ann. d. Sc. nat. 5th ser, 13, 1870/1, p. 1, esp. p. 88; R. J. HARVEY-GIBSON, Contributions towards a knowledge of the anatomy of the genus *Selaginella*, IV, The root, Ann. of Bot. 16, 1902, p. 449; J. C. TH. UPHOF, Contributions towards a knowledge of the anatomy of the genus *Selaginella*. The root. Ann. of Bot. 34, 1920, p. 493.

its shanks [1]). This structure may be taken as a simplified stem stele, but also as that of a simplified root stele; it is identical with that of the roots of *Selaginella* and of *Isoetes*.

In *S. atroviridis* the metaxylem being crescentic, a number of protoxylem strands is to be found on the concave adaxial side [2]); and in *S. Kraussiana* [1]), *S. Poulteri* and *S. delicatissima* [3]) the stele is endarch like that of *Hornea*. These cases are therefore like simple stem steles; for root steles the structures would be quite out of the normal scheme, the more remarkable as this root scheme is otherwise the same in all vascular plants.

These stelar structures, varying in about the same way as in the leafy stem, all fit in with the view that the rhizophore is a modified stem; but offer serious objections to the view that the rhizophore is a modified root [4]).

*f.* L e p i d o p h y t e s. The stelar xylem in all aerial stems is exarch with a large number of projecting or flat protoxylem strands or with a continuous protoxylem zone all around. The metaxylem may be solid, so that the stele is a protostele, or it may contain a large parenchymatous pith; in other cases there may be a "mixed pith", consisting of intermingled parenchyma and tracheids.

Little is known about the numerical relation between leaf orthostichies and protoxylem strands: HIRMER reports that in *Sigillaria elegans* their numbers seem to be equal [5]). The leaf-traces insert either at the top of the ridges or laterally, at their flanks.

In the subterranean stem organs, the stigmarian axes, the xylem sometimes is exarch too [6]), but in most cases the xylem is endarch, with a great number of anastomosing protoxylem strands. The rootlet-traces, which the present author considers to be leaf-traces, in order to reach the protoxylem strands, in an endarch xylem have to pierce the secondary and the metaxylem. As they are accompanied on their course by parenchyma strands, medullary commissures are present, in a phyllotactical arrangement, like leaf-gaps.

*g.* I s o e t e s. In the stunted stem of this remarkable group the anatomical differentiations are more or less reduced. According to LANG [7])

---

[1]) NÄGELI und LEITGEB, l.c. p. 127.

[2]) HARVEY-GIBSON, l.c. 1902 (f.n. p. 72) p. 457.

[3]) HARVEY-GIBSON, l.c. 1902, p. 460.

[4]) NÄGELI und LEITGEB, l.c. p. 126; VAN TIEGHEM l.c. (f.n. p. 72) p. 97 devotes two pages to explain the anomalous case of *S. Kraussiana* and terminates by saying: "L'anomalie signalée par MM. NAEGELI et LEITGEB disparaît ainsi en s'expliquant." As far as I see the facts are not so easily to be reasoned away.

[5]) HIRMER op.c. (f.n. p. 66), p. 271.

[6]) F. E. WEISS, A *Stigmaria* with centripetal wood, Ann. of Bot. 22, 1908, p. 221.

[7]) W. H. LANG, Studies in the morphology of *Isoetes*. II. The analysis of the stele of the shoot of *Isoetes lacustris* in the light of mature structure and apical development. Mem. & Proc. Manchester Lit. & Phil. Soc. 59, 1914/15, no. 8.

the zones inside the anomalous secondary tissue are to be homologized
to an outer primary phloem inside which a parenchymatous xylem sheath
and a strongly parenchymarized central xylem follow. For our comparison
the case is of little value.

*h*. E q u i s e t u m. The sporeling axis at first is protostelic, at least
partly; very soon however the peculiar adult structure sets in, consisting of
a number of leaf-traces, arranged on a cylinder surface and connected at
the nodes to the lower leaf-traces by commissures.

The stelar leaf-traces consist of an internal protoxylem strand, occurring in the internodes in the form of a canal with loose rings, the carinal
canal, and further somewhat more towards the stem periphery two lateral
metaxylem strands; at the nodes a somewhat larger amount of metaxylem
surrounds the protoxylem.

These conditions are to be found in the monostelic as well as in the
siphonostelic or the polystelic forms; the distribution of the sheaths does
not influence the vascular construction.

In the strobilus axis the protoxylem strands are rather irregularly
connected, forming sympodial strands in various ways. The amount of
metaxylem being much greater here, only irregularly distributed gaps in
the xylem cylinder remain.

*i*. C a l a m i t e s. The stelar structure of *Calamites* is hardly different
from that of *Equisetum*; the main difference is that the leaf-traces in some
forms descend over two internodes instead of over one, so that the number
of vascular bundles in any node is about twice that of the entering
traces.

*j*. S p h e n o p h y l l u m. The stem of *Sphenophyllum* has a protostele
with three large protoxylem strands at the periphery and a triangular
metaxylem in between. In some cases the three protoxylem strands are
double, so that there is an approach to hexarchy.

The leaves of *Sphenophyllum* always being placed in superposed trimerous whorls or in whorls with a multiple of three as number, the leaf-traces
enter before the protoxylem strands and insert on them.

*k*. C o m p a r i s o n   o f   s t e l a r   s t r u c t u r e   i n   m i c r o-
p h y l l o u s   P t e r i d o p h y t e s.

1. R e d u c t i o n   o f   c e n t r a l   p a r t   o f   x y l e m. A comparison
of the primitive *Rhyniaceae* with most higher developed microphyllous
Pteridophytes (*Psilotaceae, Lycopodium*, Lepidophytes, *Isoetes, Equisetum,
Calamites*) as well as a comparison of the sporeling structure with the
adult condition in these plants reveals us the fact that the central part of
the xylem is often more or less reduced.

This may begin with a development of living parenchyma cells between
the tracheids, so that a mixed pith is formed. No doubt this phenomenon
is connected with what BOWER has termed the size factor, bringing for
increasing xylem structures the necessity to increase their limiting surfaces

in proportion to the cube of the linear dimension, not to the square [1]).

Finally a wholly parenchymatous pith may be formed, a fact which may be connected too with the same size factor, at least in part of the cases, and moreover with the mechanical needs of the stem, a hollow cylinder being more stable than a solid core of the same weight, by its greater diameter.

2. A r r a n g e m e n t   o f   p r o t o-  and  m e t a x y l e m. The position of the protoxylem strands in the stem stele is remarkably varying. In a number of cases these strands are situated at the xylem periphery (*Gosslingia, Psilotum, Lycopodium, Sphenophyllum*), often even as outwardly projecting ridges.

In other cases the protoxylem is immerged in the metaxylem (*Asteroxylon, Tmesipteris*) and in a very few cases it occupies a central position (*Rhynia major, Hornea*). Moreover some plants have exarch and endarch conditions realized in different parts or in different kinds of stems (*Selaginella spinosa*, Lepidophytes).

It has often been held that the metaxylem inside the protoxylem, the so-called centripetal wood, should belong to another morphologic and phylogenetic category than the centrifugal wood outside the protoxylem [2]).

In view of the above facts the distinction seems to lose much of its importance, and perhaps the more simple suggestion may be advanced that when for some unknown reason the protoxylem changes its position, the metaxylem having to put up with the space left, is obliged to develop in another direction.

3. I n f l u e n c e   o f   l e a f-t r a c e s. The present author has already, on several occasions, pleaded the view that leaf-traces are not given off from the stele, but that on the contrary the stimulus inducing the traces is decurrent from the leaf into the stem. This even holds good for the case often occurring in *Psilotum* where a leaf-trace is not developed in the cortex, but the protoxylem ridge of the stele begins some mm under the leaf insertion.

In support of this view which has been proved by von Nägeli in 1858 [3]), many further facts may be alleged [4]) and it has never been refuted in literature. Now we may ask whether entering leaf-traces give rise to changes in the stem stele. As a rule the protoxylem strands of the trace are

---

[1]) F. O. Bower, Size and form, London 1930.

[2]) D. H. Scott, The old wood and the new, New Phytol. 1, 1902, p. 25.

[3]) Nägeli, l.c. (f.n. 4 on p. 71) p. 45.

[4]) J. C. Schoute, Beiträge zur Blattstellungslehre II. Über verästelte Baumfarne und die Verästelung der Pteropsida im allgemeinen, Rec. trav. bot. néerl. 11, 1914, p. 93, see p. 156; O. Posthumus, On some principles of stelar morphology, ibid. 21, 1924, p. 111, see p. 145, J. C. Schoute, On the foliar origin of the internal stelar structure of the *Marattiaceae*, ibid. 23, 1926, p. 269, see p. 271; J. C. Schoute, On pleiomery and meiomery in the flower, Ibid. 29, 1932, p. 164, see p. 214.

to be followed over a certain distance in the stele, and usually they unite with the stem protoxylem strands.

In examining the influence of these leaf-traces the first question to be raised may be whether the stellate xylem of *Asteroxylon* owes its form to leaf-trace influence.

The fact that there, where leaf-traces are absent, the xylem in *Asteroxylon* is cylindrical (rhizome, peculiar axes) points in that direction; moreover the fact that the xylem flanges always contain leaf-trace protoxylem strands in their distal parts is in favour of that view.

A solution of this point will remain impossible as long as insufficient data are available on the relation between phyllotaxis and flange distribution.

Unlike the case of the *Filicinae*, where often large leaf-gaps are to be observed, the traces of microphyllous plants usually give no further changes. This may partly be due to the fact that the leaf-traces are so much weaker than in the *Filicinae*; small traces in ferns do not cause gaps either. It may be connected, moreover, with the peripheral position of the stem protoxylem in many forms, the union of trace and protoxylem occurring at the outside.

When we look at the rare cases of endarch xylem we find that in the trailing stem of *Selaginella spinosa* the small traces simply pierce the metaxylem without causing any gaps [1]). In the stigmarian axes with centrifugal wood on the other hand the rootlet-traces as described above cause gaps.

JEFFREY once enounced the view [2]) that the microphyllous Pteridophytes only form ramular gaps, whereas the Ferns and Spermophytes moreover in all cases form leaf-gaps. It has been long recognized since that this generalization was too wide [3]).

4. S t e m  x y l e m  a n d  l e a f-t r a c e  x y l e m. POSTHUMUS makes a theoretical distinction between the original cauline xylem as present in *Rhynia* and the xylem which is formed under the influence of the entering leaf-trace, categories which he distinguishes as the stem xylem and the leaf-trace xylem [4]).

The question rises whether and in how far this distinction may be carried through, and whether is has a biological meaning. If we, for instance, could show that the stem xylem was induced acropetally and the leaf-trace xylem basipetally, then the distinction would be well established.

---

[1]) HARVEY-GIBSON, l.c. (f.n. 2 on p. 72) p. 172, Pl. 9 fig. 37.

[2]) E. C. JEFFREY, The structure and development of the stem in the Pteridophyta and Gymnosperms, Phil. Trans. 195, 1902, p. 119, see p. 144; Are there foliar gaps in the Lycopsida? Botan. Gaz, 46, 1908, p. 241,

[3]) BOWER, op.c. 1935 (f.n. p. 67) p. 329.

[4]) POSTHUMUS l.c. (f.n. p. 75) p. 155.

For the protoxylem strands this question amounts to the alternative whether the stem protoxylem strands are sympodia or not. When they are cauline, like the central protoxylem is in *Rhynia major*, their number may be regulated in harmony with the size factor by acropetal splitting or union, or perhaps by insertion of new strands between the others and by discontination.

In such a way *Lycopodium* might satisfactorily be taken, the more or less independent cauline protoxylem strands receiving the numerous leaf-trace protoxylems from various sides. In other cases, however, there are objections, namely when the distribution of these strands is regulated by phyllotaxis.

When in *Equisetum* the number of strands is always equal to that of the teeth of the next higher sheath, there is no doubt; in *Equisetum* all protoxylem strands are decurrent from leaf-traces, and the same holds good for *Calamites*. When in *Sphenophyllum* the stem invariably is triarch and the protoxylem strands always lie exactly under the leaf orthostichies, the same conclusion presents itself. The root of *Sphenophyllum* is diarch or tetrach, rarely triarch; this variation might return in the stem in case of independency of the stem protoxylem strands.

Having thus on one hand *Rhynia* with undoubtedly acropetal protoxylem and on the other hand *Equisetum* and some others with a basipetal leaf-trace protoxylem, most Pteridophytes are difficult to incorporate into either of these groups; perhaps intermediate cases may also occur. Detailed investigations about the existing conditions might possibly bring more light.

Turning now to the metaxylem we may begin with the remark that its differentiation always proceeds from the protoxylem "poles", as has even been stated in *Asteroxylon*. But as differentiation and induction are quite different processes this does not prove that the metaxylem is induced by stimuli emanating from the protoxylem.

Yet there are indications that at least in some plant groups this latter condition really occurs. Since in *Equisetum* the metaxylem is only to be found in small quantities in a definite position with regard to the protoxylem, or since in some *Lycopodium* species the metaxylem occurs in longitudinal plates always extending between protoxylem strands, we cannot escape the conclusion that the metaxylem distribution depends on that of the protoxylem.

For *Sphenophyllum* and for *Asteroxylon* the same conclusion may be drawn for the outer part of the metaxylem, the flanges. Nothing can be said, however, about the central core.

We come therefore to the general conclusion that the stem xylem which phylogenetically no doubt was once independent and which was induced acropetally, gradually may have come under the influence of the entering leaf-traces, or in other terms that it may have become dependent for its realization on stimuli emanating from these traces.

In this way the distinction between stem and leaf-trace xylem, though still being possible, looses much of its pregnancy.

One remark only remains to be made, namely that the stem stele, even when it forms no xylem or any other tissues independently, yet keeps its anatomical independence and importance; it acts as a recipient for leaf-trace stimuli. These stimuli in their centripetal course on having reached the stele, bend downward so as to induce the stelar vascular strands in their typical arrangement. This stele still develops in an acropetal way, just as it does in *Rhynia*.

5. P o l y s t e l y. From the above related facts it may be inferred that the polystelic condition is only an unimportant departure from the normal monostely. The term, framed by VAN TIEGHEM from the wrong idea of a splitting of the stele, may be used as before, notwithstanding its origin.

*l.* F i l i c i n a e.

1. R e d u c t i o n   o f   c e n t r a l   p a r t   o f   x y l e m. The phenomena of reduction of the central part of the protostele xylem in *Filicinae* are the exact counterpart of those in the microphyllous Pteridophytes; in a number of cases a mixed pith has been described [1]), and this may lead towards a complete medullation. There is, however, one complication, namely that a medulla may also be arrived at by fusion of leaf-gaps, as will be discussed below; a medullation independent from leaf-gaps is to be met with in most *Osmundaceae*, in *Gleichenia pectinata* and in the *Hymenophyllaceae* [2]).

2. A r r a n g e m e n t   o f   p r o t o-   a n d   m e t a x y l e m. Protoxylem strands are not always present in the stem stele. As TANSLEY and LULHAM write: "spiral protoxylems of many Ferns are confined to the leaf and are absent in the stem altogether" [3]); this is the case in the *Hymenophyllaceae*, the *Schizaeaceae* and according to POSTHUMUS for instance in *Platyzoma* [4]).

In many other cases the strands from the leaf-traces run down in the stem stele over a certain distance and then die out (e.g. in most *Osmundaceae* [5])). In ferns with an internal system the protoxylem strands may run down in the inner meristeles, as in *Cyathea moluccana* (= *Brunonis*) [6]) and

---

[1])  For examples see BOWER op.c. 1935 (f.n. p. 67), p. 334.

[2])  POSTHUMUS l.c. 1924 (f.n. p. 75), p. 155, 163; O. POSTHUMUS, On the anatomy of the *Hymenophyllaceae* and the *Schizaeaceae* and some additional remarks on stelar morphology, Rec. trav. bot. néerl. 23, 1926, p. 94, see p. 102.

[3])  A. G. TANSLEY and R. B. J. LULHAM, A study of the vascular system of *Matonia pectinata*, Ann. of Bot. 19, 1905, p. 475, see p. 503.

[4])  POSTHUMUS l.c. 1924 p. 166.

[5])  POSTHUMUS l.c. 1924, p. 115–139.

[6])  D. T. GWYNNE-VAUGHAN, Observations on the anatomy of solenostelic ferns, II, Ann. of Bot. 17, 1903, p. 689, see p. 708.

in *Matonia pectinata* [1]). In the normal outer dictyostele they often fuse with strands from lower traces and so form sympodial strands (e.g. in *Gleichenia pectinata* [2])).

The position of these protoxylem strands in the stem xylem is described by TANSLEY and LULHAM in the following words: "The exact position of the spiral protoxylems in relation to the metaxylem of the vascular strands of the stem is decidedly variable. Exarchy, endarchy, and mesarchy are all found within comparatively small groups, and the actual course of evolution seems to affect the position of the protoxylems much more freely and rapidly than in the other great groups of vascular plants"[3]).

Whether the stem stele sometimes may form protoxylem-strands other than those from leaf-traces has not been established. TANSLEY and LULHAM write: "spiral protoxylems in the stems of Ferns are always.... continuous with those of the petiole. In those cases in which there is no spiral protoxylem in the stem there may be a localized non-spiral protoxylem-band round the stele, which is exarch (*Loxsoma*, various *Davalliae*, &.) or endarch (*Schizaea malaccana*), but which has no connection with the leaf protoxylem; or the differentiation of tracheids may be more or less irregular (*Gymnogramme*, *Lindsaya*, etc.)" [3]).

So we see that, notwithstanding the fact that distinct protoxylem strands always seem to be foliar, there may be a more vague band of protoxylem belonging to the stem proper.

About the distribution of the metaxylem little need be said after the above remarks. Some factors influencing this distribution are the medullation, the development of a stellate xylem and the leaf-gap formation; for the last-named factors see below sub 3.

3. I n f l u e n c e   o f   l e a f - t r a c e s. In the first place we may again ask whether a stellate stem xylem may owe its form to the influence of leaf-traces. Stellate xylems occur in the family of *Clepsydropsidaceae*. As in the case of *Asteroxylon* the xylem flanges in their distal parts contain protoxylem strands on which the leaf-traces insert.

In *Ankyropteris Grayi* the correspondence between the phyllotaxis and the flange distribution is very conspicuous [4]); evidently the flange formation is ruled here by the leaf distribution.

In *Asterochlaena* and in *Asteropteris* both phyllotaxis and stele construction being more complicated, the relation between the two phenomena is not clear.

In these plants not only the stem xylem is stellate, but the whole stele assumes the same form, in contrast with the condition in *Asteroxylon*.

---

[1]) TANSLEY and LULHAM, l.c. (f.n. p. 78), p. 505.

[2]) POSTHUMUS l.c. 1924, p. 163.

[3]) See foot-note 3 p. 76.

[4]) D. H. SCOTT, Studies in fossil botany, 3rd ed., I, London 1920.

In the second place the leaf-gap formation has to be considered. The small traces of the first leaves in the sporeling usually insert on the stem stele without causing any gaps. The higher and larger traces however gradually begin to do so, as has already been described in great detail by DE BARY [1]).

About the processes going on during the formation of a leaf-gap, the following set of suppositions has been made by POSTHUMUS, suppositions which may be held to be the most satisfactory in the field [2]).

The leaf-trace in the *Filicinae* is composed of the united downward extensions of a number of elementary pinna-traces, each originally consisting of an endodermis, a pericycle, a cylindrical phloem and inside a mesarch xylem strand. These bundles may be simply united by juxtaposition, but in their further course their constituent parts usually fuse, up to the protoxylems, so that in the petiole the number of protoxylem strands may be small.

At the adaxial side of any protoxylem strand the formation of metaxylem is inhibited by some unknown influence over a certain area, dependent on the strength of the protoxylem strand; the procambial elements which might have formed the metaxylem then differentiate as parenchyma, casu quo as phloem or as sheath tissue.

A small protoxylem strand in a mesarch xylem therefore will be accompanied by a downward tapering parenchyma funnel inside the tubular metaxylem; a larger protoxylem strand will lie in a gutter-shaped metaxylem, the concavity falling towards the adaxial side.

When in a leaf-trace the parenchyma strand accompanying the protoxylem ends before the stem stele is reached, no interruption in the latter is formed (*Thamnopteris Schlechtendali* [3])); when it is continued further down, a depression in the stem xylem ensues, or a funnel-shaped pocket. The pockets caused by different leaf-traces may remain separate, but when extending far enough they will unite and a pith will be formed in this way (*Polypodiaceae, Cyatheaceae* [4]), *Schizaeaceae* [5])). The xylem being reduced in this way unto a hollow cylinder, the free parts of the pockets will appear as gaps in the cylinder, as leaf-gaps.

When in other forms a pith has been formed already by reduction of the central xylem part, the funnels will pierce the hollow xylem cylinder and cause analogous leaf-gaps [6]).

These are the views of POSTHUMUS which will be adopted here. For a different conception of the relation between leaf-trace and stem stele the

---

[1]) DE BARY, op.c. (f.n. p. 66), p. 294.

[2]) POSTHUMUS l.c. 1924, Ch. I.

[3]) POSTHUMUS l.c. 1924, p. 118, fig. 1.

[4]) POSTHUMUS l.c. 1924, p. 280.

[5]) POSTHUMUS l.c. 1926 (f.n. p. 78), p. 102.

[6]) POSTHUMUS l.c. 1924, p. 278.

reader may be referred to the excellent survey given by TANSLEY [1]). It may be remarked however that TANSLEY could not fail to recognize "the fact, which seems to be indisputable, that the leaf-trace leads, and the stele follows, in the course of evolution" [2]).

In a number of ferns where a pith is present, the stelar structure is complicated by the formation of an "internal structure", i.e. of additional vascular strands in the pith [3]). This internal structure is due to the influence of leaf-traces too, but in the details the mode of origin may be brought about in two different ways.

In the first place the internal structure may be due to the fact that the leaf-traces only partly unite to the dictyostele, part of their strands which already in the petiole occupy a more or less internal position directly running into the pith through the leaf-gap. This condition is realized in the *Cyatheae* [4]) were the leaf-traces are provided with two lateral inward folds and the strands of these folds pass into the pith, in which they may either form an inextricable network (*Hemitelia capensis* [5])) or they may end blindly (*Cyathea moluccana* (= *Brunonis* [6])). Such a leaf-trace course may be explained by assuming that the peripheral strands of any leaf-trace being accompanied by a hollow continuous parenchyma cylinder inside the strands, the central petiolar strands are forced to remain inside this parenchyma cylinder and so have to enter into the pith.

In the second place the leaf-traces may unite wholly to the dictyostele, but on their downward course in the dictyostele they may abut on lower leaf-gaps and may be forced by the parenchyma strands passing through these leaf-gaps to turn inward. These continued leaf-traces, as I have termed them [7]), when long enough, may unite and form a complex similar to that in the sporeling axis, a replica of the first protostele; the cylindrical form of this inner protostele being due to the fact that the inhibition zone of the outer dictyostele constitutes a parenchyma tube inside which the continued leaf-traces are compelled to run. When still stronger the continued leaf-traces begin to cause gaps in this "inner stele" and so a second dictyostele is built up, in which a third system may be formed, and so on.

---

[1]) A. G. TANSLEY, Lectures on the evolution of the Filicinean vascular system, New Phytol., 6, 1907, p. 25, 53, 109, 135, 148, 187, 219, 253; 7, 1908, p. 1, 29. Issued in book-form as New Phytologist Reprint 1908.

[2]) TANSLEY, New Phytologist, 1908, p. 2.

[3]) SCHOUTE l.c. 1926 (f.n. p. 75), with quotations of literature; see moreover TANSLEY l.c. 1907, p. 193; F. O. BOWER, The Ferns, Cambridge I, 1923 (p. 151), II, Cambridge 1926 (p. 101).

[4]) TANSLEY l.c. 1907, p. 223; POSTHUMUS l.c. 1924, p. 258.

[5]) G. METTENIUS, Über den Bau von *Angiopteris*, Abh. math. phys. Cl. K. Sächs. Ges. d. Wiss., 6, 1864, p. 501 see p. 525.

[6]) D. T. GWYNNE-VAUGHAN, l.c. (f.n. p. 78), p. 708.

[7]) SCHOUTE, l.c. 1926, p. 274.

TANSLEY proposes the term polycycly [1]) for the presence of internal systems, whether consisting of concentric cylinders or of scattered strands. As the term is very useful for the former case and does not apply very well to the scattered strands of the *Cyatheae*, it will be used here in the limited sense of only indicating the presence of concentric vascular structures.

This polycycly is to be found in several groups of *Filicinae*, in *Eu-* as well as in *Leptosporangiatae*. Evidently it is of a polyphyletic origin and the similarity of the phenomena is due to the development along the same lines; it is obvious that its importance for the plant, like that of the internal system of the *Cyatheae*, is to be found in the increase of the internal exchanging surface between vascular tissue and parenchyma, according to BOWER's size factor.

It occurs in all *Marattiaceae* [2]) and in *Psaronius* [3]); in some *Polypo-*

---

[1]) TANSLEY l.c. 1907, p. 193.

[2]) TANSLEY, l.c. 1907, p. 227; SCHOUTE, l.c. 1926.

[3]) Our detailed knowledge of the stelar structure of *Psaronius* has been founded by K. G. STENZEL (Über die Staarsteine, Nova Acta 24, 1854, p. 753; Psaronius in: "H. VON MEYER, Palaeontographica, Beiträge zur Naturgeschichte der Vorwelt, 12, H. R. GOEPPERT, Die fossile Flora der Permischen Formation, Cassel 1864/'65", p. 46; Die Psaronien, Beobachtungen und Betrachtungen, in: Beiträge zur Paläontologie und Geologie Österreich-Ungarns und des Orients, 19, 1906, p. 85); R. ZEILLER (Bassin houiller et permien d'Autun en d'Épinac, 2, Flore fossile 1, in: Études des gites minéraux de la France, 1890, see p. 178); K. RUDOLPH (Psaronien und Marattiaceen, Denkschr. k. Akad. d. Wiss. Wien, math. naturw. Kl. 78, 1906, p. 165).

HIRMER in his hand-book (op.c. 1927 (f.n. p. 66) see p. 545) gives an elaborate and thorough description of the anatomical facts known about *Psaronius*. In his text as well as in his diagrams, however, some misconceptions have crept in which may tend to cause confusion.

The least harmful of these errors is that in *Psaronius infarctus* a difference is assumed between two kinds of dictyosteles, as every first, third, fifth etc. dictyostele is supposed to have no relation to the continued leaf-traces which only attach to the second, fourth etc. (op.c. p. 559 and fig. 680).

As *Ps. infarctus*, described and pictured from one single specimen by ZEILLER (op.c. p. 208) and afterwards elucidated from ZEILLER' figures with perfect clearness by RUDOLPH (l.c. p. 177, Pl. 2, fig. 1), belongs to the best known *Psaronii*, we are in the position to state that *Ps. infarctus* in no way differs from the ordinary scheme and that HIRMER's fig. 680 should be cancelled.

More important are two independent errors, expressed in fig. 674, representing the bundle course in *Ps. Ungeri*, after STENZEL's drawings. HIRMER gives a radial section through a cylindrical stem part, and delineates a number of obconical dictyosteles, the one within the other. All these dictyosteles are represented as abutting free on the stelar surface; in a superficially tangential section through the stele we therefore should have met with all dictyosteles.

From the description given above it must be clear however, that the outer bundles of the whole stele, from its formation in the sporeling up to its last stages

*diaceae* (*Saccoloma elegans*, *S. domingense*, *S. inaequale* [1]); *Dennstaedtia*

in the vegetative cone, form one single mesh-work. This is not only the case in all *Marattiaceae*, and not only must it be so on theoretical grounds, but it has been stated directly by ZEILLER for *Ps. infarctus* (op.c. p. 183, Pl. 15, fig. 2).

The dictyosteles never are obconical in a cylindrical part of a stem; an obconical shape of the dictyosteles, as drawn for instance by METTENIUS for *Angiopteris evecta* (l.c. Pl. 2, fig. 1) only occurs in an obconical part of a stem; in general all dictyosteles are parallel to the stem surface, as has been fully realized by ZEILLER (op.c. p. 178) and by RUDOLPH (l.c. p. 178).

In the second place the dictyosteles in HIRMER's diagram reach farther down at one side than at the other side, all longer ends being situated at the same stem side, between two leaf orthostichies. For such an assumption which presents great difficulties for mental picturization, there is no evidence.

A last point to be mentioned here does not concern a diagram but only a statement in the text: HIRMER assumes that in *Psaronii* with a spiral phyllotaxis the dictyosteles together constitute a conical spiral, a conical winding-staircase. This again is a mistake; in a case of spiral phyllotaxis the dictyosteles are parallel to the surface as in other cases. In *Marattia* with a main series phyllotaxis the dictyosteles are quite independent, and the same is to be found in other *Marattiaceae*.

The opinion of HIRMER is based on *Psaronius Demolei* of which ZEILLER gave drawings (op.c. Pl. 24) of transverse sections; these drawings do not allow a sufficient analysis of the stelar structure.

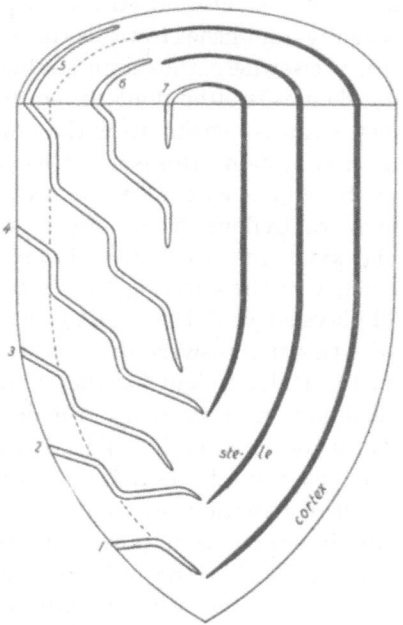

FIG. 1. Diagram of basal part of longitudinally halved polycyclic stem. 1—7 = leaf-traces and continued leaf-traces of seven leaves, all placed in a single orthostichy.

For the sake of clearness it may be advisable to substitute a new diagram of a longitudinal section through a stele with internal structure of the kind treated here. Our fig. 1 may be taken as a trial; it has been kept as simple as possible so as to be applicable not only to all *Psaronii* but also to all other polycyclic *Filicinae*. It represents a stem part, obconical in its lower and cylindrical in its upper half; at the top a half transverse section has been added. The leaves are supposed to be arranged in one single orthostichy, a case never realized in nature, and the number of dictyosteles is limited to three. The leaf-traces are represented in white, the "stem xylem" in black. The description of the general features in the text will further sufficiently explain the diagram.

[1] POSTHUMUS l.c. 1924 (f.n. p. 75) p. 219. H. KARSTEN, quoted there, has been the first to understand the structure of a polycyclic stele (1847).

*cornuta, D. rubiginosa, D. rufescens* [1]); *Pteris Kunzeana* [2]); *Acrostichum aureum* [3]), in the latter case combined with the direct entering of some trace strands through the leaf-gap into the pith as in the *Cyatheae*); in the *Cyatheaceae* in *Thyrsopteris* [4]) and finally among the *Matoniaceae* in *Matonia pectinata* [5]).

In literature the internal structures nearly always are described as running in an acropetal direction. The fact that the regularity of distribution of the internal strands is perfect in the outermost dictyostele and gradually diminishes towards the innermost dictyostele clearly proves that here, as elsewhere, the stimuli inducing the strands must have been basipetal; when as is often done the continued leaf-traces are described as reparatory strands, arising from the inner systems and closing the leaf-gaps in the outer systems, this is not in harmony with their actual origin.

4. S t e m  x y l e m  a n d  l e a f-t r a c e  x y l e m. As in the case of the microphyllous Pteridophytes we have to face the question whether all stelar xylem masses are to be taken as foliar xylem, as originated under the influence of the entering leaf-traces, or whether the original stem xylem still plays a rôle in the xylem formation.

POSTHUMUS answers this question in the latter sense [6]); according to him the stelar xylem for the greater part is stem xylem, the leaf-trace xylem only being present in the downward continuation of the leaf-traces.

On a former occasion [7]) I advocated the reverse view, that only leaf-trace xylem in the fern stem may be left, especially on account of the curious distribution the stem xylem remnants, postulated by POSTHUMUS, would be subject to in polycyclic stems.

So I am led to the following suppositions. In the Ferns the original Pteridophyte stele with its external sheaths, its phloem and its central solid xylem has been reduced into a merely topographical tissue column, acting as a recipient for leaf-traces, but without any tissue differentiation of its own. In this stele the leaf-traces once having entered bend downwards, so that all traces are arranged on a cylinder surface, indicating the stelar form.

The traces further act like organisers in this organization field, inducing the differentiation of protophloem and protoxylem, and also of all other vascular tissues next to these strands.

When a certain amount of xylem is to be formed in a narrow stele, a

---

[1]) POSTHUMUS l.c. 1924, p. 224.

[2]) POSTHUMUS l.c. 1924, p. 240. .

[3]) POSTHUMUS l.c. 1924, p. 247.

[4]) POSTHUMUS l.c. 1924, p. 258.

[5]) TANSLEY l.c. 1907, p. 196.

[6]) POSTHUMUS l.c. 1924, p. 155.

[7]) SCHOUTE l.c. 1926 (f.n. p. 75), p. 294.

protostele ensues; when the stele is larger, xylem will only be differentiated in the neighbourhood of the leaf-traces and a pith will remain in the central part (reduction of central xylem part).

In polycyclic ferns the traces abutting on lower leaf-gaps are continued into the pith; here they may only be formed inside the parenchyma tube accompanying the dictyostele and therefore are arranged anew on a cylinder surface. Here a second vascular cylinder may be formed by their inducing power and further cylinders may be formed in the same way.

**IV. Origin of primary organization.** The main stem of the plant arises exogenously in the embryo; its epidermis and cortex are connected to the outer cell layer and to the ground tissue of the embryo, its stele is formed in connection with that of the primary root.

Branches formed by dichotomy always derive their epidermis, cortex and stele from the corresponding tissues of the podium, the stele dividing in the same way as the whole stem; in lateral branches the tissue systems are nearly always of the same origin. It is only in small lateral and in adventitious branches that departures from this scheme may be found, though this is not general.

In *Rhynia Gwynne-Vaughani* the adventitious branches arising under a stoma form only their epidermis and cortex from the corresponding tissues of the main stem; the stele is formed anew somewhat higher up [1]). In the bulbils of *Lycopodium*, where the bud-trace is of about the same size as a leaf-trace [2]) and in *Dryopteris Filix-mas* where the connection between the bud stele and the leaf-base stele is brought about in very different ways [3]), we get the impression that the bud stele is formed independently and afterwards is connected to the vascular strands of the parent organ.

One point remains to be dealt with here, the occurrence of the so-called ramular gaps. In equally or unequally dichotomizing stems the stele of both shanks may sometimes show a large interruption in its vascular tissues at the side turned towards the other shank. This interruption has been termed ramular gap by JEFFREY, as a counterpart to the old term "Blattlücke".

These ramular gaps may be present in microphyllous as well as in megaphyllous plants, but they may equally be absent in both groups. They have been pictured for instance in *Lepidodendron* [4]) and in *Osmunda cinna-*

---

[1]) R. KIDSTON and W. H. LANG, On old red sandstone plants showing structure, from the Rhynie Chert Bed, Aberdeenshire. I. *Rhynia Gwynne-Vaughani*, Kidston and Lang, Trans. Roy. Soc. Edinburgh, 51, p. 761, see p. 775.

[2]) R. WILSON SMITH, Bulbils of *Lycopodium lucidulum*, Bot. Gaz. 69, 1920, p. 426.

[3]) K. G. STENZEL, Untersuchungen über Bau und Wachsthum der Farne. II, Über Verjüngungserscheinungen bei den Farnen, Nova Acta 28, 1861, p. 3.

[4]) W. C. WILLIAMSON, On the organization of the fossil plants of the Coal-measures, 3, *Lycopodiaceae* (continued), Phil. Trans. London, 162, 1873, p. 283, see Pl. 43, fig. 19.

*momea* [1]), but they are lacking in *Lycopodium* [2]), in *Todea barbara* and, in some cases, in *Osmunda cinnamomea* too [3]). A closer investigation of these phenomena and if possible of their underlying causes might be of great advantage for our knowledge of stelar structures.

### B. Secondary tissue formation.

The distinction between primary and secondary tissues in most cases is quite clean-cut and nobody will have any difficulty in determining whether a tissue is primary or secondary.

In transitional cases it appears that the distinction, however natural it may be, is not so easily formulated. On a former occasion [4]) I came to the conclusion that the best division principle is to be found in the origin of the tissue cells before or after the close of the longitudinal growth; on the whole I still cling to that view, though recognizing that the diffuse thickening growth of palm stems would require another division principle if we want to take it as secondary growth [5]).

In Pteridophytes undeniable secondary tissue formation takes place: secondary cork tissue formation in the Lepidophytes, *Sphenophyllum*, *Calamites*, *Botrychioxylon*, *Botrychium* [6]) and *Helminthostachys* [6]), and formation of secondary phloem and xylem in the same cases [7]) with the addition of *Isoetes*.

Moreover the formation of secondary xylem has been described by BOODLE in *Psilotum* [8]) and by CORMACK in *Equisetum*. With respect to *Psilotum* I might remark that from BOODLE's decription it is clear that the so-called secondary tracheids are only late in their differentiation, but that they have been formed simultaneously with the surrounding cellular' elements; according to the above definition they are therefore primary. In *Equisetum* the secondary nature of the nodal wood, claimed by CORMACK, has been rightly refuted by BARRATT [9]).

### I. Secondary cork tissue formation. The secondary cork tissue formation in *Sphenophyllum*, *Calamites*, and where

---

[1]) JEFFREY, l.c. 1902, (f.n. p. 76), p. 123.

[2]) CRAMER, l.c. (f.n. p. 71), Pl. 31, fig. 1–10; MEYER l.c. 1926 (f.n. p. 71), Pl. 12.

[3]) JEFFREY, l.c. 1902, p. 124, 125.

[4]) J. C. SCHOUTE, Über Zellteilungsvorgänge im Cambium, Verh. Kon. Ak. v. Wet. Amsterdam, 2nd sect. 9, 4, 1902, see p. 56.

[5]) J. C. SCHOUTE, Über das Dickenwachstum der Palmen, Ann. d. Buitenzorg 26, 1912, p. 1, see p. 203.

[6]) PH. VAN TIEGHEM, Sur quelques points de l'anatomie des Cryptogames vasculaires, Bull. Soc. bot. de France, 30, 1883, p. 169, see p. 170.

[7]) VAN TIEGHEM, l.c. p. 171.

[8]) L. A. BOODLE, On the occurrence of secondary xylem in *Psilotum*, Ann. of Bot. 18, 1904, 0. 505.

[9]) K. BARRATT, A contribution to our knowledge of the vascular system of the genus *Equisetum*, Ann. of Bot. 34, 1920, p. 201, see p. 217.

present in the *Filicinae*, has hardly any remarkable features: it is formed by a phellogen, arising in the epidermis in *Botrychium* [1]), in the outer cortex in *Botrychioxylon*, in the innermost cortex, or afterwards in the secondary phloem, in *Sphenophyllum* and in *Calamites*.

It always consists of radially arranged phellem cells, being suberized periderm in *Botrychium* and probably also in the fossils. A formation of phelloderm has as far as I know only been described in *Botrychium* by VAN TIEGHEM.

It is only in the Lepidophytes that the secondary cork tissue shows a much higher development, following a course of evolution not known to exist in any other recent or extinct plant.

The phellogen in these plants arose in the outer cortex, often immediately under the leaf-cushion zone, or else somewhat deeper in the outer cortex [2]). Its products consisted either of fibres only, or of fibres intermixed with parenchyma, as a consequence of the chambering of part of the fibres by transverse walls.

Whether real periderm occurred at the outside in small quantity is unknown; in any case the bulk of the tissue consisted of living cells. And as the fibres were more or less thick-walled, sometimes in a way reminiscent of collenchyma [3]), the tissue no doubt had a mechanical function; by its bulk and position it is generally even held to have constituted the chief mechanical tissue of the stem. On account of its peculiar differentiation it has been termed peridermoid by KUBART [4]); this rather meaningless term might perhaps better be replaced by secondary collenchyma.

The determination of the exact place of the phellogen in the secondary tissue has given much trouble to a number of investigators (WILLIAMSON, SOLMS-LAUBACH, BERTRAND, HOVELACQUE, ARBER and THOMAS, SEWARD, KISCH [5]), for quotations of titles see the last mentioned author). The phellogen itself usually being not very clear in the slides, the authors tried to determine its position from the occurrence of a zone with thin tangential walls and radially short cells, or from the occurrence of zones of least resistance, as illustrated by lines of split.

As a result the common opinion nowadays is that the phellogen was situated near the outer periphery of the secondary tissue, in other words

---

[1]) VAN TIEGHEM, l.c. p. 170.

[2]) M. H. KISCH, The physiological anatomy of the periderm of fossil *Lycopodiales*, Ann. of Bot. 27, 1913, p. 281, see p. 287.

[3]) KISCH l.c. p. 301, J. WALTON, Scottish lower carboniferous plants: the fossil hollow trees of Arran and their branches (*Lepidophloios Wünschianus* Carruthers), Trans. Roy. Soc. Edinburgh 58, 1935, p. 313, see p. 327.

[4]) B. KUBART, *Stigmaria* Bgt., Mitth. naturw. Ver. Steiermark, 71, 1934, p. 33, see p. 35.

[5]) KISCH, l.c.

that little phellem was formed to the outside, but a good deal of phelloderm to the inside [1]); the stigmarian part of the plants only were excepted, by having an internal phellogen.

The thorough survey of the facts as given by KISCH, and several observations by former investigators, make it probable however that this opinion must be revised, and it is not impossible that the development in all cases was centripetal. It has namely been recognized in several cases that the phellogen was not initial-celled but of polygenous origin [2]), i.e. arose in more than one tangential row of primary cells. This has been clearly demonstrated by KISCH [3]).

In later stages the phellogen evidently often followed the same line, in so far as it continued to absorb fresh primary cortex cell rows, i.e. was often a transition cambium. This has been observed very clearly by HOVELACQUE in *Lepidodendron selaginoides* [4]), and is to be seen in several drawings in literature as for instance by WILLIAMSON [5]) and by KISCH [6]); it may account for the very irregular inner outline of the secondary tissue which often invaded much further into the cortex on one side than on another side next to it [7]). For an initial-celled phellogen such an irregular boundary might not have been expected.

When actually the phellogen in most cases was a transition cambium, it is quite natural that it often is not clear in the slides, as the rows of dividing cells may have stopped their activity before new rows began. The controversies in literature about the position of the phellogen may be partly due to this circumstance; moreover it seems certain that in many cases the adult tissues gave rise to renewed growth and renewed divisions [8]).

---

[1]) HIRMER introduces the new terms exophelloderm for phellem and endophelloderm for phelloderm (HIRMER op.c. 1927, p. 216). As phellem and phelloderm are already neutral terms indicating the tissues formed to the outside and to the inside of the phellogen, irrespective of their differentiation, there seems no reason for this introduction.

[2]) For the terms polygenous origin and transition cambium see SCHOUTE in J. J. BEYER, Die Vermehrung der radialen Reihen im Cambium, Rec. trav. bot. néerl. 24, 1927, p. 631, see p. 649.

[3]) KISCH, l.c. p. 291, fig. 5.

[4]) M. HOVELACQUE, Recherches sur le *Lepidodendron selaginoides* Sternb., Mém. Soc. Linn. de Normandie, 17, 1892, p. 1, see p. 158.

[5]) W. C. WILLIAMSON, A monograph on the morphology and histology of *Stigmaria ficoides*, The Palaeontographical Soc., Volume for the year 1886, London 1887, see Pl. 6, fig. 9, 45; Pl. 8 fig. 22.

[6]) KISCH, l.c. Pl. 24, fig. 2, zone c.

[7]) W. C. WILLIAMSON, On the organization of the fossil plants of the Coal-measures, II, *Lycopodiaceae*: *Lepidodendra* and *Sigillariae*, Phil. Trans. London, 162, 1873, p. 197, see Pl. 30. fig. 41; WILLIAMSON l.c. 1887 (f.n. 5 above), Pl. 8, fig. 15.

[8]) KISCH, l.c. p. 299.

The principal argument for the assumption of a transition phellogen however is furnished by the dilatation phenomena in the secondary tissue, a topic which has rarely been mentioned in the pertaining papers. Only HIRMER remarks [1]) that "bei mehr und mehr zunehmenden Stelendurchmesser von innen heraus ein allmählich unerträglicher Druck gegen die gesamten fester gebauten Rindenpartien stattfinden [musste], wenn nicht tangential-dehnungsfähige Elemente in radialer Anordnung entsprechend eingefügt wären". HIRMER gives an elaborate diagram of the stem of *Lepidodendron vasculare* (= *selaginoides*) to explain the prevailing conditions.

Against this view of HIRMER I might remark, that a dilatating action by the expanding stele could only influence the secondary cork tissue after the soft inner cortex had been wholly crushed. Moreover the influence of this action would be strongest in the oldest innermost phelloderm parts and gradually diminish towards the peripheral phellogen.

The dilatation phenomena actually observed are quite different; what we observe is a strong dilatation of the outer secondary cork tissue zones, diminishing and finally disappearing towards the inner edge and giving rise in the outer zones to the so-called Dictyoxylon structure, i.e. the formation of a network of radially undulating fibre rows, filled up by parenchyma.

This condition has been clearly described and pictured by RENAULT for *Lepidodendron Rhodumnense* [2]) and by KISCH for *Lepidodendron brevifolium* [3]); in the latter case the dilatation wedges even reached into the cortex outside the secondary tissue. It is moreover to be seen perfectly clearly in some figures by WILLIAMSON [4]).

In other cases the secondary cork tissue evidently was dilatated as a whole by the stele, after the crushing of the inner cortex; the tissue dilatation then, though stronger in the outer layers, is clearly visible in the internal parts too [5]).

KISCH supposes [6]) that the strong dilatation in the outer zones might be

---

[1]) HIRMER op.c. 1927, p. 216.

[2]) B. RENAULT, Structure comparée de quelques tiges de la flore carbonifère, Nouv. Arch. du Mus. d'Hist. Nat. 2nd ser, 2, 1879, p. 213, see p. 252 and especially Pl. 10, fig. 6, 9–14.

[3]) KISCH, l.c. p. 308.

[4]) WILLIAMSON op.c. 1887 (f.n. p. 88), Pl. 8, fig. 23, 24.

[5]) Described for *Sigillaria spinulosa* in: B. RENAULT, Recherches sur les végétaux silicifiés d'Autun, Étude du *Sigillaria spinulosa*, par MM. B. RENAULT et GRAND'EURY, Mém. prés. par divers savants à l'Acad. d. Sc. de l'Inst. Nat. de France, 22, 9, 1876, see especially the half schematic drawings Pl. 1, fig. 5, Pl. 4, fig. 20–22 for the general dilatation, and fig. 23 and its explanation for the gradual difference between inner and outer part.

[6]) KISCH, l.c. p. 308.

explained by the assumption of a formation of small meshes in the inner zones by the cambium in its first stages, and of large meshes in the outer zones afterwards; all observation of the analogous phenomena in recent plants however militates against such a view.

A secondary zone with a cambium at its outer boundary can never be subject to dilatation by its own growth; when it shows dilatation this must be due to a pressure from within. In such a case the phenomena must be greatest in the inner strata, and must gradually diminish towards the outside.

A cambium inside a hollow cylinder on the other hand, even if the cylinder were empty, will give rise to dilatation phenomena by the mutual pressure which the turgescent cambium elements exert on each other in a tangential direction. The consequences of such a dilatation must be strongest in the oldest external strata and gradually diminish, finally disappear, in the youngest inner strata.

Of course the conditions may have been different in the different genera and species: yet we may be sure that at least in part of them the secondary cork tissue was a transition tissue with addition of new parts at its inner boundary.

## II. Secondary vascular tissue.

*a*. Normal formation. The place of origin of the cambium in the Lepidophytes, in *Spenophyllum*, in *Botrychium* [1]) and in *Helminthostachys* [2]) is directly outside the xylem cylinder, in the form of a continuous cambium cylinder; in *Sphenophyllum* its origin is described as not being simultaneous all around, it first parts being formed outside the concave metaxylem flanks [3]).

In *Calamites* with its separate primary vascular bundles the cambium arises in the bundles right outside the xylem; between these fascicular cambium arcs interfascicular, arcs are formed completing the cambium cylinder.

In all normal cases dealt with here the cambium forms secondary phloem to the outside and secondary xylem to the inside. Only the xylem will be considered here.

We may begin with the remark that in all cases with an exarch primary xylem the secondary xylem, abutting on the protoxylem, is sharply delimited towards the interior (*Lepidodendron*). But in forms with an

---

[1]) E. Russow, Vergleichende Untersuchungen betreffend die Histiologie der vegetativen und sporenbildenden Organe und die Entwickelung der Sporen der Leitbündel-Kryptogamen, etc. Mém. Acad. Imp. d. Sc. St. Pétersbourg, 7th ser. 19, 1, 1872, see p. 119.

[2]) Van Tieghem, l.c. (f.n. p. 86), p. 171.

[3]) Scott, l.c. 1920 (f.n. p. 79), p. 80; Hirmer l.c. 1927, p. 352; Van Tieghem on the other hand writes that the secondary xylem first appears outside the protoxylem (l.c. p. 173).

endarch primary xylem there is no sharp limit at all, the metaxylem insensibly merging into the secondary xylem.

The activity of the cambium may be homogeneous all around, for instance in *Lepidodendron*; the different parts of the cambium however may produce different xylem parts as well.

In *Sphenophyllum* the cambium outside the protoxylem always forms narrow xylem elements, the cambium outside the metaxylem much wider, but otherwise similar, elements.

In *Calamites* the fascicular and the interfascicular cambium parts behave differently. The fascicular cambium in all cases forms rows of tracheids, with small medullary rays in between.

The interfascicular cambium may either only form parenchyma, constituting therefore a large medullary ray (*Arthropitys bistriata*), or it may begin with parenchyma formation on which from both sides, the tracheid formation encroaches either gradually, or suddenly after some time (both conditions in *A. communis*). In these cases the short cambium elements producing the large medullary ray must have been replaced by cambium fibres as SCOTT remarks [1]); in the mature condition it is very remarkable that the radial seriation is not disturbed. It is not improbable that this phenomenon was accompanied by a tangential widening of the cambium, perhaps in consequence of the elongation of the cambium elements; in this way a dilatation of the large medullary ray, the medullary commissure and of the entire medulla was brought about.

HIRMER at least pictures two transverse section of *A. communis* of different size [2]) in which the larger one shows evident signs of dilatation in the internal parts of the large medullary rays [3]).

---

[1]) SCOTT, op. c. 1920, p. 25.

[2]) HIRMER op.c. 1927, p. 389.

[3]) HIRMER overrates both the frequency and the amount of this dilatation when he supposes (op. c. 1927, p. 392) it to be a general feature of *Arthropitys* and believes it to be correlated with a striking increase of the pith cavity.

In *A. bistriata* the whole phenomenon evidently failed (see B. RENAULT, Bassin Houiller et Permien d'Autun et d'Épinac, IV, Flore fossile, 2, Paris 1893–'96, in: Études des gîtes minéraux de la France; see especially Pl. 45, fig. 1, 3, Pl. 46, fig 2, 3) and in *A. communis* it may almost have been absent in some cases (ibid. Pl 48, fig, 2, 6); *A. gigas* on the other hand perhaps shows slight indications by the extension of the parenchyma cells perpendicular to the earliest wood (ibid. Pl. 50, fig. 2).

The amount of the dilatation no doubt is exaggerated by HIRMER because he bases it on a comparison between morphologically incomparable objects: a strong branch with a great number of stout primary vascular bundles and a large pith cavity on one hand, and a weaker branch on the other hand where everything has been laid down on a smaller scale.

Cf. moreover the related phenomena in *Astromyelon* in § 3, sub B II.

In still other cases the interfascicular cambium of *Calamites* may form only libriform fibres with small medullary rays in between (*Arthrodendron*) or finally it may display a combination of conditions, in so far as it forms some radial plates of libriform, alternating with radial plates of parenchyma, the latter always occupying the central line of the large medullary ray (*Calamodendron*).

About the construction of the secondary xylem only a few remarks may be made. In nearly all cases parenchymatic medullary rays are present between the seriated tracheary elements, as in the wood of the Spermophytes. In some *Lepidodendra* these medullary rays may contain spirally or reticulatedly thickened elements, being no doubt ray tracheids like those of *Pinus*.

It is only in *Sphenophyllum* that medullary rays of the common description may be lacking. Not in all species: in *Sph. insigne* the medullary rays are quite normal. In all other species as far as is known, instead of medullary rays there is a wholly different arrangement of the parenchyma elements. Here we find small vertical strands of narrow and erect parenchyma cells, in the angles between four tracheids; these strands are connected in some places by small short radial strands of a few procumbent cells, not forming continuous rays but discontinued very soon.

As *Sphenophyllum* is also remarkable by the fact that the secondary tracheids are not only arranged in radial rows but at the same time in tangential rows, a renewed investigation seems very desirable.

The secondary tissue formation in some cases not only occurs in the stele, but extends also into the leaf-trace in the cortex; this has been observed in *Lepidodendron* and in *Sigillaria*, and in the stigmarian rootlets. The tissues formed in these cases may even extend to a slight extent into the free part of the leaf; they are of the same kind as in the stele.

*b*. A b n o r m a l   f o r m a t i o n   o f   s e c o n d a r y   v a s c u l a r   t i s s u e. The case of *Isoetes* differs from that of all other Pteridophytes in many respects; it has been investigated by several investigators, LANG and WEST and TAKEDA [1]) being the most recent authors on the topic.

As *Isoetes* represents a plant with a very uncommon mode of growth and with special adaptations, the homologization of the different parts of its stem to those of other plants presents considerable difficulties.

The cambium originates in an anomalous position, outside the primary phloem and gives rise externally to secondary cortex. To the inside layers are formed which are usually called the prismatic tissue, described by LANG as consisting of parenchyma with tracheids and sieve tubes, considered by WEST and TAKEDA to be secondary phloem.

---

[1]) LANG, l.c. (p. 73); C. WEST and H. TAKEDA, On *Isoetes japonica*, A. Br., Trans. Linn. Soc. London, 2nd ser. 8, Botany, 1915, p. 333, see p. 343.

## § 3. Anatomy of the root. — A. Primary organization [1]).

**I. E p i d e r m i s.** The outer cell layer of the root body is not so well ·differentiated from the cortex cells as is the case in the stem. It never forms a cuticle or stomata. Its root hairs are merely processes of the epidermis cells; only in the *Marattiaceae* and in *Psaronius* are the root hairs multi-cellular.

In the latter genus the densely developed very long hairs of the basal part of the roots and of the stem surface give rise to a highly remarkable pseudocortex, by being braided into a pseudoparenchyma, as has been proved especially by SOLMS-LAUBACH [2]). This unique tissue is formed after the leaf fall and forms a thick coating of the stem in which the roots are imbedded; the more distal parts of the roots do not form this pseudocortex and are free [3]).

**II. C o r t e x.** The cortex may consist of a uniformly developed pa-renchyma, but usually it forms two or three concentric zones differing in organization: in many cases an internal cylinder of sclerenchyma is pro-duced (many *Polypodiaceae*).

A frequent feature is the development of very large intercellular spaces in the inner or in the middle cortex [4]); these may even fuse into a single cavity surrounding the stele. In *Isoetes* where the root from the beginning is somewhat dorsiventral, the intercellular spaces are biggest on the side turned away from the stem furrow. By the fact that, at this side and laterally, all intercellular spaces fuse, the stele remains attached to the outer cortex only at the furrow side. In *Phyllo-glossum* the stele in the same way occupies an excentral position,

---

[1]) Special treatments of the root being rare for all plant groups, the works of VAN TIEGHEM l.c. 1870/71 (f.n. p. 72) and of G. POIRAULT, Recherches anatomiques sur les Cryptogames vasculaires, Ann. d. sc. nat., 7th ser. 18, 1893, notwithstanding their early date may be quoted here; the latter contains an extensive chapter on the root (p. 114–158) with full quotation of the older literature.

[2]) H. Graf ZU SOLMS-LAUBACH, Der tiefschwarze *Psaronius Haidingeri* von Manebach in Thüringen, Zeitschr. f. Bot. 3, 1911, p. 721.

[3]) In a recently published preliminary communication (B. SAHNI, The roots of *Psaronius*, Intra-cortical or extra-cortical? — A discussion, Current Science 1935, p. 555) the author returns to the old view of STENZEL, that this tissue may have been secondary cortex, developing pari passu with the roots after the leaves had fallen. His arguments as given in the preliminary note are that at the outside of the pseudocortex a periderm was developed, and further that in a recent Liliacea *Asphodelus tenuifolius* numerous roots grow down through the cortex of the main root, so as to distend the main root very strongly.

As far as I see at present neither of these facts can invalidate in the least the evidence brought forward by SOLMS-LAUBACH.

[4]) NÄGELI und LEITGEB, l.c. (f.n. p. 72) p. 82.

and after METTENIUS is located "der unteren oder inneren Seite der Wurzel genähert" [1]).

The inner cortex at its inner side always is limited by an endodermis.

**III. S t e l e.** Between the endodermis and the vascular tissues a parenchymatic tissue, the pericambium, is found.

NÄGELI and LEITGEB [2]) and later on VAN TIEGHEM [3]) laid considerable stress upon the fact that in *Equisetum* endodermis and pericambium are formed very late during the ontogeny by a cell division of one single cell layer; according to them *Equisetum* therefore lacks a pericambium and only has a double endodermis, of which the outer layer alone forms the Casparian strips. This strained conception is not to be accepted [4]).

The vascular tissue in the stele occurs in the form of longitudinal xylem and phloem strands, separated by parenchyma. As in the roots of all other plants, the protophloem and protoxylem strands are alternately arranged in a single series at the outside of the vascular tissue column, all xylem and phloem strands being exarch.

The xylem strands usually consist of a radial plate, extending more or less far into the stele towards the centre. In most cases the xylem strands extending to the centre meet with their metaxylem, a single large central vessel often being present [5]). Less frequently the xylem plates being shorter a parenchymatic medulla is developed of a smaller [6]) or larger [7]) size. In *Lycopodium* the central union of the xylem strands is often less regular, so that the transverse section may show a certain likeness to that of a stem stele.

The number of strands is variable: in *Marattiaceae* there are some 10 to 20 strands of both kinds, in *Lycopodium* there may be 10 or more; lower numbers however are customary, for instance 2 in the *Polypodiaceae*.

In *Phylloglossum, Selaginella, Isoetes* and in some *Ophioglossaceae* the roots only are monarch. Moreover smaller roots in all plants generally have lower numbers than big roots. In *Lycopodium* the last thin roots are monarch too.

**IV. O r i g i n   o f   p r i m a r y   o r g a n i z a t i o n.** The root

---

[1]) G. METTENIUS, Über *Phylloglossum*, Bot. Ztg 25, 1867, p. 97, see p. 99.

[2]) NÄGELI und LEITGEB, l.c. p. 84, 109.

[3]) PH. VAN TIEGHEM, Traité de botanique, 2nd ed. I, Paris 1891, p. 681.

[4]) SCHOUTE, op.c. 1902/3 (f.n. p. 68), see p. 142.

[5]) See sor instance the specimen of *Equisetum* pictured by VAN TIEGHEM l.c. 1870/'71 (f.n. p. 72), Pl. 5, fig. 22.

[6]) See sor instance the specimens pictured by VAN TIEGHEM l.c. 1870/'71 on Pl. 5, fig. 18 (*Lastraea*) and 27 (*Botrychium*).

[7]) See sor instance the specimens pictured by VAN TIEGHEM l.c. 1870/71, Pl. 4, fig. 11 (*Marattia*); a much larger and moreover a fistular pith may be found in large specimens of *Astromyelon*, the root of *Calamites*, the smaller specimens having a solid pith or no pith at all.

organization to a large extent is a parallel to that of the stem; the absence of leaves however tends to simplify it greatly. As the question whether there is a stem xylem and a foliar xylem does not exist here, it is clear that the stele as a whole as well as the vascular strands themselves are continued acropetally.

We may ask further how the root tissue systems are first induced. In the primary root the stele is always a continuation of the stem stele. In dichotomizing roots the stele of the podium bifurcates too; lateral roots and roots borne on stems are formed endogenously, immediately outside the vascular tissue in the pericambium or in the innermost cortex. So we get the impression that the root stele is always a branch of an existing stem or root stele, whereas its epidermis may be formed anew in many cases.

The individual xylem and phloem strands probably are induced by existing strands in a previous stele too, with the exception of course of the strands in the primary axis and in wholly adventitious roots; their distribution however is governed by the additional rule of their mutual alternation, and moreover we may be sure that in any stele the number of strands is adapted to the available space. This may be deduced from the following set of facts.

In equally dichotomizing roots of *Lycopodium* the strands of the podium are continued in the shanks; in the saddle of the dichotomy some new strands arise, probably by bifurcation of existing strands. In such a way a root with 8 xylem strands may give rise to two shanks with 5 strands each [1]).

In unequally dichotomizing roots of *Lycopodium* the course of events is described by VAN TIEGHEM as follows [1]): in a root with six xylem strands two neighbour strands split radially into two: the two components adjoining the phloem strand in between and this phloem strand itself move outwards, the two xylem strands rearranging so as to become opposed; the phloem strand bifurcates too and in such a way a diarch stele of the lateral root is formed out of the mother stele which remains hexarch.

Lateral roots in other Pteridophytes in the case of a low number of strands have a fixed orientation of these strands towards the mother root; in the case of diarchy the xylem strands lie transversely, the phloem strands longitudinally, the same orientation therefore as in the lateral root of *Lycopodium* just mentioned.

Exceptions to this rule are very rare: THOMSON found [2]) that in *Marattia alata* the two xylem strands are placed longitudinally, while in *Angiopteris evecta* and *Danaea elliptica* the same condition was met with in some roots, other roots having an oblique position of the xylem plate.

---

[1]) VAN TIEGHEM l.c. 1870/'71, p. 86.

[2]) R. B. THOMSON, A seed-plant feature of the root in *Marattiaceae*, New Phytol. 33, 1934, p. 96.

These exceptions are the more remarkable as it has been generally acknowledged that in this respect there is an essential difference between Pteridophytes and Spermophytes, a difference known already to NÄGELI and LEITGEB in 1868 [1]) and confirmed by VAN TIEGHEM [2]), the Spermophytes always having a longitudinal xylem plate in diarch lateral roots. About the meaning of these exceptions see THOMSON in the quoted paper.

In monarch roots the orientation of the single strand has not been described for all cases. In *Selaginella* the roots arising from the rhizophore tip turn their xylem strand towards the rhizophore centre [3]), and in the same way dichotomizing roots, deriving their xylem strand from the splitting of the single podium strand, rotate their strands 90° so as to have them turned towards each other.

In *Lycopodium* the monarch root shanks, when formed by the dichotomy of a diarch root, in contrast to *Selaginella* turn their xylem strands right away from each other [4]); a monarch root by dichotomizing however produces two shanks with facing xylem strands as in *Selaginella* [5]).

In *Isoetes* the protoxylem strand is turned towards the weaker cortex side and towards the stem furrow; with respect to the stem it is therefore placed laterally. When the roots dichotomize the xylem strand splits too, and by a rotation the two strands in the shanks face each other as in the former cases [6]).

In monarch roots of *Ophioglossum*, if I understand correctly the descriptions given in literature [7]), the xylem strand is placed laterally.

The root structure once organized is continued in the root apex. The number of xylem and phloem strands however is not invariable in the course of the same root: NÄGELI writes (about roots in general, not particularly about those of Pteridophytes) that changes in number are absent only when the number of strands is a very small one: "Die Veränderungen im Längsverlaufe sind um so häufiger, je complicirter der Bau ist .... die Holz-, Bast-, sowie die ganzen Fibrovasalstränge können sich der Zahl nach vermehren oder vermindern, wobei bald eine Vereinigung oder Theilung, bald auch ein allmähliges Aufhören oder Verschwinden beobachtet wird" [8]).

**B. Secondary tissue formation.**

**I. S e c o n d a r y   c o r k   t i s s u e   f o r m a t i o n.** The forma-

---

[1]) NÄGELI und LEITGEB, l.c. 1868, (f.n. p. 72) p. 143.

[2]) VAN TIEGHEM l.c. 1870/'71, Pl. 3, fig. 5, 6.

[3]) NÄGELI und LEITGEB, l.c. 1868, p. 129.

[4]) VAN TIEGHEM, l.c. 1870/'71, p. 88.

[5]) NÄGELI und LEITGEB, l.c. 1868, p. 121.

[6]) NÄGELI und LEITGEB, l.c. 1868, p. 134, Pl. 19, fig. 8–10.

[7]) RUSSOW l.c. (f.n. p. 90), p. 122; VAN TIEGHEM l.c. 1870/'71, p. 108; POIRAULT l.c. (f.n. p. 93), p. 142.

[8]) NÄGELI l.c. 1858 (f.n. 1 on p. 71), p. 52.

tion of secondary cork tissue in Pteridophyte roots is a rather rare phenomenon. Only the following cases have come to my knowledge: *Calamites* (*Astromyelon*); *Sphenophyllum* (origin of phellogen in the latter genus in the pericambium or the phloem), *Botrychium*, *Helminthostachys*, *Angiopteris*, *Marattia* (place of origin according to VAN TIEGHEM in these four genera the epidermis or the outer cortex layers [1]), and also the remarkable case of *Psaronius* described by SAHNI [2]), where the place of origin of phellogen is the outer layers of the pseudocortex.

In all these cases the tissue formed is evidently a suberized periderm.

**II. Secondary vascular tissue.** The only cases of Pteridophytes producing secondary vascular tissues in their roots, seem to be *Calamites* and *Sphenophyllum*.

In *Sphenophyllum* the secondary xylem is constructed in the same way as that in the stem, with radial and tangential seriation of the tracheides and with the same anomalous parenchyma distribution.

In *Calamites* the origin of the cambium has been excellently observed and described by RENAULT [3]). From his researches it is clear that the cambium was formed as in seed plants, outside the exarch primary xylem strands and inside the phloem strands. But whereas in seed plants the tissues produced by the cambium towards the inside at first chiefly consist of wood inside the phloem strands only, so that the corrugated form of the cambium cylinder is flattened out, and whereas outside the xylem strands, when the formation sets in, usually a large medullary ray is formed, in *Astromyelon* the cambium cylinder, though either being smooth from the beginning or being flattened out very soon, formed its first tracheary parts right outside the primary xylem strands and no doubt a medullary ray inside the phloem strands.

As the tracheary parts of the xylem gradually broadened, probably by the longitudinal sliding growth of the short cambium cells elongating into fibres (see § 2 on p. 91), the cambium was widened and the large medullary rays were dilated [4]).

This dilatation evidently soon stopped, probably when the large medullary rays were so much invaded by the tracheary wood formation as to leave only small medullary rays [5]). The phenomenon itself however is not to be doubted in the younger stages; as proofs we may point out the radiating parenchyma cell arrangement around the inner primary wood flanges [6]).

---

[1]) VAN TIEGHEM, l.c. 1883 (f.n. p. 86), p. 170.

[2]) SAHNI, l.c. (f.n. p. 93), p. 556.

[3]) RENAULT op. c. 1893/'96 (f.n. p. 91).

[4]) RENAULT op.c. Pl. 56, fig. 6, reproduced in most text-books.

[5]) RENAULT op.c. Pl. 53, fig. 2, 5, 6, 7.

[6]) RENAULT op.c. Pl. 56, fig. 6, just quoted, Pl. 57, fig. 1. and for the dilatation in the lacunar cortex Pl. 57, fig. 6.

## § 4. Anatomy of the leaf. — A. The microphyll.

**I. Epidermis.** The leaf epidermis in most respects is similar to that of the stem. As there, the differentiation of its cells may vary according to the kind of underlying tissues (*Equisetum*); the epidermis of both leaf surfaces may be very different in some cases [1]; in other cases both surfaces may be similar.

The epidermis forms stomata which in conformity with the scanty differentiation of the mesophyll layers may be equally frequent on both surfaces (*Lycopodium inundatum, L. Selago* and other spp. [2]). They may even be present in such leaf parts where the mesophyll is wanting, intercellular spaces being present between the epidermis cells (*Selaginella* leaf margins, very strongly in *S. molliceps* [3])). In other cases the stomata may be localized according to the above rule of correlation with the underlying tissues: in the terrestrial *Isoetes* spp. they are formed only over the four large air canals, usually on both leaf surfaces [4]); in *Lepidodendron* they are restricted to the two furrows on the leaf under surface.

**II. Mesophyll.** A mesophyll is always present, al least along the midrib; in the marginal parts it may be lacking in rare cases (*Selaginella* spp. [3])). It may consist of a homogeneous tissue (*Psilotum*) or it may be differentiated more or less into different layers; a regular differentiation into a palisade and a spongy chlorenchyma is attained in some *Selaginella* spp. (*S. Lyallii, S. concinna* [3])), and in *Sigillariopsis sulcata* [5]).

A development of large intercellular spaces is not rare; in *Isoetes* four large longitudinal air canals are present, chambered into numerous compartments by transverse septa. In the stigmarian rootlet which according to the facts related in Ch. I § 11 will be taken as a transformed leaf, and consequently has to be dealt with here, confluent air-spaces are highly developed especially on the lateral and abaxial side, so as to cause the stele, with a thin surrounding parenchyma coating, to be suspended on small trabeculae near the adaxial side of the tubular outer cortex; at the rootlet insertion these air-spaces are lacking in a transverse zone of the rootlet cushion. At the points of dichotomy of the rootlet the air-spaces are evidently also lacking locally, so as to cause the well-known constriction of the two shank bases.

---

[1] R. J. HARVEY-GIBSON, Contributions towards a knowledge of the anatomy of the genus *Selaginella*, Spr. III, The leaf, Ann. of Bot. 11, 1897, p. 123, see description of *S. Martensii*, p. 130.

[2] HEGELMAIER l.c. 1872 (f.n. p. 67), col. 817–818.

[3] HARVEY-GIBSON, l.c. 1897, p. 151.

[4] A. BRAUN, Über die *Isoetes*-Arten der Insel Sardinien, Monatsber. k. pr. Akad. d. Wiss. Berlin a. d. J. 1863, Berlin 1864, p. 554, see p. 587.

[5] D. H. SCOTT, On the occurrence of *Sigillariopsis* in the lower coal-measures of Britain, Ann. of Bot. 18, 1904, p. 519.

As peculiar tissue strands in some microphylls we may mention the parichnos and the transfusion tissue.

The parichnos occurs in the leaves of *Lycopodium inundatum* and some other spp. [1]), sometimes only in the sporophylls (*L. clavatum*); further in the Lepidophytes, and in the sporophylls of *Isoetes Hystrix* [1]).

The parichnos may form either a double strand, situated at both sides of the leaf and running parallel to the midrib, or a single strand at the abaxial side of the midrib; in *Lycopodium* HILL stated [1]) that the single strands originate from the fusion of two original lateral strands.

In the case of recent plants the parichnos consists of a lysigenous cavity, lined by mucilage secreting cells.

Its longitudinal extension may vary; in the foliage leaves where present in *Lycopodium*, the single strand traverses the whole leaf and extends over a certain length into the cortex; in the sporophylls of *Lycopodium* its distal extension is limited, as the single strand does not extend much beyond the sporangium; in *Isoetes* the double strand is entirely confined to the sporophyll base, to the sporangium region; in the Lepidophytes, the main part of the parichnos is present as a single strand in the stem cortex, from which a double strand extends into the leaf cushion and more or less into the leaf base.

The second peculiar tissue, the transfusion tissue, consists of cellular elements with a tracheidal mode of wall thickening, but occurring outside the regular xylem. It is met with in the leaves of *Lepidodendron* and of *Sigillaria* [2]) and further in the stigmarian rootlets. Is is paralleled by all authors to the transfusion tissue of Conifers; of course it is not known whether it shared the remarkable property of the transfusion tissue of Conifers, Gnetum and Cycads, of containing a living protoplast in its cells, notwithstanding the tracheid-like wall sculpture.

In *Lepidodendron* it surrounded the foliar stele by a homogeneous zone of several cell layers [3]), in *Sigillaria* it was specially developed at the abaxial and lateral sides of the stele and evidently consisted of transfusion cells, alternating with parenchyma [4]). In the stigmarian rootlets the transfusion tissue was present in the middle zone of the tubular outer cortex, and was connected by delicate strands with the stelar protoxylem [5]); as WEISS remarks, its presence may be taken as a further argument in favour of the view of the homology of stigmarian rootlets and leaves. We might

---

[1]) T. G. HILL, On the presence of a parichnos in recent plants, Ann. of Bot. 20, 1906, p. 267.

[2]) R. GRAHAM, An anatomical study of the leaves of the Carboniferous Arborescent Lycopods, Ann. of Bot. 49, 1935, p. 587.

[3]) RENAULT, op.c. 1893/'96 (f.n. p.91), pl. 34, fig. 4–8 (*L. esnostense*); SCOTT, op.c. 1920 (f.n. p. 79), p. 143 (*L. Hickii*).

[4]) RENAULT, op.c. 1893/'96, Pl. 41 fig. 7, 19 (*S. spinulosa*), 23 (*S. Brardi*).

[5]) F. E. WEISS, The vascular branches of Stigmarian rootlets, Ann. of Bot. 16, 1902, p. 559; 18, 1904, p. 180.

add that the position in both categories of organ is essentially the same, as soon as we assume that the intercellular spaces in the stigmarian rootlets have been formed between the transfusion tissue and the stele.

**III. S t e l e.** In some cases a stele in the microphyll may be entirely lacking (*Asteroxylon*; *Psilotum* partly). In nearly all cases however a stele is present, and in simple leaves it almost invariably [1]) is a simple stele too; in dichotomously divided leaves there are usually as many steles as there are leaf lobes, as the stele dichotomizes too (*Sphenophyllum*); in some cases only the stele may be limited to the unbranched leaf part and remain simple (*Tmesipteris*, sporophyll).

The stele is not always surrounded by a well-marked endodermis and pericambium; in *Equisetum* however these tissues are well developed.

Usually every stele only contains a single vascular strand; we may however have two collateral vascular strands instead, fusing into one single strand in the upper part of the leaf (*Sigillaria* spp.) or in another case the vascular strand may be accompanied by two separate lateral xylem strands [2]).

The phloem may surround the xylem (*Tmesipteris*), but usually it is located at the under side; sometimes it only occurs in the basal part of the leaf, the apical part containing a xylem strand alone (*Lycopodium*).

The xylem strand, which is often quite small, is not always clearly differentiated into a proto- and a metaxylem. Accordingly the statements in literature about the exarch, endarch or mesarch character of the xylem often diverge; probably the distinction is not very important for these small bundles.

**IV. S e c o n d a r y   t i s s u e   f o r m a t i o n.** Secondary tissues hardly ever occur in microphylls. Cases of secondary cork tissue formation have not come to my notice, and the secondary vascular tissue formation in leaf bases in connection with secondary growth in leaf-traces is always unimportant.

**B. The megaphyll.**

**I. E p i d e r m i s.** The epidermis of petioles, rachides and large veins resembles more or less the stem epidermis; on the leaf blade the epidermis is often more specialized than in microphylls, in agreement with the greater elaboration of the leaf, for instance by undulating cell membranes.

Moreover the epidermis not only forms stomata (lacking in the *Hymenophyllaceae*) and numerous hair forms, but it may form other special elements too, like spicular cells, a kind of sclerotic cells (*Vittaria*), nectaries (*Pteridium aquilinum*) and hydathodes (*Polypodium vulgare*).

---

[1]) A polystelic condition in the simple sporophyll stalk of *Equisetum* is described by A. SANTSCHI in Contribution à l'étude anatomique du système vasculaire d'*Equisetum*, Mém. Soc. vaudoise Sc. nat. 5, 1935, p. 103.

[2]) HARVEY-GIBSON l.c. 1897, (f.n. p. 98), p. 152 (*Selaginella Lyallii*).

**II. M e s o p h y l l.** The mesophyll may be lacking in parts of the leaf blade (*Hymenophyllaceae, Leptopteris, Asplenium myriophyllum* [1])), but usually it is present in all leaf parts.

It may show layers of different organization, in the petiole more resembling the stem cortex tissues, in the blade often constituting a palisade and a spongy chlorenchyma.

In *Angiopteris* spp. vein-like structures may be present, the so-called venulae recurrentes, being strands of tissue having the structure of the leaf-margin and extending from an angle between two leaf teeth inward, running between two small true veins.

Mucilage ducts are present in the mesophyll of *Marattiaceae*.

**III. S t e l e.** Owing to the complex character of nearly all Fern leaves the stelar structure may become highly complicated.

Beginning with leaf forms, such as those of the *Coenopterideae* with their most frequently terete pinnae without any lamina, we may remark that in such leaves petiole, rachis and pinnae all have one single stele, a stele which is nearly always dorsiventral in the pinnae (*Stauropteris* excepted) but radially symmetrical in petiole and main rachis. The dorsiventral pinna steles have a gutter-shaped xylem strand, the convex side being turned towards the abaxial side which at the same time is the physiological underside. The protoxylem strands are to be found either at the concave xylem margin, or they are immersed in the metaxylem, accompanied by a parenchyma strand at their adaxial side.

The union of the pinna-traces to the rachis stele, as POSTHUMUS in particular has shown [2]), is brought about according to the same principles as the union of leaf-traces to the stem stele, the distribution of metaxylem and parenchyma notably deing due to the same causes. The peculiar spatial relations between the pinnae and the rachis, not being arranged in one single plane, find their expression in a peculiar stelar structure in the rachis and petiole.

From these facts I might come to the following theoretical view, in harmony with the views expounded above p. 79–85.

The megaphyll representing a phylogenetically differentiated branch system, every branch originally possesses one stele. This branch stele like that of the unchanged stem has been reduced into an acropetally formed tissue strand without differentiation of its own, but acting as a recipient for basipetal stimuli for vascular tissue induction.

The induction (not the differentiation) of vascular tissue starts from the distal ends of all pinnae; according to the spatial distribution of pinnae the lowest lateral pinna of the leaf is likely to finish its own branching first, so that it may be the first in which induction begins.

---

[1]) POIRAULT l.c. (f.n. p. 93), p. 198.

[2]) POSTHUMUS l.c. 1924 (f.n. p. 75), p. 170–193.

The basipetal inductions give rise to simple protophloem and proto-xylem strands and in their turn act as organizers for the further vascular tissues, in the same way as in the stem.

Turning now to the other ferns with a laminar development, we may first consider the simple or compound leaves with an open venation, without any commissural veins. For such leaves, of which many instances might be enumerated, the above views may be held without any difficulty. Petiole, midrib and blade veins are always dorsiventral here, and the xylem part of their stele, however complicated in particular cases, can always be related to the gutter-shaped strand. An admirable description of the stelar structures in pinnae, rachis and petiole has been given by THOMAE [1]).

The main complication principles of the xylem distribution are: the gutter-like strand may have incurved margins; it may close at the adaxial side unto a cylinder; it may show two lateral inward folds; it may be perforated by numerous gaps not being trace-gaps, so that in a transverse section it seems to be broken up into a number of meristeles; finally in the *Marattiaceae* an internal system is elaborated in the same way as in the stem, forming a second cylinder inside which others may be formed, giving up to seven concentric cylinders [1]).

In the last place we have to consider the leaves in which a reticulate venation has been developed by the formation of numerous commissural veins. In these cases the stelar structures in the larger veins, in pinnae, rachides and petiole are exactly the same as in leaves with an open venation, the commissural veins evidently being a secondary complication.

The whole complex of facts referred to may easily be brought into line with the above views. An important fact brought forward by POSTHUMUS is that the mode of union of a pinna-trace to a rachis stele may differ considerably, according to the casual spatial relations between trace strands and rachis stele strands. If the old view of the origin of the trace by branching of the rachis strands were correct, we might have expected, as POSTHUMUS remarks [2]), a more uniform mode of departure of the traces.

The inner systems of the *Marattiaceae* are evidently due to similar causes as in the case of the stems: THOMAE describes how the strands of the inner cylinders, when followed upwards, at the adaxial side successively run towards the next outer cylinder, so that acropetally the inner cylinders all disappear [3]). This may be read inversely as meaning that in any cylinder the newly entering strands first fill up the abaxial and lateral sides of the stele, and that when the cylinder is filled up, further strands, not finding a

---

[1]) K. THOMAE, Die Blattstiele der Farne, Jahrb. f. wiss. Bot., 17, 1886, p. 99.
[2]) POSTHUMUS l.c. 1924, p. 146.
[3]) THOMAE l.c. p. 119.

place in the same cylinder, run inwards at the adaxial side to start a new inner cylinder.

**IV. Secondary tissue formation.** In *Botrychium* a periderm may be present in the petiole bases, in continuation of the stem periderm [1]. In the same way a periderm and lenticels have been described by COSTERUS in *Angiopteris* [2] and by POTONIÉ in *Marattia* [3].

Groningen, May 1936.

Botanical Laboratory of<br>the Government University.

POSTSCRIPT: The above chapters being already in type, the valuable text-book by A. J. EAMES on the morphology and anatomy of Pteridophytes [4] appears, dealing to a large extent with the same matter as treated above.

I greatly regret not having been able to use the book in the composition of my contribution, as I might have greatly profited by it, and would certainly have quoted it in many places.

Yet, as the scope of the present Manual is quite different from that of a textbook, the two treatments of the same field differ so much as not to make useless the labour of the second in arrival.

On many points I am happy to state that my conclusions run parallel to those drawn by EAMES; in some points however they are diverging.

Of the latter category I might mention one single topic, the interpretation the *Isoetes* stock base [5].

EAMES fully realizes the difference between the lobes of *Isoetes* and the stigmarian axes. Taking the latter as a form of specialized stems (for which he introduces the term rhizomorph [6])), he follows FITTING in treating the stelar arms of *Isoetes* as their homologon.

Elaborating this view, EAMES concludes that there is a very near relation between *Pleuromeia* and *Isoetes*, and tries to establish this near relation by

---

[1]) VAN TIEGHEM l.c. 1883 (f.n. p. 86), p. 171.

[2]) J. C. COSTERUS, Het wezen der lenticellen, Thesis Utrecht 1875, see p. 35.

[3]) H. POTONIÉ, Anatomie der Lenticellen der Marattiaceen, Jahrb. K. bot. Garten Berlin, 1, 1881, p. 307.

[4]) A. J. EAMES, Morphology of vascular plants. Lower groups (*Psilophytales* to *Filicales*), New York and London, 1936.

[5]) op. c. p. 355.

[6]) op. c. p. 354. As the term rhizomorph might give rise to confusion with rhizophore, and especially as the term rhizome may be used equally well, the introduction of the new term does not seem to be necessary.

the supposition that in *Pleuromeia* up to the present only the fossilized stelar arms have been observed and that in the living plants these arms, as in *Isoetes*, were buried in a mass of soft cortical tissues which however have not been preserved.

This cannot be granted. The *Pleuromeia* stem lobes have been found more than once in situ, already as early as 1854 by SPIEKER [1]), the rootlets radiating in the matrix. Moreover SOLMS–LAUBACH observed and described the stele of these lobes [2]).

On the arguments given above I must therefore hold to my views on both *Isoetes* and *Pleuromeia*.

Finally I might make a general remark on the aim and scope of morphological botany, in connection with what has been said by EAMES on that topic in his Preface.

Morphology being fundamentally descriptive, it may be put according to EAMES on a broader basis by treating it as a comparative science, in the light of evolutionary modification and development.

So far this will be gladly granted by any biologist. But when EAMES writes that the establishment of a natural classification is the goal of most morphological study to-day, I want to remark that if so, this is not an ideal state of things.

For morphology (including anatomy) being a science in itself, not an auxiliary science, its goal should be found in its own field, not outside it.

When taxonomy is enabled to reach its aims to a larger extent by the aid of morphology, this will surely be a great satisfaction for the morphologist; the high aim of morphology itself is however, as far as I can see, to explain how a plant, recent or fossil, elaborates or elaborated its structure; to determine the laws which are or have been ruling the morphogenetic processes in the living protoplasm.

From this point of view, however imperfectly realized, the above chapters have been written.

Groningen, August 1936.

---

[1]) TH. SPIEKER, l.c. (f. n. p. 62), see p. 186.
[2]) SOLMS–LAUBACH, l.c. (f. n. p. 34), see p. 228.

CHAPTER III

EXPERIMENTAL MORPHOLOGY

by

S. WILLIAMS (Glasgow)

**§ 1. Introduction and historical account.** — Experimental Morphology is a branch of General or Causal Morphology [1]) which seeks by means of experiment to provide data for a consideration of problems dealing with the factors underlying the normal structure and development of plants. The question as to whether it can also contribute to the solution of evolutionary problems is an open one which will be discussed later. Such experimental data must always be taken in conjunction with the results of wide comparative examination of both the normal development and structure and of such deviations from the normal as occur naturally.

The point of view adopted here is that the individual development to the adult structure is "the manifestation of the properties of the specific substance under certain conditions" (LANG, loc. cit.). The investigation of the nature and mode of operation of these conditions is the main aim of experimental morphology; the nature of the specific substance (or the "spezifische erbliche Struktur" as KLEBS terms it) is at present inaccessible to experiment so that in the following account its existence will not always be stressed although it must be recognised throughout that it is the factor of paramount importance.

It is difficult to give an historical account of the development of this branch of morphology for it has shown no orderly development such as has characterised phyletic morphology and pure physiology. The problems which experimental morphology seeks to solve were already recognised in the period of developmental morphology. HOFMEISTER was the pioneer worker in this field as is evidenced in particular by his "Allgemeine Morphologie", published in 1868 [2]). There may be mentioned also the works of

[1]) For the scope of General Morphology see W. H. LANG, Pres. Address to Section K, British Assoc., 1915.

[2]) For a critical account of HOFMEISTER's contributions see GOEBEL, Wilhelm Hofmeister, English Translation, Ray Society, 1926, Chap. V.

LEITGEB, SACHS, KNY, STAHL, PRANTL and KLEBS to which reference will be made later. The onset of the phyletic period, some little time after the publication of the "Origin of Species", resulted in causal problems being relegated to the background and relatively few workers have since been attracted to them. The pioneer work of HOFMEISTER was, however, carried on and vastly extended by his pupil GOEBEL, who as BOWER [1]) says "extended the study of form from methods of record and comparison to questions of causality." From 1887, when one of the first of his contributions to experimental morphology appeared, until his death in 1932, GOEBEL contributed greatly to the advancement of this branch of Botany. His results and conclusions were included in the three editions of his great "Organographie der Pflanzen" and special mention may also be made of the "Einleitung in die experimentelle Morphologie der Pflanzen" published in 1908. LANG has also made important contributions to the experimental morphology of the Pteridophytes, notably with regard to induced deviations from the normal life-cycle. His Presidential Address to the Botanical Section of the British Association in 1915 is particularly valuable since it gives a clear and critical account of the scope of causal morphology and an indication of the importance of experiment in this branch of the science.

The following account is not intended to be a complete resumé of the experimental work carried out on the Pteridophytes but is intended rather to indicate the nature of the problems and the type of result so far attained. So far as the development and structure of the plant is concerned the various stages in the life-cycle will be considered in sequence. A separate section will deal with the bearing of experimental morphology on evolutionary problems.

## The Gametophyte

**§ 2. Factors influencing germination.** — A considerable amount of experimental work has been carried out on the factors influencing the germination of Pteridophyte spores. Little is known, however, of these factors in the case of the *Psilotales* and homosporous *Lycopodiales* except that they require special conditions including the presence of a suitable mycorhizic fungus [2]). So far as the *Equisetales* are concerned, the most complete account of germination is that of BUCHTIEN [3]). In the presence of sufficient moisture and oxygen, germination is most successful in light of medium intensity; it is slowed down somewhat both in darkness and direct sunlight.

---

[1]) F. O. BOWER, Obituary Notice of Karl Ritter von Goebel, Roy. Soc., 1933.

[2]) DARNELL–SMITH (Trans. Roy. Soc. Edin., Vol. 52, 1917) gives an account of various culture methods by means of which he was able to obtain the germination of the spores of *Psilotum*.

[3]) BUCHTIEN, O., Bibliotheca Bot., Bd. 11, 1887–89.

Little is known concerning the germination of the mycorhizic types of Fern prothallus although CAMPBELL [1]) was able to obtain the early stages of the prothalli of *Ophioglossum pendulum* and *Botrychium virginianum* in experimental cultures. The conditions influencing the germination of the Leptosporangiate types have, however, been repeatedly examined [2]). In general, apart from the presence of water and oxygen which is always essential, the presence of light is necessary and KLEBS states that *Pteridium aquilinum* is the only fern of the many he examined which gives general germination in the dark. Other Ferns give sporadic germination in the dark and it seems probable that all may be induced to do so to a greater or less extent by the employment of special cultural conditions, e.g. the spores of *Osmunda regalis* will germinate in the dark when grown on agar to which has been added 0.5% Knop's solution and 0.01% grape sugar. The qualitative nature of the light is important and KLEBS has shown that, in general, the yellow rays stimulate while the blue rays definitely retard germination. There is, however, very considerable variation in different Ferns as regards the effect of the various spectral regions on germination and KLEBS's work should be consulted for an account of this. Temperature is also an important factor and although there is again considerable variation the majority of Ferns conform to the results obtained by KLEBS in *Pteris longifolia* where the minimum temperature for germination is 12° C, the optimum between 25° C and 30° C and the maximum is 40° C.

The conditions governing the onset of germination are thus variable and, so far, there has been no analysis of the mechanism involved beyond the suggestion of KLEBS that the yellow rays stimulate a photochemical reaction in which enzymes are concerned.

## § 3. Factors influencing the Development and Form of the Pro-thallus.

— Following germination the development of the gametophyte proceeds along diverse paths in the various groups of the Pteridophytes. The nature of the factors underlying this further development has been investigated in the Equisetales and Filicales but not in the other groups.

(*a*) *Equisetinae*. The development of the gametophyte of *Equisetum* has been studied by various writers and particular mention may be made of the contributions of STAHL [3]), BUCHTIEN (loc. cit.), LUDWIGS [4]) and WAL-

---

[1]) CAMPBELL, D. H., Mosses and Ferns, 1918, p. 234.

[2]) See G. KLEBS, Sitz. d. Heidelberger Akad. d. Wiss., 1916 and 1917 and J. STEPHAN, Jahrb. für wiss. Bot., Bd. 70, 1929. Earlier literature is fully quoted here.

[3]) E. STAHL, Einfluss der Beleuchtungsrichtung auf die Theilung der *Equisetum*-Sporen, Ber. d. deutschen bot. Gesells., III Bd., 1885.

[4]) K. LUDWIGS, Untersuchungen zur Biologie der Equiseten, Flora, N.F. Bd. III, 1911.

KER [1]). When placed in water the spores swell while retaining their spherical form; they then divide into two unequal cells, a larger prothallial cell and a smaller lens-shaped rhizoid cell. STAHL investigated the question as to whether the position of this first division wall is determined by external factors such as light and gravity. For this purpose spores of *E. variegatum* were sown on watery gelatine or on moist filter paper in order to fix the spores in position. Gravity was shown to have no effect on the direction of the division since spores germinated in a vertical position in the dark showed the first wall orientated in all possible directions. Light was, however, shown to be the determining factor. Spores subjected to bright unilateral illumination showed the division wall constantly placed so that the prothallial cell faced towards the light and the rhizoid cell away from it. NIENBURG [2]) has made a cytological investigation of spores germinated in unilateral illumination and has shown that the first effect is the aggregation of the chloroplasts on the illuminated side and that this is then followed by the establishment of the nuclear spindle parallel to the rays of light and the laying down of the curved wall (Fig. 1, A and B). NIENBURG also demonstrated that the operative factor is the difference in light intensity on the two sides of the spore rather than the direction of the light rays. This first division takes place in less than twenty-four hours but STAHL found that a constant change in the direction of illumination retards the rate of division. In one experiment in which the spores were placed on a klinostat through the day and cooled to a little above 0° C during the night (division would otherwise have proceeded in the absence of light), spores were still undivided after five days although a few of them were abnormally divided into two equally large cells. BUCHTIEN confirmed this effect of bright light on the direction of the first division but also demonstrated that weak light does not so operate. The direction of growth of the first rhizoid is also to some extent determined by the influence of light. In bright light the rhizoid is negatively phototropic but, as various writers have pointed out, it is positively phototropic in diffuse light and since this is particularly shown when the atmosphere is very moist the possible influence of hydrotropism cannot be excluded.

The influence of external conditions on the further development of the prothalli was also investigated by BUCHTIEN. The young prothalli are very variable in form, the main factors concerned being light intensity and the nature of the medium upon which they are grown. Cultures subjected to one hour of direct sunlight each day first develop a filament and then proceed to the formation of a flat expanse; this is apparently the normal and most successful type. Cultures exposed to direct sunlight throughout

---

[1]) E. R. WALKER, The Gametophytes of three species of *Equisetum*, Bot. Gaz., Vol. 92, 1931.

[2]) W. NIENBURG, Die Wirkung des Lichtes auf die Keimung der *Equisetum*-Sporen, Ber. d. deutschen bot. Gesells., XLII Bd., 1924.

the day omit the filamentous stage and proceed directly to the formation of a cell plate. If grown in a 3% solution of mineral salts the spores divide by two walls at right angles to one another and proceed to the formation of a bulky cell mass, gorged with starch and often unprovided with rhizoids. In weak light, or when the spores are sown thickly, various abnormal developments occur usually involving a continuation of the filamentous stage in anomalous forms. The filamentous stage may also be retained by prothalli grown under conditions of poor nutrition, as for instance on sand.

The mature prothallus is a dorsiventral structure of rather irregular form differentiated into a massive basal region poor in chlorophyll and a number of distal green lobes. There are few data available as to the factors conditioning the dorsiventrality, the experiments of LUDWIGS (loc. cit.) being inconclusive.

The distribution of the sexual organs has been the subject of much discussion. It seems clear from the culture experiments of various workers that the prothalli are potentially monoecious but that there is a very close relation between nutritional conditions and the type of sexual organ produced. BUCHTIEN, LUDWIGS and WALKER are agreed that crowded conditions, or growth on substrata poor in mineral salts, lead to the production of depauperate male prothalli. On the other hand favourable conditions lead to the production of large prothalli which are either female or monoecious. WALKER's cultures demonstrate that well nourished prothalli of *E. Kansanum, Telmateja* and *arvense* are normally monoecious but that at any one period only antheridia or archegonia are produced; archegonia appear first and are later followed by antheridia, the sequence being repeated unless a sporophyte is produced. It seems probable that the production of archegonia depletes the prothallus of food supplies and that the formation of antheridia follows in consequence of this. Such an interpretation is supported by BUCHTIEN's experiment in which he transferred prothalli bearing archegonia from a rich substratum to sand or water and obtained the development of antheridia only. It may be noted that the reverse operation, i.e. the conversion of male prothalli into larger ones bearing archegonia by improving the conditions of growth is, according to LUDWIGS, achieved only with difficulty. In nature, the crowded sowings consequent on the interlocking action of the elaters lead, as observation has shown, to the production of some small male prothalli intermingled with larger ones bearing archegonia. Fertilisation may occur at this stage but in its absence it would seem probable from WALKER's observations that these larger prothalli would then produce successive crops of antheridia and archegonia until such time as an embryo was produced.

(b) *Filicinae.* The development and structure of the cordate type of Fern prothallus has been the subject of a considerable volume of experimental investigation. A brief outline of this may be given under the headings of (1) the factors governing the form of the prothallus, (2) the factors governing the direction of growth of the prothallus and (3) the factors governing the origin and position of the sexual organs.

1. The form of the prothallus is plastic and is readily influenced by a number of factors such as the supply of mineral salts, the intensity and spectral constitution of the light and variations in temperature.

The place of emergence of the young gametophyte from the spore, whether the latter is of the tetrahedral or the wedge-shaped type, is largely governed by the organisation of the spore wall; the wall is split at the apex of the triradiate ridges in the former type and along the ridge representing the line of contact of the four spores in the latter, i.e. on the side facing the interior of the spore group in each case. The early stages of development are very variable but the majority of Fern prothalli pass through a filamentous stage. KLEBS (loc. cit.) has shown that the characteristics of this stage are influenced by a complex set of factors of which the quantitative and qualitative nature of the light are the most important. Working with the prothalli of *Pteris longifolia*, KLEBS found that in very weak light the prothallus grows out as an undivided tube (similar to the one shown in Fig. 1, *G*); in rather stronger illumination the filament becomes segmented by cross walls, the number of such divisions increasing with increasing light intensity. Red and yellow light rays retard nuclear division and KLEBS describes how the nutrition rich spores of *Lygodium* gave rise to an undivided filament 10—12 mm long when germinated in red light. Fig. 1, *G* shows a similar undivided filament of *Pteris longifolia*. The filamentous stage may persist for a considerable period under conditions of feeble illumination or illumination by red or yellow rays. In some of the Hymenophyllaceae the filamentous condition is normal for the mature prothallus but, so far as I am aware, there are no experimental data available relating to these types. On the other hand, the filamentous stage may be omitted, as is normally so in the *Osmundaceae* and *Marattiaceae*. In *Osmunda regalis*, for instance, KLEBS found that even in very weak light or in darkness isolated longitudinal divisions occur in the filament.

Normally after the filamentous condition has persisted for a length of time varying with the nature of the illumination and temperature, longitudinal divisions take place in the distal cells of the filament leading to the formation of a cell plate. Fig. 1, *C–E* show prothalli of *Pteris longifolia* from cultures grown in successively more intense illumination. In *C*, the first longitudinal division occurs in the seventh cell of the filament; in *D*, in the fourth cell and in *E*, in the second cell. The early onset of longitudinal division is also stimulated by growth in blue light (Fig. 1, *F*). KLEBS has given measurements of the intensities of illumination under which the filament passes over to the cell plate condition in a variety of Ferns. This intensity varies within very wide limits, *Dryopteris Filix-mas*, for instance, forming cell plates at an intensity of only one tenth of that required by *Aneimia Phyllitidis*.

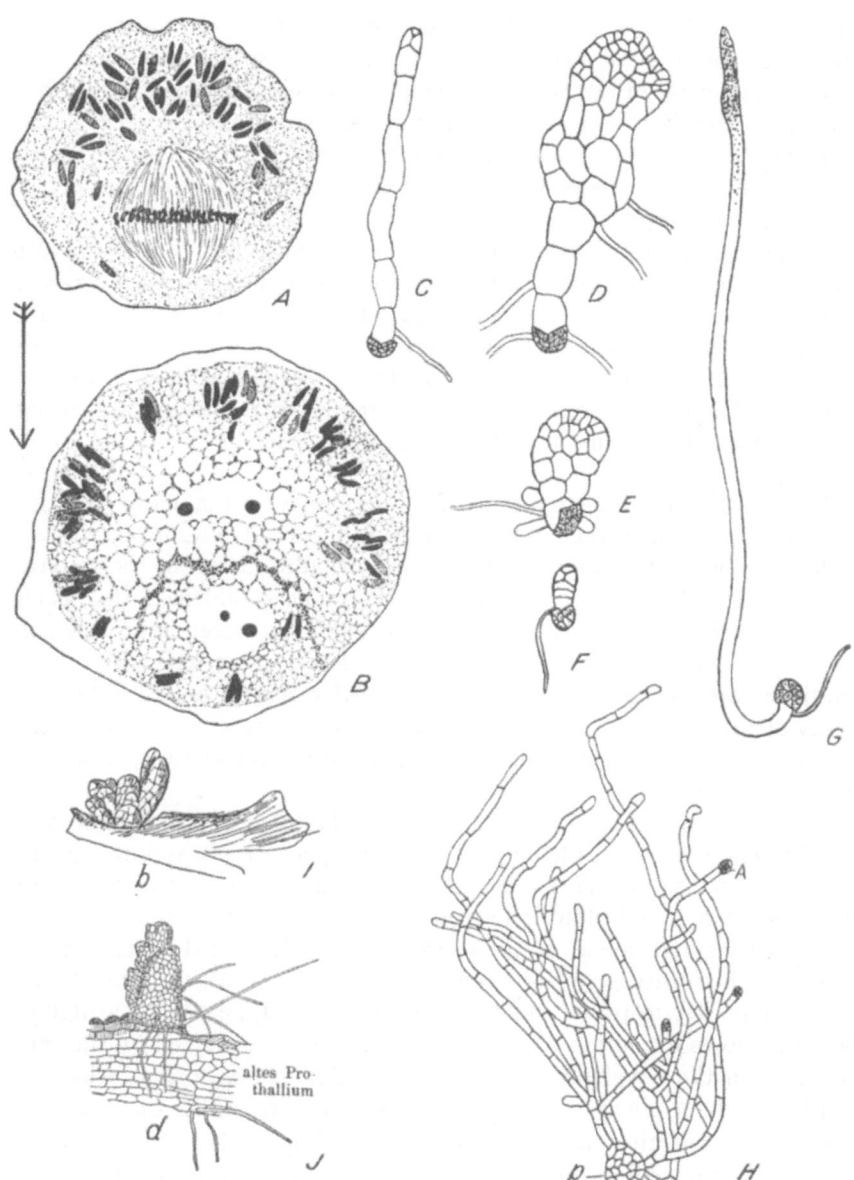

FIG. 1. *A*, *B*, sections through germinating spores of *Equisetum* fixed after 6 and 12 hours respectively; direction of unilateral illumination indicated by arrow. *C–G*, prothalli of *Pteris longifolia*; *C–E* from cultures grown in successively stronger illumination; *F*, grown in blue light; *G*, grown in red light. (X56). *H*, *Woodsia ilvensis*, prothallus (*p*) grown in weak light, showing filamentous development with terminal antheridia (*a*). *I*, *Equisetum palustre* and *J*, *E. Schaffneri;* pieces of prothallus showing regenerative growths (X14). (*A*, *B*, after NIENBURG; *C–G*, after KLEBS; *H*, after SCHLUMBERGER from GOEBEL; *I*, *J*, after LUDWIGS.)

PRANTL [1]) has shown that the development of the prothallus may be arrested at the stage of a cell plate without meristem by cultivation on a substratum devoid of nitrates but normally the cell plate soon establishes a meristem and divisions in the third direction give rise to the archegonial cushion. This further development only takes place if the light intensity is sufficiently high (KLEBS, GOEBEL and other writers) and the supply of nitrogen adequate (PRANTL, loc. cit.).

The normal adult prothallus thus established may be induced to revert to earlier stages by altering the growth conditions. For instance. PRANTL induced adult prothalli of *Osmunda regalis* to revert to the condition of a cell plate without meristem by transferring them from a full culture solution to one from which nitrogen was omitted. A reduction in the intensity of illumination or transference to darkness frequently results in the growing out of marginal cells in the form of filaments, e.g. in *Woodsia ilvensis* (Fig. 1, *H*).

The above account deals briefly with only a few of the many data obtained by numerous workers relating to the influence of external conditions on the development and form of the Fern prothallus. So far as the cordate type of prothallus is concerned it appears to be established that each growth form, in the presence of an adequate supply of water and oxygen, is the result of the interaction of a number of factors of which the most significant are the quantitative and qualitative nature of the light, the supply of mineral salts and temperature. The nature of the mechanism intervening between the stimulating factors and the response is, however, entirely unknown.

2. The direction of growth of the Fern prothallus is determined largely by light. The young cell filament is positively phototropic and negatively geotropic (PRANTL [2])). When the filament passes over to the cell plate stage, or where a cell plate is formed immediately, the flattened surface places itself at right angles to the incident rays of light if the latter are of medium or high intensity (LEITGEB [3]), PRANTL (loc. cit.), GOEBEL [4]) and other writers). Hydrotropic and geotropic influences appear to be negligible The only known exception to this latter statement is provided by the earliest stage in the development of the prothallus of *Ceratopteris* which is described by LEITGEB as having its apical cell orientated under the influence of gravity so that alternate segments are placed upwards and downwards and thus the prothallial surface stands vertically at first although it soon

---

[1]) K. PRANTL, Beobachtungen über die Ernährung der Farnprothallien und die Vertheilung der Sexualorgane, Bot. Zeit., 39 Bd., 1881.

[2]) K. PRANTL, Über den Einfluss des Lichtes auf die Bilateralität der Farnprothallien, Bot. Zeit., 37 Bd., 1879.

[3]) H. LEITGEB, Über Bilateralität der Prothallien, Flora, 1877; Studien über Entwicklung der Farne, Der Sitz. d. k. Akad. d. wiss. Wien, LXXX Bd., 1879.

[4]) K. GOEBEL, Organ. der Pflanzen, Dritte Aufl., Erster Teil, p. 601.

takes up a position at right angles to the incident rays. The adult prothallus in most Ferns is also diaphototropic in light of medium or high intensity.

The rhizoids of the prothallus are negatively phototropic but are not influenced by gravity.

3. The production of the sexual organs and the position of these on the prothallus depend on the working of a number of interrelated factors. The antheridia are produced very early in the development of the prothallus and they may appear almost immediately after the germination of the spore, particularly in examples where the spores are rich in food material. As pointed out above, the development of the prothallus may be arrested at the filamentous stage or at the a-meristic cell plate stage by unfavourable conditions of illumination or food supply and such arrested prothalli are without exception purely male. The antheridia are not localised as regards their position on the prothallus and the influence of external conditions only appears in so far as that superficial antheridia are found on the shade side of the prothallus.

Archegonia are only found on relatively older prothalli and their appearance is correlated with the establishment of a meristem and the formation of a thickened cushion, developments which, as pointed out above, depend on a suitable light intensity and other factors. Nutritional factors are obviously important but the question as to whether an increased ratio of carbohydrates to mineral salts is effective as a stimulus to fertility has not been investigated [1]).

The position of the archegonia is determined by the position of the cushion and this is dependent on the direction of the incident light rays. The cushion with its rhizoids is always formed on the shaded side of the prothallus and consequently the archegonia are also found in this position. This dorsiventrality of the prothallus can be reversed as LEITGEB, PRANTL and other workers have shown by a series of experiments on a large variety of Ferns. One or two of LEITGEB's experiments may be briefly described. Prothalli of *Ceratopteris* floating on culture solution were illuminated from below. There was a complete reversal of the dorsiventrality, archegonia and rhizoids appearing on the previously dorsal side and the production of these structures completely ceasing on the previously ventral surface. In another experiment spores of *Matteuccia Struthiopteris* were sown inside a clay cylinder suitably moistened and plugged at the top, the illumination being solely from below. The prothalli grew horizontally and in all of them the lower surface was developed as the morphological upper surface devoid of archegonia and rhizoids. It is clear from the numerous experiments which have been described that the bilaterality of the prothallus is

---

[1]) Cf. p. 130, where the factors governing the fertility of the sporophyte are discussed.

determined by the direction of the illumination when the latter is of medium or high intensity. LEITGEB failed, however, to induce the formation of archegonia on both sides of the prothallus, or alternatively their inhibition on both sides, by growing prothalli on a klinostat and thus illuminating them equally from all sides; such prothalli still showed the normal dorsiventral structure, possibly, as LEITGEB pointed out, owing to slight inequalities of illumination. In very weak illumination isobilateral prothalli may arise; PRANTL (loc. cit., 1879) mentions that prothalli of *Polypodium vulgare* growing in weak light produced archegonia on both sides and ATKINSON [1]) found archegonia on both sides of prothalli of *Pteris serrulata* which were crowded and growing almost perpendicularly so that they were illuminated from both sides.

§ 4. **Regeneration of the Prothallus.** — The prothalli of the Pteridophytes, so far as they have been investigated, possess considerable powers of regeneration. No data relating to the gametophytes of the *Psilotales* and *Lycopodiales* are, however, available.

In *Equisetum*, LUDWIGS (loc. cit., p. 434) has demonstrated that isolated portions of prothallus tissue will give rise to regenerative growths which ultimately become detached as self supporting prothalli (Fig. 1, *I*, *J*). WALKER has also noted that when prothalli degenerate any cell or group of cells may divide actively and give rise to new lobes.

In the *Filicales* the only type of prothallus which has been investigated as regards regenerative capacity is the cordate type. The experiments of STANGE [2]), HEIM [3]) and GOEBEL [4]) all indicate that virtually every cell of the cordate type of prothallus is capable of regenerative activity. A few examples of the experimental results relating to this may be described. GOEBEL states that if a prothallus is cut through longitudinally so as to damage the apex and remove a lobe a new growing point is established. The activity of the latter leads to the production of a new lobe so that the anterior portion of the prothallus again becomes cordate; the posterior portion of the damaged prothallus is not, however, restored. In a similar experiment, HEIM confirmed this result and, for the prothalli of *Osmunda regalis* and *Doodia caudata*, added the fact that adventitious prothalli were produced on the older parts. In another of HEIM's experiments the prothalli were cut through transversely; the anterior portion with the apex continued its growth in normal fashion while the basal region gave rise to numerous adventitious prothalli. Removal of the growing point also led to the production of adventitious prothalli, as many as twenty or thirty being formed on a single prothallus. Finally it may be mentioned that if prothalli

---

[1]) G. F. ATKINSON, The Biology of Ferns, 1894, p. 13.
[2]) STANGE, Ber. über die Sitz. d. Gesell. für Bot. zu Hamburg, 1886.
[3]) C. HEIM, Untersuchungen über Farnprothallien, Flora, 82 Bd., 1896.
[4]) K. GOEBEL, Organography of Plants, Part 1, p. 43.

are cut into small pieces each of these is capable of producing adventitious prothalli or, if it includes a portion of the apical region, of growing on into a normal prothallus.

The production of adventitious prothalli thus appears to be correlated with a removal of the influences exercised by the apex. Experimental removal of the apex leads to their development from any region of the prothallus, while in undamaged prothalli they are formed only on the older basal regions which presumably are no longer influenced by the growing point.

**§ 5. Fertilisation.** — Although the physiology of reproduction hardly falls to be considered in this chapter a brief account may be given of the experiments conducted by various workers on the nature of the attraction exerted by Pteridophyte archegonia upon the corresponding sperms [1]. A convenient method for investigating this was elaborated by PFEFFER. He used glass capillaries 0.1—0.15 mm in diameter, 10—15 cm long and sealed at one end. These were filled with a 0.1% aqueous solution of the substance to be investigated. The open ends of the capillaries were then dipped into water containing the motile sperms and the entrance or non-entrance of the latter noted under the microscope. By this method PFEFFER determined that the sperms of Ferns and *Selaginella* are attracted by malic acid. SHIBATA, by similar means, has shown that the sperms of *Isoetes*, *Equisetum* and *Salvinia* are also attracted by malic acid. BRUCHMANN found that the sperms of *Lycopodium clavatum* are attracted by citric acid and some of its salts. So far as I am aware, there is no information regarding the nature of the attraction in the *Ophioglossaceae* or in the *Psilotales*.

## The Sporophyte

**§ 6. Embryology.** The factors governing the development of the fertilised egg inside the Pteridophyte archegonium have been the subject of much discussion but they have not, except in a few instances, been subjected to experimental analysis. An interesting account of this subject, mainly based an a wide comparative survey of the embryology of the various Archegoniate groups, has recently been given by BOWER [2]. Suggestions are made here regarding the nature of the factors concerned, reference being made to the possible influence of external factors such as light and gravity, and internal ones such as pressure exerted upon the egg by the archegonial wall and the directive influence of the prothallial food supply.

---

[1] A convenient summary of these together with a full citation of the literature is given by BRUCHMANN, Von der Chemotaxis der *Lycopodium*-Spermatozoiden, Flora, 99 Bd., 1909, p. 193.

[2] F. O. BOWER, Primitive Land Plants, 1935, Chapter XXVI.

So far as the *Psilotales*, *Equisetales* and *Lycopodiales* are concerned there is a complete absence of experimental data relating to the causes underlying the various embryological developments exhibited by these groups.

The embryology of the Ferns has attracted more attention but even here experiment has been confined to *Marsilia* and a few of the higher Leptosporangiate types.

LEITGEB [1]) was able to demonstrate the influence of gravity on the orientation of the first segmental wall of the embryo of *Marsilia*. Megaspores were fixed to wax plates so that they could be orientated at will; these were then fertilised by the addition of microspores and thereafter microscopic examination was made of the resultant embryos. The first segmental wall was constantly found to include the archegonial axis, but, except when the latter was vertical, gravity influenced the plane of the wall so that the embryo was always divided into an upward facing half, giving rise to stem and leaf, and a downward facing half giving the foot and root. The megaspores normally lie horizontally so that the orientation of this first wall under the influence of gravity gives a suitable placing of the primary organs of the embryo for the avoidance of awkward curvatures in their further development.

So far as the higher Leptosporangiate Ferns are concerned, HEINRICHER[2]) and LEITGEB (loc. cit., 1879) have shown that the orientation of the basal wall, which constantly includes the archegonial axis, and the disposition of the primary organs of the embryo are entirely independent of the influence of light and gravity, being conditioned entirely by internal factors. For instance, embryos arising from archegonia produced on the upper side of the prothallus under the influence of illumination from below show the normal disposition of parts, as do also embryos produced on prothalli rotated on a klinostat. It seems clear that the important factor is the position of the fertilised egg in the venter of the archegonium in relation to the apex of the prothallus. A further analysis of the working of this factor has not been attempted.

**§ 7. Factors underlying Variations in External Form.** — Very little experimental work has been carried out with regard to the factors underlying the manifold variations in external form exhibited by the Pteridophytes. These differences are due fundamentally to differences in the "specific substance". There is nevertheless some possibility of examining the nature of the internal and external factors which, by acting upon this inherited constitution, influence external form. The problems are perhaps most easily formulated in relation to such genera as *Selaginella*,

---

[1]) H. LEITGEB, Zur Embryologie der Farne, Sitz. d. k. Akad. d. wiss. Wien LXXVII, 1878.

[2]) E. HEINRICHER, Beeinflusst das Licht die Organanlage am Farnembryo?, Mitt. a. d. bot. Inst. zu Graz, Jena, 1888.

*Lycopodium* and *Equisetum* where the differences between the various species are readily recognisable as variants upon a clearly defined plan of construction, but even here relatively little progress has been made.

In the genus *Selaginella* the main difference in form is that between the radial type and the much more prevalent type showing dorsiventrality accompanied by anisophylly. The difference in organisation between the *S. spinulosa* type and the dorsiventral type is a profound one and there is no indication that experiment can help to explain it. So far as the dorsiventral species are concerned, however, the specific differences are all variations on a common plan. The essential unit is a branching together with its angle meristems. Differences in tropistic behaviour and morphological develop-ment of these units in different regions of the plant give the varying configuration of the different species. There are, as yet, very few data relating to the factors underlying these various developments but the experiments of various workers have indicated the equipotentiality of these units in all parts of the plant. The presence of a system of correlating influences may be inferred but this system has not been subjected to further analysis.

The problem presented by the factors underlying the onset of anisophylly is, however, more easily attacked. The majority of the anisophyllous species show little, if any, plasticity in their response to external factors and the anisophylly once established in the ontogeny cannot be altered. GOEBEL [1]) has established the fact, however, that a few species exhibit a relatively high degree of plasticity and this is particularly so in *S. san-guinolenta*. In this latter species anisophyllous shoots become isophyllous if grown in darkness. When grown in the light, the form and arrangement of the leaves varies according as to whether the plant is grown in a dry or a wet situation; in the former the shoots are isophyllous, in the latter aniso-phyllous to a varying degree. Thus various factors, not fully analysed, underlie the anisophyllous condition of this species. The importance of light in influencing the onset of anisophylly is also indicated by GOEBEL's experiments with plants of *S. caracensis* grown from bulbils [2]). When grown in the dark these are isophyllous but when grown in the light they become anisophyllous. The presence of light rather than its direction is the im-portant factor, for plants grown in the light on a klinostat so that they were equally illuminated on all sides, still developed the anisophyllous condition.

Although the influence of external conditions can thus be demonstrated in these two species, anisophylly depends in the majority of the forms investigated on internal factors.

It may be mentioned here that in *Lycopodium complanatum* the aniso-phyllous condition is also dependent on the influence of light and shoots

---

[1]) K. GOEBEL, Organ. der Pflanzen, Dritte Aufl., Erster Teil, p. 333.
[2]) *Ibid.* p. 618.

grown in the dark, as in GOEBEL's experiments [1]), show a radial construction.

Specific differences within the genus *Equisetum* depend largely on the differential development of the shoot and root rudiments present at each node of the plant. These differences are mainly inherited characters but experiment has shown that most species possess a certain degree of plasticity which throws some light on the normal construction.

The diageotropic rhizome is characterised by the absence of chlorophyll, by the growing out of the root rudiments and by the dormancy of the majority of the lateral buds. LUDWIGS (loc. cit.) has shown that in *E. limosum* the apex of such a rhizome can be induced to grow on as a typical aerial shoot. On the other hand, axillary buds on aerial shoots, which would normally grow out as aerial branches, may be induced to form rhizomes. In LUDWIG's experiments with cuttings the majority of the buds at a node grew out as aerial shoots but one or more of the buds developed as typical rhizomes. It is clear, therefore, that there is no essential difference between the subterranean parts of the axis and those which are aerial.

The various species show differences in the degree of development of the side shoots on their aerial axes. The lateral buds may remain more or less dormant, as in *E. limosum* and *hiemale*, while in others, e.g. *E. arvense*, all the lateral shoots develop and themselves exhibit further branching. The difference between the two types depends on internal factors but it is possible to induce the development of branching in the *E. limosum* type by cutting off the apex of the shoot. It may be assumed that the potentialities of the aerial axes are the same in all species but that internal correlating factors lead to the differential developments which give rise to specific distinctions.

Further specific differences are afforded by the characters of the fertile as compared with those of the sterile shoots. The essential morphological structure is the same in both and experiment has shown that they possess similar potentialities of development even when the realisation of these is restricted by the operation of internal factors. For instance the normally unbranched and yellow-green fertile shoots of *E. arvense* and *Telmateja* may be induced by culture in a moist chamber to produce green lateral shoots and thus approach to the structure of the sterile axes (LUDWIGS, loc. cit.).

So far as I am aware, little experimental work has been carried out on the other groups of Pteridophyta. In the *Filicales* the only data refer to the conformation of sporeling leaves. GOEBEL [2]) has shown that if sporelings whose latest formed leaves have attained to the pinnate condition are placed under unfavourable conditions of growth the next formed leaves show a regression in complexity of form and are similar to the primary leaves. The conclusion is drawn that the primary leaves are potentially

---

[1]) *Ibid.*

[2]) K. GOEBEL, Organography of Plants, Eng. Edit., Pt. 1, p. 152.

similar to adult leaves but that they are arrested at an early stage of their development. GOEBEL [1]) was also able to induce regression in leaf form in *Ceratopteris* by culturing isolated stem apices. The first leaves produced by the isolated tips were of the juvenile type and GOEBEL was able to show that the degree of regression was correlated with the age of the plant used for the experiment. There was probably, as in the previous experiment, a correlation between the amount of nutrition available and the form of leaf produced.

§ 8. **Correlation of Parts.** — The phenomenon of correlation, that is, to use GOEBEL's definition, "the reciprocal influence of organs upon one another", is demonstrated in many of the experimental results recorded in this chapter and it need not be considered in any detail here. There may be instanced as clear examples of this correlation the growing on of axillary buds in *Botrychium* as a result of decapitation or damage to the stem apex. the increased formation of root borne buds following damage to the stem apex of *Ophioglossum* and the growing out of lateral branches on the aerial axes of *Equisetum limosum* following the removal of the tip. These and many other examples all indicate the existence of a system of correlating influences in every plant. So far as the Pteridophytes are concerned nothing is known of these influences beyond the observed results of their activity. Recent discoveries relating to the activity of plant hormones [2]) indicate, however, the possibility of analysing, at least in part, the nature of the mechanism involved in correlation phenomena.

§ 9. **Factors influencing the Vascular System.** — The stress laid on vascular construction in discussions of phyletic problems lends a special interest to attempts to investigate the factors influencing this. Two closely inter-related problems require to be considered, viz. (*a*) the nature of the factors leading to the differentiation of vascular tissue and (*b*) the nature of the factors underlying the distribution of vascular tissue.

(*a*) Factors underlying the differentiation of vascular tissue. Various suggestions have been made as to the nature of the factors influencing the differentiation of vascular tissue [3]) and these have been conveniently summarised by LANG [4]) as follows: 1. functional stimuli, 2. the inductive influence of older preformed parts on the developing region and 3. formative stimuli of unknown nature proceeding from the developing region. LANG

---

[1]) K. GOEBEL, Experimentelle Morphologie, p. 20.

[2]) See R. SNOW, Growth-Regulators in Plants, New Phyt., Vol. XXXI, 1932 for an account of these.

[3]) Investigations have been concerned almost exclusively with the differentiation of xylem elements, the phloem and other tissues being ignored, possibly on account of the difficulty of recognising them in the earlier stages of their development.

[4]) W. H. LANG, Pres. Address, Brit. Assoc., Section K, 1915.

pointed out, however, that the functional stimuli do not come into play at the time of the laying down of the vascular tract, though they may have importance for their maintenance later; also that the inductive influence of older parts cannot be operative in relation to the vascular system of embryos. The importance of formative stimuli of unknown nature proceeding from the developing region was therefore stressed.

One or two examples of experimental work relating to this question may be described. LANG (loc. cit., p. 10) injured or destroyed the apices of young plants of *Osmunda regalis* and as a result branching was induced in some of these. In relation to these branchings a vascular connection was established backwards with the adaxial face of the subtending leaf-trace. Comparable results to these were obtained in *Lycopodium Selago* [1]) where damage to the apex results in the development of adventitious axillary buds (Fig. 2, *A*). In these latter the first vascular tissue to be differentiated is a basal mass of tracheidal cells; then follows, in relation to the establishment of a stem apex, the formation of the procambial strand of the bud. Finally there is a backward differentiation of a chain of tracheidal cells to make contact with the trace of the subtending leaf. These two examples may be correlated with what has been observed in "natural experiments" in *Botrychium* and *Helminthostachys*. These plants possess dormant axillary buds which are at first devoid of vascular tissue and are without influence on the vascular strand of the main shoot. Such buds may be stimulated to further development by damage to the apex of the plant and in the course of this they differentiate vascular tissue within themselves and then exert a backward influence through the permanent tissue of the cortex, leading to the formation of a branch trace which makes contact with the adaxial face of the subtending leaf-trace. In *Helminthostachys* the main stele is also influenced to the extent of forming more xylem both behind and before the place of insertion of the branch.

The production of tracheides in examples of induced apogamy, particularly in those examples where isolated sporophytic organs or tissues are produced by the prothallus, are of interest in relation to the problem under consideration. The fullest account of the occurrence of such tracheides is that of LANG [2]) and the facts mentioned here are taken from this memoir. The appearance of tracheides is constantly associated with the development of apogamous buds and isolated sporophytic organs such as roots and sporangia (Fig. 6, *B*); they also occur in relation to the indeterminate sporophytic structures known as "middle shoots". Finally it may be mentioned that tracheides were observed in outgrowths from archegonial

---

[1]) S. WILLIAMS, A Contribution to the Experimental Morphology of *Lycopodium Selago*, Trans. Roy. Soc. Edin., Vol. LVII, 1933.

[2]) W. H. LANG, On Apogamy and the Development of Sporangia upon Fern Prothalli, Phil. Trans. Roy. Soc., Ser. B, Vol. 190, 1898.

projections which were mainly gametophytic in nature and even in regions of the prothallus showing no modification in external form.

This development of tracheides in such a variety of circumstances makes it difficult to identify the factors underlying their appearance. Apart from the last mentioned examples, however, it would appear possible that the establishment of a sporophytic apex or of an unorganisated sporophytic meristem is the necessary prelude to the development of vascular tissue. It is clear, however, that these examples of anomalous tracheide formation provide little or no help towards the solution of the problem presented by normal vascular differentiation, except, perhaps, support for the view that sporophytic meristems produce formative substances inducing the development of vascular tissue.

The above described experiments, and particularly those relating to the normal sporophyte, point fairly clearly to "the reality of influences proceeding backwards from developing structures", these influences leading to the differentiation of vascular tissue. The nature of this influence, or formative substance, is, however, entirely unknown. By its action, some parenchymatous cells are induced to alter their metabolism so that lignification of the wall and the other processes involved in the formation of a tracheide are set in train; others are induced to form phloem elements while still others remàin unaffected as parenchyma. This differential action of the formative substance raises a difficult problem to the solution of which no approach has yet been made.

(*b*) Factors underlying the Distribution of Vascular Tissue.

The differentiation of vascular strands proceeds along definite tracts and it is necessary to consider the nature of the directive factors concerned. One of the examples which illustrates the problem most simply is that afforded by the vascular arrangements in relation to the axillary buds of *Botrychium* described above. Here there is the establishment of a vascular connection through the cortical tissue backwards to the subtending leaf-trace. There may be related to this and similar examples the facts derived from experiments carried out on various Angiosperms [1]) and particularly those in which the formation of vascular bridges following the severance of bundles has been studied. With regard to these latter it has been shown that the new vascular connections differentiate from the basal ends of severed bundles and proceed downwards towards upper severed ends. It is obvious that here, as in the experiments on Pteridophytes, there is not merely a formative influence at work but that this, or some other factor, is directive in its action. Simon has suggested that the directive influence is a water gradient, the new vascular tissue forming first in tissues depleted of water and proceeding to those with an adequate supply. Such an explanation

---

[1]) See D. THODAY, Some Physiological Aspects of Differentiation, New. Phyt., Vol. XXXII, 1933, where recent literature is cited and briefly discussed.

might be applied to the formation of vascular connections in the Pteridophytes mentioned above. Nevertheless, the problem cannot be regarded as
solved and many more experimental data are required.

The nature of the factors underlying the varied stelar arrangements
presented by the Pteridophytes presents an even more difficult problem
than the relatively simple instances just considered. The normal progression
in stelar structure which occurs in the ontogeny of the more advanced
Leptosporangiate Ferns clearly indicates the nature of the problem and the
difficulties involved. At successive stages the apex lays down a protostele,
a solenostele and a dictyostele. As LANG has pointed out, these changes are
determined in the region between the apical meristem and the fully differentiated region behind, a region which has been little studied even from
the purely morphological
aspect and one which is
difficult to investigate by
experiment. The influence
of the leaf-traces appears
to be one of the factors
concerned but it seems
highly probable that a
complex system of factors
is operative in the formative region.

BOWER [1]), on the basis
of wide comparison, has
stressed the importance of
the Size-Factor. This is described as a "morphoplastic
factor", connoting "that
influence which tends to
secure by modification of
Form a due levelling up of
the proportion of surface to
bulk as the Size increases."
So far as the stele is concerned the important ratios
are those between endo-
dermal surface and stelar
bulk and between xylic
surface and xylic bulk. The

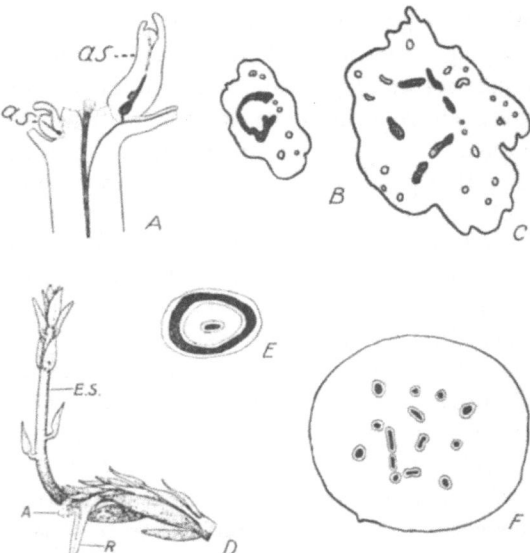

FIG. 2. *A*, longitudinal section of decapitated plant
of *Lycopodium Selago;* as, axillary adventitious buds.
( × 18). *B, C, Deparia,*(=*Aspidium) Moorei,* transverse sections of stem in starved region (*B*) and normal region (*C*). ( × 6). *D–F, Selaginella Lyallii; D,*
continued growth of distal branch laid in a horizontal position; *a,* apex which grew on as a rhizome;
e.s., erect stem; *r,* root. ( × 8). *E,* transverse section
of rhizome stele; xylem, black; endodermis, plain
line. *F,* transverse section of erect stem. (*E, F,* × 22).
(*B, C,* after THOMPSON; *D–F,* after WARDLAW).

clear effect of increasing size on the form of the stele has been amply demonstrated by BOWER (loc. cit.) and cannot be further discussed here; nor

¹) F. O. BOWER, Size and Form in Plants, 1930.

can the difficulties involved in the analysis of the mechanism of this factor be entered into. The only experimental data relating to the incidence of the Size-Factor are provided by starvation experiments. METTENIUS [1]) has shown for *Angiopteris* and THOMPSON [3]) for *Deparia* (*Aspidium* sp.) that starvation leads to a diminution in the size of the stem and that there is a simplifi- cation of stelar structure in relation to this. In the latter example for instance, the normal stele is dictyostelic, the diameter of the stele being 2.35 mm, while the stele in the diminished starved region is solenostelic with a diameter of 1.16 mm. (Fig. 2, *B* and *C*). Such experiments indicate the plasticity of the stele and the possibility of further experimental investigation of the operation of the Size-Factor.

Finally a brief reference may be made to the stelar structure of *Selaginella Lyallii* since it gives some indication of the effect of external conditions on the form of the stele. The rhizome of this plant possesses a solenostele but the upright growing branches are dialystelic (Fig. 2, *E* and *F*). It has been shown by BRUCHMANN [3]) and WARDLAW [4]) that if the tip of an erect branch is laid down horizontally on the soil the apex grows on as a rhizome which becomes solenostelic (Fig. 2, *D*). The change of position has some effect, not further analysed, on the apex and region of differentiation behind it so that a complete ring of vascular tissue is formed rather than a number of discrete strands. The operation of the Size-Factor in this example has been discussed by WARDLAW (loc. cit.).

It is clear that a large amount of experimental work, properly correlated with the results of comparative morphology, will have to be carried out before further progress can be made in the recognition and analysis of the factors influencing the vascular system.

## § 10. Organs of Indeterminate Nature. — An interesting field of inquiry which lends itself to experimental investigation is that dealing with the morphological nature of the various so-called indeterminate organs which occur in the Pteridophytes and particularly in the Lycopodiales. The most interesting of these are the rhizophores of *Selaginella*, the bulbils of *Lycopodium* spp. and the protocorm developed by the embryos of some species of the latter genus. These may be considered in turn.

The problem presented by the morphological nature of the rhizophore has been the subject of repeated discussion ever since the introduction of the term "Wurzelträger" by NÄGELI and LEITGEB. The historical aspect of

[1]) METTENIUS, Abhandl. Königl. Sachs. Ges. d. Wiss., VI.

[2]) J. McL. THOMPSON, The Anatomy and Affinity of *Deparia Moorei* Hook., Trans. Roy. Soc. Edin., Vol. L, 1915.

[3]) H. BRUCHMANN, Von den Vegetationsorganen der *Selaginella Lyallii* Spring, Flora, Bd. 99, 1909.

[4]) C. W. WARDLAW, The Vascular System of *Selaginella*, Trans. Roy. Soc. Edin., Vol. LIV, 1925.

this discussion need not be entered into nor can the evidence derived from the normal morphology be discussed. The bearing of experimental work upon this subject is, however, important and may be briefly reviewed. The main interest lies in experiments which demonstrate potentialities of the rhizophore for developments other than the normal. It has been shown by numerous workers, including BRUCHMAN [1]), BEHRENS [2]) and WILLIAMS [3]), that rhizophore rudiments can be induced artificially to grow out as normal leafy shoots. A few such experiments may be described. BRUCH-MANN was able by growing plants of *Selaginella Kraussiana* in the open air from the Summer on till Autumn to induce the production of strobili on every shoot tip, thus bringing vegetative development to a close. When such plants were brought into more favourable growth conditions, all the rhizophore rudiments in the more distal branchings grew out as vegetative shoots. In this example the interruption of vegetative growth by the formation of cones appears to be the stimulus resulting in the transformation of rhizophore rudiments into shoots. *S. grandis* shows a similar transformation in the distal cone bearing branchings and the presence of "middle shoots" is a normal feature of this species. BRUCHMANN has also described experiments on sporelings of *S. Kraussiana* in which he was able to induce the transformation of the first formed rhizophores into shoots. The sporelings normally produce three basal rhizophores. If the stem apex of an embryo is damaged the first rhizophore rudiment becomes converted into a shoot (Fig. 3, *A*), while if the hypocotyls of older sporelings are cut across the rudiments of the second and third rhizophores show a similar transformation (Fig. 3, *B*, *C*). Experiments with *S. grandis* have demonstrated similar potentialities of the rhizophore rudiments of adult shoots. If branchings, not including an apex, are laid down on moist sand, the rhizophore rudiments grow out as anisophyllous shoots (Fig. 3, *D*, *E*).

In all these experiments the stimulus which causes the rhizophore rudiment to develop as a leafy shoot is the interruption of vegetative growth, either as a result of cone formation, or of the cutting off of stem apices. The mechanism of this stimulus is not understood but an investigation of the activity of hormones in *Selaginella* might lead to interesting results. It is clear that the meristem constantly produced in the branch angles of many *Selaginella* spp., and which normally gives rise to a rhizophore, is at its inception indifferent in nature. When removed from the influence of vegetative shoot apices it always develops as a shoot. No full argument can be entered into as to the bearing of these facts on the problem presented by the morphological nature of the rhizophore but it may be briefly stated

---

[1]) H. BRUCHMANN, Von den Wurzelträgern der *Selaginella Kraussiana* A.Br., Flora, 95 Bd., 1905.

[2]) BEHRENS, J., Ueber Regeneration bei den Selaginellen, Flora, 84 Bd., 1897.

[3]) S. WILLIAMS, An Analysis of the Vegetative Organs of *Selaginella grandis* Moore, Trans. Roy. Soc. Edin., Vol. LVII, 1931.

that they support the view that the rhizophore is an organ of indifferent nature, standing, as GOEBEL says, "between root and shoot".

Another structure about the morphological nature of which there has been considerable discussion is the vegetative propagative organ found in some species of *Lycopodium*, e.g. *L. Selago* and *lucidulum*, known as the bulbil. This has been regarded on purely morphological grounds as being a transformed sporangium, an arrested shank of a dichotomy, a lateral branch or a modified leaf. This last interpretation is supported by the facts that the bulbils stand in the place of leaves in the phyllotactic spiral and that their vascular supply is similar to the leaf trace in anatomical construction and in its relation to the axial stele. Experiments on *L. Selago*[1]) have shown that if shoots of young plants are decapitated immediately below the apex the young leaf rudiments are induced to form regenerative buds (Fig. 3,

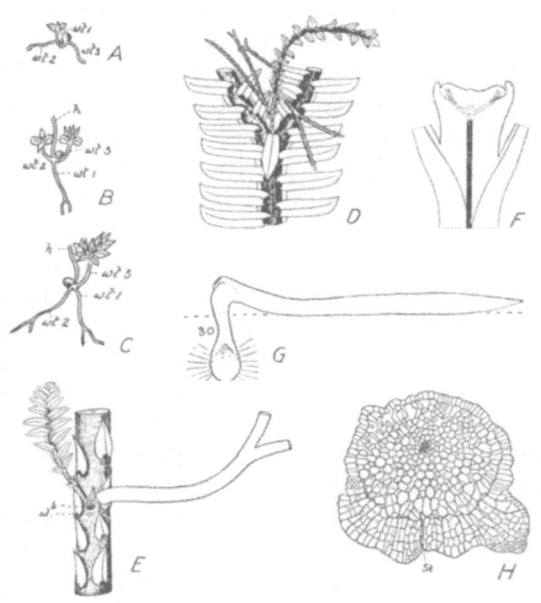

FIG. 3. *A–C*, sporelings of *Selaginella Kraussiana* showing transformation of rhizophore rudiments into shoots; *wt* 1, 2 and 3, rhizophores; *h*, hypocotyl. *D, E, Selaginella grandis;* isolated branchings showing transformation of rhizophore rudiments into shoots; *al*, scar of angle-leaf; *s*, scar of excised lateral bud. (× 3). *F, Lycopodium Selago*, longitudinal section of apex of decapitated plant showing leaf rudiments forming regenerative buds. (× 30). *G, H, Phylloglossum;* *G*, isolated leaf with adventitious tuber (× 4); *H*, transverse section of leaf showing primary cell masses. (× 30). (*A–C*, after BRUCHMANN; *G, H*, after OSBORN).

*F*). This experimental demonstration of the transformation of a leaf rudiment into a bud adds to the likelihood of the correctness of the view that the bulbil is really a modified leaf.

Finally there may be considered that curious organ produced in the ontogeny of certain species of *Lycopodium*, the protocorm. The morphological interpretation of this structure is extremely difficult and it has been regarded either as a primitive organ preceding in evolutionary origin true roots or as a mere biological adaptation of no great morphological signific-

---

[1]) S. WILLIAMS, A Contribution to the Experimental Morphology of *Lycopodium Selago*, Trans. Roy. Soc. Edin., Vol. LVII, 1933.

ance [1]). Additional data for the consideration of this problem have been provided by experimental work on *Phylloglossum* [2]), *Lycopodium Selago* [3]) and by "natural experiments" observed by HOLLOWAY [4]) in *L. ramulosum*. All these experiments concern the formation of adventitious buds on roots, stems or leaves of the plants named. Such buds arise from the proliferation of epidermal, or occasionally sub-epidermal, cells and there is formed in the early stages of their development a "primary cell mass", composed of parenchyma which in most of the examples (but not in *L. Selago*) produces rhizoids (Fig. 3, *H* and Fig. 4, *D*, *E*). Protophylls are then frequently developed by the cell mass and finally the stem apex of the adventitious bud is differentiated. HOLLOWAY pointed out with reference to *L. ramulosum* that the parenchymatous swellings produced in the formation of these buds are similar in all essentials to the protocorms of sexually produced plants. OSBORN has also stated that the primary cell mass which precedes the development of adventitious tubers on detached leaves of *Phylloglossum* is exactly comparable to the protocorm of *Lycopodium* spp. in developmental history and in function. This correspondence also holds, though perhaps not so closely, for the primary cell masses produced in the regenerative activity of *L. Selago*.

The experimental results thus indicate that the homosporous Lycopods tend to show the development of a large parenchymatous mass, frequently bearing rhizoids and protophylls, before they proceed further with the differentiation of the organs of adventitious buds; this mass is comparable in structure and function to the protocorm of sexually produced sporelings. It is not intended to enter into any discussion of the significance of the protocorm. Such parenchymatous swellings, whether produced by the normal embryo of some species or by adventitious buds, are clearly of considerable biological importance as primary storage and absorptive organs. The biological importance of the protocorm does not, however, necessarily militate against the idea that it is a primitive structure of importance in a consideration of the evolutionary origin of plant organs.

§ 11. **Regeneration.** — All that is attempted here is a brief survey of selected examples of regeneration phenomena recorded in the various groups of the Pteridophytes. No attempt will be made to apply these data to the general problems of regeneration such as the nature of the factors

---

[1]) See F. O. BOWER, Primitive Land Plants, p. 272 et seq., and W. H. LANG, On Old Red Sandstone Plants, Trans. Roy. Soc. Edin., Vol. LII, p. 620 and p. 845 for recent discussions of this question.

[2]) T. G. B. OSBORN, Some Observations on the Tuber of *Phylloglossum*, Ann. Bot., Vol. XXXIII, 1919.

[3]) S. WILLIAMS, loc. cit.

[4]) HOLLOWAY, J. E. Studies in the New Zealand Species of the Genus *Lycopodium*, Trans. New Zeal. Inst., Vol. XLIX, 1916.

influencing the character and position of the regenerative growths, the problems relating to the polarity of the plant, and the nature of the factors underlying correlation phenomena. A consideration of these would be too lengthy and would necessarily involve a consideration of groups other than the Pteridophytes.

1. *Psilotinae.* No examples of regeneration are recorded in this group.

2. *Lycopodiinae.*

(a) *Lycopodium Selago.* Experiments with bulbils and young plants grown from these show that regenerative growths may be induced to form on isolated bulbil leaves, on the stem and on partially isolated foliage leaves. In the examples of regeneration from leaves epidermal prolifer-ations occur near the base of the adaxial surface and these eventually be-come differentiated as buds. Regenerative de-velopments were obtained on the stem by decapi-tating at various levels behind the apex. When decapitation was effected immediately below the apex leaf rudiments were involved in the formation of the adventitious buds (Fig. 3, *F* and Fig. 4, *A*) but decapitation at lower levels resulted in epi-dermal, and sometimes cortical, proliferations which formed cell masses, sometimes bearing proto-phylls, before proceeding to the differentiation of the stem apex (Fig. 4, *B–F*). The significance of these facts has been dis-cussed above in the section on indeterminate organs.

(b) *L. inundatum.* GOE-BEL states that isolated first leaves of sporelings are capable of producing adventitious buds.

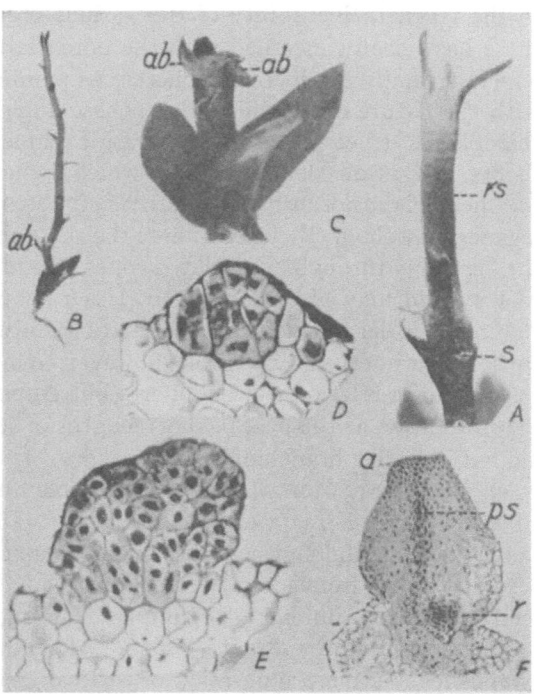

FIG. 4. Development of adventitious buds by plants of *Lycopodium Selago* grown from bulbils. *A*, re-generative shoot (*rs*) developed from leaf rudi-ment; *s*, scar of stem apex. (× 7). *B*, formation of bud in axil of leaf below a transverse incision; *ab*, bud. (× 1½). *C*, decapitated plant showing two adventitious buds of epidermal origin (*ab*). (× 7). *D, E*, stages in development of primary cell mass (× 250). *F*, adventitious bud in longitudinal section; *a*, apex; *ps*, procambial strand; *r*, first root. (× 60).

(c) *Phylloglossum*. OSBORN (loc. cit.) has described the regeneration of new tubers on detached leaves. When such leaves are laid down on moist soil, one or several epidermal cells near the proximal end on the abaxial surface proliferate to form a primary cell mass (Fig. 3, *H*). From a growing point differentiated on this cell mass a tuber develops which resembles that formed by the normal plant (Fig. 3, *G*). The results obtained were applied to the problems presented by the tuber and the protocorm. OSBORN concluded that the primary cell mass is homologous to the protocorm and that the tuber is an entirely different structure of great biological value to the plant but having little morphological significance.

(d) *Selaginella* [1]). Many experiments have been carried out with reference to the regenerative ability of the various organs of *Selaginella*. Many of these have been concerned with the potentialities of the angle meristems which normally produce rhizophores. As pointed out in the section dealing with the nature of the rhizophore, these angle meristems and also young rhizophores developed from them are capable of being transformed into shoots, so that any length of stem which includes a branching where both rudiments have not formed mature rhizophores is capable of giving rise to a regenerative shoot. Regeneration of the shoot may also be accomplished by the "growing through" of isolated cones (Fig. 5, *C, D*). Regeneration of the root system may also be induced. If a rhizophore bearing roots or root rudiments is decapitated callus formation and the production of new roots ensues. It is not usually possible, however, to induce the formation of roots other than in relation to rhizophores but BRUCHMANN was able to induce root formation at the base of short lengths of stem of *S. Kraussiana* which did not include a branching.

(e) *Isoetes* [2]). SOLMS–LAUBACH was able to induce the formation of adventitious buds in *Isoetes lacustre* and *I. Duriei*. The tuberous stocks were cut in half longitudinally and the apical regions removed. After periods varying from six to twelve months adventitious buds appeared in a relatively small number of examples. The morphological relations were complicated and the exact origin of the new growing points was not established. Active proliferation of cortical tissues was also observed. The facts relating to the adventitious buds were applied by SOLMS–LAUBACH to the problem presented by various abnormal specimens found under natural conditions.

3. *Equisetinae* [3]). — The development of true adventitious structures has not been recorded in *Equisetum*. LUDWIGS endeavoured to induce regeneration from isolated internodes of a number of species but entirely without success. On the other hand, the normal construction of the nodes, pos-

---

[1]) See S. WILLIAMS, (loc. cit., 1931), where the literature is cited.

[2]) SOLMS-LAUBACH, *Isoetes lacustris*, seine Verzweigung usw., Bot. Zeit., 1902, p. 179.

[3]) K. LUDWIGS, loc. cit.

sessing as they do the rudiments of buds and roots, makes it theoretically possible for any node under favourable conditions to regenerate a complete plant with roots, rhizomes and aerial stems. The experiments of LUDWIGS showed that it is difficult to induce this development in some species but in others, e.g. *E. Schaffneri, limosum* and *arvense*, such a result is easily achieved. Similar experiments have been carried out by PRAEGER [1]).

4. *Filicinae.* — (a) *Ophioglossum* [2]). Buds are normally formed close to the apex of a few roots of each plant but more prolific bud formation may be induced by cutting off the stem apex. Isolated root tips, or any other portion of a root, are capable of producing regenerative buds.

(b) *Marattiaceae* [3]). GOEBEL states that the stipules remain fresh and succulent after the fronds have fallen and adventitious buds may be produced by them.

(c) *Polypodium* spp. [4]). Regeneration of the leaf tip has been recorded by GOEBEL. The tip of the frond was slit longitudinally and branching ensued. Pinnae developed on the side facing the slit on the newly formed regions but no regeneration occurred on the older part of the cut surface. FIGDOR obtained similar results with the fronds of *Phyllitis Scolopendrium* [5]).

(d) *Cystopteris* spp. [6]). HEINRICHER obtained the development of adventitious buds on isolated basal parts of the fronds of *Cystopteris montana, fragilis, alpina* and *bulbifera*. Such buds appear in varying numbers on the adaxial surface of the petiole and PALISA [7]) later demonstrated that they arise from the proliferation of epidermal cells whose position is not pre-determined. HEINRICHER has also demonstrated that the isolated lower leaves of bulbils will give rise to adventitious buds.

(e) *Diplazium esculentum* and *Platycerium* spp. [8]). The tips of isolated roots of these Ferns are capable of direct transformation into buds.

(f) Sporelings, and particularly their leaves, are easily induced to produce adventitious regenerative structures. GOEBEL [9]) was able to obtain the development of adventitious buds on the isolated primary leaves of a number of Ferns and in *Ceratopteris* he was able in addition to obtain

---

[1]) R. L. PRAEGER, Propagation from Aerial Shoots in *Equisetum*, Journ. of Bot., Vol. LXXII, 1934.

[2]) K. GOEBEL, Organ. der Pflanzen, 3 Aufl., 2 Teil, p. 1221.

[3]) K. GOEBEL, Organ. of Plants, Eng. Edit., Pt. 1, p. 46.

[4]) K. GOEBEL, Einleitung in die experimentelle Morphologie, p. 215.

[5]) FIGDOR, Über Regeneration der Blattspreite bei *Scolopendrium*, Ber. d. deutschen bot. Gesell., Bd. XXIV, 1906.

[6]) E. HEINRICHER, Nachträger zu meiner Studie über die Regenerationsfähigkeit der *Cystopteris*-Arten, Ber. d. deutsch. bot. Gesell. Bd. 18, 1900.

[7]) J. PALISA, Regenerationsknospen etc., Ber. d. deutsch. bot. Gesell., Bd. 18, 1900.

[8]) GOEBEL, K., Organ. of Plants, Eng. Edit., Pt. II, p. 227.

[9]) K. GOEBEL, Einleitung etc., p. 197 et seq.

similar buds on the stem. In this last example the stem apex was removed and this resulted in the formation of an adventitious bud which carried on the growth of the axis.

**§ 12. Factors underlying the Production of Sporangia.** — The factors underlying the incidence of fertility have been investigated to some small extent in the Ferns but not in the other groups of Pteridophytes. GOEBEL has put forward the view that the production of sporangia depends upon nutritional factors and in particular on the accumulation of a large supply of carbohydrates and the raising to a high level of the ratio between organic and inorganic substances in the plant. The following experiments and "natural experiments" are put forward in favour of this view [1]. Experiments with *Matteuccia Struthiopteris* and *Onoclea sensibilis* gave the following result. When the sterile leaves were removed from the annual crown at a time when the sporophylls were still at the earliest stage of development the latter developed as sterile foliage leaves or as structures showing intermediate characters between these and the sporophylls (Fig. 5, *E*). GOEBEL

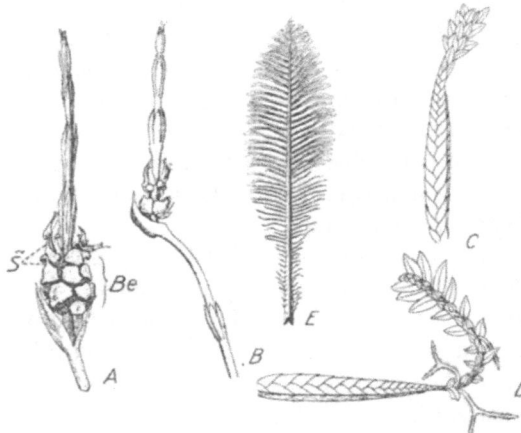

suggests that the removal of the sterile fronds leads to the sporophyll rudiments being provided with more inorganic and less organic foodstuffs than they would receive under normal conditions. Such a lowering of the ratio between organic and inorganic materials would lead on GOEBEL's view to an interference with the onset of fertility. GOEBEL has also noted that plants of *Asplenium Ruta-muraria* growing in a habitat poor in mineral supplies but providing very favourable conditions for photosynthesis produce sporangia

FIG. 5. *A, B, Equisetum Telmateja*, cones "growing through" as vegetative shoots; *s*, transition forms between sporangiophores and foliage leaves. ( × 3). *C, D, Selaginella grandis*, proliferating cones. ( × 3). *E, Matteuccia Struthiopteris*, frond showing characters intermediate between those of a sporophyll and a sterile foliage leaf. (*A, B* and *E*, after GOEBEL).

very early in their ontogeny. A comparable example is provided by the adventititous buds of *Asplenium* spp. which also become fertile at an early stage. In this example the cause of the early onset of fertility may be the fact that such buds receive abundant organic supplies from the parent plant.

---

[1] K. GOEBEL, Einleitung in die experimentelle Morphologie, p. 9 and p. 65.

The importance of the supply of organic foodstuffs is thus clear but more precise experimental work is required. The adventitious buds last mentioned would appear to supply favourable experimental material.

The production of sporangia on prothalli (Fig. 6, *A*, *B*) has been regarded by GOEBEL as an extreme example of "Frühreife" but nothing is known concerning the factors underlying this curious phenomenon.

## § 13. Experiments bearing on the Morphology of Sporophylls and Sporangiophores. — Conclusions as to the morphological nature of sporophylls must be based in the main on the results of wide comparison but experiments provide useful supplementary data. So far as the Lycopodiales are concerned the only available experimental data relate to the "growing through" of the cones of *Selaginella* spp. which has been induced by various workers including BRUCHMANN, BEHRENS and WILLIAMS. When detached cones are laid down on moist soil, or when the ends of fertile shoots are pinned down without removal from the plant, the apex of the cone grows on to produce an anisophyllous shoot whose structure is similar to that of the vegetative part of the plant (Fig. 5, *C*, *D*). Abortive sporangia are found at the junction of the cone and the new shoot. The experiment shows that there is no fundamental difference between the sporophylls and the foliage leaves.

The only experimental work on the Ferns is that of GOEBEL on *Onoclea* spp. mentioned above. These experiments show that if the foliage leaves are removed the rudiments of the sporophylls are capable of developing into vegetative fronds. The homology of the vegetative and fertile fronds is thus demonstrated.

The morphology of the sporangiophores found in some of the microphyllous types has been the subject of much discussion but little experimental work has been carried out in relation to this problem. So far as I am aware, the only recorded experiments are those of GOEBEL [1]) and LUDWIGS (loc. cit.). The former was able to induce young cones borne on side shoots of *Equisetum Telmateja* to become "vergrünte" and grow on as vegetative shoots (Fig. 5, *A*, *B*). These artificially induced deviations from the normal exhibit all stages of transition between sporangiophores and foliage leaves and GOEBEL concludes that these structures are homologous. BOWER [2]) points out, however, that the fossil evidence is against this view but a full discussion of the morphology of the cone of the *Equisetinae* cannot be entered into here since any conclusions must be based on a wide comparative treatment of living and fossil forms as well as on such few experimental data as are available.

---

[1]) K. GOEBEL, Organ. der Pflanzen, 3 Aufl., 2 Teil, p. 1240.
[2]) F. O. BOWER, Primitive Land Plants, p. 191 et seq.

**§ 14. Deviations from the Normal Life–cycle.** — The factors under-lying the rhythm of the normal life-cycle are as yet entirely obscure and it may at once be admitted that the experimental work outlined above, illuminating as it does something of the causal morphology of isolated stages in the life-history, provides no data for a solution of this major problem. It might have been expected that a study of deviations from the normal life-history would be helpful in relation to this but, as LANG points out: "So far, however, though numerous cases of these deviations from the normal development have been studied, they have only added to our knowledge of alternative processes of development that can be described but as little explained as the normal development." In spite of the truth of this general statement the experimental work which has been carried out in relation to the study of apogamy and apospory has yielded interesting results, some of which may be summarised here. All the available data relate to the Ferns, the other groups of Pteridophytes not having been subjected to experimental investigation.

(a) *Apogamy*. — In some species and varieties of Ferns apogamy is apparently obligatory, being due to the inherited constitution of the plant. On the other hand there have been described many instances of apogamous development which occur in Ferns whose prothalli are capable of bearing sexually produced embryos. Such instances may be spoken of as induced apogamy although the nature of the factors underlying their appearance are as yet but little understood. The experiments of LANG [1]) indicated that the prevention of fertilisation by avoiding watering the prothalli from above is the most significant factor leading to apogamous developments. Under these conditions various apogamous growths were obtained including the production of apogamous buds, of isolated sporophytic organs and appendages such as roots, sporangia (Fig. 6, *A*, *B*), and ramenta, and of isolated sporophytic tissues such as tracheides. Culture experiments carried out by other workers also indicate that the failure of normal fertilisation is a factor predisposing prothalli to the production of apogamous growths. The literature relating to this has been summarised by BROWN [2]). Other con-ditions have been suggested as stimulating apogamous development such as culture in bright light, culture at higher temperatures than the normal [3]) the lowering of the vitality of the prothalli by fungal and algal attack and

---

[1]) W. H. LANG, On Apogamy and the Development of Sporangia upon Fern Prothalli, Phil. Trans Roy. Soc. London, Ser. B., Vol. 190, 1898.

[2]) E. D. W. BROWN, Apogamy in *Phegopteris polypodioides*, Bull. Torrey Bot. Club, Vol. 50, 1923.

[3]) The value of the experiments of NATHANSOHN (Ueber Parthenogenesis bei *Marsilia*, Ber. d. bot. Gesell., Bd. 18, 1900), indicating that the tendency of spores to form apogamous embryos is increased by growing them at a higher temperature than the normal, is largely vitiated by the cytological facts as disclosed by STRAS-BURGER.

the failure to produce functional sexual organs under various unfavourable conditions of nutrition. For instance, in BROWN's experiments with *Phegopteris polypodioides* and other Ferns the prothalli were grown on liquid media. In full culture solutions normal sporophytes were formed; in modified solutions from which one or more of the essential mineral elements were omitted, the sexual organs either failed to develop or were non-functional and in relation to this apoga-mous developments ensued.

It is therefore fairly clear that the failure of normal fertilisation is an important factor influencing the inci-dence of apogamy but no further analysis of the mechanism in-volved has been at-tempted. The pre-vention of fertilisation does not, however, always lead to the ap-pearance of apogamy. In many Ferns old un-fertilised prothalli merely show an in-crease in size, the pro-duction of adven-titious prothalli and peculiar developments of the sexual organs. There may be in-stanced the experi-ments of MOTTIER [1] in which prothalli of *Matteucia Struthiopte-ris* and *Osmunda regalis* were grown under conditions preventing fertil-isation for a period of eight years. The prothalli formed large "clones" approximating in size to plants of *Marchantia* and continued to produce sexual organs, archegonia in favourable and antheridia in unfavourable growth periods. No apogamous developments were, however, observed. It

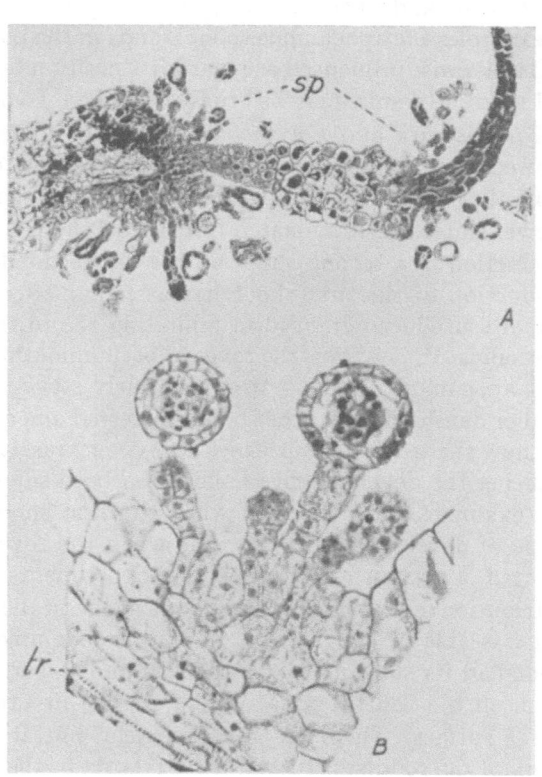

FIG. 6. Prothalli of *Phyllitis Scolopendrium* var. *A*, longitudinal section of a prothallus with two sporangial groups (*sp*) developed in succession. (× 60). *B*, sporangia and tracheides (*tr*). (× 300). (After LANG).

[1] D. M. MOTTIER, Development of Sex Organs of Fern Prothallia under Pro-longed Cultivation, Bot. Gaz., Vol. 92, 1931.

is clear therefore that in addition to the presence of external factors favour-
ing apogamous developments there must also be internal factors concerned,
possibly in the nature of an inherent susceptibility due to an anomalous
nuclear composition.

The position of apogamous growths upon the prothallus is very varied.
In examples of direct apogamy the apogamous bud normally arises from
the cushion and LEITGEB [1]) has shown experimentally that in the majority
of examples the apogamous shoot stands in the position of an archegonium
and is likewise influenced as regards its position by the effect of light. Some
of LEITGEB's results may be briefly described. Experiments on the prothalli
of *Pteris cretica* and *Polystichum falcatum*, where apogamy is obligatory,
showed that the apogamous shoot always arises on the shade side of the
prothallus. If a prothallus with an apogamous shoot already developed on
the ventral side is illuminated from below there does not usually follow the
production of a second shoot on the now shaded upper surface since the
production of the first shoot has a similar effect to the formation of a
sexually produced embryo in inhibiting the further apical growth of the
prothallus. If, however, the reversal of illumination is carried out when the
first apogamous shoot is at a very early stage of development then the
further development of the shoot is arrested and a second shoot appears on
the now shaded upper surface [2]). Only very rarely did LEITGEB succeed in
inducing the development of shoots on both sides of the prothallus but a
few examples were obtained when the first shoot had reached a middle
grade of development at the time when the reversal of illumination was
effected. LEITGEB also carried out experiments designed to induce the
distribution of the primary organs of a shoot on the two prothallial surfaces,
and was able in *Pteris cretica* to produce examples where the apogamous
shoot had its stem apex and the first and second leaves in the normal position on the ventral surface but with the first root growing out from the
dorsal surface. The facts are not clear but it seems probable that the
origin of the root was still normally related to the other shoot parts on the
ventral surface and that the illumination from below had resulted in the
root exhibiting negative phototropism and breaking through to the dorsal
surface.

It is thus clear that light is an important factor in determining the
position of apogamous shoots in these examples of direct apogamy. Their
relationship to the establishment of the prothallial meristem and the
formation of a cushion appears to be strictly comparable to that exhibited
by the archegonia on a normal prothallus. The examples of induced

---

[1]) H. LEITGEB, Die Sprossbildung an apogamen Farnprothallien, Ber. d. deutsch.
bot. Gesells., Bd. III, 1885.

[2]) This sequence of events may give the false impression that the organs of one
shoot are distributed on the two sides of the prothallus, as in examples described
by DE BARY.

apogamy in which isolated sporophytic organs and tissues are produced present, however, great difficulty in any attempt to analyse the factors influencing the place of their origin and, in the absence of further data, no useful purpose would be achieved by further discussion.

(b) *Apospory.* — Repeated attempts to induce apospory from the leaves of adult Ferns have so far been unsuccessful and BOWER states that "There appears to be a marked disability in the adult leaf for bridging over the limit between the generations in any other way than by spores." On the other hand, various workers have found it possible to induce aposporous developments from the leaves of sporelings.

An extensive series of experiments was carried out by GOEBEL[1]) in which detached juvenile leaves of various Ferns were laid down on moist soil. Various regenerative growths were obtained, including sporophytic buds, aposporously produced prothalli and structures designated by GOEBEL as "Mittelbildungen". The development of aposporous prothalli from primary leaves of *Pteris longifolia* is shown in fig. 7, *D* and *E*; similar prothalli were obtained in the case of other Ferns. The production of „Mittelbildungen" was ob-served in *Alsophila Van Geertii* (Fig. 7, *F*) and in *Ceratopteris thalictroides* (Fig. 7, *G*). These structures show partly gametophytic characters, such as the presence of antheridia and rhizoids and the absence

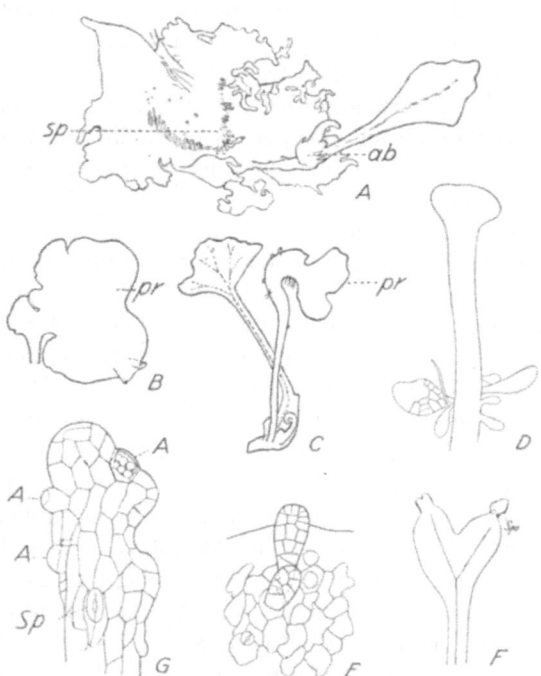

FIG. 7. *A*, prothallus of *Phyllitis Scolopendrium* var.; *sp*, group of sporangia; *ab*, apogamous bud. (× 6). *B*, *C*, aposporous growths from leaves of bud of *Osmunda regalis*; *pr*, prothallus. (*B*, × 4; *C*, × 6). *D*, *E*, primary leaves of *Pteris longifolia*; *D*, prothalli growing from petiole; *E*, origin of prothalli from epidermis. *F*, primary leaf of *Alsophila Van Geertii*, "Mittelbildungen" growing out from tip of leaf; *Spa*, stomata. *F*, "Mittelbildung" from primary leaf of *Ceratopteris*; *Sp*, stoma; *A*, antheridia. (*A–C*, after LANG; *D–G*, after GOEBEL).

of intercellular spaces, and partly sporophytic characters such as the

---

[1]) K. GOEBEL, Exper.-morph. Mitteilungen, Sitz. d. math.-phys. Klasse d. Kgl. Bayer. Akad. d. Wiss., München 1907.

presence of stomata and, in some examples, the development of vascular tissue.

In the above described experiments of GOEBEL the development of adventitious growths is clearly correlated with the isolation of the juvenile leaves from the plant. LANG [1]) has, however, succeeded in inducing the formation of aposporous prothalli from juvenile leaves of *Osmunda regalis* still attached to the parent plant. Fig. 7, *B, C* indicate the nature of the aposporous developments, which, along with other deviations from the normal, occurred on the buds developed by a plant whose apex had been damaged. Fig. 7. *C* shows one of these buds bearing two leaves, the right hand one of which consists of a cylindrical dark green stalk expanding upwards and terminating in a prothallus bearing sexual organs. Fig. 7, *B* shows a leaf from another bud where only one of the two lobes has become prothalloid. Such aposporous prothalli will, if laid down on soil, develop normally and produce sporophytes which show no tendency to aposporous development.

As pointed out in the first paragraph of this section, such induced deviations from the normal merely add "to our knowledge of alternative processes of development" and have, as yet, contributed little to the solution of the general problem presented by the normal sequence of the life-cycle. They may indeed be regarded as mere abnormalities devoid of any significant bearing on this problem [2]). Nevertheless, the study of the factors underlying the production of isolated sporophytic organs on pro-thalli may reasonably be expected to throw some light on the factors underlying the production of these structures in their normal position. GOEBEL has also pointed out that the variable nature of the regenerative growths from detached juvenile leaves offers the possibility of analysing the factors leading to the development of sporophytic shoots on the one hand and prothalli on the other No experimental work along these lines has yet been carried out but GOEBEL suggests that the conditions under which the leaf has been grown prior to its detachment may be an important factor.

## Evolutionary Problems

**§ 15. The Mechanism of Evolution.** — The only contribution which experiment can make to the investigation of the mechanism of evolution lies in the province of genetics, which is in fact a special branch of ex-perimental morphology. As this subject is dealt with elsewhere in this book, it will not be further mentioned here.

---

[1]) W. H. LANG, On Some Deviations from the Normal Morphology of the Shoot in *Osmunda regalis*, Mem. and Proc. Manchester Lit. and Phil. Soc., 1924.

[2]) See F. O. BOWER, Primitive Land Plants, p. 495.

**§ 16. The Course of Evolution.** — The question as to whether experiment can provide data useful for a consideration of evolutionary problems is an open one. It has frequently been stated that "experiment cannot reconstruct history" but this statement would appear to be true only in a limited sense. It is possible, as has been shown above, to induce many deviations from the normal structure of the plant. Any one of these deviations may in point of fact represent a normal feature of some previous evolutionary stage of the plant and, if it does, then experiment will have reconstructed history in respect of that feature. The difficulty lies, however in deciding which of the deviations. if any, does represent an earlier phase in the development of the plant under consideration. If, however, palaeobotanical data and the results of comparative morphology are taken in conjunction with the data derived from experiment, than it may be found that the latter will form a useful contribution to a probable solution of evolutionary problems.

(a) *Phyletic Problems*. It has to be admitted that experiment has so far provided very few data of service in the consideration of phyletic problems. A few examples which will illustrate the possibilities and limitations of such experimental contributions may be considered.

As described above (p. 120) LANG was able to induce the development of axillary branches in young plants of *Osmunda regalis*. Such a deviation from the normal morphological construction appears to strengthen the conclusions reached by KIDSTON and GWYNNE-VAUGHAN as to the derivation of the *Osmundaceae* from a Zygopterid source.

Experiment has also contributed to an understanding of the incidence of isophylly and anisophylly in the genus *Selaginella* and such data are useful in any consideration of the interrelationships of the various species.

Such examples are however few and their significance always open to doubt and it must be admitted that experiment seems likely to be of very limited service in the study of phylogenetic problems.

(b) *The Evolution of the Gametophyte*. Although, as shown above, there is a considerable volume of experimental evidence as to the factors underlying the form of the cordate type of Fern prothallus, this evidence throws little, if any, light on the evolutionary relationships of the various types of Pteridophyte prothallus. It is, for instance, known that under certain conditions a filamentous form of prothallus may be induced and maintained but such knowledge gives no indication of the phyletic relationship between the filamentous type of prothallus, e.g. that of the *Hymenophyllaceae*, and the cordate type. Nor is any experimental evidence available to indicate the origin of the subterranean types seen in the *Ophioglossaceae* and the homosporous *Lycopodiales* [1]).

---

[1]). The observations of LAND (Bot. Gaz., Vol. 75, 1923) on prothalli of *Angiopteris* which had been buried by falls of friable soil are interesting in relation to this

(c) *The Differentiation of the Sporophyte Plant Body.* The most recent and comprehensive attempt at a reconstruction of the historical development of the differentiated sporophyte from an undifferentiated source is that made by BOWER (loc. cit.) on the basis of a wide comparative survey and, in particular, on a consideration of the Psilophytales. The question arises as to whether experiment can contribute anything to the solution of such a problem. There may be mentioned the experiments of LANG on *Osmunda regalis* in which the development of cylindrical rudimentary structures was induced. LANG points out in relation to these that they may represent to a certain extent "the indeterminate and almost thalloid structures that underlie the differentiation of stem and leaf, however this came about." The formation of indeterminate structures possibly of comparable nature and significance, was induced by WILLIAMS in *Lycopodium Selago*.

Facts bearing on the above problem are also provided by experiments in which the replacement of leaves by shoots has been induced, e.g. those of LANG on *Osmunda regalis*. Such experimental results stress the fundamental equivalence of leaf and shoot, a fact of considerable importance in the discussion of the origin of the differentiated plant body.

Nevertheless, it is clear that the contribution of experimental morphology to the solution of this problem is not likely to be a decisive one and experiment, even if extended over a much wider range than has hitherto been the case, is likely to provide merely a check on the conclusions based on the classical comparative method.

(d) *The Origin of Heterospory.* It has been suggested on the basis of evidence derived from the fossil Equisetales that the heterosporous habit arose as a consequence of the abortion of a certain number of the spores and the consequent better nutrition of the remainder; "this process, going on more freely in some sporangia than in others, may ultimately have rendered possible the excessive development of those spores that survived, at the expense of the others, and may thus have led to the development of specialised megaspores." (SCOTT, Fossil Botany, 2nd Edit., p. 58).

Experiment has shown that nutritional factors are actually concerned in the manifestation of heterospory. GOEBEL has stated that if the photosynthetic activities of plants of *Selaginella* are retarded by growing them under conditions of feeble illumination the cones produce only microsporangia. The influence of nutrition on the heterospory of *Marsilia* has been investigated by SHATTUCK[1]). It was found possible to kill all the megaspores

---

question. They showed a cylindrical development of the cushion and the wings were inconspicuous; an endophytic fungus was present. Such observations suggest a possible mode of origin for the subterranean, cylindrical type of prothallus and BOWER (Prim. Land Plants, p. 439) mentions the possibility of experimental work in relation to this problem.

[1]) SHATTUCK, C. H. The Origin of Heterospory in *Marsilia*, Bot. Gaz., Vol. 49, 1910, p. 19.

and a varying number of the microspores in the sporocarps by suddenly lowering the temperature by means of a cold water spray. The plants were then placed under favourable growth conditions and the sporocarps allower to mature. Examination showed that in sporocarps in which the mega-spores and not less than half the microspores had aborted, some of the remaining microspores, presumably as a result of the increased supply of food materials, showed increases in size up to sixteen times the normal. The largest of these approached in size and some other features the structure of normal megaspores.

The influence of nutrition on the incidence of heterospory appears there-fore to be established and the experimental results may be regarded as supporting the view already expressed on the basis of palaeobotanical evidence.

**§ 17. Conclusion.** — The above survey will have indicated the relative paucity of experimental data available for a consideration of the morpho-logical problems presented by the Pteridophytes. It would appear that further advance must await on the critical re-examination of some of the results already obtained and the extension of experiment over a much wider range of forms.

The possible lines of attack are various. One of these is the extension of observation along lines similar to those of the experiments described in this chapter. Many of these experiments require, as GOEBEL has said, only "a plant, a pot of soil and a question". From them would come an accumu-lation of data relating to induced deviations from the normal life-cycle, regeneration phenomena, correlations of organs and tissues, the formation of vascular tissue, the onset of fertility whether in the gametophyte or the sporophyte, and many other questions. Such data would undoubtedly possess great intrinsic interest and would probably contribute, when taken in conjunction with the evidence derived from wide comparison, to the solution of the major problems of morphology.

It is, however, urgently necessary that experiment should be extended to the physico-chemical aspects of the problems. That such advance is possible is evidenced by the discoveries of WENT and other workers of growth-regulators, root forming substances and other plant hormones [1]. These raise the hope that the further analysis of correlation phenomena and other morphological problems may be rendered possible in the Pteridophytes and other phyla.

Experiments along the line of morphological inquiry together with those dealing with the physico-chemical aspects may ultimately approach to a satisfactory analysis of the nature of the factors, both internal and external

---

[1] See R. SNOW, loc. cit., for a useful review of this work. See also P. BOYSEN JENSEN, Growth Hormones in Plants, 1936 and WILLIAMS, S., Nature 139: 966.

which underlie the normal life-cycle and structure. Such an analysis will, however, leave unsolved the problem as to h o w the protoplasm responds to the influence of these factors. From the physiological side it has been pointed out that attempts at purely physico-chemical explanations are "apt to involve over-simplification of the problem" [1]) and in view of our ignorance concerning the nature and activities of living protoplasm it is clear that such explanations are imposssible at the present time. Studies in the metabolism of different tissue elements and in what WEBER [2]) has recently termed "protoplasmatische Pflanzenanatomie" give an indication of possible advance in this direction.

Finally there may be mentioned the related and even more difficult problem presented by the nature of the "spezifische erbliche Struktur". The significance of the contributions from genetical and cytological investigations in relation to this problem cannot be discussed here but research in these special branches of experimental morphology is obviously of fundamental importance and offers indeed the only direct line of attack at the present time.

---

[1]) D. THODAY, loc. cit., p. 284.

[2]) F. WEBER, Protoplasmatische Pflanzenanatomie, Protoplasma, Vol. 8, 1929.

CHAPTER IV

ASSOCIATIONS WITH FUNGI AND OTHER LOWER PLANTS

by

MARY J. F. GREGOR (Edinburgh)

**§ 1. Introduction.** — The lower forms of plant life that occur in association with living Pteridophyta may be roughly classified as epiphytes, symbionts or parasites. These groups cannot, however, be regarded as clear-cut and isolated, for transitional forms do exist. In the first-mentioned group the association is frequently a matter of chance, and many other substrata would serve the epiphyte equally well. Nevertheless the host may be injured or even killed by the epiphyte if the latter is present in sufficient numbers. In the case of symbionts the association is often obligatory, though there are instances of facultative as opposed to obligate symbiosis; sometimes the symbiosis approaches very closely to parasitism. Among the parasites are some which are mainly epiphytic in habit and merely penetrate the epidermal cells here and there by means of haustoria. Some are only parasitic for part of their life and continue to grow saprophytically after the death of the host, while others are obligate parasites.

By far the majority of the lower plants that enter into intimate association with Pteridophyta belong to the Fungi. Many of these cause diseases while others have established a symbiotic relationship as mycorhiza; the latter are discussed in detail in the following chapter. Our knowledge of the pathogenic fungi dates from the beginning of the nineteenth century, when PERSOON, FRIES, FUCKEL and other mycologists described some of the fungi occurring on Pteridophyta. In general, however, these early investigators were more concerned with the fungus than with its relationship to its host, and they frequently failed to differentiate between saprophytic and parasitic species. Even now this difficulty may sometimes be encountered, but the increased use of experimental methods has resulted in a greater emphasis on the biological relationships between fungus and host.

In addition to diseases caused by fungi there are a few attributed to Myxomycetes and bacteria, while virus diseases have also been recorded on a number of ferns. A state of symbiosis exists between the water fern

*Azolla* and the blue-green alga *Anabaena*. Innumerable epiphytes representing all classes of lower plants may, under suitable conditions, grow upon Pteridophyta.

**§ 2. The fossil record.** — Before passing on to a detailed discussion of the various associations occurring at the present day, it may be of interest to consider whether any such relationships existed during earlier geological epochs. Considering the delicate nature of most fungi, algae and bacteria it is not surprising that the records of fossil forms, though numerous, are in many cases not very satisfactory. Nevertheless it is clear that fungi certainly existed during the Carboniferous period or even earlier, while evidence of bacteria extends back to pre-Cambrian times. Many of these forms have actually been found in or on fossil Pteridophytes. At this point, however, we come up against a serious difficulty, for it is well-nigh impossible to determine whether they were true parasites, symbionts or merely saprophytes.

Mycelium has been found in petrified tissues, sometimes associated with injury to the cell walls. Often such hyphae show no diagnostic features, but occasionally indications of reproductive organs, transverse septa or clamp connections may give some guide to their affinities. Many of the records of fossil fungi refer to leaf-spot fungi, and a number of these have been described on the foliage of fossil ferns. Once again the situation is complicated, for fungi, insect injuries and even normal leaf glands may easily be confused in a fossilised condition, and undoubtedly some errors have arisen from this cause. Even when the spot is unmistakably fungal in origin it is often impossible to identify or describe the structural details of the causal organism. But the fact remains that this type of fungus injury must have been common in past ages even as it is to-day.

The humid conditions of the Carboniferous period would seem to offer an ideal opportunity for the development of epiphytes, yet many observers have commented on the remarkably clear cortical patterns of the Palaeozoic plant remains which apparently indicate that such associations did not exist in the coal swamps.

Since space does not permit of the enumeration of examples here, the reader is referred to A. MESCHINELLI's "Fungorum Fossilium omnium Iconographia" 1902, which is a list of fossil fungi recorded up to the year 1900. Interesting discussions and references to other literature will be found in A. C. SEWARD's "Fossil Plants", Vol. 1, pp. 207–222, 1898, and in R. L. MOODIE's "Paleopathology", Chapter III, by E. W. BERRY, 1923.

**§ 3. Diseases of the Gametophyte.** — The prothallia of the Pteridophyta live in such intimate connection with the soil and under such moist conditions that they very readily fall victims to the damping-off fungi. *Cephalothecium roseum* CDA., which is usually a saprophyte, was recently

found in a greenhouse in Indiana causing severe damping-off of prothallia
of *Pteris longifolia*. *Pythium Debaryanum* HESSE, a species which is now
taken to include both *P. Equiseti* SADEBECK and *P. autumnale* SADEBECK,
has been recorded by SADEBECK [1]) as a parasite on the prothallia of *Equisetum arvense*, *E. palustre* and *E. limosum*. DE BARY [2]) described *P. inter-
medium* DE BARY attacking prothallia of *Equisetum*, *Todea* and *Ceratopteris*,
and he also successfully inoculated prothallia of *Todea africana* with *P. meg-
alacanthum* DE BARY and *P. Debaryanum*.

Of far greater biological interest than the above-mentioned omnivorous
*Pythium* species is *Completoria complens* LOHDE, a parasite in the prothallia
of *Gymnogramme*, *Ceratopteris*, *Pteris cretica*, *Polystichum falcatum* etc. It
belongs to the family *Entomophthoraceae*, most of whose members are
parasitic upon insects, and it has been studied in detail by LEITGEB [3]) and
ATKINSON [4]). Infected prothallia develop yellowish spots which later turn
brown, then black; ultimately the prothallia die. The vegetative body of
the fungus consists at first of a botryose cluster of oval or twisted hyphae
lying within a single host cell, which may be completely filled by it. When
mature, slender threads grow out from some of the peripheral branches,
penetrate neighbouring cells and there form new botryose clusters into
which the protoplasm of the parent body flows. Groups of thick-walled
resting spores are sometimes formed in the centres of the hyphal clusters.
Conidia, too, are produced on delicate branches which grow out on to the
surface of the prothallium; they are thrown off with considerable force and
give rise to new infections.

## § 4. Diseases of the Sporophyte. — *a*) F i l i c i n a e. — *Glasshouse
diseases*. When ferns are cultivated in greenhouses the moist atmosphere
provides ideal conditions for the development and spread of certain types
of fungous disease. Such diseases do not normally attack the ferns in their
natural habitats and they can usually be checked by improving the
ventilation of the houses. A good example is a case known personally to the
writer where *Cladosporium herbarum* (PERS.) LK. caused an outbreak of
disease among *Pteris cretica* in midwinter. The fronds became discoloured
and soon developed a greenish mould of conidia on both surfaces. Some
plants died, but as the season advanced and more light and air entered
the houses, the disease gradually disappeared. A somewhat similar epi-
demic on the same host was described by ROSTRUP [5]), but here the causal

[1]) SADEBECK, R., Verhandl. bot. Verein d. Prov. Brandenburg, XVI, 116,
1874 and Tagebl. Versamm. deutsch. Naturf. u. Ärzte in Breslau, XLIX, 100, 1876.
[2]) DE BARY, A., Bot. Zeit., XXXIX, 528 & 553, 1881.
[3]) LEITGEB, H., Sitzungsb. K. Akad. Wiss. Wien, LXXXIV, 288–324, 1881.
[4]) ATKINSON, G. F., Bot. Gaz., XIX, 467, 1894 and Cornell Univ. Agr. Exp.
Sta. Bull., 94, 252, 1895.
[5]) ROSTRUP, E., Gartn. Tidsskr., 231, 1898.

fungus was *Coniosporium filicinum* ROSTR. and hundreds of plants suc-
cumbed to the attack. *Botrytis longibrachiata* OUD. has been recorded as a
greenhouse parasite of various ferns in Germany [1]). *B. cinerea*, too, is
harmful in countries as far apart as Alaska and Germany, while *Moniliopsis
Aderholdi* RUHLAND may cause heavy damage to ferns. WHITE [2]) studied a
disease of *Cibotium Schiedei* caused by *Pestalozzia Cibotii* WHITE. Under
greenhouse conditions this fungus kills frondlets or even entire fronds of the
fern. SORAUER [3]) has described a very curious slime disease of *Cyathea
medullaris* grown under glass. Open cankers developed at the bases of the
petioles, and cut surfaces oozed slime while the vascular bundles turned
black. The fronds withered and finally the whole plant died. It is possible
that the intumescences were caused by excess atmospheric moisture, but a
*Nectria* sp. was isolated in culture. No conclusion was reached as to the
cause of the various pathological phenomena.

*Rusts*. The diseases which attack ferns growing in the open under natural
conditions are, however, of much greater interest. Among these the fern
rusts have received the greatest share of attention and the reader is referred
to the following works for details of the numerous individual species:
P. & H. SYDOW, Monographia Uredinearum, Vol. III, 1915. J. H. FAULL,
Taxonomy and geographical distribution of the genus *Milesia*, Contr. Ar-
nold Arb., No. 2, 1932. J. C. ARTHUR, Manual of the rusts in United States
and Canada, 1934. The aecidia and spermogonia, where these are known,
occur on various species of *Abies* and may cause serious damage, especially
to seedlings and young trees. The uredo- and teleuto-sori develop on ferns,
giving rise to brown or black areas on the fronds which may be so extensive
as to cause considerable injury. The most important genera are *Uredinopsis*, *Milesina* (*Milesia*) and *Hyalopsora*. In the first-named genus the
whitish uredo-sori are enclosed in a delicate peridium, and septate teleuto-
spores are formed in a sub-epidermal layer or scattered in the mesophyll of
the frond. Amphispores, which serve as a resting stage, are produced
towards the end of the season by most, if not all, of the species. *U. Osmundae* attacks various species of *Osmunda* while the other species are restricted
to the *Polypodiaceae*. In *Milesina* the uredo-sori are very similar to those
of *Uredinopsis* but the multicellular teleutospores develop within the
epidermal cells, usually one in each cell though sometimes several (Fig. 1).
By growth and septation they eventually fill the cell, taking the same form
and outline. A considerable number of species are known, widely distributed throughout the world. ARTHUR [4]) has described *Milesia* (*Milesina*)
*australis* on *Lygodium polymorphum*, also *Puccinia Lygodii* (HARIOT) ARTH.

---

[1]) ADERHOLD, R., Centralbl. f. Bakt., 2. Abt., VI, 625–626, 1900.
[2]) WHITE, R. P., Mycologia, XXVII, 342–346, 1935.
[3]) SORAUER, P., Ber. Deutsch. Bot. Ges., XXX, 42–48, 1912.
[4]) ARTHUR, J. C., Bull. Torrey Bot. Club, LI, 53 & 55, 1924.

on the same host. SYDOW [1]), too, described *Milesina Lygodii* SYD. on *Lygodium* sp. The first and third species were based on uredo-stages only, while the second had both uredo- and teleuto-spores. FAULL [2]) states that all the collections assigned to these three species show great similarity in the uredo-stage and that further study is required before their taxonomic position can be definitely decided. He therefore excludes them from his discussion of the genus *Milesia* (*Milesina*) which, he states, only attacks ferns of the family *Polypodiaceae*. *Hyalopsora* can be distinguished from *Uredinopsis* and *Milesina* by its yellow, pulverulent uredospores. These are of two kinds, differing in thickness of wall and certain other characters, but all gradations between the two may be found: they possibly represent an active and a resting condition. The septate teleutospores are formed inside the epidermal cells. These rusts infect various genera of the *Polypodiaceae*.

In addition to those mentioned above SYDOW [3]) has described two other genera of fern rusts, *Desmella* and *Calidion*. The sori of the former consist of

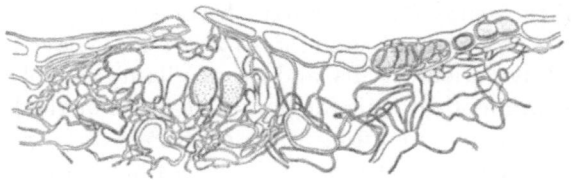

FIG. 1. *Milesina Kriegeriana* MAGN. on *Dryopteris spinulosa*. Longitudinal section showing uredo-sorus and teleutospores in epidermal cells. × 210. (Adapted from drawings by P. MAGNUS).

bunches of spore-bearing hyphae emerging through the stomata. Both uredo- and teleuto-spores are known but no aecidia or spermogonia. In *Calidion* only the uredo-stage is known, but this is distinguished from all other fern rusts by the conspicuous, thick-walled, coloured paraphyses surrounding the uredo-sorus. Four species of *Desmella* and one of *Calidion* are described, on various genera of *Polypodiaceae* and *Schizaeaceae*.

*Ferns as hosts of crop plant parasites.* It is commonly assumed that diseases of ferns, with the exception of the rusts that attack *Abies*, are of little or no economic significance. This is probably true in the majority of cases, but there are several records of dangerous parasites of cultivated crops occurring also on ferns; in this way the latter may assume some importance as perpetuators of the disease. *Corticium koleroga* (CKE.) v.H. has been found on *Polypodium lineare* and *Cyclophorus acrostichoides* (*Niphobolus fissus*) in Mysore [4]). This fungus causes a serious disease of coffee, and reciprocal cross-inoculations with isolations from the two ferns and from

[1]) SYDOW, H., Mycologia, XVII, 255, 1925.

[2]) FAULL, J. H., l.c., p. 122.

[3]) SYDOW, H. & P., Ann. Myc., XVI, 241 & 242, 1918.

[4]) VENKATARAYAN, S. V., Journ. Mysore Agr. Exp. Union, VII, 23–28, 1925.

diseased coffee plants gave positive results. *Omphalia flavida* MAUBL. et RANG. is another virulent coffee parasite which attacks many other plants, including ferns. *Pteris longifolia* has been shown by inoculation to be a host of *Sclerotium Rolfsii* SACC., a fungus which infects many different crops in warm countries. *Corticium Solani* BOURD. et GALZ. causes a disease of strawberries in North America and is also frequently found on *Pteridium aquilinum* var. *pubescens* in infected areas [1]). Bracken, again, is listed in Transvaal as a host of *Pythium aphanidermatum* (EDS.) FITZ. [2]), a species that attacks tobacco, sugar-cane, tomatoes and other valuable crop plants. STEINMANN[3]) has recorded a thread blight of the *Marasmius* group on *Nephrolepis hirsutula* in Java. A similar fungus occurred also on tea, coffee, pepper and other economic plants and was capable of causing serious damage. In the absence of fructifications, however, it was impossible to be absolutely certain that the parasites on the different hosts were all identical.

*Experiments on bracken control by parasitic fungi.* In certain parts of the world ferns of the genus *Pteridium* have become so abundant as to constitute a menace to agriculture and forestry. Attempts have been made to check the weed by various means, and the possibility of biological control by the dissemination of parasitic fungi was one of the methods considered. In this connection an interesting disease of *Pteridium aquilinum* was investigated by GREGOR [4]). *Dryopteris Filix-mas* may also be attacked by the disease and infection experiments demonstrated the susceptibility of *D. spinulosa, Polystichum aculeatum* var. *lobatum, Asplenium Trichomanes, Polypodium vulgare, Blechnum Spicant, Cystopteris fragilis* and *Phyllitis Scolopendrium.* The causal fungus is a rather unusual species of *Corticium, C. anceps* (BRES. et SYD.) GREGOR. It does not exactly cause malformation of the fronds, but since badly affected portions become very brittle and readily break off, diseased fronds commonly have an irregular and lop-sided appearance (Fig. 2). Infection takes place on the lower surface of either pinnae or rachis, most usually by means of infection cushions, though individual hyphae may enter through stomata. The internal mycelium kills the affected portions of the frond while the superficial hyphae cover the lower surface with a white felt-like mat on which the basidia are borne. In the later stages sclerotia develop freely on this superficial mycelium. The disease is markedly affected by environmental conditions, particularly atmospheric moisture, and for that reason is not likely to be of any value for eradication of bracken in a variable climate such as that of Britain.

In New Zealand the rhizomes of *Pteridium esculentum* are sometimes

---

[1]) ZELLER, S. M., Oregon Agric. Exp. Sta. Bull., 295, 1932.
[2]) WAGER, V. A., S. African Journ. of Sci., XXX, 247–249, 1933.
[3]) STEINMANN, A., Arch. voor Cacao Nederl. Indië, Deel 2, 44–47, 1928.
[4]) GREGOR, M. J. F., Phytopath. Zeitschr., VIII, 401–419, 1935.

attacked and killed by a species of *Fusarium*. CUNNINGHAM [1]) isolated this fungus and carried out infection experiments with it. In the majority of cases negative results were obtained, but in one set of experiments five out of seventeen inoculated plants died and the *Fusarium* was reisolated from the dead rhizomes. The next year, however, no trace of the disease could be found in the infected area, so this attempt at biological control of bracken was also abandoned.

FIG. 2.  Lower surface of frond of *Pteridium aquilinum* severely attacked by *Corticium anceps*.

*Galls and malformations.* Among the most interesting fern diseases, from the point of view of the pteridologist, are those which result in galls or malformations. GIESENHAGEN [2]) has described two particularly striking examples on tropical ferns. *Taphrina Laurencia* GIES. induces the for-

---

[1]) CUNNINGHAM, G. H., New Zealand Dept. Agric. Bull., 132, 4, 1927.
[2]) GIESENHAGEN, K., Flora, LXXVI, 130–156, 1892.

mation of adventitious buds on the lower surface of fronds of *Pteris quad-riaurita*, *P. aspericaulis*, *P. Blumeana*, *P. nemoralis* and *Polystichum aristatum*. The buds develop into small, abnormal fronds which form a dense bushy outgrowth resembling a witch's broom The anatomy of these abnormal fronds is very simple, and no stomata or sori occur on them. The asci of the parasite develop in their epidermal cells and later break through the outer walls to lie exposed on the surface. *Taphrina Cornu cervi* GIES. causes a somewhat simpler type of outgrowth on the fronds of *Polystichum aristatum* and *P. carvifolium*. In this case no adven-titious buds have been observed, but on both surfaces of the frond appear short cylindrical, simple or branched, structures about 1 cm. in length which contain a small vascular strand in direct connection with the vascular system of the frond. The asci of the parasite develop, in this case, beneath the cuticle of the epidermal cells of the outgrowths, and when mature they break through this and are fully exposed. GIESENHAGEN found that many of the galls on *P. aristatum* became infected with *Urobasidium rostratum* GIES., but this fungus was in no way responsible for the outgrowths.

There are a number of other species of *Taphrina* upon ferns, some of which give rise to discoloured swellings on the fronds while others merely cause leaf-spots of varying degrees of severity. A full list of these is given in SORAUER's „Handbuch der Pflanzenkrankheiten", 5th edition, Vol. II, p. 465 & 466.

*Synchytrium Phegopteridis* JUEL and *S. Athyrii* LAGERH. are responsible for small galls on fronds and petioles of *Phegopteris polypodioides* and *Athyrium Filix-femina* respectively. The galls are due to the enormous enlargement of the epidermal cells containing the parasites.

*Cryptomyces Pteridis* (REB.) REHM and its conidial stage *Gloeosporium Pteridis* (KALCHB.) BUBAK et KABAT cause a 'leaf roll' disease of *Pteridium aquilinum* [1]). Infection takes place at an early stage through the stomata and the fungus develops most freely in the lower surface of the frond, checking the growth of that region. This results in a peculiar curling of the pinnae which at the same time become thickened and exhibit a yellowish-green discolouration. Diseased plants are somewhat stunted in growth. Brown conidial fructifications and later black apothecia develop abundantly on the lower surface, just below the stomata. The function of these stomata is disturbed, which does not seriously affect plants in moist shady positions, but those growing in exposed places readily succumb to heat or drought.

Pinnules of *Osmunda regalis* may be attacked by a smut fungus, *Ustilago Osmundae* PECK [2]) which brings about a curious deformation. The affected parts turn brown or black, and are contracted into tufts. The brand spores of the fungus develop inside the diseased pinnules and sometimes burst

---

[1]) KILLIAN, K., Zeitschr. f. Bot., X, 49–126, 1918.
[2]) PECK, C. H., Bot. Gaz., VI, 276, 1881.

out along the vein but more commonly over the whole surface. Another very interesting smut, *Entyloma Nephrolepidis* RAC., has been found by RACIBORSKI[1]) on *Nephrolepis biserrata* (*N. acuta*) in Java. The mycelium invades the meristematic portions of the plants, with the result that abnormal fronds develop. These are smaller, wider and thicker than the normal ones, paler in colour and remain sterile. The hyphae live within the cells of these fronds and there form the colourless brand spores. Later, hyphae grow out on to the lower surface of the frond and cut off oval conidia. The conidiophores continue to grow, and old fronds are usually covered on the under side with a dense white mat of mycelium up to 80 μ in thickness. While discussing smuts on ferns, mention should also be made of *E. Aspidii* (BRES.) v. HÖHNEL and its conidial stage *Entylomella Aspidii* (BRES.) v. HÖHNEL. This fungus occurs on *Dryopteris spinulosa*, but no malformation results from the attack. The infected areas dry up and entire fronds may be killed.

RACIBORSKI[2]) has described another curious disease from Java on young fronds of *Acrostichum* (=*Hymenolepis*) *spicatum*. The spores of the parasite, *Platygloea Hymenolepidis* RAC., germinate on the fronds and the germ tubes each penetrate an epidermal cell. After repeated branching within the cell they emerge again as bundles of slime-coated hyphae on which basidia develop. Mesophyll cells are rarely invaded by mycelium but they show a marked hypertrophy. Infection of the epidermis, however, is usually so heavy that both surfaces of the fronds become thickly coated with gleaming white slime. The fronds are bent and twisted and subsequently some of them rot, though others appear to outgrow the disease.

*Herpobasidium Struthiopteridis* (ROSTR.) LIND gives rise to malformation of the fronds of *Matteuccia Struthiopteris*, but the fungus has not been studied in any detail. Another species, however, *H. filicinum* (ROSTR.) LIND, has been investigated fully by JACKSON[3]). It occurs in Europe on *Dryopteris Filix-mas*, *Phegopteris Dryopteris*, *P. polypodioides*, *Cystopteris montana* and *Dryopteris pulchella*, and is also known in Canada and the United States. The mycelium is systemic and perennial, presumably overwintering in the rhizomes. It gives rise to small brownish areas on the young fronds, from the lower surface of which hyphae grow out, forming a conspicuous white mould-like growth bearing basidia. The method of infection is not known. No hypertrophy is caused by this species.

*Leaf-spot diseases*. There are a very large number of fungi that cause leaf-spot diseases of ferns or, in some cases, a die-back of the tips of the fronds. Many of these belong to well-known genera such as *Gloeosporium*, *Septoria* etc. and are of no great interest or importance. On the other hand, associated with this type of disease there are some very unusual fungi

---

[1]) RACIBORSKI, M., Parasitische Algen und Pilze Java's, III, 8, 1900.

[2]) RACIBORSKI, M., Bull. Acad. Cracovie, I, 356, 1909.

[3]) JACKSON, H. S. Mycologia, XXVII, 553–572, 1935.

which have been made the subject of detailed investigations. *Valdensia heterodoxa* PEYR. is, from a mycological point of view, one of the most curious of these. It was first described by PEYRONEL [1]) on *Vaccinium Myrtillus* in Italy, but has since been recorded on plants from many different families, including one fern, *Athyrium Filix-femina*. The hyphae within the tissues are very large, up to 14 μ in diameter. Beneath the epidermis arise round fungal cells; each of these develops a short process which pierces the epidermis, often through the stomata, and swells into a second cell from which four radiating septate arms arise. In the centre, between these arms, the cell becomes covered with small outgrowths coated with mucus. When mature it is forcibly projected into the air by means of the arms, and blown on to another leaf where it adheres by the sticky central cushion. Germination occurs rapidly from any cell in the presence of moisture. PEYRONEL regards these interesting organs as bulbils and states that no spores have been observed, nevertheless he thinks that the fungus most probably belongs to the Ascomycetes.

STEVENS and DALBEY [2]) have described a serious leaf-spot disease of *Cyathea arborea* in Porto Rico. Small black spots are produced, often in such profusion as to cover more than half the leaf area. They are due to concentration of coarse dark mycelium in the epidermal cells and the subsequent development of superficial pycnidia and perithecia. The parasite shows affinities with both the *Dothideales* and the *Phacidiales*, but inclines more towards the latter. The authors have proposed a new genus for this fungus which they name *Griggsia Cyathea*. It bears a certain resemblance to *Rhagadolobium Hemiteliae* P.HENN. et LIND. [3]) which forms thin black stromata on the lower surface of the fronds of *Alsophila samoensis* in Samoa. This genus was originally included in the *Phacidiaceae* but THEISSEN and SYDOW [4]) consider that it should rather belong to the *Dothideales*. The mycelium develops freely in the mesophyll and lower epidermis, fills the air spaces below the stomata then grows out through the latter to form the stroma and asci. The stroma is thus anchored to the frond at numerous points corresponding to the position of the stomata, nevertheless it readily falls away leaving a pale-coloured area with scattered black dots which represent the stomata blocked with dark mycelium.

RACIBORSKI [5]) has described a number of new fungi causing leaf-spot diseases of ferns in Java. Among these may be mentioned *Hymenoscypha Asplenii* RAC. on *Diplazium pallidum*, *Morenoella Nephrodii* RAC. on *Dryopteris canescens*, (= *Nephrodium heterophyllum*) *Parmularia discoidea* RAC. on

---

[1]) PEYRONEL, B., Staz. Sperim. Agrar. Italiane, LVI, 521–538, 1923.

[2]) STEVENS, F. L. and DALBEY, N., Bot. Gaz., LXVIII, 222–225, 1919.

[3]) HENNINGS, P., Engl. bot. Jahrb., XXIII, 287, 1896.

[4]) THEISSEN, F. and SYDOW, H., Ann. Myc., XIII, 240, 1915.

[5]) RACIBORSKI, M., Parasitische Algen und Pilze Java's, II & III, 1900.

*Polypodium longissimum* and *Irydyonia filicis* RAC. on *Blechnum orientale.* Certain of them have since been discussed by THEISSEN and SYDOW (l.c.) who were able to add some further details concerning their structure, development and taxonomic position. This valuable monograph of the *Dothideales* includes a number of other species on ferns but unfortunately, in the majority of cases, no indication is given as to whether the fungi are parasitic or saprophytic.

FARIS [1]) studied a wither-tip disease of *Nephrolepis exaltata* caused by *Glomerella Nephrolepis* FARIS, whose conidial stage was shown to be a species of *Colletotrichum.* AGGÉRY [2]) recently published a very detailed paper on some new diseases of ferns, which included descriptions of *Homostegia Polypodii* AGG. on *Polypodium vulgare* and *P. vulgare* var. *serratum,* and *Sphaerella subostiolica* AGG. on the same hosts and also on *P. cambricum.* The genus *Sphaerella* includes several other fern parasites, namely *S. Polypodii* RABH. on *Polypodium vulgare, Dryopteris Filix-mas, Asplenium Trichomanes* and *Pteridium aquilinum, S. callistea* SYD. on *Osmunda regalis, S. Botrychii* ROSTR. on *Botrychium ternatum,* and *S. Filicum* (DESM.) FUCK. on *Dryopteris Filix-mas, D. spinulosa* and *Asplenium Adiantum-nigrum.*

It is, however, among the Fungi Imperfecti that the largest number of species causing leaf-spots are to be found, and the following list may give some idea of the variety of genera concerned: *Phoma Botrychii* JACZ. on *Botrychium matricariae, Phyllosticta Platycerii* TASSI on *Platycerium bifurcatum* (= *alcicorne*), *Vermicularia Scolopendrii* PASS. on *Phyllitis Scolopendrium, Ascochyta Pteridis* BRES. on *Pteridium aquilinum, Camarosporium Asplenii* SIEM. on *Asplenium septentrionale, Septoria Scolopendrii* SACC. on *Phyllitis Scolopendrium, S. mirabilis* PK. on *Onoclea sensibilis, S. Asplenii* E. et E. on *Athyrium angustifolium, Melasmia imitans* PECK on *Pteridium aquilinum, Gloeosporium Nicolai* AGG. on *Phyllitis Scolopendrium* and *G. Polypodii* AGG. on *Polypodium vulgare, P. vulgare* var. *serratum* and *Polystichum aculeatum* [2]), *G. Osmundae* E. et E. on *Osmunda cinnamonea, G. Pteridis* KLCHR., *G. necans* E. et E. and *G. leptospermum* PECK on *Pteridium aquilinum, G. Phegopteridis* FRANK on *Phegopteris Dryopteris, Pestalozzia funerea* DESM. var. *typica* SACC. on *Pteridium aquilinum, Cylindrocladium Pteridis* WOLF on *Polystichum adiantiforme, Dryopteris normalis* and *Nephrolepis exaltata* [3]), *Ramularia Scolopendrii* FAUTR. on *Phyllitis Scolopendrium, R. Botrychii* LINDROTH on *Botrychium Lunaria, Cercosporella Filicis-feminae* (BRES.) HÖHNEL on *Athyrium Filix-femina, Macrosporium Scolopendrii* CKE on *Phyllitis Scolopendrium, Alternaria Polypodii* MAJOR on *Polypodium* sp. [4]), *Cercospora Asplenii* JAAP on *Asplenium Trichomanes* and

[1]) FARIS, J. A., Mycologia, XV, 89–95, 1923.

[2]) AGGÉRY, B., Bull. Soc. d'Hist. Nat. Toulouse, LXVIII, 5–201, 1935.

[3]) WOLF, F. A., Journ. Elisha Mitchell Sci. Soc., XLII, 55–62, 1926.

[4]) MAJOR, T. G., Quebec Soc. Protection Plants Ann. Rept., 14, 59–61, 1922.

*Brachysporium Crepini* (WEST.) SACC. on *Ophioglossum vulgatum* [1]). *Sclerotium deciduum* DAVIS has been recorded as a parasite on *Matteuccia Struthiopteris*, but no details were given as to the symptoms produced on the host plant. Finally we might add here *Leptostroma filicinum* FRIES which forms spots on the petioles of various ferns. This list does not pretend to be complete, and there are many species omitted which may perhaps attack living plants but about whose parasitism no definite information could be found.

*Diseases of sori and spores.* There are two interesting fungi that attack the sori of certain species of *Polypodium*. *Sorica maxima* (B. et C.) GIES. is a Pyrenomycete which has been discussed in detail by GIESENHAGEN [2]). It is recorded on *P. crassifolium* and *P. punctatum* from tropical America. The mycelium forms a dense layer closely adpressed to the surface of the placenta and enveloping the bases of the sporangial stalks. From this grow out long thin stromata which swell slightly near the tip to form a perithecium containing asci with round brown spores. Small hyaline conidia are cut off from specialised hyphae borne on the surface of the stroma. Round pycnidia are also formed on similar though much shorter stromata. The fructifications of the fungus can be seen with the naked eye as fine black threads, up to 2 mm. in length, radiating from the reddish brown sori. The effect upon the host depends upon the stage of development of the sorus at the time of infection. In very young sori the superficial cells of the placenta are killed and no sporangia are produced. The fungus forms a dense mat but does not appear to reach its full development under such circumstances. If the sorus is somewhat older, sporangia are still suppressed but now a few fructifications of the parasite appear. All gradations can be seen between such cases and those where the placenta has suffered no apparent injury and sporangia and stromata occur abundantly side by side. COUCH [3]) has described another parasite, *Septobasidium Polypodii*, on *Polypodium* sp. in Jamaica. This fungus at first grows only on the sori, covering the developing sporangia and their stalks with hyphae. Penetration of the pro-sporangial cells occurs but in spite of this a good many sporangia mature. Later the mycelium spreads over the lower surface of the fertile fronds, enveloping them in a white mat of mycelium which, however, does not enter the epidermal cells but remains entirely superficial. VON SCHENK [4]) recorded a parasite, *Chytridium subangulosum* A. BRAUN, attacking spores of *Dryopteris parasitica* (=*Asp. violascens*) which had been laid out to germinate. The fungus formed spherical sporangia on the surface of the spores.

---

[1]) MAGNUS, P., Hedw., XLII, 222–225, 1903, and XLIV, 16–18, 1904.
[2]) GIESENHAGEN, K., Ber. Deutsch. Bot. Ges., XXII, 191–196 & 355–358, 1904.
[3]) COUCH, J. N., Journ. Elisha Mitchell Sci. Soc., XLIV, 255, 1929.
[4]) SCHENK, J. A. VON, Über das Vorkommen contractiler Zellen im Pflanzenreiche, p. 8, 1858.

*Diseases attributed to Myxomycetes.* SMITH [1]) has described and illustrated a disease of *Phyllitis Scolopendrium* which he said was caused by a parasitic Myxomycete, *Didymium effusum* LINK. His account was based on a single specimen sent from Cornwall. The frond was curiously forked and distorted and more or less covered on both sides with the little greyish-white sporangia which occurred on sori as well as on the vegetative parts of the frond. No mention was made of any pathological symptoms apart from the malformation. This Myxomycete is common as a saprophyte and SMITH himself points out that it is unlikely to attack perfectly healthy ferns. It seems, therefore, doubtful whether the Myxomycete was really responsible for the symptoms observed.

The plasmodia of certain Myxomycetes, for example *Spumaria alba* DC. and *Physarum gyrosum* ROST., are said to be harmful to fern fronds under moist conditions. They should probably, however, be regarded as epiphytes rather than as parasites, for there is no record of any penetration of the host tissues. Nevertheless, if present in large quantity, they do cause disease by cutting off supplies of moisture, light and air from the underlying fronds. JAROCKI [2]) has listed the following species as parasites on ferns: *Craterium minutum* FR., *Leocarpus fragilis* ROST., *Diachaea leucopoda* ROST. and *Didymium difforme* DUBY all on *Pteridium aquilinum*, also *Stemonitis flavogenita* JAHN on *Phegopteris Dryopteris*. The above remarks on the relationship between Myxomycete and fern probably apply also to these records.

*Bacterial diseases.* There is a well-known disease of ferns which has always been attributed to Nematodes. Brown discoloured areas appear on the fronds, the exact form varying considerably according to the species attacked, but the spots are often sharply delimited by the veins of the frond. A recent paper by AGGÉRY [3]) maintains that this disease is caused by bacteria which are introduced into the tissues by the Nematodes. Infection experiments with cultures of these bacteria showed that they produced all the typical macroscopic and microscopic symptoms without the aid of the Nematodes. The worms enter the fronds through the stomata in order to lay their eggs in the intercellular spaces; they bring with them bacteria which normally live as saprophytes in the soil. Four different types of bacteria were observed but were not identified, there appears therefore to be no one specific organism which is the sole cause of the disease.

Another bacterial disease has been described by the same author on *Polypodium vulgare, P. vulgare* var. *serratum* and *P. cambricum*. The earliest symptom is the appearance of yellowish spots on the fronds. These gradually turn brown, commencing at the centre. Often the midrib be-

---

[1]) SMITH, W. G., Gard. Chron., II, 72–74, 1882.
[2]) JAROCKI, J., Acta Soc. Bot. Polon., II, 183–199, 1924.
[3]) AGGÉRY, B., Bull. Soc. d'Hist. Nat. Toulouse, LXVIII, 5–201, 1935.

comes infected which results in the death of the distal portion of the frond. Bacteria are abundant in the diseased tissues. They attack chloroplasts and nuclei, ultimately destroying these. The infected cells and intercellular spaces become filled with a brown gummy substance. In severe cases the bacteria pass down the petiole to the rhizome, and fronds developed subsequently are dwarfed and malformed as well as being affected with the yellow spots. A rod-shaped bacterium was isolated which produced typical symptoms when artificially inoculated into healthy ferns. The bacteria enter the frond by the stomata or through wounds and insect punctures. A fungus, *Sphaerulina Polypodii* AGG., often developed on the dead spots but this organism was shown to be purely saprophytic.

*Virus diseases.* Although no decision has yet been reached by pathologists as to the parasitic nature of virus diseases, it may be of interest to conclude this section with a brief reference to certain records of virus infection in ferns. HINO [1]) recently published a list of Japanese plants susceptible to "mosaic and mosaic-like diseases" which included ten species of the *Polypodiaceae*, namely *Asplenium incisum, A. Wrightii, Athyrium coreanum, Dryopteris erythrosora, D. sophoroides, Coniogramme japonica, Nephrolepis cordifolia, Pteridium aquilinum* var. *japonicum, Pteris longipinnula* and *Woodwardia radicans* var. *orientalis.* There was a brief introductory note on the economic aspects of mosaic diseases but no mention of experimental work. It seems possible, therefore, that he was only dealing with some of the many types of non-infectious variegation known to occur in ferns. A second paper by the same author [2]) described and illustrated certain malformations of ferns said to be due to virus infection. Unfortunately the present writer has been unable to obtain either the paper, which is written in Japanese, or a summary of it, thus it is impossible to give details of this record.

*b)* **E q u i s e t i n a e.** — Comparatively few parasitic fungi have been recorded on *Equisetum,* the only living genus of this class. SCHAFFNER [3]) has published an account of the ravages of a Discomycete, *Stamnaria americana* MASS. et MORG., on *E. praealtum* in Ohio. Large lesions appeared on the lower internodes and finally became so extensive that the shoots were killed. The orange apothecia developed in more or less longitudinal rows and bore a superficial resemblance to the pustules of a rust. *S. Equiseti* (HOFFM.) SACC. is also known as a parasite on *E. arvense* and *E. hiemale.* Among the Fungi Imperfecti *Gloeosporium Equiseti* ELL. et Ev. occurs on *E. arvense, E. hiemale, E. laevigatum, E. limosum* and *E. sylvaticum,* while *G. Kriegerianum* BRES. attacks *E. arvense. Fusarium Equiseti* (CDA.) SACC., *F. avenaceum* (FR.) SACC. and *F. bulbigenum* CKE. et MASS. all invade the under-

---

[1]) HINO, I., Bull. Miyazaki Coll. Agric. and For., V, 97–111, 1933.

[2]) HINO, I., Journ. Japanese Bot., X, 377–380, 1934, No. 6.

[3]) SCHAFFNER, J. H., Amer. Fern Journ., XXI, 75, 1931.

ground parts of *Equisetum* spp. and also cause diseases of various important economic plants. POPP [1]) described *Fusarium graminearum* (*F. roseum* = *Gibberella Saubinetii*) attacking and killing *Equisetum*; this fungus is a dangerous parasite of cereals.

JAROCKI [2]) has recorded a Myxomycete, *Craterium minutum* FR., as a parasite on *E. sylvaticum*, but in all probability the organism was only an epiphyte, though possibly causing injury by purely mechanical means.

*c*) **Lycopodiinae**. — *Myiocopron Lycopodii* ROSTR. causes a disease of *Lycopodium complanatum* and *L. chamaecyparissus* in Denmark. The minute black perithecia of the fungus break out all over the affected plants. *Leptosphaeria marcyensis* (PECK) SACC. attacks living plants of *L. Selago* and *L. annotinum*, subsequently forming perithecia in the dead leaves and stems. *Leptosphaeria Crepini* (WEST.) DE NOT. has also been described as a parasite on *L. annotinum* and *L. clavatum*, blackening the sporophylls by the abundance of its perithecia. A number of other fungi have been recorded on various species of *Lycopodium* but no definite statements were made regarding their parasitism.

*Isoetes* is the host of four very interesting parasites. A smut, *Ustilago Isoetis* ROSTR., was described by ROSTRUP [3]) on *I. lacustre* from Denmark. The light brown, spherical spores develop in the bases of the leaves. *Rhizophidium Sphaerotheca* ZOPF [4]) attacks and kills microspores of *I. lacustre* and *I. echinospora* lying in water. This fungus belongs to the *Chytridiales*. A delicate branched mycelium or haustorium is formed within the spore and a sporangium on the surface. As many as twelve parasites may occur on one *Isoetes* spore, and they bring about changes in the spore content, converting it into large oil drops which are used by the fungus. The sporangia give rise to minute uniciliate swarm spores which bring about fresh infections. No resting spores are formed by this species. ZOPF also observed another parasite which formed from 2 to 4 thick-walled resting spores inside the *Isoetes* spores. He did not name the organism but stated that it showed certain affinities with the *Monadaceae*, a family of Flagellates. *Ligniera Isoetis* was described by PALM [5]) from Sweden in 1918. It is a member of the *Plasmodiophoraceae*, a group of organisms allied to if not included in the Myxomycetes. The parasite lives in the leaves of *Isoetes lacustre*, producing irregular brown spots on them. The colour is due to accumulations of spore-balls within the parenchyma cells. Each ball is hollow and almost completely fills its host cell, conforming to the shape of this. They are eventually set free by the decay of the invaded leaves. The method of germination and infection was not described, but a study of

---

[1]) POPP, M., Oldenburg Landw. Bl., XXV, 400, 1931.

[2]) JAROCKI, J., l.c.

[3]) ROSTRUP, E., Bot. Tidsskr., XXVI, 306, 1905.

[4]) ZOPF, Abhandl. Naturf. Ges. Halle, XVII, 92, 1888.

[5]) PALM, B., Svensk Bot. Tidskr., XII, 228–232, 1918.

early phases of the disease revealed the presence of small amoebae in the cells. These grew at the expense of the cell contents, forming a plasmodium which in turn gave rise to the dark-coloured spore-balls.

*Selaginella* is sometimes attacked by *Pythium Debaryanum* which causes 'damping off' under greenhouse conditions. *Synchytrium Selaginellae* Sor-OK. has been recorded on *Selaginella* sp., and *Taphrina Selaginellae* P.H. on *S. Menziesii*. OLSON [1]) has described a curious fungus, *Acrospermum urceolatum* OLSON, parasitic on *S. rupestris*. The small dark fructifications develop on the upper surface of both sporophylls and foliage leaves, either singly or in groups. They are vase-shaped with a wide apical aperture, and it is difficult to decide whether they should be regarded as apothecia or perithecia. The mycelium does not penetrate far into the tissues of the host. *Melanotaenium Selaginellae* HENN. et NYM. [2]) is a smut that grows in the stems and bases of the leaves of *Selaginella* in Java. The spores are at first brown, later black, and covered with warty outgrowths. They only germinate after being set free by the decay of the host plant. Another smut, *Entyloma polysporium* (PK.) FARL., was recorded by SINGH [3]) on *S. chrysocaulos* in India. Irregular brown or black patches developed on stems and leaves due to the presence of dark-coloured, thick-walled spores in the tissues. These occurred throughout the mesophyll of the leaf but only in the more superficial layers of the stem. Owing to an inadequate supply of material the full life-history of the fungus could not be investigated.

§ 5. **Symbiosis and Epiphytism.** — If one excludes mycorhiza, which are being discussed in the following chapter, the only remaining case of symbiosis is the well-known association of the water fern *Azolla* with the blue-green alga, *Anabaena Azollae*. Most of our information on the morphological aspect of this association is derived from the detailed account of the fern by STRASBURGER [4]). The growing point of the stem is sharply curved and in the resulting hollow on the dorsal surface lie some filaments of the alga. From here they pass into the cavities that develop in the upper segments of the leaves. Some of the epidermal cells lining these cavities bear curious hairs consisting of a stalk and a swollen club-shaped terminal portion. Similar hairs, either simple or branched, are also found among the algal filaments on the dorsal surface of the growing point. Some filaments of the alga penetrate beneath the indusium of the developing sporangia and there change to a resting condition. When the embryo begins to form these resting cells divide and give rise once more to characteristic filaments which become ensconced on the stem apex of the embryo. STRASBURGER found the alga present in every plant he examined of all the species of

[1]) OLSON, M. E., Bot. Gaz., XXIII, 367–372, 1897.

[2]) HENNINGS, P., Monsunia, I, 2, 1900.

[3]) SINGH, T. C. N., New Phytol., XXIX, 294–296, 1930.

[4]) STRASBURGER, E., Über *Azolla*, 1873.

*Azolla*, and since the fern showed no symptom of disease he decided that it must be a case of symbiosis. That view was generally accepted, and it has been suggested that the alga is capable of assimilating atmospheric nitrogen while the swollen hairs of the fern absorb some of the resulting nitrogenous products. The alga, on the other hand, benefits by the protection derived from the fern and perhaps also receives some supply of carbohydrates from it. All this, however, was pure hypothesis until OES [1]) experimented by growing *Azolla* on a nitrogen-free nutrient solution and found that the fern could indeed thrive under such conditions. Later LIMBERGER [2]) made an important contribution to our knowledge of this association by cultivating plants of *Azolla* freed from the alga. He found that growth and vegetative reproduction proceeded as usual and that the plants could scarcely be distinguished macroscopically from "normal" ones containing the alga. He therefore concluded that under the ordinary conditions of cultivation the fern is not dependent on the alga. After a few months, however, the alga-free ferns developed some curious abnormalities in the swollen hairs lining the now-empty leaf cavities. Sometimes several club-shaped cells appeared one above the other on the stalk cell, while occasionally the swollen apical cell never formed at all. But after prolonged cultivation without the alga these abnormal hairs disappeared. There was, nevertheless, a gradual increase in the length of the normal hairs, the terminal cells of which became densely filled with some albuminous material. Marked accumulations of starch were also observed in the leaves. Following up the line of investigation suggested by OES, LIMBERGER cultivated "normal" and alga-free ferns both on Knop's solution and on a nitrogen-free nutrient solution. On Knop's solution both types of fern grew well and reproduced vegetatively, but on the nitrogen-free solution the alga-free *Azolla* gradually died out while the "normal" ferns flourished. Thus proof was furnished that the algal association is obligatory for the fern when it is growing on a medium which is deficient in nitrogen but is not a necessity if an adequate supply of combined nitrogen is available. The association of the two organisms may therefore be regarded as a conditional symbiosis.

As pointed out in the introduction to this chapter, representatives of all classes of lower plants may be found growing epiphytically upon Pteridophytes. This type of association is much commoner in moist tropical climates than in the temperate regions, and it is sometimes difficult to distinguish between epiphytism and parasitism since the former may on occasion also result in direct injury to the tissues of the host plant. More commonly, however, the injurious effect, if present at all, is general rather

---

[1]) OES, A., Zeitschr. f. Bot., V, 145, 1913.

[2]) LIMBERGER, A., Akad. Wissensch. Wien, Math.-naturw. Kl., CXXXIV, 1–5, 1925.

than localised and is due to restriction of light and gaseous exchange. Some epiphytes are confined to particular hosts or groups of hosts and it is possible that they may have a chemical as well as a mechanical relation to their supporting plant. Many others, however, grow equally well on a wide range of plants or even on dead wood and stones.

To this latter group belong the innumerable algae that are found as epiphytes upon ferns. Sometimes the prothallia are overrun to such an extent that their development is checked, they remain sterile and are ultimately killed. MAURIZIO [1]) has given an interesting account of the nature and extent of the injury to ferns caused by heavy infestation with epiphytic algae. Fronds with a thin epidermis are more seriously affected than those with thick-walled epidermal cells. Frequently only the upper surface is invaded but under favourable conditions the algae may extend also to the lower surface and block the stomata. Sometimes they even burst open the guard cells by purely mechanical pressure and enter the air space below, but they cannot be regarded as in any way parasitic.

The Myxomycetes which have already been discussed in the section on diseases (p. 153) should probably be regarded as epiphytes rather than as parasites, but it was thought preferable to treat them under the heading of diseases since they had been recorded as parasites by the various authors quoted.

There are many epiphytic fungi, some of which represent conditions intermediate between true epiphytism and parasitism. FISHER [2]) has described a sooty mould of the tree fern *Dicksonia* caused by an epiphytic fungus, *Teichospora salicina* (MONT.) GAU. It forms a dense black film on the fronds and produces an abundance of perithecia and pycnidia, but remains entirely superficial. Another interesting epiphytic fungus has been described by SYDOW [3]) from Venezuela. It formed tiny white flecks on the lower surface of some fronds of *Diplazium expansum* whose upper surfaces were covered with epiphytic algae, mosses and lichens. The fungus formed white membranaceous stromata on which were borne the asci. SYDOW created a new genus and species for this peculiar form which he named *Nipholepis filicina*. He stated that it bore some resemblance to *Myxotheca hypocreoides* FERD. et WINGE, a species which has, however, been studied in detail by THAXTER [4]) and found to be a lichen; it occurs on *Trichomanes pinnatum* and various other ferns and flowering plants in Trinidad, and is now known as *Arthonia candida* var. *hypocreoides* (FERD. et WINGE) VAINIO.

Lichens, mosses and liverworts may all occur as epiphytes on Pteridophyta, but no useful purpose would be served by considering them more fully in this connection, since the association is not in any way an intimate one.

[1]) MAURIZIO, A., Flora, LXXXVI, 113–142, 1899.
[2]) FISHER, E. E., Proc. Roy. Soc. Vict., XLVII (N.S.), 387–388, 1935.
[3]) SYDOW, H., Ann. Myc., XXXIII, 93, 1935.
[4]) THAXTER, R., Mycologia, XIX, 160, 1927.

CHAPTER V

## MYCORHIZA

von

H. BURGEFF (Würzburg)

**§ 1. Einleitung.** — Farne sind im allgemeinen Bewohner feuchten und humösen Substrats und tragen selbst in stärkstem Masse zur Humificierung des Bodens bei; sei es dass sie mit ihren feinverzweigten Wurzeln Mineralböden auflockern und durchlässig machen und absterbend als Humus zurückbleiben, sei es dass sie sich als Epiphyten in der Traufe der Bäume als Humussammler ausgebildet haben, die ihn mit den mannigfaltigsten Organen festzuhalten und als Nahrung zu verwenden wissen.

Humöse Substrate zeigen sauere Reaction; Farne sind im allgemeinen acidophile Pflanzen. In saueren Böden herrschen Pilzmycelien vor; sie mineralisieren die absterbende organische Substanz und machen sie wieder für die grüne Pflanze aufnahmefähig. Sie leben in der Rohhumuszone; der eigentliche unter Sauerstoffabschluss gebildete echte Humus wird von ihnen kaum angegriffen, wenn nicht von neuem Sauerstoffzutritt und Abbau dieses fossilen Materials erfolgt.

Die Wurzeln aller Farne leben somit in engstem Contact mit den Pilzmycelien und haben sich mit ihnen auseinanderzusetzen. Beziehungen, die seit Jahrmillionen bestehen, haben feste Form angenommen; sie sind über die Stufe des Parasitismus hinausgewachsen zur Symbiose in weitester Fassung des Begriffs. Die nurparasitischen Verhältnisse sind im vorigen Kapitel ausführlich behandelt.

Scheiden wir den eigentlichen oder kteinotrophen und den symbiotischen Parasitismus aus, so bleiben Formen der Symbiose aller Grade zwischen den unterirdischen Teilen, Wurzeln und Rhizomen der Farne mit Pilzmycelien übrig, Vereinigungen, die wir gewohnt sind als *Mycorhiza* zu bezeichnen.

Nach der Auffassung der meisten heutigen Botaniker schliesst der Ausdruck Mycorhiza Fälle aller Art von Wurzel- oder Rhizomverpilzung ein. Ein Nutzen für die grüne Pflanze wird aus dem Bestehen einer solchen Verbindung nicht von vornherein vorausgesetzt. Es gibt aber viele Fälle, wo die

Verpilzung der Pflanze Vorteil bedeutet und ebensoviele, wo sie nicht weg-
zudenken und als *obligate Symbiose* Voraussetzung für die Existenz der
Pflanze wird, indem sie sie periodisch unabhängig von der Photosynthese
macht und ihr ein Leben als Saprophyt gestattet.

Alle Stufen kommen bei den Farnen vor. Schon bei Farnpflanzen der
Steinkohle finden wir die charakteristischen Mycelien in den Zellen der
Wurzelrinde und die nach deren Resorption entstandenen klumpigen
Excrete. (WEISS [1]), OSBORNE [2])).

Wie E. STAHL [3]) feststellte sind manche Stämme wie die Equisetales und
die Hydropteridales pilzfrei. Spätere Untersucher haben das bestätigt. Bei
anderen nimmt die Mycorhiza niedere Form der Ausgestaltung an und entspricht einer weit verbreiteten Form der Samenpflanzen und der Bryophyten; so bei den leptosporangiaten Farnen und den heterosporen *Lycopodiales*.

Nur bei 3 systematischen Gruppen hat sie höhere Ausgestaltung bei auf-
fallender Leistung erfahren: Unter den Eusporangiaten schliessen sich die
*Marattiales* noch eng an die *Filicinae leptosporangiatae* an, zwar obligat
verpilzt, aber ohne deutliche Abhängigkeit von der Verpilzung. Anders die
*Ophioglossales*, welche in sämtlichen Formen saprophytisch-mycotrophe
Prothallien führen; ebenso wie die *Psilotales*. Die merkwürdigste Ausgestaltung hat die Mycorhiza bei den isosporen *Lycopodiales* genommen.
Ihre Sporophyten sind meist autotroph, ihre Gametophyten zeigen aber
verschiedene Grade fortschreitender Compliciertheit der Mycorhiza, die bei
vielen zum Holosaprophytismus führt.

Die  g e s c h i c h t l i c h e  E n t w i c k l u n g  unserer Kenntnisse
über die Farnmycorhiza lässt sich nicht ohne Zusammenhang mit der Geschichte der Mycorhiza selbst schildern. M. C. RAYNER hat in ihrem Buch
„Mycorhiza" (London 1927) diese Entwicklung eingehend behandelt, viel
eingehender als es mir der hier zur Verfügung stehende Raum gestattet.
Zudem ist die Aufklärung der mycorhizalen Verhältnisse bei den Farnen in
engstem Zusammenhang mit der entwicklungsgeschichtlichen Erforschung
der verschiedenen Pteridophytenstämme erfolgt und nur gelegentlich Object besonderer Studien gewesen. So hat sie auch stets in der Aufdeckung
anatomischer und cytologischer Befunde bestanden, auf grund deren man
über die Beziehungen physiologischer Art zwischen den Symbionten die
verschiedensten Vermutungen, meist sehr allgemeiner Art, aussprach.
Während bei den Samenpflanzen, bei Orchideen, Ericaceen, Pirolaceen und
Monotropaceen, auch bei der Baummycorhiza schöne Ergebnisse experi-

---

[1]) WEISS, F. E., A mycorhiza from the lower coal-measures, Ann. of Bot. **18**,
255–266 (1904).

[2]) OSBORNE, T. G. B., The lateral roots of *Amyelon radicans* Will. and their
mycorhiza. Ann. of Bot. **23**, 603–611 (1909).

[3]) STAHL, E., Der Sinn der Mycorhizenbildung, Jahrb. f. wiss. Bot. **34** (1900).

mentell physiologischer Untersuchung vorliegen, steckt die Analyse der Farnmycorhiza noch in den allerersten Anfängen. Das hängt damit zusammen, dass es bisher niemals gelungen ist die Pilzsymbionten der Farne in freier Kultur zu erhalten und die Synthese der Symbiose in Reinkultur durchzuführen.

Die Art der Pilzsymbionten ist hieran schuld.

Alle Farne führen die *Phycomycoide Mycorhiza*. Die Pilzmycelien sind vielkernig, querwandlos, bilden blasige „Vesikel", eine Form der Verpilzung, die auch bei Samenpflanzen sehr verschiedener Familien der Gymnospermen und Angiospermen beobachtet ist, JANSE [1]), GALLAUD [2]), PEYRONEL [3]), JONES [4]) u.a.

Der zytologische Bau der Mycorhiza, insbesondere die Form der Organe des Austausches geben heute — unter Verwendung der Analogien mit Fällen bei den Samenpflanzen — die Möglichkeit der Klassificierung.

Vier Typen bestehen bei den Farnen: *Thamniscophage, tolypothamniscophage, ptyophage* und *chylophage Mycorhiza*.

Die von den Orchideen, den Pirolaceen und Ericaceen bekannte tolypophage Form kommt bei den Farnen nicht vor.

## § 2. Thamniscophage [5]) Mycorhiza.

— Es handelt sich bei diesem Typus um einen ausserordentlich weit verbreiteten, dem gewisse primitive Merkmale eigen sind. JANSE's Beobachtungen an einer *Selaginella*-Art des javanischen Urwaldes mögen hier als Beispiel dienen und in die Frage nach der Beurteilung der Mycorhiza einführen [1]):

Querwandlose Pilzhyphen dringen aus dem umgebenden Erdboden in die äusseren Zellen der Wurzelrinde ein, gelangen sich verzweigend in die tiefer liegenden Zellschichten und treten hier in die Intercellularräume über, denen sie hauptsächlich der Länge nach verlaufen. Hier entstehen zahlreiche terminale Anschwellungen oder „Vesikel", die Reservestoffe speichern (Fig. 1 a, b). Zweige der interzellularen Hyphen dringen nun nach Art von Haustorien in das Innere der angrenzenden Zellen, wo sie sich fein verästeln. An diesen Ästen entstehen dann die sogenannten „Sporangiolen", angenähert kugelige oder unregelmässig gestaltete Körper mit warziger Oberfläche, die nach JANSE auf kugelige Inhaltsmassen („sphérules")

---

[1]) JANSE, J. M., Les endophytes radicaux de quelques plantes javanaises, Ann. Jard. Bot. Buitenzorg **14**, 53–212, pl. V–XV (1897).

[2]) GALLAUD, J., Etudes sur les mycorhizes endotrophes Rév. gén. Bot. **17**, (1905).

[3]) PEYRONEL, B., Prime richerche sulle micorize endotrifiche e sulla microflora radicicola normale delle Fanerogame. Rivista di Biol. **5**, 463–485 (1923), l.c. **6**, 17–53 (1924).

[4]) JONES, F. R., A mycorhizal fungus in the roots of Legumes and some other plants. Journ. of Agric. Res. **29**, 459–470 (1924).

[5]) Anmerkung: θαηνίςχος = kleiner Strauch = arbusculus.

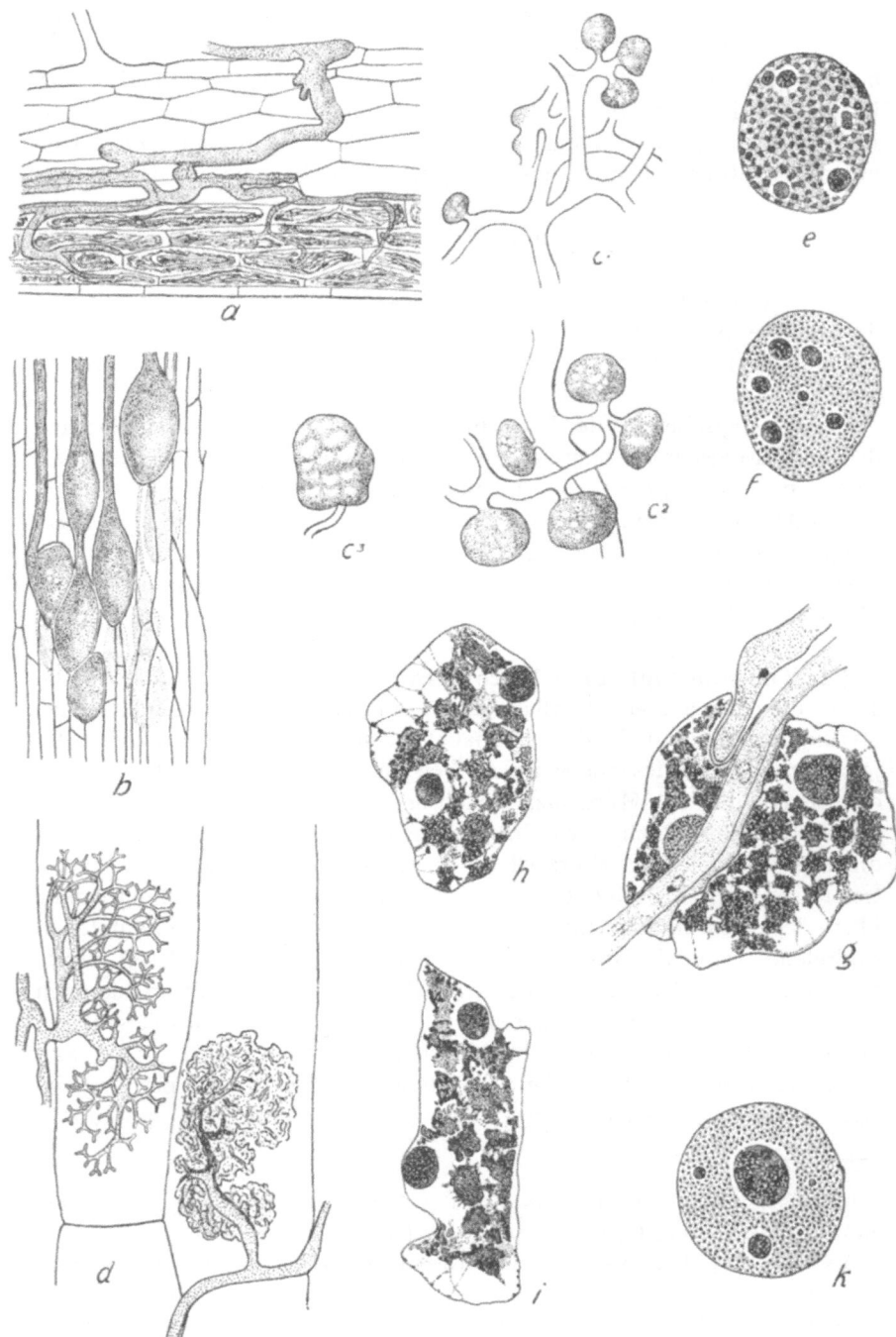

FIG. 1. — *a, Selaginella* spec. Infektion und Verpilzung der Wurzelrinde (nach JANSE). — *b, Selaginella* spec. Terminale und intercalare Vesikel an intercellularen Hyphen der inneren Wurzelrinde (nach JANSE). — *c, Selaginella* spec. „Sporangiolen in drei Entwicklungsstadien (1–3) (nach JANSE). — *d. Allium sphaerocephalum,* zwei Zellen mit zwei Arbuskeln, das eine im Beginn der Einschmelzung (nach GALLAUD). — *e–k, Psilotum triquetrum.* Fortlaufende Stadien der Zustandsänderung in den Zellkernen der Verdauungszellen (nach SHIBATA).

zurückzuführen ist. (Fig. 1c). Die Spherulae sollen feine granulae („granules") enthalten, die frei werden und die Zelle mit einer granulierten Masse und Oeltröpfchen erfüllen.

Die Natur der „Sporangiolen" hat JANSE nicht erklärt, doch schliesst er die Möglichkeit aus, dass es sich um irgendwelche Fortpflanzungsorgane des Pilzes handeln könne.

J. GALLAUD [1]) untersucht zahlreiche Pflanzen, so Lebermoose, Mono- und Dikotylen und findet wieder querwandlose oder — arme Pilze mit zusammengesetzten oder einfachen Haustorien, für welche er den generellen Ausdruck Arbuskeln („arbuscules") prägt. An ihnen entstehen die „Sporangiolen" JANSE's teils terminal („arbuscules simples") (Fig. 1d), teils lateral („arbuscules composés)".

GALLAUD deutet die Sporangiolen als Reste verdauten Pilzmaterials, die sich je nach ihrem Gehalt an organischen Stoffen verschieden färben.

Bei meinem Referat über Symbiose im Handwörterbuch der Naturwissenschaften IX, S. 812 (1934) rechnete ich JANSE's und GALLAUD's „Sporangiolenmycorhiza" zu dem 1932 [2]) aufgestellten Typus der Ptyophagen Mycorhiza, allerdings unter Hinweis auf gewisse Übergänge zur tolypophagen Form. Nachdem ich nun die Farnmycorhiza nachuntersucht habe, komme ich zu dem Ergebnis, dass dies nicht möglich ist. Bei der ptyophagen Mycorhiza werden junge noch wachsende, meist unverzweigte Hyphen zum Platzen und zum Plasmaerguss gezwungen. Das Plasma wird von der Zelle resorbiert, die Reste werden durch Celluloseauflagerung aus der Zelle ausgeschieden. Manchmal werden ausgestossene Plasmaballen auch nach kurzem selbständigem Leben und eigener Wandbildung verdaut (Ptyogene Vesikel).

Bei den meisten der von GALLAUD untersuchten Pflanzen und ebenso bei vielen Farnen findet dagegen eine Agglutination der ganzen Haustorien oder Arbuskeln statt, also eine Einschmelzung von älteren ausgewachsenen Hyphenverbänden (gelegentlich auch von Vesikeln), die der tolypophagen Mycorhiza viel näher steht und hier als *Thamniscophage Mycorhiza* bezeichnet werden soll.

## a) Primitive Formen

In der Einleitung zu diesem Kapitel ist bereits der Fall der von JANSE [3]) beschriebenen *Selaginella* geschildert. Auch BRUCHMANN [4]) hat bei der europäischen *Selaginella spinulosa* Verpilzung der Wurzeln festgestellt.

---

[1]) GALLAUD, l.c., S. 161

[2]) BURGEFF, H., Saprophytismus und Symbiose, Studien an tropischen Orchideen, Jena (1932).

[3]) JANSE, l.c., S. 161

[4]) BRUCHMANN, H., Untersuchungen über *Selaginella spinulosa*. Gotha: Perthes (1897).

Andere Arten schienen pilzfrei — wenigstens in Kultur (*Selaginella Lyallii* SPRING nach BRUCHMANN [1]).

Pilze ähnlicher Art sind nachgewiesen bei grünen Prothallien von Gleicheniaceen und Osmundaceen (CAMPBELL [2])) Genauere Untersuchungen liegen nicht vor.

Wahrscheinlich hierher gehörig ist auch die merkwürdige Verpilzung der rhizoidtragenden Kugelzellen an dem Fadenprothallium der *Schizaea pusilla* (BRITTON & TAYLOR [3])).

Einer besonderen Behandlung bedarf die obligate Mycorhiza der Marattiaceen, die eigener Züge nicht entbehrt.

### b) Marattiaceenmycorhiza

RUSSOW [4]) findet in den Zellen der Wurzelrinde von Marattiaceenwurzeln „grumöse" Massen, von denen er nach Beobachtungen an *Ophioglossum* glaubt, dass sie durch Pilzhyphen verursacht werden. KÜHN [5]) beobachtet die Entstehung der Excretklumpen aus Vesikeln und umgebenden Hyphen. Seine Isolierungsversuche des Pilzes in Rosinensaft-Hängetropfen auf dem Objectträger geben Mycel mit verschiedenerlei Conidienformen. Auch CAMPBELL [2]), und CHARLES [6]) beschäftigen sich mit diesen Objecten.

Eine sehr eingehende vorzügliche Bearbeitung der Maratticeenmycorhiza stammt von C. WEST [7]). Der Autor findet sämtliche Gattungen und Arten mit seltenen Ausnahmen einzelner Pflanzen verpilzt, so *Angiopteris*, *Marattia, Kaulfussia, Archangiopteris* und *Danaea*.

Das querwandlose kräftige ungleichmässig dicke Mycel des Pilzes dringt ohne Scheidenbildung in die Zellen der Epidermis und der äusseren Wurzelrinde ein und verzweigt sich spärlich in diesen Zellschichten. In den tieferen Schichten wird es interzellulär und entsendet, den Intercellularen entlang wachsend, zahlreiche Haustorien in die angrenzenden Zellen. Die Haustorialhyphen verzweigen sich ungemein stark und werden in dieser Weise

---

[1]) BRUCHMANN, H., Von den Vegetationsorganen der *Selaginella Lyallii* Spring. Flora **99**, 436–464 (1909).

[2]) CAMPBELL, D. H., Symbiosis in Fern-Prothallia. The Amer. Nat. **42**, 154–165 (1908).

[3]) BRITTON, E. G., and TAYLOR, A., Life history of *Schizaea pusilla*. Bull. Torrey Bot. Cl. **28**, 1—19 (1901).

[4]) RUSSOW, J., Vergleichende Untersuchungen der Leitbündel-Kryptogamen. Mém. Akad. Imp. Sc. de St. Pétersbourg **19**, (1872).

[5]) KÜHN, V. R., Untersuchungen über die Anatomie der Marattiaceen und andere Gefässkryptogemen. Flora **72**, 457–504 (1899).

[6]) CHARLES, G. M., The anatomie of the sporeling of *Marattia alata*. Bot. Gaz. **51**, 81 (1911).

[7]) WEST, C., On *Stigeosporium Marattiacearum* and the mycorhiza of the *Marattiaceae*. Ann. of Bot. **31**, 77–99 (1917).

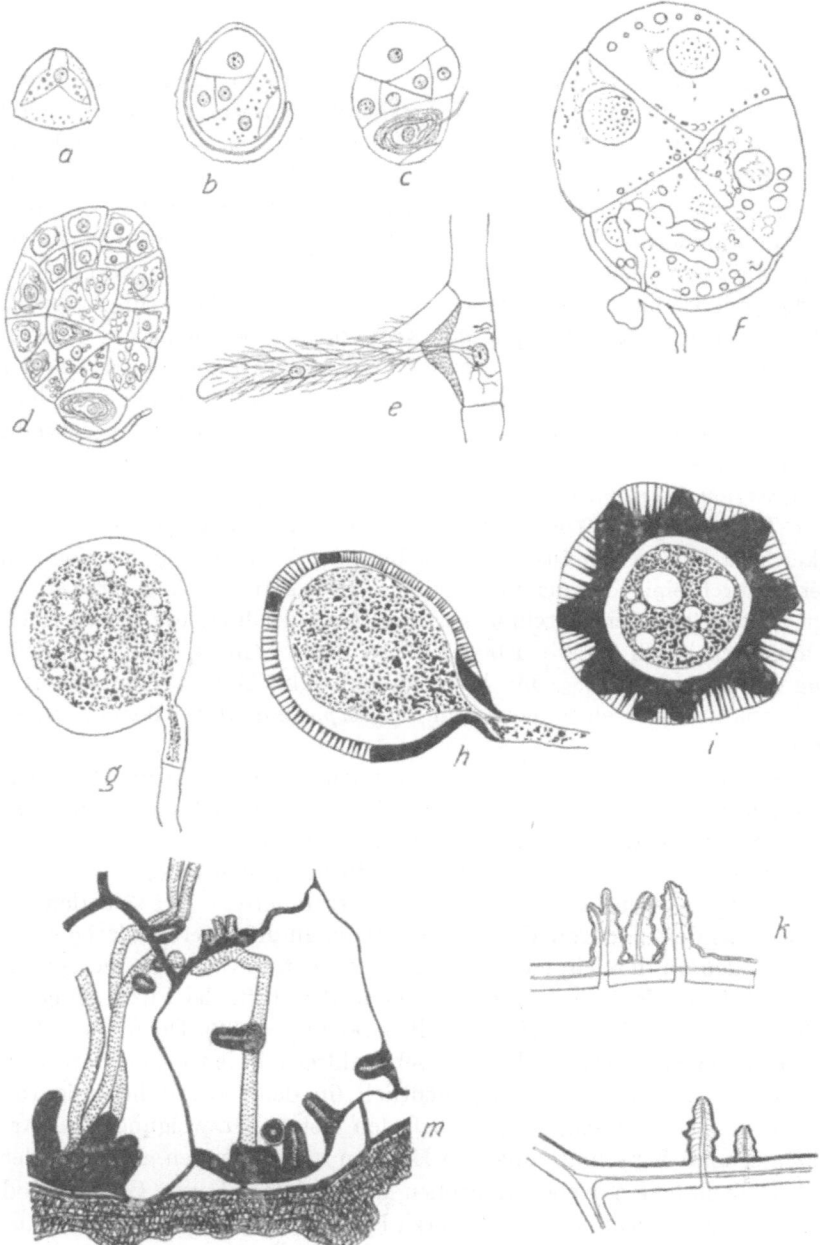

FIG. 2. — *a, Lycopodium Selago,* Spore zu Beginn der Keimung. — *b*, 5 zelliges Empfängnisstadium. — *d*, Stadium mit entwickelter Mycorhiza. — *e*, Haarfussferzenzelle. — *c, L. clavatum,* Erste Infektion des Sporenkeimlings (nach BRUCH-MANN). — *f. Ophioglossum pendulum,* zwei Monate alter Keimling, aus 4 Zellen bestehend, die basale Zelle durch eine Pilzhyphe inficiert, die bereits eine „Deformation'' erleidet (nach CAMPBELL). — *g-i, Stigeosporium marattiacearum* WEST. „Ruhesporen'' aus den Geweben der Pflanzen (nach WEST). — *k, Galeola hydra,* Structur der Röhrentüpfel. — *m, Gastrodia callosa* Röhrentüpfel auf Zellwänden und Pilzhyphen (nach BURGEFF).

zu Arbuskeln. Die Arbuskeln verlieren ihre Form und gehen in granuläre, structurlose, klumpige, stark lichtbrechende Massen („Sporangiolen") über, die bereits mit blossem Auge in den Schnitten sichtbar sind. Sie färben sich mit Jodlösung gelb und bleiben in $SO_4H_2$ und KOH unverändert. Schleimkanäle und Gerbstoffzellen werden nicht vom Pilz berührt. Ausserdem bildet der Pilz die üblichen Vesikel als „temporary reserve organs", die später entleert werden, und „resting spores": kugelige dickwandige, mit dreifacher Membran versehene ölerfüllte Gebilde (Fig. 2$g$_$i$), die an Oogonien erinnern, mangels der Antheridien aber nicht mit solchen identisch sein können. WEST vermutet, dass sie nach dem Absterben der Wurzel den Pilz regenerieren. Isolierungsversuche sind vergeblich. W. hält den Pilz für nahe verwandt mit der Gattung *Phytophthora* und beschreibt ihn zum Unterschied von dieser als conidienfreie neue Gattung und Art *Stigeosporium Marattiacearum*.

Die Mycorhiza von *Danaea* weicht nicht unbeträchtlich ab. Sie entwickelt sich hier in den äusseren Schichten der Wurzelrinde, die von den inneren durch eine sclerenchymatische Scheide getrennt sind. Das Mycel zeigt gleichmässigeren Durchmesser und ist intracellular (da Intercellularen augenscheinlich fehlen). Ebenso fehlen die „resting spores". Die Zellkerne sollen sich nicht vom Hypheninhalt abheben. West vermutet das Vorhandensein eines von dem *Stigeosporium* abweichenden Symbionten.

Um einen eigenen Standpunkt zu gewinnen, habe ich die Marattiaceenmycorhiza an *Marattia alata* und *Angiopteris evecta* nachuntersucht. *Danaea*-Pflanzen standen mir leider nicht zur Verfügung.

Der intercellulare Verlauf der Hyphen in der Innenrinde, — die Zellschichten der Aussenrinde haben keine Interzellularen — ist sehr demonstrativ, besonders dadurch, dass die inficierten an die Intercellularen grenzenden Zellen an den Tüpfeln nach innen gerichtete Celluloseemergenzen aufsetzen, die als Reaction auf eindringende Reizstoffe des Pilzes gedeutet werden müssen („Röhrentüpfel" vgl. BURGEFF [1]), S. 178). Die in die Zellen eindringenden Haustorialhyphen sind sehr zahlreich in jeder Zelle, ihre Verzweigung zu Arbuskeln ist wenig deutlich (in den WEST-schen Figuren auch nicht nachzuweisen). Schon nach den ersten Verzweigungen zu Arbuskeln scheint Einschmelzung von Hyphen vorzukommen. Klare Bilder von Arbuskeln fehlen. Neben amorphen Massen, die sich mit „Glycogenjod nach A. MEYER" grünlich, mit Chlorzinkjod an Microtomschnitten bläulich färben (Amyloid) und nur spärliche Hyphenreste enthalten, finden sich bedeutende Massen „lebenden Plasmas" mit ungemein zahlreichen Mitochondrien vermischt. Die ganzen Vorgänge der Resorption des Pilzmaterials weichen von den üblichen Bildern ab, und bedürfen noch genauerer Untersuchung. Fig. 3 oben mag hier eine Vorstellung der *Marattia alata*-

---

[1]) BURGEFF, H., Saprophytismus und Symbiose, Jena 1932.

Mycorhiza geben. Die intracellularen Vesikel werden nicht zerstört, sondern zu „resting spores". Der Materialgewinn der Pflanze bleibt problematisch.

*c.* Die hochentwickelte Form der Ophioglossaceen

MILDE [1]) beobachtete in den Wurzelrindenzellen von *Botrychium* „teigähnliche" Massen, RUSSOW [2]) fand daneben Pilzhyphen. ATKINSON [3]) untersucht ein Dutzend Arten von *Botrychium* und *Ophioglossum* und findet alle Arten verpilzt. GREVILLIUS [4]) beobachtet die Mycorhiza bei allen untersuchten scandinavischen, europäischen und tropischen *Botrychium*-arten und macht genaue Angaben über die Ausdehnung der Wurzelinfection bei den einzelnen Arten. Isolierte Pilznester hinter dem Meristem können sich weiter hinten zum Pilzmantel schliessen. „Hyphenknäuel" in den Zellen sind mit den „interzellularen" Längshyphen durch Stielhyphen verbunden. Die Knäuel sollen der Resorption durch die Pflanze verfallen. JANSE (1897) beobachtet gut die Infection von *Oph. pendulum*; wenig glücklich ist die Darstellung der Sporangiolenbildung. LANG (1902) beschreibt die Verpilzung der Prothallien von *Oph. pendulum* und *Helminthostachys zeylanica* und findet in jungen Prothallienteilen Vesikel, in alten „small shriveled bodies". BRUCHMANN (1904) beobachtet bei *Ophioglossum vulgatum* die Infection der Zellen, das Verschwinden der Stärke, die „Entartung" der Hyphen, das Zurückbleiben von „Klumpen". Der Pilz soll Humusstoffe in Baustoffe für die Pflanze verwandeln. Seine Untersuchung von *Botrychium* (1906) bringt über die Mycorhiza nichts wesentlich neues. CAMPBELL [5]) beobachtet in sehr bedeutenden Studien Keimung und Infection der Sporen von *Oph. pendulum* und *Oph. moluccanum* und beschreibt an erwachsenen Prothallien von *Oph. pendulum* die Zellinfection, den Abbau der Stärke, die Ausbildung verschiedener Arten von Vesikeln, auch charakteristische Unterschiede bei dem Prothallium von *Botrychium virginianum*, von dem er Zellen mit unregelmässigen Mycel neben „digestiv" Zellen abbildet. BÖNIKE L. [6]) gibt als erster Zeichnungen des Verdauungsvorgangs. Leider war mir der russische Text der Arbeit nicht zugänglich.

Nach eingehendem Studium der vorher aufgezählten Literatur ist es mir

---

[1]) MILDE (1869) cit. in GREVILLIUS (1895) siehe unten.

[2]) RUSSOW, l.c. S. 164.

[3]) ATKINSON, G. F., Symbiosis in the roots of *Ophioglossaceae.* Proc. Am. Ass. f. the Advanc. of Sc. 254/55 (1894).

[4]) GREVILLIUS, A. Y., Über Mycorhizen bei der Gattung *Botrychium* nebst einigen Bemerkungen über das Auftreten von Wurzelsprossen bei *B. virginianum* Swartz. Flora 80, 445–453 (1895).

[5]) CAMPBELL, l.c. S. 164.

[6]) BÖNIKE, L., Sur les mycorhizes endotrophes des Orchidées, Pirolacées et *Ophioglossaceae.* Trav. de Nat. à l'Un. Imp. de Kharkow **43**, 1–32 (1909).

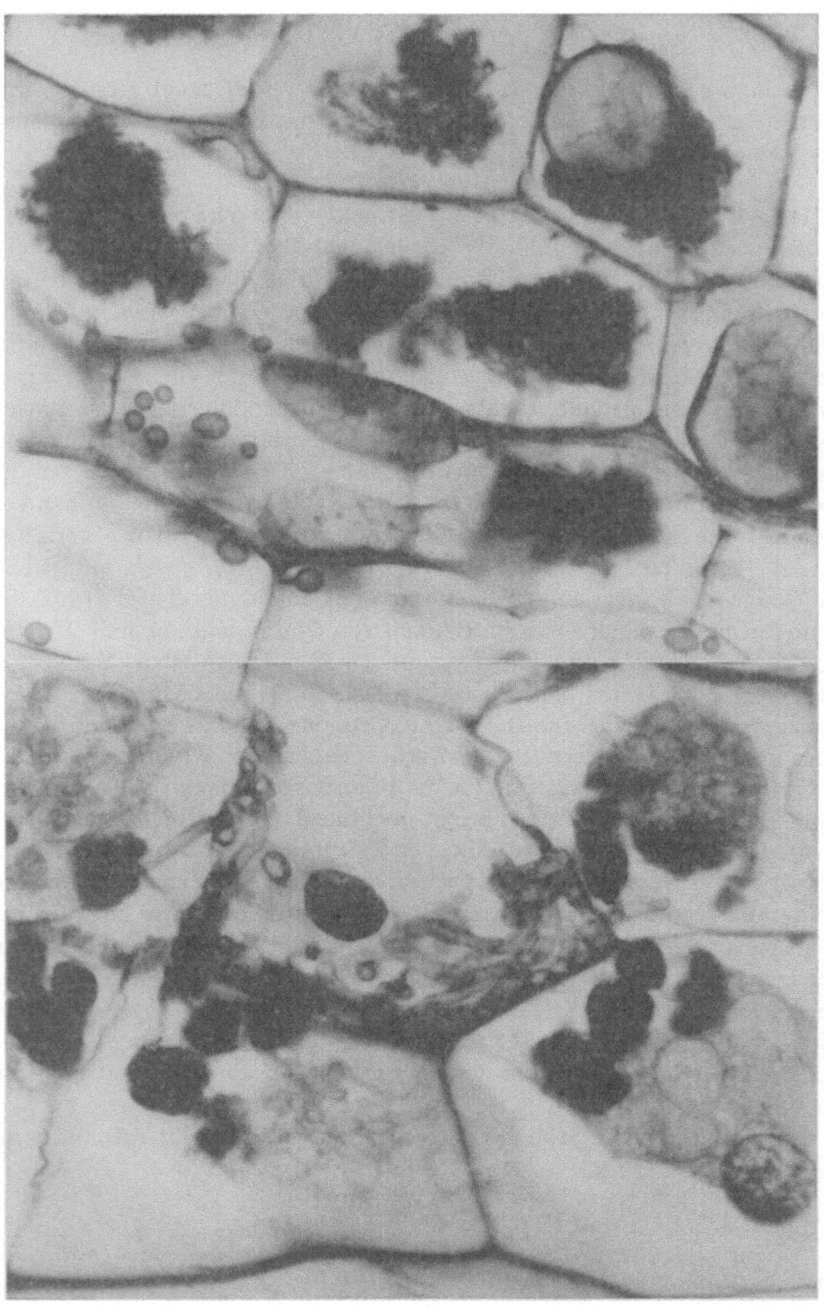

Fɪɢ. 3. — Oben, *Marattia alata*, Radialer Schnitt durch die Wurzelrinde; unten am Rand Zellen des innersten stärkereichen Speichergewebes, nach oben drei Zellschichten mit in Agglutination befindlichen Arbuskeln. In den Zellen auf der rechten Seite zwei intracellulare, in der Mitte ein intercellulares Vesikel. Die intracellularen werden zu dickwandigen „Ruhesporen". Tanninbeize-Eisenfärbung (524 : 1). — Unten, *Ophioglossum pendulum*, Oben Mitte: „Vorgeschobene" Pilzwirtzelle; in die angrenzenden Verdauungszellen nach links unten und rechts verlaufen Hyphen, die Sternarbuskel tragen. Ein Teil dieser Arbuskeln ist bereits in dunkel gefärbte Excretkörper verwandelt. Haematoxylin-Eisenalaun – Safranin (524 : 1).

nicht möglich gewesen, manche Widersprüche und Ungenauigkeiten in den Angaben und Zeichnungen der Autoren zu klären. Nun hatte ich aber das Glück in Java nicht nur reichlich Pflanzen von *Ophioglossum pendulum* und *O. moluccanum*, sondern „auf Campbells Pfaden", d.h. in einem *Asplenium nidus*-Nest bei Tjibodas auch die Prothallien des *O. pendulum* aufzufinden. Die Untersuchung erfolgte an Microtomschnitten nach den bei ptyophager Orchideenmycorhiza 1932 geschilderten Methoden. Es ergab sich folgendes:

Im Sporophyten von *Ophiogl. pendulum* sind nur die dünnen Wurzeln regelmässig verpilzt. Die Infection beschränkt sich auf die äussere interzellularenlose Wurzelrinde, derentwegen die durch die Epidermis eindringenden querwandlosen vielkernigen Hyphen intracellular verlaufen. Bei der Durchbohrung der Wände treten keinerlei Reactionen der Zelle durch Abscheidung von Cellulosescheiden ein, auch sind keine Appressorien sichtbar. Die Zellwand wird vom Pilz gelöst.

Zellen der vierten, der fünften und der sechsten Schicht können als Pilzwirtzellen fungieren. Zellen der fünften oder sechsten Schicht als Pilzverdauungszellen. Wird eine Zelle der vierten Schicht besiedelt, so wachsen sämtliche Hyphen in dieser Zelle angedrückt an den an die fünfte Zellschicht angrenzenden Zellwänden. Sie scheinen interzellular, sind aber wie dünne Wurzelquerschnitte ausweisen intracellular und augenscheinlich durch von der Verdauungszelle zudiffundierende Stoffe zum Appressionswachstum gezwungen. Fungiert eine Zelle der fünften Zellschicht als Pilz wirtzelle, so werden neben denen der 6ten auch die nicht infizierten der 5ten Schicht zu Pilzverdauungszellen. Fig. 3 unten gibt ein Bild einer solchen vorgeschobenen Pilzwirtzelle. Differenziert sich in seltenen Fällen eine Zelle der 6ten Schicht zur Pilzwirtzelle so wird das Bild der ganzen Mycorhiza invers. Appressionswachstum der Hyphen erfolgt jetzt an den nach aussen gerichteten Wänden.

Die Infection der Verdauungszellen ist höchst merkwürdig; eine kräftige aus der Pilzwirtzelle eindringende Hyphe wird zu einem dünnwandigen viele blasige Aussackungen treibenden Haustorium, das ich Sternarbuskel nennen will. (Fig. 4 links oben). Die Stärke wird nun langsam abgebaut. Der Abbau führt jedoch nicht immer zu völligem Schwund der Stärke. Vielfach lassen sich sehr kleine Stärkekörner in den Amyloplasten der Verdauungszellen noch nachweisen. Der Sternarbuskel füllt schliesslich mit seinen Aussackungen den Querdurchmesser der Zelle aus. Man beobachtet an weiter fortgeschrittenen Stadien ein Knittern der Wände, die immer mehr zusammenschrumpfen. Der Inhalt zeigt noch gefärbte Kerne und dunkle Schollen anderer Substanz (Fig. 4 links unten). Die zerknitterte Wand wird immer mehr zusammengeballt und liefert einen unregelmässig rundlichen, noch mit der Zuleitungshyphe verbundenen, scharf berandeten, stark lichtbrechenden Excretkörper, die „Sporangiole" JANSE's. (Fig. 3 unten, 4 links oben). Die Zelle wird nun nach und nach bei immer wiederholtem Einbruch neuer

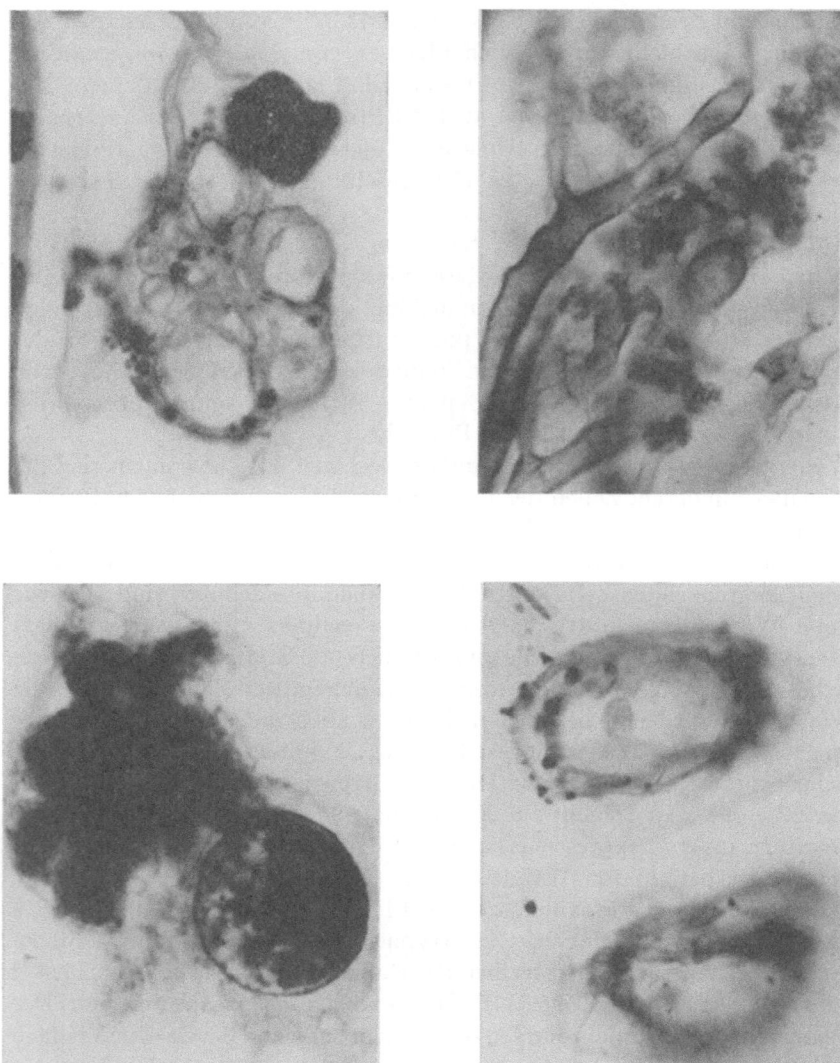

Fig. 4. — L. oben, *Ophioglossum pendulum*, Junger Sternarbuskel umgeben von Amyloplasten, deren zusammengesetzte Stärke noch nicht abgebaut wurde, daneben geschwärzter aus einen Arbuskel enstandener Excretkörper. Tannin Eisenpräparat (1154 : 1). — L. unten, Daselbe; Sternarbuskel in der Verdauung mit anliegendem Zellkern. Hämatoxylin-Eisenalaun-Safranin (1154 : 1). — R. oben, *Psilotum triquetrum* Ausschnitt aus Fig. 6 R. oben: Arbuskeln bei stärkstmöglicher Vergrösserung. Die zarten Einzelhyphen sind im Schnitt sichtbar. Tannin-Eisen (1154 :1).— R. unten, *Lycopodium complanatum*, Pilzknäuel aus dem Rindengewebe des Prothalliums mit den auf die Hyphen aufgesetzten Röhrentüpfeln (1154 : 1).

Hyphen und wiederholter Coagulation neuer Sternarbuskeln mit den Excret-körpern angefüllt. In Fig. 3 unten sind in jeder Verdauungszelle die haema-toxylingeschwärzten Excretkörper neben intakten Sternarbuskeln zu sehen. Junge Excretkörper färben sich mit Glycogenjod und Jodjodkali dunkel gelb und entfärben sich bei höherer Temperatur. Beim Erkalten kehrt die Farbe zurück. In alten Wurzeln unterbleiben die Reactionen. Trotzdem scheint es sich nicht allein um Glycogen zu handeln. Das in den Ophioglossumwurzeln aller untersuchten Arten massenhaft vorhandene fette Oel gibt dieselben Reactionen mit Jod. Es lässt sich durch Färbung mit Sudan III leicht nachweisen. Eine ähnliche Substanz tritt auch in dicken Tropfen in Hyphen der Pilzwirtzellen auf [1]). Mit Aether aus frischen Wurzeln von Ohp. vulga-tum extrahiert, ist das Oel in Menge zu erhalten, von gelber Farbe, löslich zu gelber Lösung in absolutem Aethylalkohol, u n l ö s l i c h in abs. Me-thylalkohol aber rasch verseifbar mit $KOH + NH_3$. Ein Erwerb dieses Oels durch die Pilzverdauung scheint nicht ausgeschlossen. Die Frage be-darf genauerer Untersuchung als sie jetzt im Winterzustand der Pflanzen möglich ist. Alte Excretkörper färben sich mit gewöhnlicher Jodjodkalium-lösung bläulich, nach Auswaschen des überschüssigen Jods blau. Die Reak-tion ist auf die geschrumpften Wände des Sternarbuskels zurückzuführen, die zum Unterschied von den gewöhnlichen Hyphen in ihren Chitinwän-den Amyloid enthalten. Die stark lichtbrechende wahrscheinlich aus Plas-ma bestehende Umhüllung der Exkretkörper verschwindet bei Behandlung mit Javellescher Lauge, es bleiben die reinen Pilzmembranen übrig. Die Excretkörper werden also nicht mit Cellulose umhüllt wie die Excret-klumpen der tolypophagen Mycrorhiza.

Das vollsaprophytische Prothallium von *Ophioglossum pendulum* ist sehr klein. Von LANG [2]) und CAMPBELL [3]) ist es genau beschrieben. CAMPBELL hat auch die Sporenkeimung geschildert. Im Alter von 2 Monaten wird die basale Zelle des vierzelligen Stadiums inficiert (Fig. 2f). „The branching mycelium of the mycorhiza was closely applied to the surface of the cell, and a haustorium was sent down through the cell wall into the basal cell." Das Bodenmycel entspricht hier dem Mycel der Pilzwirtzelle und zeigt Ap-pressionswachstum; die Basalzelle wird gleich zur anlockenden Ver-dauungszelle. „In the cell infected with the fungus, the contents show the peculiar aggregated appearance characteristic of the infected cells of the older Prothallium." An der Verdauung des ersten Haustoriums ist nach CAMPBELL's Worten wohl nicht zu zweifeln.

Das erwachsene Prothallium ist verzweigt. Der Pilz ist intracellular in

---

[1]) BRUCHMANN findet nicht nur die verpilzten Prothalliumzellen des *Botrychium Lunaria* sondern auch die „Ausstülpungen" der Pilzhyphen mit fettem Oel erfüllt.

[2]) LANG, W. H., On the prothalli of *Ophioglossum pendulum* and *Helminthos-tachys zeylanica*. Ann. of Bot. 16: 2–56 (1902).

[3]) CAMPBELL, D. H., Studies on the *Ophioglossaceae*. Ann. Jard. Bot. Buiten-zorg 21: 138–194 (1907) desgl. 1908 l.c. S. 164.

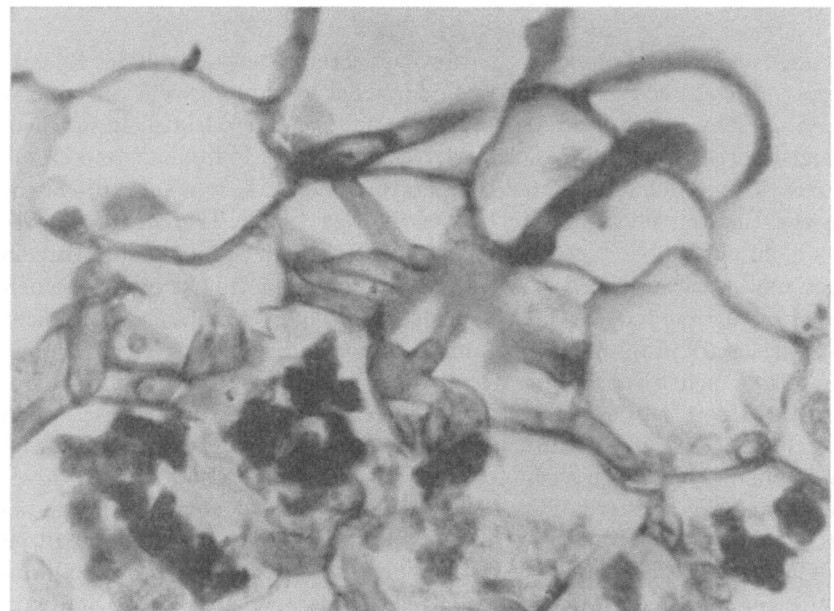

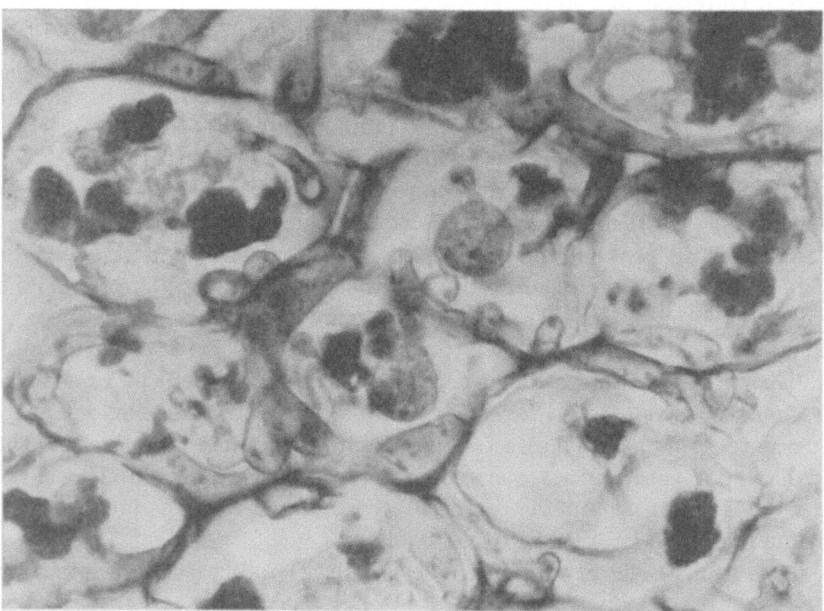

FIG. 5. — Oben, *Ophioglossum pendulum*, Längsschnitt durch das Prothallium.
Rechts oben: Pilzhyphe durch die kurze Rhizoidzelle eingedrungen, in den äusseren
Zellen Mycel, in den inneren Sternarbuskel in allen Stadien der Verdauung. Hae-
matoxylin-Eisenalaun-Eosin (524 : 1). — Unten, Zellen aus dem centralen Teil
der Prothalliums. Dickwandige grosskernige Hyphen durchbohren die Wände der
lebenden Zellen und erzeugen an Seitenzweigen Sternarbuskel, die der Verdauung
verfallen. Färbung und Vergrösserung wie vorher.

den ganzen basalen Partien verbreitet; jung inficierte Zellen befinden sich hinter den wachsenden Meristemen der Prothalliumzweige. Neuinfectionen finden von aussen durch die kurzen Haare statt (LANG 1902). Die cytologischen Verhältnisse der Mycorhiza sind nach der Schilderung der Autoren nicht eindeutig. Ich bringe das folgende nach eigenen Beobachtungen:

Die Zellen des Gametophyten haben etwa den halben Durchmesser der Wurzelrindenzellen des Sporophyten; das bedingt ein etwas geändertes Verhalten des Pilzes. Die Infection erfolgt durch die Haarzelle, die Hyphe verzweigt sich in der subepidermalen Schicht (Fig. 5 oben), deren Zellen zu Pilzwirtzellen werden. In den tiefer liegenden Zellen beobachtet man alle Stadien der Verdauung hier wesentlich kleiner ausfallender Sternarbuskeln, daneben typische Pilzwirtzellen. Weiter nach innen hört die säuberliche Differenzierung in Pilzwirt- und Verdauungszellen auf. Immer dickere derbwandige grosskernige Hyphen wachsen nach allen Richtungen durch die Zellen unter starker Zerstörung der Zellwände (Fig. 5 unten) und bilden, anscheinend jeweils in Nachbarzellen hinein, ihre Haustorien oder Sternarbuskeln, die in der gleichen Art wie in den Zellen des Sporophyten verdaut und zu etwas weniger regelmässig gestalteten Excretkörpern werden. Die derbwandigen Hyphen unterliegen keiner Schädigung seitens der Pflanze. Auch die vielfach ohne Scheidenbildung durchbohrten Zellen besitzen gesunde, schwach hyperchromatische, nur wenig vergrösserte Zellkerne. Abweichend von der Sporophytenmycorhiza ist also im wesentlichen nur der Verlust der Differenzierung in Pilzwirt- und Verdauungszellen in den centralen Teilen des Prothalliums. Die bei der Sporophytenmycorhiza fehlenden von den Zellen nicht angegriffenen, derbwandigen Riesenhyphen übernehmen hier augenscheinlich die Zuleitung des Materials im dichteren Gewebe des Prothalliums. Was sie aus dem umgebenden Erdboden herbeischaffen verfällt der Pflanze auf dem Weg der Thamniscophagie.

Nach längerem vegetativen Wachstum beginnt das Prothallium mit der Anlage der Geschlechtsorgane. Sie und ihre Nachbarzellen werden vom Pilzmycel nicht berührt. Bei Helminthostachys besteht nach LANG (1902) eine ausgesprochen vegetative Phase, auf die nach einer Streckung die Anlage der ♂ und ♀ Organe an getrennten Prothallien erfolgt. Die Verpilzungsart scheint von der der Ophioglossum-Prothallien nicht verschieden. Für die von BRUCHMANN [1]) untersuchten Prothallium von *Oph. vulgatum* und *Botrychium Lunaria* ergeben sich aus den BRUCHMANNschen Beschreibungen der Endophyten keine wesentliche Abweichungen.

Die gesamten Ophioglossaceen haben augenscheinlich den gleichen Verpilzungstypus. Unterschiede bestehen in der Dicke der vegetativen Hyphen, in der Art des Appressionswachstums in den Pilzwirtzellen, das ich in den

---

[1]) BRUCHMANN, H., Über das Prothallium und die Keimpflanze von *Ophioglossum vulgatum* L. Bot. Ztg. **62**, 227–248 (1904) ders. Über das Prothallium und die Sporenpflanze von *Botrychium lunaria*. Flora **95**, 203–230 (1906).

Wurzeln von *O. vulgatum* viel ausgesprochener fand als bei *O. pendulum* und *moluccanum*. Die Hyphen liegen hier völlig flachgedrückt der Wand der Verdauungszellen an. Die Art der Ausbreitung des Mycels in der Wurzelrinde wechselt nach GREVILLIUS [1]), der die meisten Arten untersuchte, bedeutend. Bis zu 7 Schichten können inficiert sein (*Botrychium lanceolatum*). Bei anderen Arten nimmt die Dicke der Pilzschicht auf 5–4–3–2 Schichten ab. Bei *B. australe* finden sich nur unzusammenhängende Pilznester.

### § 3. Die tolypothamniscophage [2]) Mycorhiza der Psilotales. —

Bei *Psilotum* wurde die Verpilzung von SOLMS-LAUBACH [3]) entdeckt, auch die Infection der Brutknospen durch den Pilz beobachtet und abgebildet. Kurz damit beschäftigt haben sich noch JANSE [4]) und BERNATZKY [5]). Der letztere glaubte sich mit der Isolierung der Endophyten in freier Kultur erfolgreich, hat aber sicher einen falschen Pilz in Kultur gehabt.

SHIBATA [6]) hat dann eine eingehende Untersuchung vorgenommen, von deren Ergebnissen meine eigenen Befunde in einzelnen Punkten abweichen.

Der Bau der Pflanze ist von SOLMS-LAUBACH (1884) geschildert worden. Sie besitzt ein stark verzweigtes unterirdisches Rhizom, dessen mit zweizelligen Rhizoiden besetzte Äste sich verzweigend abwärts wachsen, während andere aus der oberen Zone an das Licht tretend zu aufrechten Assimilationssprossen werden. Die unterirdischen Rhizomteile sind in einiger Entfernung vom Meristem verpilzt, die dünnen Abzweigungen stärker als die Hauptachsen. Der Pilz dringt mit querwandlosen Hyphen durch die Haare in die Rhizomrinde ein und inficiert einen Teil der Zellen in 6–7 Schichten der Aussenrinde. In den inficierten Zellen verschwindet die Stärke aus den Amyloplasten. Zuerst erscheinen in der Zelle wenige Windungen allmählig perlschnurartig anschwellender, dicker Hyphen, von denen dünnere und verzweigte Äste abgehen, die das Innere der Zelle nach und nach anfüllen (Fig. 6 links oben). Andere Zellen werden von Anfang an durch dünnere Hyphen besiedelt. SHIBATA sah diesen Unterschied und spricht von „Pilzwirtzellen" und „Verdauungszellen". Da, wie auch er beobachtet, der Pilz in allen Zellen schliesslich verdaut wird, glaube ich die Unterscheidung nicht gerechtfertigt. Vor allem in den dünnhyphigen Zellen bildet der Pilz häufig grosse blasige plasma- und reservestoffreiche Vesikel. Der Pilz-

---

[1]) GREVILLIUS, l.c. S. 167.

[2]) τολύπη Knäuel, ϑαμνίςϰος = arbusculus.

[3]) SOLMS-LAUBACH, H., Der Aufbau des Stockes von *Psilotum triquetrum* und dessen Entwicklung aus der Brutknospe. Ann. Jard. Bot. Buitenzorg **4**, 139–194 (1884).

[4]) JANSE, l.c. S. 161.

[5]) BERNATZKY, J., Beiträge zur Kenntnis der endotrophen Mycorhizen. Természetrajzi Füzetek **22**, 88–110 (1899).

[6]) SHIBATA, K., Cytologische Studien über die endotrophen Mycorhizen. Jahrb. f. wiss. Bot. **37**, 641–684 (1902).

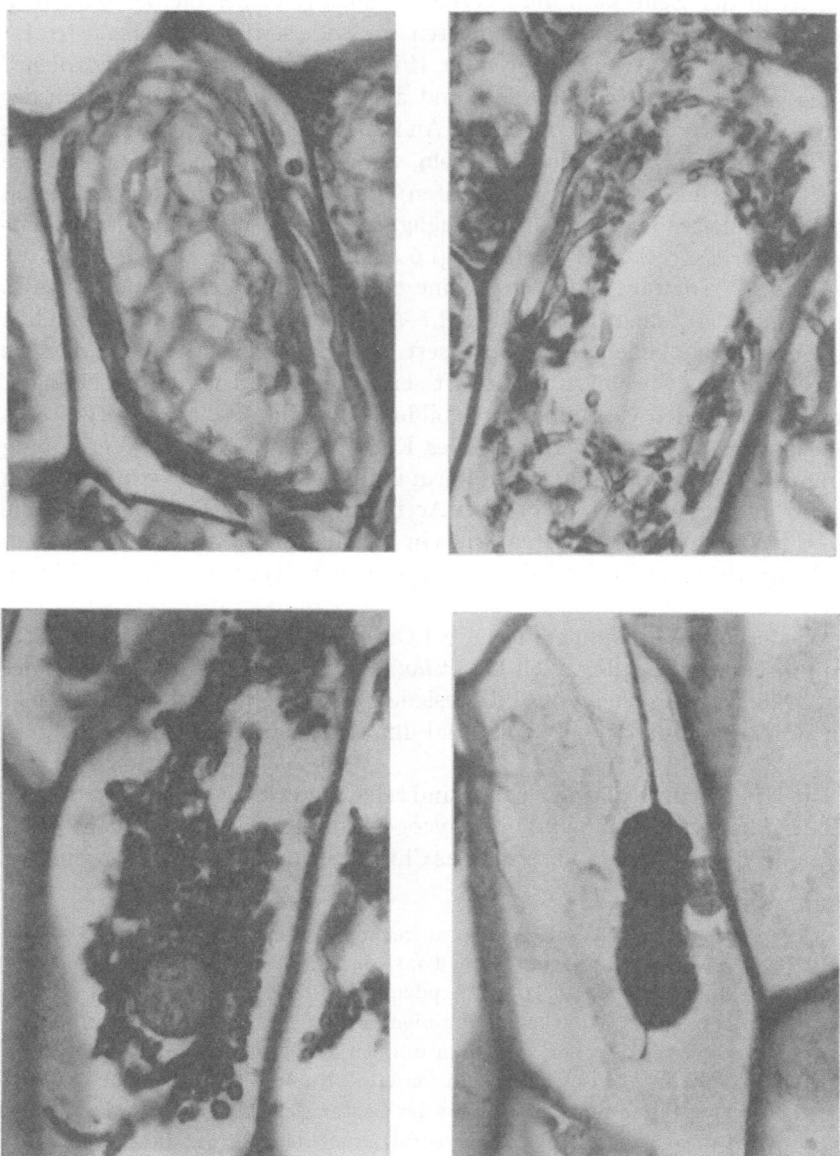

FIG. 6. — L. oben, *Psilotum triquetrum*. Rhizomrindenzelle mit groben, perlschnur-
förmig angeschwollenen Hyphen, die dünne verzweigte Äste in d. Innere der Zelle
senden. Tannin-Eisenfärbung (393 : 1). — R. oben, *Psilotum triquetrum*. Arbuskelbildung an den Hyphen. Tannin-Eisen (393 : 1). — L. unten, *Psilotum triquetrum*.
Hyphen und eins der Vesikel im Beginn der Verdauung. Sehr stark färbbar. Tannin-
Eisen (393 : 1). — R. unten, *Psilotum triquetrum*. Excretkörper oder „Klumpen"
an Hyphenresten hängend, daneben der Zellkern. Tannin-Eisen.

knäuel in der Zelle kann nun verhältnissmässig locker bleiben oder auch sehr dicht werden, in jedem Fall treten am Ende seitliche Hyphenäste auf, welche zu kurzen Arbuskeln werden. (Fig. 6 rechts oben und 4 rechts oben).

Sie sind SHIBATA entgangen [1]) und auch recht schwer und nur mit der stärksten Optik gut zu beobachten. An ihnen beginnt die jetzt einsetzende Verdauung, die zuerst die Arbuskeln, dann die Hyphen und zuletzt die Vesikel einschmilzt. (Fig. 6 links unten). Es bleibt ein fester, an den Resten ehemals zuleitender Hyphen aufgehängter, gelblich gefärbter, stark lichtbrechender Excretkörper übrig. (Fig. 6 rechts unten).

Mit der Verdauung einher geht eine charakteristische Veränderung des Zustandes der Zellkerne. (vgl. Fig. 1 e–k). Schon in den pilzerfüllten Zellen haben sie sich bedeutend vergrössert. Vor der Verdauung werden sie amoeboid. Das Chromatin vermehrt seine Masse und wird zu scholligen, durch feine Fäden verbundenen Gebilden. Auch die Nucleolen vergrössern sich und treten an die Peripherie des Kerns. Nach beendigter Verdauung nimmt der Kern wieder normale Form und Grösse an. Die Veränderungen sind wohl der Ausdruck besonderer Activität gewisser Kernfunctionen, die mit der Verdauung des Pilzmaterials in Verbindung stehen, sie kommen in noch drastischerer Weise bei der tolypophagen Mycorhiza der Orchideen vor.

Wiederholte Verdauung wie sie bei Orchideen die Regel ist, beobachtet man in einzelnen Zellen auch bei *Psilotum*. Die Zelle wird dann nach der ersten Phagocytose neu vom Pilz besiedelt und eine zweite Verdauung legt einen neuen Mantel von Pilzmaterial um den schon vorhandenen Excretkörper [2]).

Die Hyphen im Psilotumrhizom sind reich an fettem Oel, das der Pflanze bei der Phagocytose anheimfällt. Glycogen konnte ich indessen nicht nachweisen. Die Pilzmembran besteht aus Chitin (SHIBATA), trotzdem enthalten

---

[1]) Diese Behauptung kann ich nicht aufrecht erhalten.

An *Psilotum*-Pflanzen, die auf einem Topf von *Hura crepitans* aufgegangen waren, fand ich eine sehr regelmässige Verpilzung der Rhizoms. Der Pilz ist in allen Merkmalen bis auf die Arbuskelbildung identisch, er hat die gleichen perlschnurförmigen peripheren Hyphen mit ihren dünnen Abzweigungen und die gleichen Vesikel. Er wird in ebensolcher Weise verdaut. Nur die Arbuskelbildung fehlt. Zwischen beiden Pilzen muss mindestens der Gattungscharakter gewahrt sein. Den Namen der tolypothamniscophagen Mycorhiza möchte ich trotz dieser Beobachtung für *Psilotum* beibehalten.

[2]) Bilder wie SHIBATA's Fig. 33 und 34 sind nicht anders zu deuten. Sie haben SHIBATA den Anlass gegeben, einen unmittelbaren localen Einfluss des Kerns auf die Verdauung und Klumpenbildung anzunehmen der m.A. nach nicht besteht. (SHIBATA ist hier durch eine entsprechende Beobachtung von MAGNUS an *Neottia* beeinflusst worden, vgl. BURGEFF 1936, S. 14) [3]).

[3]) BURGEFF, H., Samenkeimung der Orchideen und Entwicklung ihrer Keimpflanzen mit einem Anhang über praktische Orchideenanzucht. Jena (1936).

die verdauten Klumpen Amyloid und färben sich mit Jodjodkali grünlich, nach Behandlung mit Javellescher Lauge blau. SHIBATA konnte mit Schweizers Reagens das Amyloid herauslösen und sieht in ihm eine von der Pflanze ausgeschiedene Kittsubstanz. Ich halte es auch für möglich, dass ein bei der Verdauung veränderter Bestandteil der Pilzmembran vorliegt. Die microchemische Seite der Psilotum-Mycorhiza erinnert damit stark an die der Ophioglossaceen, der sie auch durch die neu aufgefundenen Arbuskeln nach der morphologischen Seite ähnelt.

Die Prothallien von *Psilotum* und *Tmesipteris* sind von LAWSON [1]), das von *Tmesipteris* auch von HOLLOWAY [2]) aufgefunden und beschrieben worden. Habituell erinnern sie stark an die von Ophioglossum pendulum. Als Holosaprophyten sind sie viel regelmässiger inficiert als die hemisaprophytischen Sporophyten. Die Form der Mycorhiza scheint in keinem Punkt von der für das Psilotumrhizom geschilderten abzuweichen. Die Infection geht auch hier durch die Haare. Alte Zellen der Prothallien sollen absterben. LAWSON macht den Pilz verantwortlich für die unbekannte Sporenkeimung.

Ich habe seit 1909 zahllose Versuche der Isolierung des *Psilotum*-Pilzes unternommen; in den letzten Jahren unter Verwendung der neuen Wuchsstoff- und Vitamintechnik. Seit 1935 gelingt es mir regelmässig den Pilz aus sterilen Rhizomausschnitten mit einzelnen Hyphen herauswachsen zu lassen. Sein Wachstum ist ganz ausserordentlich langsam. Schliesslich stellen die Hyphen das Wachstum ein. Einer Übertragung auf neuen Nährboden nach Trennung von dem zugehörigen Gewebestück des Psilotum-Rhizomes folgte stets der Tod der Hyphe. An der Identität der auswachsenden querwandlosen perlschurförmig angeschwollenen Hyphen mit dem *Psilotum*-Endophyten besteht kein Zweifel. Sehr ähnliche, wenn nicht identische Hyphen erhielt ich aus Marattia alata-Wurzeln. Es ist immerhin zu hoffen, dass die Kultur schliesslich gelingt und neue Wege zur Analyse und Synthese der Farnmycorhiza eröffnet.

Die Mycorhiza der Lycopodienprothallien bleibt noch zu besprechen. Man hat ihr seit TREUB's und BRUCHMANN's klassischen Untersuchungen stets das allergrösste Interesse entgegengebracht, zumal der Stoffaustausch bei diesen abweichenden Formen der Verpilzung in völliges Dunkel gehüllt war. Verdauungsvorgänge waren nicht beobachtet. TREUB rechnete trotzdem mit der Möglichkeit der Übertragung von Glycogen und Oel aus dem Pilzmycel in die Pflanze, die deren saprophytischen Wuchs gewährleisten konnten. BRUCHMANN lehnte diese Hypothese

---

[1]) LAWSON, A. A., The prothallus of *Tmesipteris tannensis*. Trans Roy. Soc. Edinburgh **51**, 775–794 (1916); The Gametophyte Generation in the *Psilotaceae* l.c. **52**, 93–113 (1921).

[2]) HOLLOWAY, J. E., The prothallus and young plant of *Tmesipteris*. Trans. New Zealand Inst. **50**, 1 (1901).

ab, sprach der Pflanze die gesamte (nur bei *L. Selago* von Pilz unterstützte) Aufnahmefähigkeit für organische Bodenstoffe zu und sah die Pilzfunction lediglich in einer das Material umformenden. Die von dem Prothallium aufgenommenen „Humus"-Stoffe sollten gewissermassen durch die tiefen peripheren Pilzschichten filtriert und verändert und in eine für die Pflanze assimilierbare Form gebracht werden.

Schon 1909 hatte ich mit Prothallien, die ich der Freundlichkeit Herrn BRUCHMANN's verdankte, Pilzisolierungsversuche angestellt, die erfolglos blieben. Ich habe sie nicht wiederholt, da es mir an Material fehlte. BRUCHMANN's Kunst des Findens dieser verborgenen Schätze ist mit ihm ins Grab gesunken. Erst in Java auf dem Gipfel des Pangrango (3000 m) gelang mir ein Fund von zwei Prothallien des *Lycopodium clavatum–divaricatum* mit jungen Embryonen. Diese und eine Reihe mir von Herrn Collegen P. CLAUSSEN (Marburg) liebenswürdiger Weise zur Verfügung gestellten Microtomschnitte von *Lyc. clavatum, complanatum* und eine Schnittserie von *L. Selago* ermöglichten mir eine eigene Untersuchung, an die ich mit grösster Spannung heranging. Was sie ergab, und wieweit es mir gelang das Austauschproblem zu klären, darüber mögen die beiden folgenden Kapitel berichten:

### § 4. Die ptyophage Mycorhiza der Lycopodien des Phlegmaria-Typus.

— Wie schon früher gesagt erfolgt der Materialaustausch zwischen Pilz und Pflanze bei der ptyophagen Mycorhiza unter der Erscheinung der Plasmoptyse. Einige der von JANSE beschriebenen Mycorhizatypen gehören hierher. Kräftige junge Hyphen dringen in besondere Verdauungszellen ein. Die Hyphenspitze platzt und die entstehenden Plasmaergüsse werden von der Pflanze verdaut, wobei Reste als Excretkörper zurückbleiben [1]). JANSE nennt die Ptyosomen gleich den aus eingeschmolzenen Arbuskeln der tamniscophagen Mycorhiza entstandenen Excretkörpern „Sporangiolen". PETRI [2]) findet die richtige Erklärung; er bezeichnet die durch Hyphenbruch in Freiheit gesetzten resp. in die Zelle ergossenen Plasmamassen als „prosporoidi".

Ich selbst [3]) habe an tropischen Orchideensaprophyten der Gastrodiinae hochcomplicierte Formen solcher Sporangiolen untersucht und dann diesen nicht auf die phycomycoide Mycorhiza beschränkten Typus als *Ptyophage Mycorhiza* (πτύειν speien, φαγεῖν essen), die „Spor-

---

[1]) Excretkörperlose Übernahme des Pilzinhalts ist nur durch FRANKE), bei *Monotropa hypopitys* aufgefunden worden.

[2]) PETRI, L., Ricerche sul significata morfologico e fisiologico dei prosporoidi (sporangioli di Janse) nelle micorize endotrofiche. Nuov. Giorn. Bot. Ital. n. ser. **10**, 541 (1903).

[3]) BURGEFF, H., l.c. S. 166.

[4]) FRANKE, H. L., Beiträge zur Kenntnis der Mycorhiza von *Monotropa hypopitys* L. Flora N.F. **29**, 1–52 (1934).

angiolen" oder „prosporoidi" zur Vermeidung der wenig glücklichen an Fortpflanzungsorgane erinnernden Namen als *Ptyosomen* bezeichnet.

Bei den isosporen Lycopodiaceen sind mit zwei Ausnahmen (*L. inundatum* und *L. cernuum*) nur die Gametophyten [1]) inficiert. Halb-und ganzsaprophytische Typen kommen vor. Unter den letzteren fallen als besonderer Typus die sehr zarten gestreckten cylindrischen Prothallien des *Lycopodium Phlegmaria, Billardieri-gracile* und *L. Selago* aus dem Rahmen. Trotz vorzüglicher Beschreibung des Baus und der Mycorhiza dieses Typus durch die bedeutenden Autoren TREUB [2]), BRUCHMANN [3]) [4]) und HOLLOWAY [5]) hätte ich die grundlegende Entscheidung der Stoff-Übertragungsfrage nicht ohne eigene Untersuchung an *Lyc. Selago* entscheiden können. Ich stelle diese am besten untersuchte Art in den Vordergrund.

Die Sporenkeimung ist von BRUCHMANN (1910) beobachtet. Die Spore keimt selbständig und erzeugt ein 5-zelliges Stadium, das in diesem Zustand etwa 1 Jahr lang auf die Infection durch den Pilz warten kann. Inficiert wird die mit einer rudimentären Rhizoidzelle versehene Basalzelle (Fig. 2b) worauf sich das Prothallium rasch vergrössert. (2d) Bereits in den Nachbarzellen der Basalzelle entstehen die „Sporangiolen". In den Knäuel- oder Pilzwirtzellen verschwinden die Inhaltsstoffe, dafür treten in den Sporangiolen- oder Pilzverdauungszellen Stärke und Fetttropfen auf. Der Pilz wächst hinter dem Meristem her, ohne dies zu berühren und inficiert den basalen rübenförmigen Teil des Prothalliums.

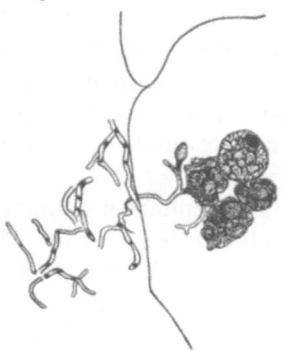

Fig. 7 — *Lycopodium Selago*, Pilzwirt- und Verdauungszelle. In letztere ist eine Hyphe eingedrungen und hat drei Ptyosomen erzeugt. Neben ihnen der Zellkern.

Wie ich feststellen konnte, sind die „Sporangiolen" echte Ptyosomen, die sich aber wegen der sehr geringen Grösse schwer erkennen lassen. An einer in die Zelle eingedrungen Hyphe platzt die Spitze und erzeugt einen Plas-

---

[1]) Nur bei *L. salakense* sollen sie pilzfrei sein.

[2]) TREUB, M., Etudes sur les Lycopodiacées. Ann. Jard. Bot. Buitenzorg **5**, 87–133 (1886).

[3]) BRUCHMANN, H., Über die Prothallien und die Keimpflanzen mehrerer europäischen Lycopodien. Gotha (1898).

[4]) ders., Die Keimung der Sporen und die Entwicklung der Prothallien von *Lycopodium clavatum* L., *L. annotinum* L. und *L. Selago* L. Flora N.F. **1**, 220–267 (1910).

[5]) HOLLOWAY, J. E., A comparative study of the anatomy of six New Zealand species of *Lycopodium*. Trans. New Zeal. Inst. **42**, 356 (1909); Studies in the New Zealand species of the genus *Lycopodium* part IV. The structure of the prothallus in five species, l.c. **52**, 193 (1920).

[6]) BURGEFF (1932), l.c. S. 166.

maerguss (Fig. 7) der grössere Vacuolen aufweist, die im Leben mit Fett, Glycogen oder Zellsaft angefüllt sein dürften. Basalwärts wachsen an der Hyphe Seitenhyphen aus, die dasselbe Schicksal erleiden. So erklärt sich die scheinbar traubenförmige Anordnung dieser Gebilde. An Microtomschnitten, die ich aus einem BRUCHMANNschen Originalpraeparat, das in Formalin eingelegt war, herstellte, gaben die Ptyosomen noch eine ungemein starke Färbung mit Glycogenjod, die beim Erwärmen schwand und beim Erkalten zurückkehrte. Das ist verwunderlich, da Formaldehyd Glycogen nicht fällt. Die Erklärung gab die Behandlung einer Schnittserie mit Javellescher Lauge. Der Inhalt der Ptyosomen war darnach restlos verschwunden, es blieb eine sich mit Chlorzinkjod bläuende Cellulosewand zurück, an der hie und da noch ein Stück der Traghyphe ebenfalls gebläut daransass. Die Plasmaergüsse werden also wie bei der ptyophagen Mycorhiza der Orchideen von der Zelle mit Cellulose umkleidet, ebenso wie die zuführenden Hyphen (vgl. BURGEFF 1932, 175/176). Die Hyphen der Pilzwirtzellen zeigen keine Cellulosereaction. Das Glycogen wird vermutlich durch die tüpfellosen Cellulosewände der Ptyosomen zurückgehalten.

Im terminalen dünneren dorsiventralen Teil des fructificierenden Prothalliums besiedelt der Pilz die den Geschlechtsorganen abgewandte Seite. Hier sind die subepidermalen Zellen meist als Pilzwirtzellen die inneren als Phagocyten ausgebildet.

Eine sehr auffällige Funktion haben die Rhizoidzellen. BRUCHMANN (1910) und HABERLANDT[1]) haben sie eingehend untersucht. Von dem Fussteil jeden Rhizoids wird durch eineWand der der Basis des Prothalliums zugekehrte Teil abgeschnitten. (Fig. 2e). Die entstandene „Haarfussfersenzelle" wird nach aussen mit einer starken Celluloseverdickung versehen, die unterhalb des Rhizoids ein halbkugeliges oder linsenförmiges vorspringendes Membranpolster trägt. BRUCHMANN nennt die Zelle „Animier"- oder „Provocations"- oder „Pilzexpeditionszelle", HABERLANDT einfacher: „Durchlasszelle". Die Function der plasmareichen Zelle besteht darin, den Pilz von innen anzulocken und zum Austritt in den Boden zu veranlassen. Bei dem Durchgang der Hyphen verschleimt der innere Teil des Membranpolsters. Nach dem Durchgang umwächst das Pilzmycel die Schleimschicht des Rhizoids, sodass sich die Bodensubstanzen lösenden Functionen von Pilz und Pflanze hier unterstützen. Das Rhizoid bleibt also zum Unterschied von denen der Lycopodienprothallien der anderen Typen vom Pilzdurchgang verschont, was auf selbständige Leitung von Material in Haar und Hyphen zu deuten scheint. Erst durch die Ptyophagie in den Zellen werden die vom Pilz gewonnenen Stoffe der Pflanze zugänglich.

Sehr ähnlich wie bei *Lycopodium Selago* liegen die Dinge bei *L. Phleg-*

---

[1]) HABERLANDT, G., Die Pilzdurchlasszellen des Prothalliums von *Lycopodium Selago*. Beitr. z. Allg. Bot. **1**, 293–300 (1918).

*maria*. Auch hier sind nach TREUB [1]) Fusszellen entwickelt, durch die der
Pilz nach aussen tritt und das Rhizoid mit lockerem Mycel umwächst. Die
Haarzelle steht aber völlig auf der mit Mycel erfüllten Fusszelle. Ptyopha-
gie am Mycel der Prothalliumzellen kann nach TREUBS Beschreibung und
Zeichnung nicht sicher festgestellt werden, doch deutet die starke amoeb-
oide Veränderung der Kerne in den Pilzzellen auf Verdauungsvorgänge.

Bezüglich des Stoffwechsels zwischen Pilz und Pflanze hat TREUB eine
Reihe wichtiger Beobachtungen gemacht. Die Leitfähigkeit des Pilzmycels
muss sehr gross sein, denn die Pilzknäule benachbarter Zellen sind nur
durch eine einzige Hyphe verbunden. Microchemische Reactionen, die
TREUB in gewandter Weise anwendet, zeigen den hohen Gehalt des gesamten Pilzmycels an Glycogen. Er sagt: „L'endophyte se trouve exactement
dans ces cellules du prothalle ou l'huile abonde. Le champignon se nourrit-
il en partie de cette huile, la retirant au prothalle et la transformant en
glycogène dans ses filaments? Ou bien l'huile du prothalle provient-elle en
partie du glycogène élaboré par le champignon et cédé à son hôte? Je ne
suis pas à même de décider ce point litigieux." — — Ich glaube dass das
Prothallium seine gesamten organischen Stoffe vom Pilz erhält, und falls
sich bei *L. Phlegmaria* die Ptyophagie bestätigt n e b e n  d e m  G l y-
c o g e n  a u c h  O e l.

Auch *Lycopodium Billardieri* v. *gracile* gehört nach HOLLOWAY's Beschreibung zum Phlegmaria-Typus. Localinfectionen durch die Haare
werden an den Zweigen beobachtet, ähnlich wie sie auch TREUB an den
gestielten Bulbillen bei L. phlegmaria festgestellt hat. Im Innern des Prothalliums findet HOLLOWAY zwei Sorten von Zellen, solche mit Knäueln
und solche, in denen die Knäuel verschwinden und "clusters of spores"
zurückbleiben, die wohl mit Ptyosomen identisch sind. Einige Besonderheiten seien noch erwähnt, so das Fehlen von Oel und Stärke in den Zellen,
das Fehlen von Vesikeln und das Intercellularwerden des Pilzes in den über
der Basalzone gelegenen Geweben des Prothalliums, dazu die Umscheidung
der in die Rhizoiden austretenden Hyphen. Es ergiebt sich kein einheitliches
Bild. Nachuntersuchung erscheint bei dieser Art ebenso am Platze, wie
bei *L. Phlegmaria*.

**§ 5. Die chylophage [2]) Mycorhiza der Lycopodien.** — *a*) T y p u s
d e s  L y c o p o d i u m  c o m p l a n a t u m. —BRUCHMANN [3]) hat das
Prothallium von *Lycopodium complanatum* bereits seines eigenartigen Baues
wegen als Sondertypus charakterisiert. Das rübenförmige Gebilde (Fig. 8
u. r. und 9 r.) trägt auf dem Scheitel die Geschlechtsorgane. Das Gewebe des

---

[1]) TREUB, l.c. S. 179.
[2]) χυλός = Saft.
[3]) BRUCHMANN, l.c. S. 179.

Scheitels ist von dem vegetativen rübenförmigen Teil durch ein intercalares Meristem (M) getrennt, das die vegetativen Teile nach unten, die reproduktiven nach oben erneuert. Hier interessieren die ersteren, die sich aus einem centralen Gewebe und einer Mantelschicht zusammensetzen. Die Mantelschicht besteht unter der mit Rhizoiden besetzten Epidermis aus einer bis zu 8 Zellen starken Rindenschicht (R), die nach innen an eine einfache Schicht sehr lang werdender radial gestreckter Pallisadenzellen (P) anstösst. Das Prothallium wird von der Basis aus inficiert. Der Pilz wächst hinter dem Meristem her. Austritt von Pilzhyphen erfolgt durch die Haare in acropetaler Reihenfolge (A). Die Rindenzellen enthalten regelmässig gebaute Mycelknäuel, die aus nahezu unverzweigten aufgewickelten Hyphen bestehen, und nach Art von Taustapeln in den einzelnen Zellen liegen. Die Knäuel sind — was schon TREUB an anderen Arten beobachtete — nur durch einzelne Hyphen verbunden. Gelegentlich finden sich Vesikel in den Zellen. Von diesem Mycel der Aussenrinde, das sehr nahe an das Meristem heranreicht, wachsen Hyphenstränge aus und dringen z w i s c h e n die noch sehr jugendlichen Pallisadenzellen ein. Ihr Wachstum findet aber mit der Abgrenzung der Pallisaden an das weitlumige Centralgewebe sein Ende.

Die Zellen des Pallisadengewebes führen reichlich grosse um den Zellkern gelagerte stärkegefüllte Amyloplasten, deren Stärkeinhalt während der Infection der Intercellularen etwas abzunehmen scheint, sich aber hinter dieser Zone wieder herstellt und erst in der Basis älterer Prothallien abnimmt. Auch die Rindenzellen enthalten Stärke, die nach der hier intracellularen Pilzinfection nur teilweise verschwindet (im Gegensatz zu BRUCHMANNs Beobachtungen). Auch das Centralgewebe ist stärkereich. Fettartige Reservestoffe waren an den Microtomschnitten Herrn CLAUSSENS natürlich nicht mehr nachzuweisen.

Das merkwürdigste am ganzen Bild der Mycorhiza ist nun folgendes. Sämtliche Hyphen mit Ausschluss der in den basalen Prothalliumteilen scheinen in gesundem Zustand zu sein. V o n  e i n e r  P i l z v e r - d a u u n g  z e i g t  s i c h  k e i n e  S p u r. Nach BRUCHMANN soll der Pilz die von den Rhizoiden des Prothalliums aufgenommenen Humusstoffe umwandeln und der Pflanze vermitteln, die ja ringsum von Pilzgewebemantel umschlossen ist.

Hier versagt jede Erklärung nach den alten Vorstellungen. Es besteht auch keine Analogie mit irgend einer anderen Mycorhizaform, ausser der der Baummycorhiza; diese vermittelt ohne Zweifel dem Baum allerlei Stoffe, worunter Salze, Kalium Phosphorsäure und besonders Stickstoff eine Rolle spielen. Saprophytismus, d.h. Kohlenhydraterwerb wird aber in keinem Fall bei der Mycorhiza der Holzgewächse beobachtet.

Der BRUCHMANNschen Hypothese der selbständigen Aufnahme der organischen Stoffe aus dem Boden fehlt nach neueren Vorstellungen die Basis. Solche Stoffe sind im Boden bei der Concurrenz unter den Microorganismen nicht verfügbar, sie können nur von diesen selber, d.h. vom Pilz

geliefert werden (vgl. HOLLÄNDER in BURGEFF 1936, S. 261 und BUR-
GEFF l.c. S. 176. Hier bleibt nur die Möglichkeit, dass
der Pilz aus seinen Verbindungen mit dem Bodenmycel lösliche Stoffe direkt in die Zelle hinein abscheidet oder guttiert.

1932 gelang mir zuerst der Nachweis der Guttation an Pilzhyphen in der
Pflanzenzelle. Hyphen in den Rindenzellen der saprophytischen Orchidee
*Gastrodia callosa* scheiden Stoffe in die Zelle ab. Die Zelle reagiert mit einer
Ablagerung von Wandstoffen auf die guttierende Stelle. Der Ausfluss des
Guttationssekrets verhindert das Festwerden der Wandstoffe in der Achse
der Hydathode; es entsteht eine röhrenförmige Scheide oder ein „Röhren-
tüpfel'' der hier nicht einer Zellwand sondern (Fig. 2*k*) einer Pilzhyphe
aufgesetzt ist. (Fig. 2*m*). Diese Röhrentüpfel färben sich mit Haematoxylin
ungemein stark und enthalten ausser Cellulose korkähnliche Auf- oder
Einlagerungen.

Genau entsprechende Gebilde lassen sich nun an den
Hyphen der Aussenrindenzellen und auch — wenn auch weniger typisch —
an dem interzellularen Mycel der Pallisaden-
schicht des *L. complanatum* nachweisen. Sie
fallen sogar ganz ausserordentlich stark auf und
geben — nur bei *L. complanatum* vorkommend
— dieser Mycorrhiza die besondere Note (Fig. 4
rechts unten). Freilich sind die Bildungen sehr
klein und entsprechen den geringen Hyphen-
durchmesser von etwa 1,2 µ. Betrachtet man
die kegelförmigen Gebilde von oben, so sieht man
eine centrale Aufhellung, die wohl dem nicht fest
gewordenen Porus des Tüpfels entspricht. Chlorzinkjod ergibt nach vorheriger Behandlung mit
Javellescher Lauge eine wenn auch schwache,
doch deutliche Bläuung der kegelförmigen oder
knopfförmigen Gebilde. Im Pallisadengewebe
sind die Wandstoffauflagen weniger specifisch
ausgebildet, mehr flächen- oder rippenförmig.
Das intracellulare Mycel entwickelt sich vom
Meristem aus mit gesunden gut differenzierten,
besonders in den Vesikeln sichtbaren Zellkernen.
Gleich hinter der ersten aus nur wenigen Zellschichten bestehenden Zone beginnen die hae-
matoxylingeschwärzten Auflagerungen. Die Vesikel bleiben hier ohne weitere Entwicklung,
während sie im abgestorbenen Basalgewebe alter

Fig. 8 — *Lycopodium*-Pro-
thallien: Links *Lyc. clavatum* var. *divaricatum* WALL.,
R. oben, *Lyc. clavatum*, R.
unten *Lyc. complanatum*
(die letzten nach BRUCH-
MANN).

erwachsener Prothallien zu sehr grossen dickwandigen Gebilden heranwachsen können. Augenscheinlich sind in dem Pallisadengewebe die

intercellularen Pilzhyphen stark geschädigt, was mit der starken Stoffab-
gabe der Hyphen zusammenhängen mag, die den Pallisaden die Speiche-
erungsfunction ermöglicht. Das Pallisadengewebe wäre also das eigentliche Absorptionsgewebe der Pflanze.

Über die Art der vom Pilz abgegebenen löslichen Stoffe lässt sich nach
den fertigen Schnitten nichts aussagen. Eine Entfärbung mit nachheriger
Behandlung mit dem Jodreagenz auf Glycogen (A. MEYER) ergab keine
charakteristische Färbung (wie sie etwa beim Prothallium von *L. clavatum*
deutlich auftritt). Genauere Untersuchung ist nur an lebendem Material
möglich.

Es scheint mir soviel festzustehen, dass *Lycopodium complanatum* einen
völlig abweichenden Mycorhizatypus darstellt, bei dem die in den peripheren Geweben gelagerten lebenden Pilzhyphen lösliche Stoffe in die Pflanzenzellen oder in Intercellularräume guttieren. Nachdem ein eigentlicher
Verdauungsvorgang nicht nachweisbar ist, schlage ich hierfür die Bezeichnung einer C h y l o p h a g e n  M y c o r h i z a vor.

Zum Typus des *Lycopodium complanatum* gehört dem Bau nach ein von
LANG (1894) beschriebenes neben *Psilotum* aufgefundenes *Lycopodium*-
Prothallium unbestimmter Art.

*b*) T y p u s  d e s  L y c o p o d i u m  c l a v a t u m  u n d  L.  a n n o-
t i n u m. — *Lycopodium clavatum* ist in seinen Gametophyten eingehend
beschrieben von BRUCHMANN (1898) und LANG [1]. Die Entwicklung und
Sporenkeimung beobachtete BRUCHMANN 1910 [2]. Die Infection durch den
Pilz erfolgt im neunzelligen Stadium des Keimlings. (Fig. 2c) Wieder ist es
die über der Rhizoidzelle gelegene Basalzelle in welche die erste Hyphe
eindringt und in der sie zu einem Knäuel heranwächst. Von hier aus werden
die anschliessenden peripheren Zellen besiedelt. An älteren Prothallien
werden dann die Epidermiszellen nicht mehr vom Pilz berührt. Die Form
des Prothalliums ist zuerst die eines rübenförmigen Körpers, später die
eines abgeflachten Kreisels, dessen starkes Randwachstum zuletzt zu vielen
Einfaltungen des Randes führt, der die auf der breiten mittleren Fläche im
reproductiven Gewebe liegenden Geschlechtsorgane umgiebt. Oberhalb des
Randes zwischen vegetativem und reproductivem Gewebe liegt wieder das
intercalare Meristem. Von diesem aus entwicklen sich eine bis 9 Zellen starke
Rinde, dann eine einzellige aus radial gestreckten Zellen bestehende Pallisadenschicht und eine nach innen anschliessende aus isodiametrischen Zellen
gebildete Speicherschicht.

Die Rindenzellen sind mit Mycelknäueln angefüllt, die auch hier wieder
in schiffstau-ähnlichen Rollen beisammenliegen und etwa 25 bis 30 Windungen einer einzelnen Hyphe enthalten; ähnliche Knäuel finden sich hier
auch in der Pallisadenschicht. Aus ihr führen aber je mehrere Hyphen in

[1]) LANG, W. H., The prothallus of *Lycopodium clavatum* L. Ann. of Bot. **13**,
279–317 (1899).

[2]) BRUCHMANN, l.c. S. 179,4.

den jeweils über der Pallisadenzelle gelegenen Zwickel der Speicherge-
webezellen und werden hier intercellular. Sie bilden glatte Hyphenzöpfe,
die die Zellen jedoch nur in den Interzellularen selbst berühren und die be-
rührenden Wände nicht auseinanderdrängen. Nach aussen hat das Mycel
Verbindung mit dem Bodenmycel durch die Rhizoiden mittleren Alters.

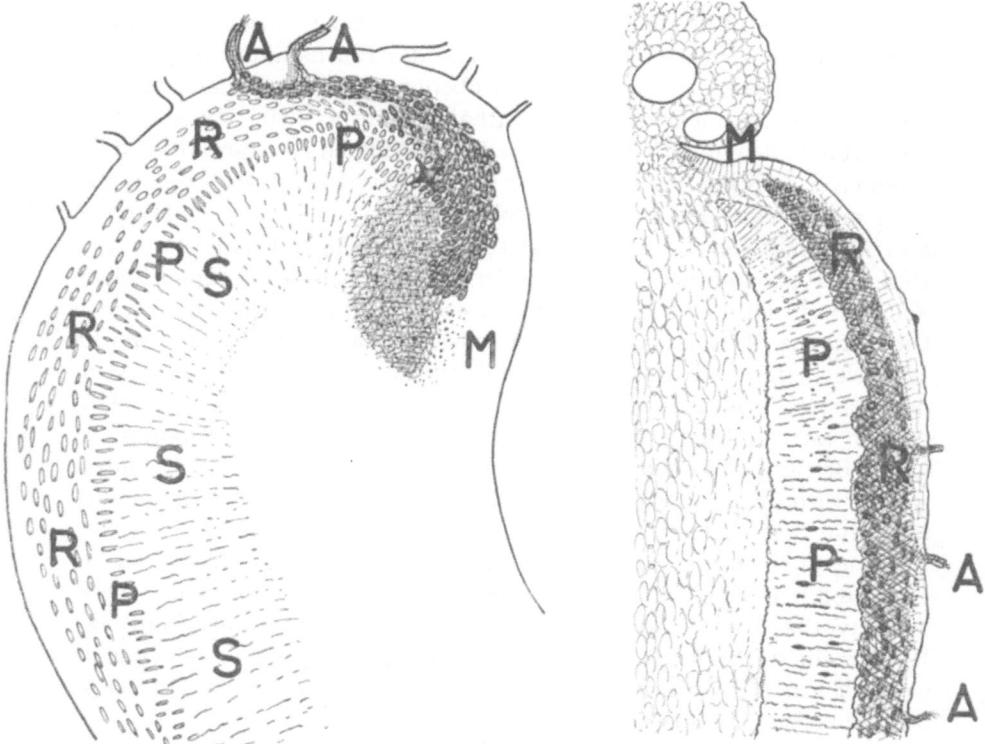

FIG. 9. — Links, *Lycopodium clavatum* v. *divaricatum*. Schematische Zeichnung der Verteilung
des Pilzmycels in einem radialen Längschnitt der Hälfte einer Prothalliumfalte. Nur Mycel
und Stärkeinhalt der Zellen sind angedeutet. Der „Glycogenbogen" ist dunkler gezeichnet.
Die Hyphen ausserhalb dieses Bogens sind abgestorben. R. Rindenknäuel, P Pallissadenknäuel,
S Interzellulares Mycel der Speicherschicht. M Meristemzone mit Stärke. A Auswandernde
Hyphen. Rechts, *Lycopodium complanatum*. Schematische Zeichnung der Rindenschichten.
R Rindenschichten, P Pallisadenschicht, A Auswandernde Hyphen.

Ich konnte an meinem Material von *L. clavatum* v. *divaricatum* von Java
nicht sicher feststellen ob die Bodenverbindungen des Mycels Infections-
oder Emissionshyphen sind. Für die Auffassung der Function der Mycorhiza ist das gleich, da Pilzhyphen in beiden Richtungen als Leitorgane wirken können. Fig. 9 (links) gibt eine schematische Skizze des Pilzkörpers
auf der einen linken Hälfte eines Schnitts durch eine nach der Seite gerichteten Membranfalte eines grossen Prothalliums, das einen bereits grünen Sporophyten trug, aber noch in vollem Wuchs war. Die Skizze ist nach einem

dünnen Handschnitt, der in A. MEYER's Glycogenreagenz lag, hergestellt. Reservestoffe finden sich nur um das Meristem (M) herum und im anliegenden Speichergewebe (S) in Form von Stärke. Glycogen, kenntlich an der intensiv rötlichbraunen Färbung ist in sämtlichen intra-wie intercellularen Mycelien des oberen dunkeler gezeichneten Bogens enthalten, dessen Gewebe sich ausfaltend später zu den peripheren Teilen des Prothalliums werden. Aus dem ganzen Bild der Mycorhiza ebenso aus der Beobachtung zahlreicher nach allen Richtungen durch Stücke des Prothalliums hergestellter Microtomschnitte folgt, dass nur jener obere mit Jod Glycogenfärbung gebende Mycelbogen lebendes Mycel enthält. Die in nächster Meristemnähe inficierten Zellen und die Hyphen der Interzellularen des stärkereichen jugendlichen Teils der Speicherschicht stehen über den „Bogen" in Verbindung mit den aus den beiden Rhizoiden (bei A) austretenden Hyphen, also mit dem Bodenmycel. Da die glatten dünnhyphigen Knäuel kaum verzweigte Hyphen aufweisen und die Verbindungen zwischen den einzelnen Zellen meist durch nur eine Hyphe aufrecht erhalten werden ergeben sich beträchtliche Mycellängen, die sich aus der Zahl der Windungen des Knäuels in ihrer Grössenordnung schätzen lassen. Nimmt man die Windungslänge einer mittleren Zelle mit etwa 120 μ an und rechnet 25 Windungen im Durchschnitt, so erhält man für die Länge eines Knäuels den hohen Betrag von 3 mm für den ganzen Verlauf der Hyphen mit 17 hintereinandergeordneten Knäueln 51 mm Hyphenlänge für die Verbindung des Pilzes mit dem Boden.

Die etwa einen mm lange Wegstrecke wird also für den Pilz durch die Knäuelbildung zu einer um das zweiundfünfzigfache verlängerten. Wenn durch vorhandene Verzweigungen der aus dem Boden eingedrungenen Hyphen die wahrscheinlich vorkommen, dieser Wert auch um einen ganz wesentlichen Teil verringert werden mag, so bleiben doch recht lange Hyphenschläuche, die sich in den meristemnahen Geweben ausbreiten u n d  d a n n  d u r c h  i h r  d i r e k t  h i n t e r  d e m  M e r i s t e m  e i n - s e t z e n d e s  A b s t e r b e n  i s o l i e r t  i n  d e n  auf Fig. 9 L. mit einem X bezeichneten  G e w e b e n  d e r  P f l a n z e  z u r ü c k b l e i - b e n.  Man sieht zuerst in den Zellen der Pallisadenschicht dass die Hyphen ihr pralles Aussehen verlieren, also in Zellen, die noch nicht ihre normale Grösse gewonnen haben. Durch das Absterben werden die intercellularen Hyphen isoliert und verlieren gleichfalls ihr Aussehen und ihre Reaction auf Glycogen. Auch die Knäuel im Rindenge- webe sterben ab, so dass sich der lebende Körper des Pilzes auf das Areal des „Glycogenbogens" beschränkt.

Die Ursachen des Absterbens lassen sich nicht deuten. Es könnte sich um eine chemische Einwirkung der Pflanzenzellen auf das Mycel handeln. Bei dem Fehlen aller Verdauungserscheinungen bei anderen Typen der Lycopodiaceenmycorhiza ist es aber vielleicht wahrscheinlicher, dass die Pilzverbindungen zwischen den Pilzknäueln der einzelnen Zellen teilweise durch das Wachstum der Zellen zerrissen werden. Die hier zugrunde liegende Auffassung nimmt an, dass der Pilz von der Pflanze nur mit einem Teil seines Bedarfs an Kohlenhydraten oder anderen Stoffen ernährt wird, in der Hauptsache sein Material aus dem Boden mitbringt, angelockt von irgend welchen Reizstoffen. Mit dem Abreissen der knäuelverbindenden Hyphen verfällt das gesamte Material der Pflanze.

In den älteren Teilen der Speicherschicht liegen die Hyphenzöpfe in den Intercellularen in einer schleimartigen mit Safranin färbbaren Masse eingebettet. Die Wände der anschliessenden Zellen beginnen sich basalwärts immer mehr mit Haematoxylin zu schwärzen.

Ob es sich um einen allmähligen Korkabschluss der Intercellularen handelt ist fraglich. Wichtig erscheint dagegen das rasche Verschwinden der Glycogenreaction hinter der Zone des Absterbens der Hyphen. Die Schwierigkeit der Erklärung des Glycogendurchgangs durch die Pflanzenmembran, auf die TREUB hinweist, scheint zu bestehen. Enzyme können jedoch dabei eine Rolle spielen.

Röhrentüpfel und Emergenzen auf den sehr dünnen Pilzhyphen lassen sich bei *Lycopodium clavatum* nicht nachweisen. Natürlich kann eine direckte Ernährung der Pflanze durch Pilzguttation durch deren Fehlen nicht ausgeschlossen werden.

Das Gesamtbild, das sich bei *L. clavatum* ergibt, ist jedenfalls ein völlig anderes als das von *L. complanatum*. Active Mycorhiza, die man hier vielleicht eine „necrophage" nennen könnte wenn man nicht den Ausdruck der chylophagen beibehalten will, besteht nur unmittelbar hinter dem Meristem, a l s o a m R a n d e d e s P r o t h a l l i u m s. Da jede fördernde Verpilzung eine gewisse Gewebemasse voraussetzt, die ihr unterliegt, wird hier das ungemein starke Randwachstum der Prothallien des Clavatum-Typus verständlich (fig. 8). Die Massenentwicklung des Randes ist Voraussetzung für eine genügende Masse verpilzten Gewebes.

*Lycopodium complanatum* mit seiner schlanken Rübenform (Fig. 8 rechts) bedarf mit seinem lebend guttierenden Mycel dieser Masse nicht; noch weniger das ptyophage *L. Selago*, das sich mit der bei vielen Saprophyten vorkommenden Cylinderform begnügt.

Eine endgültige Klärung der Dinge können meine Beobachtungen in keiner Weise bedeuten. Hierzu braucht es lebendes Material, das erlaubt, den Hyphenzustand genauer festzustellen, den Fettstoffwechsel aufzuklären nach glycogenspaltenden Enzymen zu suchen und manche andere Specialuntersuchung vorzunehmen, unter welchen die Isolierung und Beobachtung des Pilzes natürlich die wichtigste wäre. Die Kenntnis seiner Art des Wachstums, seiner Fähigkeiten der Stoffleitung wären Vorbedingungen für die Ausarbeitung oben angeführter Arbeitshypothesen bei allen Lycopodienprothallien.

*Lycopodium clavatum* fand ich von seiner tropischen Varietät *divaricatum* nicht verschieden; nur ist die Faltenbildung des Prothalliumrandes bei ihm noch stärker.

*c*) T y p u s   d e s   L y c o p o d i u m   i n u n d a t u m   u n d   c e r-
n u u m. — Die Keimung der Sporen bei *Lycopodium inundatum* hat schon
DE BARY [1]) beobachtet, später hat GOEBEL [2]) die Entwicklung der jungen
Keimpflanzen geschildert. Die ersten Stadien waren grün und frei von der
Verpilzung. Erwachsene Prothallien die seit FRANKENHAUSER (1858) be-
kannt geworden waren, hat GOEBEL entwicklungsgeschichtlich untersucht.
Sie bestehen aus einem basalen rübenförmigen oder knolligen Teil, der mit
Rhizoiden am Boden gefestigt ist. Eine Mycorhiza tritt regelmässig auf.
Epidermis und Subepidermis enthalten Pilzknäuel, die durch einzelne
Hyphen verbunden sind. Von ihnen aus werden die Intercellularen des basalen und centralen Gewebes vom Pilz besiedelt, der die Zellen stark aus-
einanderdrängt und local verschieben kann. Die Zellen dieses Speichergewebes führen fettes Oel, aber keine Stärke. Sporophyten, Keimpflanzen
wie erwachsene, können in besonderem intercellularenreichen „Polsterge-
webe" (BRUCHMANN und GOEBEL) intercellulare Verpilzung zeigen,doch
bleibt diese auf besondere Teile der Stämmchen beschränkt.

Ganz analog verhält sich das von TREUB [3]) untersuchte *L. cernuum* aus
Java; kaum verschieden sind *Lycopodium laterale* und *ramulosum* sowie
*cernuum* aus Neuseeland (HOLLOWAY [4])). Verpilzt ist bei diesen Formen die
Epidermis und zwar intracellular, die darunterliegenden Schichten der
basalen, bei *cernuum* rhizoidarmen Knolle führen intercellulare Verpilzung,
bei welcher die Hyphen die Zellen „runden". Bei *L. ramulosum* treten im
Basalgewebe auch intracellulare starkwandige Vesikel auf.

Von einer beobachteten Degeneration der Mycelien in oder zwischen den
Zellen ist nirgends bei den Autoren die Rede. GOEBEL äussert sich sehr vor-
sichtig über die Function des Pilzes. TREUB hält ihn für einen Commen-
sualen und führt die Rhizoidarmut auf die Infection der Epidermis zurück.
HOLLOWAY hält den Pilz für einen Symbionten. Er beobachtet locale Infec-
tionen und bildet das Austreten von umscheideten Hyphen aus den Rhi-
zoidbasen ab.

Nach Analogie mit den vorhergeschilderten Typen, nach dem Fehlen von
Excretkörpern aller Art, glaube ich nicht fehlzugehen wenn ich den inun-
datum-cernuum-Typus bei der chylophagen Mycorhiza einreihe. Das Fehlen der Pallisadenschicht ist das einzige charakteristische Merkmal im anatomischen Bau, das diesen Typus etwa vom clavatum-Typus unterscheidet.
Die Mycorhiza scheint vereinfacht, sie gewährt auch nur eine Beihilfe zur

---

[1]) BARY, A. DE, Sur la germination des Lycopodées, Ann. Sc. Nat. Bot. **4**, 30
(1858).

[2]) GOEBEL, K., Über die Prothallien und Keimpflanzen von *Lycopodium inundatum*. Bot. Zeitung **45**, 161–168, 177–188 (1887).

[3]) TREUB, M., Etudes sur les Lycopodiacées, Ann. Jard. Bot., Buitenzorg **4**,
107–128 (1884).

[4]) HOLLOWAY, l.c. S. 179.

Entwicklung der ja grünen und höchstens hemisaprophytischen Prothallien.

Auch der junge Sporophyt von *L. cernuum* ist nach TREUB (1890) intercellular verpilzt und zwar soll derselbe Pilz hier in etwas abweichender Form die Pflanze besiedeln. Das Central-Gewebe der embryonalen Knolle besteht aus unregelmässig sternförmigen Zellen, die weite Intercellularen besitzen. Hier breitet sich der Pilz aus, wobei die von ihm umsponnenen Zellen zwar keine Reservestoffe speichern aber lange am Leben bleiben. Gelegentlich kommt es in den Zellen zu einer Bildung dickwandiger „Sporen", die TREUB abbildet. Die peripheren Gewebe der Knöllchen führen Stärke. Auch an der erwachsenen Pflanze kommen Knöllchen vor, die auf eine entsprechende Art verpilzt sind (JANSE 1897). Der Pilz dringt hier schon bei der Infection durch die Intercellularen der Epidermis in das Innere, wo er sich stark verzweigt und alle Intercellularen anfüllt. Material dieser Knöllchen brachte ich aus Java mit und konnte es genauer untersuchen.

Die kräftigen und stark färbbaren Infectionshyphen finden sich schon nahe hinter dem Meristem; sie verzweigen sich in dünnere Äste, ihr Durchmesser sinkt von 8 auf 6 auf 4 μ herab. Auch ihre Färbbarkeit nimmt ab. Nach und nach füllen sie in dichten Massen die weiten Intercellularen des Sternparenchyms des Knöllchens und gewinnen wieder stärkere Färbbarkeit. Etwas weiter basalwärts sieht man zuerst die feineren, dann die gröberen Hyphen collabieren. S c h l i e s s l i c h  i s t  d i e  g a n z e  P i l z m a s s e  a b g e s t o r b e n  und füllt — jetzt mit stärkster Färbbarkeit — die Intercellularen in der ganzen Knöllchenbasis an. Die Zellen des Knöllchens bleiben am Leben, haben gesunde, Zellkerne und inhaltslose Plasten. Stärke tritt nicht auf. Ob das Pilzmaterial verdaut wird, oder der Autolyse verfällt, lässt sich nicht entscheiden. Inhaltsstoffe der Hyphen sind nicht nachzuweisen,doch enthalten die dicken Hyphenwände der collabierten Masse Stoffe, die sich intensiv mit allen Jodlösungen färben. Fügt man Wasser hinzu, so geht ihre braune Farbe in eine rötlich-violette über, die Wände der Knöllchenzellen färben sich blau, sie enthalten Amyloid. Glycogen lässt sich in den Pilzhyphen nicht nachweisen. Der in der Hyphenmasse färbbare Stoff zeigt keinerlei Eiweissreactionen, ist beständig gegen Schwefelsäure, Kalilauge, $H_2O_2$, wird aber von Javellescher Lauge sofort zerstört. Nach Behandlung mit dieser färben sich die Hyphenwände mit Jod gelb. Die Stoffübertragungsfrage kann also an dem Alkoholmaterial nicht geklärt werden. Es besteht aber kaum ein Zweifel, dass man den Fall an die Seite des *Lycopodium-clavatum* Falles stellen kann. Der Verfall der intercellularen Hyphen ist hier noch viel demonstrativer, es handelt sich wahrscheinlich um eine necro-chylophage Mycorhiza, über deren Beziehung zur Verpilzung der Prothallien keine Aussagen gemacht werden können.

Die kräftigen Pilzhyphen, die hier auftreten, sind völlig verschieden von denen der Lycopodien-Prothallien, deren Durchmesser den Wert von 1–1.5 μ kaum überschreitet. Abweichend sind sie auch durch ihre Wandstärke und Verzweigung.

**§ 6. Rückblick.** — Versucht man sich nach unseren heutigen Kenntnissen von den Beziehungen der Farne zu den Pilzen ein Bild zu machen und das am meisten wesentliche herauszustellen, so muss vor allem folgendes in die Augen fallen.

Farne sind Pflanzen uralten Stammes, die vor Millionen von Jahren in viel bedeutenderer Zahl und in viel reicherer Formentwicklung die Erde besiedelt haben. So müssen auch ihre Beziehungen zu Pilzen seit langem bestehen, es müssen auch diese Pilze zu altertümlichem Typus gehören.

So sind die Endophyten der Farne querwandlose oder -arme nicht fusionierende Mycelien, polyenergid wie viele Algen; ihre besonderen Organe lediglich Absorptionshyphenbüschel (Arbuskeln) und als Reproductionsorgane blasige Auftreibungen, die als ephemere Reservestoffspeicher dienen oder zu „Ruhesporen" mit mehrfacher stark verdickter Wand werden können. Geschlechtsorgane sind nie beobachtet, doch kommen in den Vesikeln gelegentlich anscheinend einkernige sporenähnliche Gebilde vor, die ihnen den Charakter von Sporangien geben können.

Die Leistung der Pilze in der phycomycoiden Mycorhiza ist eine geringe. Das Pilzwachstum ist ein sehr langsames. Die Prothallien der Ophioglossaceen und der Lycopodiaceen sind auch im warmen Klima sehr schlechte Wachser und producieren eine sehr geringe Vegetationsmasse.

Ein Vergleich mit Orchideen- und Monotropaceen-Sapropyten ist hier am Platz. Diese sind mit augenscheinlich erdgeschichtlich jüngeren und sehr leistungsfähigen Hymenomyceten, oder doch diesen nahestehenden Imperfectenmycelien (*Rhizoctonia*) associiert und producieren vor allem in den Tropen recht beachtliche Vegetationsmassen, die bei holzzerstörenden Pilzen der *Gastrodiinae* und *Vanillinae* ihre Symbionten zu ausgesprochen grossen Gewächsen, bei Galeola bis zur 40 m hohen Liane heranwachsen lassen. Andere Gruppen wie die Burmanniaceen, Triuridaceen, gewisse Gentianaceen und Polygalaceen bleiben freilich Kleingewächse — aber diese führen ebenfalls die phycomycoide Mycorhiza, wenn auch in einer anscheinen leistungsfähigeren Form. Der Vergleich der phycomycoiden mit der „hymenomycoiden" Mycorhiza zeigt somit die erstere als die weitaus primitivere Form.

Erstaunlich ist die grosse Mannigfaltigkeit in der Function der Mycorhiza, die ungemein verschiedene Ausbildung bei den *Ophioglossales*, den *Psilotales*, den *Lycopodiales*, welche letzere mit ihren drei oder vier Typen ganz verschiedener Form des Erwerbs von Pilzmaterial eine ganz unerwartet reiche functionelle Differenzierung zeigen.

Wer mag hier verantwortlich sein, der Pilz oder die Pflanze. Die Frage wäre vielleicht lösbar, wenn sich herausstellt, dass bei verschiedenen Mycorhizatypen ähnliche oder gleiche Pilzarten gefunden werden. Bedeutsam schein mir bereits, dass aus sterilen Wurzelausschnitten von *Marattia* und *Psilotum* genau entsprechende Hyphen in meinen Kulturen zum Auswachsen kamen. Man könnte versucht sein nach den zahlreichen Beobachtungen und Vermutungen der Autoren eine Entscheidung über Gleichartigkeit und Verschiedenartigkeit der Endophyten verschiedener Gruppen zu treffen. Bei der Unsicherheit aller dieser Vermutungen — sie sind neuster Zeit von VAN DER PIJL [1]) zusammengestellt — erscheint ein solcher Versuch aussichtlos. Ob Phycomyceten oder chitinbildende Oomyceten, ob Endogonaceen, Phythophtoraceen, Blastocladiaceen oder Saprolegniaceen vorliegen, wird erst die Isolierung und Reinkultur aufzeigen können.

---

[1]) PIJL, L. VAN DER, Die Mycorrhiza von *Burmannia* und *Epirrhizanthes* und die Fortpflanzung ihres Endophyten. Rec. des Trav. Bot. Néerlandais. 31, 741–779 (1934).

# CHAPTER VI

## ZOOCECIDIA

by

W. M. DOCTERS VAN LEEUWEN (Leersum)

The Pteridophytes possess the faculty of forming galls under the influence of mites and insects, just as well as the Anthophytes. Therefore one would expect relatively more galls on this ancient group of plants than on the Phanerogams. The contrary is true. Relatively few galls are known of the Pteridophytes, and these are mainly caused under the influence of animals belonging to the phylogenetically older groups: 1. by gall-mites, 2. by *Diptera*, and 3. by *Thysanoptera*. Of the latter it is the more primitive *Terebrantia* which cause galls on ferns. On the Gymnosperms too, the tripses cause galls, but just as with the Phanerogams it is not only the *Terebrantia*, but also the *Tubulifera*.

With the *Pteridophyta* most galls are found on the younger group of the *Polypodiaceae*, which is very rich in species; very few galls occur on the phylogenetically older groups, such as *Marattiaceae*, *Equisetaceae*, and *Lycopodiaceae*. The faculty of forming galls has come to its full development with the younger plant groups, viz. with the Phanerogams, and this particularly so with the *Dicotyledonae*. Confer KARNY (1926, p. 38). Tracing the system backwards, one perceives that the number of galls is very small with the lower plants: *Algae*, *Fungi*, *Musci*, and *Hepaticae*.

In Europe and round the Mediterranean galls have been collected in sufficient number to justify a comparison in figures. HOUARD (1908, 1909, and 1913) describes 7494 zoocecidia of Anthophytes, and only 26 of Pteridophytes (21 of which of *Polypodiaceae*, 1 on a *Selaginella*, and 4 on *Equisetum*), i.e. not yet 0.4%. Of the Netherlands Indies 1534 zoocecidia have been described (cf. DOCTERS VAN LEEUWEN, 1926), 35 of which occurred on Pteridophytes, i.e. 2.3%. The greater wealth of these plants in the tropics consequently is attended with a greater richness in galls. At any rate, these figures show that by comparison the number of galls on Pteridophytes is only small.

Most galls occur on the leaves, less so on the stalks, and, as far as I am aware, galls on roots have not yet been described.

The leaf-galls are partly rollings of the leaf-edge, which sometimes are able to take up the entire half of the leaf-blade, partly they are swellings of the leaf-blade. These may either be protuberances, equally strongly formed on both sides of the leaf, or be developed on one side only. The galls on the stalks may be spindle-shaped when the gall-causer lives in the centre, or lopsidedly developed, or globular excrescences, which are fixed on the stalk by a thin stem.

The gall-causers are still little-known, especially of the territories outside Europe. Representatives of the following groups of animals are gall-causers.

1. *Eriophyidae* or gall-mites. These are known in all parts of the world. Only a few species are described, viz. *Eriophyes pterides* MOLL (a dubious species), and *Eriophyes pauropus* NAL., which has been described for the first time by NALEPA (1908, p. 530) as occurring in Samoa, but which afterwards appeared to be common in many parts of the tropics of the Old World. This gall-mite causes galls on numerous species of *Nephrolepis* (cf. DOCTERS VAN LEEUWEN, 1921, p. 83). Gall-mites form rollings of the leaf-margin or excrescences on the leaf-blade. *Eriophyes pauropus* NAL. causes galls also from the sori. On *Angiopteris evecta* HOFFM. a gall-mite causes a kind of erineum of a callus nature on the underside of the leaves.

2. *Thysanoptera*. Galls caused by these animals are only known as occurring in the Netherlands Indies. They all consist of rollings of the leaf-margin, and are formed on *Asplenium nidus* L., *Pleopeltis superfacialis* BEDD., and *Pleopeltis pteropus* MOORE. The latter gall is caused by *Physothrips pteridicola* KARNY.

3. *Cecidomyidae* or gall-midges. A great part of the galls is caused by these insects. Of Europe a few species of these gall-midges have been described, but of other parts of the world all of them are still to be examined. Usually the galls develop on the leaves, and form swellings of the leaf-blade, which develop equally strongly on both sides, or grow out on one side only. A unilocular stalk-gall is known as occurring on *Selaginella pentagona* SPRING., which has been studied by STRASBURGER (1873, p. 105).

4. *Muscidae* or flies. A gall caused by a fly, *Chortophila signata* BRUSCHK., on various European ferns, which consists of a rolling and accumulation of small leaves at the end of the leaf-tops.

5. *Coccidae* or scale-insects. A coccid causes a gall on the stalks of *Psilotum nudum* GRISEB. It consists of an accumulation in the form of a witches' broom of small twigs, in anatomy resembling the rhizome. This gall is only known as occurring in Java.

6. *Hymenoptera*. On *Pteridium aquilinum* KUHN a stalk-gall has been found, which is described as caused by a Cynipid, but of which nothing else is known. A procecidium is formed on the leaves of the same fern by a sawflie: *Selandria stamineipes* KL.

On the following groups of *Pteridophyta* galls have been found.

1. *Selaginellaceae*. A stem-gall, mentioned above, is known as occurring on a representative of this family. Besides, during the expedition, made in 1926, to the Central Mountain Territory of North New Guinea, an other gall has been found, consisting of a transformed bud. The cavity was inhabited by a larva of a gall-midge.

2. *Psilotaceae*. On *Psilotum nudum* GRISEB. the above mentioned stalk-gall.

3. *Equisetaceae*. A few galls are known as occurring in Europe on representatives of the genus *Equisetum*, probably caused by *Diptera* and *Coleoptera*.

4. *Marattiaceae*. On *Angiopteris evecta* HOFFM. the above mentioned mite-gall is known as occurring in Java.

5. *Hymenophyllaceae*. A few galls are known as occurring in South America, and in Java. Besides, I myself found a stem-gall on a *Hymenophyllum* in the Central Mountain Territory of North New Guinea in 1926, which has not yet been described.

6. *Polypodiaceae*. The greatest number of galls is known as occurring on this family which is very rich in species. These galls occur in all parts of the world, and belong to the highest developed galls of the Pteridophytes.

Only very few data are at our disposal about the biology of the gall-causers, that of the galls, and their anatomy. Almost the entire field is still open for investigation. GIESENHAGEN (1909, p. 327) described the anatomy of two dipterocecidia on representatives of the family of the *Hymenophyllaceae* in Brazil, viz. *Hymenophyllum lineare* Sw. and *H. ciliatum* Sw. The former is a leaf-gall, the latter a stem-gall. The leaf-gall arises by the folding of the two leaf halves, while the wall swells. This wall consists of a thin-walled parenchym, and part of the cells gets more or less the aspect of stone-cells. The stem-gall is a top-gall, more or less globular, with a very pilose surface. The wall consists of a multicellular parenchym, the cells of which are thin-walled.

More is known about the gall, caused by a gall-mite: *Eriophyes pauropus* NAL. on various species of *Nephrolepis*. The development and the anatomy of the gall has been described by DOCTERS VAN LEEUWEN (1910, p. 142), and examined more extensively by GIESENHAGEN (1919, p. 66). For the first investigation galls were at our disposal which were only developed on the leaf-margin. GIESENHAGEN found on the materials which were at his disposal also galls on the actual leaf-blade. Afterwards we found such galls on numerous species of *Nephrolepis*, and even galls which had developed from the sori, (cf. DOCTERS VAN LEEUWEN, 1921, p. 85).

In case of strong infection, the entire leaf may be transformed into a gall. The principal matter appears to be the formation of emergences and branched hairlets, which is usually attended with the formation of a hollow swelling, a narrow opening giving access to the outer world. The wall of these galls consists of thin-walled parenchym-cells, which grow larger as they are situated more inwards, and which may grow out to giant-cells in the innermost layer. In these cells large quantities of amylum are deposited. The cells forming the inner wall of the gall-chamber are thin-walled, and do not contain amylum, but a dense, granular protoplasm, and a very large nucleus. The gall-chamber is divided unto smaller cavities by excrescences of the gall-wall. Many cells of the inner wall grow out into multicellular, branched, brown hairlets. According to GIESENHAGEN, the gall arises by a wound of one or more leaf-cells, made in the young leaves by the gall-mite. Round these wounded cells a wound-cambium arises, from which the wall and its covering of excrescences and hairlets develop.

## LITERATURE:

W. und J. DOCTERS VAN LEEUWEN–REIJNVAAN. Über die Entwicklung einiger Milbengallen. Ann. d. Jardin bot. d. Buitenzorg. Vol. XXIII. 1910, p. 142.

W. und J. DOCTERS VAN LEEUWEN–REIJNVAAN. Über die unter Einfluss eines Cocciden entstandene Umbildung der oberirdischen Triebe von *Psilotum triquetrum* in dem Rhizom ähnlich gebauten Gebilden. Ber. d. deutschen bot. Gesellsch. Vol. XXIX. 1911, p. 166.

W. und J. DOCTERS VAN LEEUWEN–REIJNVAAN. Über die von *Eriophyes pauropus* NAL. an verschiedenen Arten von *Nephrolepis* gebildeten Blattgallen. Ann. d. Jardin bot. d. Buitenzorg. Vol. XXXI. 1921, p. 83.

Mrs. J. DOCTERS VAN LEEUWEN–REIJNVAAN and W. M. DOCTERS VAN LEEUWEN. The Zoocecidia of the Netherlands East Indies. Batavia, 1926.

K. GIESENHAGEN. Über zwei Tiergallen an Farnen. Ber. d. deutschen bot. Ges. Vol. XXVII. 1909, p. 327.

K. GIESENHAGEN. Entwicklung der Milbengalle an *Nephrolepis biserrata* SCHOTT. Jahrb. f. wiss. Bot. Vol. LXIII. 1919, p. 66.

C. HOUARD. Les Zoocécidies des Plantes d'Europe et du Bassin de la Méditerranée. Paris. Vol. I, 1908; Vol. II, 1909; Vol. III, 1913.

C. HOUARD. Les Zoocécidies des Ptéridophytes de l'Ancient Continent; leur Histoire. Notes Ptéridologiques, Fascicule XI. Paris, 1920.

C. HOUARD. Les Zoocécidies des Plantes d'Afrique, d'Asie et d'Océanie. Paris, Vol. I, 1922; Vol. II, 1923.

C. HOUARD. Les Zoocécidies des Plantes de l'Amérique du Sud et de l'Amérique Centrale. Paris, 1933.

H. H. KARNY. Phylogenetic considerations. Chapter XI of DOCTERS VAN LEEUWEN, The Zoocecidia of the Netherlands East Indies. Batavia, 1926, p. 57.

A. NALEPA. Botanische und Zoologische Ergebnisse einer wiss. Forschungsreise nach den Samoa-Inseln, den Neu-Guinea-Archipel und den Salomons-Inseln. VI. *Eriophyiden*. Denkschr. d. Math. Naturwiss. Klasse d. Kaiserl. Akad. d. Wissensch. Wien. Bd. LXXXIV. 1908, p. 523.

E. STRASBURGER. Einige Bemerkungen über *Lycopodiaceae*. Die Bulbillen und Pseudobulbillen der *Selaginella*. Bot. Ztg. Vol. XXXI. 1873, p. 105.

Chapter VII

CYTOLOGY

by

Lenette Rogers Atkinson (Amherst, Mass.)

**§ 1. Historical sketch.** — The field of cytology, as treated in this chapter, comprises a study of the cytoplasm and its inclusions, both living and non-living, exclusive of the nucleus. Until recently, close inspection of these elements of the cell in the *Pteridophytes* at least, was neglected, their phenomena being greatly overshadowed by the more striking ones exhibited by the nucleus. Nor has this fern group been popular in cytological study, for it does not lend itself well to observation. Its widely differing members are not universally distributed and its tissues are not easily prepared for investigation. Impregnation by the fixatives is difficult, probably largely because of the presence of a heavy cuticle, in the gametophyte tissue especially. Good sections, once obtained, are, in handling, easily washed off the mount and must be treated in some special way to prevent this. Vital observation cannot be carried to any extent, for, either the organs are too deeply embedded to be seen without sectioning, or the chloroplasts mask the other elements in the cell. The literature is, therefore, fragmentary but the pieces can be put together to give an account not too inadequate, in spite of the gaps still present in our knowledge.

The spermatoids, by reason of their motility, were objects early investigated. Savi (1834) [1] saw them in *Salvinia*, Nägeli (1844) [2] in the Filicales and (1846) [3] in *Pilularia*, Mettenius (1850) [4] in *Isoetes*, Thuret (1851) [5] in

---

[1] R. Savi, Continuazione delle ricerche sulla fecondazione della *Salvinia natans*. Nuov. Giorn. d. lett. Pisa 28. 1834.

[2] Nägeli, Bewegliche Spiralfäden an Farnen. Zeitschr. f. wiss. Bot. 1, 168–188. 1844.

[3] Nägeli, Ueber die Fortpflanzung der Rhizocarpeen. Zeitschr. f. wiss. Bot. 1. 1846.

[4] G. Mettenius, Zur Fortpflanzung der Gefässkryptogamen. Beit. z. Bot. 1. 1850.

[5] M. G. Thuret, Notes sur les Anthéridies des Cryptogames. Ann. Sc. Nat. Bot. Sér. 3. 16. 1851.

*Equisetum* and HOFMEISTER (1851) [1]) in *Selaginella*. Among the subsequent discussions concerning the component parts of the spermatozoid, the sources of its nucleus, cilia and vesicle, or speculations as to the presence of centrosomes and their origin, the remarkably accurate descriptions of the male gamete by BELAJEFF (1885–1899) [2]) stand out in high relief. In *Isoetes, Selaginella, Equisetum* and various members of the *Filicinae*, he observed the transformation of the reticulate nucleus of the mother cell (spermatid) [3]) into the seemingly homogeneous spiral band, enclosed in the cytoplasm of that cell. He stated that the mature spermatozoid is composed of a spiral nucleus surrounded by a thin plasmatic envelop which becomes an appendage at the posterior end (vesicle of later authors). At the anterior end it is bounded on one side by the delicate tip of the nucleus and on the other by an intensely staining thread which takes the same stain as the cytoplasm, a reaction confirmed by various workers, among them JACKSON and RICKETT(1928) [4]). This thread was traced by BELAJEFF to a rounded body (blepharoplast of later authors) in the cytoplasm of each spermatid cell, where, by elongation and expansion, it comes to lie about the nucleus of the spermatozoid. Both SHAW (1898) [5]) and BELAJEFF (1898) [2]) were able to find this body in the grandmother cell of the spermatozoid (spermatng mother cell) where it undergoes division, the daughter granules migratidle to opposite sides of the nucleus. During mitosis they lie near the spindie poles and after cell division is completed, each becomes transformed in the manner previously described by BELAJEFF. YAMANOUCHI (1908) [6]) observed these granules in *Dryopteris parasitica* and described their elongation in the spermatid cell first to a wedge-shaped body, then to a band, more or less

---

[1]) W. HOFMEISTER, Vergleichende Untersuchungen. Leipzig. 1851.

[2]) WL. BELAJEFF, Antheridien und Spermatozoiden der heterosporen Lycopodiaceen. Bot. Zeit. 43, 793–802, 808–819. 1885. — Über Bau und Entwicklung der Spermatozoiden bei den Gefässkryptogamen. Ber. d. Deut. Bot. Ges. 7, 122–125. 1889. — Über den Nebenkern in spermatogenen Zellen und die Spermatogenese bei den Farnkräutern. Ber. d. Deut. Bot. Ges. 15, 337–339. 1897. — Über die Spermatogenese bei den Schachtelhalmen. Ber. d. Deut. Bot. Ges. 15, 339–342. 1897. — Über die Aehnlichkeit einiger Erscheinungen in der Spermatogenese bei Tieren und Pflanzen. Ber. d. Deut. Bot. Ges. 15, 342–345. 1897. — Über die Cilienbildner in den spermatogenen Zellen. Ber. d. Deut. Bot. Ges. 16, 140–143. 1898. — Über die Centrosome in den spermatogenen Zellen. Ber. d. Deut. Bot. Ges. 17, 199–205 1899.

[3]) The cell which by metamorphosis becomes the spermatozoid.

[4]) JACKSON and RICKETT, Staining reactions of Fern Gametes. Science 68 (1752), 89–90. 1928.

[5]) W. R. SHAW, Über die Blepharoplasten bei *Onoclea* und *Marsilia*. Ber. d. Deut. Bot. Ges. 16, 177–184. 1898.

[6]) SH. YAMANOUCHI, Spermatogenesis, oogenesis and fertilization in *Nephrodium*. Bot. Gaz. 45, 145–175. 1908. (*Nephrodium molle = Dryopteris parasitica*).

closely associated with the nucleus [1]). This manner of development was confirmed by RANKIN (1934) [1]) [2]). YASUI (1911) [3]) found a granule in the cytoplasm of *Salvinia natans* and described its elongation to a thread attached to the nucleus.

SHAW (1898) [4]) described a variation of this process in *Marsilia* [5]). In this plant there are only four spermatogenous mitoses and, during the second of these, SHAW observed a small granule in the cytoplasm of the cell. It divides to form two but they disappear during the third mitosis. He called this body a blepharoplastoid, for, as it disintegrates, a new one appears near each spindle. At the completion of cell division, this granule divides in turn and each daughter granule migrates to occupy a polar position during the final mitosis. Thus one of these bodies is received by each spermatid cell. SHAW called them blepharoplasts. Each soon fragments into granules which form an elongating thread in close union with the nucleus. SHARP (1914) [6]) confirmed this report, showing however, a strong achromatic figure about the dividing blepharoplasts.

More recently ALLEN (1911) [7]) found in the spermatid of *Adiantum*, comma-shaped or irregular, blepharoplasts, lying in the cytoplasm between the reticulate nucleus and the cell wall. Cell and nucleus grow and the blepharoplast elongates to a narrow band, running half-way around the nucleus but not in contact with it. Projecting beyond the nucleus anteriorly, it remains in this relation even in the mature spematozoid. Nucleus and blepharoplast migrate to one side of the cell and assume a spiral shape. This is accompanied by a condensation of the chromatin into an apparently homogeneous mass and a shrinkage of the cytoplasm to form a close-fitting sheath, the bulk of which, lying back of the middle spiral, shows a series of vacuoles.

Until 1930, the term blepharoplast was variously used to indicate the deeply-staining band described above, or the granule from which it arose, or even for the entire cilia-bearing, anterior portion of the male gamete. Then MÜHLDORF [8]), after a long study of the mature spermatozoid, carried

---

[1]) See figs. 9–30 (after YAMANOUCHI, ALLEN, RANKIN and YUASA).

[2]) D. E. RANKIN, The life history of *Polypodium polypodioides*. Journ. Elisha Mitchell Sci. Soc. 49, 303–328. 1934.

[3]) YASUI. On the life history of *Salvinia natans*. Ann. Bot. 25, 469–483. 1911.

[4]) W. R. SHAW, Über die Blepharoplasten bei *Onoclea* und *Marsilia*. Ber. d. Deut. Bot. Ges. 16, 177–184. 1898.

[5]) See figs. 40–52 (after SHAW and SHARP).

[6]) L. W. SHARP, Spermatogenesis in *Marsilia*. Bot. Gaz. 58, 419–431. 1914.

[7]) R. F. ALLEN, Studies in spermatogenesis and apogamy in ferns. Trans. Wis. Acad. Sci. Arts and Letters. 17, 1–56. 1911.

[8]) A. MÜHLDORF, Berichtigungen und Ergänzungen unserer Kenntnisse über die Morphologie und Histologie pflanzlicher Spermien. Biologia Generalis 6, 457–482. 1930.

his terminology for the Bryophytes over to the Pteridophytes. In an attempt to simplify terms and avoid confusion, he divided the body of the spermatozoid into three parts, without regard to origin, namely, the nucleus, the "plasma-part" ("Plasmastück") and the motor apparatus ("Bewegungsorgan"), consisting of cilia-bearing band, or "stalk" of the motor apparatus ("Stamm des Bewegungsorgan"), and the cilia.

The "plasma-part" in the Hepaticae is a constituent of the spermatozoid at the moment of fertilization, but in the Pteridophytes it is transient in nature, taking the form of a vesicle hanging at the posterior end of the spermatozoid and dropping off before the archegonium is entered. DRACINSCHI (1930 [1]), 1932) [2]), adopting this terminology, was forced to modify it when, in her investigation of the *Selaginella* spermatozoid, she found present both the Hepatic and the fern type of "plasma-part". She, therefore, called the vesicle the "plasma-remainder" and named four constituent parts of the body for the mature gamete, namely, the nucleus, the "plasma-part", the vesicle ("Plasmarest") and the motor apparatus composed of "stalk" and cilia.

A second cell constituent which for a long time has held the attention of botanists is the plastid. It is very conspicuous and its role in photosynthesis gives it great importance. SCHIMPER (1883, 1885) [3]), investigating the *de novo* origin of plastids, observed certain transition stages among the "active chlorophyll grains" of spores of *Osmunda*, root tips of *Azolla* and chromatophores of fertile stalks of *Equisetum*. He concluded that the same plastid may exhibit various phases, e.g., a leucoplast, becoming a chloroplast, then returning to its first state, or a chloroplast becoming a leucoplast and then a chromoplast.

HABERLANDT (1888) [4]), attracted by the diversity in form of the chloroplasts of *Selaginella* gave a description of their development. In a funnel-shaped, assimilating cell of the vegetative leaves, there is a green, trough-shaped body, thick where it lies at the bottom of the cell and thinner at the margins. This is the chloroplast. Its contour is irregular and the nucleus lies in close connection with it at the bottom of the trough. In some half a dozen species, this solitary chloroplast shows a constriction between two, or more, lobes. The lobes move apart but remain connected by a thin plas-

[1]) M. DRACINSCHI, Über das reife Spermium der Filicales und von *Pilularia globulifera* L. Ber. d. Deut. Bot. Ges. 48, 295–311. 1930.

[2]) M. DRACINSCHI, Über die reifen Spermatozoiden bei den Pteridophyten. Bul. Fac. de Stiinte din Cernăuti 6, 63–134. 1932.

[3]) A. F. W. SCHIMPER, Über die Entwickelung der Chlorophyllkörner und Farbkörper. Bot. Zeit. 41, 105–112, 121–131, 137–146. 1883. (*Azolla, Osmunda regalis*).
— Untersuchungen über die Chlorophyllkörner und die ihnen homologen Gebilde. Jahrb. f. wiss. Bot. 16, 1–247. 1885. (*Equisetum arvense*, p. 108).

[4]) G. HABERLANDT, Die Chlorophyllkörper der Selaginellen. Flora 46 (71), 291–308. 1888. (*S. Martensii, S. caesia, S. grandis, S. cuspidata*).

matic bridge. Both fixed and fresh material of the stem parenchyma revealed in each cell of the meristem, one single chloroplast somewhat larger than the nucleus and clinging to it. A dumb-bell-shaped object arises by constriction between whose halves the plasmatic bridge soon becomes colorless. The resulting bodies may divide again and again as the parenchyma tissue differentiates, but each division is incomplete in the sense that the plasmatic connection does not break. In the older cells chains may reach a length of 0.15 mm., and consist of as many as 30 spindle-shaped members with plasmatic bridges of variable length. This chain HABERLANDT considered as one chloroplast, an opinion substantiated by the observed contraction, under adverse conditions, of the plasmatic connections, even to the extent of bringing the separated members again into close contact. Each guard cell of a stoma possesses, when young, two somewhat elongated chloroplasts connected by a filament. In the older guard cells, the chloroplasts are usually four-membered. HABERLANDT also pointed out that starch is formed in the absence of pyrenoids as in the Phanerogams and the other Pteridophytes and that it is present in the form of inclusions in the chloroplast.

By 1907 it was rather generally accepted that the chloroplast consisted of a colorless ground substance, protein in nature, through which the chlorophyll was in some way distributed. PRIESTLY and IRVING [1]), cut at 1μ frozen *Selaginella* plastids measuring about 0.02 mm. and found the sections to be in the form of uninterrupted rings, and had no doubt that the chlorophyll is restricted to the outer layer. ZIRKLE (1926) [2]) described the plastids of *Adiantum* as hollow spheroids. His conclusions however have met with some criticism. We look, therefore, to future investigation for a more adequate knowledge as to the structure of the chloroplast in the Pteridophyte group [3]).

During the years from 1910-1920 investigations of another nature were going on, stimulated by the discovery in plant cells (MEVES, 1904) [4]) of granular and rod-shaped bodies which were called chondriosomes or mitachondria [5]). Exhaustive searches led to the detection of these bodies in many

[1]) J. H. PRIESTLY and A. IRVING, The structure of the chloroplast considered in relation to its function. Ann. Bot. 21, 407–413. 1907. (*Selaginella Martensii, S. Kraussiana*).

[2]) C. ZIRKLE, The structure of the chloroplast in certain higher plants. Amer. Journ. Bot. 13, 301–320, 321–341. 1926.

[3]) HEITZ (1936) thinks that the chlorophyll is distributed in discs placed parallel to the surface of the flattened lenticular chloroplasts. The striations hitherto considered as pathological may be merely the discs seen in side view. Ber. d. Deut. Bot. Ges. 54, 362–368.

[4]) FR. MEVES, Über das Vorkommen von Mitochondrien bezw. chondromiten in Pflanzenzellen. Ber. d. Deut. Bot. Ges. 22, 284–286. 1904.

[5]) Elongated threads are often called chondrioconts and chains of granules chondriomites. The entire mitochondrial content of a cell constitutes the chondriome.

types of cells. Observations were made principally among the Phanerogams, Algae and Fungi, where the material is particularly favorable for *in vivo* study. The mitochondria were characterized as capable of division, able to pass from one form to another, staining more or less rapidly, not with the usual vital stains, but with Dahlia violet, methyl violet 5B or Janus green. These bodies are dissolved or greatly altered by acetic acid or alcohol, thus making necessary special treatment in fixation. The so-called mitochondrial methods were introduced into botanical research, of which REGAUD's combination of formol and bichromate of potassium is a great favorite, especially among European botanists.

Theories and hypotheses had been developed during this period and an attempt to extend them to all groups of the plant kingdom led to investigations of the Pteridophytes. Because of this tendency to look to them for further substantiation of concepts originating elsewhere, it is thought advisable to review briefly the theories concerned.

PENSA, LEWITSKY, MEVES and ALVARADO and at first GUILLIERMOND[1])[2]), having observed the apparent transformation in cells of granules and rod-shaped bodies into chloroplasts, were led to believe that the elements of the chondriome are all alike, some of them remaining as mitochondria whose role is unknown, others being transformed, as the cells differentiate, into plastids, specialized for photosynthesis.

RUDOLPH, SAPEHIN, SCHERRER and MOTTIER[3]), however, following the chloroplasts through all the stages in the life history of certain Cryptogams, and observing these chloroplasts persisting side by side with mitochondria, held to the old theory of the individuallity of the plastids. They declared that one had to do here with distinct genetical lines for the plastids and the mitochondria, explaining that in the Phanerogams there is apt to be confusion, as the two lines show similar forms in the embryonic cells.

---

[1]) These papers are cited not because they were the first to appear on the subject, but rather because the theory of the author is stated therein.

[2]) A. PENSA, Osservazioni di morfologia e biologia cellulare nei vegetali (mitochondri, cloroplasti). Arch. f. Zellforsch. 8, 612–662. 1912. — G. LEWITSKY, Über die Chondriosomes in pflanzlichen Zellen. Ber. d. Deut. Bot. Ges. 28, 538–546. 1910. — FR. MEVES, Die Chloroplastenbildung bei den höheren Pflanzen und die Allinante von A. MEYER. Ber. d. Deut. Bot. Ges. 34, 333–345. 1916. See also Arch. f. mikr. Anat. 89. 1916. — S. ALVARADO, Die Entstehung der Plastiden aus Chondriosomen in den Paraphysen von *Mnium cuspidatum*. Ber. d. Deut. Bot. Ges. 41, 85–96. 1923.

[3]) K. RUDOLPH, Chondriosomen und Chromatophoren. Ber. d. Deut. Bot. Ges. 30, 605–629. 1912. — A. A. SAPEHIN, Untersuchungen über die Individualität der Plastide. Ber. d. Deut. Bot. Ges. 31, 14–16, 321–324. 1913. Also, Arch. f. Zellforsch. 13, 319–398. 1915. — A. SCHERRER, Die Chromatophoren und Chondriosomen von *Anthoceros*. Ber. d. Deut. Bot. Ges. 31. 1913. — D. MOTTIER, Chondriosomes and the primordia of chloroplasts. Ann. Bot. 32, 19–114. 1918.

There followed further investigation by GUILLIERMOND [1]) on the Phanero-gams and Fungi, and by his students EMBERGER [1]) and MANGENOT [2]) on the ferns and Algae, respectively, which led to the conception of the duality of the chondriome. There are, according to these botanists and their supporters, two classes of bodies in plants, resembling the animal chondriome (Fig.141), which present common morphological and histo-chemical properties but which differ functionally. The one, composed of active chondriosomes or plastids which are colorless in embryonic cells but are capable of producing pigment, play a role in photosynthesis and may undergo appreciable modifications, due to the products elaborated. The second class, those inactive in photosynthesis are called ordinary chondriosomes and their rôle remains obscure. The chondriosomes are considered as independent and permanent parts of the cell, just as are the nucleus, vacuoles and cytoplasm. These with the wall, lipoid granules and other inclusions con-stitute the elements of a plant cell.

Variations in this classification and differences in nomenclature have appeared. P. A. DANGEARD [3]) described for the plant cell a plastidome, composed of plastids, a vacuome composed of vacuoles and a spherome in which seems to be included the ordinary mitochondria of GUILLIERMOND and certain globules which he at first called microsomes. BOWEN [4]) pro-posing in a work left uncompleted, to go to the problem with no pre-conceived notions of facts or nomenclature, agreed on the presence of a plastidome and vacuome in the plant cell, but was impelled by caution to call the ordinary chondriosomes of GUILLIERMOND, the pseudochondriome as he could not prove them comparable to the animal chondriome. He found also disc-like bodies in *Equisetum* cells which he called "osmiophilic platelets" and which he thought were elements as yet undiscovered. It seems more probable that they represent a chondriome altered by the fixative as stated by GUILLIERMOND [5]) who has clearly demonstrated the

[1]) See pp. 204–205.

[2]) G. MANGENOT, Sur l'évolution des chromatophores et le chondriome chez les Floridées. C. R. Acad. Sci. 170 (26), 1595–1598. 1920. — See also, C. R. Acad. Sci. 170, 63–65. 1920.

[3]) P. A. DANGEARD, La structure de la cellule végétale dans ses rapports avec la théorie du chondriome. C. R. Acad. Sci. 173, 120–123. 1931. — Sur la reproduction sexuelle chez le *Marchantia polymorpha* dans ses rapports avec la structure cellu-lare. C. R. Acad. Sci. 178, 267–271. 1924. — See also, C. R. Acad. Sci. 170, 301–306, 709–714. 1920. and, C. R. Acad. Sci. 171, 69–74. 1921.

[4]) R. H. BOWEN, Studies in the structures of plant protoplasm. II The plastiome and pseudochondriome. Zeitschr. f. Zellforsch. u. mikros. Anat. 9, 1–65. 1929. — See also, ibid. 6, 689–725. 1928 (*Equisetum arvense*). — Bull. Tor. Bot. Club 53, 179–196. 1927. (*E. arvense*). — Ann. Bot. 53, 309–327. 1929 (*E. arvense*).

[5]) A. GUILLIERMOND, A propos des recherches récentes de M. BOWEN sur l'appa-reil de Golgi. C. R. Biol. 98, 368–371. 1928. Ibid. 101, 567–572. 1929. — See also, S. R. BOSE, Ann. Bot. 45, 303–314. 1931.

inconstancy of osmic acid in impregnation. BOWEN himself, said that it is impossible to predict what will be blackened by osmic methods [1]).

In opposition to the above hypotheses is that of the school of A. MAYER[2]), which is supported by SHARP[3]), that the chondriosomes are merely visible products of protoplasmic activity or temporary physical states of the colloidal substances composing the protoplasm. In substantiation of this we are shown figures made artificially with emulsions by LÔWSCHIN [4]) and WALKER [5]) which resemble more or less closely these chondriosomal forms.

Whatever may be the name by which the forms described above are known or whatever functions and origins may be attributed to them, the fact remains that they are almost universally demonstrable under certain given conditions when they take a prominent and conspicuous place among the constituents of the plant cell.

There is some literature to be found concerning the spindle figure [6]) but its nature and origin during somatic and reduction divisions, its rôle in the cell and relation to the phragmoplast and new cell wall, are still matters needing further investigation. A similar state of affairs exists in regard to investigation of the plasmodesmen [7]).

---

[1]) R. H. BOWEN, The use of osmic-impregnation methods in plant cytology. Bull. Tor. Bot. Club 56, 33–52. 1929.

[2]) A. MAYER, Bemerkungen zu G. Lewitsky, Über die Chondriosomen in pflanzlichen Zellen. Ber. d. Deut. Bot. Ges. 29, 158–160. 1911. — Ibid. 34, 168–173. 1916.

[3]) L. W. SHARP, Introduction to Cytology. McGraw-Hill. New York. 1934. p. 92.

[4]) LÔWSCHIN, „Myelinformen" und Chondriosomen. Ber. d. Deut. Bot. Ges. 31, 203–209. 1913.

[5]) C. E. WALKER, Artefacts as a guide to the chemistry of the cell. Roy. Soc. Proc. London (Biol. Sci.) Ser. B. 103, 397–403. 1928. — See also, C. E. WALKER and M. ALLEN, ibid. 101, 468–482. 1927.

[6]) W. J. V. OSTERHOUT, Über Entstehung der karyokinetischen Spindel bei *Equisetum*. Jahrb. f. wiss. Bot. 30, 159–168. 1897. — R. W. SMITH, The achromatic spindle in the spore mother cells of *Osmunda regalis*. Bot. Gaz. 30, 361–377. 1900. — P. DENKE, Sporenentwicklung bei *Selaginella*. Beih. Bot. Zentralbl. 12, 182–199. 1902. (*S. pallescens, S. Martensii*). — J. E. GRAUSTEIN, Evidences of hybridism in *Selaginella*. Bot. Gaz. 90, 46–74. 1930. p. 52 (six exotic sp. of *Selaginella* and *S. apus, S. rupestris*). — Y. GIHEI, Einige Beobachtungen über die Zellteilung in den Archesporen und Sporenmutterzellen von *Psilotum triquetrum* Sw., mit besonderer Rücksicht auf die Zellplattenbildung. Bot. mag. (Tokio) 34, 117–129. 1920. — V. JUNGERS, Mitochondries, chromosomes et fuseau dans les sporocytes d'*Equisetum limosum* La Cellule 43, 323–340. 1934. — W. A. BECKER and J. H. SIEMASZO, Über das Verhalten der Cytoplasmaeinschlüsse der *Equisetum*-sporen während der Zellteilung. La Cellule 45, 29–42. 1936. (*E. arvense, E. limosum*).

[7]) P. TERLETZKI, Anatomie der Vegetationsorgane von *Struthiopteris germanica* und *Pteris aquilina*. Jahrb. f. wiss. Bot. 15, 452–501. 1884. — A. MEYER, Das Irrthümliche der Angaben über das Vorkommen dicker Plasmaverbindungen zwischen den Parenchymazellen einiger Filicinen und Angiospermen. Ber. d. Deut. Bot. Ges.

## BIBLIOGRAPHY

These papers have been drawn upon very freely in the preparation of the chapter or have been added to the list because they contain extensive bibliographies.

DRACINSCHI, M. 1930. Über das reife Spermium der Filicales und von *Pilularia globulifera*. Ber. d. Deut. Bot. Ges. 48, 295–311. (*Osmunda regalis, Dryopteris Filix-mas, D. Thelypteris, D. austriaca, Phegopteris Robertiana, Athyrium Filix-femina, Phegopteris polypodioides, Asplenium Adiantum-nigrum, A. Trichomanes*).

—— 1932. Über die reifen Spermatozoiden bei den Pteridophyten. Bull. Fac. Stiinte Cernauti 6, 63–134. 1932 (Filicales: *Polystichum Braunii, Ceratopteris thalictroides, Alsophila* sp., *Aneimia phyllitidis, Athyrium Filix-femina* var. *carymbiferum, Salvinia natans, Selaginella Martensii Watsonii, S. pulcherrima, S. plumosa, S. Emmeliana, S. caulescens, S. inaequalifolia; Equisetum Telmateja, E. silvaticum; E. arvense; Isoetes lacustris, I. japonica*.

EMBERGER, L. 1920. Evolution du chondriome chez les Cryptogams vasculaires. C. R. Acad. Sci. 170, 282–285. (*Athyrium Filix-femina*).

—— 1920. Evolution du chondriome dans les Fougères. C. R. Acad. Sci. 170, 469–471. (*Phyllitis Scolopendrium, Asplenium Ruta-muraria, Pteridium*).

—— 1920. Etude cytologique de la Sélaginelle. C. R. Acad. Sci. 171, 263–266.

—— 1921. Etude cytologique des organes sexuels des Fougères. C. R. Acad. Sci. 171, 735–737. (Polypodiaceae).

—— 1921. Recherches sur l'origine et l'évolution des plastides chez les Ptéridophytes. Arch. Morph. Gen. et Exp. 1, 1–186. (*Athyrium Filix-femina, Dryopteris Filix-mas, Phyllitis Scolopendrium, Asplenium Ruta-muraria, Pteridium aquilinum, Adiantum capillus-Veneris; Selaginella pallescens, S. Watsoniana; Equisetum arvense, E. limosum*) Bibliography.

—— 1921. Contribution à l'étude cytologique du sporange chez les fougères. C. R. Acad. Sci. 173, 1485. (Polypodiaceae).

—— 1922. Sur l'évolution des plastides chez les végétaux. Assoc. Fr. Adv. Sci. Montpellier 46, 330–334. (*Equisetum ramosissimum, E. limosum*).

—— 1922. Sur la cytologie des Lycopodinées homosporées. C. R. Soc. Biol. 87, 1394–1396. (*Lycopodium Selago, L. alpinum, L. clavatum*).

—— 1922. A propos des résultats de Sapehin sur la cytologie des Lycopodinées homosporées. C. R. Soc. Biol. 87, 1396–1398. (*Lycopodium Selago*).

—— 1922. Nouvelle contribution à l'étude cytologique des Sélaginelles. C. R. Soc. Biol. 87, 1398–1400.

—— 1923. Sur le système vacuolaire de Sélaginelles. C. R. Soc. Biol. 88, 218–221.

—— 1923. Remarque sur la cytologie des Sélaginelles. C. R. Soc. Biol. 88, 225–226.

—— 1923. Observations sur les chloroplastes des Sélaginelles. C. R. Soc. Biol. 88, 513–517. (*Selaginella Kraussiana, S. Martensii* var. *compacta, S. pulcherrima*).

—— 1923. Recherches sur le protplasme des Lycopodinées. Arch. d'Anat. Micros. 19, 309–349. (*Lycopodium Selago, L. clavatum, Selaginella Kraussiana, S. Martensii* var. *compacta*) Bibliography.

—— 1924. Contribution à l'étude de la formation des plastes chez les végétaux. C. R. Acad. Sci. 179, 420–422. (*Phyllitis*).

—— 1925. Le chondriome des végétaux. C. R. Acad. Sci. 181, 226–228.

—— 1925. Sur la réversion des plastes chez les végétaux. C. R. Acad. Sci. 181, 879–880.

—— 1927. Nouvelles recherches sur le chondriome de la cellule végétale. Rev. Gen. de Bot. 39, 341–363, 420–448.

—— 1928. A propos du chondriome: réponse à M. PAVILLARD. Bull. Soc. Bot. de France 75, 696–698. (J. PAVILLARD, A propos du chondriome. Bull. Soc. Bot. de France 75, 42–45. 1928).

MANGENOT and EMBERGER. 1920. Sur les mitochondries dans les cellules animales et végétales. C. R. Soc. Biol. 83, 418–420. (*Athyrium Filix-femina*).

------

14, 154–158. 1896. (*Matteuccia Struthiopteris, Polypodium vulgare*). — E. STRASBURGER, Über Plasmaverbindungen pflanzlicher Zellen. Jahrb. f. wiss. Bot. 36, 493–610. 1901 (*Selaginella Martensii, S. lepidophylla*).

GUILLIERMOND, A., MANGENOT, G., PLANTEFOL, L. 1933. Traité de Cytologie Végé-
    tale. Le François, Paris. Extensive bibliography.
SHARP, L. W. 1934. Introduction to Cytology. McGraw-Hill. New York.
SHÜRHOFF, P. N. 1924. Die Plastiden. Handbuch der Pflanzenanatomie. Cytologie
    I. Berlin. Bibliography.
YUASA, A. 1932. Studies in the Cytology of Pteridophyta. I. On the spermatozoid
    of *Pteris cretica* L. var. *albo-lineata* Hk. Bot. Mag. (Tokyo) 46, 4–12.
——— 1933. II. The morphology of the spermatozoids of some ferns. Cytologia 4,
    305–337. (*Polypodium aureum, Adiantum capillus-Veneris, Pteris cretica* var.
    *albo-lineata, Leptogramme totta* (= *Dryopt. africana*), *Polystichum varium, Dry-
    moglossum microphyllum, Leptochilus zeylanicus*).
——— 1933. III. The morphology of spermatozoids in eight species of ferns.
    Bot. Mag. (Tokyo) 47, 681–696. (*Athyrium nipponicum, A. rigescens, Blechnum
    nipponicum, Davallia solida, Polystichum Standishii, P. tripteron, Pteris cretica,
    Notogramme* (= *Coniogramme*) *japonica.*
——— 1933. IV. On the spermatozoids of *Selaginella, Isoetes* and *Salvinia*. Bot.
    Mag. (Tokyo) 47, 698–709. (*Selaginella involvens, Isoetes japonica, Salvinia
    natans*).
——— 1934. V. Spermatoteleosis in *Notogramme* (=*Coniogramme*) *japonica* Presl and
    *Pteris multifida* Poiret with special reference to the development of border-
    brim, lateral bar and cilia-bearing band. Journ. Fac. Sci. (Sec. 3) 4, 389–397.
——— 1935. VIII. The morphology of spermatozoids in seven species of ferns. Bot.
    Mag. (Tokyo) 49, 375–382. (*Alsophila Bongardiana, Asplenium incisum, A.
    tripteropus, Doodia media, Dryopteris triphylla, Matteuccia orientalis, Osmunda
    cinnamomea*).

## § 2. The Filicinae. — T h e  g a m e t o p h y t e. — In the mature

spore of the *Osmundaceae* (Fig. 1) there is present a centrally-placed nucleus,
vacuoles which are large and numerous and a chondriome consisting of
many large chloroplasts and ordinary mitochondria in the form of granules
and rods. KIRBY (1928) [1]) finds that the spores are of two types, those with
a high number, 80-100, of lens-shaped plastids arranged outwards from
the nucleus in radiating lines, and those with as few as 14-30 which are
peripheral in position. These chloroplasts contain 6-9 starch grains, except
in the case of *O. gracilis* where there is no starch.

In the spores not containing chlorophyll, the chondriome consists of
granules and the short rods. Some of these rods are larger than others and
are probably the elements which later develop chlorophyll just before the
membranes of the germinating spore break.

In living prothallia the cells are packed with large green lenticular
plastids (Fig. 2) measuring about 5 μ, [2]) which make observations of the
other elements of the cell difficult. Both ZIRKLE [3]) and RANKIN [4]) describe
these chloroplasts as hollow, flattened spheroids whose outer stroma sur-

---

[1]) K. S. N. KIRBY, The development of the chloroplasts in the spores of *Osmunda*. Journ. Roy. Micros. Soc. 48 (3), 10–34. 1928 (*O. palustris aurea, O. regalis, O.
gracilis*).

[2]) M. MOEBIUS, Über die Grösse der Chloroplasten. Ber. d. Deut. Bot. Ges. 38,
224–232. 1920 (*Cyclophorus adnascens, Dryopteris Filix-mas, Platycerium divergens,
P. iridioides*). — See also Chap. VIII, p. 279, this Manual.

[3]) P. 200, note 2 (*Adiantum*).

[4]) P. 198, note 2 (*Polypodium polypodioides*).

rounds a central vacuole containing 2-3 minute starch grains. Fixed material reveals rod-shaped and granular mitochondria, often clinging to the chloroplast, and a large vacuole (Fig. 2) filled with phenol compounds which show tannoid reactions.

During the formation of trichomes, such as the glandular hairs of *Dryopteris*, and on the outgrowth of rhizoids from prothallial cells, there is a degeneration of the plastids (Fig. 3) described by GRATZY-WARDENGG [1]).

**Spermatogenesis.**

The cells of the lower epidermis which give rise to the antheridia in the *Filicinae* [2]) are first distinguishable from the other epidermal cells by their denser cytoplasm and slightly smaller chloroplasts (Fig. 4). As the cells round out and divide to form the young antheridium, there is a gradual absorption of starch, the chloroplasts decrease in size, become spindle-shaped and stain more deeply (Fig. 5). The mitochondrial rods and granules about them seem to increase. The wall cells of the antheridium contain a sparse cytoplasm, small green plastids, mitochondria in the form of

---

FIG. 1–30. (*Filicinae*). — Fig. 1. Young spore of *Todea barbara* enclosing vacuoles, mitochondria and plastids (*P*). (After PY.) — Fig. 2. Prothallial call of Polypodiaceous type containing mature plastids (*P*) and precipitated phenol products (*C*) in a large vacuole. (After EMBERGER.) — Fig. 3. Stages in degeneration of chloroplasts in a prothallium of *Dryopteris Filix-mas*. (After GRATZY-WARDENGG.) — Fig. 4–7. Antheridial cells of *Adiantum capillus-Veneris*. (After EMBERGER.); 4. Antheridial initial and fragment of adjacent cell showing relative sizes of plastids; 5. Later stage than fig. 4. showing spindle-shaped plastids (*P*); 6. Cell of antheridial wall enclosing vacuoles (*CP*), ordinary mitochondria (*M*) and small plastids (*P*); 7. Spermatogenous cell with mitochondria (*M*), chondrioconts derived from former plastids, and vacuoles. — Fig. 8. Spematid of *Polypodium polypodioides* enclosing mitochondria and plastids. CHAMPY–KULL fix., ALTMANN's stain. (After RANKIN.) — Fig. 9–10. Spermatid mother cell of *Dryopteris dentata* with a granule (blepharoplast) at each pole. Flemming's fix. (After YAMANOUCHI). — FIG. 11. Spermatid of *Polypodium polypodioides* with mitochondria in the cytoplasm and blepharoplast body near the nuclear membrane. CHAMPY–KULL fix. (After RANKIN.) — Fig. 12–14 Spermatid cells of *Dryopteris dentata* enclosing an elongated blepharoplast. FLEMMING's fix. (After YAMANOUCHI.); 12. Rounded blepharoplast body in cytoplasm; 13. Surface view of wedge-shaped blepharoplast; 14. Cross section of stage shown in fig. 13. — Fig. 15. Blepharoplast half-way around slightly elongated nucleus in spermatid of *Polypodium polypodioides*. Granular mitochondria in the cytoplasm. CHAMPY–KULL fix. (After RANKIN.) — Fig. 16–17. Further elongation of the blepharoplast in spermatid of *Dryopteris dentata*. (After YAMANOUCHI.); 16. Surface view; 17. Cross section of stage of fig. 16. — Fig. 18–20. Spermatid cells of *Adiantum capillus-Veneris*. FLEMMING's fix. (After R. F. ALLEN); 18. View perpendicular to those of figs. 19, 20 showing blepharoplast band (*B*) very close to nucleus in the contracted cytoplasm of the cell. — Fig. 21–23. Stages in elongation of the blepharoplast body in the spermatid of *Notogramma japonica*. (After YUASA.); 21. Elongated blepharoplast showing dark-staining thread and lighter staining band; 22. Blepharoplast attached to the nucleus; 23. Blepharoplast and nucleus coalesced. — Fig. 24–25. Lateral and anterior view, respectively, of metamorphosing spermatid of *Adiantum capillus-Veneris*. (After

---

[1]) E. GRATZY-WARDENGG, Degeneration von Chloroplasten an Farnprothallien. Protoplasma 14, 52–63. 1932. (*Dryopteris Filix-mas, Matteuccia Struthiopteris*).

[2]) EMBERGER.

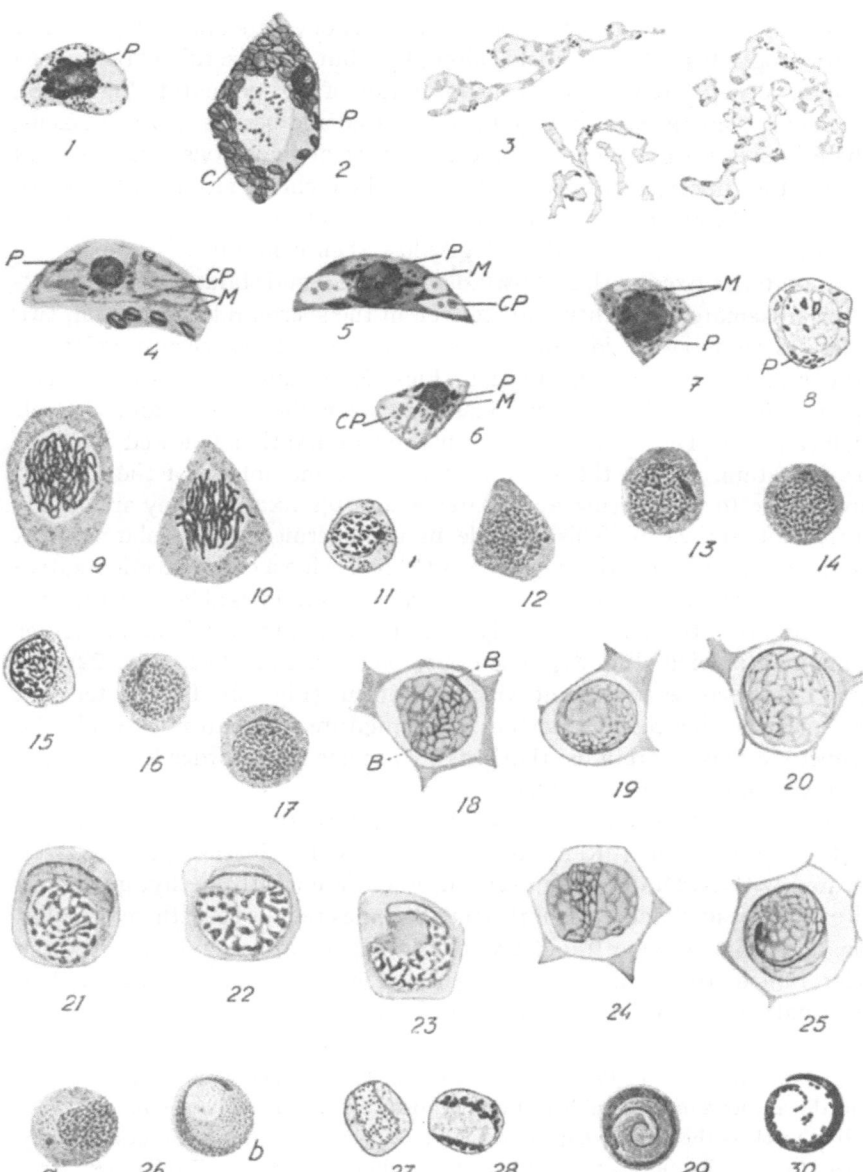

ALLEN.) — Fig. 26. Coiled posterior end (*a*) and anterior end (*b*) of spermatid of *Dryopteris dentata* FLEMMING's fix. (After YAMANOUCHI.) The blepharoplast extends beyond the nucleus at the anterior end. — Fig. 27–28. Migration of the mitochondria toward the nucleus in the spermatid of *Polypodium polypodioides*, a stage parallel to that of fig. 26, *a*. REGAUD's fix. (After RANKIN.) — Fig. 29. Nearly mature spermatozoid within spermatid of *Dryopteris dentata*. FLEMMING's fix. (After YAMANOUCHI.) — Fig. 30. Same stage as fig. 29 but showing mitochondrial granules along the inner surface of the nucleus of *Polypodium polypodioides*. REGAUD's fix. (After RANKIN.)

rods and granules and vacuoles full of phenol compounds (Fig. 6). The spermatogenous cells lose their chlorophyll but the plastids may be seen in some forms up to the time of the formation of the spermatid (Figs. 7, 8). The cytoplasm appears dense in fixed material and besides a large nucleus, there is present a rich chondriome consisting of short rods, granules and long rather thick chondrioconts (Fig. 7). These chondrioconts EMBERGER does not hesitate to designate as the former chloroplasts. The vacuoles seem to decrease in size and gradually lose their contents.

In material fixed with FLEMMING's fixative [1]) and stained with HEIDEN-HAIN's haematoxylin, there can be seen in the spermatid mother cell, two granules on opposite sides of the resting nucleus. These occupy polar positions during the ensuing division (Figs. 9, 10) and one is seen in each spermatid, situated in the cytoplasm between the nucleus and the cell wall (Fig. 12). The CHAMPY-KULL method of fixation followed by ALT-MANN's stain, reveals these bodies during the metaphase of the division giving rise to the spermatid [1]). After REGAUD's fixative, they are apparently not stained [1]). This granule in the spermatid cell enlarges in a manner variously described [2]). YUASA (1934 [3]) using a chrom-acetic fixative or R. F. ALLEN's modification of FLEMMING's fluid, is able to distinguish two regions in this enlarging body, a deeply-staining margin and a lighter remaining portion (Fig. 21). The entire body extends itself (Fig. 22) and gradually becomes coalescent with the nucleus (Fig. 23). The lighter area becomes the cilia-bearing portion in the mature spermatozoid, while the deeper-staining margin is thought to become the border-brim [4]) (the deeply-staining band of BELAJEFF).

DRACINSCHI[5]) finds that it is possible to establish the chondriosome nature of this border brim ("Randsaum") and says that in the spermatid cell treated with ALTMANN's stain and differentiated with methyl green, the already thread-shaped chondriosome becomes red along with the smaller plastids. Both RANKIN [1]) and EMBERGER, on the other hand, describe the migration of the mitachondria to the nucleus (Fig. 27, 28) where they eventually form a double row of granules along its inner surface (Fig. 30).

---

[1]) RANKIN, note 2, p. 198; ALLEN, note 7, p. 198; YAMANOUCHI, note 6, p. 197 and the author's own unpublished observations on *Lygodium palmatum*.

[2]) As well as the wedge-shaped, or round, or comma-shaped bodies described by YAMANOUCHI, RANKIN and ALLEN, the author has observed relatively large granular, oval or vesiculate bodies each with threads extending from it in opposite directions parallel to the surface of the nucleus.

[3]) A. YUASA, Studies in the cytology of the Pteridophyta. V. Spermatoteleosis in *Notogramma (Coniogramme) japonica* Presl and *Pteris multifida* Poiret with special reference to the development of border-brim, lateral bar and cilia-bearing band. Journ. Fac. Sci. Sec. III, 4 (4), 389–397. 1934.

[4]) YUASA, Studies in Cytology of Pteridophytes I.

[5]) DRACINSCHI, notes 1, 2, p. 199.

DRACINSCHI also finds granules in or on the plasmatic membrane surrounding the nucleus when CAJAL's silver impregnation method has been employed. They are arranged in a single row (Fig. 38) or in groups (Figs. 35, 137) but their nature has not been determined.

We are, therefore, confronted with incomplete and conflicting evidence in this matter. Are we to conclude that the chondriome of the spermatid cell is distributed in the spermatozoid in three places, first along the surface of the body in the form of the border brim, secondly over the inner surface of the nucleus as variously arranged granules, and thirdly in the vesicle in the form of starch-forming plastids? And if the deep-staining band in the spermatozoid, and therefore the granule from which it arose, be of a chondriosomal nature, then how shall we explain their presence in material fixed with FLEMMING's fluid or their absence after REGAUD's[1])?

The spermatozoids differ in size, in the amount of coiling and in the number and attachment of the cilia[2])[3]) (Fig. 33). In the *Polypodiaceae* [3])[4]) the male gametes are discharged from the antheridium coiled within a membrane which is invisible unless stained [3]). They free themselves suddenly and then they appear conical in shape (Fig. 31), for the body, consisting of from 2.5-3 spirals rests on a vesicle („Plasmarest"). This vesicle is composed of a homogeneous ground substance in which can be detected fat globules, 4-6 oval or round plastids containing starch, and small granules staining with neutral red which represent the vacuome [5]). As the spermatozoid rotates along an irregularly spiral path, the vesicle slips backwards and is finally discarded. The spermatozoid elongates (Fig. 32), becomes coiled into 4-5 spirals, is very flexible and is is now ready to enter the neck of the archegonium. It consists of a nucleus and a motor apparatus. The nucleus [6]) extends the whole length of the spermatozoid, beginning as a narrow thread which becomes broader until it occupies the entire width of the body and then ends in a blunt point. It is surrounded by a plasmatic sheath which runs out to a thread at the posterior end, but exhibits granules along the inside length of the nucleus (Fig. 35). The stalk of the motor apparatus (cilia-bearing band) occupies the greater part of the anterior end of the spermatozoid, running back to a point at a region midway along the body. It is resistant to maceration and shows only two visible struc-

---

[1]) Page 201, line 6.

[2]) YUASA.

[3]) DRACINSCHI.

[4]) MÜHLDORF, Page 198, note 8.

[5]) WENT could not see the vacuoles in the male gamete but suggested that they had possibly escaped notice because of their smallness: — F. A. F. C. WENT, Die Vermehrung der normalen Vacuolen durch Teilung. Jahrb. f. wiss. Bot. 19, 296–356. 1888. (*Salvinia natans, Polypodium paradiseae, Alsophila australis, Cyathea medullaris*).

[6]) Chap. VIII, p. 242; p. 264, this manual.

tures, a dark-staining thread (p. 208) at one edge and 40-50 granules scattered over its outer surface [1]). The granules are the basal structures for the cilia and their number corresponds to that of the cilia. Each cilium runs out to a slender end-piece.

The mature spermatozoids of *Pilularia* [2]) (Fig. 34) are ovoid in shape. The spiral body makes about two and a half turns about a relatively large vesicle. The vesicle contains fat globules of varying sizes and 10-16 plastids containing starch (Fig. 36). The starch grains are often found in decomposition. There is no localization of stain as in the ferns to indicate the vacuome, but the great amount of slime surrounding the spermatozoid makes the results of this test uncertain. The body of the gamete is composed of a nucleus and the motor apparatus. The border-brim extends the whole length of the body and at the anterior end projects like a hook (Fig. 37) beyond the nucleus and cilia-bearing band. The 40-50 cilia are

---

FIG. 31–52 (*Filicinae*). — Fig. 31. A recently liberated Polypodiaceous spermatozoid. (After MÜHLDORF.) — Fig. 32. Mature spermatozoid without the vesicle. *Thelypteris palustris.* (After DRACINSCHI.) — Fig. 33. Diagram of spermatozoids of Pteridophytes, showing relative lengths of body and cilia, position of border-brim („Randsaum") and nucleus and arrangement of basal granules in the cilia-bearing band („Stamm"); — a, *Selaginella;* b, *Isoetes;* c, *Filicinae;* d, *Pilularia globulifera;* e, *Salvinia natans.* (Modified from DRACINSCHI.) — Fig. 34. Mature spermatozoid of *Pilularia* before the shedding of vesicle. (After DRACINSCHI.) — Fig. 35. Fragment of the nucleus of a spermatozoid of *Dryopteris* showing granules of unknown origin. (After DRACINSCHI.) — Fig. 36 Plastids with starch from the vesicle of a spermatozoid of *Pilularia globulifera.* (After DRACINSCHI.) — Fig. 37. Upper portion of the spermatozoid of *Pilularia globulifera*, showing the hook-shaped border-brim („Randsaum"), the linear series of basal granules from which the cilia arise and anterior point of nucleus. (After DRACINSCHI.) — Fig. 38. Granules of unknown nature along nuclear surface of spermatozoid of *Pilularia globulifera.* (After DRACINSCHI.) — Fig. 39. Spermatozoid of *Salvinia natans* coiled about its vesicle. (After DRACINSCHI.) — Fig. 40–52. Spermatogenesis in *Marsilia.* — Fig. 40. Anaphase of the first spermatogenous mitosis of *M. quadrifolia* with no granules appearing at the poles. (After SHARP.) — Fig. 41. Second spermatogenous mitosis of *M. quadrifolia* showing polar granules and surrounding cytoplasmic radiations. (After SHARP.) — Fig. 42. Prophases of third spermatogenous mitosis of *M. vestita* showing already-divided granules ("blepharoplastoids") in the cytoplasm. (After SHAW.) — Fig. 43. Metaphase of the third spermatogenous mitosis of *M. quadrifolia* showing granules at the poles. (After SHAW.) — Fig. 44. Anaphase of the third spermatogenous mitosis of *M. quadrifolia* showing the granules of the second mitosis disorganizing at the side of the spindle and the granules of the third mitosis at the poles. (After SHARP.) Fig. 45. Resting stage of the spermatid mother cell of *M. vestita* showing a divided granule in the cytoplasm. (After SHAW.) — Fig. 46. Metaphase of fourth spermatogenous mitosis of *M. quadrifolia* with an irregular body (blepharoplast) at each pole. (After SHARP.) — Fig. 47. Late telophase of a spermatid mother cell in *M. vestita* showing blepharoplasts in polar positions. (After SHAW.) — Fig. 48. Fragmented blepharoplasts in spermatid cells of *M. vestita.* (After SHAW.) — Fig. 49. Elongated blepharoplast lying near the nucleus in the spermatid cell of *M. vestita.* A mass of starch is seen at the right of the nucleus. (After SHAW.) — Fig. 50. Lateral view of a spermatid cell of *M. quadrifolia* showing the blepharoplast

---

[1]) See also A. YUASA, Feulgen's nucleal-staining applied to Pteridophyta. Proc. Imp. Acad. *22*, 226–268. 1936. for negative reaction of border-brim, lateral bar, cilia-bearing band and cilia.

[2]) DRACINSCHI, p. 199, note 1.

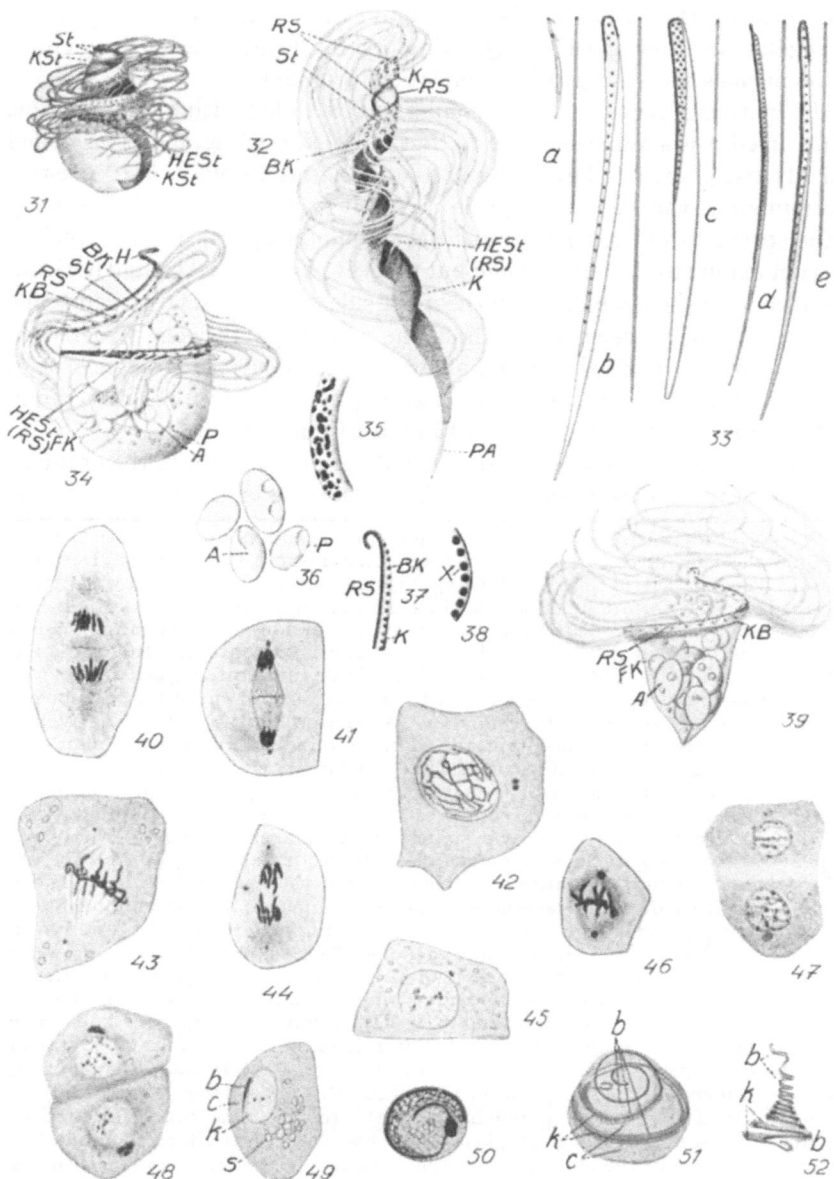

extending beyond the anterior end of the nucleus. (After SHARP.) — Fig. 51. A three-dimensional drawing of the spermatid of *M. vestita* whose nucleus and the dark-staining thread associated with the nucleus, is coiled about the cytoplasmic portion of the cell. (After SHAW.) — Fig. 52. The body of the spermatozoid of *M. vestita* after the vesicle has been shed. (After SHAW.) A, starch; *BK*, basal granule, point of origin of cilium; *FK*, fat globule; *HESt*, posterior end of "stalk" of motor apparatus; *K*, nucleus; *KB*, beginning of nucleus; *P*, plastid; *PA*, plasmatic sheath; *RS*, border-brim („Randsaum"); *St*, "stalk" of motor apparatus.

more delicate than those of the ferns and can be traced to basal granules arranged in one row on the cilia-bearing band. The granules in the vicinity of the nucleus are arranged in a single row (Fig. 38).

The mature spermatozoid of *Salvinia* [1]) is extruded within a membrane. When freed it is similar in shape to that of *Pilularia* (Fig. 39). It is coiled scarcely twice about a large vesicle containing fat globules, 6–10 plastids and numerous vacuoles which stain intensely with neutral red. The body of the spermatozoid consists of nucleus and motor apparatus. The nucleus does not extend to the anterior end but tapers to a point 16μ behind it. Granules can be stained which lie on the inner side of the plasmatic membrane surrounding the nucleus. The stalk of the motor apparatus is bordered on one side by the border-brim and shows along its surface a single row of basal granules, 25–30 in number, each giving rise to a relatively long cilium.

## Oogenesis and fertilization.

In the formation of the egg EMBERGER describes a return to the granular chondriome in a cytoplasm containing numerous small vacuoles (Figs.

FIG. 53–75 (*Filicinae*). — Fig. 53. A young archegonium of *Dryopteris dentata* in which the ventral cell is dividing. (After YAMANOUCHI.) — Fig. 54. Detail of dividing ventral cell of fig. 53 showing "thread structure of unknown origin" described by YAMANOUCHI. Modified FLEMMING's fix. (After YAMANOUCHI.) — Fig. 55. Immature egg of *Polypodiaceae* containing, besides ordinary mitochondria and chondrioconts, small vacuoles (*V*) with precipitated contents. REGAUD's fix. (After EMBERGER.) — Fig. 56. Granular chondriome in mature egg cell of *Polypodiaceae*. REGAUD's fix. (After EMBERGER.) —Fig. 57. Coiled spermatozoid nucleus in cytoplasm of egg of *Lygodium palmatum*. (After ROGERS.) — Fig. 58. Egg cell of *L. palmatum* shortly after one spermatozoid has entered the cytoplasm; functional spermatozoid resting on the nucleus of the egg while its cilia are in the periphery of the cytoplasm; supernumery spermatozoids crowded together at the surface of the egg cell. Arrow indicates direction of the neck of the archegonium. (After ROGERS.) — Fig. 59. Cilia found at periphery of cytoplasm after one spermatozoid has penetrated the egg nucleus. *L. palmatum* (After ROGERS.) — Fig. 60. Polyspermy in *L. palmatum*. One spermatozoid nucleus within the egg nucleus, cilia and a second spermatozoid in the vacuolate cytoplasm; supernumary spermatozoids at surface of the egg cytoplasm. (After ROGERS.) — Fig. 61. Four-celled embryo of *Adiantum capillus-Veneris* with its large vacuoles. Plastids, chondrioconts, granular and rod-shaped mitochondria are seen in the cytoplasm. REGAUD's fix. (After EMBERGER.) — Fig. 62. Detail of the chondriome of an apical cell of the root of *Athyrium Filix-femina;* granular and rod-shaped mitochondria (*M*) and some vesiculate elements (*P*). REGAUD's fix. (After EMBERGER.) — Fig. 63. Detail of chondriome of the cortical parenchyma of the root of *Athyrium Filix-femina;* ordinary mitochondria (*M*), plastids (*P*). REGAUD's fix. (After EMBERGER.) — Fig. 64. Cells of the cortical parenchyma of the root of *Athyrium Filix-femina;* ordinary mitochondria (*M*), amyloplasts (*A a*). REGAUD's fix. (After EMBERGER.) — Fig. 65–67. Stages in the elongation of the mitochondrial elements in cells of the central cylinder from the root of *Athyrium Filix-femina*. REGAUD's fix. (After EMBERGER.)— Fig. 68. Oblique section of the apical cell in the stem of *Pteridium aquilinum* with the elements of the chondriome and the vacuoles clearly defined in the cytoplasm. REGAUD's fix. (After EMBERGER.) — Fig. 69. Cells of young cortical parenchyma of the stem of *Pteridium aquilinum* enclosing granular and rod-shaped mitochondria (*M*), vacuoles (*V*) and a few plastids (*P*). REGAUD's fix. (After EMBERGER.) — Fig. 70. Detail of plastids in fig. 71. — Fig. 71. Older cortical parenchyma cells of

---

[1]) See note 2, page 210.

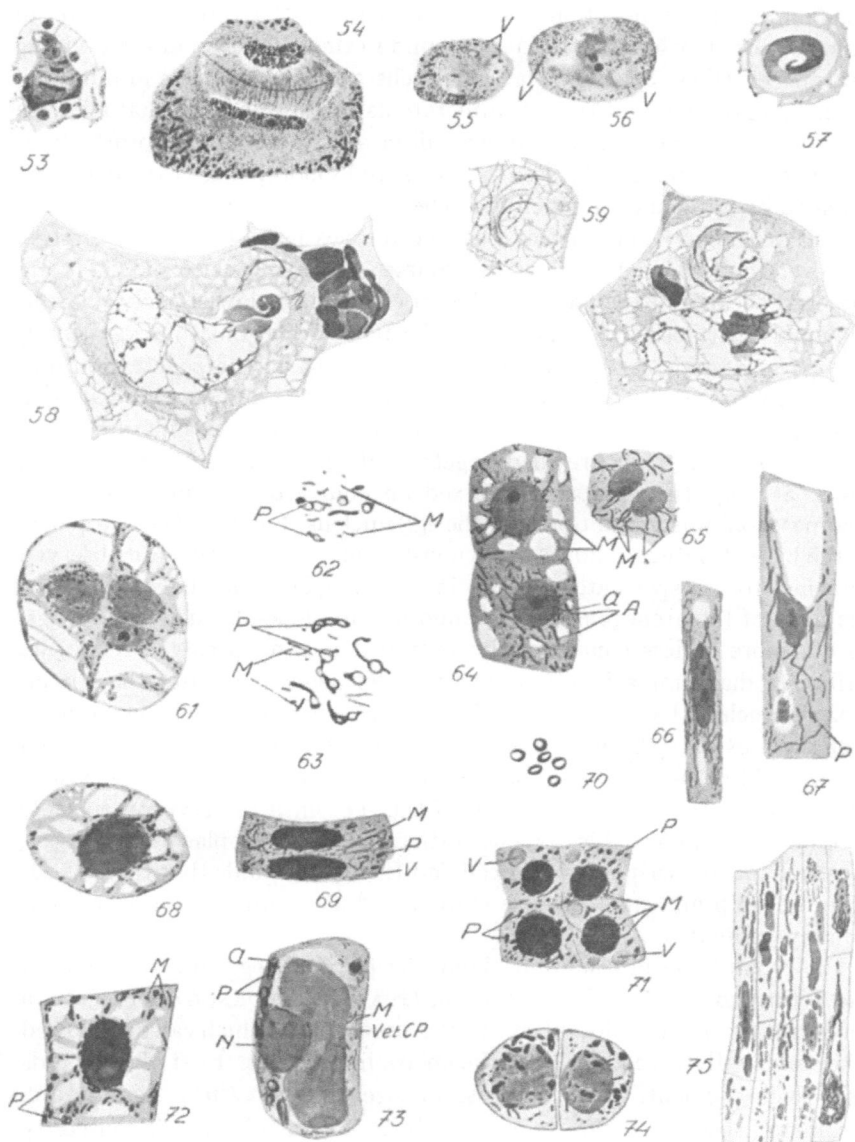

*Pteridium aquilinum* enclosing ordinary mitochondria (*M*), vacuoles (*V*) and plastids (*P*) the last being larger than those in fig. 69. (After EMBERGER.) — Fig. 72. Stelar parenchyma of stem of *Pteridium aquilinum* showing mitochondria (*M*) and plastids (*P*). (After EMBERGER.) — Fig. 73. Epidermal cell from a leaf of *Phyllitis Scolopendrium* enclosing a large central vacuole (*Vet CP*), small mitochondria in the form of granules and rods (*M*), large starch-forming plastids (*Pa*) and nucleus (*N*). REGAUD's fix. (After EMBERGER — Fig. 74. Ordinary mitochondria and developing plastids in an immature stomatal apparatus of the leaf of *Phyllitis Scolopendrium*. (After EMBERGER.) — Fig. 75. Vacuolar system in the vascular meristem of a young leaf of *Athyrium Filix-femina*. (After EMBERGER.).

55, 56). In this connection one might mention the thread stucture of unknown origin which YAMANOUCHI found in the cytoplasm of the dividing central cell (Figs. 53, 54). The canal cells also undergo this granulation of the chondriome and then disintegrate as the egg reaches maturity. In the cells of the neck of the archegonium and those of the prothallium forming the venter can be seen vacuoles and chloroplasts, the latter surrounded by ordinairy mitachondria. The egg cell at the time of fertilization is well distended [1]), not concave as so often described (Fig. 56, 58). It is, however, impossible to follow the course of the spermatozoid in living material because of the large number of chloroplasts in the neck cells and the embedded position of the venter. In fixed archegonia of *Lygodium* [1]) the spermatozoids can be found at various stages in the penetration into the cytoplasm of the egg. They are coiled (Fig. 57) and surrounded by a hyaline area interpreted as the cytoplasmic sheath. The supernumerary spermatozoids remain crowded together at the surface of the egg cell (Figs. 58, 60). In a preparation fixed an hour after flooding, the coiled spermatozoid was found touching the egg nucleus (Fig. 58). The cytoplasm at the base of the egg is now conspicuously vacuolate. No cilia are observed attached to the spermatozoid, but in the cytoplasm on the side toward the neck of the archegonium are found definite hair-like structures (Figs. 58–60) more or less concentrically arranged in semi-circles [2]). They are strikingly diagrammatic, often persisting for some time after the spermatozoid nucleus has penetrated that of the egg. One is inclined to interpret them as cilia thrown off as the spermatozoid makes its way toward the egg. After penetration, the spermatozoid nucleus undergoes a slow process of expansion in the nucleus of the egg, until the two constituents become indistinguishable. Vacuolization of the cytoplasm progresses, especially at the periphery of the cell and EMBERGER finds that the plastids again develop and manufacture starch. Ordinary mitochondria are continually present.

The cells of the prothallium about the fertilized egg lose their starch and return to the chondriocont form. GRATZY-WARDENGG [3]) observed a degeneration of the chloroplasts in the prothallium which can be followed successively from cell to cell as the sporophyte grows. First the plastids lose their pigment, then decrease in size and sometimes agglutinate, becoming speckled and vacuolate (Fig. 3), a change in viscosity

---

[1]) L. ROGERS (ATKINSON), Development of the archegone and studies in fertilization in *Lygodium palmatum*. La Cellule 37, 327–352. 1926. Fertilization is also described by CAMPBELL. Ann. Bot. 2, 233–264. 1888 (*Pilularia*); by SHAW, p. 198, note 4; by YAMANOUCHI, p. 197, note 6. — Also Chap. VIII § 7.

[2]) See also SHAW's fig. 3. The fertilization of *Onoclea*. Ann. Bot. 12, 262–285. 1898 (*O. sensibilis, Matteuccia Struthiopteris*).

[3]) P. 206, note 1.

often being accompanied by amoeboid activity. The final result is a "chondriocont-like leucoplast", whose return to the state of a chloroplast can be brought about in most cases by the removal of the sporophyte.

Polyspermy occurs, at least in *Lygodium* (Fig. 60).

**The sporophyte.** — The four-celled embryo contains large vacuoles and a chondriome which consists of short rods, granules and somewhat heavily-stained, starch-bearing elements, the young plastids. These elements are about the nucleus (Fig. 61). The wedge-shaped intitial cell of the leaf of *Gymnogramme sulphurea* at the momentwhen that organ first pierces the archegonial envelope, contains a weakly-staining cytoplasm with large vacuoles [1]. In the neighboring cells there are small vacuoles present, enclosing precipitates which stain heavily with haematoxylin. The mitochondria are few in number and in the shape of granules and short rods, but in the superficial cells they develop rapidy into starchbearing chloroplasts.

In the independant sporophyte, the apical cell of the root is large, its cytoplasm containing numerous vacuoles without visible content and a chondriome made up of granules and rods (Fig. 62). Some of the rods are vesicular in shape indicating the presence of starch [2]. [3]. The segments nearest the apical cell show essentially the same characteristics. Toward the root cap the vacuoles in the cells are filled with phenol compounds which make observation difficult but ordinary mitochondria as well as longer starch-containing rods can be observed. These begin to disintegrate in the older cells of the root cap and eventually disappear with the cytoplasm. In the meristematic tissue behind the apical cell, there is a considerable elongation of the mitochondria (Fig. 64) and under high magnification starch inclusions may be observed. In the external region of the cortex there are large amyloplasts, so packed with starch that it appears to be in chains (Fig. 63). These are accompanied by ordinary mitochondria. The vacuoles are small. They are round or elongated and, swelling rapidly, form the typical large vacuole filled with phenol products. In the central cylinder, progressing from the apex of the root to the older cells there is a gradual lengthening of the chondrioconts accompanied by a loss in staining capacity (Figs. 65, 66). These chondrioconts form a confused bundle of fliaments about the nucleus (Fig. 66), an element here and there showing a grain of starch. Existing side by side with them are granular mitochondria and short or thin rods. Even in the members of the pericycle the elongation is considerable. The vacuoles of the central cylinder develop

---

[1] A. VLADESCO, Recherches morphologiques et expérimentales sur l'embryogénie et l'organogénie des fougères leptosporangiées. Dissertation. Paris 1934.

[2] EMBERGER.

[3] MOTTIER, p. 201, note 3 (*Adiantum pedatum*).

slowly. They are pseudomitochondrial [1]) in form at first but eventually increase in size and fill with phenol compounds (Figs. 66, 67).

In the stem of the *Filicinae* [2]) the apical cell is large and its vacuoles [3]) have no visible contents (Fig. 68). The chondriome consists of short rods and granules packed about the nucleus, some among them containing starch while others of similar form do not. As the cells of the cortical parenchyma differentiate there is, in general, an elongation of the ordinary mitachondria to chondrioconts (Fig. 69). In the older parenchyma cells these become gradually less chromatic until only the plastids are distinguishable (Fig. 70, 71). The stelar parenchyma is made up of long cells which when young contain vacuoles without visible contents and a chondriome consisting of rods, granules and lightly elongated chondrioconts, with or without starch (Fig. 72). In the procambial strands the starch is absorbed and the plastids, as well as the ordinary mitochondria, undergo elongation.

The young leaf cells have large nuclei, small elongated or round vacuoles, containing phenol compounds, and granular and rod-shaped mitochondria in which all the elements seem to be of the same value. Soon some of them elongate and manufacture starch and as the protoplasm becomes more vacuolate, some of the mitochondria appear markedly thicker and plastid-like, as in the root parenchyma. The small ordinary mitochondria remain as earlier described and the large chondriosomes fragment to form plastids. Inactive chondrioconts are rare. In the epidermal cells of the differentiated leaf there are granules and rods present as well as large lenticular plastids

FIG. 76–95 (*Filicinae*). — Fig. 76. Central cell of a young sporangium of *Dryopteris dentata* enclosing mitochondria in the form of short rods and granules. Formol-chormic-and-osmic-acid fix. (After SENJANINOVA.) — Fig. 77. Portion of tapetal layer and enclosed spore mother cells in a sporangium of *Asplenium Ruta-muraria;* chondriome of the tapetum containing somewhat elongated elements but that of the spore mother cells is granular. REGAUD's fix. (After EMBERGER.) — Fig. 78. Spore mother cell of *Todea barbara* just before reduction division; a few chondrioconts persist about the nucleus. Sublimate-formol fix. of MANGENOT. (After PY.) — Fig. 79. Tapetal cell at the stage of the spore mother cell. *Todea barbara.* (After PY.) — Fig. 80. Section of a sporangium of *Asplenium Ruta-muraria* at the spore mother cell stage. REGAUD's fix. (After EMBERGER.) — Fig. 81. Diakinesis in *Todez barbara* of the first division in the spore mother cell showing the mitochondria scattered indiscriminately in the cytoplasm. Sublimate-formol of MANGENOT. (After PY.) — Fig. 82. Same stage as that for fig. 81; mitochondria arranged in groups in the cytoplasm of *Dryopteris dentata.* Formol-chromic-and-osmic-acid fix. (After SENJANINOVA.) — Fig. 83–86. Reduction divisions in *Dryopteris dentata.* Formol-chromic-and-osmic-acid fix. (After SENJANINOVA); 83. End of first reduction division; invasion of the equatorial region of the cell by the mitochondria; 84. The

[1]) EMBERGER uses this term to designate the filamentous form developing from the early spherical elements. See classical description of the vacuole by the DANGEARDS. — C. R. Acad. Sci. 169, 1005–1011. 1919. and C. R. Acad. Sci. 170, 474–478. 1920.

[2]) EMBERGER.

[3]) WENT described the vacuole in the apical cell of the stem in the species listed on p. 209, note 5.

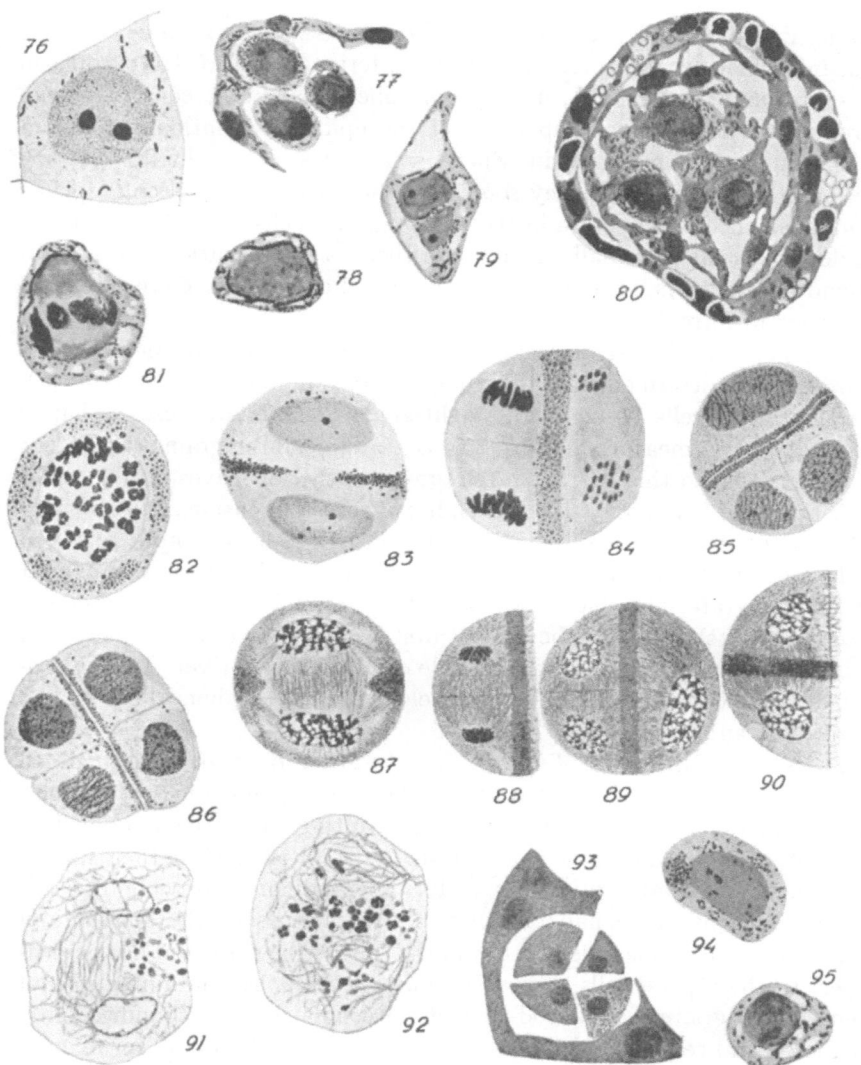

mitochondrial band separates the spindles of the second reduction division;
85. Young tetrad showing splitting of the mitochondrial plate; 86. Division of the
mitochondrial body into four nearly equal parts by the new cell walls of the tetrad.
Fig. 87–90. Stages comparable to those of figs. 83–86, in the spore mother cells of
*Dryopteris dentata* following fixation with FLEMMING's fluid. (After YAMANOUCHI.)
— Fig. 91. First reduction division in the spore mother cell of *Marsilia quadrifolia;*
starch mass moving to an equatorial position. FLEMMING's fix. (After MARQUETTE.)
Fig. 92. Same stage as that of fig. 84 and of fig. 88; starch masses separating the two
spindles of the second reduction division. *M. quadrifolia*. (After MARQUETTE.) —
Fig. 93. Young tetrad surrounded by tapetal tissue showing granular chondriome.
REGAUD's fix. (After EMBERGER.) — Fig. 94. Young spore of *Dryopteris dentata;*
certain mitochondrial elements have enlarged. (After SENJANINOVA.) — Fig. 95.
Small plastids and rod-shaped and granular mitochondria in a young spore of a
tetrad. *Todea barbara*. (After PY.)

containing starch (Fig. 73). The parenchyma cells show large and numerous plastids which decrease in size toward the interior of the leaf. The cytoplasm is sparse and the vacuoles, usually only one in each cell, contain phenol products. The stomatal apparatus of the epidermis contains at first a homogeneous chondriome in which small plastids gradually become differentiated (Fig. 74). They soon take the form of typical chloroplasts and mask the mitochondria in the mature guard cells. The plastids in the under epidermis are small. In the stele there is an elongation of the mitochondrial elements as in the root and the vacuoles are distinctly filamentous in form (Fig. 75).

The hairs of the water leaves of *Salvinia* present a pnenomenon of degeneration described in both living and fixed material by CHOLODNJ [1]). In the young cells the typical, bright-green plastids are biscuit-shaped and very large, measuring $7\mu$ by $20\mu$, containing starch grains whose size is comparable to that of the starch grains in the air leaves, but in older cells there are found only colorless bodies of small dimensions ($0.5\mu$ by $10\mu$) whose contours are irregular. In a hair consisting, for example, of seven cells, this change from large to small plastids does not go on throughout at the same rate. The chloroplasts at the tip gradually decrease in size and lose their starch, those at the base at first become larger and greener, then suddenly, they too, lose starch, become colorless and shrivel. The chloroplasts are accompanied by the other elements of the chondriome in the shape of granules and rods.

The sporangium of the *Filicinae* [2]) [3]) [4]), arises from an epidermal cell in which there is an absorption of the phenol compounds in the vacuoles and a reduction to a size which is maintained during the subsequent development of the organ. The large chloroplasts of the initial cell lose their starch, become fusiform and divide actively. The ordinary mitochondria are present as rods and small granules. As the tissues of the sporangium differentiate small plastids are seen in the cells of the stalk and mechanical layer. In the tapetal cells there are numerous vacuoles containing phenol compounds, granular and rod-shaped mitochondria and chondrioconts. In the central cells, the plastids lose their chlorophlyll and all forms may be observed from lenticular chloroplasts to chondrioconts but they are not as numerous as the ordinary mitochondria (Fig. 76). In the spore mother cell there is a large nucleus in a dense cytoplasm containing small vacuoles, granular mitochondria, rods and beautiful chondrioconts

[1]) N. CHOLODNJ, Über die Metamorphose der Plastiden in den Haaren der Wasserblätter von *Salvinia natans*. Ber. d. Deut. Bot. Ges. 41, 70–79. 1923.

[2]) EMBERGER.

[3]) G. PY, Recherches cytologiques sur l'assise nouricière des microspores et les microspores des plantes vasculaires. Dissertation. Paris 1932. (*Todea barbara*).

[4]) M. SENJANINOVA, Chondriokinese bei *Nephrodium molle*. Zeit. f. Zellforsch u. Mikros. Anat. 6, 493–508. 1928.

which seem to surround the nucleus (Figs. 77, 78, 80). During the heterotypic prophases as the cell increases in size [1]) there is a segmentation of the chondriome and only rare chondrioconts are seen (Fig. 82). Some of these are thicker and more deeply-staining than the others and probably represent the plastids. The vacuoles are well distributed in the cell. At metophase, PY (Fig. 81) does not observe the polarization of the chondriome described by SENJANINOVA (Fig. 82). The mitochondrial elements remain granular throughout the reduction divisions in the spore mother cells but in the plurinuclear tapetum surrounding them which contains many chondrioconts, rods and granules and numerous vacuoles, there is a development of plastids which begin to manufacture starch (Figs. 79, 80). The chondriome of the young spores (Fig. 93) is granular. As they mature and the tapetum disappears there appear chondrioconts and then plastids among the mitochondria of the ordinary form (Figs. 94, 95). The vacuoles may be rather large and are found to contain phenol compounds.

SENJANINOVA describes for *Dryopteris* (Figs. 82–86) the formation of a compact mitochondrial mantle which in behavior is very like that described for *Equisetum* by LEWITSKY [2]). At the metaphase of the spore mother cell, the mitochondria surround the nucleus either as a compact mass or in groups (Fig. 82) but during the telophases, the entire chondriome accumulates at the equator (Fig. 83-85) where it forms a zone which, although it contains vacuoles, is still more dense than the surrounding cytoplasm. The plate soon shows a cleavage at its center. When after the homoeotypic division the new wall is laid down between the pair of sister nuclei, this wall reaches from the outer membrane of the spore mother cell on the one hand and to the above-described mitochondrial plate on the other and when all the walls are complete each spore receives approximately a quarter of the mitochondria of the spore mother cell (Fig. 86). One is struck by the close resemblance of the figures of SENJANINOVA [3]) to those of YAMANOUCHI [4]) who described for *Dryopteris* a granular zone which formed at the equator during the reduction divisions (Figs. 87–90). This is also reminiscent of MARQUETTES observations [5]) on *Marsilia* where in the spore mother cell he saw an aggregation of starch grains in close contact with the nucleus. These occupy a position at one

[1]) R. DE LITARDIÈRE, Variations de volume du noyau et de la cellule chez quelques Fougères durant la prophase hétérotypique. C. R. Acad. Sci. 156, 562–564. 1913. (*Polystichum aculeatum, Dryopteris Filix-mas, Asplenium Trichomanes, A. Adiantum-nigrum*).

[2]) P. 231, note 3.

[3]) P. 218, note 4.

[4]) SH. YAMANOUCHI, Sporogenesis in *Nephrodium*. Bot. Gaz. 35, 1–30. 1908. (*N. molle*).

[5]) W. MARQUETTE, Concerning the organization of the spore mother-cells of *Marsilia quadrifolia*. Trans. Wis. Acad. Sci., Arts and Letters. 16, 81–106. 1908.

side of the spindle at the beginning of division but at the end of the first division move in between the disappearing spindle fibres where they remain until the nuclei of the second divison are reconstructed (Figs. 91 92). During the cell division to form the spores, the starch is so distributed that each spore receives an equal quantity.

## § 3. Lycopodiinae. — The gametophyte. — The gametophyte of this group has been investigated in its morphological rather than in its cytological aspects.

**Spermatogenesis.** — Cytological accounts of the developing spermatozoid are not available but DRACINSCHI [1]) has given a description of the mature gamete. In *Isoetes* (Fig. 96) the free-swimming cell consists of motor apparatus, nucleus and two vesicles which are soon lost. These vesicles first appear as two discs but soon after the release of the spermatozoid they absorb water to such an extent that they double their original diameter. Each contains fat and vacuome elements but shows no positive reaction for starch. The nucleus, pointed at each end, is enclosed in a spongy plasmatic structure and does not extend to the most anterior tip

FIG. 96–112. (*Lycopodiinae*). — Fig. 96. Mature spermatozoid of *Isoetes lacustre*. (After DRACINSCHI) — a, detail of cilia. — Fig. 97. Spermatozoid of *Selaginella* coiled about the vesicle. (After DRACINSCHI.) — Fig. 98. Spermatozoid of *Selaginella* after the vesicle has been liberated. (After DRACINSCHI.) — Fig. 99. Spermatozoid of *Selaginella involvens* at the moment of extrusion from the antheridium. (After YUASA.) — Fig. 100. Detail of anterior portion of the spermatozoid of *Selaginella involvens* showing the common point of attachment of the cilia, a, b. (After YUASA.) — Fig. 101. Mature spermatozoid of *Selaginella involvens* enclosing granules (b) in the posterior portion of the body; *a*) sharp posterior tip (After YUASA.) — Fig. 102. Chloroplasts in a cell at the apical region in the stem of *Lycopodium*. (After EMBERGER.) — Fig. 103. Elongated elements of the chondriome in a cell of the stelar tisue of *Lycopodium*. (After EMBERGER.) — Fig. 104. Young spore mother cell of *Lycopodium* showing the somewhat elongated chondrioconts. (After EMBERGER.) — Fig. 105. Chondrioconts in a spore mother cell of *Lycopodium*. (After EMBERGER.) — Fig. 106. Living cells of the apical region of the stem containing each a long filament. *Selaginella Kraussiana*. (After EMBERGER.) — Fig. 107. Cells in the living cortical parenchyma of the stem of *S. Kraussiana* showing different types of chloroplasts. (After EMBERGER.) — Fig. 108. Chloroplast chain in a living cell of the cortical parenchyma of the stem. *S. Kraussiana*. (After EMBERGER.) — Fig. 109. Two chloroplasts in living pericycle cells of *S. Martensii*. (After EMBERGER.) — Fig. 110. A tooth from a living cell of the leaf of *S. Martensii*. (After EMBERGER.) — Fig. 111. Guard cells from a living leaf of *S. Martensii*. (After EMBERGER.) — Fig. 112. *a*) Healthy chloroplast from a cell on the under side of a variegated leaf of *S. Martensii*. (After SCHWARZ.); *b*) Beginning of degeneration of the chloroplasts in a variegated leaf of *S. Martensii*. (After SCHWARZ.) — Fig. 113. A state of nearly complete degeneration of the chloroplasts in a leaf cell of *S. Martensii*. (After SCHWARZ.) — Fig. 114. Stomatal apparatus in leaf of *S. Martensii;* healthy chloroplast in the guard cell at the left, degenerating chloroplasts in the cell at the right. (After SCHWARZ.) — Fig. 115. Cell of the central cylinder of the stem of *Selaginella* showing elongated chondriome; *P*, plastids; *M*, ordinary mitochondria. (After EMBERGER.) — Fig.116. Young sporangium of *Selaginella*. (After EMBERGER.) — Fig. 117. Portion of the apical region of a median section through the mature macrospore of *S. rupestris;* protoplasm containing much protein matter. (After LYON.) — Fig. 118. Late diaster

[1]) P. 199, note 1. (*Isoetes lacustre, I. japonica*).

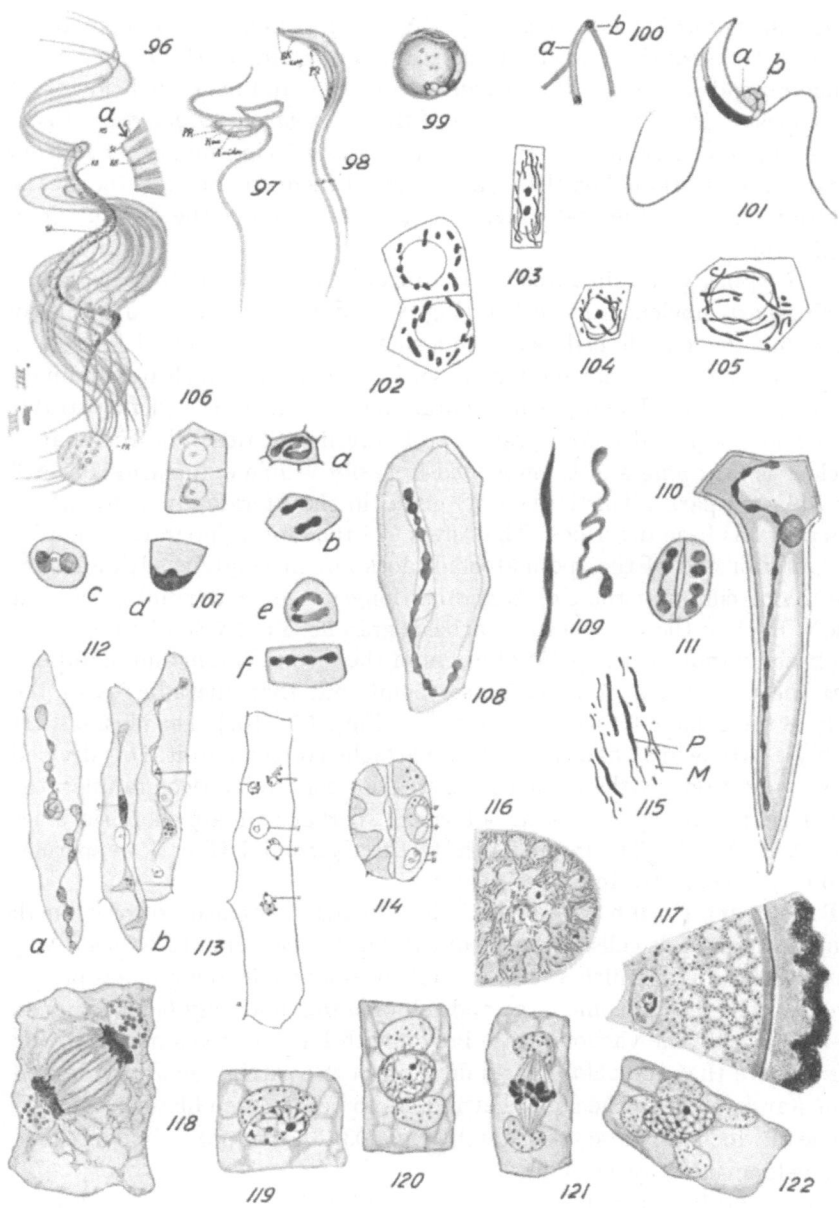

in a young leaf cell of *Isoetes lacustre* showing polar positions of starch grains. (After MARQUETTE.) — Fig. 119. Young leaf cell of *I. melanopoda* enclosing a plastid at one side of the nucleus. (After MA.) — Fig. 120. Young leaf cell of *I. melanopoda* with two daughter plastids on opposite sides of the nucleus. (After MA.) — Fig. 121. Young leaf cell in *I. melanopoda;* metaphase with two plastids occupying polar positions in the cell. (After MA.) — Fig. 122. Four plastids in a cell from the upper part of the leaf of *I. melanopoda.* (After MA.)

of the antherozoid. The "stalk" [1]) of the motor apparatus is homogeneous and transparent, traversed on its outer margin by the border-brim,
which extends almost to the nucleus on one side and back far below it on
the opposite side (Fig. 33). Like that of the *Filicinae* and *Equisetum* it is to
be regarded as a chondriosome. The cilia, about 25 in number, arise from
basal granules situated on the "stalk," four of them grouped together at the
anterior end and the rest extending backward along the "stalk" in a
single  row.

In *Selaginella* the liberated spermatozoid consists of motor apparatus,
plasma-part, nucleus and vesicle (Fig. 97). When first released the body
surrounds the lens-shaped vesicle in a single coil. Then the vesicle absorbs
water, becomes spherical and is soon set free. It consists of clear, seemingly
homogeneous, fluid cytoplasm, containing fat substances demonstrable
with Sudan III and starch grains, probably in plastids. The rod-shaped
nucleus is $2.5\mu$ long and is embedded in easily visible cytoplasm (Fig. 98)
the "plasma-part". This is the only genus in the Pteridophytes for which
this part has been described. The "stalk" of the motor apparatus occupies
the anterior end of the spermatozoid, does not macerate easily and gives
rise to one cilium at the tip, to another longer one at a point somewhat
below. Each of these is traced to a basal granule. YUASA [2])did not observe
the basal granule at the point of origin of the second cilium and maintains
that the two cilia arise at the same point, but that one adheres to the
body of the gamete for a certain distance (Figs. 100, 101). The plasma-part
is only a little wider than the stalk and attached directly to it. It is divided
into two portions by the nucleus, the anterior containing two granules, the
posterior containing 4–5, some of which, perhaps all, are chondriosomes
(Figs, 98, 99, 101). The resemblance to the *Chara* and Hepaticae spermatozoid [3]) is very striking.

**T h e   s p o r o p h y t e.** — In the viscous cytoplasm of the small
living meristematic cells at the tip of the vegetative stem of *Lycopodium* [4])
may be seen a large nucleus and several rod-shaped or round chloroplasts.
In older cells they become large and full of starch and may be as many as
15–20 in a cell. The vacuoles stain brown with I-KI. In fixed material, one
can observe that the chloroplasts develop in the parenchyma, divide and
form starch, while in the stele they regress to become fine filaments, in the
course of which they lose their starch (Fig. 103). In the root a similar course
of development may be followed.

The young leaf, forming early in the bud, exhibits the same cytology
as those cells from which it has come. The meristem cells at the base of an
older leaf contain somewhat elongated chloroplasts which are larger than

---

[1]) P. 199.

[2]) YUASA (*Selaginella involvens.*).

[3]) MÜHLDORF, p. 198, note 8.

[4]) EMBERGER.

those of the stem. Near the stele, the plastids are filamentous but at the tip of the leaf they assume their mature form and become pigmented.

Except that the sporangium in *Lycopodium*[1]) arises from cells containing small plastids and not from colorless cells as in *Selaginella,* one might superimpose the two stories of development. The primordia of the future sporangium do not differ in cellular content from those leaf cells giving rise to them, but as the sporogenous cells develop, the chlorophyll in them disappears and during the entire process of sporogenesis, no chlorophyll can be seen. It is a well known fact that the spores of *Lycopodium* do not give a reaction for starch. In the stalk and dehiscence layer the plastids continue to grow, remaining as rods more or less elongated in the stalk but becoming round at the summit of the sporangium. Here they reach their maximum development at the time the spore mother cells form and then they become absorbed during the maturation of the sporangium. The tapetal cells contain well-developed plastids but in the sporogenous cells there are only chondrioconts to be seen in living material. Fixed preparations reveal the second class of mitochondria in the form of filaments, rods and granules. They are especially visible in the peripheral cells of the stalk of the young sporangium. The vacuoles do not present striking particulars but in certain cells their contents may turn yellow brown with I–KI in the manner of glycogen [1]). Oil globules are everywhere abundant. The spores of *Lycopodium* in living material do not appear green but SAPEHIN [2]) reported the presence in the spores of fat globules which reduce osmium. The non-photosynthetic chondriome has the classic form and chlorophyll appears at germination.

*In vivo* observations of the exceedingly small meristem cells of the stem tip in *Selaginella* are difficult by reason of the metabolic products which mask even the rather large nucleus and the fine undulous filament [1]) [3]) whose index of refraction is only slightly greater than that of the cytoplasm. This filament (Fig. 106) changes position in the cytoplasmic currents, is sensitive to osmic acid and presents the fixing and staining characteristics of a chondriosome. In older cells it takes on pigment and prepares for a division which is so seldom completed that chains of chlorophyll bodies are formed (Figs. 107, 108) as described by HABERLANDT [4]). Fixed preparations confirm this and bring out the second line of mitochondria in the form of numerous granules, rods and rare chondrioconts. In fixed material of the apical cells of *S. Kraussiana,* EMBERGER found a chondriome in which it was impossible to distinguish the filament seen in living material except for its proximity to the nucleus. The vacuoles of

---

[1]) EMBERGER.

[2]) 1915. P. 201, note 3.

[3]) P. A. DANGEARD, Plastidome, vacuome et sphérome dans *Selaginella Kraussiana.* C. R. Acad. Sci. 170, 301–306. 1920.

[4]) P. 199, note 4.

these young cells are small and round, filled with reserves which stain a dirty green with haematoxylin following REGAUD's fixative. In older cells they enlarge but have not been observed to pass through the pseudomitochondrial stage observed in the Phanerogams[1]). Fatty globules are present in all cells.

In the young vascular tissue the rather rigid plastids are accompanied by rods, granules and numerous filamentous, more or less elongated, chondrioconts (Fig. 115). In the central cylinder all mitochondrial elements elongate, the plastids lose their pigment and in the future vessels, the entire chondriome is absorbed. In the pericycle the chloroplasts elongate and present diverse and beautiful forms. In the cortical parenchyma the chlorophyll bodies are usually in chains and are accompanied by chondrioconts, rods and granules of the usual type.

The cytology of the root [2]) of the *Lycopodiinae* is said to be similar to that of the stem.

The leaf inherits the constitution of the tissue from which it was derived. In the young living organ, each cell presents a large nucleus and a single pale green plastid near or over it. This plastid divides to form two but rarely more, while in those cells nearest the point of insertion of the leaf, there is less chlorophyll, the plastids being replaced by filaments of little or no color. These filaments are homologous to the chondrioconts of the apical cells. Approaching the tip of the undifferentiated leaf tissue [2]) [3]) the filamentous plastid thickens, rounds up and becomes lenticular or dumbbell shaped. These may break apart and in the cells at the tip of the leaf the chloroplasts may appear round but they do not long remain so, for by a series of incompleted divisions there is formed an infinite variety of extraordinary forms as described by HABERLANDT[3]). EMBERGER describes a second method whereby the chains are formed. A long filament becomes nodulate (Fig. 109) its aspect varying with the nature of the cell enclosing it. The young teeth possess plastids, round in some species, in the form of chains in others (Fig. 110) but as the teeth develop, the assimilation apparatus disorganizes, the starch is absorbed and the plastids lose their pigment. In fixed material of the teeth cells which grow near the insertion of the leaf, there is shown a return to a chondriome of rods and grains. In differentiated leaves the epidermal cells of the two leaf surfaces do not exhibit similar chloroplasts for those which receive the sun's rays have fewer chains and contain more chlorophyll. The guard cells are as described by HABERLANDT (Fig. 111).

In the leaves of green house plants of *Selaginella* kept at cold temper-

---

[1]) GUILLIERMOND, A. MANGENOT, PLANTEFOL, Traité de Cytologie végétale. 1933. p. 297–307 and fig. 157.

[2]) EMBERGER.

[3]) HABERLANDT, p. 199, note 4.

atures, Schwarz [1]) finds a degeneration of the chloroplasts which results in the appearance on the leaves of white areas, more or less large. Each cell of the leaf receives a normal plastid (Fig. 112 a), but in some regions they lose their pigment and become granular (Fig. 112 b). Then their contours seem to dissolve and in the end the chloroplast disappears (Fig. 113). Even sister cells may behave differently (Fig. 114), characteristic plastids being visible in one of them, degenerated plastids in the other. A similar process is described in the ligule.

In the ligule at an early stage there can be seen in living material of *Selaginella* a crescent-shaped, flexuous body adhering to the nucleus. It is similar to that found in the apical region of the stem. In material treated with Regaud's fluid, this body, or plastid, is accompanied by numerous mitochondria of the ordinary form. The filament can not be seen in living cells of the organ at a later stage, but in fixed material its degeneration can be followed. The plastid becomes thin and loses its affinity for stains. There is a fragmentation of all the elements of the chondriome during which the ordinary mitochondria are first affected. Eventually all the elements become vesiculate and disappear.

The vacuoles of this group are small and round in young cells but become very large in the adult stages without passing through the filamentous form described for the other ferns [2]). Fat granules are almost always present and are carried more rapidly than the mitochondria in the cytoplasmic currents.

One should not leave the subject of the leaf without mentioning, for what it is worth, a certain tendency toward polarity manifested by the plastids of *Isoetes*. Marquette [3]) and Ma [4]) describe a single plastid in very young leaves (Fig. 119) lying in close contact with the nucleus. This plastid may divide, its two halves moving to opposite sides of the nucleus (Figs. 118, 120, 121). This may be in response to some sort of polarity in the cell. On the other hand, the plastids may take these polar positions because they can not very well be anywhere else by reason of the vacuolate condition of the cell, or because of the large area occupied at the equator of the cell by the spindle.

In cells where there are four (Fig. 122) or more chloroplasts, they obviously can not occupy polar positions, nor can they in leaves of plants growing in full light where the number of plastids may be as high as 30 per cell.

---

[1]) W. Schwarz, Über die Ursachen und das Zustandkommen der Panaschierung bei einer Form der *Selaginella Martensii*. Protoplasma 10, 427–451. 1930.

[2]) P. 215, 216.

[3]) W. Marquette, Manifestations of polarity in plant cells which are apparently without centrosomes. Beih. z. Bot. Centralbl. 21, 281–303. 1907 (*I. lacustre*).

[4]) R. M. Ma, The chloroplasts of *Isoetes melanopoda*. Amer. Journ. Bot. 15, 277–284. 1928.

FITTING [1]), however, after studying living material of 19 species of *Isoetes* and 9 of *Selaginella*, reported that in the macrospore mother cell a mantle of starch grains forms about the nucleus. It was observed to elongate, divide into two parts, each of which migrates to opposite sides of the cell and there divides again. The amount of starch increases for a time but disappears before the spore is mature.

In the sporangium of *Selaginella* pigment is found only in the external layers of the young organs. In the primordial tissue, DANGEARD [2]) distinguished 1–2 long plastids about the nucleus. In the young sporangium EMBERGER finds them more delicate than those in the stem at the same level and, except for their position about the nucleus, indistinguishable from other members of the chondriome (Fig. 116). The history from this point is identical with that for *Lycopodium* [3]). In the spore mother cells, the plastids enlarge somewhat and become distinguishable from the other chondriosomes. In the pedicel the chloroplasts remain small, in the dehiscence layer they grow at first and then are absorbed at the maturation of the layer. They are present in the nutritive layer but remain thin. One or two are transmitted to the spores and are visible in living cells but they do not become green until the moment of germination. The spores contain little or no starch but a large number of oil drops and small bodies which are probably protein bodies [4]) (Fig. 117).

## § 4. Equisetinae. — T h e  g a m e t o p h y t e. — The spores of *Equisetum* are round with a centrally-placed nucleus surrounded by oval chloroplasts and mitochondria in the form of rods and granules which are difficult to differentiate (Fig. 123). The large central vacuole of the young spore has been replaced by smaller ones, localized at the periphery of the spore. JOYET-LAVERGNE (1926–'34) [5]) finds two categories based on the physico-chemical reactions of the spores. Spores A stain more heavily with neutral stains and their leuco-derivatives than spores B, bleach more markedly, reduce more strongly osmic acid and are richer in glutathion. By analogy with similar reactions in the nucellus and pollen of Phanerogams, this author would ascribe a female tendency to spores A and a male tendency

---

[1]) H. FITTING, Bau und Entwicklungsgeschichte der Makrosporen van *Isoetes* und *Selaginella* und ihre Bedeutung für die Kenntniss des Wachsthums pflanzlicher Zellmembranen. Bot. Zeit. 58, 107–165. 1900.

[2]) P. 223, note 3.

[3]) P. 223.

[4]) F. LYON, The spore coats of *Selaginella*. Bot. Gaz. 40, 285–295. 1905. (*S. rupestris, S. apus, S. Emmelliana*).

[5]) PH. JOYET-LAVERGNE, La sexualisation cytoplasmique et les caractères physico-chimiques de la sexualité. Protoplasma 3, 357–390. 1928. (Contains also bibliography of earlier papers). — Le changement de sexe et la sexualisation cytoplasmique. Ibid. 11, 320–348. 1930. — See also K. GOEBEL, Organographie Pt. II. 1930. p. 1075 where 'an inner unlikeness of the spores' is discussed.

to spores B, formulating laws concerning this inferred sexualization of the cytoplasm. This work has met with some criticism based on observed reversal in sex brought about by changes in physiological factors [1]).

The first division wall separating the rhizoid cell from the prothallial cell, orientated in a direction perpendicular to the light rays falling on it engenders a polarity of the plastids (NIENBURG 1924) [2]) so that those plastids lying with their longitudinal axes directed toward the centrally placed nucleus (Fig. 124), migrate gradually toward the lighted side of the spores. When there are only a few straggling chloroplasts left, the nucleus takes up its position on the side away from the light (Figs. 125–127) and proceeds to division [3]).

The prothallial cells contain 45–60 plastids which more or less resemble each other. REINHARD (1933) [4]) observing their division *in vivo* finds that there is a six hours lapse of time from the first indication of the dumbbellshaped division-form to the completion of the process. Often chains of four or more plastids persist, similar to but always shorter than, those of *Selaginella* [5]).

**Spermatogenesis.** — We must go back to 1912 (SHARP) [6]) for the fullest account of the formation of the male gamete of *Equisetum*. This history has been confirmed by SETHI (1928) [7]) whose interest, however, is morphological rather than cytological. In the sporogenous cells of the embedded antheridium, there is a gradual regression of plastids (Fig. 132), although they may still be observed as late as the spermatid mother cell stage (Fig. 135). In the spermatid mother cell there appears near the nucleus, a minute granule of unknown origin. (Fig. 128). It stains intensely with haematoxylin and shows faint cytoplasmic radiations. It divides, the halves grow larger, migrate to opposite sides of the nucleus and remain

---

[1]) E. SCHRATZ, Untersuchungen über die Geschlechterverteilung bei *Equisetum arvense*. Biol. Zentralbl. 48, 617–639. 1928. — See also abstracts of JOYET-LAVERGNE papers by NEEDHAM and NEEDHAM in Protoplasma 3, 259. 1928. — Ibid. 4, 174. 1928.

[2]) W. NIENBURG, Die Wirkung des Lichts auf die Keimung der *Equisetum*-spore. Ber. d. Deut. Bot. Ges. 42, 95–99. 1924. See also, STAHL, Ber. d. Deut. Bot. Ges. 3. 1885.

[3]) W. A. BECKER and T. H. SIEMASZO (1926) substantiate this, noting that the plastids never invade the region of the phragmoplast. Granules staining with Dahlia violet, showing no displacement during mitosis, do not invade that region either, but can be found on its outer surface. La Cellule 45, 29–42. (*E. arvense, E. limosum*).

[4]) H. REINHARD, Über die Teilung der Chloroplasten. Protoplasma 19, 541–564. 1933. (*Equisetum arvense*).

[5]) P. 224.

[6]) L. W. SHARP, Spermatogenesis in *Equisetum*. Bot. Gaz. 54, 89–109. 1912.

[7]) M. L. SETHI, Contributions to the life history of *Equisetum debile*. Ann. Bot. 42, 729–737. 1928.

at the poles during division (Figs. 129–132). During the anaphases the granule becomes less chromatic and might be neglected unless greatly overstained, when it appears as a body of considerable size. In the spermatid cell it is seen again, appearing as an uneven ring of granules (Fig. 133) which fragments, becomes arranged in a row (Figs. 134, 135) and in the mature spermatozoid becomes the darkly staining thread so clearly seen at the anterior end (Figs. 136, 138).

An antheridium may contain as many as 512 spermatozoids[1]) which are discharged singly or in pairs in a membrane from which they soon free themselves. The mature spermatozoid (Fig. 138) consists of a nucleus, a motor apparatus and a vesicle which at first is hardly visible but by absorption of water rapidly increases to a considerable size. This vesicle contains 3–10 vacuoles staining with neutral red and 20–30 starch grains arranged in plastids in groups of two or three. The nucleus begins as a thread about $8\mu$ behind the "stalk" and gradually comes to occupy the entire band. Over its membrane can be stained the refractive granules which have already been mentioned for other groups [2]) but whose nature has not been determined (Fig. 137).

---

FIG. 123–148 (*Equisetinae*). — Fig. 123. Portion of mature spore of *Equisetum arvense* containing large plastids and ordinary mitochondria. (After Py.) — Fig. 124. Radially arranged chloroplasts about a centrally-placed nucleus in a spore of *Equisetum* (After NIENBURG.) — Fig. 125. Migration of chloroplasts toward lighted side of an *Equisetum* spore at the time of germination. (After NIENBURG.) — Fig. 126. First division in an *Equisetum* spore. (After NIENBURG.) — Fig. 127. First wall in *Equisetum* spore cutting off a rhizoid cell with few plastids and a prothallial cell with many plastids. (After NIENBURG.) — Fig. 128–136. Spermatogenesis in *Equisetum* (After SHARP.); 128 Spermatozoid mother cell with a dark-staining body in the cytoplasm; 129. Division of the body in the cytoplasm of the spermatid mother cell; 130. Separation of these bodies of fig. 129; 131. Metaphase of the division to form the spermatid; bodies of fig. 130 lying at the poles; 132. Spermatid cells in whose cytoplasm are seen ring-shaped bodies, the blepharoplasts, and degenerating plastids; 133, 134. Fragmentation of the blepharoplast in the spermatid cell; 135. Blepharoplast granules arranged in a series in the spermatid cell; plastids in the cytoplasm; 136. Spermatid cell in which the nucleus and the blepharoplast have become spirally coiled. — Fig. 137. Portion of nuclear surface of the spermatozoid of *Equisetum Telmateja*. (After DRACINSCHI.) — Fig. 138. Mature spermatozoid of *E. Telmateja; RS*, border-brim („Randsaum"); *BK*, basal granule, attachment point of cilium; *KB*, beginning of nucleus; *HEST*, posterior end of the „stalk" of the motor apparatus; *K*, nucleus; *F*, fat globules; *P*, plastid; (After DRACINSCHI.) — Fig. 139. Detail of the chondriome of a cell near the tip of the root in *E. limosum; P*, plastids, *M*, ordinary mitochondria. (After EMBERGER.) — 140. Detail of the chondriome from a cell of the cortical parenchyma of *E. arvense; P*, plastids *M*, ordinary mitochondria. (After EMBERGER.) — Fig. 141. a) Cells of the vegetative tip of the aerial stem of *E. arvense* whose chrondriome is more or less homogeneous. (After EMBERGER.) — b) Cells from the liver of a chicken, figured for comparison of the chondriome with that shown in fig. 141a. (After EMBERGER.) — — Fig. 142. Vegetative tip of the stem of *E. arvense* in which it is impossible to distinguish plastids from mitochondria. (After EMBERGER.) — Fig. 143. Large lenticular plastids and small spherical pseudochondriosomes in a cell from the inner layer of a leaf of *E. arvense*. Plastids cluster in the polar region of the dividing cell.

[1]) DRACINSCHI, p. 199, note 2.
[2]) P. 210.

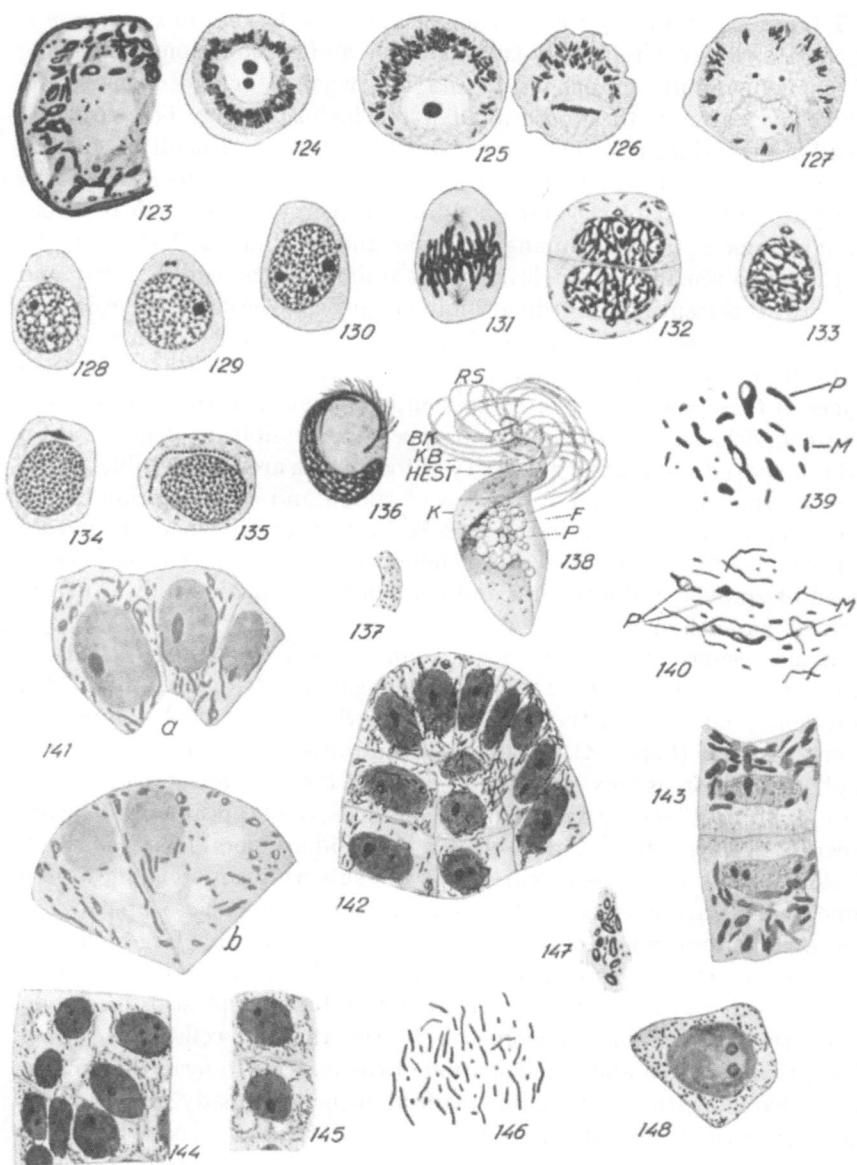

CHAMPY fix. (After BOWEN.) — Fig. 144. Portion of a figure given by EMBERGER of a young sporangium of *E. limosum*, in which the plastids are not easily distinguishable. — Fig. 145 Portion of a figure given by EMBERGER of the sporogenous cells of *E. limosum*, in which the chondriome is homogeneous. — Fig. 146. Detail of homogeneous chondriome of fig. 145. (After EMBERGER.) — Fig. 147. Fragment of the cytoplasm of the tapetal layer of *E. arvense* enclosing very large plastids and smaller mitochondrial granules and rods. (After PY.) — Fig. 148. Spore mother cell of *E. arvense* with its homogeneous chondriome. (After PY.).

**The sporophyte.** — The apical cell of the root of *Equisetum* is vacuolate [1]) with a relatively small nucleus and a chondriome consisting of smaller granular mitochondria and somewhat elongated chondriconts which by reason of their occasional starch inclusions may be considered small plastids (Fig. 139). These two categories are very difficult to separate since they stain alike. As differentiation of the tissue begins, there is an elongation of most of the mitochondrial elements of the cortical parenchyma, some of them becoming vesicular and retaining a dark stain, the rest of them remaining thin, losing their staining capacity (Fig. 140), and probably representing the ordinary mitochondria of the cell. The vesiculate elements enlarge in the outer regions of the cortex and become typical periform or dumb-bell shaped plastids with starch in each. Transition stages of these bodies may often be seen in a single cell. In these cells the long chondrioconts seem to have disappeared, possibly by segmentation but rod-shaped and granular ordinary mitochondria are again visible. In the central cylinder there is an absorption of starch and an elongation of the mitochondrial elements as in the ferns [2]). In the root cap where the phenol compounds are less dominant, the chondriome is easily seen. It is composed of granular mitochondria and more or less elongated rods as in the cortical parenchyma.

In the vegetative tip of the aerial stem of *Equisetum* [1]) the apical cell contains small, pale green, rounded or elongated chloroplasts, embedded in a dense, fairly refractive cytoplasm, in which the vacuolar limits are clearly defined (Figs. 141, a, 142). Those mitochondria having no rôle in photosynthesis are invisible *in vivo*, being of the same refractive index as the cytoplasm. Fixed by mitochondrial methods, they appear more delicate, generally stain less readily than the plastids and are notable for their lack of starch inclusions. The apical cell of the subterranean stem shows the same general arrangement. The young lateral buds enclose apical cells whose chondriome is made up of rods, granules and somewhat larger, slightly more heavily stained chondrioconts which can be traced in the maturing cells through all the stages to mature starch-forming plastids. The bud scales show the same phenomena. In the vascular cells, there is an absorption of starch and an elongation of the plastids to form the tangled mass characteristic of the mitochondrial elements already described for the other vascular cryptogams. In the maturation of a conducting cell from the procambial strand (LENOIR 1926 [3]) the cytoplasm becomes granular in appearance. The cells elongate as this granulation progresses, the cytoplasm becomes plasmolized, the nuclear chromatin thickens and

---

[1]) EMBERGER.

[2]) P. 215.

[3]) M. LENOIR, La nécrobiose dans les éléments du cambium chez *Equisetum arvense*. Rev. Gen. Bot. 38, 615–631. 1926.

the nuclear membrane ruptures as if division were to take place. Then all the contents of the cell disappear and leave the vessels empty.

The leaves of the sporophyte, as might be expected contain the same cellular elements as the stem from which they have developed. BOWEN mentioned and figured an arrangement of the plastids in the neighborhood of the poles during division (Fig. 143). In cells of the young sporangium there are to be found more or less elongated chondrioconts, rods and granules (Fig. 144). As one passes from the shield to the small sac which will be the sporangium these elements lose their staining capacity, become thinner and in the primary sporogenous cells (Figs. 145, 146) are reduced to a remarkable homogeneity in which it is impossible to recognize the chloroplasts which were present in earlier cell generations. In this genus the dehiscense layer of the sporangium does not, as in the ferns, contain plastids at the time of maturation. Phenol compounds are lacking except in the external cell layers of the shield. In the young nurse cells [1]) the chondriome is formed of chondrioconts, rod-shaped and granular mitochondria. The vacuoles are at first small and not numerous but soon increase in number. By a disappearance of the already existing cell walls and a failure to form new ones after nuclear division there is formed a multi-nucleate tissue in which the spore mother cells begin to round up. This tissue is characterized by small vacuoles and a chondriome formed of granules and narrow chondrioconts, some of which are shaped like tadpoles, indicating the beginning of starch formation.

The spore mother cell (Fig. 148) in fixed material shows a fragmented chondriome which in aspect is very like that of a meristematic cell. The cytoplasm is dense, the vacuoles are small and the chondriome is composed of rods and granules among which PY sees certain ones whose staining capacity is greater. EMBERGER does not hesitate to call them plastids. This is borne out by the fact that in living spore mother cells extruded in all stages of division from pricked sporangia (LENOIR 1934) [2]) green plastids can be clearly seen. During the metophase and anaphases of the division, LEWITSKY[3]) described an irregular chondriome mantle which can be observed about the nucleus and PY later confirmed this [1]) [4]). The orientation is lost during interkinesis and the chondriome returns to a state of freely dividing

[1]) PY, p. 218, note 3 (E. arvense).

[2]) M. LENOIR, Etude vitale de la sporogénèse et des phénomènes d'apparence électro-magnetiques concomitants chez l'Equisetum variegatum. La Cellule 42, 355–408. 1934. (Contains also bibliography for earlier papers).

[3]) G. LEWITSKY, Die Chondriosomen in der Gonognese bei Equisetum palustre. Zeit. f. wiss. Biol. Planta Arch. f. wiss. Bot. 1, 301–316. 1925.

[4]) V. JUNGERS (1934) describes a similar mitochondrial grouping for E. limosum but considers it as transitory and not as representing a morphological entity of the cell. La Cellule 43, 323–340.

elements. These elements soon migrate to the equator of the cell where they remain until the formation of the walls which separate the daughter cells. The vacuoles are distributed in that part of the cell not occupied by this mantle. LENOIR [1]) also observes in living material a cap-like arrangement of the chloroplasts in the early prophases of the heterotypic division and again in the early differentation of the spore but it is lost by the time the spore matures. This phenomenon, with certain nuclear responses, he would like to represent as evidence of electromagnetic polarity. During the reduction divisions the periplasmodium [2]) contains small vacuoles, long chondrioconts and small plastids which, although they preserve the shape of the chondriocont, are manufacturing starch. The spores also contain small plastids which continue to develop and manufacture starch and appear green in living material. There are present in the cell ordinary chrondriosomes in the form of granules, short chondrioconts, a large centrally placed vacuole and a lateral nucleus. The periplasmodium is now literally packed with mitochondria and large starch forming plastids (Fig. 147) while the vacuoles are less numerous than before. Degeneration soon sets in and at the time of liberation of the spores, the sporangium is occupied by the spores and their elators.

§ 5. Conclusions. — In considering the results of investigations of the Pteridophytes as a whole, without reference to the genera studied, the types of tissue under consideration or the methods employed, there are to be noted recurring cytological data which make it possible to follow certain tendencies in the group. The chondriosomes exhibit, in general, a more or less homogeneous appearance in the cytoplasm of meristematic cells, in the mature and fertilized egg (the granular chondriome of EMBERGER), sometimes likewise, in the spermatid and occasionally in the spore mother cell. These bodies are almost universally elongated in stelar tissue and exhibit diverse forms in assimilating tissue where the plastids are the dominating elements. The vacuoles in meristematic cells tend to be round and then to become larger as the cells enclosing them differentiate until the condition of the large central vacuole is reached in the mature cell. This development is reversible, and may or may not, pass through a filamentous stage. There is a tendency to a polarization of the cytological elements in cells undergoing division, whether it be shown by the plastids or by the elaborated products enclosed in the cytoplasm. Our knowledge, however, is not complete and all these tendencies must be investigated further before the laws underlying the behavior of the elements in question can be fully understood and their rôle in the cell clearly interpreted.

---

[1]) See foot-note 2, p. 231.
[2]) See foot-note 1, p. 231.

CHAPTER VIII

KARYOLOGIE

von

W. Döpp (Marburg a. Lahn)

**§ 1. Geschichtlicher Überblick, Literatur, Methodik.** — Die Pteridophyten bieten der Untersuchung ihrer Kernverhältnisse vielfach erhebliche Schwierigkeiten, die z.T. in den hohen Chromosomenzahlen vieler Vertreter, z.T. in der Kleinheit der Kerne bei gewissen Gattungen (*Lycopodium, Selaginella*) begründet sind. Diese Schwierigkeiten sind ein Grund dafür, dass bei den Pteridophyten eine Verknüpfung von Cytologie und Genetik noch nicht in genügendem Maße vorgenommen worden ist.

Anderseits sind zahlreiche Vertreter infolge ihres recht erheblichen Kerndurchmessers zur Bearbeitung sehr geeignet, und *Osmunda* und *Psilotum* sind infolge ihrer grossen Kerne zu klassischen Untersuchungsobjekten geworden. Trotz der oft nicht leichten Aufgabe ist daher eine grosse Anzahl von Arbeiten über die grosskernigen Formen erschienen, während die kleinkernigen nur wenig Berücksichtigung gefunden haben. Beim Studium derselben fällt jedoch auf, dass gerade in den letzten Jahren eingehendere Untersuchungen über die Chromosomenstruktur, — Grösse und — Form nicht gemacht worden sind, im Gegensatz zu den Phanerogamen, im Gegensatz auch zu den Moosen, bei denen das Heterochromatin und die Geschlechtschromosomen sowie die Genetik den Anlass zu einer Reihe neuerer Arbeiten gaben.

Unter den früheren Arbeiten verdienen die Untersuchungen von YAMANOUCHI (1908 a, b) hervorgehoben zu werden. YAMANOUCHI studierte bei *Dryopteris dentata* eingehend die Kernverhältnisse auf verschiedenen Stadien der Entwicklung und stellte dabei einerseits die Z a h l e n - k o n s t a n z der Chromosonen, andererseits die durch die B e f r u c h - t u n g bewirkte Verdoppelung der Chromosomenzahl sowie deren R e - d u k t i o n in den Sporangien einwandfrei fest.

Wichtige Tatsachen waren allerdings schon weit früher ermittelt worden. Der erste, der — unter Anwendung der damals noch neuen Paraffin- methode — den männlichen Kern in der Eizelle (von *Pilularia*) und damit also den Befruchtungsvorgang beobachtete, war CAMPBELL (1888). — Eine Reduktion hatten bereits STRASBURGER [1]) an *Osmunda regalis* und ROSEN (1896) an *Psilotum triquetrum* festgestellt.

Sehr eingehend beschäftigt sich die Arbeit von DE LITARDIÈRE (1921), die gegenüber den bis damals erschienenen einen wesentlichen Fortschritt bedeutet, mit den Kernverhältnissen, besonders mit der gewöhnlichen Kernteilung und der Chromosomenstruktur bei zahlreichen *Filicinae*.

Die Reifeteilungen, die zu verschiedenartigen Deutungen bezüglich Meta- und Parasyndese Anlass gegeben haben, wurden untersucht von STRASBURGER [1]), ROSEN, CALKINS, FARMER und MOORE, GRÉGOIRE, YAMANOUCHI, BEER, DIGBY, SARBADHIKARI, SZAKIEN u.a.

Die Spermatogenese und der Bau der Spermatozoiden sind seit den Arbeiten von BELAJEFF, BUCHTIEN, GUIGNARD und SCHOTTLÄNDER bis in die jüngste Zeit hinein Gegenstand von zahlreichen Untersuchungen gewesen (YAMANOUCHI, YUASA, DRACINSCHI); auch die Befruchtung beanspruchte lebhaftes Interesse (CAMPBELL, SHAW, THOM, YAMANOUCHI, ROGERS).

Ferner haben die Abweichungen vom normalen Kernphasenwechsel (Parthenogenesis, Apogamie, Aposporie) hinsichtlich der Chromosomenverhältnisse wiederholt die Aufmerksamkeit auf sich gezogen (FARMER u. DIGBY, ALLEN, STEIL, DÖPP u.a.).

Die Aposporie gibt bei den Farnen die Möglichkeit, experimentell zu polyploiden Formen zu gelangen An solchen haben HEILBRONN, LAWTON und MANTON zytologische Untersuchungen angestellt.

Erst in letzter Zeit wurde der Ansatz dazu gemacht, experimentell die Kernteilungen in den Sporangien zu beeinflussen (FRIEBEL 1933) und dadurch abweichende Prothallien oder Sporophyten zu erhalten.

Am Schluss dieses Paragraphen ist die wichtigste Literatur zusammengestellt. Die bis 1896 bekannt gewordenen Tatsachen hat ZIMMERMANN (1896) zusammen gestellt. Zahlreiche Angaben sind ferner in den Werken von BOWER und CAMPBELL enthalten. Da die Kernverhältnisse der Pteridophyten in allen wesentlichen Punkten mit denen der Phanerogamen übereinstimmen, können allgemeinere Lehr- und Handbücher mit Erfolg benutzt werden, die ja überdies z.T. auch spezielle Angaben über die Pteridophyten enthalten. In erster Linie erwähne ich die Pflanzenkaryologie von TISCHLER, ferner die Werke von SHARP–JARETZKY, BĚLAŘ, GEITLER und KÜSTER und schliesslich die zusammenfassende Darstellung von HEITZ (1936). — Auf die Aufstellung einer Liste der Chromosomenzahlen habe ich verzichtet und verweise auf die Tabulae Biologicae und die Karyologie von TISCHLER (1. Auflage und die demnächst erscheinende II. Hälfte der 2. Auflage). — Im übrigen sei auf die Spezialliteratur hingewiesen.

---

[1]) Biol. Centralbl. 14, 1894.

Bezüglich der M e t h o d i k der Untersuchung möchte ich kurz darauf aufmerksam machen, dass ausser den altbewährten Untersuchungsmethoden in den letzten Jahren die Karminessigsäurefärbung und auch die Nuklealfärbung nach FEULGEN Anwendung gefunden haben. Die erstere haben vorgenommen HEITZ, FRIEBEL, YUASA; ferner habe ich selbst mit gutem Erfolg teils in der ursprünglich von HEITZ (1926) angegebenen, teils in der von LORBEER [1]) modifizierten Weise Karminessigsäure verwendet. Die Nuklealfärbung ist von FRIEBEL und WESTBROCK [2]) benutzt worden. — Einige Angaben über Lebendbeobachtung [3]) sind auf S. 241, 255 und 280 zu finden.

## LITERATUR

ALLEN, R. F. Studies in spermatogenesis and apogamy in ferns. Transactions Wisconsin Acad. Sc. 17, 1911, 1–56.

ANDERSSON–KOTTÖ, IRMA, The genetics of ferns. Bibliographia genetica 8, 1931, 269–294.

ANDERSSON–KOTTÖ, I. and GAIRDNER A. E., The inheritance of apospory on *Scolopendrium vulgare*. Journal of Genetics 32, 1936, 189–228.

BARANOV, P., Entwicklungsgeschichte des Sporangiums und der Sporen von *Lycopodium clavatum* L. Ber. D. Bot. Ges. 43, 1925, 352–360.

BARTOO, D. R., Development of sporangium in *Schizaea rupestris*. Bot. Gaz. 88, 1929, 322–331.

BEER, R., On the development of the spores of *Helminthostachys zeylanica*. Ann. of Bot. 20, 1906, 177–186.

BEER, R., Studies in spore development III. The premeiotic and meiotic nuclear divisions of *Equisetum arvense*. Ann. of Bot. 27, 1913, 643–659.

BELAJEFF, W., Ueber Bau und Entwicklung der Spermatozoiden der Gefässkryptogamen. Ber. D. Bot. Ges. 7, 1889, 122–125.

BELAJEFF, W., Über den Nebenkern in spermatogenen Zellen und die Spermatogenese bei den Farnkräutern. Ber. D. Bot. Ges. 15, 1897, 337–339.

BELAJEFF, W., Über die Spermatogenese bei den Schachtelhalmen. Ber. D. Bot. Ges. 15, 1897, 339–342.

BELAJEFF, W., Über die Cilienbildner in den spermatogenen Zellen. Ber. D. Bot. Ges. 16, 1898, 140–144.

BELAJEFF, W., Über die Centrosome in den spermatogenen Zellen. Ber. D. Bot. Ges. 17, 1899, 199–205.

BĚLAŘ, K., Die cytologischen Grundlagen der Vererbung. Handb. d. Vererbungswiss. I, B. Berlin 1928.

BERGHS, J., Les cinèses somatiques dans le *Marsilia*. La Cellule 25, 1907, 73–84.

BOWER, F. O., The ferns (*Filicales*). Cambridge 1923–1928.

BUCHTIEN, O., Entwicklungsgeschichte des Prothalliums von *Equisetum*. Bibl. Botanica 8, 1887, 1–49.

BURLINGAME, L. L., The sporangium of the *Ophioglossales*. Bot. Gaz. 44, 1907, 34–56.

CALKINS, G. N., Chromatin-reduction and tetrad-formation in Pteridophytes. Bull. Torr. Bot. Club 24, 1897, 101–115.

CAMPBELL, D. H., Zur Entwicklungsgeschichte der Spermatozoiden. Ber. D. Bot. Ges. 5, 1887, 120–127.

---

[1]) Jahrb. f. wiss. Bot. 80, 1934.

[2]) Ann. of Bot. 44, 1930.

[3]) Vgl. auch LENOIR, Comptes rend. Soc. Biol. Paris 113, 1933, 1221–1223 (*Equisetum*).

CAMPBELL, D. H., The development of *Pilularia globulifera* L. Ann. of Bot. 2, 1888, 233–264.

CAMPBELL, D. H., The structure and development of Mosses and Ferns. New York 1905.

DENKE, P., Sporenentwicklung bei *Selaginella*. Beih. Bot. Centralbl. 12, 1902, 182–199.

DIGBY, L., On the archesporial and meiotic mitoses of *Osmunda*. Ann. of Bot. 33, 1919, 135–172.

DÖPP, W., Die Apogamie bei *Aspidium remotum* AL. BR. Planta 17, 1932, 86–152.

DÖPP, W., Weitere Untersuchungen an apogamen Farnen. I. *Aspidium Filix-mas* Sw. var. *crist. hort.* Ber. Deutsch. Bot. Ges. 51, 1933, 341–347.

DRACINSCHI, M., Über das reife Spermium der *Filicales* und von *Pilularia globulifera.* (Vorl. Mitteil.) Ber. D. Bot. Ges. 48, 1930, 295–311.

DRACINSCHI, M., Über das reife Spermium von *Equisetum*. Bul. Fac. Stiinte d. Cernăuti 5, 1931, 84–95.

DRACINSCHI, M. (1932), Über die reifen Spermatozoiden bei den Pteridophyten (*Selaginella, Equisetum, Isoetes, Filicinae leptosporangiatae*). Bul. Fac. Stiinte d. Cernăuti 6, 1932/33, 63–134.

EKAMBARAM, T. and VENKATANATHAN, T. N., Studies on *Isoetes coromandelina*. I. Sporogenesis. Journ. Indian Bot. Soc. 12, 1933, 191–225.

EKSTRAND, H., Über die Mikrosporenbildung von *Isoetes echinosporum*. Svensk bot. Tidsk. 14, 1920, 312–318.

ERNST, A., Bastardierung als Ursache der Apogamie im Pflanzenreich. Jena 1918.

FARMER, J. B. and MOORE, J. E. S., New investigations into the reduction phenomena of animals and plants. Proc. Roy. Soc. London 72, 1904, 104–108.

FARMER, J. B. and MOORE, J. E. S., On the maiotic phase (reduction divisions) in animals and plants. Quart. Journ. Micr. Sci., N.S., 48, 1905, 489–557.

FARMER, J. B. and DIGBY, L., Studies in apogamy and apospory in ferns. Annals of Bot. 21, 1907, 161–199.

FARMER, J. B. and DIGBY, L., On the cytological features exhibited by certain varietal and hybrid ferns. Ann. of Bot. 24, 1910, 191–212.

FRIEBEL, H., Untersuchungen zur Cytologie der Farne. Beitr. z. Biol. d. Pfl. 21, 1933, 167–210.

GEITLER, L., Grundriss der Cytologie. Berlin 1934.

GEORGEVITCH, P., Aposporie und Apogamie bei *Trichomanes Kaulfussii*. Jahrb. f. wiss. Bot. 48, 1910, 155–170.

GRAUSTEIN, J. E., Evidences of hybridism in *Selaginella*. Bot. Gaz. 90, 1930, 46–74.

GRÉGOIRE, V., La formation des gemini hétérotypiques dans les végétaux. La Cellule 24, 1907, 369–420.

GRÉGOIRE, V., Les cinèses de maturation dans les deux règnes. La Cellule 26, 1910, 223–422.

GREGORY, R. P., Spore formation in leptosporangiate ferns. Ann. of Bot. 18, 1904, 445–458.

GUIGNARD, L., Développement et constitution des anthérozoides. III. Fougères. Rev. Gén. Bot. 1, 1889, 71–78.

HANNIG, E., Über die Bedeutung der Periplasmodien (I, II. u. III). Flora 102, 1911, 209–278; 335–382.

HANNIG, E., Über das Vorkommen von Perisporien bei den Filicinen nebst Bemerkungen über die systematische Bedeutung derselben. Flora 103, 1911, 321–346.

HAYES, D. W., Some studies of apogamy in *Pellaea atropurpurea* (L.) LINSB. Transact. Microsc. Soc. 43, 1924, 119–135.

HEILBRONN, A., Über experimentell erzeugte Tetraploidie bei Farnen. Ztschr. f. ind. Abst. u. Vererbungs.-Lehre, Suppl.-Bd. 2, 1928, 830–844.

HEILBRONN, A., Polyploidie und Generationswechsel. Ber. D. Bot. Ges. 50, 1932, 289–300.

HEILBRONN, A., Über die genetische Lokalisation Stoffwechsel und Generationsfolge regulierender Anlagen. Biol. Zentralbl. 53, 1933, 431–441.

HEITZ, E., Der Nachweis der Chromosomen. Zeitschr. f. Bot. 18, 1926, 625–681.

HEITZ, E., Chromosomenstruktur und Gene. Zeitschr. f. indukt. Abst.- u. Vererbungslehre. 70, 1935, 402–447.

Jørgensen, C. A. The microsporangia of *Pilularia globulifera*. Dansk. Bot. Ark. Kopenhagen 5, 1928, 1–9.

Küster, E., Die Pflanzenzelle. Jena 1935.

Lawton, E., Regeneration and induced polyploidy in ferns. Amer. Journ. of Bot. 19, 1932, 303–333.

Litardière, R. de, Formation des chromosomes hétérotypiques chez le *Polypodium vulgare* L. C. R. Ac. Sc. Paris, 155, 1912, 1023–1026.

Litardière, R. de, Recherches sur l'élément chromosomique dans la caryocinèse somatique des *Filicinées*. La Cellule 31, 1921, 255–473.

Litardière, R. de, Observations cytologiques sur le *Salvinia natans*. Arch. de Botan. 2, 1928, 47–52.

Manton, Irene, Contributions to the cytology of apospory in ferns. I. A case of induced apospory in *Osmunda regalis*. Journ. Genetics 25, 1932, 423–430.

Marquette, W., Concerning the organization of the spore mother cells of *Marsilia quadrifolia*. Trans. Wisconsin Acad. of Science, 16, 1908, 81–106.

Okabe, S., Über eine tetraploide Gartenrasse von *Psilotum nudum* Palisot de Beauvois (= *P. triquetrum* Sw.) und die tripolige Kernteilung in ihren Sporenmutterzellen. Sci. Rep. Tohoku Imp. Univ. 4, 1929, 373–380.

Okuno, S., Chromosome numbers in some sporophyll-bearing ferns. Bot. Mag . Tokyo 50, 1936, 332–337.

Osterhout, W. J. S., Über Entstehung der karyokinetischen Spindel bei *Equisetum*. Jahrb. wiss. Bot. 30, 1897, 159–168.

Rogers, L. M. (1927), Development of the archegone and studies in fertilization in *Lygodium palmatum*. La Cellule 37, 1926/27, 315–352.

Rosen, F., Kerne und Kernkörperchen in meristematischen und sporogenen Geweben. Beitr. z. Biol. d. Pfl. 7, 1896, 225–312.

Rosenberg, O., Apogamie und Parthenogenesis bei Pflanzen. Handb. d. Vererbungswiss. II, L. Berlin 1930.

Sarbadhikari, P. C., Cytology of *Osmunda* and *Doodia* I. On the somatic and meiotic mitoses of *Doodia*. Ann. of Bot. 38, 1924, 1–26.

Sarbadhikari, P. C., Cytology of *Osmunda* and *Doodia*. II. On the gametophyte and post-meiotic mitoses in the gametophytic tissue of *Doodia*. Ann. of Bot. 41, 1927, 1–35.

Schottländer, P., Beiträge zur Kenntnis des Zellkerns und der Sexualzellen bei Kryptogamen. Beitr. z. Biol. d. Pfl. 6, 1893, 267–304.

Sharp, L. W., Spermatogenesis in *Equisetum*. Bot. Gaz. 54, 1912, 89–119.

Sharp, L. W., Spermatogenesis in *Marsilia*. Bot. Gaz. 58, 1914, 419–431.

Sharp, L. W., Einführung in die Cytologie. Deutsche Übersetzung von R. Jaretzky. Berlin 1931.

Shaw, W. R. (1898a), The fertilization of *Onoclea*. Annals of Bot. 12, 1898, 261–285.

Shaw, W. R. (1898b), Über die Blepharoplasten bei *Onoclea* und *Marsilia* V. M. Ber. D. Bot. Ges. 16, 1898, 177–184.

Smith, R. W., The achromatic spindle in the spore-mother-cells of *Osmunda regalis*. Bot. Gaz. 30, 1900, 361–377.

Steil, W. N., Apogamy in *Nephrodium hirtipes*. Bot. Gaz. 59, 1915, 254–255.

Steil, W. N., A study of apogamy in *Nephrodium hirtipes*. Ann. of Bot. 33, 1919, 109–132.

Steil, W. N., Incomplete nuclear and cell division in the tapetum of *Botrychium virginianum* and *Ophioglossum vulgatum*. Am. Journ. of Bot. 22, 1935, 409–425.

Stevens, W. C., Über Chromosomenteilung bei der Sporenbildung der Farne. Ber. D. Bot. Ges. 16, 1898, 261–265.

Stevens, W. C., Spore-formation in *Botrychium virginianum*. Ann. of Bot. 19, 1905, 465–474.

Strasburger, E., Apogamie bei *Marsilia*. Flora 97, 1907, 123–191.

Szakien, Br. (1927), La formation des chromosomes hétérotypiques dans l'*Osmunda regalis*. La Cellule 37, 1926/27, 367–395.

Tabulae Biologicae. Bd. IV, VII und XI (Chromosomenzahlen).

Thom, C. The process of fertilization in *Aspidium* and *Adiantum*. Transact. Acad. Sc. St. Louis 9, 1899, 285–314.

Tischler, G., Allgemeine Pflanzenkaryologie. Berlin 1921–1922 (Handbuch d.

Pflanzenanatomie, herausgeg. v. K. LINSBAUER). — 2. Aufl., I. Hälfte: „Der Ruhekern". Berlin 1934.

WINKLER, H., Parthenogenesis und Apogamie im Pflanzenreich. Jena 1908.

YAMAHA, G., Einige Beobachtungen über die Zellteilung in den Archesporen und Sporenmutterzellen von *Psilotum triquetrum* Sw. mit besonderer Rücksicht auf die Zellplattenbildung. (V.M.) Bot. Mag. Tokyo 34, 1920, 117–129.

YAMANOUCHI, SH., (1908a), Sporogenesis in *Nephrodium*. Bot. Gaz. 45, 1908, 1–30.

YAMANOUCHI, SH., (1908b), Spermatogenesis, oogenesis and fertilization in *Nephrodium*. Bot. Gaz. 45, 1908, 145–175.

YAMANOUCHI, SH., Chromosomes in *Osmunda*. Bot. Gaz. 49, 1910, 1–12.

YEATES, J. S., The nucleolus of *Tmesipteris tannensis* BERNH. Proceed. R. Soc. London, Ser. B. 98, 1925, 227–224.

YUASA, A., Studies in the cytology of Pteridophytes. I. On the spermatozoid of *Pteris cretica* L. var. *albo-lineata* HK. Bot. Mag. Tokyo, 46, 1932, 4–12.

YUASA, A. (1933a), Studies in the cytology of Pteridophyta. II. The morphology of spermatozoids of some ferns. Cytologia 4, 1933, 305–337.

YUASA, A. (1933b), Studies in the cytology of Pteridophyta. III. The morphology of spermatozoids in eight species of ferns. Bot. Mag. Tokyo 47, 1933, 681–696.

YUASA, A. (1933c), IV. On the spermatozoids of *Selaginella, Isoetes* and *Salvinia*. Ebenda 47, 1933, 687–709.

YUASA, A. (1934a), Studies in the cytology of the Pteridophyta. V. Spermatogenesis in *Notogramme japonica* PRESL and *Pteris multifida* POIRET. Journ. of the Fac. of Sc. Imp. Univ. Tokyo. Sect. III, 4, 1934, 389–397.

YUASA, A. (1934b), Studies in the cytology of Pteridophyta. VI. Reduction division in *Ophioglossum ellipticum* HOOKER et GREVILLE. Bot. Mag. Tokyo, 48, 1934, 567–570.

YUASA, A. (1935a), Studies in the cytology of Pteridophyta. VIII. Reduction division in *Isoetes japonica* AL. BR. (Prel. note). Bot. Mag. Tokyo 49, 1935, 27–31.

YUASA, A. (1935b), The morphology of spermatozoids in seven species of ferns. Bot. Mag. Tokyo 49, 1935, 375–382.

ZIMMERMANN, A., Über die Proteinkrystalloide. I. Tübingen, Beitr. z. Morph. u. Phys. d. Pfl.-Zelle, 1, 1893, 54–79.

ZIMMERMANN, A., Die Morphologie und Physiologie des pflanzlichen Zellkernes. Jena 1896.

## § 2. Der Ruhekern. Terminologie, Morphologie, Physiologie. —

Wichtigste Literatur: FARMER und DIGBY (1907), DIGBY (1919), DE LITARDIÈRE (1921), ROSEN (1896), SARBADHIKARI (1924, 1927), SCHOTTLÄNDER (1893), YAMANOUCHI (1908 a, b), YEATES (1925).

Den Ausdruck „Ruhekern" möchte ich nicht, wie es vielfach geschieht, auf d i e Kerne beschränken, die sich nicht mehr teilen, sondern auch auf das Stadium zwischen zwei Kernteilungen, auf die Interphase, ausdehnen.

Der Kern besteht aus Kernwand, einem oder mehreren Nukleolen und dem lebenden Kerninhalt, dem Karyoplasma. Bezüglich der für das Karyoplasma in Frage kommenden Terminologie schliesse ich nich an BĚLAŘ (1928, S. 13–15), GEITLER (1934) und HEITZ (1936) an, die in diesem zweierlei Substanzen unterschieden: 1. Die Kerngrundsubstanz (Karyolymphe, Kernsaft), 2. das Chromatin. Dabei ist die Bezeichnung „Chromatin" ein Sammelbegriff für das, was sich bei der Teilung in Chromosomen verwandelt.

Hinsichtlich der äusseren Gestalt des Kernes gilt für die Pteridophyten nichts besonderes; es sei hier nur auf die abweichenden Formen hingewiesen, wie sie sich einerseits bei Spermatozoiden (schraubige Form), andererseits bei Restitutionskernen (§ 8) finden. Solche im Extremfall oft hantelförmige Kerne sind z.B. in den Sporenmutterzellen gewisser apogamer *Polypodiaceen* (Fig. 98), sowie, z.T. in noch auffallenderer Form, in der Tapete von *Botrychium* und *Ophioglossum* (Fig. 101) anzutreffen (STEIL 1935). — Erwähnt seien an dieser Stelle noch die amöbenartigen Kerne in den Sporenzellen (S. 244) der Makroprothallien von *Marsilia quadrifolia* und *Pilularia globulifera* [1]).

Ueber die G r ö s s e der Kerne gibt uns eine Tabelle in TISCHLERS Karyologie (1934, S. 36) Aufschluss. Besonders grosse Kerne enthalten beispielsweise die Sporenmutterzellen und die Eizellen, ferner auch die Scheitelzellen in Wurzel, Stamm und Blatt bei den leptosporangiaten Farnen. Vergleichende Messungen sind allerdings bisher nur in geringem Umfang ausgeführt worden. Der Durchmesser schwankt von etwa 30μ (höchster gefundener Wert für die Sporenmutterzellen von *Psilotum triquetrum*; Eizelle von *Marsilia Drummondii*) bis herab zu etwa 3μ (Blattzellen von *Selaginella Martensii*). Während die *Filicinae*, ferner auch *Psilotum*, *Isoetes* und *Equisetum* verhältnismässig grosse Kerne besitzen, ist die Kerngrösse bei den *Selaginellaceae*, z.T. auch bei den *Lycopodiaceae* beträchtlich geringer. Aus der Arbeit von BARANOV (1925) ergibt sich ein Kerndurchmesser (durch Ausmessen der Figuren erhalten) von ca. 8μ, entsprechend bei *Selaginella* (DENKE 1902) ein Wert von ebenfalls 8μ (Prophase der ersten Reifeteilung).

TAB. 1

*Unterschiede in den Dimensionen* [2]) *bei dem normalen Athyrium Filix-femina und dessen aposporen Varietäten* (nach FARMER u. DIGBY 1907).

| | Athyrium Filix-femina Normalform | A. F.-f. var. clarissima Bolton | A. F.-f. var. clarissima Jones | A. F.-f. var. unco-glomeratum Stansfield |
|---|---|---|---|---|
| *Prothallium* | | | | |
| junge Zellen | 100 | 135 | 180 | 250 |
| ältere Zellen | 100 | 140 | 170 | 300 |
| junge Kerne | 100 | 140 | 190 | 250 |
| ältere Kerne | 100 | 160 | 220 | 280 |
| Spermatozoiden | 100 | 135 | 160 | 250 |
| Epidermiszellen | | | | |
| des *Blattes* | 100 | 110 | 180 | 200 |
| Chromosomenzahl | 38–40 (haploid) | 84 | 90 | 100 |

Was die K e r n p l a s m a r e l a t i o n betrifft, so liegen Messungen vor für die Zelldimensionen gewisser *Polypodiaceae* und deren experimentell erzeugte polyploide Abkömmlinge (HEILBRONN 1928, LAWTON

---

[1]) A. SCHULTZ, Planta 25, 1936.

[2]) Genauere Angaben werden leider nicht gemacht.

TAB. 2

*Grössenverhältnisse bei Dryopteris Filix-mas var. pseudo-mas und Varietäten* (nach
FARMER u. DIGBY 1907).

| | Dryopteris Filix-mas var. pseudo-mas | desgl. var. polydactyla Wills | desgl. var. polydactyla Dadds | desgl. var. cristata apospora |
|---|---|---|---|---|
| Prothalliumzellen am Scheitel . . | 100 | 85 | 70 | 60 |
| Kerne dieser Zellen | 100 | 60 | 70 | 75 |
| Spermatozoiden . | 100 | — | 85 | 90 |
| Epidermiszellen des Blattes . . | 100 | 90 | 100 | 70 |
| Chromosomenzahl Gametophyt . . | 72 | 66 | 90 | 60 |
| Sporophyt . . . | 144 | 132 | 132 | 66 |

1932); ferner sind die Grössenverhältnisse verschiedener Varietäten
einer Art im Hinblick auf ihre Chromosomenzahl vergleichend untersucht
worden (FARMER und DIGBY 1907; Tab. 1 und 2). Bezüglich des ersten
Punktes verweise ich auf den § 9 und die dort befindlichen Tabellen und
bemerke dazu, dass eine genau durchgeführte Analyse, wie sie in den
Untersuchungen von F. VON WETTSTEIN bei den Laubmoosen zur Auf-
findung bestimmter Gesetzmässigkeiten geführt hat, bei den Farnen
noch nicht vorliegt. Volumenmessungen sind kaum ausgeführt, sondern
fast nur Längen-und Flächenbestimmungen. Immerhin ergaben sich bei
den polyploiden Formen grössere Dimensionen als bei der Ausgangsform.
Die vergleichenden Messungen von FARMER und DIGBY (1907) an Zellen
und Kernen für verschiedene Varietäten der gleichen Art ergaben teils bei
höherer Chromosomenzahl grössere Dimensionen (Tab. 1, 2), teils aber
auch kleinere Werte gegenüber der Normalform, trotz höherer Chromosomenzahl bei der Varietät. Es erscheinen mir dabei allerdings auch gewisse Ungenauigkeiten in den Zählungen und Messungen nicht ausgeschlossen.

Wie weit entsprechende Beziehungen für verschiedene Arten oder
Gattungen gelten, können wir nicht mit Sicherheit sagen, da Grössen-
bestimmungen kaum gemacht worden sind. STRASBURGER [1]) hat Messungen an embryonalen Zellen und deren Kernen bei einigen Farnen sowie
bei zahlreichen anderen Pflanzen vorgenommen und gefunden, dass das
Verhältnis von Kern- und Zelldurchmesser stets angenähert 2 : 3 ist.
Von DÖPP (1932) ausgeführte Messungen ergaben folgendes (Werte in µ
angegeben):

---

[1]) Histol. Beitr., Heft 5, 1893.

TAB. 3

|  | Dryopteris remota | Dryopteris Filix-mas | Dryopteris spinulosa |
|---|---|---|---|
| Sporen |  |  |  |
| Länge . . . . . . . . . . . | 68 | 52 | 56 |
| Breite . . . . . . . . . . | 43 | 36 | 36 |
| Sporenmutterzellen, Durchmesser . . . . . . . . . . | 29,7 | 24,8 | 25,8 |
| Sporenmutter zellen, Kerndurchmesser . . . . . . | 21,3 | — | 18,1 |
| Chromosomenzahl |  |  |  |
| Gametophyt . . . . . . | etwa 130 | etwa 80 | 82 |
| Sporenmutterzellen . . . . | etwa 260 | etwa 160 | etwa 160 |

Weitere Messungen an Kernen hat FRIEBEL (1933) ausgeführt.

Ueber Beziehungen zwischen Chromosomenzahl und Zahl und Grösse der Chloroplasten vgl. HEILBRONN (1928).

Die K e r n s t r u k t u r ist im Leben gut zu beobachten an Prothallien von *Polypodiaceae*, sowie an jungen Indusien von *Polypodiaceae*. An gefärbten Präparaten ist im Ruhekern die schwache Färbung des Chromatins oft sehr auffällig, z.B. an Eikernen und in Sporenmutterzellen.

Das Verhalten des C h r o m a t i n s gegenüber sauren und basischen Farbstoffen untersuchten SCHOTTLÄNDER (1893), ROSEN (1896) und VIDAL [1]) (1912). Die Ergebnisse waren die gleichen wie die bei Phanerogamen gewonnenen. SCHOTTLÄNDER fand, dass der Eikern „erythrophil", der männliche Kern dagegen „cyanophil" ist. ROSEN (1896) sah Blaufärbung des Chromatins in Meristemkernen. Entsprechend ergab sich in den Untersuchungen VIDALS [1]) (1912) an Meristemkernen von *Equisetum* Cyanophilie, während die nicht mehr teilungsfähigen Kerne der Internodien sich als erythrophil erwiesen.

Die Kerne zeigen bei zahlreichen *Filicinae* nach YAMANOUCHI (1908 a,b) und DE LITARDIÈRE (1921) ein stark aufgelockertes C h r o m a t i n, das eine fädige Struktur erkennen lässt (Fig. 20); in den Fäden sind kleine Chromatinkörperchen („C h r o m o m e r e n"?) zu erkennen (vgl. auch SZAKIEN (1927) und § 3 u. 6). Ob H e t e r o c h r o m a t i n (HEITZ 1929, 1933, 1936) [2]) in Form von kleineren oder grösseren Stücken (Chromocentren) vorhanden ist, lässt sich noch nicht mit Sicherheit sagen, da hierüber genauere Angaben fehlen. Jedenfalls bildet DIGBY (1919) für *Osmunda* und SARBADHIKARI (1924, 1927) für *Doodia* stark gefärbte Chromatinkörnchen in den Ruhekernen ab, deren Natur näher zu prüfen wäre. Ferner wissen wir aus den Untersuchungen von DE LITARDIÈRE, dass die (sehr kleinen!) Chromosomen von *Salvinia* und *Azolla* in dem neugebildeten Tochterkern keine strukturellen Verände-

---

[1]) Ann. Sc. nat. Sér. IX, Bot., 15, 1912.

[2]) Berichte D. Bot. Ges. 47, 1929; Planta 18, 1933.

rungen erfahren und daher, abgesehen von der etwas geringeren Färbbar-
keit, in der Interphase genau so aussehen wie in der Telophase (Fig.27—33).
Nun ist ja das Heterochromatin dadurch charakterisiert, dass es in der
Telophase keine Rückbildung erfährt wie das übrige (Eu-) Chromatin und
stark färbbar bleibt. Es wäre daher zu untersuchen, inwieweit am
Aufbau dieser kleinen Chromosomen Heterochromatin beteiligt ist. Bei
*Dryopteris Filix-mas, D. spinulosa* und *Woodwardia radicans* beschreibt
DE LITARDIÈRE stark färbbare Gebilde, die er als Karyosomen bezeichnet
(Fig. 3); sie sollen bei der Teilung ins Cytoplasma gelangen (?). Auch hier
wäre festzustellen, ob es sich nicht etwa um Chromozentren handelt. —
In dem Falle van *Salvinia* und *Azolla* ist das Erhaltenbleiben der
I n d i v i d u a l i t ä t  der Chromosomen während der Ruhe besonders
markant; aber auch an anderen Objekten konnte DE LITARDIÈRE ebenso
wie SZAKIEN an *Osmunda* die Chromosomen als solche in der Interphase
identifizieren. — Neuerdings haben YUASA (1932) und DRACINSCHI (1932)
gezeigt, dass auch in den Kernen von Spermatozoiden, die früher als
homogen beschrieben wurden, eine Struktur nachweisbar ist (bisher noch
nicht bei *Selaginella*). DRACINSCHI konnte sogar an lebenden Spermatozoiden von *Isoetes* Chromatingranula beobachten (Fig. 90).

Der Ruhekern enthält einen bis mehrere N u k l e o l e n. Die Zahl
derselben wechselt in ein und demselben Gewebe, da mehrere miteinander
verschmelzen können. Von 62 Kernen von *Tmesipteris* (YEATES 1925) hatten
10 je 6, 45 je 5, 6 je 4 und einer 3 Nukleolen. Auch die Maximalzahl ist bei
den einzelnen Arten verschieden (§ 5). Die Form ist kugelig oder mehr
länglich bis langgestreckt und dann verschiedenartig gewunden (Fig. 1 a,
b), in anderen Fällen sanduhrförmig (BEER 1913) oder schliesslich ganz
unregelmässig. Infolge der Verschmelzungen entstehen Nukleolen von
grösserem Durchmesser. YEATES (1925) hat festgestellt, dass das Gesamt-
volumen der Nukleolen in ganz verschiedenartigen Geweben von Gametophyt und Sporophyt bei *Tmesipteris tannensis* um so grössere Werte
annimmt, je grösser der Kerndurchmesser ist. Ob noch andere Faktoren
die Grösse des Nukleolus bestimmen, ist noch nicht untersucht [1]. DE
LITARDIÈRE fand in den Wurzeln von *Azolla caroliniana* zweierlei Kerne:
Im Periblem und Plerom war der Nukleolus gross und das Chromatin nur wenig gefärbt (Fig. 31); umgekehrt zeigten die Kerne im Kalyptrogen und Dermatogen kleine Nukleolen, aber stärker gefärbtes Chromatin (Fig. 30). Er
schliesst daraus, dass färbbare Substanz aus den Chromosomen herauswan-
dert und für den Aufbau von Nukleolen verwendet wird. Etwas tatsächliches wissen wir aber hierüber nicht. Die für den Nukleolus angegebenen
Strukturen (Fig. 1 c) dürften wohl häufig auf Fixierungsfehlern beruhen.
In den Kernen einer ganzen Anzahl von *Filicinae* haben ZIMMERMANN[2]) und

---

[1]) Vgl. Planta 18, 1933 (HEITZ).
[2]) Beitr. z. Morphologie und Physiologie der Pflanzenzelle. Tübingen, 1, Heft 1,
1893.

andere Autoren (Zusammenstellung bei TISCHLER 1934, S. 163) E i w e i s s- k r i s t a l l o i d e festgestellt (Fig. 2).

Es sei hier kurz hingewiesen auf die Fälle, wo normalerweise stets m e h r e r e K e r n e im g e m e i n s a m e n P l a s- m a ohne Zwischenwände vorhanden sind: 1. Für den Halskanal zahlreicher *Filicinae* sowie auch von *Isoetes*[1]) (nicht dagegen von *Lycopodium*) werden 2 oder 4 Kerne ohne Zwischenwände angegeben (Fig. 5). (Nach YAMA-NOUCHI (1908) wird zwar bei *Dryopteris dentata* eine Zellplatte gebildet, die jedoch wieder verschwinpet.) In dieser Hinsicht ist also das Archegon der *Filicinae* weiter abgeleitet als das der Moose. 2. Zum Beginn der Makroprothallienentwicklung bei *Selaginella*, z.B. bei *S. Kraussiana* (CAMPBELL 1902)[2]) und *Isoetes* (CAMPBELL 1890)[3]) findet Vielkernbildung statt. Erst später werden zwischen die Kerne Wände eingeschaltet. Das Makroprothallium bei *Selaginella* zeigt also hinsichtlich des Vorhandenseins von freien Kernteilungen Verhältnisse, wie sie im Prinzip im Makro-

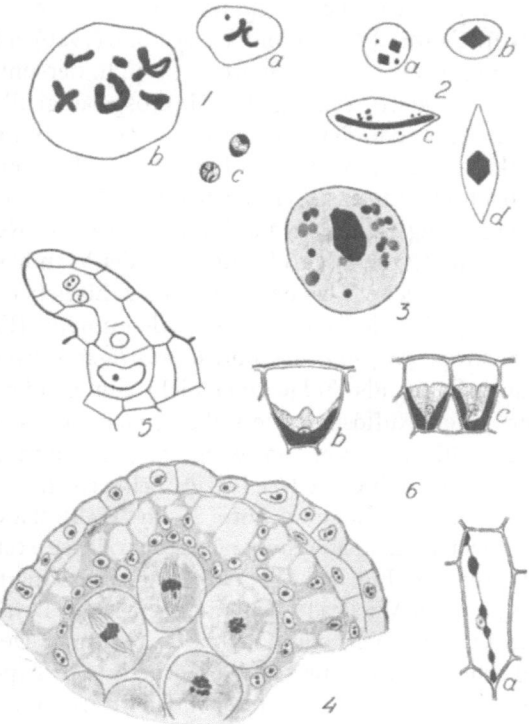

FIG. 1. Nukleolen. *a.* Kern einer ausgewachsenen Prothalliumzelle von *Pityrogramma chrysophylla*. Vergr. 800 ×. Nach SCHOTTLÄNDER 1893. *b.* Kern einer jungen Prothalliumzelle von *Athyrium Filix-femina* var. *clarissima* JONES. Vergr. 800 ×. Nach FARMER und DIGBY 1907. *c.* Nukleolen aus Archesporzellen von *Tmesipteris tannensis*. Vergr. 900 ×. Nach YEATES 1925. — Fig. 2. Kerne mit Eiweisskristallen aus dem Mesophyll. *a, b.* von *Polypodium punctatum, c.* von *P. caespitosum, d.* von *Woodwardia radicans*. Nach ZIMMERMANN 1893. — FIG. 3. Kern einer älteren Dermatogenzelle aus der Wurzel von *Woodwardia radicans* mit „Karyosomen". Nach DE LITARDIÈRE 1921. — FIG. 4. Periplasmodium mit darin eingebetteten Sporenmutterzellen von *Pilularia globulifera*. Vergr. 150 ×. Nach JØRGENSEN 1928. — FIG. 5. Archegon von *Dryopteris dentata*. Im Halskanal 2 Kerne ohne Zwischenwand. Nach YAMA-NOUCHI 1908*b*. — FIG. 6. *a.* Zelle aus der innersten Rindenschicht der Achse von *Selaginella Kraussiana;* Kern der „Chlorophyllkette" anliegend. *b, c.* Assimilationszellen von *S. Martensii* bezw. *S. caesia;* Kerne den Chloroplasten dicht angelagert. *a–c.* nach HABERLANDT 1924. — 1–3, 5, 6 umgezeichnet; 4 photogr. Wiedergabe.

[1]) LYON, Bot. Gaz., 37, 1904.
[2]) Ann. of Bot., 16, 1902.   [3]) Ann. of Bot., 5, 1890/91.

prothallium der Phanerogamen wiederkehren. — Während nach ARNOLDI (1910), YASUI( 1911)[1]) und K. GOEBEL (1930)[2]) im Makroprothallium von *Salvinia natans* in der Sporenzelle (d.h. der unteren, grösseren der beiden durch die erste Teilung der Makrospore gebildeten Zellen) freie Kernteilungen stattfinden, ist dies nach den neueren Untersuchungen von K. H. GÖBEL (1935)[3]) nicht der Fall. 3. Im Rhizom von *Psilotum triquetrum* beobachtete SHIBATA (1902)[4]) einige Male zwei Kerne in solchen Zellen („Verdauungszellen"), die klumpenförmige Reste des symbiontischen Pilzes enthalten. 4. Durch Auflösung der Wände von Tapetenzellen entsteht eine einheitliche Plasmamasse mit zahlreichen Kernen (*Filicinae, Equisetum*)[5]), die die Sporenmutterzellen umgibt (Fig. 4). 5. In der Tapete von *Botrychium virginianum* und *Ophioglossum vulgatum* entstehen 2- und 4-kernige Zellen als Folge einer nicht genügenden Entwicklung der Spindel oder einer Auflösung derselben oder der schon gebildeten Zellplatte (STEIL 1935). — Vgl. ferner S. 267, 274 und 275.

Ueber die P h y s i o l o g i e des Zellkerns liegen nur wenige Angaben vor, von denen ich 'einige anführe. Nach HABERLANDT[6]) liegt bei *Selaginella* der Kern in den Meristemzellen der Achse stets dicht an den Chromatophoren an (Fig. 6). Die Wandung der Schleimkanäle der *Marattiaceae* besteht nach WEST (1915)[7]) aus typischen Drüsenzellen mit den für diese charakteristischen, stark färbbaren Kernen. Den Charakter von Drüsenzellkernen besitzen ferner die Kerne in der Tapete und auch noch in dem in vielen Fällen daraus entstehenden Periplasmodium (CARDIFF 1905, BEER 1906, HANNIG 1911, KUNDT 1911, SENJANINOVA 1927, TIPPMANN 1928, JØRGENSEN 1928 u.a.)[8]). Ob und inwieweit dabei eine Beteiligung des Kerns bei der Ernährung der Sporenmutterzellen und Sporen oder bei der Perisporbildung stattfindet, wissen .wir nicht. Ich kann nur sagen, dass mehrfach zahlreiche Tapetenkerne in unmittelbarer Nähe des in Bildung begriffenen Perispors festgestellt worden sind. Die Untersuchungen von FITTING[9]) (1900) über die Bildung der Sporenhaut bei *Selaginella* haben ergeben, dass Sporenhäute, die nicht nur weit vom Kern, sondern überhaupt vom Plasma der Spore entfernt sind, zu wachsen vermögen. Im übrigen verweise ich noch bezüglich der Frage des Membranwachstums

---

[1]) Flora 100, 1910; Ann. of Bot. 25, 1911.

[2]) Organographie der Pflanzen.

[3]) Dissertation Marburg 1935.

[4]) Zitiert auf Seite 245.

[5]) Eine Ausnahme bildet nach TIPPMANN (Planta 6, 1928) *Angiopteris*.

[6]) Physiolog. Pflanzenanatomie, 6. Aufl. 1924, S. 252.

[7]) Ann. of Bot. 29, 1915.

[8]) J. D. CARDIFF, Bot. Gaz. 39, 1905. — A. KUNDT, Beihefte Bot. Centralblatt, 27, 1. Abt., 1911. — M. SENJANINOVA, Zeitschrift f. Zellforschung und mikroskopische Anatomie, 6, 1927. — T. TIPPMANN, Planta, 6, 1928.

[9]) Bot. Zeitung, 58, 1900.

auf das bei Tischler (1935, I) auf S. 308 und 312–316 Gesagte. — Shibata (1902) [1]) macht einige interessante Angaben über das Verhalten des Kerns in den „Verdauungszellen" von *Psilotum*. Der Kern ist stark vergrössert und amöboid geformt, die Chromatinsubstanz ist zu grossen Klumpen zusammengeballt. Auch der Nukleolus ist bedeutend grösser geworden. In Zellen mit verdauten Pilzmassen ist der Kern wieder abgerundet und hat normales Aussehen angenommen. In anderen Fällen von Mykorrhiza (*Lycopodium clavatum* und *annotinum* [2]), *Botrychium*) konnte keine Veränderung in der Gestalt und im Verhalten des Zellkerns festgestellt werden (Bruchmann 1910, Campbell 1921) [3]). — Einige Angaben über D e g e n e r a t i o n des Kernes in den Gefässanlagen von *Equisetum* machen Vidal (1912) und Lenoir (1926) [4]).

**§ 3. Die typische Kernteilung.** — Wichtigste Literatur: Rosen (1896), Beer (1913), Berghs (1907), de Litardière (1921), Digby (1919), Sarbadhikari (1924, 1927), Yamanouchi (1908 a, b), Yeates (1925). — Die bis jetzt vorliegenden Untersuchungen beschäftigen sich hauptsächlich mit den *Filicinae*; über die Kernteilung bei den *Lycopodiinae* und den *Equisetinae* sind wir weniger gut unterrichtet.

Ueber den Einfluss äusserer Faktoren (Schwerkraft, Licht) auf die Richtung der Teilungsspindel vgl. Zimmermann (1896, S. 84–86). — Ob eine P e r i o d i z i t ä t der Kernteilung vorliegt, ist nicht genau bekannt. Beobachtungen hierüber liegen vor von Gerhardt [5]) (1927) an *Osmunda regalis*. Er findet zu den verschiedenen Tageszeiten nur geringe Schwankungen. Da er aber nur die Gesamtzahl der in Teilung begriffenen Kerne anführt, ohne bei seinen Angaben die einzelnen Kernteilungsphasen hinsichtlich ihrer Häufigkeit zu berücksichtigen, so erscheint doch eine Periodizität keineswegs ausgeschlossen. — Ueber die D a u e r einer gewöhnlichen Kernteilung (Reifeteilungen vgl. § 4) ist meines Wissens nichts näheres bekannt. Ich habe (Döpp 1932) lediglich in einem Fall, Prothallium von *Aspidium* (*Dryopteris*) *remotum*, festgestellt, dass die Chromosomen etwa 5 Minuten brauchen, um von der Aequatorialplatte zu den Polen zu gelangen.

Die Fig. 18–25 geben einen Ueberblick über den Verlauf einer Mitose.

Im einzelnen sind bisher die Angaben sowohl für verschiedene Spezies wie auch diejenigen verschiedener Autoren für die gleiche Spezies bezüglich der Art der Umbildung der Chromosomen in der Telophase und ihrer Entwicklung in der Prophase sehr verschieden.

---

[1]) Pringsheims Jahrb. f. wiss. Bot., 37, 1902.

[2]) Bei *Lycopodium Selago* beobachtete Bruchmann Vergrösserung und Formveränderung des Kerns.

[3]) H. Bruchmann, Flora 101, 1910, 220–267. — D. H. Campbell, Ann. of Bot. 35, 1921, 141–158.

[4]) Ann. Sc. nat. Sér. IX, Bot., 15, 1812; Rev. gén. Bot. 38, 1926, 615–631.

[5]) E. Gerhardt, Dissertation Marburg 1927.

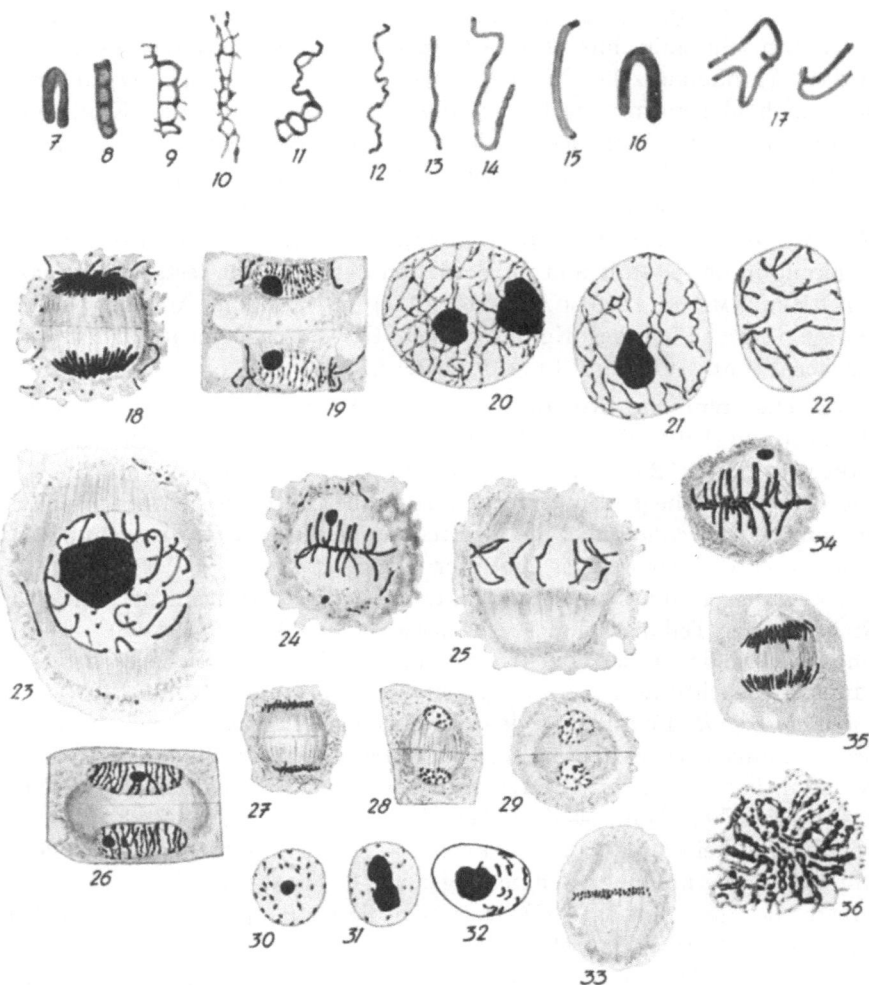

FIG. 7–17. *Hymenophyllum tunbridgense*. Chromosomen im Verlauf der Kernteilung. 7–9 Telophase; 10 Interphase; 11–15 Prophase; 16 Metaphase; 17 Anaphase. — FIG. 18–25 *Pteris cretica*, Kernteilungsstadien. 18, 19 Telophase; 20 Interphase; 21–23 Prophase; 24 Metaphase; 25 Anaphase. — FIG. 26 *Blechnum occidentale*, Telophase. — FIG. 27–33. *Azolla caroliniana*, Kernteilungsstadien. 27–29 Telophase; 30 Interphase aus dem Dermatogen, 31 aus dem Plerom der Wurzel; 32 Prophase; 33 Metaphase. — FIG. 34. Metaphase aus dem Prothallium von *Dryopteris dentata*. Nukleolus am Rand der Spindel. — FIG. 35. Anaphase aus dem Archespor von *Dryopteris dentata*. — FIG. 36. *Osmunda regalis*, Telophase in Polansicht. Von DIGBY als Längsspaltung der Chromosomen gedeutet. — 7–33 nach DE LITARDIÈRE (1921); 34 nach YAMANOUCHI (1908*b*); 35 nach YAMANOUCHI (1908*a*); 36 nach DIGBY (1919). — 7–17 und 32 umgezeichnet, die übrigen Fig. photogr. Wiedergaben.

Hinsichtlich der sich in der Telophase abspielenden Veränderungen der Chromosonen unterscheidet DE LITARDIÈRE (1921) vier Typen, die auch durch verschiedenes Verhalten in der Prophase charakterisiert sind. (Benutzt wurden zur Untersuchung in der Hauptsache Wurzelspitzen, im übrigen Rhizome und junge Blätter.)

1. Der erste Typ findet sich bei Arten mit grossen, dicken Chromosomen (*Hymenophyllum tunbridgense, H. demissum, H. asplenioides, H. fucoides, Trichomanes reniforme, T. Colensoi, Leptopteris superba, Osmunda cinnamomea, O. regalis*). Das Verhalten der Chromosomen in der Telophase ist ähnlich dem bei zahlreichen Angiospermen: An den Polen angelangt, rücken die Chromosomen vorübergehend sehr eng aneinander (,,tassement polaire"). Sie bekommen ,,Alveolen", d.h. Hohlräume, deren Natur nicht bekannt ist (Fig. 7–11). Diese werden grösser, und mit zunehmender Kerngrösse verlängern sich die Chromosomen und entfernen sich voneinander, wobei zwischen ihnen feine Anastomosen entstehen. Der Ruhekern ist charakterisiert durch eine gleichmässige Verteilung der Chromosomensubstanz; trotzdem sind die Chromosomen aber immer noch als solche zu erkennen. Zu Beginn der Prophase zeigen sich die Chromosonen zunächst als alveolosierte Gebilde, die denen der Telophase ganz ähnlich sind. Durch Verschwinden einzelner Randstücke infolge anderer Verteilung der Chromosomensubstanz entstehen Zickzackfäden von einfacher Struktur (Fig. 11, 12); gleichzeitig gehen auch die Anastomosen verloren. Dann (nicht früher!) wird in diesen Fäden ein Längsspalt sichtbar (Fig. 14), und durch Verkürzung, Dickerwerden und gleichmässigere Ausgestaltung am Rand wird die für die Metaphase typische Chromosomenform herausgebildet (Fig. 12–16). — Die Längsspaltung kann nach DE LITARDIÈRE, wie er an seinen Abbildungen belegt, nur in der Prophase vor sich gehen; seiner Auffassung nach hat sie nichts mit der Alveolosierung in der Telophase zu tun.

2. Der zweite Typus (Beispiel: *Pteris cretica*) ist dadurch charakterisiert, dass die Chromosomen, die hier viel dünner sind, keine Alveolosierung zeigen (Fig. 18–25). Sie werden zu langen, feinen, durch Anastomosen (die auch hier in der Prophase wieder verschwinden) verbundenen Fäden ausgezogen. Dementsprechend entstehen in der Prophase keine Zickzackbänder, sondern das Chromatin liefert dünne, gekrümmte Fäden, die sich spalten, verkürzen und verdicken. Diesen Typus fand DE LITARDIÈRE bei zahlreichen Vertretern der Polypodiaceen (*Dryopteris-, Asplenium-, Adiantum-, Polypodium*-Arten u.a.), ferner in den Familien der *Cyatheaceae, Parkeriaceae, Schizaeaceae, Gleicheniaceae, Marsiliaceae, Ophioglossaceae*.— Der Längsspalt, der in der frühen Prophase auftritt, ist bei vielen Formen in der späteren Prophase bis zum Ende derselben infolge sehr engen Kontakts nicht mehr sichtbar (Fig. 22, 23).

3. Bei *Blechnum occidentale, Angiopteris evecta* und *A. d'Urvilleana, Marattia fraxinea* und *Trichomanes radicans* lassen einige Chromosomen

in der Telophase Alveolen erkennen (gelegentlich kann dies auch beim Typus 2 vorkommen), im gleichen Kern befindliche andere dagegen bleiben frei davon und verhalten sich nach dem 2. Typus (Fig. 26). Die alveolisierten Chromosomen zeigen in der Prophase ganz das entsprechende Aussehen wie bei 1, liefern also Zickzackfilamente, während die einfachen Chromosomen in der Prophase sich offensichtlich wie die von *Pteris cretica* (2. Typus) verhalten.

4. Der vierte Typ (Fig. 27–33), dem *Azolla* und *Salvinia* angehören, zeichnet sich durch eine nur geringfügige Veränderung der sehr kleinen Chromosomen in der Telophase aus (vgl. § 2). Sie treten zwar auch hier durch Anastomosen miteinander in Verbindung, erfahren aber keine Alveolisierung. In der Telophase erscheinen sie als kurze Stäbchen, während der Interphase als kugelige oder ellipsoidische Gebilde. In der Prophase erfolgt Verlängerung und Längsspaltung.

Allen Fällen gemeinsam ist eine Verlängerung der Chromosomen während der Stadien zwischen Telophase und Interphase; eine solche geht auch während der Prophase vor sich. Auf die Verlängerung in der Prophase folgt bei den meisten Arten wieder eine Verkürzung, die allerdings zuweilen nur gering sein kann.

*Equisetum* · scheint, soweit bekannt (BEER 1913), mit *Pteris cretica* übereinzustimmen, während die *Lycopodiaceae* und *Selaginellaceae* mit ihren kleinen Chromosomen sich vielleicht wie *Azolla* und *Salvinia* verhalten könnten. Auch die *Psilotaceae* und *Isoetaceae* wären noch näher zu untersuchen.

Es liegen offenbar gewisse Beziehungen zur systematischen Zugehörigkeit vor, denn einige Familien (*Salviniaceae, Hymenophyllaceae*, nicht so sehr dagegen die *Polypodiaceae*) zeigen einheitliche, für sie charakteristische Verhältnisse. Allzu grosse systematische Bedeutung kann man aber diesem verschiedenen Verhalten nicht beimessen, denn nahe verwandte Arten brauchen nicht das gleiche Verhalten zu zeigen, und sogar bei derselben Pflanze kann sich in verschiedenen Teilen ein verschiedenes Bild ergeben (vgl. auch *Blechnum occidentale*, Typus 3).

Die verschiedenen von DE LITARDIÈRE gefundenen Möglichkeiten treffen übrigens nicht nur für die Pteridophyten zu, sondern auch bei den Phanerogamen sind ganz ähnliche Verhältnisse beschrieben worden. Es fragt sich nun freilich, ob sich nicht durch weitere Untersuchungen ein einheitlicheres Bild ergeben würde.

Aehnliche Strukturen wie DE LITARDIÈRE bilden YAMANOUCHI[1]) (1910) bei *Osmunda cinnamomea* sowie SHARP (1914, Fig. 19, 26, 33 daselbst) in

---

[1]) Für die Archesporzellen von *Dryopteris dentata* gibt YAMANOUCHI (1908a) Vakuolisierung an. Aus seinen Abbildungen geht eine solche jedoch nicht einwandfrei hervor. Was er als Alveolen auffasst, scheinen mir eher Zwischenräume zwischen den die Chromosomen verbindenden Anastomosen zu sein.

den spermatogenen Zellen von *Marsilia quadrifolia*[1]) und SZAKIEN (1927), ein Schüler GRÉGOIREs, bei *Osmunda regalis* (Fig. 37) ab.

YAMANOUCHI und DE LITARDIÈRE beobachteten die *Längsspaltung* der Chromosomen erst in der Prophase; nach ihnen kommt also der Vakuolisierung in der Telophase nicht die Bedeutung zu, dass durch sie eine Spaltung des Chromosoms in zwei Hälften erreicht wird. Für Objekte wie *Pteris cretica*, *Equisetum* (BEER) u.a., die in der Telophase keine Vakuolisierung zeigen, konnte erst recht nicht die Behauptung aufgestellt werden, dass die Längsspaltung schon in der Telophase erfolge[2]). Dagegen liessen diejenigen Formen, die Alveolisierung zeigen (*Osmunda*, *Doodia*) bei anderen Autoren die Meinung aufkommen, dass die Längsspaltung bereits in der Telophase oder gar schon in der Anaphase durchgeführt werde (Miss DIGBY 1919, SARBADHIKARI 1924, 1927). Besonders *Osmunda* ist (auch bezüglich der Reifeteilungen) häufig der Gegenstand von Untersuchungen gewesen, die zu verschiedenen Deutungen geführt haben.

DIGBY (1919) sieht entsprechend wie in ihren Arbeiten über *Primula* und *Galtonia* in den sich in der Telophase abspielenden Vorgängen bei *Osmunda regalis* (Fig. 36) eine Teilung in zwei Tochterchromosomen, die sich weit von einander entfernen sollen, sodass im Ruhekern nur einfache Fäden wahrgenommen werden können. In der Prophase soll wieder eine Annäherung der beiden Fäden erfolgen. Im wesentlichen zu den gleichen Resultaten gelangte SARBADHIKARI bei *Doodia* (1924, 1927). Die Fig. 12 bei SARBADHIKARI zeigt eine Spaltung sogar schon in der Anaphase.

Demgegenüber stehen also die Auffassungen von DE LITARDIÈRE (1921) und SZAKIEN (1927), die während der Telo- und der Interphase nichts erkennen konnten, was nach einem Auseinanderweichen von Fäden ausgesehen hätte. Auf Grund eigener Beobachtungen bin ich zu dem gleichen Ergebnis gekommen.

Dass verschiedene Autoren, z.B. YAMANOUCHI (1908) und DIGBY (1913), der Auffassung von einem k o n t i n u i e r l i c h e n  S p i r e m in der Prophase zuneigen, dürfte in der Schwierigkeit des Erkennens freier Enden begründet sein. Das gleiche gilt für die Prophase der 1. Reifeteilung[3]).

---

[1]) Diese Angabe steht nicht im Gegensatz zu den Befunden von DE LITARDIÈRE, die an vegetativen Teilungen gewonnen wurden.

[2]) Der Zeitpunkt, den die Autoren angeben, ist allerdings sehr verschieden. Während DE LITARDIÈRE schon auf früheren Prophasestadien einen Längsspalt beobachten konnte, der allerdings später in manchen Fällen (z.B. *Pteris cretica*) vorübergehend wieder verschwinden kann, verlegen andere, z.B. YAMANOUCHI (1908a), den Zeitpunkt der Spaltung erst in die letzten Stadien der Prophase oder gar (BEER 1913) in die Metaphase. Der Spalt dürfte aber auch in diesen Fällen schon auf einem früheren Stadium aufgetreten, jedoch übersehen worden sein.

[3]) DE LITARDIÈRE (1921) und BEER (1913) geben an, dass ein kontinuierliches Spirem nicht vorliegt.

Die A n a s t o m o s e n , die nach DE LITARDIÈRE durch Aneinander-
haftenbleiben der Chromosomen während des Auseinanderweichens in der
Telophase entstehen, verschwinden in der Prophase wieder und zwar infolge der Verlängerung und Verkürzung der Chromosomen; das körnige
Aussehen der Prophasechromosonen soll von den abgerissenen Anastomosen herrühren. Zuweilen beobachtete DE LITARDIÈRE auch Anastomosen am Ende der Prophase oder in der Metaphase, die erhalten geblieben oder evtl. auch durch Neubildung in der gleichen Weise wie in der
Telophase entstanden sein könnten.

Die Stelle der „ I n s e r t i o n ” der Chromosomen an die „Spindel-
fasern” während der Metaphase ist nach DE LITARDIÈRE meistens „terminal” [1]) (Beispiel: *Pteris cretica*, Fig. 25), häufig auch intermediär, seltener
median (*Hymenophyllum tunbridgense*, Fig. 16, 17). Die Chromosomen sind
an der Ansatzstelle fadenförmig zugespitzt. In ein und derselben Aequatorialplatte ist die Insertion oft verschieden. Die Trennung der Tochterchromosomen beginnt meistens an der Insertionsstelle.

Die Chromosomen in der Ana- und Telophase sind entsprechend ihrer
Insertion entweder alle stäbchenförmig (nahezu gerade gestreckt oder etwas
gekrümmt wie bei *Pteris cretica* u.a. *Polypodiaceae*, Fig. 35), oder eine Anzahl von ihnen ist *V*-förmig (*Hymenophyllum tunbridgense*, Fig. 7, 16)
mit gleichen oder ungleich langen Schenkeln.

Während der Wanderung zu den Polen findet eine Verkürzung und Verdickung der Chromosomen statt.

Die N u k l e o l e n , die während der Interphase recht grosse Gebilde
darstellen, sind in der späten Prophase oder in der Metaphase (Fig. 24)
nur in Form kleiner Körperchen vorhanden. Während der Metaphase
oder später sind (besonders schön bei den *Psilotaceae*!) an der Grenze der
Spindelfigur Nukleolen oder Reste davon gelegentlich zu beobachten
(Fig. 34, 74), die offenbar ins Plasma gelangen und dort aufgelöst werden (ROSEN 1896, DE LITARDIÈRE 1921, YEATES 1925) [2]).

Nach DE LITARDIÈRE (1921) erscheint die S p i n d e l bei *Hymeno-*

---

FIG. 37–51. *Osmunda regalis*. 37 a, b Telophase der letzten gewöhnlichen Mitose im
Archespor; 38 Interphase; 39 ff. Prophase der 1. Reifeteilung; 39, 40 Bildung der
Leptonemafäden; 41 Leptonema; 42 Beginn des Zygonemastadiums; 43 Zygonema
(Bukettanordnung); 44, 45 Pachynema; 46, 47 Strepsinema. 48–50 Verkürzung und
Verdickung der Fäden; 51 Gemini mit Aequationsspalt in den univalenten Chro-
mosomen. Nach SZAKIEN (1927). — FIG. 52. Zygonemastadium von *Osmunda
regalis*. Nach GRÉGOIRE (1907). — FIG. 53. *Osmunda regalis* mit "looping over".
Vergr. 1200 ×. Nach DIGBY (1919). — FIG. 54. Das gleiche (*Osmunda regalis*) nach
FARMER und MOORE (1905). — FIG. 55. *Doodia*. "Second contraction". Vergr. ca.
1300 ×. Nach SARBADHIKARI (1924). — Fig. 56. *Osmunda regalis*. Heterotypische

---

[1]) Vgl. hierzu jedoch die Angaben von HEITZ (1936).

[2]) Wie ich aus einer Anmerkung in der Arbeit von DE LITARDIÈRE (1921, S. 379)
entnehme, soll nach PAMPALONI (Annali de Bot. 1, 1903) der Nukleolus des alten
Kerns in den Tochterkern hineingelangen. Alle anderen Angaben besagen aber
demgegenüber, dass der Nukleolus stets neu gebildet wird (Fig. 74).

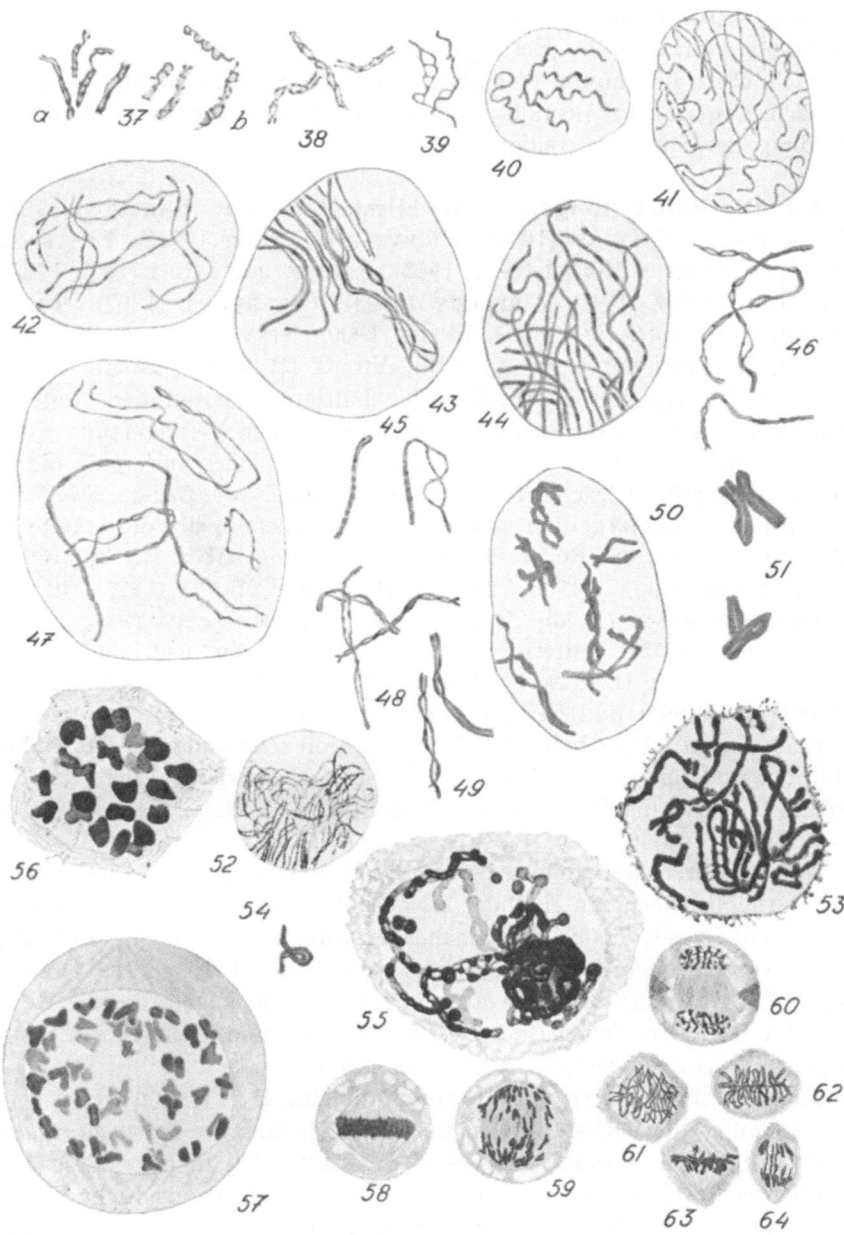

Metaphase in Polansicht. Nach DE LITARDIÈRE (1921). — Fig. 57. *Dryopteris remota*, Diakinese. Vergr. ca. 1600 ×. Aus DÖPP (1932). — FIG. 58–60. Meta-, Ana- und Telophase der heterotypischen Teilung von *Dryopteris dentata*. Nach YAMANOUCHI (1908a). — FIG. 61–64. Desgl. Aufeinanderfolgende Stadien der homoeotypischen Teilung. Nach YAMANOUCHI (1908a). — 37–51, 54 umgezeichnet; 57 Original, die übrigen Fig. photograph. Wiedergaben.

*phyllum tunbridgense* in Gestalt zweier Polkappen. Mit Vergrösserung derselben verringert der Kern bei dieser Art ebenso wie bei *Pteris cretica* sein Volumen und nimmt eine abgeplattete Gestalt mit grosser Aequatorialachse an; DE LITARDIÈRE schliesst daraus, dass der Kernsaft an der Bildung der Spindel beteiligt sei.

**§ 4. Die Reifeteilungen.** — Wichtigste Literatur: BARANOV (1925), BEER (1913), CALKINS (1897), DIGBY (1919), DÖPP (1932), EKSTRAND (1920), FARMER u. MOORE (1904, 1905), FARMER und DIGBY (1907, 1910), GRÉGOIRE (1907, 1910), GREGORY (1904), OSTERHOUT (1897), ROSEN (1896), SARBADHIKARI (1924), SMITH (1900), STEVENS (1898), SZAKIEN (1927), YAMANOUCHI (1908 a, 1910), YUASA (1934 b, 1935 a). — Die meisten der erwähnten Arbeiten beschäftigen sich mit den Reifeteilungen bei den *Filicinae*; allerdings sind manche Vertreter und sogar ganze Familien (z.B. die *Hymenophyllaceae*) nicht oder nur wenig untersucht. Wenig Berücksichtigung haben auch die *Lycopodiaceae, Selaginellaceae* und *Equisetum* gefunden. Mit *Lycopodium* befasst sich die Arbeit von BARANOV (1925); die Reifeteilungen von *Psilotum* haben ROSEN (1896), FARMER und MOORE (1905), YAMAHA (1920) und MEYER (1928), die von *Isoetes* EKSTRAND (1920), EKAMBARAM und VENKATANATHAN (1933) und YUASA (1935 a) untersucht. Sehr wenig sind wir noch über *Selaginella* unterrichtet (DENKE 1902, GRAUSTEIN 1930). An *Equisetum* haben OSTERHOUT (1897) und BEER (1913) gearbeitet.

Nach SMITH (1900) beträgt bei *Osmunda* die Zeitdauer für das Ruhestadium und für die Prophase bis zur Synapsis etwa 14 Tage, für die Synapsis selbst ungefähr 3–4 und für die beiden Reifeteilungen ungefähr 2–3 Tage.

Ueber die erste Reifeteilung bei den Pteridophyten gehen die Meinungen noch erheblich mehr auseinander als bezüglich der gewöhnlichen Mitose, wie sich dies besonders bei dem am meisten untersuchten Objekt, *Osmunda*, zeigt. Auf der einen Seite stehen die Vertreter der Parasyndese (YAMANOUCHI, GRÉGOIRE, SZAKIEN), auf der anderen diejenigen, die die Konjugation nach dem Schema der Metasyndese ablaufen sehen (FARMER und MOORE, GREGORY, DIGBY, SARBADHIKARI).

Nach Auffassung derjenigen Autoren, die eine P a r a s y n d e s e vertreten, laufen die Vorbereitungen zur 1. Reifeteilung bei *Osmunda regalis* (GRÉGOIRE 1907, YAMANOUCHI 1910, SZAKIEN 1927) in der Weise ab, dass in der Prophase die Chromosomen, die sich zunächst als dünne Fäden herausdifferenziert haben (Leptonemastadium, Fig. 39-41), paarweise miteinander in Verbindung treten (Zygonemastadium, Fig. 42, 43) und konjugieren. Die so entstandenen Doppelfäden zeigen eine sehr charakteristische „Bukett"-Anordung, wie schon die bekannte Abbildung GRÉGOIRES zeigt (Fig. 52). Die Vereinigung geht soweit, dass die Doppelchromosomen paarweise zu je einem einzigen Faden verschmolzen erscheinen (Pachynema, Fig. 44, 45); höchstens an einigen Stellen sind die Konjugationspartner etwas voneinander entfernt und daher als gesonderte Fäden zu erkennen. Zunächst ist hierbei noch Bukettanordnung vorhanden; sie macht jedoch bald einer unregelmässigen Anordnung der Doppelfäden Platz, wobei der innige Kontakt, der vorher bestand, gelockert wird (Fig. 46, 47). Nach Durchlaufung eines Strepsinemastadiums (Fig. 46, 47) verkürzen und verdicken sich die Chromosomen (Fig. 48–50), bis schliesslich die verschiedenartig gestalteten Chromosomen der Diakinese (Fig. 51, 56, 57, 66–70) zu beobachten sind. — Ganz ähnlich wird die Konjugation bei *Dryopteris dentata* (YAMANOUCHI 1908 a) und *Dryopteris remota* (DÖPP 1932) beschrieben. Sie scheint auch bei *Ophioglossum* (YUASA 1934 b) und *Isoetes* (YUASA 1935 a) in dieser Weise zu verlaufen.

Demgegenüber behaupten die Anhänger der M e t a s y n d e s e, dass die Gemini aus einem Pachynema-Faden, in dem zwei homologe Chromosomen „end to end" verbunden sein sollen, während der „second contraction" (Fig. 55) durch Umbiegung der Enden (looping, Fig. 53, 54) und darauffolgende Parallellagerung der Einzelchromosomen entstehen. Das, was bei Annahme einer Parasyndese als Längsspalt zwischen zwei ganzen univalenten Chromosomen angesehen wird, betrachten die Vertreter der Metasyndese als eine Spaltung, die ein univalentes Chromosom längsteilt. So beschrieben FARMER und MOORE bei *Osmunda* den Konjugationsmodus. BEER (1913) schildert in ganz ähnlicher Weise die Konjugation bei *Equisetum arvense*. Bei Miss DIGBY (1919) und entsprechend bei SARBADHIKARI (*Doodia*, 1924) kommt als neuer Gesichtspunkt hinzu, dass diese Autoren in den Vorgängen der meiotischen Prophase eine Wiedervereinigung der in der vorhergehenden Telophase getrennten „Längshälften" je eines Chromosoms sehen, wie sie in jeder Prophase erfolgen soll. Die Doppelfäden (Fig. 43), die GRÉGOIRE u.a. als z w e i g a n z e u n i v a l e n t e Chromosomen ansehen, halten sie also für die vorher voneinander getrennten, dann aber wieder vereinigten Längshälften j e e i n e s u n i v a l e n t e n Chromosoms.

Dazu ist folgendes zu sagen:

1. Eine Längsspaltung der Chromosomen von *Osmunda* in der Telophase scheint mir nach der Untersuchung von SZAKIEN nicht erwiesen. Die Chromosomen erfahren, wie aus SZAKIENs Abbildungen hervorgeht, in der Telophase charakteristische Veränderungen (Fig. 37), bleiben aber in der Interphase als einheitliche Gebilde vorhanden (Fig. 38); es gehen also aus einem Chromosom nicht zwei Fäden hervor, die während der Interkinese voneinander getrennt sind. Zu Beginn der Prophase stellen nach SZAKIEN die Chromosomen Zickzackfäden von einfacher Struktur wie in einer gewöhnlichen Prophase dar (Fig. 39). Diese einfachen Fäden können also offenbar keine Längshälften von Chromosomen sein. Auch eine Spaltung von univalenten Fäden auf einem Prophasestadium kann nicht vorliegen. Denn dann müssten doch wohl Stadien

beobachtet werden, in denen die Einzelfäden eines Paares nur halb so dick sind wie ein in den vorhergehenden Stadien beobachteter einzelner Faden vor dem Auftreten der Doppelstruktur, was nicht der Fall ist. Es treten also g a n z e  u n i v a l e n t e  Chromosomen paarweise zusammen. Zu dem gleichen Ergebnis konnte ich, u.a. durch Berücksichtigung der Fadendicke auf den einzelnen Stadien, bei *Dryopteris remota* (1932) und *D. spinulosa* und *D. Filix-mas* kommen.

2. Von den englischen Autoren (DIGBY, SARBADHIKARI) wird der ,,second contraction" eine besondere Bedeutung beigelegt. Diese besteht darin, dass die Fäden sich auf einem gewissen Stadium der Synapsis von der Kernwand ablösen und sich im Kerninnern eng aneinanderlagern (Fig. 55), wobei das Umbiegen der Fäden und die (angenäherte) Parallellagerung der vorher ,,end to end" verbundenen Einzelchromosomen erfolgen soll (Fig. 53). Während also in der Synapsis die Parallellagerung der Hälften der univalenten Chromosomen erfolge, werde in der ,,second contraction" die Paarung der ganzen univalenten Chromosomen durchgeführt. Demgegenüber ist in Uebereinstimmung mit BĚLAŘ (1928, S. 182) und GEITLER (1934) zu sagen, dass die ,,second contraction" ebenso wie die Synapsis (s.unten) zweifellos ein Artefakt darstellt.

3. Aus den Figuren von DIGBY, ebenso aus denen von SARBADHIKARI, lässt sich eine Bildung von Gemini in der von ihnen geschilderten Weise nicht feststellen. — Die grossen Zwischenräume zwischen den Schenkeln des *V*-förmigen Gebildes, die sich die englischen Autoren durch eine Umbiegung entstanden denken, werden von SZAKIEN durch die Angabe erklärt, dass nach dem Pachynemastadium ein Strepsinemastadium, in dem die Konjugationspartner stellenweise sich voneinander trennen, durchlaufen werde (Fig. 47).

4. Die Bilder in Fig. 53, 54 lassen sich durch eine während oder nach der Parallelkonjugation erfolgte Verklebung zweier Enden erklären. Für ringförmige Gemini bleibt auch für DIGBY keine andere Erklärung übrig, als dass freie Enden auf einem späteren Stadium miteinander vereinigt sind. Es ist daher nicht einzusehen, warum man die Vereinigung der Enden im Sinne der Metasyndese auffassen soll.

5. Bei genauer Aneinanderreihung der einzelnen Stadien ergibt sich meines Erachtens kein anderer Schluss als der, dass der Spalt, der in den Gemini die univalenten Chromosomen trennt (Fig. 66–68) und auch nach DIGBYs Ansicht der Konjugationsspalt ist, dasselbe darstellt, wie der früher in der Prophase beobachtete Zwischenraum (Fig. 43).

Es seien noch einige Einzelfragen erörtert. Während der heterotypischen Prophase vergrössert sich das K e r n v o l u m e n beträchtlich. BEER (1913) gibt für *Equisetum arvense* eine Vergrösserung des Durchmessers von 14 auf 17μ an. Während der Diakinese scheint eine geringe Abnahme des Volumens einzutreten (vgl. Prophase der gewöhnlichen Mitose, § 3).

In der Synapsis erscheint bei *Dryopteris* der Kern in Form eines etwas eingedrückten Gummiballes, wobei das „Bukett" an der konkaven Stelle des Kerns sitzt. Später wird er wieder kugelig.

Die synaptische Zusammenballung der Chromatinmasse an einer Stelle des Kernraums, die man an schlecht fixierten Präparaten (Fig. 65 b, c) beobachtet, ist ein Kunstprodukt (DÖPP 1932). An lebenden Sporenmutterzellen (Fig. 65 a) sowie bei guter Fixierung (Fig. 65 d) ist eine Verklumpung nicht zu beobachten. Die Chromosomen füllen den ganzen Kernraum aus.

Dass ein kontinuierliches Spirem vorhanden ist, erscheint mir unwahrscheinlich.

In den Leptonema – sowie in den Pachynemafäden von *Polypodiaceae* kann man kleine Körperchen, die C h r o m o m e r e n, erkennen (vgl. auch die Abbildungen von *Osmunda* bei DIGBY, SARBADHIKARI und SZAKIEN).

In der D i a k i n e s e können die beiden Chromosomen eines Geminus parallel gelagert sein, oder die Gemini nehmen andere Formen an (Fig. 51, 56, 57, 66–70, 79, 81). Wie CALKINS (1897), YAMANOUCHI (1908) u.a. ausführen, entstehen die verschiedensten Formen ($H, I, L, O, T, U, V, X, Y$). Entweder sind die Konjugationspartner an einem oder an beiden Enden verbunden, oder beide Enden sind frei.

YAMANOUCHI (1908 a) findet, dass bei *Aspidium molle* der A e q u a t i o n s s p a l t, der die univalenten Chromosomen der Länge nach halbiert, erst in der Anaphase der heterotypischen Teilung auftritt. SZAKIEN (1927) dagegen beobachtete bei *Osmunda regalis* diesen Spalt schon in der Diakinese. Auch bei *D. remota* konnte ich (DÖPP 1932) schon während der Diakinese den Längsspalt erkennen. Dieser teilt das univalente Chromosom längs, nicht quer; eine Querteilung wurde auf Grund unrichtiger Deutung von CALKINS (1897) angegeben [1]), von STEVENS (1898) jedoch wieder bestritten.

Für die Reifeteilungen werden von HUMPHREY (1894) [2]) Centrosomen beschrieben (*Osmunda, Psilotum*), ebenso von CALKINS (1897) für *Pteris tremula*. Hier liegt sicher ein Irrtum vor. Weder SMITH (1900) noch DIGBY (1919) haben bei *Osmunda* Centrosomen gefunden, ebenfalls nicht OSTERHOUT (1897, *Equisetum*).

Die S p i n d e l der heterotypischen Teilung, deren Bildung, wie in einigen Fällen beobachtete wurde, etwa während der Synapsis beginnt (SMITH, DÖPP), ist häufig zunächst multipolar polyarch (vgl. jedoch FARMER und DIGBY (1910)), später dagegen bipolar. Von OSTERHOUT (1897) und BEER (1913) wird bei *Equisetum* zunächst eine multipolare Spindel angegeben, aus der eine multipolar diarche und schliesslich eine bipolare Spindel hervorgeht.

---

[1]) CALKINS beschreibt „Tetraden", die durch eine Längs- und eine darauf folgende Querteilung eines Spiremsegmentes entstanden seien.

[2]) Ber. D. Bot. Ges. 12, 1894.

Soweit genauere Beobachtungen vorliegen, scheint die Trennung der beiden univalenten Chromosomen in der ersten Reifeteilung stattzu-finden, sodass also die e r s t e  T e i l u n g  die R e d u k t i o n bewirkt und daher als Reduktionsteilung bezeichnet werden kann.

Während der A n a p h a s e (Fig. 59) ist die charakteristische *V*-Form zu erkennen, die durch Auseinanderweichen der durch den Aequations-spalt gebildeten Hälften an deren der Aequatorialebene zugewandten Enden zustandekommt. Die Chromosomen wandern nicht in gleicher Front zu den Polen, sondern einige bleiben hinter den anderen zurück. Die *V*-Form bleibt offenbar während der Interkinese erhalten.

Die Chromosomen erscheinen in der Prophase der z w e i t e n  R e i f e-t e i l u n g (Fig. 61–64) im Gegensatz zu einer gewöhnlichen Teilung mit weit voneinander gerückten Spalthälften; entweder sind die beiden Teilstücke parallel oder an einem Ende verbunden, in *V*-förmiger Anord-nung, die noch von der vorhergehenden Anaphase herrührt. In der Me-taphase sind sie dagegen parallel gelagert. Die homoeotypische Spindel bei *Equisetum* ist nach BEER zunächst entweder multipolar polyarch oder multipolar diarch. Später ist sie bipolar. Auch in anderen Fällen scheint die Spindel sich so zu verhalten.

Ueber andere Inhaltsbestandteile in den Sporenmutterzellen und die Vorgänge bei der Wandbildung s. bei MARQUETTE (1908)[1]) und SENJA-NINOVA (1928)[2]).

## § 5. Die Nukleolen in Beziehung zu Kernteilung und Chromoso-men. — Wichtigste Literatur: BEER (1913), DE LITARDIÈRE (1921), YEATES (1925). — Schon ROSEN (1896) bildet für *Psilotum* die s y m m e t r i-s c h e  L a g e der Nukleolen in den beiden Tochterkernen ab (Fig. 71). Ferner machte DE LITARDIÈRE die Beobachtung, dass die Nukleolen in g l e i c h e r  Z a h l und in s y m m e t r i s c h e r  L a g e in den jungen Tochterkernen entstehen (Fig. 72). Die Symmetrie erstreckt sich auch auf die F o r m und G r ö s s e der Nukleolen. Er gibt dafür aber nur eine ganz allgemein gehaltene Erklärung ab. Die gleichen Festellungen machte YEATES (1925) für *Tmesipteris tannensis* (Fig. 73). Nun hat HEITZ (1931, 1935)[3]) festgestellt, dass die Nukleolen von Blütenpflanzen an genau festliegenden achromatischen Stellen einzelner bestimmter Chromosomen (SAT-Chromosomen) entstehen. Da diese in einer für die betreffende Pflanze konstanten Zahl vorhanden sind, so ergibt sich eine bestimmte Anzahl und eine gesetzmässige, in Bezug auf die Aequatorialebene sym-metrische Anordnung der in den Tochterkernen entstehenden Nukleolen. Wenn diese Ergebnisse auch für die Pteridophyten zutreffen würden, so wäre auch hier für die Zahl- und Symmetrieverhältnisse eine Erklärung

[1]) Trans. Wisconsin Ac. 16, 1908.
[2]) Arch. f. mikrosk. Anatomie und Zellforschung 6, 1928.
[3]) Planta 12, 1931.

gefunden. Es sind nun natürlich bei den Farnen zunächst genaue Untersuchungen erforderlich, die an Objekten einsetzen müssten, die vielleicht SAT-Chromosomen besitzen (*Hymenophyllum*? Vgl. § 6).

Einige Beobachtungen liegen aber durchaus schon im Sinne der HEITZschen Ergebnisse. In den Ruhekernen ist die während der Telophase vorliegende Nukleolenzahl infolge der Verschmelzung von zwei oder mehr Nukleolen meistens nicht mehr vorhanden. Nun zeigte aber YEATES (1925), dass im Sporophyten von *Tmesipteris tannensis* eine Maximalzahl von 6 Nukleolen nicht überschritten wird. Ich habe mich gleichfalls bei mehreren *Polypodiaceen* davon überzeugen können, dass die Nukleolenzahl über ein bestimmtes Höchstmass nicht hinausgeht. Eine weitere Konsequenz würde sein, dass die Höchstzahl im Gametophyten halb so gross ist wie im Sporophyten und dass die letztere gerade ist. Dies hat sich bei einigen Arten tatsächlich als zutreffend herausgestellt. YEATES erhielt als Höchstzahl in Sporen und im Gametophyten 3, im Sporophyten 6. Im

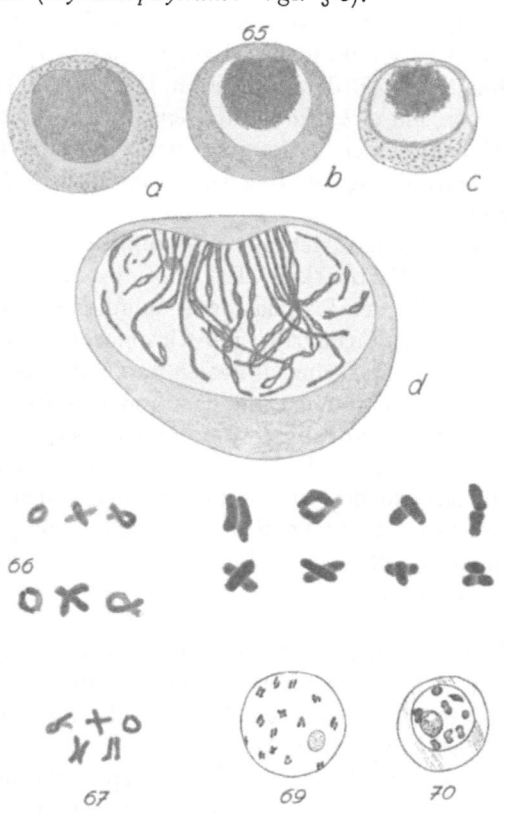

FIG. 65 *a–d*. Sporenmutterzellen von *Dryopteris remota* in „Synapsis". *a*. nach lebendem Material, *b*. gleiche Zelle wie *a*, sofort nach Fixierung mit BOUIN gezeichnet, *c*. nach einem Mikrotomschnitt (schlecht fixiert), *d*. bei stärkerer Vergrösserung (gut fixiert). *a–c* Vergr. 650 ×, *d* Vergr. 1800 ×. Aus DÖPP (1932). — FIG. 66–68. Geminiformen. 66 *Osmunda regalis*; 67 *Psilotum triquetrum*; 68 *Dryopteris remota*; 66, 67 nach FARMER und MOORE 1905 (umgezeichnet); 68 aus DÖPP (1932; Orig.). — FIG. 69. *Lycopodium clavatum*, Diakinese. Vergr. ca. 900 ×. Nach BARANOV (1925) umgezeichnet. — FIG. 70. *Selaginella apoda*, Mikrosporenmutterzelle in Diakinese. Nach GRAUSTEIN (1930) umgezeichnet.

Archespor von *Dryopteris spinulosa* fand ich 6, im Gametophyten nur 3 Kernkörperchen im Höchstfall; entsprechend zählte ich im Prothallium von *D. Filix-mas* 1–2, in Wurzelspitzen und Sporangien 1–4 Nukleolen. *D. Filix-mas* besitzt im haploiden Zustand etwa 80

Chromosomen; die gleiche Zahl hat die apogame Varietät *D. Filix-mas* var. *cristata* hort. in beiden Generationen (DÖPP 1932, 1933). Die Varietät besitzt dementsprechend auch im Sporophyten nur 1–2 Nukleolen. Eine aus dieser experimentell von mir erzeugte Pflanze (vgl. § 8 und 9) mit der doppelten Chromosomenzahl zeigte dagegen 1–4 Kernkörperchen[1]). Zu erwähnen ist noch, dass FARMER u. DIGBY (1907) für die Prothallien einiger Varietäten von *Athyrium Filix-femina*, die mindestens die doppelte Zahl von Chromosomen gegenüber der Normalform besitzen, eine höhere Nukleolenzahl angeben (bei der Normalform 1–2, bei den Varietäten dagegen mehr).

In der Telophase nehmen die Nukleolen an Grösse zu, während andererseits die Färbbarkeit der Chromosomen abnimmt; umgekehrt verringert sich in der Prophase das Volumen der Nukleolen bei gleichzeitiger Zunahme der Färbbarkeit der Chromosomen. Daraus schliesst DE LITARDIÈRE (1921), dass die Nukleolen durch Ausfliessen von Substanz aus den Chromosomen entstehen und dass andererseits in der Prophase Nukleolensubstanz wieder auf die Chromosomen übertragen werde. Zu der gleichen Auffassung ist auch BEER (1913) gekommen. BEER und YEATES (Fig. 75) fanden, dass in den Sporenmutterzellen in früheren Prophasestadien von den Nukleolen tröpfchenförmige Gebilde abgegeben werden. Wir wissen jedoch sonst nichts Tatsächliches über diese Dinge.

Vgl. noch § 3.

### § 6. Die Chromosomen (Individualität, Zahl, Grösse, Form, Bau).

— Wichtigste Literatur: DIGBY (1919), FARMER and DIGBY (1907), GRÉGOIRE (1907), HEITZ (1926), DE LITARDIÈRE (1921), SARBADHIKARI (1924, 1927), SZAKIEN (1924, 1927).

I n d i v i d u a l i t ä t. — Die Interphasekerne von *Salvinia* und *Azolla* lassen sehr schön die einzelnen Chromosomen erkennen, da diese seit der Telophase fast far keine Veränderung erfahren haben (Fig. 30, 31). DE LITARDIÈRE konnte auch bei zahlreichen anderen Objekten, sowie SZAKIEN bei *Osmunda regalis* (1927), Schritt für Schritt durch die Interphase hindurch die Chromosomen verfolgen (Fig. 10, 20, 38).

C h r o m o s o m e n z a h l. — Vorausschicken möchte ich, dass einerseits bei zahlreichen Arten die Chromosomenzahlen noch nicht ermittelt sind, andererseits manche früheren Zählungen einer Nachprüfung bedürfen.

Die bisher angegebenen Chromosomenzahlen schwanken zwischen sehr hohen und sehr niedrigen Werten. Bezüglich der bei vielen Pflanzen sehr hohen Chromosomenzahlen nehmen die Pteridophyten eine besondere Stellung ein (Fig. 76—81). Nur wenige Angiospermen besitzen so hohe Haploidzahlen von 60 Chromosomen und darüber, wie sie bei den Farnen recht häufig sind. Besonders hohe Werte finden sich bei den

---

[1]) Vgl. hierzu auch GEITLER (Planta 17, 1932).

*Polypodiaceen, Ophioglossaceen, Psilotaceen* und bei *Equisetum*, während andererseits bei den *Hymenophyllaceen, Osmundaceen, Marsiliaceen, Salviniaceen, Lycopodiaceen* und *Isoetaceen* erheblich niedrigere Zahlen vorliegen.

Die geringste Haploidzahl [1]) (4) hat ARNOLDI (1910) [2]) für eine von ihm untersuchte Form von *Salvinia natans* angegeben. Andere Forscher fanden dagegen bei *Salvinia natans* n=8 Chromosomen (KUNDT 1911 [3]), YASUI 1911) [4]). Vermutlich handelt es sich hier um verschiedene Rassen (vgl. § 9). Die Zahl n=8 wird ferner angegeben für *Selaginella*-Arten (DENKE 1902, HEITZ 1926). Dann folgen *Isoetes*-Arten (TAKAMINE [5]) 1921) mit 11, *Pilularia* mit 13 (DE LITARDIÈRE 1921), *Lycopodium clavatum* mit 14 (BARANOV 1925), *Polypodium aureum* (HEILBRONN 1928) sowie *Marsilia*-Arten (STRASBURGER 1907, DE LITARDIÈRE 1921, MARSCHALL [6]) (1925) mit 16 und *Osmunda*-Arten mit etwa 20–22 Chromosomen im haploiden Zustand (YAMANOUCHI 1910, DIGBY 1919, DE LITARDIÈRE 1921, OKUNO 1936). Schon recht hohe Werte weisen die Arten der Gattung *Dryopteris* auf (60–80, gewisse Varietäten z.T. mehr). Die meisten Chromosomen wurden bisher aufgefunden bei *Tmesipteris tannensis* (100, nach YEATES 1925), *Psilotum triquetrum* (Varietät (104; OKABE 1929), *Equisetum arvense* (ca. 115; BEER 1913), *Ophioglossum vulgatum* (100–120; BURLINGAME 1907), und schliesslich bei *Ophioglossum ellipticum* 170 als bisher h ö c h s t e Haploidzahl (YUASA 1934 b). Auch bei *Dryopteris remota* (DÖPP 1932) ergibt sich für die Sporenmutterzellen eine recht hohe Zahl (130 Gemini = 260 Einzelchromosomen). Auffallend hoch ist ferner der von FRIEBEL (1933) für gewisse abnorme Sporenmutterzellen von *Polybotrya cervina* gefundene Wert (bis über 600 Chromosomen). Zahlen um 100 herum und darüber sind bei den Angiospermen Ausnahmen; im übrigen sind diese Grössenordnung sowie darüber liegende Zahlenwerte nur für gewisse Flagellaten und andere Protisten angegeben. — Bemerkenswert ist, dass in der Familie der *Polypodiaceen* die Zahl n=32 bei verschiedenen Gattungen wiederkehrt (z.B. *Cystopteris fragilis, Adiantum capillus-Veneris, Phyllitis Scolopendrium, Pteridium aquilinum* und *Pteris*-Arten). — Soweit bisher bekannt ist, können die Arten einer Gattung entweder die gleiche Zahl besitzen (*Marsilia*) oder aber erhebliche Unterschiede aufweisen (z.B. in der Gattung *Salvinia* (niedrigste Zahl n = 4, höchste n = 32) und bei *Selaginella* (niedrigste Zahl n = 8, höchste

---

[1]) Bei den hier angegebenen Zahlen, soweit nichts anderes vermerkt, handelt es sich stets um die Haploidzahl, z.T. nicht um direkt gefundene Werte, sondern um solche, die sich aus der von den Autoren angegebenen Diploidzahl errechnen lassen.

[2]) Flora 100, 1910.

[3]) Beih. Bot. Centralbl. 27, I. Abtlg., 1911.

[4]) Ann. of Bot. 25, 1911.

[5]) Bot. Mag. Tokyo 35, 1921.

[6]) Bot. Gaz. 79, 1925.

n = 28), aber auch Varietäten einer Art können erhebliche Unterschiede
in der Chromosomenzahl sowohl gegenüber der Normalform wie auch untereinander zeigen. Als Beispiel nenne ich die von FARMER und DIGBY (1907)
untersuchten Varietäten von *Dryopteris Filix-mas* und *Athyrium Filix-
femina*, ferner *Dryopteris Filix-mas* aus der Marburger Gegend (ca. 80 Chromosomen, DÖPP 1932) und *Dryopteris Filix-mas* ssp. *eu-Filix-mas* var. *crenata* und var. *Borreri* (65 Chromosomen, DE LITARDIÈRE 1921), ferner
(ebenfalls bei Marburg gefunden) *Dryopteris spinulosa* ssp. *dilatata* (82 Chromosomen, DÖPP 1932) und *D. spinulosa* ssp. *dilalata* var. *oblonga* (65 Chromosomen, DE LITARDIÈRE 1921), ausserdem die von verschiedenen Standorten stammenden, morphologisch aber offenbar durchaus gleichen Pflanzen von *Salvinia natans*. Die Erscheinungen der Polyploidie sollen jedoch
in einem besonderen Abschnitt besprochen werden.

G r ö s s e n v e r h ä l t n i s s e. — Bezüglich der Grössenverhältnisse der Chromosomen (DE LITARDIÈRE 1921, FRIEBEL 1933) ist zunächst
zu sagen, dass solche Längen wie bei den Phanerogamen (z.B. bei *Trillium
grandiflorum*) niemals erreicht werden. Durch besondere Länge zeichnen
sich nach DE LITARDIÈRE die Chromosomen von *Ceratopteris thalictroides*
aus (ca. 12μ). — Im einzelnen ergeben sich Unterschiede nach der Zugehörigkeit zu Familie, Gattung, Art oder Varietät. Besonders grosse Dimensionen hinsichtlich Länge und Dicke finden sich bei *Hymenophyllum tunbridgense* und den übrigen Arten der Gattung *Hymenophyllum*, ferner auch
bei den *Osmundaceen*, während andererseits die Chromosomen der *Salviniaceen*, *Lycopodiaceen* (Fig. 69) und *Selaginellaceen* [1]) recht kleine Gebilde
sind. Was die Verschiedenheiten zwischen Arten einer Gattung betrifft,
so sind z.B. die Chromosomen von *Hymenophyllum demissum* dünner als
die von *Hymenophyllum tunbridgense*, und entsprechend diejenigen von
*Adiantum cuneatum* feiner als die von *A. capillus-Veneris*. Der Querdurchmesser schwankt jedoch nicht nur bei verschiedenen Familien, Gattungen oder Arten, sondern auch für Varietäten ein und derselben Art.
Beispielsweise fand DE LITARDIÈRE die Chromosomen von *Phyllitis Scolopendrium* var. *ramo-marginatum* dünner als die der Normalform oder der
Varietät *crispum*; ferner ergaben sich Längenunterschiede zwischen *Pteris
cretica* und der Varietät *albo-lineata* (Tab. 4) [2]). Darüber hinaus zeigen sich
aber Unterschiede in Bezug auf Länge und Dicke im gleichen Kern oder in
der gleichen Platte, wie aus Tab. 4 und Fig. 82 hervorgeht. Ausser den in
Tab. 4 angegebenen Zahlen scheinen auch Zwischenwerte vorzukommen.

---

[1]) Bei den *Lycopodiaceen* und *Selaginellaceen* ergeben sich Dimensionen von 1–2 ·
μ (BARANOV 1925, HEITZ 1926).

[2]) Bei Material von *Salvinia natans* von verschiedenen Standorten, das morphologisch keinerlei Unterschiede zeigte, fand DE LITARDIÈRE (1921, 1928) erhebliche
Verschiedenheiten nicht nur bezüglich der Zahl, sondern auch der Länge der Cho_
mosomen.

TAB. 4a

Verschiedene Längen von somatischen Chromosomen in μ aus der gleichen Platte.
Nach DE LITARDIÈRE (1921).

| | |
|---|---|
| *Pteris cretica* . . . . . . . . . . . | 2,25; 3 |
| *Pteris cretica* var. *albo-lineata* . . . . | 2,8; 4 |
| *Nephrolepis exaltata* forma *Piersonii* . | 2; 2,5; 4; 4,5; 5,5 |
| *Dryopteris setigera* . . . . . . . . | 1; 1,5; 2; 2,5; 3; 4 |
| *Dennstaedtia cornuta* . . . . . . . . | 3; 5 |
| *Azolla caroliniana* . . . . . . . . | 0,5; 1; 5 |

TAB. 4b

Dickenmessungen von Chromosomen (in μ).

| | | | |
|---|---|---|---|
| *Polypodium aureum* | Metaphase der hetero-typischen Teilung | 1–1,6 | FARMER und DIGBY 1914[1] |
| | homoeotypische Tlg., Beginn der Anaphase | 0,6–1,1 | |
| *Osmunda regalis* | Metaphase bei Archesporteilungen | 0,6–1 | FARMER und DIGBY 1914[1] |
| | Anaphasechromosomen d. homoeotypischen Teilung | 0,7–1 | |
| *Hymenophyllum tunbridgense* | somatische Metaphase | 1,25–1,5 | DE LITARDIÈRE 1921 |

Soweit es sich um grössere Unterschiede handelt, dürften diese als erblich konstant anzusehen sein. Bei geringeren Unterschieden wird jedoch die Grenze zwischen einer auf Modifikation beruhenden Variabilität und erblicher Bedingtheit schwer oder gar nicht zu ziehen sein. — Vielleicht liegen auch bei *Psilotum triquetrum* verschiedene Chromosomenlängen vor (vgl. DE LITARDIÈRE 1921, S. 344, 345). Bei dem als Bastard aufgefassten *Polypodium Schneideri* sind im Ruhekern Chromosomen von verschiedener Dicke zu erkennen. DE LITARDIÈRE nimmt an, dass sich in den Bastardkernen die Chromosomen der mutmasslichen Eltern (*P. vulgare* und *P. aureum*) durch ihre Dimensionen unterscheiden.

C h r o m o s o m e n f o r m. — Die in der Aequatorialplatte angeordneten Chromosomen sind mehr oder weniger lange, entsprechend ihrer Insertion fast gerade oder schwach gekrümmte Stäbchen (Fig. 24, 34, 76, 78, 80) oder stark gebogene Gebilde (Fig. 16). Bei zahlreichen *Filicinae*, besonders bei den *Polypodiaceen* und *Cyatheaceen* sind die Chromosomen lang gestreckt; die äusseren liegen vielfach mit ihrer ganzen Länge in der Aequatorialebene, während die zentral gelegenen aufgerichtet sind und sich nur mit einem kleinen, oft deutlich umgebogenen Stück in der Aequatorialebene befinden (DE LITARDIÈRE 1921). Bei *Equisetum* herrschen die gleichen Verhältnisse. Bei *Azolla* und *Salvinia* (Fig. 27—33) dagegen sind die Chromosomen im Verhältnis zu ihrer Dicke sehr kurz; in der Seitenansicht der Aequatorialplatte erscheinen sie als kleine runde oder eiförmige Gebilde, da der grössere Teil des Chromosoms in der Aequatorialebene liegt. Auffallend ist der Unterschied in Grösse und Form zwi-

---

[1] Phil. Trans. Roy. Soc. London, Ser. B, 205, 1914.

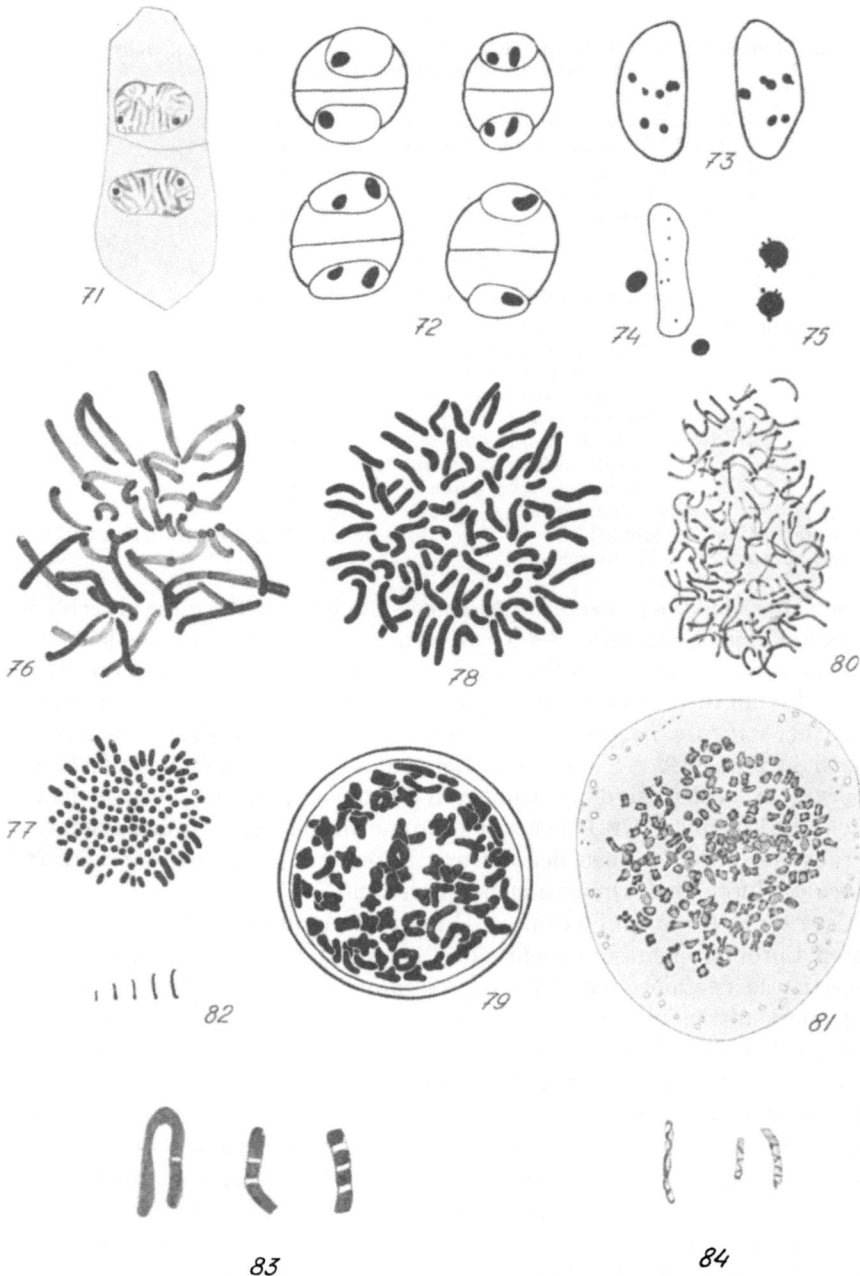

FIG. 71–73. Symmetrieverhältnisse der Nukleolen in jungen Tochterkernen. Fig. 71. *Psilotum triquetrum* (Vergr. ca. 600 ×). Nach ROSEN (1896); Fig. 72. *Pteris cretica*. Nach DE LITARDIÈRE (1921); Fig. 73. *Tmesipteris tannensis*. Nach YEATES (1925). — FIG. 74. Junger Tochterkern von *Tmesipteris tannensis* mit 6 neugebildeten Nukleolen; 2 alte Nukleolen ausserhalb des Kerns. Nach YEATES (1925). —

schen *Selaginella*-Arten (HEITZ 1926). Während bei *Selaginella grandis, S. Ouvrardii* und *S. Vogelii* in der Metaphase stäbchenförmige Chromosomen vorliegen, erscheinen sie bei *S. Martensii* als kleine kugelige Gebilde. HEITZ nimmt an, dass diese durch Zerfall von ehemals stäbchenförmigen Chromosomen (Querteilung) entstanden sind (vgl. S. 276).

Die Chromosomen der Anaphase oder frühen Telophase sind je nach ihrer Insertionsstelle entweder alle stäbchenförmig (gerade gestreckt oder etwas gekrümmt) wie bei *Pteris cretica* und zahlreichen anderen *Polypodiaceen*, oder die Chromosomen sind z.T. V-förmig (z.B. *Hymenophyllum, Pilularia globulifera*).

Durch Aetherbehandlung konnte FRIEBEL (1933) bei *Matteuccia orientalis* statt der Interkinesekerne Kerne mit diakineseartigen Geminis und Einzelchromosomen erhalten (Fig. 110). Auch in den Tetradenkernen bekam er nach Aetherbehandlung Chromosomen, die wie in einer Diakinese gestaltet und angeordnet waren (Fig. 111).

C h r o m o s o m e n b a u. — Nach DE LITARDIÈRE (1921) stellt ein Chromosom von *Hymenophyllum* oder von einer anderen Art mit grossen Chromosomen in der Telophase ein Gebilde dar, das nicht mehr homogen erscheint wie in der Metaphase, sondern eine Anzahl von ,,Alveolen'' im Inneren enthält. Ueber die Natur derselben kann DE LITARDIÈRE nichts Genaues angeben. — Während YAMANOUCHI (1908) eine einheitliche Chromosomensubstanz annimmt, unterscheidet DE LITARDIÈRE eine ,,nicht färbbare'' Grundsubstanz und eine ,,färbbare'' Substanz, die die erstere durchtränkt. — Aehnliche Abbildungen wie DE LITARDIÈRE gibt SZAKIEN (1927).

Wie schon früher gesagt, hält im Gegensatz zu DE LITARDIÈRE und SZAKIEN Miss DIGBY (1919) die Telophase-Chromosomen von *Osmunda* für längsgespalten. Ich kann mich jedoch dieser Ansicht nicht anschliessen.

Einige Abbildungen von DE LITARDIÈRE und SZAKIEN (Fig. 37, 84) lassen auf das Vorhandensein eines Schraubenfadens ( C h r o m o n e m a) und daher auf eine Schraubenstruktur der Chromosomen schliessen. Es ist

---

FIG. 75. Nukleolen von *Tmesipteris tannensis* während der Synapsis. Vergr. ca. 1000 ×. Nach YEATES (1925). — FIG. 76. Somatische Aequatorialplatte von *Leptopteris superba* mit 42 Chromosomen. Nach DE LITARDIÈRE (1921). — Fig. 77. Spätere Anaphase in Polansicht aus einem Prothallium von *Dryopteris remota*. Ca. 130 Chromosomen. Vergr. ca. 1700 ×. Aus DÖPP (1932). — FIG. 78, 79. *Dryopteris spinulosa* ssp. *dilatata*. Fig. 78 Prothallium, Aequatorialplatte; 82 Chromosomen. Vergr. ca. 1700 × ; Fig. 79 Diakinese, ca. 80 Gemini. Vergr. ca. 1700 ×. Aus DÖPP (1932). — FIG. 80. Somatische Aequatorialplatte von *Dryopteris Filix-mas* ssp. *eu-Filix-mas* var. *crenata*. Ca. 130 Chromosomen. Nach DE LITARDIÈRE (1921). — FIG. 81. Metaphase der heterotypischen Teilung von *Ophioglossum ellipticum* mit ca. 172 Gemini. Vergr. ca. 700 ×. Nach YUASA (1934*b*). — FIG. 82. Verschieden lange Chromosomen aus einer Aequatorialplatte von *Dryopteris setigera*. Nach DE LITARDIÈRE (1921). — FIG. 83. Chromosomen von *Hymenophyllum tunbridgense*. Nach DE LITARDIÈRE (1921). — FIG. 84. *a*. Chromosomen von *Osmunda regalis*, *b*. von *Hymenophyllum tunbridgense*. Schraubenstruktur? Nach DE LITARDIÈRE (1921). — FIG. 77–79 Orig., Fig. 76, 80, 91 photogr. Wiedergaben, die übrigen umgezeichnet.

daher anzunehmen, dass auch die Chromosomen der Farne entsprechend dem allgemeinen Schema (GEITLER 1934, HEITZ 1936) gebaut sind.

Weitere Meinungsverschiedenheiten bestehen hinsichtlich der Angaben über C h r o m o m e r e n. Nach DE LITARDIÈRE und SZAKIEN sollen solche Angaben auf Beobachtungsfehlern beruhen. Bei einigen *Hymeno-phyllum*-Arten fand nun aber DE LITARDIÈRE (1921) häufig Chromosomen (Fig. 83), die an die Satellitenchromosomen von höheren Pflanzen, z.B. von *Vicia faba* [1]), sowie einiger Insekten erinnern. Es können eine oder mehrere achromatische Stellen vorhanden sein. Leider ist jedoch nichts darüber angegeben, ob diese Erscheinung konstant und erblich fixiert ist. (Da DE LITARDIÈRE an den achromatischen Stellen eine Art Scheide beobachtete, nimmt er an, dass jedes Chromosom von einer Hülle umgeben sei.) DE LITARDIÈRE gibt nun allerdings zu, dass solche Chromosomen mit einer oder mehreren achromatischen Stellen zugunsten der Auffassung angeführt werden könnten, dass das Chromosom aus einer Anzahl von Untereinheiten zusammengesetzt ist. — Auf Grund von Beobachtungen an *Polypodiaceen* sowie auf Grund der in der Literatur vorliegenden Abbildungen komme ich zu dem Schluss, dass wenigstens in den Fällen mit grösseren Chromosomen eine Längsdifferenzierung des Leptonema- und Pachynema-Fadens in Chromomeren und zwischen diesen liegende Stücke deutlich erkennbar vorhanden ist.

Die Chromosomenstruktur bei den *Lycopodiinae* und *Equisetinae* kennen wir noch weniger als die der *Filicinae*. Untersuchungen über den Bau der Chromosomen sind daher dringend erforderlich.

Bemerkenswert ist, dass die kleinen Chromosomen von *Salvinia* und *Azolla* in der Telophase keine Rückbildung erfahren (DE LITARDIÈRE 1921; vgl. § 3). — Ueber H e t e r o c h r o m a t i n und Chromocentren s. § 2. — G e s c h l e c h t s c h r o m o s o m e n sind bei den Farnen bisher noch nicht bekannt geworden.

**§ 7. Kernverschmelzung.** — Wichtigste Literatur: ALLEN (1911), BELAJEFF (1889, 1897, 1898, 1899), BUCHTIEN (1887), CAMPBELL (1887, 1888), DRACINSCHI (1930, (1931, 1932), FARMER und DIBGY (1907), GUIGNARD (1889), ROGERS (1927), SCHOTTLÄNDER (1893), SHARP (1912, 1914), SHAW (1898 a und b), THOM (1899), YAMANOUCHI (1908 b), YUASA (1932, 1933 a, b, c, 1934 a, 1935 b).

G e s c h l e c h t s z e l l e n. — Die S p e r m a t o g e n e s e der Pteridophyten und der B a u der S p e r m a t o z o i d e n sind öfters bis in die jüngste Zeit hinein genau studiert worden (BELAJEFF, BUCHTIEN, DRACINSCHI, GUIGNARD, SCHOTTLÄNDER, SHARP, YAMANOUCHI, YUASA). Bei den letzten spermatogenen Teilungen tritt ebenso wie bei den Moosen und Cycadeen ein centrosomenartiges Körperchen auf (Fig. 85),

---

[1]) HEITZ, Planta 12, 1931.

über dessen Herkunft sich nur Mutmassungen äussern lassen. Das nach den Untersuchungen von SHARP (1912, 1914) u.a. hieraus entstehende, längliche, stark färbbare Gebilde [1]) ist nach DRACINSCHI nicht, wie früher angenommen wurde, der Ursprungsort der Geisseln, dient also nicht als Blepharoplast, sondern es stellt (nach DRACINSCHI) ein Chondriosom dar. Die Spermatozoiden der Pteridophyten mit Ausnahme von *Lycopodium* und *Selaginella* unterscheiden sich von denjenigen der Moose ausser durch den Besitz zahlreicher Geisseln durch eine mehr oder weniger hohe Zahl schraubiger Umdrehungen ihres Kernes (Abb. 86—90). Der Kern besitzt keinen Nukleolus. Die Kernsubstanz färbt sich sehr stark und wurde früher als homogen angegeben, jedoch konnte DRACINSCHI an gefärbten, aber auch an lebendem Material Chromatingranula (Chromomeren?) feststellen (Abb. 90), ebenso YUASA an gefärbten Präparaten. Der Kern der E i z e l l e erscheint etwas abgeplattet oder auch eingedrückt (vgl. die Mitose in Abb. 91!), im übrigen aber von regelmässiger abgerundeter Form und befindet sich im Ruhezustand. Die von THOM abgebildeten, stark gelappten Eikerne sind entweder keine normalen Kerne mehr, oder es liegt ein Fixierungsfehler vor. — Vgl. Kap. VII (Cytology).

Der B e f r u c h t u n g s v o r g a n g ist hauptsächlich bei den *Polypodiaceen* untersucht worden. Das Spermatozoid dringt, mit seinem Vorderende voran, in die Eizelle ein. Dabei bleiben sämtliche Teile ausser dem Kern im Cytoplasma zurück. Nach SHAW (1898) und YAMANOUCHI (1908 b) dringt bei *Onoclea sensibilis* bezw. *Dryopteris dentata* der männliche Kern ohne Form- und Strukturänderung [2]) in den Eikern ein, was in den meisten Fällen zutreffen dürfte. Dort wird dann die Kernsubstanz allmählich aufgelockert, bis kein Unterschied mehr zwischen dem Chromatin des Eikerns und dem des Spermatozoids vorhanden ist (Fig. 94). Einige neue Nukleolen treten auf. Nach CAMPBELL (1888) erfolgt aber bei *Pilularia*, ebenso nach ROGERS (1927) bei *Lygodium palmatum* vor dem Eindringen des Spermatozoids in den Eikern an der Wand desselben eine erhebliche Gestaltsveränderung, indem der Kern sich abrundet (Fig. 92, 95). — Beobachtungen von ROGERS u.a. über Polyspermie lassen nicht ausgeschlossen erscheinen, dass gelegentlich Verschmelzung des Eikerns mit mehr als einem männlichen Kern eintritt, womit vielleicht eine Möglichkeit zur Entstehung polyploider Formen gegeben ist. — Als Ausnahmefall sei ferner die einmal beobachtete Verschmelzung zwischen Bauchkanalzelle und männlicher Geschlechtszelle erwähnt (bei *Selaginella apus* (LYON 1904)) [3]).

---

[1]) Von DRACINSCHI als „Randsaum", von YUASA als „border-brim" bezeichnet.

[2]) Das gleiche gilt nach LAWTON (1932) für *Woodwardia virginica;* die Autorin bildet in einer Mikrophotographie einen männlichen, schraubig gewundenen Kern (es handelt sich dabei übrigens um diploide Geschlechtszellen) innerhalb eines Eikerns ab.

[3]) Bot. Gaz. 37, 1904.

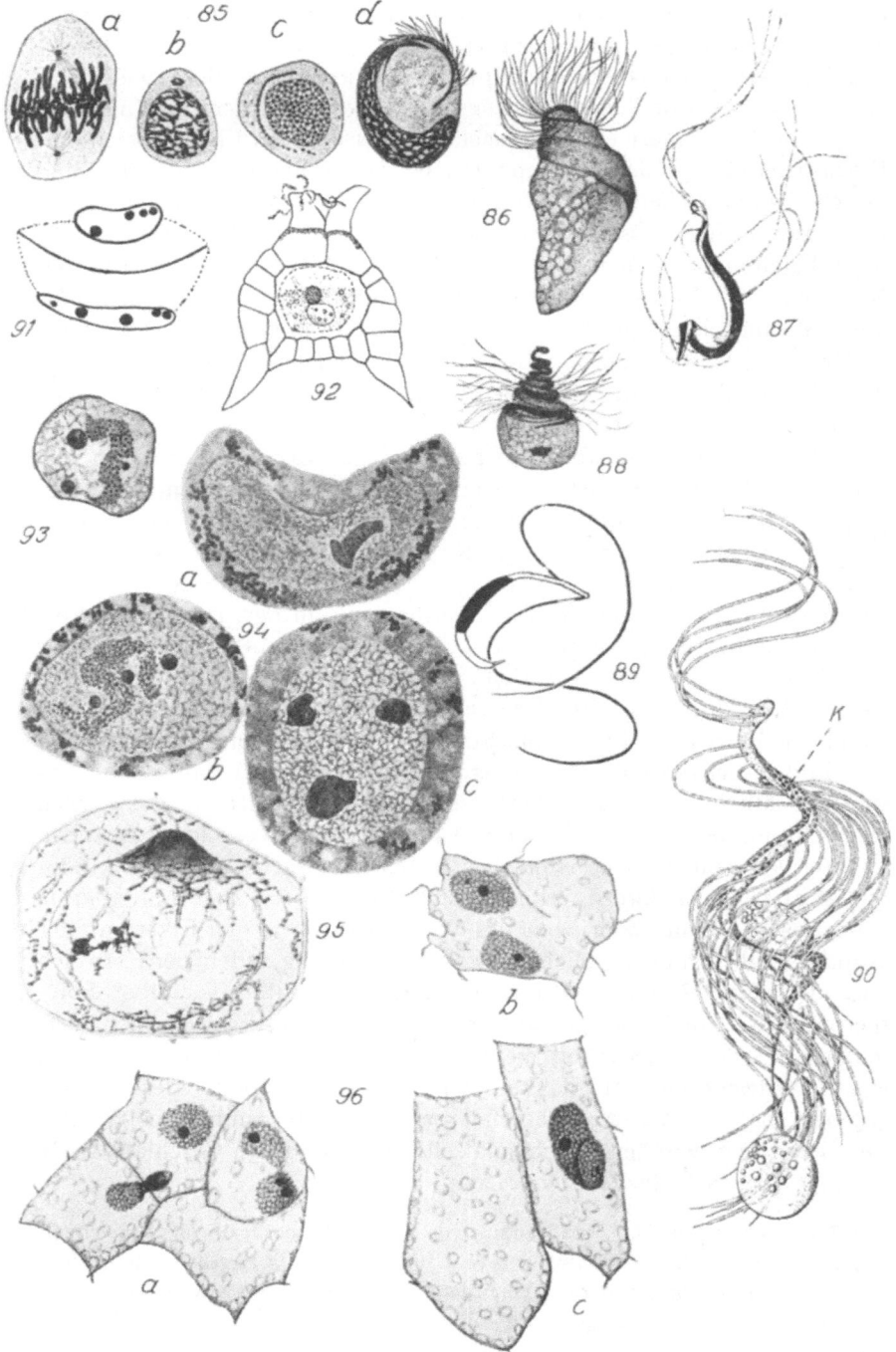

Für einige Farne ist behauptet worden, dass der normale Befruchtungsvorgang nicht mehr vorhanden sei, dafür aber an einer anderen Stelle eine Verschmelzung von Kernen in Betracht käme. FARMER und DIGBY (1907) fanden bei *Dryopteris Filix-mas* var. *pseudo-mas* var. *polydactyla* Wills., dass die Prothallien keine Archegonien erzeugen, dass aber die Kerne benachbarter Prothallienzellen sowohl auf dem mehrschichtigen Teil wie in den Flügeln miteinander verschmelzen können, nachdem der eine Kern durch eine hierfür an der Wand gebildete Oeffnung in die Nachbarzelle eingewandert ist (Fig. 96 a, c). Aehnliche Verhältnisse beobachteten sie bei *Dryopteris Filix-mas* var. *pseudo-mas* var. *polydactyla* Dadds (Fig. 96 b). Die Embryonen bildeten sich aus dem Prothalliumgewebe heraus, und zwar vermuten FARMER und DIGBY, dass die Keimpflanzen ihren Ursprung nehmen aus den in der geschilderten Weise diploid gewordenen Prothalliumzellen. Direkt beobachtet haben sie dies jedoch nicht, sondern sie vermuten es lediglich auf Grund ihrer Chromosomenzählungen. Der Wechsel der Chromosomenzahl würde demnach trotz des Fehlens der Archegonien der gleiche sein wie bei einem normalen Farn.

Diesen Modus verglichen die Autoren mit dem „BLACKMANSchen Typus" bei den Uredineen und bezeichneten ihn als Pseudo-Apogamie. STEPHENS und SYKES (1910) [1]) haben in Prothallien von *Pteris Droogmantiana* Zellen mit zwei Kernen gefunden, die miteinander verschmelzen können. Die Zweikernigkeit soll hier so zustande kommen, dass die Wandbildung bei der Kernteilung unterbleibt. Ob die auf diese Weise diploid gewordenen Zellen nun einen diploiden Sporophyten liefern, geben die Autoren nicht an. Abgesehen von diesem Befund und abgesehen von den von HAYES (1924), JUNG (1927)[2]) und FRIEBEL (1933) beobachteten zweikerni-

---

FIG. 85. *Equisetum arvense*, Spermatozoidentwicklung. *a.* letzte spermatogene Teilung mit Blepharoplast, *b–d.* Entwicklung einer spermatogenen Zelle zum Spermatozoid; Umgestaltung des Kerns und des Blepharoplasten. Vergr. ca. 1400 ×. Nach SHARP (1912) aus SHARP–JARETZKY (1931). — FIG. 86–90. Spermatozoiden. FIG. 86. *Equisetum arvense.* Vergr. ca. 1400 ×. Nach SHARP (1912). Fig. 87. *Isoetes japonica.* Vergr. ca. 1100 ×. Nach YUASA (1933c). Fig. 88. *Marsilia quadrifolia.* Vergr. ca. 1000 ×. Nach SHARP (1914). Fig. 89. *Selaginella involvens.* Vergr. ca. 2200 ×. Nach YUASA (1933c). Fig. 90. *Isoetes lacustre*, nach lebendem Material. Bei *K* Kernstruktur sichtbar. Vergr. ca. 1650 ×. Nach DRACINSCHI (1932). — FIG. 91. Kernspindel mit Eikern (unten) und Bauchkanalkern von *Dryopteris dentata.* Nach YAMANOUCHI (1908b). — FIG. 92. Archegon von *Pilularia globulifera.* ♂ Kern (der kleinere!) dicht am Eikern. Vergr. ca. 220 ×. Nach CAMPBELL (1888). — FIG. 93. ♂ Kern im Eikern von *Onoclea sensibilis*, 14 Stunden nach Wässerung der Kultur. Vergr. ca. 900 ×. Nach SHAW (1898a). — FIG. 94 a–c. Verschiedene Stadien der Auflockerung des ♂ Kerns im Eikern von *Dryopteris dentata.* Nach YAMANOUCHI 1908 b. — FIG. 95. *Lygodium palmatum.* ♂ Kern, abgerundet, am Eikern. Chromatin des Eikerns in der Nähe des Spermakerns dicht angehäuft. Nach ROGERS (1927). — FIG. 96 a, c. *Dryopteris Filix-mas* var. *pseudo-mas* var. *polydactyla* Wills. b var. *polydactyla* Dadds. Erklärung im Text. Nach FARMER und DIGBY 1907. — FIG. 87, 89–93 umgezeichnet, die übrigen photogr. Wiedergabe.

---

[1]) Ann. of Bot. 24, 1910.

[2]) Diss. Marburg 1927.

gen Prothalliumzellen sind die Angaben von FARMER und DIGBY noch nicht bestätigt und auch noch keine ähnlichen Vorgänge beobachtet worden, obgleich Prothallien apogamer oder apogam erscheinender Farne wiederholt untersucht worden sind. Ich selbst habe in zahlreichen untersuchten Fällen keinerlei Anzeichen für eine Verschmelzung vegetativer Prothalliumkerne finden können. Dass FARMER und DIGBY die Entstehung von Embryonen aus den diploid gewordenen Zellen nicht direkt beobachtet haben, wurde schon betont. Im übrigen wäre zu sagen, dass manche ihrer diesbezüglichen Abbildungen wenig beweisend sind, während andere den Eindruck erwecken, als ob auch Restitutionskerne vorliegen könnten. Auch wäre zu erwägen, ob nicht z.T. Artefakte, durch die Präparatherstellung bedingt, in Frage kommen dürften. Es muss daher eine Nachprüfung dieser Verhältnisse vorgenommen werden. — Eine ganz andere Möglichkeit zur Erreichung des diploiden Zustandes wird von R. F. ALLEN (1911) für *Polystichum falcatum* angegeben. Die Sporophyten entwickeln sich apogam an den haploiden Prothallien und sind selbst auch haploid. Erst in den Sporangien erfolgt nach R. F. ALLEN ein Sexualvorgang, indem die 16 Zellen des sporogenen Zellkomplexes sich paarweise unter Kernverschmelzung zu 8 Zellen vereinigen, die nach Reduktionsteilung die Sporen liefern. Nach den Erfahrungen FRIEBELS (1933), der bei Sporenmutterzellen von *Polybotrya cervina* eine Plasmaverschmelzung und darauf erfolgende Fusion der Teilungsspindeln beobachtete, sind von vornherein solche Verschmelzungsvorgänge bei Sporenmutterzellen nicht ausgeschlossen; es ist jedoch zu untersuchen, ob nicht auch hier ebenso wie bei *Dryopteris hirtipes* (STEIL 1919) und *D. remota* sowie *Dryopteris Filixmas.* var. *cristata* (DÖPP 1932, 1933) Restitutionskernbildung vorliegt. Auch Miss STEIL (1915) hatte ja ursprünglich angenommen, dass in den Sporangien von *D. hirtipes* Zell- und Kernverschmelzung entsprechend den Angaben ALLENS stattfände; sie kam jedoch in ihrer späteren Arbeit zu anderen Ergebnissen (§ 8).

**§ 8. Abweichungen vom normalen Kernteilungsmodus.** — Wichtigste Literatur: BARTOO (1929), DÖPP (1932, 1933), FRIEBEL (1933), GRAUSTEIN (1930), HANNIG (1911). — Abweichungen gegenüber dem normalen Verhalten bei der gewöhnlichen Mitose sowie bei den Reifeteilungen sind bei den Pteridophyten, besonders bei den *Polypodiaceen*, sehr häufig, teils als für die Sporenbildung wesentliche Erscheinung im Sporangium bestimmter Arten oder Varietäten (*Dryopteris hirtipes* u.a.; vgl. unten), bei anderen Arten als Ausnahmefall oder auch als häufigere Abnormität während der Reifeteilungen bei sonst normalen Verhalten, oder schliesslich als regelmässig auftretender Vorgang an solchen Stellen, denen für die Fortpflanzung keinerlei Bedeutung zukommt (z.B. in der Tapete). In verschiedenen Fällen ist eine experimentelle Beeinflussung der gewöhnlichen und der allotypischen Mitosen gelungen.

Eine wichtige Rolle spielt die R e s t i t u t i o n s k e r n b i l d u n g, die bei bestimmten, apogamen [1]) Vertretern im sporogenen Gewebe auf dem 8-Zellstadium auftritt. Zunächst von STEIL (1919) bei *Dryopteris hirtipes* untersucht, wurde sie näher analysiert von DÖPP (1932) bei *Dryopteris remota* und ausserdem bei *Dryopteris Filix-mas* var. *crist.* hort. festgestellt (DÖPP 1933). Sie tritt ferner nach meinen Beobachtungen noch bei einer ganzen Anzahl anderer apogamer *Polypodiaceen* auf. Während die Mitose im 8-zelligen Archespor (als Chromosomenzahl sei 2n angenommen) auf dem Stadium der Prophase und der Metaphase noch ein durchaus normales Aussehen besitzt, zeigen in der darauffolgenden Anaphase die Chromosomen ein abweichendes Verhalten. Die Spalthälften wandern nicht nach den Spindelpolen auseinander, sondern bleiben im allgemeinen paarweise in paralleler Anordnung nebeneinander liegen und bilden einen oft sehr unregelmässig gestalteten Knäuel, aus dem nur ein einziger Kern mit 4 n Chromosomen hervorgeht (Fig. 97, 98). Es entstehen auf diese Weise Kerne von meistens sehr abweichender Gestalt (Fig. 98). Sehr häufig sind Hantel- und Nierenform. Zusammen mit diesem Vorgang erfolgt „unvollständige Zellteilung" mit nur stellenweise ausgebildeten Wänden (Fig. 98). Die Restitutionskerne in den 8 nunmehr als Sporenmutterzellen aufzufassenden Archesporzellen verhalten sich wie die Sporenmutterzellkerne bei normalen Arten; die einzelnen Stadien der Reduktionsteilung sowie die zweite Reifeteilung werden in der gewöhnlichen Weise durchlaufen, sodass die Sporen wiederum 2 n Chromosomen enthalten. Bei der apogamen Entstehung des Sporophyten aus dem Gametophyten findet dagegen keine Aenderung in der Chromosomenzahl statt. Es hat sich bei einigen der genannten Arten gezeigt — das gleiche gilt sicherlich auch in anderen Fällen — dass die Restitutionskernbildung von entscheidender Bedeutung für die Erzeugung keimfähiger Sporen ist; denn solche Sporangien von `Dryopteris remota* z.B., die anstelle der Restitutionskernbildung eine gewöhnliche Teilung ausführen, wobei 16 Sporen mutterzellen[2]) entstehen, liefern nur sterile Sporen. — Es ist aus verschiedenen Gründen sehr wahrscheinlich, dass die Konjugationspartner im Zygotän-, Pachytän- und Diakinese-Stadium die beiden vorher gebildeten Tochterchromosomen darstellen, die ja nicht auf zwei Kerne verteilt wurden, sondern in den gleichen Kern, nämlich den Restitutionskern, hineingelangen. Findet anstelle der Restitutionskernbildung die Teilung in der gewöhnlichen Weise statt, so erfolgt so gut wie keine Geminibildung. — Während der Entwicklung zur Diakinese nimmt der Kern übrigens wieder normale Gestalt an, und die unvollständigen Wandstücke verschwinden.

Restitutionskernbildung verbunden mit unvollständiger Zellteilung kann ferner stattfinden bei der 1. oder 2. Reifeteilung (Fig. 100, 103); sie

---

[1]) Ich benutze die Terminologie von WINKLER (1908).
[2]) Dies ist bei geschlechtlichen Formen die Normalzahl.

dürfte hierbei Anlass geben zur Entstehung von in Form und Chromosomenzahl abweichenden Sporen, die in Triaden oder Dyaden angeordnet sind. Solche abweichenden Teilungen und Sporentypen sind bei den genannten apogamen Formen und bei der apogamen *Selaginella rupestris* (GRAUSTEIN 1930; Fig. 113), bei gewissen Varietäten (Fig. 100), aber auch bei vermutlichen Bastarden (*Polypodium Schneideri* (FARMER und DIGBY 1910) und *Polybotrya cervina* (FRIEBEL 1933)) und sogar bei gewöhnlichen Arten mit normaler geschlechtlicher Fortpflanzung (*Polypodium vulgare* (FARMER und DIGBY 1910), *Aneimia Phyllitidis* (Fig. 112 b, c) und *Onoclea sensibilis* (FRIEBEL (1933)) anzutreffen. Die in seltenen Fällen vorhandenen grossen kugeligen Sporen, die in Einzahl aus einer Sporenmutterzelle entstehen, sind wohl z.T. auch hierher zu rechnen; andererseits könnten sie auch durch Kernteilung ohne Wandbildung entstanden sein.

Sehr unregelmässige Teilungsvorgänge beobachtete STEIL (1935) in der Tapete von *Botrychium virginianum* und *Ophioglossum vulgatum*. Hier entstehen durch wiederholte Restitutionskernbildung grosse, polyploide Kerne von abweichender Form (Fig. 101).

Restitutionskernbildung lässt sich auch experimentell durch Anwendung geeigneter Methoden erzielen und zwar durch Kälte (DÖPP, noch unveröffentlich; vgl. § 9 und Fig. 120) sowie durch Einwirkung von Chloroform und Röntgenstrahlen auf Sporangien (Fig. 115, 116).

Eine weitere Abweichung besteht in der Bildung von K e r n e n m i t g e r i n g e r e r C h r o m o s o m e n z a h l. Bei *Dryopteris remota* und *D. Filix-mas* var. *crist.* hort. kann anstelle der Restitutionskernbildung eine Teilung in Tochterkerne und -zellen (Sporenmutterzellen) erfolgen, die aber von einer gewöhnlichen Mitose sehr stark abweicht. Es erfolgt wohl eine Trennung der Chromosomen in zwei Gruppen; dabei wandern aber die Spalthälften eines Chromosoms nicht zu den beiden Polen hin auseinander, sondern sie gelangen beide in den gleichen Kern. Die Verteilung geschieht ganz unregelmässig, sodass der eine Tochterkern dabei fast stets mehr Chromosomen erhält und daher grösser ist als der andere (Fig. 99). Beispielsweise entstehen bei *Dryopteris remota* aus einer der 8 Archesporzellen zwei Sporenmutterzellen mit 2 × 50 und 2 × 80 Chromosomen anstatt einer Sporenmutterzelle mit 2 × 130 Chromosomen. Häufig degeneriert der kleinere Kern und die kleinere Zelle (Fig. 99 rechts); in vielen Fällen erfolgt dagegen durchaus normale Reduktionsteilung und Sporenbildung. Die so gebildeten Sporen scheinen nach meinen Beobachtungen auch keimfähig zu sein, jedoch ist sicheres noch nicht festgestellt. Ebenso wie bei der Restitutionskernbildung konjugieren offenbar die vorher gebildeten Spalthälften eines jeden Chromosoms paarweise miteinander. Trennen sich nur wenige Chromosomen von der Hauptmasse ab, so entsteht ein Zwergkern, der aber, ohne dass er durch eine Wand abgetrennt würde, resorbiert wird. In diesem Fall entsteht nur e i n e Sporenmutterzelle, die natürlich eine geringere Chromosomenzahl haben muss.

A b t r e n n u n g von C h r o m o s o m e n von der Hauptmasse sowie als häufige Folgeerscheinung Bildung von Z w e r g k e r n e n, oft in grosser Zahl, kann ferner eintreten während der 1. und 2. Reifeteilung und zwar s p o n t a n (Fig. 100, 102) wie auch (FRIEBEL 1933) nach e x - p e r i m e n t e l l e r B e e i n f l u s s u n g (Fig. 108, 114–117). Der erste Fall ist verwirklicht bei apogamen Farnen wie auch bei solchen, die als Bastarde angesprochen werden (*Polypodium Schneideri*, FARMER und DIGBY 1910; Fig. 102), aber schliesslich auch bei gewissen Varietäten (Fig. 100) und ganz normalen Formen (*Polypodium vulgare*, FARMER und DIGBY 1910). Nach Einwirkung von Narkotika und Behandlung mit Röntgenstrahlen entstandene Zwergkerne zeigen die Fig. 108 (Sporen von *Matteuccia orientalis*, nach Ätherbehandlung entstanden (FRIEBEL 1933)) und 114 (Sporen von *Dryopteris spinulosa*, nach Einwirkung von Röntgenstrahlen entstanden; Orig. ).

Auf Kernteilung ohne Wandbildung bei der Sporogenese sind die in Fig. 109 und 112 dargestellten m e h r k e r n i g e n S p o r e n zurückzuführen. Diese Mehrkernigkeit kann nach FRIEBELs Untersuchungen (1933) spontan (Fig. 112 a *Aneimia Phyllitidis*; Fig. 112 d *Onoclea sensibilis*) wie auch nach Aetherbehandlung (Fig. 109, Riesensporen von *Matteuccia orientalis*) auftreten.

Es können also in verschiedener Weise h e t e r o p l o i d e Sporen entstehen. FRIEBEL hat aus solchen von *Aneimia Phyllitidis* und *Onoclea sensibilis* abnorme Prothallien erhalten; diese lieferten Sporophyten, die von normalen Pflänzchen wesentlich verschieden waren (§ 9). Sonst wissen wir bisher über die Keimung dieser Sporen leider so gut wie nichts. Ich glaube aber, trotzdem sagen zu können, dass die bei den *Polypodiaceae* häufig gegebene Möglichkeit der Entstehung polyploider sowie aneuploider Sporen während der Reifeteilungen sicherlich von grosser genetischer Bedeutung ist.

An weiteren Abweichungen, die in Sporangien vorkommen können, wären u.a. zu nennen:

1. Bildung von Sporenmutterzellen bei *Polypodiaceae* schon nach dem 3. Teilungsschritt (normalerweise erfolgt diese erst nach dem 4.). Beispiel: *Polybotrya cervina* nach FRIEBEL (1933).

2. Das Vorkommen verschiedener Kernteilungsphasen in den Sporenmutterzellen des gleichen Sporangiums, während sonst die Sporenmutterzellen sich genau auf dem gleichen Stadium befinden. Bei apogamen Formen (eigene Beobachtung) sowie bei *Polybotrya cervina* (FRIEBEL 1933).

3. Kernteilung ohne Wandbildung im Archespor führt zur Bildung grosser vielkerniger Sporenmutterzellen (BARTOO 1929; *Schizaea rupestris*).

4. Verschmelzung der Plasmen der sporogenen Zellen, besonders häufig vom 1. — 3. Teilungsschritt, beobachtet von FRIEBEL (1933) bei *Polybotrya cervina* (spontan) sowie bei *Matteuccia orientalis* nach Ätherbehandlung

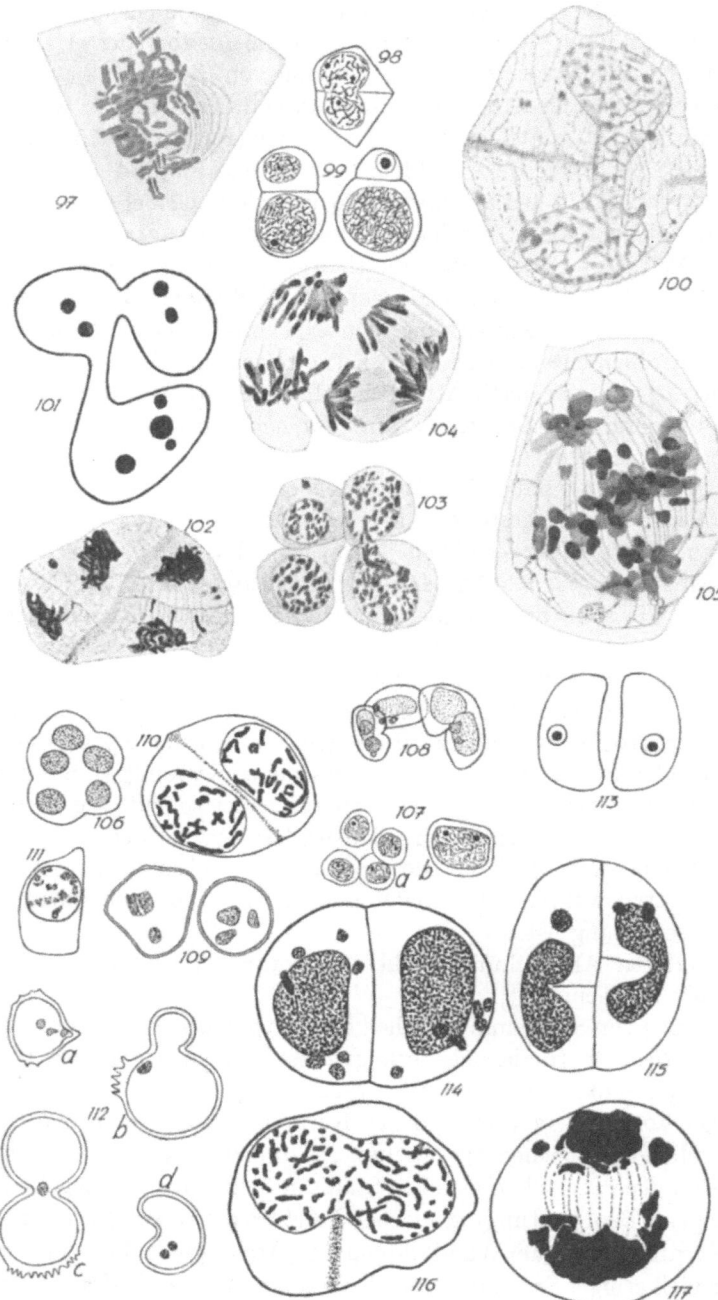

FIG. 97. Restitutionskernbildung bei *D. remota*. Vergr. ca. 1650 ×.
Aus DÖPP (1932). — FIG. 98. Sporenmutterzelle mit Restitutionskern und unvollständigen Wandstücken von *D. remota*. Vergr. ca.

(Fig. 106). Die Zahl der Kerne in einem gemeinsamen Plasma kann verschieden sein; im Höchstfall sind es 16.

5. Bei *Polybotrya* können nach FRIEBEL (1933) die in ein gemeinsames Plasma eingeschlossenen Kerne während der Teilungsstadien miteinander verschmelzen; es entstehen dabei also Verschmelzungsprodukte von 2–16 Sporenmutterzellkernen. Auf diese Weise können Diakinesekerne mit 600 und mehr (uni- und bivalenten) Chromosomen entstehen. Nach der 1. Reifeteilung wurden Ruhekerne gebildet, die keine Weiterentwicklung mehr erfuhren. Vielleicht ist auch der grosse Kern der in Fig. 107 b abgebildeten Sporenmutterzelle von *Matteucia Struthiopteris*, die nach $KNO_3$–Injection allein in einem Sporangium auftrat, auf eine solche Verschmelzung zurückzuführen (Fig. 107a zeigt normale Sporenmutterzellen zum Vergleich).

6. N i c h t p a a r u n g von Chromosomen (Fig. 105) wurde beobachtet bei *Polypodium Schneideri* = *P. aureum* × *P. vulgare* var. *elegantissimum* (FARMER und DIGBY 1910), bei *Dryopteris remota* = *D. Filix-mas* × *D. spinulosa* (DÖPP 1932) und zwar nur in denjenigen Sporenmutterzellkernen, die nicht durch Restitutionskernbildung, sondern durch eine gewöhnliche Teilung in 16-Zahl im Sporangium entstanden sind, bei *A. germanicum* = *A. Trichomanes* × *septentrionale* (DÖPP, noch nicht veröffentlicht) und schliesslich bei *Polybotrya cervina* (FRIEBEL 1933). Es handelt sich dabei also um Formen, die auf Grund ihrer morphologischen Eigenschaften mit mehr oder weniger grosser Wahrscheinlichkeit als Bastarde

---

500 ×. Aus DÖPP (1932). — FIG. 99. Je zwei durch unregelmässige Teilung entstandene, ungleich grosse Sporenmutterzellen von *D. remota*. Rechts Degeneration des Kerns der kleineren Zelle. Vergr. ca. 500 ×. Aus DÖPP (1932). — Fig. 100. Sporenmutterzelle von *Polypodium vulgare* var. *elegantissimum* mit Restitutionskern (Restitutionskernbildung ist anstelle der heterotypischen Teilung erfolgt). Im Cytoplasma Chromatin. Vergr. ca. 1200 ×. Nach FARMER u. DIGBY (1910). — FIG. 101. Restitutionskern aus der Tapete von *Botrychium virginianum*. Nach STEIL (1935). — FIG. 102. Homoeotypische Teilung von *Polypodium Schneideri* mit vereinzelt bleibenden Chromosomen und Chromatin im Cytoplasma. Vergr. ca. 1200 ×. Nach FARMER und DIGBY (1910). — FIG. 103. Junge Sporen von *Dryopteris hirtipes*. Rechts anstelle zweier Kerne ein Restitutionskern. Vergr. ca. 1150 ×. Nach STEIL (1919). — FIG. 104. *Dryopteris hirtipes*. Zwei Teilungen in einer Sporenmutterzelle, die offenbar zur Bildung von 5 Sporen führen werden. Vergr. ca. 1150 ×. Nach STEIL (1919). — FIG. 105. *Polypodium Schneideri*. Heterotypische Spindel. Bivalente und univalente (die kleineren!) Chromosomen. Vergr. ca. 1200. Nach FARMER und DIGBY (1910). — FIG. 106–112. Abnormitäten in Sporangien. Nach FRIEBEL (1933). Erklärung im Text. 106 Vergr. ca. 250 ×, 107 Vergr. ca. 200 ×, 108 Vergr. ca. 600 ×, 109 Vergr. ca. 250 ×, 110 Vergr. ca. 600 ×, 111 Vergr. ca. 600 ×, 112 a–c Vergr. ca. 120 ×, 112 d Vergr. ca. 300 ×. — FIG. 113. Makrosporendyade von *Selaginella rupestris*. Nach GRAUSTEIN 1930. — FIG. 114–117. Abnormitäten nach Behandlung mit Röntgenstrahlen (114, 115, 117) und Chloroform (116). Vergr. ca. 1350 ×. 114 Schnitt durch Tetrade (nur 2 Sporen getroffen) mit zahlreichen Kleinkernen ausser den beiden grossen Kernen. 115 Dyade mit Restitutionskernen und einigen kleinen Kernen. 116. Anstelle der heterotypischen Teilung ist Restitutionskernbildung erfolgt. 117. Mehrere isolierte Chromosomengruppen (Telophase der heterotypischen Teilung). 97–99, 114–117 Orig., 102–105 photogr. Wiedergaben, die übrigen Fig. umgezeichnet.

angesehen werden. Das Auftreten von Asyndese würde für den Hybriden-charakter der betreffenden Pflanzen sprechen. Stets kommen jedoch in den Diakinesekernen univalente Chromosomen nicht ausschliesslich vor, sondern mit bivalenten zusammen. Auffällig ist die zusammen mit der Asyndese auftretende Sterilität der Sporen; von den genannten Formen werden nämlich *Polypodium Schneideri, Asplenium germanicum* und *Polybotrya* als steril angegeben. Bei *Dryopteris remota* sind die aus den Asyndese zeigenden Sporenmutterzellen entstandenen Sporen vollkom-men steril, während die nach Restitutionskernbildung auftretenden Spo-ren gute Keimfähigkeit zeigen.

7. Drei- und mehrpolige Spindeln. Beispiel: Tetraploide Gartenrasse von *Psilotum triquetrum* (OKABE 1929) in der Metaphase der ersten Reife-teilung; Spindeln mit drei und mehr Polen von STEIL (1935) in der Tapete eusporangiater Farne (*Botrychium, Ophioglossum*) beobachtet.

8. Bildung von mehr als 4 Sporen aus einer Sporenmutterzelle (STEIL 1919, OKABE 1929, FRIEBEL 1933). (Das Vorkommen von Dyaden oder Triaden ist schon früher erwähnt.) Vgl. Fig. 104.

9. Bei *Botrychium* und *Ophioglossum* entstehen nach STEIL (1935) in der Tapete 2- und 4-kernige Zellen infolge einer nicht genügenden Entwick-lung der Spindel oder einer Auflösung derselben oder der schon gebildeten Zellplatte.

10. Die verschiedentlich für die Tapete angegebene ,,Amitose'' (*Helmin-thostachys zeylanica*, BEER 1906; *Equisetum*, HANNIG 1911) dürfte wohl in vielen Fällen in Wirklichkeit Restitutionskernbildung sein (vgl. STEIL 1935).

Aus dem Gesagten geht hervor, dass einige als Artbastarde angesehene Farne besondere cytologische Verhältnisse zeigen (vgl. auch GRAUSTEIN 1930). Da jedoch cytologische Untersuchungen an künstlich erzeugten Artbastarden noch nicht vorgenommen worden sind, einfach aus dem Grunde, weil solche Bastarde bisher kaum hergestellt, geschweige denn bis zur Bildung von Sporangien gebracht werden konnten, so lässt sich sicheres über die infolge von Bastardierung auftretenden cytologischen Veränderungen nicht aussagen; ausserdem ist ja auch zu bedenken, dass zahlreiche der genannten Abweichungen bei Nichtbastarden vorkommen.

Einer Nachprüfung bedürfen die Angaben von GRAUSTEIN (1930) über *Selaginella* (,,the contents of the nucleus pass out into the cytoplasm''; ,,the nucleus is divided without chromosome formation''.

Ausserhalb der Sporangien scheinen Abweichungen seltener zu sein. Es seien an dieser Stelle genannt die von JUNG (1927) [1]) in Prothallien von *Polystichum Braunii* beobachteten Zellen mit Doppelkernen (es handelt sich dabei offenbar um Kernteilung ohne Wandbildung; diese Prothallien trugen übrigens z.T. Spaltöffnungen!) und die 2-kernigen Zellen in gewissen

---

[1]) Diss. Marburg 1927.

(vermutlich heteroploiden) Prothallien von *Onoclea sensibilis* (FRIEBEL 1933). Im übrigen vgl. noch das auf S. 243 u. 244 Gesagte.

**§ 9. Polyploidie und Aneuploidie.** — Wichtigste Literatur: DÖPP (1932), FARMER u. DIGBY (1907), FRIEBEL (1933), HEILBRONN (1928, 1932, 1933). LAWTON (1932), MANTON (1932), OKABE (1929). — Die Untersuchungen der letzten Jahrzehnte[1]) haben gezeigt, dass Polyploidie unter den Angiospermen in der Natur eine verbreitete Erscheinung ist. Auch bei den Pteridophyten ist sie anzutreffen. Ausserdem ist es bei letzteren in mehreren Fällen geglückt, Polyploidie experimentell hervorzurufen. Sie tritt naturgemäss sowohl im Gametophyten wie im Sporophyten in Erscheinung.

TAB. 5 [2]) (Cf. unten)

| | *Haploidzahl* | |
|---|---|---|
| *Pteris cretica* | ca. 30 | DE LITARDIÈRE 1921 |
| „ „ var. *Ouvrardii* | ca. 60 | DE LITARDIÈRE 1921 |
| *Pteris tremula* | 32 | GREGORY 1904 |
| „ „ | 60–65 | CALKINS 1897 |
| „ „ | ca. 95 | DE LITARDIÈRE 1921 |
| *Polypodium aureum* | 16 | HEILBRONN 1928 |
| „ „ | 34–36 | { FARMER und DIGBY 1910, DE LITARDIÈRE 1921 |
| „ „ | 76 | FRIEBEL 1933 |
| *Marsilia quadrifolia* | 16 | { STRASBURGER 1907, DE LITARDIÈRE 1921 |
| „ „ | 50–70 | GUSTAFSON 1935/36 [3]) |
| *Marsilia Drummondii* | im Gametophyten und Sporophyten 32 | STRASBURGER 1907 |
| „ „ | ca. 105 | GUSTAFSON 1935/36 [3]) |
| *Salvinia natans* | 4 | ARNOLDI 1910 |
| „ „ | 8 | KUNDT 1911, YASUI 1911 |
| „ „ | 8 | DE LITARDIÈRE 1928 |
| „ „ | ca. 15 | HEITZ 1926 |
| „ „ | ca. 24 | DE LITARDIÈRE 1921 |
| *Salvinia auriculata* | 15 | HEITZ 1926 |
| „ „ | 32 | DE LITARDIÈRE 1921 |
| *Psilotum triquetrum* | 52 | OKABE 1929 u.a. |
| „ „ , Varietät | 104 | OKABE 1929 |
| *Equisetum arvense* | ca. 14–15 | LENOIR 1926 |
| „ „ | ca. 29 | LENOIR 1922 |
| „ „ | ca. 115 | BEER 1913 |
| *Equisetum limosum* | 12 | STEINECKE 1932 |
| „ „ | ca. 45–50 | v. BOENICKE 1911 |

---

[1]) Vgl. insbesondere die Zusammenstellung von TISCHLER (Bot. Jahrb. 67, 1934, 1 ff.).

[2]) Die auf S. 235 ff. nicht aufgeführten Literaturangaben sind in den Tabulae biologicae, Bd. IV, VII u. XI zu finden.

[3]) Hereditas 21, 1935/36, S. 56 ff.

TAB. 6 (Cf. unten)

| | | |
|---|---|---|
| *Adiantum capillus-Veneris* | ca. 31 bezw. 32 | { DE LITARDIÈRE 1921, REINHOLD 1926 |
| „          *cuneatum* | ca. 60 | DE LITARDIÈRE 1921 |
| *Polypodium aureum* | 16, 34–36, 76 | vgl. Tab. 5 |
| „          *vulgare* | ca. 90 | FARMER und DIGBY 1910 |
| *Aneimia rotundifolia* | 38 | } |
| „          *Phyllitidis* | 76 | } FRIEBEL 1933 |
| „          *obliqua* | 76 | } |
| *Salvinia natans* | 4, 8, 15, 24 | } Vgl. Tab. 5 |
| „          *auriculata* | 15, 32 | } |
| *Selaginella Emmeliana* | 8 | } DENKE 1902 |
| „          *serpens* | 8 | } |
| „          *Vogelii* | 8 | } |
| „          *Ouvrardii* | 10–11 | } HEITZ 1926 |
| „          *Martensii* | 26–28 | } |
| *Psilotum triquetrum* | 48 | { ROSEN 1896 / GUIGNARD 1898 / OKABE 1909 |
| *Tmesipteris tannensis* | ca. 100 | YEATES 1925 |
| *Isoetes echinospora* | 11 | EKSTRAND 1920 |
| „          *asiatica* | 11 | } TAKAMINE 1921 |
| „          *japonica* | ca. 22 | } |
| „          *japonica* | 33 | YUASA 1934 |
| *Equisetum arvense* | ca. 29 | LENOIR 1922 |
| | ca. 14–15 | LENOIR 1926 |
| „          *limosum* | ca. 45–50 | v. BOENICKE 1911 |
| | 12 | STEINECKE 1932 |
| „          *arvense* | ca. 115 | BEER 1913 |

Spontan auftretende Polyploidie. Die hohen Chromosomenzahlen vieler Pteridophyten legen die Vermutung nahe, dass sie polyploiden Charakter haben. Bisher liegen dafür aber nur wenige sichere Anhaltspunkte vor.

Auf einige Fälle, die — vorausgesetzt, dass die Zählungen stimmen — durch Vervielfältigung der ursprünglich vorhandenen Chromosomenzahl entstanden sein dürften, hat schon DE LITARDIÈRE (1921) hingewiesen (Tab. 5), nämlich auf die von *Pteris cretica* und der Varietät *Ouvrardii*, die er für eine tetraploide Rasse von *Pteris cretica* hält, ferner von *Pteris tremula* und *Salvinia natans*. Morphologisch verschiedene Rassen von *Salvinia natans* sind allerdings noch nicht beschrieben. In Tab. 5 sind noch einige weitere Verschiedenheiten in den Zählungsergebnissen für die gleiche Art angegeben. Aehnliche Beziehungen wie die eben genannten liegen vielleicht auch innerhalb einer Gattung für verschiedene Spezies oder gar für verschiedene Gattungen vor. Die hierfür in Frage kommenden Arten sind mit ihren Chromosomenzahlen in Tab. 6 zusammengestellt [1]). Bezüglich *Selaginella* nimmt HEITZ (1926) an, dass bei *S. Martensii* ein Multiplum von 8 vorliege, und dass dabei möglicherweise nicht eine Vermeh-

---

[1]) Vgl. hierzu die Zusammenstellungen bei TISCHLER (1921–22), S. 609 und 611.

rung des Genoms, sondern ein Chromosomenzerfall durch Querteilung eingetreten sei, da sich *S. Martensii* im Gegensatz zu den 16-chromosomigen (2 n = 16) Arten, die lange, stäbchenförmige Chromosomen besitzen, durch sehr kurze, beinahe kugelige Chromosomen auszeichnet. Ähnliches vermutet DE LITARDIÈRE (1928) bezüglich *Salvinia natans*.

Im Sporenmaterial eines jeden Farnes finden sich u n r e g e l m ä s s i g g e s t a l t e t e, sowie in der G r ö s s e a b w e i c h e n d e S p o r e n. Als Ursache kommen gewisse Unregelmässigkeiten bei den Reifeteilungen in Betracht (§ 8). Hierin liegt zweifellos eine Möglichkeit zur Gewinnung polyploider Formen. FRIEBEL (1933) fand bei *Aneimia Phyllitidis* (Fig. 112 a–c) spontan aufgetretene, z.T. 2-kernige Sporen, die ihrer Grösse nach sehr wahrscheinlich als polyploid anzusehen sind, wenn er auch keine Zählung vorgenommen hat. Diese Grossporen zeigten bei Einzelkultur schnellere Keimung, und die aus ihnen hervorgegangenen Prothallien zeichneten sich durch schnelleres Wachstum und grössere Zelldimensionen aus. An gewissen, offenbar aus 2-kernigen Sporen hervorgegangenen Prothallien von *Onoclea sensibilis* wurden in der Nähe der Meristembucht 2-kernige Zellen und in den ,,Flügeln'' grosse, amöboide Kerne festgestellt. Auch Sporophyten, die vermutlich polyploid sind, wurden erhalten.

E x p e r i m e n t e l l   e r z e u g t e   P o l y p l o i d i e.  Für die experimentelle Erzeugung von polyploiden Formen bestehen, ebenso wie bei den Laubmoosen, 3 Möglichkeiten: 1. Apospore Erzeugung eines Gametophyten aus diploiden Zellen des Sporophyten (vegetative Blattzellen oder Zellen der Sporangienwand), wobei die entstehenden Prothallien diploid sind (FARMER und DIGBY 1907, HEILBRONN 1928, 1932, LAWTON 1932, MANTON 1932). Bedürfen die so entstandenen Prothallien zur Erzeugung eines Sporophyten der Befruchtung, so entstehen auf diese Weise 3n- und 4n-Pflanzen. 2. Beeinflussung der Reifeteilungen in der Weise, dass mehrkernige oder einkernige, jedenfalls aber Sporen mit höherer Chromosomenzahl gebildet werden (FRIEBEL 1933). 3. Beeinflussung der Kernteilung in Prothallien (DÖPP, noch unveröffentlicht).

Was die erste Möglichkeit betrifft, so liegen die ersten genaueren cytologischen Untersuchungen über diploide Prothallien vor von FARMER und DIGBY (1907) an Varietäten von *Athyrium Filix-femina*, die sich durch apospore Entstehung der Prothallien, verbunden mit Sterilität der Sporangien, auszeichnen, sowie an Varietäten einiger anderer Farne. Die Prothallien besitzen nach den Untersuchungen dieser Autoren die gleiche Zahl von Chromosomen wie der Sporophyt. FARMER und DIGBY stellten an diesen Prothallien, die gegenüber denen der Normalform als diploid oder annähernd diploid anzusehen sind, an den Kernen derselben und an den Spermatozoiden grössere Dimensionen fest als an den gewöhnlichen, aus Sporen entstehenden Prothallien von *Athyrium Filix-femina*, die beträchtlich weniger Chromosomen besitzen (Tab. 1). Da sich die aposporen

Prothallien als apogam oder parthenogenetisch [1]) erwiesen, war es nicht möglich, polyploide Sporophyten zu erhalten. Dies gelang jedoch HEIL-BRONN (1928) mit *Polypodium aureum* und danach MANTON (1932) mit *Osmunda regalis* (Fig. 118) und LAWTON (1932) mit *Dryopteris marginalis* und *Woodwardia virginica* (Fig. 119). Die ebenfalls apospor erzeugten Prothallien dieser Farne lieferten nur dann eine Keimpflanze, wenn Befruchtung eingetreten war. Auf diese Weise erhielten die Autoren 4 n -Sporophyten, aus denen durch Auslegen von Primärblättern oder Blattstücken auf geeignetes Substrat wieder Prothallien gewonnen werden konnten, die sicherlich — Zählungen sind an ihnen (abgesehen von HEILBRONN) allerdings nicht gemacht worden — auch 4 n Chromosomen besitzen. Ausserdem erhielt MANTON und allem Anschein nach auch LAWTON Sporophyten mit dem 3-fachen Chromosomensatz. Sporophyten von *Woodwardia virginica*, die wahrscheinlich triploid sind, traten in den Versuchen von LAWTON als Kreuzungsergebnis zwischen n- und 2 n- Prothallien auf. Durch Verschmelzung von 2 n- mit n-Gameten sind sicherlich auch

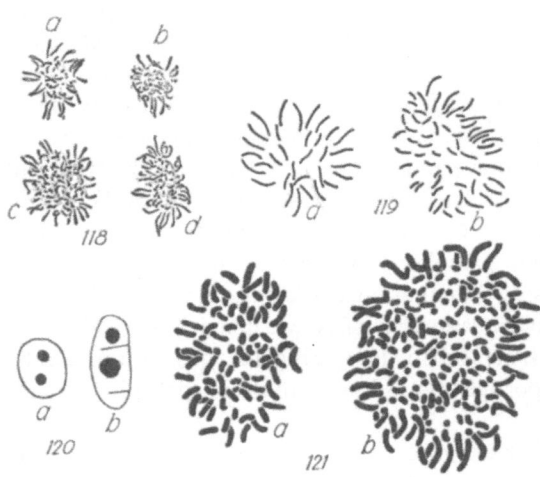

FIG. 118. *Osmunda regalis*. Aequatorialplatte *a*. einer gewöhnlichen Pflanze (44 Chromosomen),*b*. aus einem apospor erzeugten Prothallium (44 Chromosomen), *c*. aus einer tetraploiden (88 Chromosomen), *d*. aus einer triploiden Pflanze (66 Chromosomen). Vergr. ca. 500 ×. Photogr. Wiedergabe. Nach MANTON (1932). — FIG. 119. *Woodwardia virginica*. *a*. Platte aus einem haploiden, *b*. aus einem apospor entstandenen diploiden Gametophyten. Umgezeichnet. Nach LAWTON (1932). — FIG. 120–121. *Dryopteris Filix-mas* var. *crist*. hort. (Orig.). 120 *a*. Kern eines gewöhnlichen Prothalliums, 120*b* durch Kältewirkung erzeugter Restitutionskern eines jungen Prothalliums. Vergr. ca. 400 ×. 121 *a*. Platte (Wurzelspitze) einer gewöhnlichen Pflanze (82 Chromosomen), 121 *b*. Platte (Wurzelspitze) einer durch Kältewirkung erzeugten Pflanze mit doppelter Chromosomenzahl (173 Chr.). Vergr. ca. 1250 ×.

die beiden triploiden Pflanzen von *Osmunda regalis* entstanden, die Miss MANTON anführt, wenn sie auch näheres über deren Entstehung nicht angeben kann.

Die Tab. 7 gibt einen Ueberblick über die polyploiden Gametophyten und Sporophyten, die HEILBRONN, MANTON und LAWTON erhielten, mit den von diesen Autoren festgestellten Chromosomenzahlen.

---

[1]) Vgl. Anm. 1 a.S. 269.

TAB. 7

| | Gametophyten | | | Sporophyten | | |
|---|---|---|---|---|---|---|
| | aus Sporen | durch Aposporie entstanden | | | | |
| | $n$ | $2n$ | $4n$ | $2n$ | $3n$ | $4n$ |
| *Polypodium aureum* (HEILBRONN 1928, 1932) . . . . . . | 16 | 32 | 60–70 | | | 56–68 |
| *Osmunda regalis* (MANTON 1932) . . . | | 44 | | 44 | 66 | 88 |
| *Dryopteris marginalis* [1]) (LAWTON 1932) . . . | ca. 36 | ca. 65 | | | | 120–130 |
| *Woodwardia virginica* (LAWTON 1932) . . . | ca. 32 | 57–64 | | 59–65 | | 123–130 |

Die höchste erreichte Zahl ist in den genannten Fällen also 4 n. Oktoploidie ist bisher nur in e i n e m Fall [2]) erzielt worden.

Eine genauere cytologische Untersuchung der Reifeteilungen der polyploiden Sporophyten liegt noch nicht vor. OKABE (1929) bildet eine Metaphaseplatte der 1. Reifeteilung mit 104 Gemini ab; im übrigen treten charakteristische Abnormitäten auf (dreipolige Spindeln; Bildung von 5, 6, 7 usw. anstatt 4 Sporenkernen). Ferner macht HEILBRONN (1932) einige Angaben, aus denen hervorgeht, dass die tetraploide Pflanze von *Polypodium aureum* Sporen erzeugt, die, zu 90 % keimfähig, allem Anschein nach diploide Prothallien liefern.

TAB. 8. (Nach LAWTON (1932))

*Dryopteris marginalis* (Material von Long Island, N.Y.)

| | Gametophyt | | Sporophyt | |
|---|---|---|---|---|
| | Fläche der Prothalliumzellen in $\mu^2$ (Mittelwert) | Zunahme in % gegenüber dem $n$-Gametophyt | Fläche der unteren Epidermiszellen in $\mu^2$ | Zunahme in % gegenüber dem $2n$-Sporophyten |
| $n$ | 5900 | | | |
| $2n$ | 7900 | 34 | 3952 | |
| $4n$ | 19200 | 73 | 6916 | 75 |

HEILBRONN und LAWTON (Tab. 8) haben die Grössenverhältnisse der polyploiden Gametophyten und Sporophyten studiert. Während HEILBRONN (1928) bei den diploiden Prothalliumzellen von *Polypodium aureum* eine Flächenvergrösserung im Verhältnis 2 : 1 oder gar 3 : 1 erhielt, fand LAWTON (1932) nur eine Vergrösserung um 34 % bei *Dryopteris marginalis* und 53 % bei *Woodwardia virginica*. Es ergaben sich ferner im Sporophyten Grössenunterschiede bezüglich der Epidermiszellen, Schliesszellen, Sporangien und Kerne. Die Chloroplasten erwiesen sich allerdings bei 2 n-Prothallien

---

[1]) Die wirklichen Werte dürften nach LAWTON sein $n = 38$, $2n = 76$, $4n = 152$.

[2]) ANDERSSON–KOTTÖ, Svensk Bot. Tidskr. 30, 1936, S. 60.

als kleiner, aber zahlreicher als bei n-Prothallien (HEILBRONN 1928). Näher kann hier auf Einzelheiten nicht eingegangen werden. — Für die diploide und die tetraploide Rasse von *Psilotum triquetrum* gibt OKABE Messungen an (Sporenmutterzellen in Diakinese):

Diploide Rasse: Kerndurchmesser 19,73 $\pm$ 1,95 $\mu$
Tetraploide ,, : Kerndurchmesser 25,62 $\pm$ 2,00 $\mu$

Die 2 n-Prothallien verhielten sich im allgemeinen durchaus normal. Sie besassen normale Antheridien, Spermatozoiden und Archegonien. HEILBRONN gibt ferner für 2 n-Prothallien von *Polypodium aureum* schnellere Entwicklung und frühere Geschlechtsreife an. Auch die Sexualorgane des tetraploiden *Dryopteris marginalis* schienen normal zu sein. Andererseits stellten sich im Zusammenhang mit der Erhöhung der Chromosomenzahl nachteilige Erscheinungen ein. Die Regenerationsfähigkeit des Sporophyten nahm ab (HEILBRONN, LAWTON). Die Sexualorgane bei den 4 n-Prothallien von *Polypodium aureum* waren abnorm; nur durch Apogamie konnten Sporophyten gebildet werden. Andere, offenbar im Zusammenhang mit der Polyploidie stehende Erscheinungen waren ausser der von HEILBRONN beobachteten Apogamie das Auftreten von Zellen oder Organen beim Gametophyten, wie sie für die Sporophytgeneration charakteristisch sind, nämlich von Epidermiszellen, Tracheiden, Spaltöffnungen und Sporangien (LAWTON 1932). — Zu erwähnen ist noch, dass HEILBRONN (1932, 1933) vergleichende Untersuchungen über Chlorophyllgehalt, Assimilation und Atmung bei dem diploiden und dem tetraploiden *Polypodium aureum* angestellt hat. Durch ein umfassendes Studium der erwähnten Erscheinungen an Polyploiden, das bis jetzt noch im Angangsstadium steht, werden wir sicherlich tiefere Einblicke in die Fragen des Generationswechsels, der genom-und plasmonbedingten Vererbung und der Bedeutung des Kernes erhalten.

Den Gedanken, durch B e e i n f l u s s u n g der K e r n t e i l u n g s v o r g ä n g e in den Sporangien p o l y p l o i d e Sporen zu erhalten, hat bei Farnen als erster FRIEBEL (1933) in die Tat umzusetzen versucht. Es gelang ihm, bei *Matteuccia orientalis* durch Aetherbehandlung Riesensporen zu erzielen (Fig. 109), die sich als mehrkernig erwiesen. Leider konnte er nicht prüfen, ob solche Sporen als Ausgangsmaterial zur Gewinnung von Polyploiden dienen können, da er nur in einem Falle einen kümmerlichen Keimling erhielt, der sein Wachstum bald einstellte.

Bei dem apogamen *Dryopteris Filix-mas* var. *cristata* hort. gelang es mir (noch nicht veröffentlicht), in einer Zelle eines jungen fadenförmigen Prothalliums, in der ich eine Mitose zur Zeit der Metaphase und der darauf beginnenden Anaphase beobachten konnte, durch Beeinflussung mit Kälte die Chromosomenzahl zu verdoppeln und auf diese Weise zu Prothallien und zu Pflanzen mit doppelter Chromosomenzahl zu gelangen (Fig. 120, 121). Die Zählung an Aequatorialplatten in Wurzelspitzen ergab annähernd 82 und 173 Chromosomen.

Von verschiedenen Autoren vorgenommene Zählungen für die gleiche Art haben öfters Werte ergeben, die sehr viel weniger voneinander abwichen als im Falle der Polyploidie. So fand z.B. DE LITARDIÈRE für *Osmunda regalis* n = 22 Chromosomen, während DIGBY für *Osmunda regalis* var. *palustris* forma *aurea* n = 20 angibt. Man beachte ferner z.B. die verschiedenen für *Dryopteris Filix-mas* (§ 6) und *Polystichum falcatum* [1]) gefundenen Werte. Liegt hier nur ein Fehler in der Zählung vor, oder sind die Zahlen wirklich verschieden, sodass es sich tatsächlich um Aneuploidie handelt? Ich möchte annehmen, dass wenigstens in manchen Fällen das letztere zutrifft. Wie Aneuploidie zustandekommen kann, dürfte aus dem in § 8 gesagten hervorgehen.

Die weitere Erforschung der Pteridophyten auf etwa vorhandene Heteroploidie im Zusammenhang mit den Fragen der Verbreitung und der durch die Heteroploidie bedingten Erscheinungen dürfte eine wichtige Aufgabe sein.

## § 10. Abweichungen vom normalen Kernphasenwechsel. —

Wichtigste Literatur: ALLEN (1911), ANDERSSON-KOTTÖ (1931), ANDERSSON-KOTTÖ und GAIRDNER (1936), DÖPP (1932, 1933), ERNST (1918), FARMER und DIGBY (1907), GEORGEVITCH (1910), ROSENBERG (1930), STEIL (1919), STRASBURGER (1907), WINKLER (1908). — Solche treten bei den Pteridophyten häufig auf. Die wesentlichsten hier zu erwähnenden Erscheinungen sind die der P a r t h e n o g e n e s i s und der A p o g a m i e [2]) (beides wird zusammengefasst unter dem Begriff Apomixis [2])).

Hierbei sind zwei Möglichkeiten vorhanden. In dem einen Fall ist die Apomixis verbunden mit A p o s p o r i e [3]); die Sporangien sind dann steril, sodass Sporen nicht produziert werden, und die Prothallien entstehen aus vegetativen Blatt- oder Sporangienzellen. Beispiele hierfür sind nach FARMER und DIGBY (1907) das parthenogenetische *Athyrium Filix-femina* var. *clarissima* Bolton und die apogame Varietät *Athyrium Filix-femina* var. *clarissima* Jones [4]), ferner *Trichomanes Kaulfussii* (GEORGEVITCH 1910). Die Chromosomenzahl ist in diesen Fällen stets die gleiche; ein Wechsel derselben findet weder bei der apomiktischen Entstehung des Sporophyten noch bei der apoporen Entwicklung der Prothallien statt.

Demgegenüber stehen Fälle von Apogamie, bei denen es nach Restitutionskernbildung und darauffolgender Reduktion in den Sporangien zur Bildung von wenigstens zu einem grossen Teil normal aussehenden, keimfähigen Sporen kommt. Der Phasenwechsel würde sich hierbei nach den Untersuchungen von STEIL (1919) und DÖPP (1932, 1933) bei *Dryopteris hirtipes*, *Dryopteris remota* und *Dryopteris Filix-mas* var. *cristata*

---

[1]) Tabulae Biologicae, Bd. IV.

[2]) Vgl. Anm. 1 S. 269.

[3]) Nicht alle aposporen Formen sind jedoch apomiktisch.

[4]) Chromosomenzahlen sind in Tab. 1 angegeben.

hort. [1]) folgendermassen darstellen, wenn die Chromosomenzahl ohne Rücksicht darauf, ob es sich um eine als haploid oder diploid anzusehende Zahl handelt, mit n bezeichnet wird:

Spore (*n*)

↑

(R e d u k t i o n s t e i l u n g)

↑

Sporenmutterzellen (*2n*)

↑

(R e s t i t u t i o n s k e r n b i l d u n g)

↑

sporogenes Gewebe, 8-zellig, jung (*n*)

↑

Pflanze, apogam entstanden (*n*)

↑

Prothallium (*n*)

↑

Spore (*n*)

Ein Phasenwechsel findet also statt, aber nicht in der gewöhnlichen Weise. In diesem Zusammenhang sei hingewiesen auf die ganz ähnlichen Verhältnisse bei parthenogenetischen *Hieracium*-Arten (ROSENBERG 1927) [2]).

Einige Fälle von Parthenogenesis bedürfen noch der Klärung in cytologischer Hinsicht, und zwar diejenigen von *Marsilia Drummondii* (STRASBURGER 1907) und einiger *Selaginella*-Arten. Von *Marsilia Drummondii* gibt STRASBURGER an, dass in den Sporangien ein Teil der Sporenmutterzellen nach der Diakinese, in der bivalente Chromosomen zu beobachten sind, eine Reduktionsteilung erfährt, während in anderen eine typische Kernteilung ohne Reduktion stattfindet. Es werden demnach haploide und diploide Sporen gebildet. Eine genauere·Untersuchung der cytologischen Verhältnisse, wie überhaupt des gesamten Fragekomplexes wäre sehr erwünscht. — Ueber die Kernverhältnisse in den Sporangien der von BRUCHMANN (1912) [3]) untersuchten parthenogenetischen *Selaginella*-Arten (*S. rubricaulis, S. spinulosa*) ist nichts bekannt. Die apogame *S. rupestris* bildet nach GRAUSTEIN (1930) Mikro- und Makrosporendyaden (Fig. 113) aus; die Mikrosporen haben z.T. abweichende Formen. Auch bei der apomiktischen *S. anocardia* (GEIGER 1935) [4]) steht eine cytologische Untersuchung noch aus.

Als Ursache der Apomixis wird von ERNST (1918) Bastardierung vermutet. Leider hat man hierfür jedoch keine sicheren Anhaltspunkte. Andererseits machen Beobachtungen von HEILBRONN (1932) wenigstens an dem von ihm untersuchten *Polypodium aureum* Beziehungen zwischen Apogamie und Polyploidie wahrscheinlich.

---

[1]) Vermutlich auch bei *Polystichum falcatum* (vgl. STEIL 1919).

[2]) Hereditas 8, 1926/27.

[3]) Flora 104, 1912.      [4]) Flora 129, 1934–1935.

Nach FARMER und DIGBY (1907) soll im Sporophyten des apposporen und parthenogetischen *Phyllitis Scolopendrium* var. *crispum Drummondae* eine etwas höhere Chromosomenzahl vorhanden sein als im Gametophyten. Vielleicht handelt es sich hier jedoch um fehlerhafte Zählungen. — Eigentümliche Verhältnisse liegen vor bei einer als „peculiar" bezeichneten Varietät von *Phyllitis Scolopendrium* (ANDERSSON–KOTTÖ u. GAIRDNER 1936; vgl. Kap. IX (Genetics)). Diese zeigt Aposporie (ohne Sorus- und Sporangienbildung), was durch einen einfachen rezessiven Faktor bedingt ist. Die Prothallien erzeugen normale Sexualorgane, und es erfolgt Befruchtung. Die durch diese erfolgende Chromosomenvermehrung wird aber infolge der apposporen Entwicklung der Prothallien nicht kompensiert durch eine Reduktionsteilung in den Sporangien, sodass die Chromosomenzahl in den aufeinanderfolgenden Generationen zunimmt; sie vergrössert sich jedoch auffallenderweise nicht in dem Maße, wie man hätte erwarten sollen. In einer der untersuchten Generationen wurden gezählt 40–45 (60 hätten es auf Grund der Entstehungsweise sein sollen), in der darauffolgenden 60–68, in der nächsten schliesslich 90–100 Chromosomen (ANDERSSON–KOTTÖ 1931). Die Autorin nimmt an, dass in den Antheridien und Archegonien eine Herabsetzung der Chromosomenzahl stattfindet.

### § 11. Schluss.

Aus dem bisher Gesagten geht hervor, dass trotz zahlreicher Untersuchungen noch vieles in Bezug auf Zahl und Grösse der Chromosonen, Heterochromatin sowie Chromosomenstruktur überhaupt, Nukleolenverhältnisse und Kernteilung bei den Pteridophyten zu erforschen bleibt. Durch genaue vergleichende Einzeluntersuchungen über die Kern- und Chromosomenverhältnisse auch bei den zahlreichen lokalen Formen, Unterarten, Varietäten und Bastarden, die bei den Farnen beschrieben worden sind, kann die Kernforschung der Systematik der Farne wichtige Dienste leisten.

Ferner sind weitere experimentelle karyologische Studien erforderlich, vor allem in Verbindung mit genetischen Fragestellungen und Methoden.

Polyploidie ist bei den Farnen offenbar recht häufig anzutreffen. Ueber ihre Entstehung sind wir jedoch noch sehr im Unklaren. Weitere Untersuchungen an Pteridophyten könnten noch wichtige Aufschlüsse geben über die Ursachen der Polyploidie, die durch die Polyploidie bedingten Eigentümlichkeiten und die Bedeutung der Polyploidie für die Artentstehung. Ebenso kann das Studium der Abweichungen bei der Sporenentstehung (§ 8) Licht werfen auf die Fragen der Entstehung neuer Formen.

Zahlreiche Erscheinungen, die infolge der Häufigkeit ihres Auftretens bei den Polypodiaceen schon beinahe nicht mehr als Ausnahmen gelten können, wie Apogamie, Aposporie und Restitutionskernbildung, bedürfen noch der weiteren Bearbeitung durch zytologische, aber auch durch genetische Arbeitsmethoden, besonders im Hinblick auf ihre Ursachen und auf Fragen des Generationswechsels.

CHAPTER IX

GENETICS

by

I. ANDERSSON-KOTTÖ (Merton, London)

**§ 1. Historical notes.** — The first published record of fern hybrids was in 1837, when KICKX [1]) reported *Asplenium Ruta-muraria* × *A. germanicum*, and *Pityrogramma chrysophylla* × *P. calomelanos*. The latter hybrid was also described by MARTENS [2]) in the same year. Since this time numerous supposed hybrids have been described, chiefly in the leptosporangiate ferns; not only intervarietal and interspecific but also intergeneric hybrids have been reported. In the Equisetales a few species-hybrids have been described and some of the *Selaginella* species have been assumed to be of hybrid origin.

Most of these plants were described as hybrids because when found in nature or under cultivation they differed in some respects from the specific type, approached another species or were more or less intermediate between two species. Under wild conditions such aberrant forms have often been observed accompanying two species, or in an area lying between those occupied by the two species respectively. Sterility and departures from the normal in anatomy, cytology, vigour etc. have also been considered as indications of hybridity. For similar reasons, forms more or less intermediate have been described as hybrids in cases where spores of different species have been sown or prothallia planted together.

Though many of these ferns may possibly be of hybrid origin, this has by no means been proved. Many of these found in the wild, for example the species-hybrids described by COCKAYNE and ALLAN [3]) in New Zealand,

---

[1]) M. KICKX, Comm. to Bull. Acad. Roy. Sci. Brussels. 1837.

[2]) M. MARTENS, Notice sur un cas d'hybridité dans les Fougères. Bull. Acad. Roy. Sci. Brussels. 1837.

[3]) H. H. ALLAN, Illustrations of wild hybrids in the New Zealand flora. V. Genetica 9. 1927. — L. COCKAYNE and H. H. ALLAN, An annotated list of groups of wild hybrids in the New Zealand flora. Ann. Bot. 48. 1934.

are probably of hybrid origin; others are more doubtful. HEILBRONN [1]) could get no hybrids of *Asplenium Ruta-muraria* × *A. Trichomanes* or of *A. septentrionale* × *A. Trichomanes* from sowing spores together or adding spermatozoids of one parent to the female prothallia of the other. From *A. septentrionale* × *A. Ruta-muraria*, however, some plants appeared which did not resemble either parent as much as they did *A. germanicum*. This latter species has repeatedly been assumed to be a hybrid. Negative results were obtained by HOYT [2]), who attempted to cross-fertilise various species and genera. Attempts to make artificial crosses in *Selaginella* and *Equisetum* by the present author also failed.

In other cases the experimental conditions and technique employed were such that no conclusions could safely be drawn. Present-day technique in hybridisation and culture has shown that many of these supposed cases of hybridisation are very improbable. In the author's opinion, reports of intergeneric and wide interspecific crosses should be received with great caution.

More recently, CZAJA [3]) claims to have produced hybrids between *Gymnogramme (Pityrogramma) chrysophylla* ♀ and sulphurea ♂, but the presumed hybrids all died in the embryo stage. The same applies to the young plants obtained by DÖPP [4]) from *Dryopteris Filix-mas* × *spinulosa*, a cross that it would be of interest to obtain as these two species are the reputed parents of *D. remota*. *D. remota* × *Filix-mas* gave no result. On the other hand the same author appears to have succeeded in crossing *D. spinulosa* × *remota* (fig. 1), which would thus be the first case of an undoubted specieshybrid. However, hybrids may now be obtained and the progeny bred under absolutely controlled conditions; there seems no doubt therefore that where species are capable of forming hybrids, such hybrids will in due course be obtained, and where they are fertile their progeny investigated.

Ferns which by their progeny proved to be hybrids were first produced by ANDERSSON-KOTTÖ [5]) in 1927 from intervarietal crosses. They were obtained from cross-fertilisation between ♀ and ♂ gametophytes under controlled conditions. For the culture media and technique used, refer-

---

[1]) HEILBRONN, A. Apogamie, Bastardierung und Erblichkeitsverhältnisse bei einigen Farnen. Inaug. Diss., München 1910.

[2]) W. D. HOYT, Physiological aspects of fertilization and hybridization in ferns. Bot. Gaz. 49. 1910.

[3]) A. TH. CZAJA, Über Befruchtung, Bastardierung und Geschlechtsertrennung bei Prothallien homosporen Farne. I. Zeits. f. Bot. 13. 1921.

[4]) W. DÖPP, Versuche zur Herstellung von Artbastarden bei Farnen. I. Ber. deuts. Bot. Ges. 53. 1935.

[5]) I. ANDERSSON-KOTTÖ, Note on some characters in ferns subject to mendelian inheritance. Hereditas 9. 1927.

ence may be made to ANDERSSON-KOTTÖ[1]) and LAWTON[2]). Heterozygotes of unknown origin had however been investigated earlier, in as much as HEILBRONN[3]) in 1910 found segregation in the progeny of some varieties of *Athyrium Filix-femina*, and LANG[4]), 1923, first found a mendelian segregation in ferns from a natural heterozygote of *Phyllitis Scolopendrium* of unknown parentage.

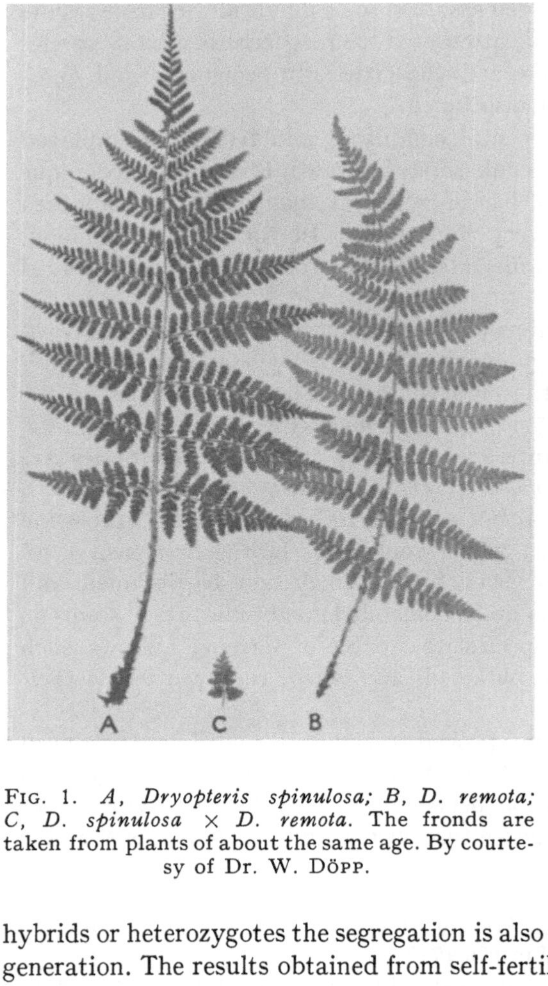

The great polymorphy shown by many fern genera indicates that hybridisation may frequently occur in uncontrolled crosses and in nature.

## § 2. Genetical experiments with structural characters.

— With the few species that as yet have been investigated among ferns, the segregation and recombination of the genes for various characters is clearly demonstrated and the inheritance is in principle the same as in higher plants and in mosses. Mendelian segregation is shown in the next generation ($F_2$) when gametophytes from experimentally obtained hybrids ($F_1$) are fertilised *inter se*. From natural hybrids or heterozygotes the segregation is also apparent in the succeeding generation. The results obtained from self-fertilisation of separate game-

FIG. 1. *A, Dryopteris spinulosa; B, D. remota; C, D. spinulosa × D. remota.* The fronds are taken from plants of about the same age. By courtesy of Dr. W. DÖPP.

---

[1]) I. ANDERSSON, The genetics of variegation in a fern. J. Genet. 13. 1923.
Loc. cit. 1927, and subsequent papers.

[2]) E. LAWTON, Regeneration and induced polyploidy in ferns. Amer. J. Bot. 19. 1932.

[3]) HEILBRONN, loc. cit.

[4]) W. H. LANG, On the genetic analysis of a heterozygotic plant of *Scolopendrium vulgare*. J. Genet. 13. 1923.

tophytes derived from a hybrid or heterozygote show the distribution of the genetical factors at reduction division in the sporangium.

Mendelian segregation and recombination is well demonstrated in *Polystichum angulare*[1]). A plant of unknown parentage proved to be heterozygous for two pairs of factors, one determining the shape of fronds, pinnae and pinnules and the other their position (i.e. length of rachis, imbrication of pinnae and pinnules). From recombination of these two pairs of factors four types of plant appeared (fig. 2) as follows:

(1) Plants similar to the parent, dominant for both factors (Type I). — (2) Plants like (1) as regards the shape of the pinnules, but differing in the imbricated position of the pinnae and pinnules (Type II) (dominant in shape, recessive in position).—(3) A type with truncated fronds, usually ending in a short hornlike protrusion of the rachis; with pinnae of varying length; pinnae more or less cuneate-flabellate, palmately lobed and toothed (Type III). This type is dominant for position but recessive for shape of pinnules. — (4) Similar to (3) but with pinnae and pinnules imbricated. This is evidently the double recessive. (Type IV).

FIG. 2. *Polystichum angulare*. The four types which appear as a result of recombination of two pairs of factors. *A*, Type I (*angulare* type); *B*, Type II; *C*, Type III; *D*, Type IV.

The four classes of plants, obtained from gametophytes fertilised inter se, appeared in an approximate ratio of 9 : 3 : 3 : 1, as would be expected if the parents were heterozygous for two factors. It was shown experimentally that plants of Type I, as expected, were either homozygous or segregating in the next generation, giving a 9 : 3 : 3 : 1 ratio; or were heterzygous for one factor, giving a 3:1 ratio. Plants of Type II gave either Type II only, or Types II and IV in a 3 : 1 ratio. Similarly Type III gave either Type III only, or Types III and IV in a 3 :1 ratio. At a later date the Type IV plants were bred from, and gave only their own type.

---

[1]) ANDERSSON–KOTTÖ, loc. cit. 1927.

From self-fertilisation of separate gametophytes the equal distribution of these factors at reduction division was proved, as indicated by the following numbers: of 78 gametophytes self-fertilised, 22 gave sporophytes of Type I, 20 of Type II, 18 of Type III and 18 of Type IV. The sporophytes from each self-fertilised gametophyte were uniform.

*Athyrium Filix-femina.* Plants were raised by HEILBRONN[1]) from a sporophyte of *Athyrium Filix-femina* var. *multifidum* Mappelbeckii, a variety with a small multifid apex of the pinnae. Seven plants were more or less like this variety, two were similar to typical *Filix-femina* and fourteen were more branched.

According to the experiments of ANDERSSON-KOTTÖ[2]), the branching of fronds and pinnae which is characteristic of the so-called crested and other more simply branched varietes is dominant over the typical *A.F.-f.* type of non-branching frond, and a segregation of 3 : 1 occurs in $F_2$. The more complicated branching types of *Athyrium* do not show a clear segregation. This applies to the var. *Fieldii* Moore, from which HEILBRONN raised a family and to other more or less cruciate forms bred by the present author.

In the variety *Frizelliae* each pinna branches immediately from the rachis, producing a semicircular, flabellate pinna. Instead of a single pinna there are thus two transformed parts, one covering the other. Heterozygotes between this variety and typical *Filix-femina* show both types of parental pinnae on the same frond, and in the next generation a segregation occurs, which gives normal *Filix-femina*, heterozygotes, var. *Frizelliae* in a ratio of 1 : 2 : 1, as found by ANDERSSON-KOTTÖ. A factor for shortened rachis and imbricated pinnae and pinnules, which is recessive, was also found to segregate simply.

Among the many distinct degrees of lacination of the pinnules occurring in this species, the more laciniated types are recessive to the less laciniated types and to the *Filix-femina* type.

Presence of anthocyanin colour in the rachis is dominant to its absence.

*Phyllitis Scolopendrium.* LANG[3]) showed that entire leaf and incised leaf in *Phyllitis Scolopendrium* are allelomorphic, a segregation of 3 entire: 1 incised occurring in $F_2$ from a natural hybrid.

The characters investigated by ANDERSSON-KOTTÖ[4]) in this species and found to behave in a simple mendelian way were as follows:

---

[1]) A. HEILBRONN, Apogamie, Bastardierung und Erblichkeitsverhältnisse bei einigen Farnen. Inaug. Diss. München. Fischer, Jena. 1910.

[2]) I. ANDERSSON–KOTTÖ, The genetics of ferns. Bibliographia Genet. 8. 1931.

[3]) W. H. LANG, On the genetic analysis of a heterozygotic plant of *Scolopendrium vulgare*. J. Genet. 13. 1923.

[4]) I. ANDERSSON–KOTTÖ, A genetical investigation in *Scolopendrium vulgare*. Hereditas 12. 1929.

(a) dwarf, which is a simple recessive to tall. — (b) Murication of the upper surface of the frond, incompletely dominant to smooth surface. — (c) Rugosity of the upper surface of the frond, correlated with undulation. The heterozygote is more or less intermediate. — (d) Bimarginate edge, which is characterised by a longitudinal fold on the under-surface, developed simultaneously with sectioning of the edge up to the fold. The portion of the upper surface corresponding to the fold becomes rugose, the rugosity forming a ridge or second margin. This character is more developed in the homozygote than in the heterozygote. — (e) The fold or ridge-like protrusion of some layers of the under-surface of the frond. The fold is developed so as to form a broad transparent membrane, which stands away from the frond and is correlated with the contraction of the whole lamina. In sporophytes heterozygous for this character the fold is discontinuous and its development is correlated with serration of the edge. — (f) Simple crenation of the edge without other reductions in size or development of the lamina is recessive to entire edge. — (g) Crenation of the edge, and a peculiar retardation in growth along the whole lamina, without transformation of upper or lower surface, is recessive to entire edge and normal development. This factor causes degrees of reduction of the lamina and various interrelated processes affecting nervature and position of sorus (see p. 292). The most extreme reduction of lamina occurs in the branched type, in which scarcely more than the branched rachis remains. (fig. 3). — (h) Terminal branching at a definite point (different for different varieties) is recessive to non-branching. — (i) Regular branching, together with entire edge, is dominant over irregular branching and crenate edge. — (k) The presence of laterally protruding lobes or side-branches is recessive to their absence. These side-lobes contain a strengthened vein which branches sympodially, or a side-branch of the costa. — (l) Basal branching, consisting of protruding basal lobes with a branch of the costa, is possibly allelomorphic to no basal branching — (m) The arrested growth at the apex of the frond which results in a more or less pocket-like structure on the underside and protrusion of the midrib into a horn (characteristic of the variety peraferum) is dominant over hornless apex, but the segregation approximates to 2 : 1 rather than 3 : 1. The factor causing this type of terminal growth affects different types in different ways. It was recombined with the factors for crenate edge, for dwarfness and for branching. In the last instance each branch possesses an apical pocket-horn structure. A similar arrest of growth may occur at other points along the frond, e.g. at the apex of side-branches or basal branches or anywhere where a more or less strengthened vein occurs. The factors involved have not so far been investigated.

The same applies to the factors causing difference in width of lamina and lobes, degree of laciniation and number of divisions of the costa or apex of the frond.

In the classification of the ferns, venation and the situation of the sorus in relation to venation, the relation of the sori to one another, to the costa and to the edge of the frond are of the gratest importance. The departures from the typical Scolopendrium sorus and venation which have been met with in these experiments are therefore of special interest. Since development of the sorus is intimately connected with the development of venation and of the interrelated processes of growth of the whole frond, it is not astonishing that deviations in position of the sorus were found to be caused by some of the factors referred to above, which primarily affect the growth of the leaf-blade and retard the growth at certain points or along certain regions.

Departures from the typical position, shape and indivivuality of the sorus thus occur together with the various developments of the other characters and are inherited together with these. Instead of the sori being

FIG. 3. *A*, extravagant type of *Phyllitis Scolopendrium*, showing extreme reduction of lamina, position and shape of sori. *B*, *Phyllitis Scolopendrium* type.

confined to the lower surface of the frond, a continuation of the sori along the margin to the upper surface may occur. This is the result of unequal development of the upper and under surface, or of the development of a marginal or intramarginal fold, or of contraction of parts of the frond. A continuous band of sori on the upper side of the shelf-like fold along the edge of the frond is thus found in some varieties. Another arrangement of the sori is due to factors which primarily cause a mode of growth that leads to reduction of the upper surface, at the same time specially affecting the part between the two branches of the primary vein. The edge of the frond thus appears pinnate, and the two sori which face one another protrude in this way in a 'pinnate' manner. They continue over the apex of the pinna down the corresponding part of the upper side. In other words, each pinna consists of one turned-over double sorus. However, sori may also continue from the lower to the upper surface, each sorus covering itself on either surface, without much transformation of the lamina; this has been found in a hybrid with a slight submarginal fold, slight irregular incision of the edge and very slight murication of the upper surface.

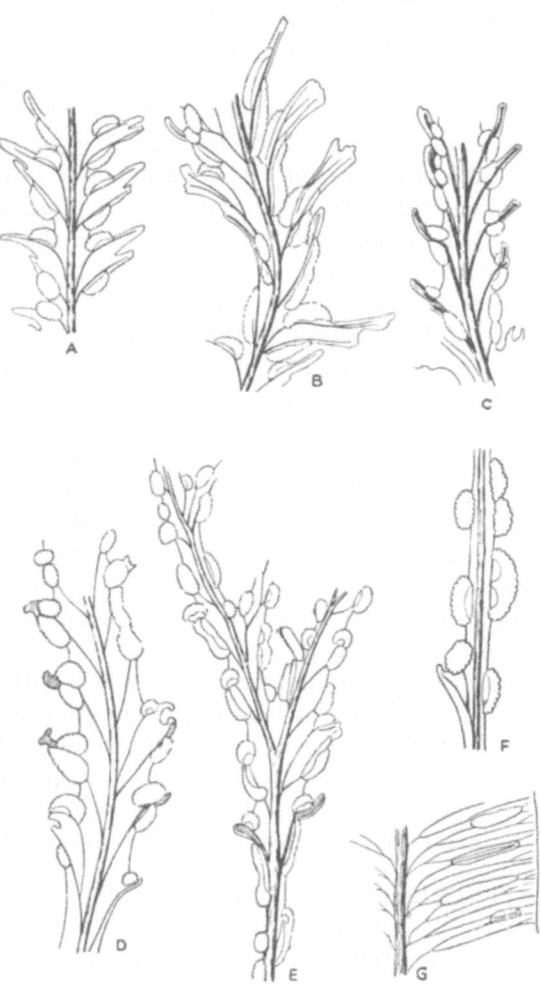

Dislocation of parts of the sorus, or spiny protrusions containing the folded-back sorus, occur when the ridges or intramarginal folds on the under-surface cross the sori. The typical position

FIG. 4. *A–F,Phyllitis Scolopendrium*. Sections of fronds, showing shape and position of sori, venation and diminution of leaf blade. *G*, normal development and position of sori in *Ph. Scolopendrium* type.

may also be altered by several types of lobing, crenation and incision of the edge of the frond.

Sori may be produced on the upper surface of the frond without corresponding sori on the under surface, as seen in some muricated types. They are then situated in depressions of the upper surface, distinctly within the margin of the frond, and are of irregular shape and without indusia.

The most interesting departure from the position and shape of sorus characteristic of *Phyllitis Scolopendrium*, combined with alteration of venation, was found in the different types (fig. 3 and 4) of a family derived from the variety *cornutum*. In this family plants may be branched or unbranched, and they differ also in the width of lamina developed. The edge is more or less deeply crenated. The factor for crenation also affects the width of the lamina and venation, and may be said to have a curious retarding influence along the whole lamina. As a consequence of the alterations which occur, the position and shape of the sorus is affected. The plants may have a comparatively broad lamina, where the double sori typical of *Ph. Scolopendrium*, which open towards one another, are seen to do so only along part of the sorus. Or they may have narrower fronds and lobes, together with reduced venation and fusion of veins; in this case the two sori are situated at an angle towards one another, following the edge and opening outwards, or the sori which originally opened towards one another have become widely separated and come to lie back to back. In this position they are unequal in size, the extrorse sorus in placed more or less parallel to the margin and costa, the introrse one is very short and inserted at right angles. Finally, certain branched types of the same family frequently develop only the extrorse sori, the frond then having the appearance of a costa bordered by a disrupted line of sori, all parallel to the costa (fig. 4 F).

The shape of the indusium corresponds with that of the sorus. Partial abortion or absence of indusium occurs in the suprasoriferous types and in some of the crenated forms with sori along the edge. Further, where a well-developed membrane-like edge occurs, the indusium is much reduced or disappears in the membrane.

Reciprocal hybrids have in all cases been alike and fully fertile.

**§ 3. Obligatory apospory.** — The only clear case of genetical determination of ·apospory in ferns occurs in *Phyllitis Scolopendrium* [1] [2] [3]. Apospory and the whole abnormal development of the so-called "peculiar" sporophytes of this species is caused by one gene which behaves as a simple recessive to normal development and no apospory. Reciprocal crosses give normal sporophytes in $F_1$ and a segregation in $F_2$ of 3 normal to 1 aposporous. The normals either breed true or segregate like the $F_1$, the 'peculiars' give only 'peculiars'.

[1] I. ANDERSSON–KOTTÖ, The genetics of ferns. Bibliogr. Genet. 8. 1931.

[2] I. ANDERSSON–KOTTÖ, Observations on the inheritance of apospory and alternation of generations. Svensk Botanisk Tidskrift 26. 1932.

[3] I. ANDERSSON–KOTTÖ and GAIRDNER, A. E., The inheritance of apospory in *Scolopendrium vulgare*. J. Genet. 32. 1936.

This gene for 'peculiar' alters the whole development of the life cycle. The sporophytes throughout are of very slender growth and only attain a height of about 8 cm. (fig. 5 E). The fronds (fig. 5 A–D) never approach the size, shape or robustness of normal plants. They are similar to the normal as regards anatomy and venation and in some other respects, though on a reduced scale, but the tissue between the veins of the leafblade is gametophytic. This tissue gradually grows out into a prothallial sheet along the edge. Sori, sporangia and spores are entirely absent. The prothallial outgrowth gradually weighs down the frond and the prothallium establishes itself on the soil as a separate gametophyte. Normal sexual organs are developed and after fertilisation a similar new 'peculiar' structure arises, which repeats the process. Since apospory is not offset by apogamy, as usually occurs in obligatorily aposporous ferns, we get a series of increasingly high polyploid 'peculiars'. Further reference is made to these 'peculiars' below (see § 7 and 8).

**§ 4. Polyploid segregation.** — The only clear case of polyploid segregation of a mendelian character in a fern to my knowledge occurrs in *Phyllitis Scolopendrium*. The data (ANDERSSON-KOTTÖ, unpub.) do not yet suffice for definite conclusions, but it is apparent that a segregation of 5 normal to 1 'peculiar', of the type described above, is obtained when separate gametophytes are self-fertilised, the gametophytes being obtained from a fertile hybrid between a 'peculiar' with 50 chromosomes and a gametophyte of normal *Scolopendrium* type with 30 chromosomes, the hybrid having 80 chromosomes.

**§ 5. Mutation.** — Though many fern genera show that mutation must be comparatively frequent, no mutation has so far been recorded in experiments under controlled conditions where the genotype of the material was previously known.

The somatic mutations in *Nephrolepis exaltata* described by BENEDICT [1] are of interest, since they parallel the variation in characters of the sporophytes obtained from spores in other fern genera. Most of the cultivated species of this genus show variation by bud-sports, and in *N. exaltata* such variation is specially frequent, as many as 200 new bud sports having arisen since 1895. They occur discontinuously and are classified by BENEDICT as progressive and regressive. The progressive variations have ap-

---

[1] R. C. BENEDICT, The origin of new varieties of *Nephrolepis* by orthogenetic saltation. I. Progressive variations. Bull. Torr. Bot. Club 43. 1916. — Evolution as illustrated by ferns. Brooklyn Bot. Gard. leaflets ser. 10. 1922. — Which Boston fern is best? J. Hered. 13. 1922. — The origin of new varieties of *Nephrolepis* by orthogenetic saltation, II. Regressive variation or reversion from the primary and secondary sports of *bostoniensis*. Amer. J. Bot. 9. 1922. — New bud sports in *Nephrolepis*. Genetics 8. 1923. — The moss-leaved fern. J. Hered. 15. 1924. —

FIG. 5. *A–D*, fronds of 'peculiar' sporophytes of *Phyllitis Scolopendrium; E*, fully grown 'peculiar' sporophyte; *F*, comparable fronds of a young sporophyte of *Scolopendrium* type. For comparison with *C* and *E* see normal frond of mature plant of *Scolopendrium* type in Fig. 3.

peared along several lines and show increased division of the leaf, ruffling of the pinnae or dwarfing. Among further variations may be mentioned dichotomy of pinnate and apex, cresting and vivipary. The characters appear separately or together. Progressive increase occurs in successive sporophyte generations, each mutation producing a higher grade of the character. The regressive mutations rarely show a complete return to the parent form.

A genetical investigation is impossible here owing to sterility. Only one fertile variety has appeared from *Nephrolepis exaltata bostoniensis* [1]). This was twice-pinnate with occasional once-pinnate fronds, and gave from spores, forms which parallel the variations of *N. e. bostoniensis* and also other forms. The succeeding generation showed decreased variability and progressive variation as regards increased leaf division.

§ 6. **Chloroplast characters.** — The study of chloroplast characters is facilitated where there is a well-developed gametophyte generation, as in the Ferns. The mode of inheritance of chloroplast characters such as size and various development of pigment, has been studied by ANDERSSON-KOTTÖ [2]). In all cases of variegation, that is the presence of cells with normal chloroplasts and cells with pale green, generally smaller chloroplasts together in one individual, the transition from one type of chloroplast to the other is sudden and evidently occurs at the division of one cell.

In *Adiantum cuneatum* it may be assumed that the pale cells are due to mutation (in the wider sense), the mutation occurring (a) somatically in sporophyte and gametophyte and (b) at spore formation or reduction. Somatic mutation in the gametophyte occurs in about one cell division in ten thousand. This should be compared with the considerably higher mutation rate at reduction, viz. 1.3 or 2.8 pale to 1 green.

In *Lastraea atrata* variegation is the phenotypic expression in the sporophyte of a factor which causes no visible effect in the gametophyte. In the variegated plants investigated both generations had the same chromosome number and the periodic change from variegation of the sporophyte to whole-colour of the gametophyte is therefore of interest.

Variegated plants of *Phyllitis Scolopendrium* were exceptional as compared with other variegated ferns investigated, in having in each sporangium only one type of gametophyte, i.e. either all green or all pale. No segregation occurs therefore at this point. Since the change from normal to pale is reversible in the sporophytic tissue, and the green gametophytes subsequently breed true, we may assume that a labile condition is created at each somatic division in the sporophyte. This condition is transmitted

---

[1]) R. C. BENEDICT, Variation among the sporelings of a fertile sport of the Boston fern. J. Hered. 15. 1924.

[2]) I. ANDERSSON, loc. cit. 1923. — I. ANDERSSON–KOTTÖ, Variation in three species of ferns (*Polystichum angulare, Lastraea atrata* and *Scolopendrium vulgare*). Z.I.A.V. 56. 1930. — ANDERSSON–KOTTÖ, I. loc. cit. 1931.

by the pale gametophytes whether selfed or crossed (as ♀ or ♂) with green.

A more complicated inheritance of variegation occurs in *Polystichum angulare*. The facts briefly are that variegated sporophytes produce four phenotypically distinguishable types of gametophytes: green, pale non-viable and two types of variegated. Variegated sporophytes are only obtained on crossing a pale region of a gametophyte of the first of the two variegated types with a green gametophyte or a green region of a varie-gated gametophyte. These variegated sporophytes breed like their varie-gated parents. The following give only green sporophytes: all green gametophytes, green parts of the first type of variegated gametophytes, and variegated gametophytes of the second type. These green sporo-phytes may then (*a*) breed true; or produce (*b*) green and non-viable pale gametophytes, (*c*) green and variegated gametophytes of the second type, (*d*) green, variegated of the second type and pale gametophytes, or (*e*) pale and variegated gametophytes of the second type. It was found that the phenotypic change from green to pale green in the gametophytes coincided with a genotypic change. From further breeding work, and the results of breeding from different parts of the gametophytes, self-fertili-sation as well as cross-fertilisation, the conclusion was reached that the phenotypical change from green to pale coincides with a change or mu-tation in a factor determining chlorophyll development. This factor was found to exist in seven different states. The factor in each state may by either stable or unstable. The instability consists in the power of mutation from one state to another specific state, at a certain stage of the life cycle. The factor in each state has thus its own characteristic mode of activity.

The pale patch of the variegation pattern of the gametophyte here usually involves the whole growing point, and later on, when the growing point has changed back to green, the pale region comes to lie more or less across the gametophyte; whereas in the *Adiantum* case referred to above, the pattern consists of pale wedges or larger areas situated on either side of the growing point.

Giant chloroplasts [1]) in *Polystichum angulare* are probably controlled by a simple mendelian factor, since a segregation of gametophytes with normal and with giant chloroplasts in the ratio 1 : 1 occured in the progeny of a sporophyte derived from a cross between a normal gametophyte as ♀ and a gametophyte with giant chloroplasts as ♂. This sporophyte had normal chloroplasts.

**§ 7. Effects of alteration of genome quantity.** — It is well known that quantitative alteration of the genome is often accompanied by alterations in cell-size, cell-number and other characters. Since the form and size of a generation and its organs depend on the form, size and number

---

[1]) ANDERSSON–KOTTÖ, loc. cit. 1931.

of cells and the relation between these characters, the effects of a quantitative alteration of the nucleus may be considered here. In *Osmunda regalis* Manton [1]) found 2 n and 4 n sporophytes indistinguishable on a casual inspection. No increase in cell-size or number, or in size of organs, or of the whole plant, was found in the increasingly high polyploid series of the 'peculiar' type of *Phyllitis Scolopendrium* [2]). In *Polypodium aureum* on the other hand, Heilbronn [3]) found that increase in quantity of nuclear material was accompanied by an increase in cell-size; 2 n gametophytes, obtained through induced apospory from 2 n sporophytes and from spores of the tetraploid plants had larger cells than the n gametophytes. The same applied to 2 n as compared with 4 n sporophytes. Width of lobes and size of sori and sporangia were also greater in the 4 n than in the 2 n sporophytes. No noticeable increase in the size of the gametophyte as a whole was found. Similarly, in *Dryopteris marginalis* and *Woodwardia virginica*, Lawton [4]) found that the 2 n prothallia obtained by induced apospory had larger cells than the n ones. The leaves of the tetraploid sporophytes were composed of larger cells than those of the diploids, and were irregularly lobed.

The higher polyploid gametophytes when fully grown are of typical shape. In some species the diploid and tetraploid gametophytes retain both antheridia and archegonia, while in others, loss of one or both may occur.

While in the 'peculiars' of *Phyllitis* the polyploid sporophytes and gametophytes are similar in all respects to the normal (diploid and haploid) forms, in *Polypodium aureum* [5]) we get various interesting aberrations. The diploid gametophytes are sexual, but under certain conditions they can be induced to develop into sporophytes; the tetraploid ones are asexual. They only possess an occasional antheridium, which becomes transformed into a growing point; and they are sooner or later, in one way or another, differentiated into sporophytes. The intermediate structures between gametophytes and sporophytes which are obtained when fronds are regenerated are of special interest. It is evident that they are at first of gametophytic nature, as indicated by the cell-shape and the presence of rhizoids and antheridia, but gradually the apical end becomes trans-

---

[1]) I. Manton, Contributions to the cytology of apospory in ferns. I. A case of induced apospory in *Osmunda regalis*. J. Genet. 25. 1932.

[2]) I. Andersson–Kottö, On the comparative development of alternating generations, with special reference to ferns. Svensk Botanik Tidskrift 30. 1936.

[3]) A. Heilbronn, Über experimentell erzeugte Tetraploidie bei Farnen. Verh. V. Int. Kong. Vererb. Berlin 1927.

[4]) E. Lawton, Regeneration and induced polyploidy in ferns. Amer. J. Bot. 19. 1932.

[5]) A. Heilbronn, Polyploidie und Generationswechsel. Ber. Deuts. Bot. Ges. 50. 1932.

formed into a primary leaf, on which a lateral growing point gives rise
to a normal sporophyte. From regeneration of the gametophytic cells
typical prothallia. are obtained. As HEILBRONN [1]) points out, a higher
degree of polyploidy favours the sporophytic generation rather than
the gametophytic generation. A haplont is always a gametophyte,
a diplont may be either a sporophyte or a gametophyte according to
external conditions, and a tetraplont tends·to be a sporophyte. The production in its turn of a gametophyte from the tetraploid sporophyte is much
more difficult than in the case of the corresponding diploid sporophyte,
and the tetraploid gametophytic generation loses its sexual function and
develops apogamously into sporophytes.

That the normal sequence of events in the succeeding generation is
upset by the experimental alteration of the sequence in the preceding
generation, or by increased genome quantity, has also been shown in
*Dryopteris marginalis* by LAWTON, where the gametophytic regenerations
from diploid sporophytes were found to bear sporangia. Sporangia on
prothallia had previously been reported by LANG [2]) but these prothallia
were grown from spores under abnormal conditions and their chromosome number was not ascertained.

The effects of a quantitative alteration of the genome on the
sporophyte as compared with the gametophyte cannot as a rule be strictly
compared in ferns owing to the dissimilarity between the alternate generations. An exception is found in the 'peculiars' of *Phyllitis Scolopendrium*,
in which the tissue of the sporophytic generation may be compared with
that of the gametophyte, inasmuch as the tissue situated between the
veins and all round the leaf-blade (replacing sporophytic tissue) consists
of cells similar to those of the gametophyte. The frond gradually grows
out into a gametophyte which gives rise to a similar new sporophytic
generation and this again behaves in the same way. The prothallial cells
in the sporophyte can thus be compared with the later stages of the gametophyte from the same frond when grown as a separate entity. From comparison of cells in the polyploid series it was concluded [3]) that gametophytes
with different quantities of genome are similar, and sporophytes with
different quantities of genome are similar, but that there is a difference
between the gametophytic and the sporophytic generations. There is a
readjustment in cell size when the sporophyte grows out into a gametophyte.
This readjustment from the larger cell-size of the sprophyte to the smaller
size characteristic of the gametophyte, is not due to a dropping-out of

---

[1]) A. HEILBRONN, Über die genetische Lokalisation Stoffwechsel und Generationsfolge regulierender Anlagen. Biol. Zbl. 53. 1933.

[2]) W. H. LANG, Apogamy and the development of sporangia upon prothallia.
Phil. Trans. Roy. Soc. 190. 1898.

[3]) I. ANDERSSON–KOTTÖ, Loc. cit. Svensk Bot. Tids. 1936.

chromosomes or to any irregularity in the division of the chromatin mass. As the sporangial stage is totally absent in these 'peculiars', it would appear that it is not the transit from one nuclear phase to the other, or reduction in chromosome number, which is responsible for the readjustment, but rather the alternation of morphological generations.

It seems probable also from other facts that the sequence of events leading to differences in the cell-size and rate of growth must be considered for the life-cycle as a whole. In other words, the growth and development of gametophyte and sporophyte is integrated in the pattern of growth and development of the whole life cycle.

§ 8. **The effects of genetical factors in sporophyte and gametophyte.** — Several mendelian genes were referred to in § 2 which show themselves in the sporophyte alone. No simple mendelian gene affecting the gametophyte only has so far been found in ferns. Lang [1]) however found a race of *Phyllitis Scolopendrium* which showed a tendency to form sporangia on old gametophytes when fertilisation was prevented. The cause of this interesting departure is unknown; it is not even known if the race was a diploid one, or how it bred in this respect. The race was otherwise normally developed as regards sex and other characters.

Among the genes which might be expected to express themselves in both sporophytic and gametophytic generations, such as those determining the size, number and contents of the cell, the mendelian gene for dwarfness is up to the present the only one for which a definite place of visible action is established. The effects of this gene on cell number are apparent in the sporophyte but not in the gametophyte. It is unfortunate that in ferns the inheritance of chloroplast characters is not dependent on genes segregating in a simple way. In *Polystichum*, the unstable gene mutates from one state to another in a definite way and mutation may take place in gametophyte or in sporophyte or in both. In *Scolopendrium*, the gametophytes may show either of the chloroplast types of the parent sporophyte but the change from one chloroplast type to the other, as happens in the sporophyte, never takes place in the gametophyte. In *Adiantum* both sporophyte and gametophyte show changes from normal chloroplasts to small whitish chloroplasts and *vice versa*. In *Lastraea* only the sporophyte shows the chloroplast variation.

It is of special interest to note that the gene for dwarfness acts visibly throughout the sporophytic generation, though not throughout diplophase. This is apparent from the result of combining this gene with 'peculiar'. We have here, in the dwarf 'peculiars', sporophyte and gametophyte within the same nuclear phase and only the sporophytic tissue, i.e. the

---

[1]) W. H. Lang, On a variety of *Scolopendrium vulgare* which bears spores on the prothallus. Ann. Bot. 43. 1929.

whole sporophyte including the veins is affected by the dwarf gene. The
prothallial tissue of the frond and the subsequent free-living gametophyte
are unaffected.

The effect of combining these two recessive genes for 'peculiar' and
dwarfness is a shift in the comparative development of the life-cycle;
the sporophytic part becomes reduced in size compared with the gametophytic part. Fronds of fully grown plants before growing out into prothallia
attain only a maximum width and length of 1 cm., usually less, and have
a stipe never exceeding 7 mm. Though slender, they have a shape and
venation similar to that of the tall peculiars, and no abnormality in development has been met with in any family. On these small fronds the
edge grows out into the usual sized large prothallial cushion around the
periphery and at a distance fully developed plants look like a slender
stalk with roots at one end and a prothallium, often many times larger,
at the other end. Owing to the slenderness of the stalk the prothallial
outgrowth weighs the fronds down towards the soil.

The gene for 'peculiar' [1]) causes a drastic alteration in the sequence of

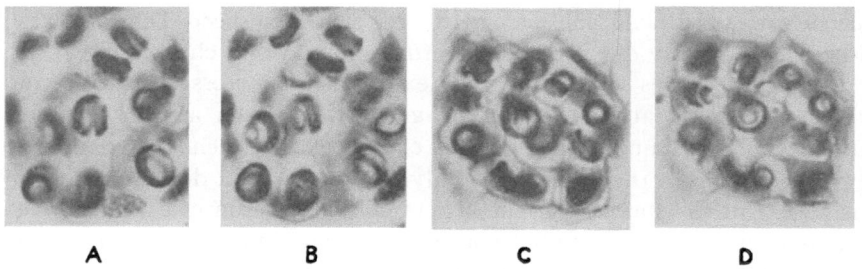

**A**                **B**                **C**                **D**

Fig. 6. *A* and *B*, spermatozoids in a sporangium; *C* and *D* spermatozoids in an
antheridium.

events in the life cycle as a whole. Some consequences of this alteration
appear in the hybrids between normals and 'peculiars' and also in succeeding generations. Some of the sporangia of the normal hybrids are
found to contain spermatozoids instead of spores (fig. 6 A and B) and the
polyploid series of the 'peculiars' and also the normal segregates from the
cross, instead of showing a doubling of the chromosome number in successive generations, often have less than the double chromosome number.

As regards expression of dominance, it might be deduced from the
occurrence of spermatozoids in sporangia that the gene for 'peculiar' has
different effects on different characters; it is clear that we must consider
the effects of the gene on the life-cycle as a whole. Long complete development (normal type) is dominant over short, incomplete development
(peculiar type); but in regard to sexuality the $F_1$ shows incomplete-

---

[1]) I. ANDERSSON–KOTTÖ, Loc. cit. 1936.

I. ANDERSSON–KOTTÖ and GAIRDNER, A. E., Loc. cit. 1936.

dominance. The differentiation of sex cells is independent of the gametophytic stage. The gene for 'peculiar' thus has a special effect on the time
of sexual activition. In both normal and 'peculiar', sex activation occurs
in the gametophyte, but the time of activation is accelerated in the 'peculiar'; the result of crossing normal and 'peculiar' is the activation of sex
at an earlier stage than occurs in the normal life cycle — in addition to its
occurrence at the usual stage. In this way a cell characteristic of the
gametophyte in ferns (a sex cell) is transferred to the normal sporophyte
and there replaces, inside the sporangium, a cell as characteristic of the
sporophyte as the spore. If a single gene, as is here the case, can thus alter
the developmental sequence and cause a characteristically gametophytic
character to be shifted to the sporophyte, it may be presumed that other
characters and even organs may similarly be shifted from one generation
to the other.

The life-cycle of the hybrids is also partly different from that of either
parent and any previously known fern, since sporangia are seen to develop
spermatozoids without the intermediate stage of spores and gametophyte.

The fact that the polyploid series of 'peculiars', and also the normals
from reciprocal crosses between normal and 'peculiar', frequently have
less than the expected chromosome number is not due to any chromosome
elimination or irregularity during ontogeny. Reciprocal crosses with the
normal type have shown that a lowering of the chromosome number
probably occurs at the sexual stage and more frequently in antheridia
than in archegonia. No gross cytological disturbance occurs and the chromosome numbers indicate that a process of a reductional nature may
occur at this stage. This is plausible since, owing to the shift of the developmental processes of the life-cycle as a whole, the sexual stage in the
'peculiars' corresponds to the sporangial stage in the normals, as indicated
by the formation of sexual cells in the sporangia of the hybrid. There is
no segregation in this respect. After the cross all gametophytes from the
hybrid and from the normal progeny as well as the 'peculiars' show the
process.

Since a single gene, which no doubt arose by mutation, causes the drastic
alterations of the normal sequence of events which we see in the 'peculiar', it may be expected that other alterations of the life cycle have
appeared or may appear in a similar way. By analogy with the mutation
process in other organisms, the reverse mutation may occur, i.e. from the
recessive to the dominant type and it is reasonable to contemplate shifts
and drastic alterations in the life-cycle as a whole arising by gene mutation
and leading to the life-cycle seen in the present day ferns.

§ 9. **General considerations.** — As shown above, the genetical analysis of ferns has only just begun in a few species, but the rich variation
in many characters which is shown by so many of the present day fern

species will furnish good material for genetical investigations; they have also the advantage over the Angiosperms that the gametophytic generation is well developed. It is possible to keep a gametophyte almost indefinitely by subdivision, and to use it repeatedly both for self-fertilisation and for cross-fertilisation. The comparatively simple mode of growth in the gametophyte makes it favourable for the investigation of the time and mode of development of such characters as may express themselves at this stage of the life cycle. In the Leptosporangiates moreover, the spore-content of a single sporangium shows by its progeny the nature of the sporophytic cell from which it was derived, and we can draw conclusions as to what has happened at the reduction-division in the 16-cell stage, which gives rise to the characteristic number of (usually 64) spores.

Now that we have a satisfactory technique for the production of hybrids and can breed the progeny under controlled conditions, we may be able to answer some questions concerning the evolution of ferns, and the causes of their different ontogenetic development. We have a means of analysing the mode of inheritance of separate characters and of drawing conclusions as to what new types may be produced by recombination of existing genetical factors. With a knowledge of the relation between phenotype and genotype we may also reach conclusions on developmental problems such as the working of the different meristems and whether they are independent, or how far correlations in development limit the possibility of the appearence of new characters. Finally, genetical experiments with ferns may contribute to a better understanding of the processes which lead to alternation of generations and alternation of nuclear phases, and the relation between the two; and of the causes which lead to the higher development of the one or the other generation.

CHAPTER X

# GROWTH, TROPISMS AND OTHER MOVEMENTS

by

H. G. DU BUY (College Park, Md.) and E. L. NUERNBERGK (Freiburg i. B.)

**Introduction and review of the main literature.** — To understand the growth and tropistic movements of the *Pteridophytae* [1]) with their great variety of forms, their presence in extremely different environments, from water to deserts, would require far more analytical data than there are now available. Furthermore, the presence of an alternation of generations means that each phase in the life history may develop under different environmental conditions: i.e., a heterotrophic generation alternates with an autotrophic generation. It is therefore necessary not to generalize from conclusions drawn from experiments on some specialized form, nor to apply indiscriminately the same growth-scheme to all plants, as has been done during the development of the theory of growth-substances.

Only by careful application of the theory of growth regulation by chemical substances (a theory originally developed for the growth of a particular plant organ, and therefore frequently opposed in its wider application) will it be possible to see if and how far the several growth types found in ferns fit into this scheme.

The literature deals (1) with the development and growth of the prothallium and sex organs, and (2) with the development of the leaves, mainly from the point of view of systematics. Where the growth of these two phases depends on different environmental factors, we shall deal with them separately.

The earlier literature is indicated in the two standard works of SACHS [2])

---

[1]) The nomenclature here followed is that of CHRISTENSEN 's "Index Filicum". In many cases cross-references indicate the equivalent scientific names employed by writers in plant physiology.

[2]) J. SACHS: Lehrbuch der Botanik, 4th ed. Leipzig (1874).

and GOEBEL [1]). Other parts of this Manual deal with later literature in which the physiological part is only of incidental importance.

Publications dealing mainly with related problems are: HOFMEISTER [2]), BANKE, DE BARY, GOEBEL [3]), KNY, LEITGEB [3]), PRANTL [3]), HEIM, PERRIN, REED, LAAGE, NAGAI [4]).

The chief recent work on the physiology of the fern gametophyte is that of KLEBS [5]), who dedicated several years to the analysis of factors influencing the growth of the gametophyte. His extensive work cannot be adequately summarized in this brief review.

## I. Growth of the gametophyte

The gametophyte is very suitable for investigations on growth and development, because the development from spore to young prothallium which forms the sex organs shows a succession of stages in which cell division takes place in three different directions:

1. the spore forms a rhizoid [6]) and a more or less elongated germination tube by transverse cell division only;

2. a prothallium plate is formed by longitudinal cell division;

3. finally cell division in the third dimension forms a many layered young prothallium.

**§ 1. Growth of prothallia.** — 1) Normally the prothallia are simple heart-shaped because all the marginal meristematic cells develop from an initial edge-shaped apical cell and continue to function during the whole development. This is called *marginal growth.*

---

[1]) K. GOEBEL: Grundzüge der Systematik und speciellen Pflanzenmorphologie, Leipzig (1882) = Outlines of classification and special morphology of plants, Oxford (1887). – Vergleichende Entwicklungsgeschichte der Pflanzenorgane, ,,Enzyklopädie der Naturwiss. p. 99–432'', Berlin (1883).

[2]) W. HOFMEISTER: Die Lehre von der Pflanzenzelle, Leipzig (1867).

[3]) See notes on p. 305, 307 and 310.

[4]) C. HEIM: Untersuchungen über Farnprothallien, Flora (Jena) 82, 35, (1896).
A. LAAGE: Bedingungen der Keimung von Farn- und Moossporangien. Diss. Halle (1906). See also: Beih. Bot. Cbl. 21, I, 76–115 (1900).
I. NAGAI: Physiologische Untersuchungen über Farnprothallien. Flora (Jena) 106, 281–330 (1914).
G. PERRIN: Influence des conditions extérieures sur le développement et la sexualité des prothalles des Polypodiacées. C.r. Acad. Sci. Paris 147, 433–435 (1908).
H. S. REED: The value of certain nutritive elements to the plant cells. Ann. of Bot. 21, 501–543 (1907).

[5]) G. KLEBS: Zur Entwicklungsphysiologie der Farnprothallien. I, II, III. Sitzsber. Heidelberg. Akad. Wiss., Math.-naturw. Kl. 1916, B4; 1917, B4, B7 (1916/1917).

[6]) The behaviour of rhizoids of this origin is not necessarily the same as that of rhizoids of the full grown prothallium.

2) Other types: In the *Hymenophyllaceae* [1]) growth is mainly apical. The apical meristem, however, increases in breadth. After some time the central part becomes adult and stops growing. Therefore two meristems remain. The first grows in the direction of the old shoot, the second forms a lateral shoot. This prothallium is then similar in shape to analogous organs of lower plants e.g., *Vaucheria*.

*Osmundaceae* develop in the same way but several other parts of the meristem stop their growth.

3) *Vittaria* and *Monogramma* [1]); *Anogramma*. Young spatula-shaped prothallia are in the beginning meristematic in their proximal region, later only on both sides. Under the meristem develops (as in *Gymnogramme*) a reproductive axis which grows in the soil and bears the archegonia [3]). (The reproductive axis of some ferns serves only for perennation.)

In *Anogramma* the prothallia when inadequately nourished form only adventitious structures. When well nourished they also form archegonia.

Similar is the behaviour of other species, e.g., *Blechnum*, *Ceratopteris*, *Gymnogramme*, *Polypodium*, and PERRIN [2]) as well as CZAJA [4]) stated that a scant nutrition or one that lacks nitrogen, yields uni-sexual male prothallia, while plenty of nitrogen produces hermaphrodite prothallia. In a solution without nitrogen the development takes no normal course at all (PRANTL [5])).

Besides this the male and female sexual organs develop better in dim light, whereas in strong light only antheridia are produced. Likewise, they are influenced by different spectral colors, e.g. red light produces exuberant, mostly unisexual cultures (PERRIN [2])).

When prothallia of *Anogramma* die, the fertilized egg cells draw nourishment from the vegetative tubers. The development of the germinating plant inhibits in this case any vegetative growth of the tuber (Cf. p. 306). The tuber can also function as a resistant stage.

*Influence of "hormones" on the development of prothallia.* — MEYER [6]) mentions that the spores of Pteridophyta contain more "growth substance" than other parts of these plants. However, at present we do not know

---

[1]) K. GOEBEL: I. Ueber epiphytische Farne und Muscineen. Ann Jard. bot. Buitenz. 7, 1–73; II. Zur Keimungsgeschichte einiger Farne. Ibidem p. 74–119 (1888).

[2]) See note p. 304.

[3]) K. GOEBEL: Organographie der Pflanzen, 2nd part, 3rd edit. p. 1101 Jena (1930).

[4]) A. TH. CZAJA: Ueber Befruchtung, Bastardierung und Geschlechtertrennung bei Prothallien homosporer Farne. Z. Bot. 13, 545–589. (1921).

[5]) K. PRANTL: Beobachtungen über die Ernährung der Farnprothallien und der Verteilung der Sexualorgane. Bot. Ztg. 39, 953 (1881).

[6]) F. MEYER: Ueber die Verteilung des Wuchsstoffes in der Pflanze während ihrer Entwicklung. Diss. Frankfurt a.M. (1936), p. 34.

anything about the possible influence of that peculiar growth substance on the development of the prothallia originating from these spores.

On the other hand, a few investigations have been made with hormones applied to fern prothallia by means of culture-media or nutrient solutions. For instance, concerning the influence of follicle-hormone (Progynon = $C_{18}H_{22}O_2$) ORTH [1]) made some experiments. It was impossible to state any effect on the ratio between the two sexes in prothallia of *Gymnogramme*, *Blechnum*, *Ceratopteris* and *Polypodium*, just as little as in Angiospermae. As it has been shown above, this ratio depends much on the nutrition and is changed in favour of the archegonia, if much nitrogen is available.

A favourable influence of Progynon is seen only in prothallia of *Equisetum*, where this hormone supports a better development of the rhizoids, especially of the first rhizoid of the spore, and besides that, a better stretching growth and generally a more exuberant development of the prothallia. Owing to this effect the ratio between the number of produced archegonia and antheridia rose to the optimal amount of 50% to 50%.

*Regeneration in prothallia.* — When a heart-shaped prothallium is cut in two parts, these parts do not regenerate, but the growing point enlarges and forms a new lobe.

When the whole growing point (all the embryonic cells) is removed, many adult prothallium cells grow out into new prothallia. This shows that embryonic tissue can be regenerated, and that the presence of a part of a growing point inhibits the outgrowth of adult prothallium cells into new prothallia [2]) (cf. p. 305 "tubers of *Anogramma*").

According to GRATZY–WARDENGG [3]) this phenomenon is related to the presence of an osmotic gradient in prothallia. A maximum of both, the osmotic value and the suction force exists in the growing point, where the strongest meristematic cell division occurs, moreover it can be observed in the neighbourhood of the embryo or of the initial cells in regeneration. Toward the base of the prothallium both osmotic factors diminish gradually, hence it may be assumed that every embryonic tissue, growing points as well as points of regeneration, acts like attracting points with regard to the substances necessary for growth. Therefore, these substances are not available any longer for other parts of the prothallium tissue.

### § 2. Development of dorsiventrality. — *Tropisms of the Gametophyte.*
— The Filicinae show very clearly that a dorsiventral organ does not need to

---

[1]) H. ORTH: Die Wirkung des Follikelhormons auf die Entwicklung der Pflanze. Z. Bot. 27, 565–607 (1934).

[2]) K. GOEBEL: Ueber Regeneration im Pflanzenreich. Biol. Zbl. 22, 385–397, 417–438, 481–505 (1902).

[3]) S. A. E. GRATZY–WARDENGG: Osmotische Untersuchungen an Farnprothallien. Planta (Berl.) 7, 307–339 (1929).

consist of several cell layers. A single, apparently symmetrical (radial) cell, or a single layer of cells may exhibit dorsiventrality. Secondly the ferns show that dorsiventrality is very dependent on environmental conditions. This confirms the saying of PRANTL [1]) that there is practically no part of a plant whose future shape is more determined by environmental conditions.

*Influence of environmental conditions, especially of light.* — Whereas the germination tubes mostly show positive phototropism, stages 2 and 3 (see p. 304) are plagio-phototropic and show labile dorsiventrality. This labile dorsiventrality is emphasized by DÖPP [2]). In *Matteuccia Struthiopteris* and other Filicinae, prothallia are often formed which bear antheridia, archegonia, and rhizoids on both sides. In most cases, however, the rhizoids and archegonia, and partly also the antheridia, are produced on the nonilluminated side of the prothallium (cf. p. 310).

*Influence of light.* — The influence of light on the prothallia begins during germination (see further below). For instance, in *Equisetum* the first cell-wall of the germinating spore always develops perpendicular to the gradient of light-absorption, the more strongly illuminated daughter cell becoming the first prothallium cell, the more weakly illuminated the rhizoid-cell (STAHL [3])). In this case, the first visible influence of light consists of an atraction of the chromatophores to the lighted side of the spore, followed by the turning of the nuclear spindle axis parallel to the direction of the light (NIENBURG [4])).

However, this cannot be applied to the germination of other fern-spores, in which the point, where the first rhizoid develops is not influenced by light (GOEBEL [5])).

LEITGEB [6]) stated that the more strongly illuminated side always turns out to be the dorsal side. GOEBEL [5]) gives a detailed description of the development in *Adiantum*. In light of sufficient intensity the positively phototropic germinating tube develops into a two-dimensional prothallium, which places its surface at right angles to the direction of the light rays. It shows so-called transverse phototropism. There are also prothallia which develop directly from the spore into a cell plate (LEITGEB [6]), GOEBEL[5])).

In *Ceratopteris* this two dimensional cell plate shows at first positive

[1]) K. PRANTL: Ueber den Einfluß des Lichtes auf die Bilateralität der Farn-prothallien. Bot. Ztg. 37, 697 (1879).

[2]) W. DÖPP: Untersuchungen über die Entwicklung von Prothallien einheimischer Polypodiaceen. Pflanzenforschung, No. 8 (Jena 1927).

[3]) E. STAHL: Einfluß des Beleuchtungsrichtung auf die Teilung der *Equisetum*-spore. Ber. dtsch. bot. Ges. 3, 334–340 (1885).

[4]) W. NIENBURG: Die Wirkung des Lichtes auf die Keimung der *Equisetum*-spore. Ber. dtsch. bot. Ges. 42, 95–99 (1924).

[5]) K. GOEBEL: Ueber die Einwirkung des Lichtes auf die Flächenentwicklung der Farnprothallien. Rec. trav. bot. néerl. 25A, 122 (1928).

[6]) H. LEITGEB: Studien über die Entwicklung der Farne. Sitzgsber. Akad. Wiss. Wien, math.-naturw. Kl. 80, I, 201–227. (1879).

phototropism and at a certain stage of development goes over to transverse phototropism. In *Osmunda regalis* the cell plate shows transverse phototropism from the beginning.

1. The dependence of developmental growth on the wavelength of light.

*A. Germination of spores.* Germination of spores under normal conditions takes place in 4–20 days. The germination of the spores of *Dicksonia* is independent of the wavelength of light. The spores of *Equisetum arvense, E. palustre* and *Pteris longifolia* do not germinate in the dark and in infra red (SCHULZ [1])), and are dependent, like the spores of *Pteris longifolia*, on the wavelength and on the intensity of the light.

According to KLEBS [2]), red and green light have a favourable effect, whereas blue and violet light bring about an inhibition, although this is dependent on the intensity (STEPHAN, fig. 4)[3]).

However, ORTH [4]) recently stated that this is not the rule for all species. All spores of ferns are capable of germinating in light, and ORTH could not find any spore germination when kept exclusively in darkness. For instance, *Pteridium aquilinum*, a sun-fern germinates well in light of every wave-length (esp. of 5500 Å) and endures high light intensities. *Athyrium Filix-femina* and *Dryopteris Filix-mas* germinate well in light of short wavelengths the intensity of which may be low or high. This group of ferns belong to the red-green-shade-plants.

In all species red, yellow and green light (esp. 7100 and 5700 Å) stimulates germination as well as growth of the germination tube, but the wavelengths about 7500 Å–10,000 Å and 4900 Å–6,000 Å are inhibiting. Out of these experiments we may conclude: light of *different* wave-length exerts a *specific* influence on the germination and differentiation of fern-spores.

Furthermore, the age of the spores plays an important role with regard to the germination, for instance, *Equisetum*spores, older than 4 weeks, do not germinate. X-rayed spores of *Polypodium aureum* seem to develop in an abnormal way (KNUDSON 1934) [5]); in any case their chloroplasts develop into giant plastids which have often a diameter of about 22 μ.

---

[1]) W. SCHULZ: Ueber die Einwirkung des Lichtes auf die Keimfähigkeit der Sporen der Moose, Farne und Schachtelhalme. Beih. Bot. Cbl. 61, 81. (1901/02).

[2]) See note p. 304. Compare also: A. BURGERSTEIN: Einfluß des Lichtes verschiedener Brechbarkeit auf die Bildung von Farnprothallien. Ber. dtsch. bot. Ges. 26a, 449–451 (1908).

[3]) J. STEPHAN: Untersuchungen über die Lichtwirkung bestimmter Spektralbezirke und bekannter Strahlungsintensitäten auf die Keimung und das Wachstum einiger Farne und Moose. Planta (Berl.) 5, 381–443 (1928).

[4]) R. ORTH: Zur Keimungsphysiologie der Farnsporen in verschiedenen Spektralbezirken. Jb. Bot. 84, p. 358–426 (1937). Cf. A. SEYBOLD: Ueber den Lichtfaktor photophysiologischer Prozesse. Jb. Bot. 82, 741–794 (1936).

[5]) L. KNUDSON: Giant Plastids in Ferns Produced from X-Rayed Spores. Am. J. Bot. 21, 712 (1934).

*B.* Klebs[1]) reports for *Pteris longifolia* that the prothallium cell plate is formed mainly under the influence of blue-violet radiation which increases transverse and longitudinal cell division. Regarding *Cystopteris fragilis*, Mussack[2]) states that the flat, thin prothallium develops perpendicularly to the direction of the light (lens-shaped cells?). In light of shorter wave-length the growth by stretching increases, while in light of longer wave-length transversal cell-division is intensified.

Stephan[3]) defines the influence of the several spectral colors on *Dick-sonia antarctica* and *Pteris longifolia* more clearly. He observes that in all wavelengths > 6000 Å there is no development of prothallia at all, and that this occurs only in parts of the spectrum > 5800 Å. Typical growth with wedge shaped apical cells at the vegetation point requires blue light of a wavelength of about 5000 Å. Like germination these processes are dependent on the intensity of the light, but also in this case plants of different types show a different behaviour. In most cases the longer wavelengths increase the stretching of the cells and the shorter wavelengths have an inhibiting effect. In *Equisetum* however, the shorter wavelengths increase the stretching of the cells. Closer investigation of this problem would be worth while in relation to phototropism.

The influence of the wavelength and the intensity of the light on the chain of processes causing cell division is even more variable, so that except for the general rule that the shorter wavelengths on the whole increase the cell division, no further generalizations can be made. There are furthermore no investigators dealing with the influence of the light on partial processes of growth, e.g., on the reversible and irreversible extensibility of the cell wall etc. (du Buy 1933, p. 879) [4]). Although Klebs has laid the foundation for the work on the development of fern prothalla, it is impossible to refer to his work briefly. This shows, however, the main value of his work because he has emphasized the complexity and the variability of the above problems from the beginning, a fact which does not make his work popular, but nevertheless deals with the phenomena in an adequate way.

2. The dependence of transverse phototropism on the wavelength of light. Development of dorsiventrality and phototropic behaviour also seem to depend on light of wavelength < 5500 Å (cf. above).

Leitgeb[5]) states that young prothallia of *Ceratopteris thalictroides* develop dorsiventrality and show positive phototropism when the cultures

[1]) See note p. 304.

[2]) A. Mussack: Untersuchungen über *Cystopteris fragilis.* Beih. Bot. Cbl. 51, I. 204–254 (1933).

[3]) See note p. 308.

[4]) H. G. du Buy: Ueber Wachstum und Phototropismus von *Avena sativa*. Rec. Trav. bot. néerl. 30, 798. (1933).

[5]) See note p. 307.

are placed under a glass jar containing $K_2Cr_2O_7$ solution or ammoniacal cupric oxide solution. Both solutions are transparent for wavelengths about 5250 Å and shorter.

KLEBS [1]) grew prothallia of *Pteris longifolia* under red glass, transparent from infra-red to 6100 Å and a little yellow and green. The prothallia develop bilateral symmetry, grow rhizoids and antheridia on both sides, and show no pronounced transverse phototropism, but grow up from their substratum. According to STEPHAN [2]), this can be explained thus: the development of the prothallium is due to the transmitted yellow-green light, but not enough of the shorter wavelengths were transmitted to cause dorsiventrality and hence the normal transverse phototropic behaviour.

3. Reversal of dorsiventrality by light. When *Osmunda* prothallia are illuminated from below, the ventral side becomes concave. When growth is rapid the curvature can be so great that the meristematic layer grows towards the light. In this case no change in dorsiventrality occurs. When growth is less rapid the curvature is not great enough, and the dorsal side remains longer in the shadow. In this case the dorsiventrality will be changed, the curvature decreases, and the prothallium forms rhizoids and archegonia at the former dorsal side [3]). This is shown by LEITGEB (1879) [4]) in *Ceratopteris* and by NĚMEC [5]) in *Gymnogramme*.

*Influence of gravity and other tropisms.* — Almost nothing is known concerning the influence of gravity and other tropisms on prothallia. Only LEITGEB [4]), [6]) assumes some influence of gravity on the development of the embryos of macrospores of *Marsilia*. The macrospore always germinates in a nearly horizontal position, hence, the small prothallium also lies horizontally. After that, the first division of the fructified eggcell takes place in such a manner that the new cell wall will be situated in the direction of the archegonium axis, the plane of the wall being placed more or less at right angles to the direction of gravity.

As the upper cell later on produces the first leaf and the shoot, and the lower cell develops the roots, the whole embryo will grow up vertically. However, it may be observed that the real influence of gravity on the whole development of the embryo is restricted solely to the germination of the macrospore, for macrospores germinated beforehand and then brought in such a position that the direction of the archegonium axes

[1]) See note p. 304.

[2]) See note p. 308.

[3]) Compare the labile polarity, p. 306.

[4]) See note p. 307.

[5]) B. NĚMEC: Die Symmetrieverhältnisse und Wachstumsrichtungen einiger Laubmoose. Jb. Bot. 43, 501 (1906).

[6]) H. LEITGEB: Zur Embryologie der Farne. Sitzgsber. Akad. Wiss. Wien, Math. naturw. Kl. 77, I, 222–242, (1878).

points vertically upwards, will develop their embryo axes likewise not vertically but horizontally.

Generally it may be stated that the influence of light on the development of prothallia is by far the most important of all, because without light, usually no development of the prothallium plate occurs, but only a more or less elongated tube is formed. In this respect the behaviour of prothallia is very different from that of the gametophyte of Hepaticae where not only light, but also gravity and other environmental factors (moisture: hydrotropism) influence shape and direction of growth to a considerable amount.

**§ 3. Growth of the rhizoids of the full grown Gametophyte.** — Concerning the growth of the rhizoids of prothallia it may be stated that they have the same physiological features as the rhizoids of the Hepaticae, i.e. they react to light, gravity, moisture etc., and have apical growth (like the root hairs of the Spermatophyta). The sole difference consists in the fact that in ferns there are no cone-shaped rhizoids. Nearly all fern rhizoids are unicellular, except for instance the rhizoids of *Danaea* according to BREBNER [1]).

*Movements of the rhizoids of the full grown Gametophyte.*

Phototropism of prothallium rhizoids of Equisetum. According to STAHL and BUCHTIEN [2]) the rhizoids of *Equisetum* are slightly negatively phototropic in low light intensities and strongly negatively phototropic in higher intensities. STEPHAN [3]) on the contrary reports that the rhizoids are always positively phototropic to light of low *and* of high intensity (sun). This discrepancy can be interpreted in accordance with the hypothesis of DU BUY-NUERNBERGK [4]). This is: "light which increases the growth of a radial organ above that which occurs in darkness will always cause negative phototropism, provided that the absorption differences of the light are great enough".

Phototropism and Geotropism of prothallium rhizoids of Filicinae. According to LEITGEB [5]) the older literature mostly states that the rhizoids of Filicinae are negatively phototropic. BISCHOFF [6]) showed that this is not proved. LEITGEB only stated

---

[1]) See G. HABERLANDT: Physiologische Pflanzenanatomie, 6th ed. Leipzig 1924, p. 209.

[2]) O. BUCHTIEN: Entwicklungsgeschichte des Prothalliums von *Equisetum*. Bibliotheca bot. No. 8. Stuttgart (1887). E. STAHL: Einfluß der Beleuchtungsrichtung auf die Teilung der *Equisetum*-Sporen. Ber. dtsch. bot. Ges. 3, 334–340 (1885).

[3]) See note p. 308.

[4]) H. G. DU BUY and E. NUERNBERGK: Phototropismus und Wachstum der Pflanzen I. Erg. d. Biol. 9, 358. Berlin (1932).

[5]) See note p. 307.

[6]) H. BISCHOFF: Untersuchungen über den Geotropismus der Rhizoiden. Beih. Bot. Cbl. 28 I, 94 1912).

that the rhizoids are always formed on the shaded side of the prothallium, and grow out "exotropically", i.e. regulated by symmetrical internal factors. The prothallium itself is transversely phototropic and since the rhizoids grow out at right angles to its surface it looks as if the rhizoids were negatively phototropic.

It remains to be proved that they are not aphototropic, particularly because BISCHOFF followed the same line of thought. He showed that prothallium rhizoids of Filicinae are not geotropic. The rhizoids grow out independent of the direction of gravity.

## II. Growth of the Sporophyte

As in other vascular plants, i.e., the Spermatophyta, we have to distinguish in the Pteridophyta between root, stem and leaves, (= fronds) all three together composing the plant body ("Kormus"). Of these three elements the leaves of ferns show the most remarkable features and therefore, their growth and behaviour have been mostly investigated. On the other hand, concerning the stem and root of the Pteridophyta, both differing little or not at all from the corresponding organs of the Spermatophyta, nearly no investigations are known, which deal particularly with mechanism and physiology of their growth. That is why we shall deal mainly with growth and tropisms of the leaves of ferns in the following paragraphs. The terms stipe (= petiole) and lamina (= blade) as parts of the frond (leaves) will be used synonymously.

**§ 4. Growth types in Filicinae.** — To begin with the growth of the roots, it may be sufficient to mention the statement [1]) that Pteridophyta roots grow similarily to the roots of *Ephedra vulgaris*, both belonging to a special kind of growth-type. According to PORODKO [2]) this growth type is distinguished from the usual growth type of roots (compare further on p. 313, 314) by the fact that the growth is nearly exclusively apical [3]), and the middle part of the root almost lacks growth.

According to GOEBEL [4]), the form and mode of development of leaves in Filicinae are much more variable. Compare for example, the small leaves of the Hymenophyllaceae with the very massive leaves of *Angiopteris*. Yet the early stages in the development of these different types of leaves have much in common as can be shown by a comparison of primary and pinnate leaves.

---

[1]) Some further informations on roots are given by K. GOEBEL: Organographie der Pflanzen, 1st part, 3rd ed., p. 91 Jena (1928).

[2]) H. PORODKO: Neue Längenwachstypen der Hauptwurzeln. Planta (Berl.) 6, 234–254 (1928).

[3]) Cf. W. PFEFFER: Pflanzenphysiologie, vol. 2, p. 6, 10. 2nd ed. Leipzig (1904).

[4]) See note p. 304 and p. 305, note 3.

In some of them the meristematic tissue is marginal. (The veins are bifurcated which is a necessary consequence of this type of growth).

In others, leaf formation is due to a meristem limited to the apex which produces the lateral pinnae by monopodial branching, when branching occurs.

The following types can be recognized.

(A) The young plant begins with a cotyledon which has marginal growth from the beginning. The leaf or only the veins of this cotyledon and of the primary leaves show bifurcational branching (*Marattiaceae, Osmundaceae*).

(B) The apex of the leaf grows constantly in one direction, and later marginal growth appears. (For instance all ferns which show long stretched leaves with many lateral parts, e.g. some *Adiantum* species).

(C) Apical growth may persist through several vegetative periods, e.g. many *Nephrolepis, Hymenophyllaceae, Gleicheniaceae, Lygodium*. (The results obtained on tissue-cultures make it probable that we are dealing in *Lygodium* with leaves which show unlimited growth).

Generally, the apical growth of *Filicinal* leaves is combined with hyponasty of the extreme part of the growth zone thus forming the well known curl of the young fronds. In the following part of the growth zone hyponasty is replaced by epinasty which on its part causes the unrolling of the leaf (see fig. 4 and 5 $D_1$; cf. also p. 320).

The hyponastic curling of the infant apex is not always to be found in fern-fronds. GOEBEL (p. 1183) [1]) mentions some cases, e.g., *Pteris serrulata*, *P. cretica* and *P. umbrosa*, where, notwithstanding the apical growth, the laminar part of the frond is stretched, and only the stipe shows a sharp curvature. We shall deal with this subject again on p. 321. To another type of an uncurled leave belongs further the frond of *Adiantum uniforme*.

A comparison with other plants is helpful in understanding the tropistic responses of ferns. The causal analysis of growth and tropisms has been worked out in the greatest detail for apical growth types of Angiosperms. As a result of this work it can be concluded that the primary step in analysis is to determine the growth type of the plant.

According to DU BUY and NUERNBERGK (1932–1935) [2]) at least four apical growth types can be distinguished:

1. *Avena*-type. Here growth is distributed over the whole coleoptile and is regulated principally by auxins, produced only in the tip. When the tip is removed, growth nearly stops. Only after some hours a temporary regeneration of auxin production occurs.

2. *Helianthus*-type. Here growth is also distributed over almost the whole seedling. When the tip is removed, however, both tip and base continue to grow. Growth is regulated by substances produced by all non-adult cells which are distributed over the whole plant. These cells preserve their embryonic state for several days.

[1]) See note p. 305.
[2]) See note p. 311.

3. *Vicia*-type. Growth takes place only at the apical part of the plant. When the tip is removed, no growth can be found. Growth is regulated by substances which are produced only by non-adult cells found only in the apical part. They remain in this embryonic state for only a short time.

4. Root-type. This type resembles more or less the *Vicia*-type, but the growth zone is a little bit more extended (cf. also p. 312).

Fern leaves of growth type C show the features of the third group, the *Vicia*-type. A few general conclusions concerning this point are given below. A more elaborate report will be given elsewhere.

*Osmunda, Polypodium, Asplenium bulbiferum and related species.* There are many papers dealing with their development from a morphological point of view, relating the shape of initial cells with the shape of the leaves, etc. (GIESENHAGEN, HELM) [1]). They occasionally mention the eventual change in shape resulting from a change in external conditions [2]). However, there is no mention made of the relation between shape and the action of growth regulators, etc. We give here some of our data.

The fern *Dryopteris*, showing the features of growth type B, has limited apical growth. Since lateral leaflets are developed later and in the lower part of the petiole, the cells of the petiole must remain longer in an embryonic state. This is shown by a longer growing region, and an occurence of curvature after application of auxin [3]) over this whole region. The whole development of the leaf takes place in 20–30 days, at about 20° C. After this time, growth stops and the application of auxin no longer has an effect [4]).

PRANKERD [5]) is the only investigator who gives more detailed data regarding the development of *Asplenium bulbiferum*, a growth type similar to *Dryopteris*. To quote the author: "Three phases may conveniently be distinguished in the life-history of a fern frond: the first or infant phase

---

[1]) K. GIESENHAGEN: *Asplenium viride* Huds. forma *daedalum*. Ein Beitrag zur Entwicklungsmechanik des Farnwedels. Flora (Jena) 123, 105–132 (1928).

J. HELM: Anlage und Entwicklung des Blattes von *Trichomanes bimarginatum* V. D. B. Planta (Berl.) 23, 442–473 (1935).

[2]) Cf. K. GOEBEL: Morphologische und biologische Studien XIII. Weitere Untersuchungen über die Gruppe der *Drynariaceae*. Ann. jard. bot. Buitenz. 39, 117–126. (1928).

[3]) Detailed reference to the growth substance "auxin" is given by DU BUY–NUERNBERGK in "Phototropismus und Wachstum der Pflanzen, I, II, III" (Ergebn. d. Biologie, vol. 9, 10, 12. Berlin 1932, 1934, 1935).

[4]) It may be mentioned here that the growing apex of *Dryopteris* moves in small circumnutations. (CH. DARWIN: The power of movements in plants, London 1880. Cf. the German translation: „Das Bewegungsvermögen der Pflanzen", 2nd ed. Stuttgart 1898, p. 54, 217).

[5]) T. L. PRANKERD: On the irritability of the fronds of *Asplenium bulbiferum* with special references to graviperception. Proc. Roy. Soc. London 93 B, 143–152 (1922).

when the apex of the frond is curled, the second or adolescent phase, from the time when the first leaflets appear beneath the apical coil, until the frond is quite uncurled, when the third or mature phase is reached". (Cf. PRANKERD, p. 147, fig. 4). As we have seen, this mode of development holds only for growth type B. PRANKERD's third stage, for instance, is never reached by growth type C.

Furthermore, she has proved definitely that there a is real difference between the first and second stages except in the development of the leaf, a morphological feature. The author gives no data concerning the distribution of growth over the whole plant during its development. This makes a further analysis of the causes of the differences in "heliotropic and geotropic irritability" etc., in the two phases more difficult.

Several experiments have been made by us on the growth of *Nephrolepis cordifolia*. Fig. 1 and 2 show that, the growth of *Nephrolepis* — and similarly of *Osmunda* etc. — either in the light or in the dark (green house or dark room; see in connection herewith fig. 3), takes place in the apical part of the plant. The zone of maximal elongation is always 1–2 cm from the tip. Therefore in young plants, the growth seems to be basal (1 cm from the tip), afterwards the zone of maximal elongation shifts to the middle part and the tip (cf. fig. 2). When the tip is removed no growth can be found. Growth is regulated by cells in an embryonic state (cf. fig. 3E) and only occurs in the apical part. Several interesting conclusions can be drawn:

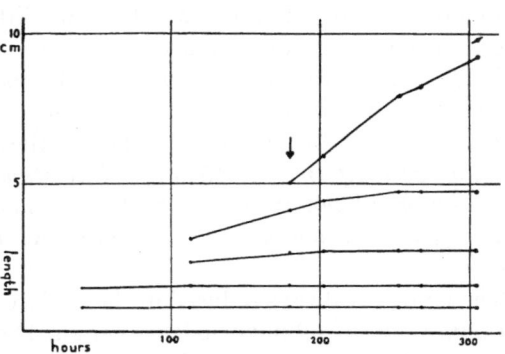

FIG. 1. — Typical apical growth curve (*Nephrolepis cordifolia*). The different lines represent the growth curves of the different zones. New zones have been added gradually. The arrow indicates the development of the first pair of laminae.

1. The Pteridophyta which belong to this physiological group show the same features of growth as higher plants, whereas the growth of the cells of Bryophyta, as far as they have been investigated, cannot be changed by auxin or by decapitation, but results from the individual growth of each cell (see Manual of Bryology, p. 222, 224, 228) [1]). In Bryophyta, therefore, the growth resembles the growth of cells of lower plants.

2. A tropistic curvature can only be produced in the apical part, depending on the properties of the growth type to which the fern belongs. This curvature will not shift to the base and is fixed easily.

3. We can expect, furthermore, that the mobility of the growth regulators is not great. This has important consequences for photo- and geo-

[1]) Cf. also DU BUY–NUERNBERGK 1935, p. 336 following (see p. 314).

tropic "sensitivity", for the validity of the theory of BLAAUW etc. After removal of one lateral half of the tip, only the cells under the undamaged part grow, due to the production of growth regulators by this part (fig. 5C). The curvature of the damaged part is partly due to exsiccation. The cells show curvature only while they are young (1–2 cm from tip). This zone of non-adult cells is also the zone where unilateral application of growth substance causes a curvature (fig. 5E, F). Application of growth substance on other parts has no effect (fig. 5G).

§ 5. **Influence of darkness (etiolation) on growth.** — As in phototropism (see further below) the general influence of light on the growth of fern leaves is comparatively small. For the most part it is effective only, when lasting for a longer period. Then, for instance, lack of light produces an etiolation which shows different aspects in the different species even of the same family.

With regard to the *Polypodiaceae* HOMMER [1]) stated that etiolated fronds of *Polystichum aculeatum, Dryopteris Filix-mas,* and *Woodwardia radicans* are longer than normal, whereas these fronds are shorter in *Polystichum adiantiforme, Asplenium lineolatum, Adiantum cuneatum, A. gracillimum, Blechnum brasiliense, B. occidentale, Doodia aspera, Dryopteris dentata (Nephrodium molle), Polypodium Reinwardtii* and *Pteridium aquilinum.*

The lamina is always reduced in the etiolated state, some species (*Doodia, Adiantum cuneatum*) have no leaves at all. The leaves often show hyponasty (cf. p. 321), e.g., the small leaves of *Adiantum gracillimum, Asplenium lineolatum, Pteridium,* and in many species the petioles develop even farther and owing to this their number is reduced.

Usually, the leaves of the etiolated plants remain green, as ferns (like conifers) are able to form chlorophyll even in darkness; only in *Pteridium* white fronds have been observed.

Regarding the stipes of etiolated fronds the size of their cells often increases (*Polystichum aculeatum*), as well as the number of cells, if the fronds have developed to a greater length (see above). Generally, in etiolation all cells represent a kind of infant stage, and on this account no development of spores occurs in etiolated leaves.

Summarizing it can be stated that the behaviour of ferns when etiolated does not differ much from that of dicotyledonous plants.

§ 6. **Phototropism of leaves.** — The phototropism of leaves growing in the light is not great. According to PRANKERD [2]) the "heliotropic irritability" in *Asplenium* begins in the "infant" stage, reaches a maximum in the

---

[1]) M. HOMMER: Ueber das Etiolement bei Farnpflanzen und die Ursachen des Etiolements im allgemeinen. Bot. Arch. 14, 1–46 (1926).

[2]) See note p. 314.

"adolescent" stage and ceases early in the "mature" stage. This is, however, only true for ferns belonging to a similar growth type. As is shown in fig. 4 the phototropic response runs more or less parallel with the growth of the reacting leaf.

The dorsiventrality of the leaves causes a different "sensitivity" to light coming from different directions, as described by PRANKERD. When the young frond is placed abaxially to the light, a curvature of about 90° is the result; when the young frond is placed adaxially, the effects of epinasty and light balance each other. As a result the backward curve of the fully developed leaf is depressed, so that the leaf remains vertical.

When the light strikes the young leaf at right angles to the plane of the leaf, only a slight positive movement takes place.

From the moment that leaflets are developed, the curvatures are no longer pronounced. By adaxial and abaxial illumination the position of the leaf in relation to the vertical is respectively somewhat increased or decreased. PRANKERD mentions also a "diaheliotropistic" movement caused by lateral illumination which induces a torsion of the stipe. No phototropic curvature occurs at this stage of development.

All these statements hold only for the ferns of the above mentioned growth type, and within this type only for certain species.

In *Nephrolepis*, for instance, the younger stages including stages with several leaves show curvatures of nearly the same extent following illumination from the abaxial and the lateral side (fig. 4, *A–D*).

After adaxial illumination they show a curvature only of somewhat less extent. The size of curvature decreases gradually in older leaves, being gradually replaced by torsion. This torsion is dependent on the presence of the leaves although the elongation of the stipe is independent of their presence (cf. fig. 4, *E–G*). Details of this work will be given elsewhere.

The behaviour of *Dryopteris* and *Nephrolepis* shows that the phototropic (and as we shall see, also the geotropic )"sensitivity" seems very variable. This is due to the fact that the size of a curvature is taken as a measure of the sensitivity. It must be emphasized that a curvature is only the last link in the complex chain of processes initiated, for example, by light.

The term "phototropic sensitivity" has therefore no analytical value at all. The case of *Dryopteris* shows the absurdity of this term: during one stage of development it has at least four different phototropic sensitivities!

We must, therefore, distinguish between the sensitivity of one or more of the parts of the whole chain reaction (the curvature). For example, the sensitivity of the photolabile system, the sensitivity of the cell wall to changes in the environment, and the sensitivity of the cell wall as measured by its ability to react. Only if these distinctions are made further analysis will be possible.

If, for example a single stage of *Dryopteris* illuminated with the same amount of light shows different degrees of curvature (depending on the direction of illumination); we must realize that the curvature never can be a proper measure for the phototropic sensitivity, although it undoubtedly has one photosensitive system.

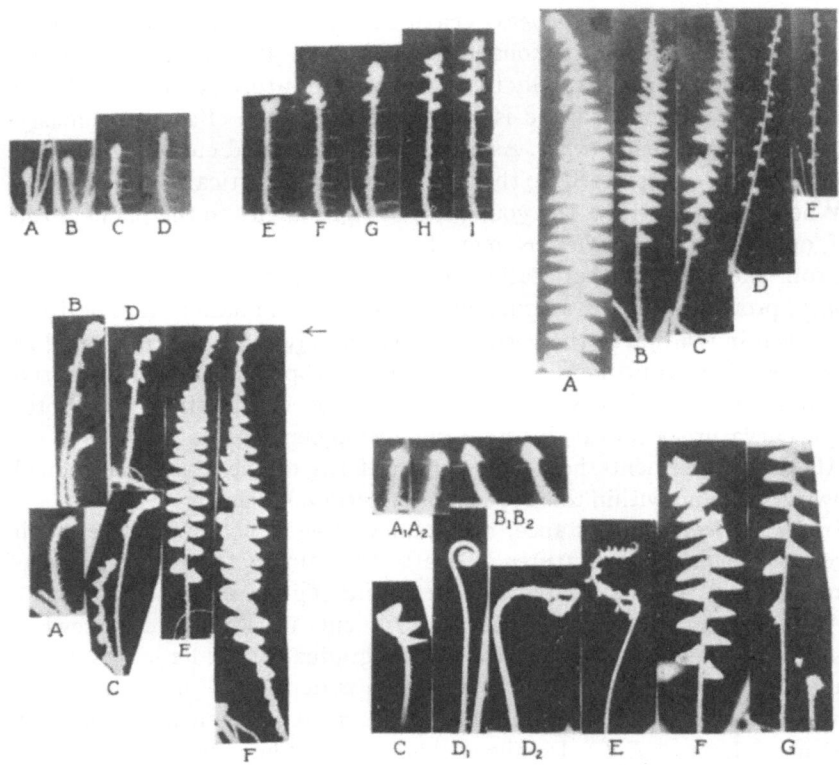

Above left: FIG. 2. — *Nephrolepis cordifolia*. A-I: Growth of leaf in daylight. In C the marks were replaced.
Above right: FIG. 3. — A. Leaf developed in daylight.
   B. developped in daylight later in darkness.
   C. same as B.
   D. Leaf developed in darkness.
   E. Leaf embryo and leaf developed in darkness (notice the small growing point).
Below left: FIG. 4. — Shadowgraphs of phototropic curvatures of *Nephrolepis cordifolia*. A–D: abaxial, adaxial and radial illumination of young fronds.
   E and F: Torsions and curvatures of fronds with laminae. The arrow indicates the direction of the illumination.
Below right: FIG. 5. — Influence of 3-indole acetic acid on fern fronds.
   $D_2$ and $D_1$: *Dryopteris dentata*, the others: *Nephrolepis cordifolia*.
   $A_1$, $B_1$: on tip of decapitated leaf.
   $A_2$, $B_2$: after 24 hours.
   C. Longitudinal removal of tip.
   $D_1$. Young petiole without growth regulator.
   $D_2$. with growth regulator.
   E, F. Results after some days of unilateral application of growth regulator.
   G. Result of unilateral application of growth regulator on adult petiole: practically no curvature.

Using always the developmental stage, only relative values for spectral sensitivity, threshold curvature, etc. can be obtained.

For measurement of phototropic sensitivity, by which is usually meant the sensitivity of the photolabile system, we must look for a measurable reaction that is independent of structural peculiarities of the plant, e.g., the growth reactions.

For the rest, the literature on phototropism of ferns does not contribute to our knowledge of growth reactions, spectral sensitivity, validity of the law of BLAAUW–FRÖSCHEL etc. [1]). The following, however, may be considered:

The change in the ability of the cell walls to react can be seen from the zonal growth curve on p. 315.

By analogy with higher plants of the same growth type it is probable that the curvature results from the difference in growth reaction of the illuminated and the non-illuminated side (theory of BLAAUW in the strict sense). Therefore the differences in "sensitivity" in very young and in older plants finds a satisfactory explanation in the decreased ability of old cell walls to react.

§ 7. **Geotropism of fern leaves.** — Geotropism of fern leaves is, like phototropism, related to growth. The only analysis of this subject is contained in the work of PRANKERD and of WAIGHT [2]).

In *Asplenium bulbiferum* the cells react to gravity except in the last stage of development. When the leaves are nearly full grown there remains a little growth and only a slight geotropism. In the other stages, however, there is a relation between the stage of development of the individual cells (and not the rate of growth of the whole plant; see further below) and the geotropic reaction. This means that plants of different length can show similar curvatures when the stage of development is the same.

All the experiments referring to this are carried out in the light because a few controls showed that growth and reactions are the same in the dark. It is interesting that the behaviour of higher plants is different in this respect. The geotropic behaviour of ferns should therefore be tested especially in

---

[1]) For details see DU BUY–NUERNBERGK 1935 l.c. (p. 314).

[2]) T. L. PRANKERD: see p. 314.

—— The ontogeny of graviperception in *Osmunda regalis*. Ann of. Bot. 34, 709–720 (1925).

—— Studies in the geotropism of Pteridophyta. Proc. Linnean Soc. of London 141, 3–4 (1928/1929).

—— Studies in the geotropism of Pteridophyta IV. On specifity in graviperception. Journ. Linnean Soc. Bot. 48, 317–336 (1929).

—— and F. M. O. WAIGHT: On the presentation time and latent time for reaction to gravity in Pteridophytes. Rep. Brit. Ass. Adv. Sci. 1922. 1923, 397 (1923).

F. M. O. WAIGHT: On the presentation time and latent time for reaction to gravity in fronds of *Asplenium bulbiferum*. Ann. of Bot. 37, 55–61 (1923).

relation to changes in "sensitivity" during development. The function of the "statoliths" which disappear in the later stages of development is not at all proved.

It has already been mentioned, that the geotropic reaction of the investigated ferns shows in general a great variability. This relates not only to the effect of different temperatures on growth and geotropic response but also to seasonal and diurnal influences which have nothing to do with the force of gravity in itself.

Regarding at first the temperature effects, PRANKERD'S [1]) investigations on *Asplenium bulbiferum* do not disclose any considerable differences between the behaviour of ferns and other plants. For instance, at 10°C she observed a presentation time double that at 20°C, which points to chemical changes acting in accordance with the VAN 'T HOFF law. The growth-rate at the two temperatures was about $1/_3$. Moreover, at high temperatures (30°C) a time factor has been shown to operate, this temperature at once increasing the presentation time by $1/_3$, and by $2/_3$ after one day's exposure. The latent time (delay in response), however, being the same at 30°C and 20°C, was not affected by the time factor. From other experiments PRANKERD inferred that an injurious high temperature affects only the initial ("perception") and not the end ("response") stages in the scheme of geotropic reaction.

Concerning the periodical changes of geotropic sensitivity in *Asplenium bulbiferum*, PRANKERD [2]) stated that there is both an annual and a diurnal variability of the presentation time. The latter decreases by 33% at night and 20% in winter, while the latent time is 10% at night and 25% in winter compared with summer value. Further, the amplitude of movement is greater in summer.

According to PRANKERD, these rhythms are not correlated *directly* with growth or any other (investigated) internal or external factor, especially as the corresponding diurnal and seasonal growth rate, measured under constant conditions, does not vary considerably if at all.

With regard to the differences between the geotropic behaviour of the two investigated species i.e. *Asplenium bulbiferum* and *Osmunda regalis* it may be mentioned that:

1. the most sensitive stage of the frond in *Asplenium* is the stage with 5–7 pairs of leaflets, while in *Osmunda* it occurs at maturity — a stage at which all geotropic irritability has ceased in *Asplenium*. Moreover, the almost mature frond of *Osmunda regalis* shows the shortest presentation time (20 sec.) recorded for any plant.

[1]) T. L. PRANKERD: Studies in the geotropism of the Pteridophyta. V. Some effects of temperature on growth and geotropism in *Asplenium bulbiferum*. Proc. Roy. Soc. London 116 B, 479–493 (1935).

[2]) T. L. PRANKERD: Studies in the geotropism of Pteridophyta VI. On rhythm in graviperception and reaction to gravity. Proc. R. Soc. Lond. 120 B, 126–141 (1936).

2. the growth ratios for certain stages can be the same where the presentation times are different.

Finally, attention may be called to the geotropism of the stalk pulvini of *Angiopteris longifolia* as there is a striking parallel to the behaviour of the pulvini of Gramineae. Those pulvini do not become visible on the stalks till late and can serve for orienting the leaf if it is put out of its proper direction. In that case a knee-like bending occurs. GOEBEL [1]) pointed out that these pulvini are spots where growth ceases last, but perhaps we encounter here the same phenomenon as in the Gramineae pulvini mentioned above, i.e. growth has already ceased in the pulvini which are in the normal position, but will start again, as soon as the pulvini are put into another position. SCHMITZ [2]) was able to show that in the case of Gramineae pulvini new growth-regulator (auxin) has been produced (unilaterally) under influence of geotropism thus causing growth and bending. It is still to be investigated whether this explanation could not explain the behaviour of the *Angiopteris* pulvini.

**§ 8. Hyponasty of fern leaves.** — In certain cases the hyponasty of fern leaves is due to the influence of light. For instance, according to HERZOG [3]) the leaves of many *Salvinia* species do not stretch in sunlight, but do in dim light.

On the other hand, there is no direct causal relation between the apical growth of fern leaves and their circinate embryonic apex [5]).

A straight leaf tip occurs in all three fern leaf growth types: in the first (*A*) as in some *Trichomanes* species; in the second (*B*) as in *Ophioglossum*, *Botrychium*; in the third (*C*) as in some *Pteris* species (see p. 313).

In these last examples, the same phenomenon occurs as in many dicotyledonous seedlings, flower stalks, etc.: several millimetres below the apex the stalk shows a curvature of about 180°, so that the apex is directed downward. Discussion on the change from positive to negative geotropism, change of internal factors, etc. can be found in DU BUY–NUERNBERGK [4]) and in RAWITSCHER [5]). A suggestion of a possible further analysis is given by DU BUY-NUERNBERGK [6]).

---

[1]) K. GOEBEL: (1930, p. 1182) see note, p. 305.

[2]) H. SCHMITZ: Ueber Wuchsstoff und Geotropismus bei Gräsern. Planta (Berl.) 19, 614 (1933).

[3]) R. HERZOG: Anatomische und experimentell morphologische Untersuchungen über die Gattung *Salvinia*. Plants 22, 490–514. (1934).

[4]) H. G. DU BUY a. E. NUERNBERGK: Phototropismus u. Wachstum d. Pflanzen I, p. 401 and foll., p. 447 foll. Erg. d. Biol. 9. Berlin (1932).

[5]) F. RAWITSCHER: Der Geotropismus der Pflanzen. Esp. p. 121–131. Jena (1932).

[6]) DU BUY–NUERNBERGK l.c., p. 529 ff.

Finally it may be mentioned that GOEBEL [1]) has denoted a certain kind of the phenomena just described, as unfolding movements ("Entfaltungs-bewegungen"). A good example illustrating these is the bending downwards of the apical part of the young leaves of *Pteris Wallichiana* which is caused by geotropism, that is to say a paratonic factor. For details concerning this kind of *epinasty* see the works of GOEBEL [1]).

**§ 9. Dorsiventrality.** — The presence of dorsiventrality in the majority of *Selaginellaceae* (BAKER: 260 species) is dependent on light.

According to GOEBEL (1915) [2]), *Selaginella caracensis* shows anisophylly only when it is grown in the light. The same is true for *Selaginella sanguino-lenta*. In this case, however, a low humidity can inhibit the action of the light (e.g. the relation between humidity and phototropism in Bryophyta, DU BUY and NUERNBERGK (1932, p. 428) [3]).

The *direction of the light* seems to have no influence; the pseudo-radial organs possess inherent dorsiventrality, (*Selaginella lepidophylla* (GOEBEL 1908, p. 115) [4]); *Selaginella sulcata*) and this dorsiventrality cannot, as far as is known, be reversed by experimental conditions.

The so-called reversed dorsiventrality of the reproductive part, e.g. in *Selaginella chrysocaulos, S. Lyallii, S. suberosa*, is the result of an increase in growth of the leaflets of the upper side and a decrease in growth of those of the lower side compared with sterile parts. This is no true reversal, for vegetative cuttings made of "reversed" fertile stems show the original dorsiventrality. This "reversal" depends on unknown internal factors, related to sporulation. (BEHRENS) [5]).

**§ 10. Heterophylly.** — The influence of environmental factors on leaf development of numerous ferns shows once more that this material in consequence of its high degree of plasticity is very suitable for the study of form determination. Numerous agents, especially atmospheric moisture and in a certain measure the intensity of illumination, are efficacious in .this respect, as may be shown by only two quotations out of a great many: according to ANGST [6]) injury to the leaves or rhizome of *Pteridium aquilinum*

---

[1]) K. GOEBEL: Entfaltungsbewegungen der Pflanzen. 2nd ed. Jena (1924), p. 229.

[2]) K. GOEBEL: Morphologische und biologische Bemerkungen 24. Die Abhängig-keit der Dorsiventralität vom Lichte bei einer *Selaginella*-Art. Flora (Jena) 108, 315–318 (1915).

[3]) DU BUY-NUERNBERGK l.c., p. 428.

[4]) K. GOEBEL: Einleitung in die experimentelle Morphologie der Pflanzen. Leip-zig–Berlin (1908).

[5]) J. BEHRENS: Über Regeneration bei den Selaginellen. Flora (Jena) 84, 159–166 (1897).

[6]) E. C. ANGST: Observations on *Pteris Aquilina*. Publ. Puget Sound Biol. Stat. 5, 261–263 (1925/1928).

may result in the production of smaller fronds. On the other hand BOODLE[1]) states that the leaves of the same fern vary according to the intensity of light in which they mature, and therefore can be classed as shade- and sun-leaves, the latter being xerophytic. (Compare herewith the existence of sun- and shade-fern species [2])).

Yet these facts do not represent the main points of heterophylly. They are shown much better by the following ones: While in most plants the factors which govern leaf development are relatively fixed, experiments on heterophylly in ferns, however, make it clear that the conditions for leaf formation which are permanent for one plant species may occur in others only during an early stage.

*Heterophylly and light.* — *Drynarieae.* In these ferns all leaves have the same shape in the early stages. The change in illumination during day and night has a certain influence on this shape which can be described as follows: light promotes the plate growth and inhibits the leaf stalk formation (cf. p. 317). There fore,when the development begins during daytime the plate growth begins directly, and bifurcation and stalk formation are suppressed. Undivided mantle leaves which grow over the substratum are formed. In darkness stalked assimilatory leaves develop. There are many transitions between these extreme and normal leaf formation, e.g., in *"Drynaria heraclea"*. This is a non-heterophyllic species. When this develops in the dark there is no suppression of leafstalk formation.

The initial leaf stages of other heterophyllic ferns have the same developmental potentialities. If their development is to be interpreted as in the case of the Drynarieae, we must suppose that the power to react to environmental conditions changes periodically.

The same is true for *Platycerium*. These ferns have also undivided mantle leaves pressed against the substratum and assimilatory leaves. Young plants, however, have only mantle leaves. Here also light has an influence on the shape. When, for example, *Platycerium bifurcatum* is cultivated in dimly lighted rooms it develops only assimilatory leaves. The development of mantle leaves depends on the presence of an adequate illumination.

Yet this is not the only factor involved. The mantle leaves can be regarded as inhibited forms of assimilatory leaves, this inhibition being furthered by more or less definite internal or external factors.

Heterophylly is also found in epiphytes, e.g., *Polypodium* species and *Asplenium Nidus*.

In the case of *Polypodium Willdenowii, P. rigidulum, P. quercifolium* [3]), a

---

[1]) L. A. BOODLE: The structure of the leaves of the Bracken (*Pteris aquilina*) in relation to environment. J. Linn. Soc. (Bot. 35, 659–669 (1901).

[2]) F. KANNGIESZER: Schattenfarne. Gartenwelt 22, 61, 284 (1918).

[3]) The three species belong to the genus *Drynaria*, according to CHRISTENSEN.

false distinction has been made between "fertile" and "sterile" leaves, by analogy with the difference between sporophyll and assimilatory leaves in *Osmunda, Helminthostachys, Botrychium, Stenosemia aurita, Matteuccia Struthiopteris, Onoclea sensibilis*, etc. The *sporophylls* of the latter group, however, show a reduction of chlorophyll.

In *Polypodium*, however, the fertile leaves ("sporophylls") are dark green. This situation can be explained by looking at the growth type of the leaves. The development of the leaves of these ferns is principally the same as in the case of *Platycerium*. The difference is that in *Polypodium* the differentiation between niche leaves and assimilatory leaves begins first within the vegetative region. The "sterile" leaves are the unstalked niche leaves, the "fertile" leaves the long (stalked) pinnate leaves. *Polypodium heracleum*, a transition form, shows both types of development in one leaf: the leaf foot forms a niche leaf, the frond an assimilatory leaf.

*Asplenium oceanicum.* Another type of heterophylly is found in "stolons" of this species (*Asplenium obtusilobum* Hook., not Desv.). These are leaves in which the blade is only slightly developed.

In accordance with the preceding discussion, it is probable that this suppression is governed by definite yet unknown factors.

**§ 11. Shoot formation by leaves.** — Shoot formation by leaves belongs to the normal course of development in many ferns. We can with GOEBEL [1]) distinguish two groups:

1. leaves on which shoots regularly develop under normal conditions (hygrophytic ferns e.g. *Diplazium celtidifolium, Asplenium viviparum* etc.);

2. leaves which develop shoots only when subjected to certain conditions (meso-xerophytic ferns).

A special example of the last group is shown by the ferns in which embryonic shoots occur only at the apex of leaves. These leaves have a more or less stolon-like shape e.g. in *Adiantum Edgeworthii, Aneimia rotundifolia, Camptosorus rhizophyllus, Asplenium oceanicum* (only special leaves), *A. Mannii*.

**§ 12. Growth of aerophores, leaf hairs of Salvinia.** — In some ferns, particularly in slime ferns, aerophores or pneumatophores (BOWER, MET-TENIUS) [2]) are described. These are pin-shaped tissues growing out from the persistent base of the older pinnae, e.g. of *Plagiogyria glauca*, or from the circinate apex of the leaf, e.g. of *P. pycnophylla*. In the latter case they alternate with the pinnae projecting through the covering of mucilaginous hairs.

---

[1]) See note p. 306.

[2]) F. O. BOWER; The Ferns, vol. 1, 204; vol. 2, 277 Cambridge (1923, 1926.) S. G. METTENIUS: Ueber einige Farngattungen II. *Plagiogyria*. Abh. Senckenbergschen Naturf. Ges. 2. (1858).

The pneumatophores occur in rudimentary form in other ferns (*Dryopteris Thomsonii, D. decussata and D. stipellata*; cf. HABERLANDT) [1]). Their whitish color is due to their air content. They are described as glands, (but have nothing to do with the slime layer: this is due to gland hairs), and as cone-shaped respiratory cells. They are proliferations of the aeration tissue which can be found in most ferns (GOEBEL) [2]). The factors involving their development and their function are still purely hypothetical (as is the function of the mucilage layer, the nectaries, etc.).

Other interesting structures are the leaf hairs of *Salvinia*. ANDREWS and ELLIS [3]) made some investigations on them and found the hairs to be positively chemotactic. Egg-yolk and raw egg, applied to the leaves, produced discolorations. The authors suggest that the insects which can always be found on the leaves, are possibly decomposed. In any case, some food is absorbed in this way.

## III. Regenerations in fern leaves.

**§ 13. Regeneration.** — Broadly speaking, there may be several types:
1. anatomical regeneration of the removed part: restitution.
2. formation of tissues or organs on parts of plants.
3. processes involved in wound healing.

In ferns particularly there is a continuous series ranging from typical regeneration to cases where development is caused in the normal plant by the action of internal or external factors.

Embryonic tissue especially is capable of regeneration, as when leaf tips are cut longitudinally, for example in *Blechnum brasiliense, Polypodium heracleum*, etc. Following regeneration, the plant shows a bifurcation, which also often occurs in fern leaves under normal conditions (GOEBEL 1902, p. 503 seq.) [4]).

Adult cells, on the other hand, were found incapable of regeneration as in the case where the tips of young and old leaves were cut. Even growth by stretching stops.

In one group shoot buds develop under any circumstances out of primordia which already can be found in the embryonic leaves (e.g. *Asplenium viviparum, Asplenium oceanicum, Adiantum Edgeworthii, Adiantum caudatum* and *Adiantum dolabriforme*).

In a second group, regeneration occurs only under certain conditions

---

[1]) G. HABERLANDT: Physiologische Pflanzenanatomie, 6th ed., p. 405, Leipzig (1924).

[2]) K. GOEBEL: Morphologische u. biologische Studien VIII. Ueber Schleimfarne und Aerophoren. Ann. Jard. bot. Buitenz. 36, 84–106 (1926).

[3]) F. M. ANDREWS and M. M. ELLIS: Some observations concerning the reactions of the leaf hairs of *Salvinia*. Bull. Torr. Bot. Club 40, 441–446 (1913).

[4]) K. GOEBEL: Über Regeneration im Pflanzenreich. Biol. Zbl. 22, 385–397, 417–438, 481–505 (1902).

(contact of the leaf tip with the soil, high humidity, etc.: *Aneimia rotundifolia, Asplenium rutaefolium* (GOEBEL 1902, p. 389) [1]).

In a third group, especially investigated by BEYERLE [2]), regeneration occurs in isolated etiolated primary leaves. Non-isolated primary leaves show no regeneration. Only in a normal culture of young *Polypodium heracleum* shoot buds are developed spontaneously. Inhibiting factors (etiolation, culture under water, ether gas) have no effect, neither have the cutting of the vascular system at the base of the leaf blade, the cutting of the roots or of one or all growing points. The only regenerated parts often formed were one or more apical buds.

In isolated primary leaves, regeneration takes place only in the young stages. Exceptions occur in *Hemionitis palmata* and *Adiantum capillus-Veneris*. This last fern shows, according to REINHOLD [3]), regeneration in leaves, stems, and in older fertile leaves. In *Hemionitis palmata* only shoot buds were regenerated. Further, PALISA [4]) found that *Cystopteris* species develop the regeneration buds in the base of the fronds.

BEYERLE [5]) [6]) observed regeneration in isolated leaves of the following species:

*Adiantum conicum*, form. *alata*, *A. fulvum* (not *Adiantum Dutrianum, elegans, reniforme, tenerum*);
*Alsophila tomentosa*;
*Aneimia densa* (not other *Aneimia* species);
*Anogramma leptophylla*;
*Asplenium Nidus, A. serratum*;
*Athyrium Filix-femina*;
*Blechnum capense* (= *Lomaria capensis*).
*Ceratopteris thalictroides*;
*Cibotium Schiedei*;
*Cyathea dealbata, C. medullaris* [3]);
*Davallia canariensis*;
*Dicksonia fibrosa* (not *D. squarrosa*);
*Diplazium melanocaulon*;
*Drynaria rigidula*;
*Thelypteris palustris*;
*Gymnogramme Laucheana* [6]), *G. Hookeri*;
*Hemionitis palmata*;

---

[1]) See page 325, note 4.

[2]) R. BEYERLE: Untersuchungen über die Regeneration von Farnprimärblättern. Planta 16, 622–665 (1932).

[3]) R. REINHOLD: Ueber regenerative Sprossbildungen bei *Adiantum capillus-Veneris* L. Mitt. Inst. allg. Bot. Hamburg 6, 309–337 (1926).

[4]) J. PALISA: Die Entwicklungsgeschichte der Regenerationsknospen, welche an den Grundstücken isolierter Wedel von *Cystopteris*arten entstehen. Ber. dtsch. bot. Ges. 18, 398–410 (1900).

[5]) See p. 328.

[6]) *Gymnogramme Laucheana* is said by K. DOMIN (Rozpr. II Tr. Ceske Akad. 38, No. 4, 1929) to be a horticultural form of *Pityrogramma chrysophylla*, and *G. Hookeri* to be *Pityrogramma adiantoides*. In addition, *Asplenium Sandwichium, Dicksonia Dayi*, and *Pellaea ternata* are likewise doubtful names employed by BEYERLE.

*Marattia alata*;
*Nephrolepis biserrata* (*Osmunda regalis* developed prothalloid outgrowth in one case);
*Polypodium aureum, P. Fendleri, P. heracleum, P. loriceum, P. musifolium*;
*Polystichum adiantiforme* (not *P. angulare*); *P. varium* (= *Lastraea opaca*);
*Pteris tremula*;
*Tectaria Maingayi*.

These ferns regenerate shoot buds, undifferentiated outgrowths, prothalloid shoots, and intermediate formations. All regenerated parts are always developed from the epidermis (exogenous or pseudo-endogenous development).

The undifferentiated outgrowths do not always die, but develop into buds or prothallia in *Aneimia densa, Hemionitis palmata, Pteris tremula*.

The intermediate formations between shoots and prothallia are planless proliferations with stomata, vascular elements, antheridia, and rhizoids, as in *Hemionitis palmata*, or more definitely shaped as in *Pityrogramma chrysophylla* (GOEBEL) [1]), *Ceratopteris thalictroides* (GOEBEL, BEYERLE) [1]), *Anogramma leptophylla, Davallia canariensis, Nephrolepis biserrata* and *Polypodium heracleum*. (BEYERLE).

The place of development of regenerated parts is variable, e.g., on the whole leaf in *Polypodium aureum* and *P. heracleum*, on the side, in *Polystichum adiantiforme, Thelypteris palustris, Ceratopteris thalictroides*, and on the tip only (GOEBEL) [2]) in *Adiantum Edgeworthii, Aneimia rotundifolia, Asplenium Mannii, A. oceanicum* and *Camptosorus rhizophyllus*.

The development of prothalloid organs on leaves shows that there is no inherent difference between the two generations, and consequently no clear relation to the chromosome number. For instance, BROWN [3]) obtained regenerations from the petiole of the leaf of *Phegopteris polypodioides*. A cellular mass resembling a prothallium was formed, from which rhizoids developed as well as true leaves and structures intermediate between prothallia and true leaves (LAWTON [4])).

There is, however, a more pronounced relation between the stage of development of the cells and the kind of regenerated organ. For instance:

(1) *Ceratopteris thalictroides* develops mainly *prothallia* on decapitated young plants with one or two leaves, and *shoot buds* on older plants.

2. In *Davallia canariensis* and in *Nephrolepis biserrata*, the shoot buds are mainly developed at the basal (older, see p. 314 foll.) part of the leaf, the prothallia at the apical part.

3. If only prothallia are regenerated, these are mainly developed at the apical (youngest) margins in the neighbourhood of the nerve endings.

---

[1]) See page 325, note 4.

[2]) See p. 306, note 2.

[3]) E. D. W. BROWN: Regeneration in *Phegopteris polypodioides*. Bull. Torr. Bot. Club 45, 391–397 (1918).

[4]) See note p. 331.

TABLE I

Regeneration of primary leaves as dependent upon different light conditions. (After Beyerle 1932)

| | Direct daylight | Daylight of slightly diminished intensity | Daylight of strongly diminished intensity | Dark or previous darkness | Potassium bichromate | Ammon. copper oxyd |
|---|---|---|---|---|---|---|
| Polypodium aureum | > 90% regener. shoot buds | 90–100% shoot buds | 50–70% | — (leaves died) | 90–100% sh. b. (etiolated shape) | slow regener. of great shoot buds |
| Polypodium Fendleri | — | — | —[1] | +[3] | — | |
| Polypodium heracleum | % | 47% regenerated sh. b. + int. f.[1] + proth. | | | | |
| Davallia canariensis | 50%–59% sh. b. + interm. form. + proth. | Older pinnated leaves = 40%; injured younger = 79%; the rest 65–76% sh. b. + proth. etc. | | — | 64% sh. b. + int. form. + proth. | 71% sh. b. + prolif. + proth. |
| Nephrolepis biserrata | | 70% regener. sh. b. + int.[1] f. + prot. | | + | + | 36% regen. sh. b. + int. f.[1] + (few) prot. |
| Aneimia densa | + | 1% indif.[2] form. | | | | |
| Hemionitis palmata | fewer new formations | many new formations | | 2% shoot buds | | |
| Ceratopteris thalictroides | mostly 10% sometimes up to 45% prothalloid formations; seldom interm. formations | | + | + | + | + |

[1] Intermediate formations.    [2] Indifferentiated formations.    [3] Leaves died.    [4] No influence was observed.

TABLE II

*Polypodium aureum*

*New formations and regenerations on regenerated and then isolated leaves (After Beyerle 1932)*

| New formations on regenerated leaves | Greater regenerated leaves > 5 mm | Smaller regen. leaves 3–6 mm | Small reg. leaves 1,5–3 mm | Regenerated leaves, 2–5 mm on moist filterpaper | | |
| --- | --- | --- | --- | --- | --- | --- |
| | on filter paper with nutrient sol. in daylight of a little diminished intensity | | | Direct daylight | Daylight of a little diminished int. | Daylight of strongly diminished int. |
| Number of leaves. . . | 133 | 160 | 34 | 104 | 40 | 86 |
| Died without regeneration. . . | 14 | 83 | 6 | 16 | 1 | 13 |
| Fresh but without regeneration. . . | 9 | 34 | 5 | 9 | 11 | 11 |
| Tubers and buds. . . | 103 } 109 | 30 } 34 | 8 } 13 | 60 } 74 | 23 } 26 | 36 } 52 |
| Non-prothalloid proliferations . . | 6 | 4 | 5 | 14 | 3 | 16 |
| Prothalloid-apical outgrowths, and pro-thallia on margin . | 1 | 9 | 10 | 5 | 2 | 9 |
| Intermediate formations. . . . . . | — | — | — | — | — | — |

4. When development is s u p p r e s s e d by internal or external factors, e.g. light, only prothallia are regenerated. This is also the case in fern leaves which normally develop shoot buds. (*Drynaria rigidula, Polypodium aureum* (Tab. I, II), *P. heracleum* and *P. musifolium*).

From these points the conclusion can be drawn that buds can be obtained only when the development of the cells has passed a certain stage. In all previous stages or in all cases where inhibition of development takes place, only sporangia can be found. An analogy with the action of organisers, however, has not yet been worked out.

**W o u n d  h o r m o n e s.** — With regard to the regenerations in the neighbourhood of wounds, not all primary leaves of ferns react alike (BEYERLE) [1]). In *Polypodium heracleum* and in *Polypodium aureum* regenerated parts are formed around the wound.

In most plants no regenerated parts are developed after wounding. *Davallia canariensis* rots around the wound, and then forms prothallia. Rotting leaves of *Polypodium heracleum* and *Polypodium aureum* also often form prothallia. Expressed sap, however, has no influence.

Concerning the influence of wounds on the underlying tissue we refer to the observations of HOLDEN [2]) on the wound reactions of the petioles of *Pteris aquilina* (= *Pteridium aquilinum*). These petioles often show wound-scars. The wounds are mostly very superficial, not penetrating the subepidermal sclerenchyma, but some are deeper seated.

Usually, these wound reactions are characterized:

1. by a compensatory local thickening and partial or complete lignification of the cortical parenchyma which may or may not be accompanied by elongation,

2. by the local delignification of the subepidermal sclerenchyma,

3. by a deposit of tannin in the cell walls in the affected area.

However, wound reactions in the tissues composing the vascular strands are rare, and where they do occur, are confined to the starch sheet and conjunctive parenchyma, which thicken and may elongate and divide.

**T a b l e  o f  r e g e n e r a t i o n s,  a r r a n g e d  i n  o r d e r  o f place  of  development.** — 1. *Regeneration on leaves.*

*Adiantum capillis-Veneris*: non-isolated (etiolated) leaves, isolated primary leaves and adult leaves, parts of leaves, stems and roots develop shootbuds, but never prothallia or prothalloid formations (REINHOLD [3])).

,,      *caudatum*: shootbuds preformed on adult leaves always develop (GOEBEL [4]), KUPPER [5]).

---

[1]) See note p. 326.

[2]) H. S. HOLDEN: Further observations on the wood reactions of the petioles of *Pteris aquilina*. Ann. of. Bot. 30, 127–133 (1916).

[3]) See p. 326.

[4]) See p. 325, note 4.

[5]) W. KUPPER: Ueber Knospenbildung an Farnblättern. Flora (Jena) 96, 337–408 (1906).

*Adiantum conicum (alatum), fulvum*: isolated leaves regenerate (BEYERLE [1])).
    ,,    *dolabriforme, rotundifolium*: shootbuds preformed on adult leaves always develop (GOEBEL [1])).
    ,,    *Edgeworthii*: shootbuds preformed on adult leaf apex always develop (GOEBEL [1]), KUPPER [2]).
*Alsophila tomentosa*: isolated leaves regenerate (BEYERLE [1])).
    ,,    *van Geertii*: primary (?) leaves develop prothallia and intermediate structures (GOEBEL [3])).
*Aneimia densa*: isolated primary leaves develop undifferentiated proliferations ($\pm$ 2% regeneration); Tab. I (BEYERLE [1])).
    ,,    *Dregeana*: detached primary leaves develop prothalloid proliferations (GOEBEL [3])).
    ,,    *rotundifolia*: shoot buds, preformed on adult leaf apex develop only in contact with soil (GOEBEL [1]), KUPPER [2]).
"*Anogramma*" *chrysophylla*: isolated leaves regenerate prothallia (BEYERLE [1]).
*Anogramma leptophylla*: isolated leaves regenerate formations intermediate between shoots and prothallia (BEYERLE [1]).
"*Aspidium capense*": young and older primary leaves develop prothalloid formations (BEYERLE [1]).
    ,,    *macrophyllum*: preformed shoot buds are developed on adult leaves under certain conditions (GOEBEL [1])).
    ,,    *marginale*: first, second and third leaf develop prothallia and intermediate structures (LAWTON [4])).
"*Asplenium*" *celtidifolium*: shoot buds, preformed on adult leaves develop always (GOEBEL [1])).
*Asplenium dimorphum*: leaves develop prothallia (KUPPER [2])).
*Asplenium Mannii*: shootbuds, preformed on apex of adult leaves, are more or less metamorphosed in stolons and develop always (GOEBEL [1]), KUPPER [2])).
*Asplenium Nidus*: primary leaves develop shootbuds on both sides. In this case large fresh leaves sometimes develop prothallia (BEYERLE [1])).
    ,,    *oceanicum*: shootbuds, preformed on apex of adult leaves always develop in more or less metamorphosed stolons (GOEBEL [1]), KUPPER [2])).
    ,,    *rutaefolium*: shootbuds, preformed on apex of adult leaves develop only in contact with soil (GOEBEL [1])).
    ,,    *serratum*: primary leaves develop prothallia on margins (BEYERLE [1])).
    ,,    *viviparum*: attached leaves always develop preformed shoot buds (KUPPER [2])).
*Athyrium Filix-femina*: $\pm$ 15% of decapitated young plants only develop a new apical bud (BEYERLE [1])).
    ,,    *filix* var. *clarissima*: leaves develop prothalloid structures (BOWER [5]), DRUERY [6]), FARMER–DIGBY [7])).
*Blechnum brasiliense*: the embryonic part of a leaf cut longitudinally develops small leaflets (BEYERINCK [8])).

---

[1] R. BEYERLE see p. 326; K. GOEBEL see p. 325, note 4 and p. 326.

[2] See page 330 note 5.

[3] K. GOEBEL (1907) see note p. 333.

[4] E. LAWTON: Regeneration and induced polyploidy in ferns. Amer. J. Bot. 19, 303–333 (1932).

[5] F. O. BOWER: On apospory in ferns, with special reference to Mr. C. T. DRUERY's observations. Journ. Linnean Soc. London (Bot) 21, 360–368 (1884).

[6] C. T. DRUERY: Observations on a singular mode of development in the Ladyfern. Further notes on a singular mode of reproduction in *Athyrium Filix-femina*. Ibidem 21, 354–357; 358–359 (1884).

[7] J. B. FARMER and L. DIGBY: Studies in apogamy and apospory in Ferns. Ann. of Bot. 21, 161–193 (1907).

[8] M. W. BEYERINCK: Over regeneratie-verschijnselen etc. (see note p. 334).

*Blechnum capense*: see *Lomaria capensis*.

*"Campylogramme Trollii"*: primary leaves develop shootbuds and prothallia ($\pm$ 2% regeneration BEYERLE [1])).

*Ceratopteris thalictroides*: primary leaves develop adventitious buds at apex. Petioles of isolated primary leaves develop prothallia and intermediate structures (BALLY [2]), BEYERLE [1]), GOEBEL 1930, p. 1225 [3]), KOEHLER [4])).

*Cyathea medullaris*: 6% of primary leaves develop undifferentiated marginal outgrowths (BEYERLE [1])).

„    *dealbata*: isolated leaves regenerate (BEYERLE [1])).

*Cibotium Schiedei*: 6% of primary leaves develop prothallia on margin (BEYERLE [1])).

*Cystopteris bulbifera*: shoot buds preformed on adult leaves always develop KUPPER [5])).

*Cystopteris fragilis*: formations of rhizoids and sporophytic buds develop on attached leaves (HEILBRONN [6])).

*Davallia canariensis*: primary leaves develop shootbuds, intermediate formations and prothallia (at the apex) (BEYERLE [1])).

*Dicksonia Dayi* forma *obtusiloba*: primary leaves develop prothalloid formations on margin (BEYERLE [1])).

*Dicksonia fibrosa*: isolated young primary leaves only develop prothalloid formations on margin (BEYERLE [1])).

*Diplazium celtidifolium*: see *"Asplenium" celtidifolium*.

„    *melanocaulon*: some primary leaves develop tubers at base (BEYERLE [1])).

*"Drynaria heraclea"*: primary leaves develop shootbuds, intermediate formations, prothallia and isolated rhizoids; large regenerated leaves develop only shootbuds, small or dying leaves develop prothallia (BEYERLE [1])).

„    *rigidula*: primary leaves develop only shootbuds (BEYERLE [1])).

*Dryopteris Filix-mas*: see *"Lastraea pseudomonas"*.

„    *hirtipes*: see *"Nephrodium hirtipes"*.

„    *marginalis*: see *"Aspidium marginale"*.

„    *Phegopteris*: see *"Phegopteris polypodioides"*.

„    *Thelypteris*: already developed primary leaves develop shootbuds on petiole; undeveloped primary leaves develop prothalloid formations at apex (BEYERLE [1])).

*Equisetum*: aerial shoots are developed (PRAEGER [7])).

*"Gymnogramme chrysophylla"*: primary leaves develop prothalloid structures (GOEBEL [8])).

„    *farinifera*: petioles of attached leaves develop prothallia (WORONIN [9])). According to DOMIN this is a variety of *Pityrogramma chrysophylla*.

„    *Hookeri*: primary leaves develop abundant prothalloid outgrowths on margin (BEYERLE [1])). See footnote No. 6, p. 326.

---

[1]) See p. 326.

[2]) W. BALLY: Ueber Adventivknospen auf Primärblättern von Farnen. Flora (Jena) 99,2 (1909).

[3]) See p. 305, note 3 and p. 334, note 8.

[4]) E. KOEHLER: Farnstudien: Flora (Jena) 113, 337 (1920).

[5]) See p. 330, note 5.

[6]) A. HEILBRONN: Apogamie, Bastardierung und Erblichkeitsverhältnisse bei einigen Farnen. Flora (Jena) 101, 12 (1910).

[7]) R. L. PRAEGER: Propagation from aerial shoots in *Equisetum*. Journ. of Bot. 72, 175–176 (1934).

[8]) See note p. 322 (K. GOEBEL 1908).

[9]) H. WORONIN: Apogamie und Aposporie bei einigen Farnen. Flora (Jena) 98, 101–162 (1908).

*"Gymnogramme Laucheana"*: primary leaves develop prothalloid outgrowths and shoots in the neighbourhood of the ends of nerves and at the leaf stalks (BEYERLE [1])). See footnote No. 6, p. 326.

*Hemionitis arifolia* ("*H. cordata*"): shootbuds preformed on base of adult leaf-blades develop only on old dying or isolated leaves (GOEBEL [1])).

,,     *palmata*: many primary leaves develop shootbuds, intermediate structures and prothallia on all parts; more developed leaves develop adventitious buds which are already present. Prothalloid structures are only developed, when adventitious bud with petiole is taken away (GOEBEL [1]), BEYERLE [1])).

*"Lastraea opaca"*: 8% of leaves develop small prothalloid apical outgrowths, both dying soon after that (BEYERLE [1])).

*"Lastraea (Nephrodium) pseudomonas* var. *cristata"*: leaves develop prothalloid structures (DRUERY [2])).

*"Lomaria capensis"*: ± 10% of primary leaves develop shortliving prothalloid structures (BEYERLE [1])).

*Lycopodium inundatum*: isolated leaves develop shootbuds (GOEBEL [3])).

*Marattia alata*: at the base of leaves and on isolated chaff scales regeneration of buds and rhizoids occurs (BEYERLE [1])).

*"Nephrodium hirtipes"*: see STEIL [1]).

*Nephrolepis biserrata*: primary leaves develop shootbuds, prothallia and intermediate structures (BEYERLE [1]), KUPPER [5])).

*Notolaena flavens*: primary leaves develop new formations (WORONIN [6])).

*Notolaena Marantae*: blade and petiole of leaves develop shootbuds (WORONIN [6]))

*Onoclea*: leaves develop shootbuds (KUPPER [5])).

*Osmunda regalis*: primary leaves develop prothallia (GOEBEL 1928, p. 535 [7])).

*Pellaea nivea*: attached primary leaves develop prothallia (WORONIN [6])).

*Phegopteris polypodioides*: isolated adult(?) petioles develop prothallia, intermediate structures and shootbuds (BROWN [8]).

*Phyllitis Scolopendrium* var. *crispa*: leaves develop prothallia (KUPPER [5])).

*Pityrogramma chrysophylla*: see "*Anogramma*" *chrysophylla* and "*Gymnogramme*" *chrysophylla*.

*Platycerium bifurcatum (alcicorne), grande, Hillii*: leaves develop shootbuds in strong light and prothallia in dim light (KOEHLER [9])).

*Polypodium heracleum*: see "*Drynaria heraclea*".

,,     *aureum*: isolated primary and secondary leaves develop shootbuds, intermediate structures and prothallia, mainly on abaxial side (Table I, II). Older leaves develop prothalloid structures (BEYERLE [1]), GOEBEL [10]), HEILBRONN [11])).

---

[1]) See p. 326 (esp. notes 1 and 2).

[2]) C. T. DRUERY: Notes upon an aposporous *Lastraea (Nephrodium)*. Journ. Linnean Soc. London (Bot) 29, (1892).

[3]) K. GOEBEL: Ueber Prothallien und Keimpflanzen von *Lycopodium inundatum*. Bot. Ztg. 45, 161, 177 (1887).

[4]) W. N. STEIL: Apogamy in *Nephrodium hirtipes* Hk. Ann. of Bot. 33, 109–132 (1919).

[5]) See p. 330, note 5.

[6]) See p. 332, note 9.

[7]) K. GOEBEL: Organographie 1st part, 3rd ed. Jena (1928).

[8]) See p. 327.

[9]) See p. 332, note 4.

[10]) K. GOEBEL: Experimentell-morphologische Mitteilungen. I. Sitzgber. Ak. Wiss. München, math.-physik. Kl. 37, 119–136 (1907).

[11]) A. HEILBRONN: Ueber experimentell erzeugte Tetraploidie bei Farnen. Z. ind. Abst. u. Vererbungsl. Suppl. 2, 830 (1928).

*Polypodium Fendleri*: primary leaves develop shootbuds and rudimentary leaves (BEYERLE [1])).

,,    *lycopodioides*: older leaves develop prothalloid structures (GOEBEL [2])).

,,    *loriceum*: primary leaves develop shootbuds, prothallia and prothalloid structures (BEYERLE [1])).

,,    *musifolium*: primary leaves regenerate (only 50–60%) shootbuds and prothalloid structures (BEYERLE [1])).

,,    spec.: primary leaves develop prothalloid structures on margin, intermediate structures at base. No regeneration in older primary leaves (GOEBEL [2])).

,,    *vulgare*: see KOEHLER [3]).

*Polystichum adiantiforme*: see "*Aspidium capense.*"

,,    *angulare* var. *pulcherrimum*: leaves develop prothallia (BOWER [3]) DRUERY [3]), KUPPER [4])).

,,    *varium*: see "*Lastraea opaca.*"

*Pteris longifolia*: blades and petioles of primary leaves develop prothallia (GOEBEL [2])).

,,    *cretica* var. *albo-lineata*: first, second and third leaves produce prothallia.

,,    *tremula*: only about 20% of isolated young primary leaves develop nothing else but prothalloid structures (BEYERLE [1])).

"*Scolopendrium vulgare* var. *crispum*": isolated leaves, in contact with soil develop prothallia (DRUERY [5])). The same as *Phyllitis Scolopendrium*.

*Selaginella denticulata, Galeottiana, inaequalifolia, laevigata, Martensii, uncinata*: cell proliferations are developed on decapitated apex (BEHRENS, BEYERINCK, PFEFFER [6])).

*Tectaria Maingayi*: see "*Campylogramme Trollii*".

,,    *martinicensis*: see "*Aspidium macrophyllum*".

*Trichomanes Kraussii*: isolated adult leaves develop prothallia (KUPPER [4]), WORONIN [7])).

*Woodwardia virginica*: first, second and third isolated leaf develop prothallia (LAWTON [7])).


2. *Regenerations on roots and growing points.* — In general, the shoot formation on roots can be explained, according to GOEBEL [8]) by a "Sprossbildende Substanz". For instance, in *Ophioglossum*, the shoot formation

---

[1]) See p. 326.

[2]) K. GOEBEL (1908): see p. 322.

[3]) See p. 332, note 4 (KOEHLER) respectively p. 331 (BOWER, DRUERY).

[4]) See p. 330, note 5.

[5]) C. T. DRUERY: On a new Instance of Apospory in *Polystichum angulare* var. *pulcherrimum*. Journ. Linnean Soc. London (Bot). 22, (1886); See also p. 41a.

[6]) J. BEHRENS: see p. 322.

M. W. BEYERINCK: Beobachtungen und Betrachtungen über Wurzelknospen und Nebenwurzeln. Natuurk. Verh. Ak. Wetensch. Amsterd. 25 (1886); see also: Verzamelde Geschr. 2, 7–122. Delft (1921)). — Over regeneratieverschijnselen aan gespleten vegetatiepunten van stengels en over bekervorming. Neder. kruidk. Arch. Nijmegen ser. 2, vol. 4, part. 1, p. 63–105 (1883); see also: Verzamelde Geschr. 1, 293 Delft (1921).

W. PFEFFER: Die Entwicklung des Keimes der Gattung *Selaginella*. HANSTEIN's Bot. Abh. 1, 67 (1871).

[7]) See p. 332, note 9 (WORONIN) respectively p. 331 (LAWTON).

[8]) K. GOEBEL: Organographie der Pflanzen, 1st ed., p. 39 Jena (1898). Cf. also 3rd ed. (see p. 305, note 3) p. 1221 (1930).

does not occur at the shoot pole, but at the root pole itself. This reaction shows itself clearly, especially when the tip of the plant is removed beforehand, or when the respective root tip has been cut off to a length of some centimetres from the plant. Other parts of the root can regenerate as well, but to a lesser extent than the tip. These phenomena are mentioned by SACHS [1]) for the first time. ROSTOWZEW [2]) described the same thing with regard to the species *Diplazium esculentum, Platycerium bifurcatum*, and FRIES [3]) mentions it for the rootbearers of *Salvinia* [4]).

We are quoting some characteristic examples for this kind of regeneration.

*Antrophyum plantagineum*: shoot buds formed on some distance of root apex, develop. (GOEBEL [5]).
*Diplazium esculentum*: shoot buds formed on apex of root, develop. (GOEBEL [5]), LACHMANN [cf. ROSTOWZEW] [2]).
*Hecistopteris pumila*: shoot buds develop on roots (GOEBEL [6]).
*Ophioglossum pedunculosum*: non preformed shoot buds are formed on apex of roots. They develop always, on other parts of roots only after isolation (GOEBEL [7])).
    ,,    *vulgatum*: same. They develop mainly after cutting (GOEBEL [7]) 1902, p. 492; POIRAULT [8])).
    ,,    in general: roots develop shoot buds (ROSTOWZEW [9])).
*Platycerium bifurcatum, grande, Hillii, Willinckii, Stemaria*: shoot buds are formed on apex of roots, they develop from presumably preformed buds (GOEBEL [7]) 1902; 1930, p. 1222; ROSTOWZEW [2]), SACHS [1]).
*Polypodium heracleum*: longitudinally cut growing point develops leaflets on both sides of the halved growing points (GOEBEL [7])).
*Selaginella Martensii*: After removal of the growing point of the leaf the rootbearers grow out (FRIES [3])).

3. *Regeneration on prothallia*. — In heart-shaped fern prothallia, e.g. *Polypodiaceae*, the halved growing point develops a new wing. If the growing point is excised then many prothallia develop.

---

[1]) J. SACHS: Vorlesungen über Pflanzenphysiologie. 1st ed. p. 29 (1882); 2nd ed. p. 27 (1887) Leipzig.

[2]) S. ROSTOWZEW: Beiträge zur Kenntnis der Gefäßkryptogamen I. Flora 73, 155–168 (1890).

[3]) R. E. FRIES: Ett bidrag till kännedomen om *Selaginella*-Rothbärarne. (Ein Beitrag zur Kenntnis der Wurzelträger von *Selaginella*). Svensk. bot. Tidskr. 5, 252–259 (1911).

[4]) The direction of these rootbearers is independent of environmental conditions, i.e., they are "exotropic".

[5]) See p. 306, note 2.

[6]) K. GOEBEL: Archegoniatenstudien VIII. *Hecistopteris*, eine verkannte Farngattung. Flora (Jena) 82, 67–75 (1896).

[7]) See note 8 p. 334 and p. 306.

[8]) G. POIRAULT: Recherches anatomiques· sur les Cryptogames vasculaires. Ann. sci. nat. bot. 7. sér. 18, 113–256; 148 (1893).

[9]) S. ROSTOWZEW: Beitrag zur Kenntnis der Ophioglosseen. Moskau (1892).

Older prothallia of *Asplenium Trichomanes, Athyrium alpestre, Polypodium punctatum* develop small adventitious prothallia at the apical notch (NAGAI, STEIL, WALDMANN) [1]).

Apogamic prothallia, e.g. *"Aspidium* (= *Dryopteris*) *remotum"*, *Hemionitis palmata, Lygodium japonicum*, develop embryos on different parts of the prothallium, especially on the midrib (GOEBEL [2]), WALDMANN [1]), REINHOLD 1926, p. 312 [3]) and the literature quoted by the latter).

*Dryopteris dilatata*: develops sporangia on prothallia (LANG [4])).
*Osmunda* spec.: same (PACE [5])).
*Phyllitis Scolopendrium*: same (LANG [4])).
According to LANG in these cases the sporangia only seem to be directly attached to the prothallia, in fact they originate from a sporophyte, though very small, which develops on the prothallium.
*Hecistopteris pumila*: prothallia develop brood buds (GOEBEL [6])).
*Hymenophyllum* spec.: same.
*Lycopodium Phlegmaria*: same (TREUB, GOEBEL 1930, p. 1067 [7])).
*Monogramma*: same.
*Vittaria:* same (GOEBEL 1930, p. 1093).

## IV. Secondary thickening

**§ 14. Secondary thickening.** — Occurs (in living ferns) only in the tuberous short axes of the *Isoetinae*, the shoot bases of the *Psilotaceae*, and is slightly indicated in stems of some *Equisetinae*, as for instance in *Equisetum maximum*. It occurs further in some *Ophioglossales* (e.g. *Botrychium*) by way of cambial activity, whereas in the main group of the Filicales it has not been observed anywhere.

On the other hand most of the fossil ferns showed secondary thickening, e.g. the *Lepidodendrales* with a meristem like a cork cambium, the stems and roots of *Calamitaceae*; the *Sphenophyllales* and the *Pteridospermae*,

---

[1]) J. NAGAI: Physiologische Studien über Farnprothallien. Flora 106, 281–330 (1914).

W. N. STEIL: Studies of some new cases of apogamy in ferns. Bull. Torr. Bot. Club 45, 93–108 (1918); Vegetative reproduction and aposporous growth from the young sporophyte of *Polypodium irioides*. Ibidem 48, 203–205 (1921): The development of prothallia and antheridia from the sex organs of *Polypodium irioides*. Ibidem 48, 271–277 (1921).

H. WALDMANN: Beiträge zur Entwicklungsgeschichte der Prothallien einheimischer Polypodiaceen. Diss. Marburg (1928).

[2]) See p. 334, note 8.

[3]) See p. 326.

[4]) W. H. LANG: On apogamy and the development of sporangia upon fern prothalli. Philos. Trans. Roy. Soc. 190 B. London (1898).

[5]) L. PACE: Some peculiar fern prothallia. Bot. Gaz. 50, 49–58 (1910).

[6]) See p. 334, note 8.

[7]) K. GOEBEL: see p. 305 (note 3).

M. TREUB: Etudes sur les Lycopodiacées. II. Le prothalle du *Lycopodium Phlegmaria* L. Ann. Jard. bot. Buitenz. 5, 87, 98. (1886).

the former having an axial vascular bundle containing secondarily formed scalariform-reticulate and bordered pit tracheids, the latter with cambial growth producing a wood structure inwards and a bark zone (phloem and cortical tissue) outwards.

## V. Movements in ferns which are not caused by growth

**§ 15. Diurnal movements of the leaves of Marsilia.** — Nastic leaf movements which are not caused by growth are seldom found in ferns. One of the most remarkable examples are the diurnal movements of the petioles of adult leaves of *Marsilia*. This marsh plant has leaves like a clover plant or *Oxalis* with 4 leaflets: a fairly long leaf stalk, bearing at its end 2 pairs of petioles originating from the bifurcation of the primary leaves. The petioles are attached by pulvini to the main stalk.

In daylight the petioles are horizontally stretched, but at night they rise, and fold together two by two, the terminal pair being enclosed by the two lower leaflets.

They also rise in full sunlight (,,Tagesschlaf''). NUERNBERGK [1]) has shown that these "parheliotropic movements" are caused by strong short wave-length radiation, especially by wave-lengths < 5000 Å, and also ultra-violet rays (mercury vapour-lamp). The long wave-lengths of the spectrum including infra-red and heat are entirely ineffective.

A change of position or shape of the chloroplasts does not seem to occur in strong light.

With regard to the motile mechanism of the pulvini no investigations are known, but we may assume that there is no significant difference between the behaviour of *Marsilia* and the behaviour of the pulvini of the Leguminosae and other plants capable of nastic response. Therefore the difference in osmotic pressure caused by a change of permeability of the cell walls in the (illuminated) upper and (non illuminated) lower layer of the parenchyma tissue is likely to cause the unequal shortening and elongation respectively of the two sides of the pulvinus, which brings about the movement [2]).

**§ 16. Mechanism of the annulus of the Polypodiaceae and related phenomena.** — Another movement not caused by growth and especially characteristic of ferns is the dehiscence of the sporangium of the *Polypodia-*

---

[1]) E. NUERNBERGK: Beiträge zur Physiologie des Tagesschlafs der Pflanzen. Bot. Abh. No. 4. Jena (1925).

[2]) The increase of permeability of the cell walls is probably due to light. Compare with respect to this: H. WEIDLICH: Die Bewegungsmechanik der Variationsgelenke, Bot. Arch. 28, 219 (1930); M. BRAUNER: Untersuchungen über die Lichtturgorreaktionen des Primärblattgelenkes von *Phaseolus multiflorus*, Planta (Berl.) 18, 288 (1932) and the discussion by H. v. GUTTENBERG in: Fortschritte d. Bot. 4, 269 (1935).

*ceae.* The sporangia of this family of the *Filicinae* are shaped like a stalked lentil and consist of only one cell layer enclosing the spores. The sporangium cells generally are flat-polyhedral and have a thin wall with exception of the edge of the "lentil"; here the cell walls are partially thickened like a horseshoe and, starting from the stalk they form a kind of ring, the so called annulus.

The inner walls of that annulus are particularly thickened while the radial walls become thinner towards the outside. When mature the cells contain water and as soon as they loose it by drying the thin outer walls draw inwards causing a shortening of the peripheral edge of the annulus. At the end of this process the annulus opens at a preformed point, the so-called "stomium" which lies at the end of the thickened cells, and consists of two thin-walled cells, the *lip cells*. Then the annulus may nearly double back upon itself, pulling with it a mass of ripe spores. Finally, the annulus suddenly returns to its original position, at the same time forcibly throwing out the spores.

According to STEINBRINCK [1]) the cause of this phenomenon is to be found in the considerable adhesion of the water in the cells to their wall membranes, and further in the great force with which the water molecules mutually cohere. Hence, as soon as the water diminishes by evaporation the outer cell walls, being more yielding than the thickened inner ones, are pulled inwards by adhesion and cohesion. Finally, the water column breaks up: this is the moment where the annulus jerks back. Now its cells appear obscure, they contain but little water distributed along their walls besides a cavity apparently filled with air (but void of air, if the process of dehiscence took place in vacuum).

From this we may conclude that it is of no importance whether the annulus cell walls generally are permeable to air or not — and in fact they are —, the matter of importance is only the coherence of the water as well as its adhesion to the walls. These factors prevent the air from entering into the cell cavity during the first stage of the dehiscence when the annulus has not yet returned, since every smal air bubble would necessarily interrupt the adhesion between water and wall.

With regard to the strength of the cohesion force it may be mentioned that according to RENNER and URSPRUNG [2]) its value can amount up to

---

[1]) C. STEINBRINCK: Der Oeffnungs- und Schleudermechanismus des Farn-sporangiums. Ber. dtsch. bot. Ges. 15, 86 (1897); — Kohäsions- oder hygroskopischer Mechanismus? Ibidem 21, 217–229 (1903); A. URSPRUNG: Der Oeffnungsmechanismus der Pteridophyten-Sporangien. Jb. Bot. 38, 635–666 (1903); — Beiträge zum Bewegungsmechanismus einiger Pteridophyten Sporangien. Ber. dtsch. bot. Ges. 22, 73–84 (1904).

[2]) A. URSPRUNG: Ueber die Kohäsion des Wassers im Farnanulus. Ber. dtsch. bot. Ges. 33, 153–162 (1915). O. RENNER: Theoretisches und Experimentelles zur Kohäsionstheorie des Wasserbewegung. Jb. Bot. 56, 617–667 (1915).

350 atmospheres, a value which also has been proved true in physical experiments in vitro.

Another phenomenon, probably related to the dehiscence of the annulus is the dehiscence and dissemination of the macrospores in *Selaginella* which also may be caused, according to GOEBEL [1]), by some mechanism, depending upon cohesion or shrinking. For details we refer to the investigations of GOEBEL.

Finally, attention may be called to the curling-mechanism of some fern-fronds (e.g., *Polypodium polypodioides, P. vulgare, Ceterach officinarum, Asplenium Trichomanes, A. Ruta-muraria*), which has been investigated by PESSIN [2]) and by SCHMIDT [3]). The leaves of *Polypodium* and *Ceterach* under dry conditions loose the greater amount of water from their lower surface with the result that a curling of the blades occurs, which leaves the lower epidermis exposed and the upper epidermis concealed on the inside. In *Asplenium*, however, the direction of movement is opposite, there the lower surface is curled concavely.

The *curling* of the living leaves is mainly the result of a cohesion (and shrinkage)-mechanism, though in the beginning of the movement also a diminution of the turgor pressure may play a little part. The *expansion* is, in the case of dead leaves, entirely due to imbibition of the cell walls, while in the living ones it may also be due to imbibition and (osmotic) turgor pressure of the cell contents.

## § 17. Hygroscopic movements of spores of Equisetum. — In close relation to the cohesion mechanism of the annulus are the hygroscopic movements of spores of *Equisetum*. The wall of those spores consists not only of endosporium (intine) and exosporium (exine) but also of a third outermost layer, the so-called perispore (perinium), which also originates from the periplasmodium. In the mature spore this perispore ruptures in the form of 2 screw-shaped, parallel, coiled straps, broad bladed at their ends. Their centre, however, remains attached to the exosporium at one place forming in such a way 4 ribbon-like appendages, the so-called elaters. These elaters when wet swell on their outside more than on their inner side and thus will bend, coiling around the spore. When drying out they reversely uncoil.

These movements occur automatically when atmospheric humidity changes (for instance also by breathing on them). They are caused by the different power of imbibition existing on the two sides of the elaters, which

[1]) K. GOEBEL: Archegoniatenstudien IX. Sporangien, Sporenverbreitung und Blütenbildung bei *Selaginella*. Flora (JENA) 88, 207–228 (1901).

[2]) L. J. PESSIN: A physiological and anatomical study of the leaves of *Polypodium polypodioides*. Amer. J. Bot. 11, 370–381 (1924).

[3]) W. SCHMIDT: Ueber den Einrollungsmechanismus einiger Farnblätter. Beih. Bot. Cbl. 26, I, 476–508 (1910).

is likely to depend on colloidal-chemical differences. Possibly they serve for a better dissemination of the air-borne spores, depositing them in moist places favourable for their germination.

**§ 18. Movements of the chloroplasts.** — The tactic movements of chloroplasts in ferns do not differ in any respect from the corresponding movements known in other plant classes [1]). The same is to be said about the change of shape of chloroplasts. Therefore, only a few statements will be given [2]).

As in nearly all other higher developed plants, the assimilatory cells of ferns contain many but small chlorophyll plastids. Exceptions are found in *Selaginella Martensii, S. grandis, S. Kraussiana* and *S. caesia.*

In *Selaginella Martensii* and *S. grandis* every cell contains but one large, trough-shaped chloroplast lining the inner part of the cell walls. In *S. Kraussiana* and *S. caesia* there are two equal chloroplasts, which correspond, however, according to their position to the two halves of the one trough-shaped chloroplast of *S. Martensii.*

As in mosses, the shape of the chloroplasts in ferns is rounded and spherical or hemispherical in strong sunlight and during long darkening, but in diffuse or dimmed light the chromatophores become flattened and more lenticular.

Usually combined herewith are tactic movements of the chloroplasts, particularly conspicuous in prothallia. Here the chloroplasts line the outer cell walls parallel to the surface of the organ when light is diffuse or weak, the side walls being then without chlorophyll (Epistrophe according to FRANK) [3]).

In prothallia with only one cell layer, but also in certain spongy parenchyma and palisade tissue the epistrophe is often replaced by the diastrophe, as SENN (p. 74 foll., 93 and elsewhere) [2]) calls the position of the chloroplasts where these border the two opposite (outer) cell walls parallel to the organ surface, the walls facing the main light being somewhat preferred.

In dimmed sunlight the chloroplasts partially leave the surface walls and pass over to the side walls (FRANKS' Apostrophe). Finally, in excessive sunlight the chloroplasts leave the surface walls as well as the side walls

---

[1]) See for instance Manual of Bryology, p. 231. The Hague (1932).

[2]) See further G. SENN: Die Gestalts- und Lageveränderung der Pflanzen-Chromatophoren. Leipzig (1908).

[3]) B. FRANK: Ueber die Veränderung der Lage der Chlorophyllkörner und des Protoplasmas in der Zelle, und deren innere und äußere Ursachen. Jb. Bot. 8, 216 (1872).

and cluster together in the inner part of the cells, forming one or more compact lumps (SCHIMPER's Systrophe) [1].

Evidently these movements serve to bring the chloroplasts into the best conditions of light with regard to assimilation and protection against too intensive radiation. Both the tactic movements and the change of shape of the chloroplasts are caused only by the blue-violet rays, the yellow-red spectral part not being efficacious and acting like darkness.

Concerning the way the tactic movements are carried out several theories have been put forward, but it would lead too far to deal with them particularly in this chapter.

**§ 19. Tactic movements of spermatozoids.** — Except the *Cycadinae* and *Ginkgoinae* the Pteridophyta are the highest class in the realm of plants developing spermatozoids for reproduction. Therefore, these are the phylogenetic latest plants, in which there is to be found anything like a tactic movement, apart from the movements of chloroplasts of course.

Generally, the behaviour of the spermatozoids of ferns developed in the antheridia of the prothallia does not differ from the behaviour of other spermatozoids, for instance in Bryophyta [2]. Therefore, it may be sufficient to deal with only the most important physiological peculiarities.

With regard to the morphological structure it may be mentioned that apart from the Lycopodiinae with only 2 cilia all other ferns have spermatozoids bearing numerous grouped cilia, especially near the anterior ends of their bodies. Often the spermatozoids are a flat, spirally coiled band, resembling a "corkscrew", e.g. in the *Polypodiaceae* and *Marsiliaceae*. The direction of the coils of this "corkscrew" points to the left as well as to the right, and similarly some sperms when moving appear to revolve counterclockwise, others in a clockwise direction.

The spermatozoids apparently show neither clear phototactic nor geotactic movements, but conspicuous chemotaxis, the direction of which usually is positive, less frequently negative (see later).

This chemotaxis is caused by very different chemicals, especially the salts of organic acids, but not by carbohydrates and ammonia salts. The different species of sperms show a diverse behaviour, as results from the following list (BRUCHMANN, BULLER, LIDFORSS, SHIBATA [3])):

---

[1] A. F. W. SCHIMPER: Untersuchungen über die Chlorophyllkörner und die ihnen homologen Gebilde. Jb. Bot. 16, 1. (1885).

[2] See Manual of Bryology, The Hague 1932, p. 225, 231.

[3] H. BRUCHMANN: Von der Chemotaxis der *Lycopodium*-Spermatozoiden. Flora (Jena) 99, 193–202. (1909).

A. H. R. BULLER: Contributions to our knowledge of the physiology of the spermatozoa of Ferns. Ann. of Bot. 14, 543–582 (1900).

B. LIDFORSS: Ueber die Chemotaxis der *Equisetum*-Spermatozoiden. Ber. dtsch. bot. Ges. 23, 314–316 (1905).

K. SHIBATA: Untersuchungen über die Chemotaxis der Pteridophyten-Spermatozoiden. Jb. Bot. 49, 1–60 (1911).

*Lycopodium* reacts to citric acid,
*Equisetum*      „     „ tartaric and malic acid,
*Isoetes*        „     „ d-tartaric, fumaric, malic and succinic acid,
*Salvinia*       „     „ malonic and malic acid,
*Filicinae*  react „    „     „     „     „
*Selaginella*  reacts „ malic acid.

The spermatozoids (*Isoetes* and Filices) are also influenced by certain other organic acids related with the above mentioned ones as to their chemical constitution (SHIBATA). They react further to alkaloids, *Equisetum* for instance to morphine-, cocaine-chlorohydrates and strychnine nitrate. But not only organic compounds are efficacious, but also some metallic ions, e.g. potassium, rubidium (BULLER) and lithium on Filices, calcium and strontium on *Salvinia* and *Osmunda*, and in the case of *Equisetum* even many other metallic ions (SHIBATA). In all these cases the different compounds, whether organic or inorganic must be ionised in order to be effective. From that we may conclude that the ions are the very agents, and not the undissociated molecules.

Moreover the degree of concentration of the chemotacticum is of importance too. For instance H- and OH-ions cause positive movements in lower concentration, but negative taxis in higher concentration. The same occurs with the malic acid, but here perhaps the degree of dissociation of the acid, that is to say the amount of H-ions, is responsible for the effect. As malic acid ($COOH-CH_2-CHOH-COOH$) or its salts are the main compounds to cause taxis of the fern spermatozoa, we may consider them to be of the nature of those substances which usually emanate from the archegonia and which attract the spermatozoids to the latter.

PFEFFER [1]) was the first to investigate the chemotactic movements from a quantitative point of view. He established the validity of the law of WEBER–FECHNER [2]) and gave values for the threshold that is the lowest effective concentration of a chemotacticum. For instance, for malic acid and Filices spermatozoa he observed a threshold of about 0,001% or less.

PFEFFERS' method consisted in bringing spermatozoids in a shallow dish containing pure water or weak solutions of malic acid. After this he put the point of a small capillary tube into the liquid, the tube also being filled with malic acid, but of higher concentration than the content of the dish. Consequently, the malic acid diffused out the capillary tube into the liquid attracting the spermatozoa like the aperture of an archegonium. He

---

[1]) W. PFEFFER: Locomotorische Richtungsbewegungen durch chemische Reize. Untersuch. Bot. Inst. Tübingen *1*, 363–482 (1884). See also Ber. dtsch. bot. Ges. 1, 524–533 (1883).

[2]) See DU BUY–NUERNBERGK 1935, p. 472 (referred to on p. 314).

then determined the lowest difference of concentration between the solution of the capillary tube and the solution of the dish, a concentration difference just sufficient to cause an accumulation around and an entering of spermatozoids into the tube. In this way he obtained the following *relative* thresholds in Filices:

If the solution of the dish contained

| 0,0005% malic acid, | the concentration in the capillary tube | 0,015% malic acid, |
|---|---|---|
| 0,001 % „ „ | had to be | 0,03 % „ „ |
| 0,01 % „ „ | | 0,3 % „ „ |
| 0,5 % „ „ | | 1,5 % „ „ |

It follows from this that the malic acid concentration in the tube must be 30 times higher than that in the dish, if spermatozoa shall be attracted by the tube.

PFEFFER already stated that this relative, constant threshold is not valid but for 0,0005–0,9% malic acid concentration in the dish. SHIBATA [1]) obtained other values, and we must assume that the spermatozoids being in the solution during a longer time become gradually adapted to a certain concentration of the agent. The resulting consequency would be that the law of WEBER-FECHNER, if at all, cannot be valid but transitorily and then only within a limited range (cf. DU BUY–NUERNBERGK [2])).

In connection with these facts, it is very interesting to discuss the problem of the manner in which spermatozoids are directed when moving towards the source of the chemotacticum. Two theories have been stated: the so-called "local action theory of tropisms" (= theory of LOEB), which includes the "topotactic" movements of PFEFFER, and the theory of JENNINGS [3]), the so-called "trial and error"-theory (also including the "phobotactic" movements of PFEFFER [4])).

According to the *"local action theory"* the positive response of a sperm takes place as follows: the spermatozoid, swimming about at random and coming to that portion of the medium in which the chemotacticum is present, has its movements directly modified by the chemotactic agent, so that the direction of its axis is turned towards the region of the chemotacticum; it then proceeds in a straight line towards the agent. The total action is caused by the u n e q u a l  d i s t r i b u t i o n (that is to say by the different concentration)  o f  t h e  c h e m o t a c t i c u m  o n  d i f-

---

[1]) See p. 341.

[2]) See p. 342 respectively p. 314.

[3]) H. S. JENNINGS: Behavior of the lower organisms. New York, Macmillan (1906). — See also: german transl. by MANGOLD: „Das Verhalten der niederen Organismen". Leipzig–Berlin (1910).

H. S. JENNINGS: The interpretation of the behavior of the lower organisms. Science N.S. 27, 698–710 (1908).

[4]) W. PFEFFER: Pflanzenphysiologie, vol. 2, p. 754 foll., 757. 2nd ed. Leipzig (1904).

ferent parts of the sperm, this acts locally on certain parts of the body, directly influencing the motor organs and modifying the action of them. Therefore, the essential part of the reaction is the orientation of the organism so that its axis is placed parallel to the direction of the gradient of the chemotacticum (HOYT [1]).

According to the *theory of* JENNINGS the spermatozoid, coming into the region of a chemotacticum to which it reacts "positively", enters this region without reacting, but upon tending to leave the region, it reacts by turning back (phobotactically); the "positive" response is thus obtained by a series of "negative" reactions whenever the organism tends to go in a direction leading away from the source of chemotacticum (HOYT [1]). The essential point in this case is the fact that the c h a n g e   o f   t h e   c o n-
d i t i o n s to which the sperm is subjected, a c t s   o n   t h e   o r g a-
n i s m   a s   a   w h o l e, and that the sperm responds as a whole, perfor-
ming complex and coordinated movements. Therefore, the nature of the reaction is that the spermatozoid shows no response in passing to-
wards the region of optimal concentration, but reacts by turning back upon tending to pass from the optimum. According to JENNINGS, the reaction obtained depends on the physiological condition of the organism, which is in turn partly determined by the past experiences of the organism.

Most authors (e.g. SHIBATA [2]) believe that the positive response of fern sperms is due to the directive action of the chemotactic substance on the motor organs, the axis being thus turned towards the agent. Most workers, however, while holding this view, have noted that many sperms do not react as thus described, but may wander about the field, or pass through it indifferently, or may avoid it. Although such reactions often have been ascribed to individual differences, they yet induced some investigators to abandon the "local action theory" in favour of the "trial and error-
theory". For instance, even HOYT [1], in his detailed investigations on the chemotaxis of fern sperms considers their movements and reactions to be due to the effect of the chemotacticum on the o r g a n i s m   a s   a   w h o l e and not to the action of different concentrations of the agent on local parts of the sperms.

METZNER [3], however, was able to settle the different contradictions mentioned above, and showed that in general the chemotactic movements of the sperms follow the "local action theory of tropisms". The sperms seem to react "as a whole" only when no local concentration difference of the agent is operating, for instance when a spermatozoid is

---

[1] W. D. HOYT: Physiological aspects of fertilization and hybridization in ferns. Bot. Gaz. 49, 340–370, 355 (1910).

[2] See p. 341.

[3] P. METZNER: Studien über die Bewegungsmechanik der Spermatozoiden. Beitr. allg. Bot. 2, 435–499 (1923).

swimming directly towards the source of the chemotacticum or away from it (see also further below).

However, as soon as a sperm moves in a direction such as to bring about a local concentration difference on the opposite sides of the organism, the latter changes its direction swimming now in a smooth curve and without vacilation towards the higher concentration of the agent.

Based on his observations METZNER supposes that an increase of the concentration of the chemotacticum also increases the frequency of vibration of the cilia, and this opinion would be compatible with the fact that a unilateral increase of the cilia frequency caused by a higher concentration leads to a curved course.

Regarding the permanent rotation of the swimming sperms on their axes some investigators believed that this fact did not lend probability to the explanation of the tactic movements according to the "local action theory". METZNER, however, has refuted this opinion. Firstly, the rotation and the lashing about of the cilia does not disorder the "gradient field" of a chemotacticum by forming vortices, as the only result of the swinging of the cilia appears to be an attraction and sucking up of the liquid towards and through the circles formed by the ends of the vibrating cilia. Consequently, vortices can arise only *within* the swinging space of the cilia and in the proximity of the sperm-nucleus, but never near the outside of the cilia, the probable places of chemotactic perception.

Secondly, the cilia show a reaction-time of about 0.1 sec., while the rotation time amounts to about 0.2–0.25 sec. per revolution. Hence, when a rotating sperm is influenced by a higher chemical concentration on one side of the cilia, the increase of the frequency of vibration (see above) does not occur until after the influenced part of the cilia has moved about half a revolution towards the opposite side. Consequently, the impulse bringing about the change of direction always occurs on the side opposite to the place of the chemical influence.

There exists also a metachronism with respect to the movement of the cilia, i.e., the cilia of a spiral row (see p. 341) do not react together at the same time, but they react one after the other, the first governing the vibrations of the resting ones. The angular velocity of this metachronism is about the same as the velocity corresponding with the rotation, and does not modify the mode of reaction in any way.

Finally, we can observe in certain cases a kind of response much resembling the phobotactic reactions in the sense of the terminology of JENNINGS.

They occur for instance, when a sperm penetrates into a concentration of the chemotacticum exceeding the optimum thereby retarding again the vibration of the cilia, or when — with the same effect on the cilia — a sperm on its course gets from a higher to a lower concentration (e.g., in the neighbourhood of the source of the chemotacticum). In all such cases the spermatozoid turns off, the direction of the new course being dependent on the direction which the sperm had when meeting the alteration.

Therefore, this turning aside (or back) does not show any uniformity but takes the character of a phobotaxis. Since, however, a real turning back or swimming back in exactly the same direction, in which the organism approached, does not seem to occur, METZNER calls this kind of response the *pseudophobotactic reaction*.

In the case of a real phobotactic response we should expect that the sperm by turning back suddenly would reverse the direction in which the cilia swing, and this with regard to the *whole* of the cilia. As METZNER could not prove any reaction of this kind, he assumes that the turning back is caused by a local slackening of the cilia action only on that side of the organism, where the higher or lower concentration of the agent first acts. Consequently, the essential point of these pseudophobotactic reactions r e m a i n s the local action of the agent on different parts of the organism, the very thing which has been learned from the normal kind of response formerly described.

Finally, recapitulating the leading features of the two theories, it is to be stated that only the "local action theory" gives an analytically elaborate and logical explanation of the mode of tactic response, whereas the theory of JENNINGS is not able to reduce the complexity of the tactic movements to analytically comprehensible and less complex processes.

# Chapter XI

# CHEMIE UND STOFFWECHSEL

von

Karl Wetzel (Berlin)

**§ 1. Einleitung.** — Kaum eine andere systematische Pflanzengruppe ist biochemisch und stoffwechselphysiologisch so mangelhaft erforscht wie die der Pteridophyten. Selbst in recht zuverlässigen biochemischen Handbüchern wie denen von Czapek, Wehmer und Klein klaffen in den Angaben zwischen Moosen und Phanerogamen gähnende Lücken. Eine planmässige Bearbeitung des Stoffbestandes und Stoffwechsels der Pteridophyten fehlt uns noch vollständig. Die heute vorliegenden einschlägigen Arbeiten lassen sich hinsichtlich ihrer Zielsetzung in 2 Gruppen einteilen. Die einen befassen sich mit einigen besonders auffälligen Erscheinungen, wie dem Kieselsäure- und Aluminiumgehalt der Pteridophyten oder dem Auftreten eigenartiger Stoffe, wie der Farnsäuren, harzartiger Körper oder der Sporopollenine der Sporenwände. Die andere Gruppe von Arbeiten knüpft an die Verwendung einiger Pteridophyten in der Pharmazie an und steht vorzugsweise im Dienst einer Förderung der Drogenanalyse. Dagegen ist der Stoffwechsel und seine Dynamik bei Pteridophyten nur in wenigen Stichproben untersucht, die wohl etwas vorzeitig und generalisierend zu der Auffassung geführt haben, dass die Gefässkryptogamen stoffwechselphysiologisch sich unmittelbar an die Blütenpflanzen anschliessen. Es erscheint indessen kaum zweifelhaft, dass eine eingehendere Kenntnis auch hier zu einer wesentlich differenzierteren Auffassung führen und dem Phylogenetiker manch wertvollen Hinweis auf die verwandtschaftlichen Beziehungen innerhalb dieser systematischen Gruppe geben würde.

Die folgende Abhandlung kann daher kein befriedigendes Bild über Chemie und Stoffwechsel der Pteridophyten geben, sondern muss sich mit einer Zusammenfassung der in der Literatur weit zerstreuten Angaben und bescheidenen eigenen Versuchen einer Abrundung der sich daraus ergebenden Vorstellungen begnügen, die wir mit dem Wunsche niederlegen, dass sie bald durch umfassendere Untersuchungen ihre notwendige Erweiterung und Ergänzung finden möchten.

**§ 2. Die Zellmembran.** — Der Hauptbestandteil der Zellmembran der Pteridophyten ist die Zellulose; daneben aber treten schwer hydrolysierbare Hemizellulosen in grösserer Menge auf. Wenn ältere Autoren (vgl. MERKELBACH) [1]) als Hydrolysenprodukte der Zellulose von Pteridophytenzellwänden fast regelmässige Mannane angeben, so beruht dieser Irrtum auf unvollständiger Hydrolyse der Hemizellulosen. Es ist bekannt, dass gerade Mannane der Hydrolyse mit 5% iger HCl länger widerstehen: so mussten HÄGGLUND und KLINGSTEDT [2]) $2^1/_2$ Tage mit 72%iger $H_2SO_4$ bei Zimmertemperatur hydrolysieren, um alle Hemizellulosen in Lösung zu bringen. Hieraus erklärt sich auch die angebliche Mannosearmut der Pteridophyten-Hemizellulose. Als Hydrolysenprodukte der Hemizellulosen ergeben sich erhebliche Mengen von Galaktose, Arabinose und Mannose, neben wenig Methylpentosen. Bemerkenswert erscheint noch, dass sich die Hemizellulosen fertiler und steriler Sprosse von Equisetumarten chemisch unterscheiden durch die fehlende Galaktose in den ersteren, wo sie nur in den Sporophyll tragenden Köpfen gefunden wurde.

Der Pentosengehalt scheint in einzelnen Arten derselben Gattung stark zu schwanken. So wird angegeben für: *Lycopodium clavatum* 9,35% Pentosen, *L. annotinum* 8,1%, *L. alpinum* 17,0%, *L. Selago* 18,6%. Ob dieses Mehr an Pentosen auf einen Überschuss an Hemizellulosen oder auf einen solchen an Pektinen zurückzuführen ist, bedarf noch näherer Untersuchung.

**§ 3. Die Sporopollenine** [3]). — Eine völlig andersartige chemische Zusammensetzung weisen die Membranen der Sporen auf, die mit derjenigen der Mikrosporen der Phanerogamen weitgehend übereinzustimmen scheint. Man bezeichnet diese Membransubstanzen daher als Sporopollenine. Mikrochemisch haben diese Stoffe manche Reaktionsweise mit Cutin und Suberin gemein, chemisch aber haben sie mit diesen Stoffgruppen nichts zu tun. Sie gehören ihrer Zusammensetzung nach vielmehr zu den polymeren Terpenen, mit denen sie auch in ihrem Verhältnis von C : H (5 : 8) übereinstimmen. Dagegen ist der Sauerstoffgehalt sehr variabel und für einzelne Sporopollenine recht charakteristisch. Nach den bisherigen Untersuchungen darf man annehmen, dass der O weder in carboxylatischer noch in carbonylartiger Bindung vorliegt; neben Hydroxylen wird man vielleicht noch ätherartige Bindungen annehmen müssen. Ihr besonderes Charakteristicum besitzen die Sporopollenine in ihrer chemischen Beständigkeit: verdünnte Mineralsäuren vermögen sie überhaupt nicht, konzentrierte nur langsam zu zerlegen. Alkalien vermögen praktisch nur in der

---

[1]) MERKELBACH: Die chem. Zusammensetzung der Zellwände bei einigen Gefässkryptogamen. Diss. Freiburg 1907.

[2]) HÄGGLUND und KLINGSTEDT: Ann. d. Chem. 459, 26, 1927. Cellulosechemie 5, 58, 1924.

[3]) Vgl. ZETZSCHE: KLEIN, Handb. d. Pflanzenanalyse II, S. 205.

Schmelze anzugreifen. Selbst Acetylbromid das sogar Zellulose und Lignin in lösliche Substanzen überzuführen vermag, bleibt ohne erkennbare Wirkung auf die Sporenmembranen. Diese Eigenschaft und die andere mit Halogenen oxydationsbeständige Verbindungen einzugehen, sind für die Abtrennung dieser Substanzen von andern Zellstoffen von grosser praktischer Bedeutung.

Im allgemeinen besteht nur das derbere Exospor aus Sporopolleninen, während am Aufbau des zarteren Endospors Zellulose massgeblich beteiligt ist. Mengenmässig überwiegt das erstere sehr stark (90%). Auf die hohe Resistenz der Sporopollenine gegen chemische, physikalische und biologische Einflüsse ist die wundervolle Erhaltung der Sporen und Pollen in abgelagerten Kohle- und Torfschichten zurückzuführen.

Bekanntlich macht die Pollenanalyse praktischen Gebrauch von dieser Erscheinung, um die Entstehungsgeschichte organogener Ablagerungen besonders der Moore aufzuhellen. Es scheint indessen, dass die hohe chemische Resistenz der Pollenmembranstoffe hierbei noch nicht methodisch voll ausgenützt ist. Man könnte sich hiervon noch eine bedeutsame Erleichterung und Präzisierung der pollenanalytischen Methode versprechen. Hinsichtlich der Isolierung der Sporopollenine aus Ölschiefer vergl. KLEIN. Handbuch d. Pflanzenanalyse S. 339 ff, aus Torf S. 309 und aus Braunkohle S. 313.

## § 4. Verkieselung der Membran.

— Besonders reichliche Einlagerung von Kieselsäure findet sich bekanntlich in den Membranen der Aussenschichten von Equiseten. Ihre Isolierung erfordert die vorangehende Entfernung der Alkalisalze durch Säuren und der organischen Substanzen durch Oxydantien. Nach weiterer Behandlung des Rückstandes mit Ammoniak und fettlösenden Agenzien, bleibt, ein Rest, der zu 97–98% aus $SiO_2$ besteht und unter dem Mikroskop deutlich die Strukturen der Aussenmembran zeigt. LANGE[1]) hat Kieselsäureverbindungen auch im Zellsaft von Equisetum hiemale nachweisen können. In welcher Form, ob als Gel oder kristallisiert, ob in organischer oder anorganischer Bindung das Si vorkommt, scheint noch nicht bekannt zu sein. Erhebliche Kieselsäuremengen hat KEEGAN[2]) auch in *Pteridium aquilinum* nachgewiesen, die im Herbst von 17 auf 53% anstiegen.

Reich an Kieselsäure sind *Equisetum Telmateja*, *E. arvense* und *E. hiemale*, die aus diesem Grunde zum Scheuern von Metall und Holzgefässen Verwendung finden. In der Heilkunde werden die Schachtelhalme von alters her als Diureticum verordnet.

Ein Maximum an Kieselsäure findet sich in der Asche von *Equisetum maximum*, wo CHURCH 63% des Aschengehalts an Kieselsäure feststellte.

---

[1]) LANGE: Ber. chem. Ges. 11, 822, 1878.
[2]) KEEGAN: Chem. News 11, 289, 1915.

Die folgende Tabelle gibt weiteres Zahlenmaterial wieder (vgl. ([1—3])).

| Pflanze | Kieselsäure in % des Aschengehalts | |
|---|---|---|
| *Equisetum Telmateja* . . | 31,083% | $SiO_2$ |
| „ *hiemale* . . . | 8,75 % | „ |
| „ *arvense* . . . | 6,188% | „ |
| *Lycopodium alpinum* . . | 10,24% | „ |
| „ *cernuum* . . | 30.25 % | „ |
| „ *clavatum* . . | 6,40 % | „ |
| „ *Billardieri* . | 3,14 % | „ |
| „ *Selago* . . . | 2,53 % | „ |
| *Selaginella spinulosa*. . . | 6,67 % | „ |
| *Psilotum triquetrum* . . . | 3,77 % | „ |
| *Cyathea serra* . . . . | 12,65 % | „ |
| *Ophioglossum vulgatum* . | 5,32 % | „ |
| *Salvinia natans*. . . . | 6,71 % | „ |
| *Marsilia quadrifolia* . . | 0,88 % | „ |
| *Pteridium aquilinum* . . | 49,85 % | „ |

GILLOT und DURAFOUR [4]) geben für unter bestimmten Bedingungen (Ca-arm und Mg-arm) gezogene Pflanzen von Adlerfarn sogar einen $SiO_2$-Gehalt von 68,8% der Asche an. Weitere Angaben über den Kieselsäuregehalt bei *Marattiaceen* s. POIRAULT [4]) und bei *Selaginella* s. HARVEY-GIBSON [6]).

**§ 5. Lignin.** — Auf Grund der ausführlichen Untersuchung LINS-BAUERS [7]) wurde bis in die neuere Zeit angenommen, dass bei den Pteridophyten die Verholzung der Zellmembran zum erstenmal im ganzen Pflanzenreich einsetze. WAKSMAN und STAVENS [8]) geben demgegenüber zwar zu, dass die Farbreaktionen auf Holz in Moosen tatsächlich ausbleiben, jedoch gelang es ihnen auf *chemischem* Wege aus *Sphagnum* neben 35,2% Zellulose immerhin 9,2% Lignin zu isolieren. Der niedrige Methoxylgehalt des Mooslignins lässt übrigens erkennen, dass sich dasselbe auch im chemischen Aufbau von dem Holzstoff der Pteridophyten und höheren Pflanzen unterscheidet.

Völlig holzfrei unter den Pteridophyten wurde bisher nur *Isoetes lacustris* befunden (BURGERSTEIN [9])). Bei *Salvinia* dagegen ist nur mehr

---

[1]) CHURCH: Chem. News 1874, S. 137; Journ. of Bot. 1875, S. 169.

[2]) MARIANI: Justs Jahresber. I, 58, 1888.

[3]) HORNBERGER: Landw. Vers. Stat. 32, 371, 1886.

[4]) GILLOT und DURAFOUR: Bull. Soc. de Natur. de L'Ain, 1904. Just II, 1052, 1904.

[5]) POIRAULT: Ann. Sci. Nat. (7), 18, 113, 1893.

[6]) HARVEY–GIBSON: Ann. of Bot. 7, 355, 1893.

[7]) LINSBAUER: Oesterr. Bot. Zeitg. 49, 317, 1899.

[8]) WAKSMAN und STAVENS: Soil Science 26, 133. Vgl. auch STADTNIKOFF: Brennstoffchemie 11, 21, 1930.

[9]) BURGERSTEIN: Sitz. Ber. Kaiserl. Akad. Wiss. Wien 70, 1874.

das Xylem ligninfrei, während im mechanischen System auffallend dunkel gefärbtes Lignin auftritt.

Bei den *Equiseten* sind die mechanischen Systeme der oberirdischen Organe ebenfalls unverholzt, bei *Selaginella* geben nur die Mittellamellen Ligninreaktion, während *Lycopodium* normal verholztes Festigungsgewebe besitzt. Bei einer ganzen Anzahl von Arten dieser Gattung kommen Verholzungen sogar im Mesophyll vor (*L. annotinum, L. clavatum, L. volubile, L. complanatum*). Selbst die Epidermis und partiell sogar die Schliesszellen sind hier verholzt. Auch *Selaginella* hat ähnlich wie die Cycadeen und Coniferen verholzte Schliesszellenwände. Stark verholzt sind vor allem auch die Aussenwände der Farnsporangien. In vielen Fällen zeigen auch die Endodermisradialwände Lignineinlagerung.

Im ganzen also ist die Verholzung bei den Pteridophyten weit verbreitet. LINSBAUER deutet diese Erscheinung als Ausdruck einer bei den Pteridophyten bereits weitgehend durchgeführten Arbeitsteilung der Zellen.

**§ 6. Kohlehydrate.** — Wir verweisen hier zunächst auf das Kapitel über den Membranaufbau und die daran beteiligten Kohlehydrate. Im allgemeinen weisen die Pteridophyten hinsichtlich der Kohlehydrate ähnliche Verhältnisse wie die höheren Pflanzen auf. Das wichtigste Reservekohlehydrat ist die S t ä r k e, die vor allem in Rhizomen z.T. auch im Stengelmark der baumförmigen tropischen Farne gespeichert wird. RIPPERTON [1]) untersuchte im Hinblick auf die Möglichkeit einer technischen Ausbeutung den Stärkegehalt der Stämme hawajanischer Baumfarne (*Cibotium Chamissoi*), die in Hawai ein Areal von etwa 400.000 acres bedecken. Danach scheint der Stärkegehalt dieser Farnstämme demjenigen der Cannarhizome wenig nachzustehen; obwohl neben der Stärke noch erhebliche Mengen reduzierender Zucker isoliert werden konnten, musste doch von einer technischen Gewinnung der Farnstärke aus Rentabilitätsgründen Abstand genommen werden. In den Rhizomen der Farne scheint die Stärke wenn nicht der einzige, so doch immer der wesentlichste Reservestoff zu sein. (CZAPEK 1, 395). Über die quantitativen Verhältnisse ist auch hier wenig bekannt. Wir selbst fanden in den nachmittags abgenommenen Blättern von *Blechnum brasiliense* etwa 1% des Frischgewichts an Stärke, also Mengen, wie man sie aus Blättern höherer Pflanzen auch kennt. Der Gehalt der Rhizome an Stärke (15, 4–28,2% nach KRUSE) wird demjenigen der Kartoffel wenig nachstehen, wie aus dem mikroskopischen Bild zu vermuten ist, das dicht mit Stärkekörnern erfüllte Rindenparenchym-, Stärkescheide- und Holzparenchymzellen zeigt. Die Rhizome von *Pteridium aquilinum* sind so stärkereich, dass sie früher die Hauptnahrung der Maoris liefern konnten.

Zucker: An Zuckern hat LIEBER [2]) in den Knollen einer *Nephrolepis-*

---

[1]) RIPPERTON: Hawaii exp. Stat. Bulletin No. 53, 1924.

[2]) LIEBER: Ber. naturw. med. Ver. Innsbruck 1910, XXXVIII, 34.

art neben d-Glukose, noch d-Fruktose und Maltose, insgesamt etwa 16%
Zucker nachgewiesen. In den Blättern von *Blechnum brasiliense* fanden wir
etwa 0,87 g vergärbare Zucker pro 100 g Frischgewicht. Auffallenderweise
soll nach Bäsecke Saccharose in den Blättern von *Phyllitis Scolopen-*
*drium* gänzlich fehlen. Es ist nicht ausgeschlossen, dass der Rohrzucker
gelegentlich durch T r·e h a l o s e ersetzt ist. Anselmino und Gilg [1])
berichten über erhebliche Trehalosemengen in *Selaginella lepidophylla*
(2,5%), doch sind andere *Selaginella*-Arten trehalosefrei befunden worden.
Inwieweit auch in Farnen stoffliche Beziehungen zwischen Trehalose und
Mannit wie etwa bei Pilzen bestehen, bedarf noch der Aufklärung. Anselmino und Gilg haben in ihrer Versuchspflanze keinen Mannit gefunden.

**§ 7. Glykoside.** — Unsere Kenntnisse über das Vorkommen von
G l y k o s i d e n in Pteridophyten ist ganz besonders dürftig. Weder
Czapek noch Wehmer machen hierüber irgendwelche Angaben. Als erster
hat wohl Greshoff [2]) B l a u s ä u r e im Adlerfarn, ferner in *Gym-*
*nogramme aurea* und *G. cordata*, sowie in einer nicht näher identifizierten
*Lastraea-* und *Athyrium-* Art festgestellt. Später hat sich Mirande um die
Entstehung der Blausäure in Farnen bemüht. *Pteridium aquilinum* enthält danach ein cyanogenes Glykosid von Amygdalin-Charakter, das bei
der Hydrolyse Benzaldehyd abspaltet. Mirande [3]) stellte ein solches Glykosid auch bei *Cystopteris fragilis alpina*, einem in der Dauphinée und in
Savojen häufigen Farn fest. Das Glykosid ist wie das emulsinartige Ferment zu seiner Spaltung in allen grünen Pflanzenteilen enthalten; wie es
indes bei Glykosiden üblich ist, sind Substrat und Ferment trotzdem lokal
getrennt, so dass die Fermentwirkung erst bei Schädigung oder Zerstörung der Plasmastruktur deutlich in Erscheinung tritt. Das einige
Zeit der Autolyse überlassene Pflanzenmaterial gibt bei der Destillation
Benzaldehyd ab. Es wird vermutet, dass das Glykosid hauptsächlich
im Zellsaft, das Ferment dagegen im Plasma lokalisiert ist.

**§ 8. Organische Säuren.** — *a)* Aliphatische Säuren: Es ist u.W. nie
versucht worden, das Vorkommen von organischen Säuren in Pteridophyten einer planmässigen Untersuchung zu unterziehen. Die meisten Angaben beruhen vielmehr auf Nebenbefunden und sind daher im höchsten
Grade lückenhaft. Fettsäuren: So entnehme ich die einzige Angabe über
das Auftreten von E s s i g s ä u r e einer Analysentabelle für *Equisetum*
*maximum* von Braconnet. Über die Entstehung und das physiologische
Schicksal der Säure werden keinerlei Angaben gemacht. P r o p i o n-

---

[1]) Anselmino und Gilg: Ber. Deutsch. Pharm. Ges. 23, 1913.
[2]) Greshoff: Pharm Weekbl. voor Nederl. 45, 770, 1908.
[3]) Mirande: C. r. d. séances de l'Acad. des Sciences 167, 695, 1918.

s ä u r e soll sich angeblich im Rhizom von *Dryopteris Filix-mas* sowie in dem extrahierten Wurmfarnöl gefunden haben.

Viel sicherer und auch häufiger ist das Auftreten der B u t t e r s ä u r e. Ihr Vorkommen scheint an das gleichzeitige Auftreten von P h l o r o - g l u c i n gebunden, denn alle Farne, in denen der eine Stoff nachgewiesen werden konnte, enthielten auch den andern. Beide Stoffe kommen indes wahrscheinlich nicht frei, sondern als Komponenten der F a r n s ä u r e n (s. S. 356) vor. Die Buttersäure ist in ihnen als Phloroglucinbutanon gebunden. Manche dieser Farnsäuren wie Aspidin und Filmaron spalten im Organismus die Komponenten Phloroglucin und Buttersäure ab. Weitere Angaben siehe unter Farnsäuren.

C a p r y l - und C e r o t i n s ä u r e sind Bestandteile des Rhizomöls und des Sporenfettes von *Dryopteris Filix-mas*, während die Sporen von *Lycopodium clavatum* die L y c o p o d i u m s ä u r e glyceridartig gebunden enthalten. Im Fett des Stengels von *Equisetum arvense* L. konnte die ungesättigte L i n o l e n s ä u r e festgestellt werden. Die Dicarbonsäure A z e l a i n s ä u r e wurde aus Sporenextrakt des Wurmfarns, die ungesättigte Tricarbonsäure A c o n i t s ä u r e als Ca-Salz aus *Equisetum fluviatile* und *E. palustre* isoliert.

Verbreiteter scheint nach den Angaben der Literatur die O x a l s ä u r e in den Farnen zu sein, da sich die Mehrzahl der Angaben über Säurevorkommen in Pteridophyten auf diese Säure bezieht. In der bereits citierten Analyse für *Equisetum maximum* werden Spuren von Oxalat erwähnt. Im übrigen aber herrschte seit den Untersuchungen KOHLS [1]), der nur in 3 von 34 untersuchten Farnarten Oxalatvorkommen nachweisen konnte, die Ansicht vor, dass Kryptogamen im Gegensatz zu den höheren Pflanzen arm an Oxalsäure seien, ein Standpunkt den auch DE BARY in seiner Vergleichenden Anatomie vertrat.

POIRAULT [2]) konnte indes zeigen, dass den früheren Forschern die Oxalatkristalle offenbar nur wegen ihrer besonderen Kleinheit in den Farnen entgangen waren. Er führt eine recht respektable Reihe von Oxalat führenden Farnen auf und weist auch bereits auf Oxalatkristalle aus verschiedenen Kristallsystemen hin, wobei ihm nicht verborgen blieb, dass die Kristallform besonders vom Schleimgehalt und der Reaktion des Zellsafts abhängig war: in schleimreichen Zellen mit stärker saurem Zellsaft herrschten die monoklinen Kristalle, in säurearmen Zellen dagegen Formen des quadratischen Systems vor. In einer ausführlichen Studie hat neuerdings FREY diese Angaben auch für die höheren Pflanzen bestätigt und näher präzisiert. Hinsichtlich der L o k a l i s a t i o n der Oxalatkristalle wird von allen Autoren einheitlich hervorgehoben, dass nur die Epidermis und das unmittelbar darunter liegende hypodermale Gewebe Oxalat führen, während die Bündel stets frei davon befunden wurden.

---

[1]) KOHL: Kalk- und Kieselsalze. Marburg 1890.

[2]) POIRAULT: Journ. de Bot. 1893, 72.

Von den Wasserfarnen führt nach KOHL *Marsilia* Oxalat, während es bei den *Lycopodiinae* den Gattungen *Selaginella* und *Psilotum* fehlt [1]).

Spärlicher noch sind die Angaben über das Vorkommen der A p f e l- s ä u r e in Pteridophyten, obwohl diese Säure nach den eigenen orientierenden Untersuchungen in Farnen analog den Verhältnissen in den höheren Pflanzen und in scharfem Gegensatz zu den Pilzen viel verbreiteter ist als die Oxalsäure. BELZUNG und POIRAULT [2]) beschreiben Ca-Malat-Vorkommen in Blättern von *Angiopteris evecta* und REGNAULT [3]) weist auf Äpfelsäure bei *Equiseten* hin.

Im Folgenden geben wir eine Tabelle wieder, die die Ergebnisse eigener Säureuntersuchungen an Blättern einiger Farne enthalten. Die Säure wurde nach den im Leipziger Botanischen Institut ausgearbeiteten Methoden quantitativ bestimmt und zwar in beiden Formen, der optisch aktiven und der racemischen Komponente. Als optisch aktive Säure trat stets die l'Form auf, ein weiterer Analogiefall zu den höheren Pflanzen. Die Angaben beziehen sich auf 1 g Trockengewicht; um eine Bezugsmöglichkeit auf das Frischgewicht zu ermöglichen, ist bei allen Arten noch der % uale Anteil des Trockengewichts am Frischgewicht angegeben. Säurezahlen in Millimol 2 basischer Säure.

| Pflanze | Gesamt-säure | l' Äpfel-säure | i Äpfel-säure | Trock. Gew. in % des Frischgew. |
|---------|--------------|----------------|---------------|--------------------------------|
| Platycerium. . . . . . | 0,247 | 0,075 | 0,017 | 16,7 |
| Polypodium aureum . . | 0,221 | 0,075 | 0,077 | 28,6 |
| Blechnum brasil. (alt). . | 0,127 | 0,025 | 0,057 | 29,6 |
| Blechnum brasil. (jung) . | 0,257 | 0,014 | 0,081 | 23,7 |
| Polypodium aureum f. ra-mosum. . . . . . . | 0,213 | 0,006 | 0,111 | 16,1 |
| Dicksiona antarctica . . | 0,142 | 0,003 | 0,077 | 45,6 |
| Dryopteris Filix-mas . . | 0,114 | 0,033 | 0,047 | 20,1 |

Es ist ersichtlich, dass in allen untersuchten Farnen die Äpfelsäure mehr als die Hälfte der Gesamtsäure ausmacht, wohingegen von Oxalsäure nur Spuren gefunden wurden. Quantitativ ist also nicht die Oxalsäure, sondern die Äpfelsäure die wichtigste organische Säure der Farne.

Mit den höheren Pflanzen haben die Farne offensichtlich auch den diurnalen Säurestoffwechsel gemein: wenn auch keine ausführlichen Unter-

---

[1]) Vergl. auch MONTEVERDE: Justs Jahresber. 1889. Giesenhagen: Flora 76. TERLETZI: Jahrb. wiss. Bot. 15, 491. VINGE: Bidrag til kännedomen om ormbunkarnes bladbyggnad. Lund 1889.

[2]) BELZUNG u. POIRAULT: Journ. de Bot. 1892 (p. 286).

[3]). REGNAULT: Ann. Chim. et Phys. (2) 62, 208, 1836.

suchungen darüber vorliegen, so spricht doch eine Angabe von LANGE über *Dryopteris Filix-mas*, die einen nächtlichen Säureansteig um 30% des Abendwertes feststellt, durchaus für eine solche Annahme, wenn sie auch einer Erweiterung und Bestätigung auf breiterer Grundlage bedarf.

Hinsichtlich der übrigen Säuren fanden wir nur noch eine Bemerkung über das Auftreten von Weinsäure in *Lycopodium complanatum*.

### § 9. Aktuelle Azidität und Pufferungskapazität. — Die in Ermangelung von Literaturangaben vorgenommenen orientierenden eigenen Versuche lassen erkennen, dass die Farne nicht zu den typischen Säurepflanzen zu zählen sind. Dafür spricht auch das reichliche vorkommen von Amiden in allen untersuchten Organen der Pflanzen (s. S. 377). Wir geben im folgenden eine kleine Übersicht der gemessenen aktuellen Azidität bei einzelnen Farnen wieder; die angegebenen Werte beziehen sich auf den Zellsaft der Blätter: *Pteridium aquilinum* ph 5,66, *Nephrolepis cordifolia* ph 5,55, *Platycerium* spec. ph 5,83, *Dryopteris Filix-mas* ph 5,83, *Athyrium Filix-femina* ph. 6,14, *Polypodium punctatum* ph 5,82, Die Werte sind mit der Chinhydronelektrode gemessen. Sie sind möglicher weise etwas zu hoch, da die Reaktion des Zellsafts sich in kurzer Zeit unter Braunfärbung (Oxydation) nach der alkalischen Seite hin verschiebt. Man darf vermuten, dass hierbei die Oxydation phenolischer Hydroxylgruppen eine Rolle spielt.

Trotz des geringen Säuregrads ist die P u f f e r u n g s k a p a z i t ä t des Zellsafts ansehnlich. Die folgende Kurve (Fig. 1) gibt die β Werte des Zellsafts von *Polypodium punctatum* wieder:

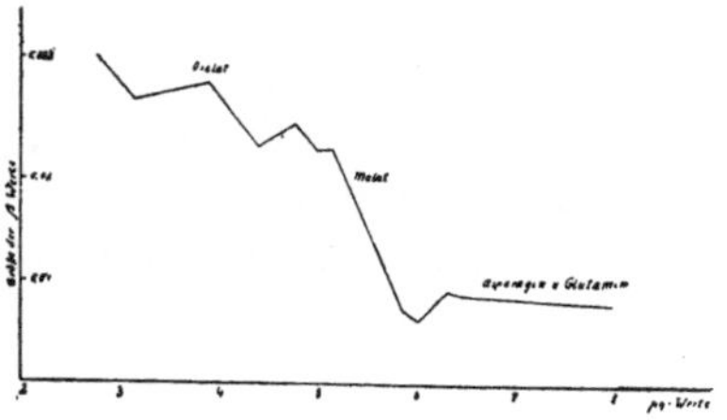

FIG. 1. — Pufferungskapazität des Zellsafts aus Blättern von *Polypodium punctatum*.

Die Kurve lässt erkennen, dass der Zellsaft in der Hauptsache nur gegen Säuren gepuffert ist. Im Bereich von ph. 5,26 bis 2,76 schwankt der

$$\beta \text{ Wert} \left( \frac{\text{Gramm-Aequivalent Säure} \times \text{Volumen zugesetzter Säure}}{\text{ph. Verschiebung} \times \text{Volumen Pufferlösung}} \right) \text{zwischen}$$

0,032 und 0,023). Gegen Lauge dagegen ist der Zellsaft nur ganz unbedeutend gepuffert; zwischen ph. 5,86 und ph 8,02 liegt der β Wert um 0,009.

Der Verlauf der Pufferungskurve lässt mit einiger Wahrscheinlichkeit schon auf die Natur des Puffergemisches schliessen. Es ist nicht zu bezweifeln, dass den ersten Ansturm der H· das Malatgemisch abdämmt. Der Verlauf der Kurve zwischen ph 4,5 und 2,7 lässt auf die Anwesenheit nicht unerheblicher Oxalatmengen schliessen.

Die geringe Pufferung nach dem neutralen Gebiet hin dürfte auf Amide und auf ein Bicarbonatsystem zurückzuführen sein. Die Anwesenheit beträchtlicher Carbonatmengen konnte analytisch ebenso wie diejenige der Amide bestätigt werden.

**§ 10. Die Farnsäuren.** — Diese Säuren verdienen insofern ein erhöhtes Interesse als sie offenbar ausschliesslich in Farnen vorkommen. Das gemeinsame chemische Merkmal dieser Stoffe ist das Phloroglucingerüst, das bald mehr oder weniger methyliert bez. mit Buttersäure zu Butanonen verknüpft ist. Der einfachste Körper, von dem sich die übrigen unschwer ableiten lassen, ist die *Filicinsäure*:

$$
\begin{array}{c}
\text{C.(CH}_3\text{)}_2 \\
\text{OC} \diagup \quad \diagdown \text{COH} \\
\text{HC} \diagdown \quad \diagup \text{CH} \\
\text{COH}
\end{array}
$$

Ebenfalls im Rhizom von *Dryopteris Filix-mas* fand BOEHM [1]) das *Aspidinol*, ein Methylphloroglucin-Monomethylester-Butanon, dessen von BOEHM angegebene Konstitutionsformel durch KARRER [2]) bestätigt wurde.

$$
\begin{array}{c}
\text{C.CH}_3 \\
\text{HOC} \diagup \quad \diagdown \text{C.OCH}_3 \\
\text{C}_3\text{H}_7\text{.OC.C} \diagdown \quad \diagup \text{CH} \\
\text{COH}
\end{array}
$$

Die Substanz ist alkohollöslich und wird durch Eisenchlorid grün gefärbt. Schmelzpunkte 156 bis 161°.

*Flavaspinsäure* ist nach BOEHM:

$$
\begin{array}{c}
\text{C.CH}_3 \qquad\qquad\qquad\qquad\qquad \text{C.CO.C}_3\text{H}_7 \\
\text{OC} \diagup \quad \diagdown \text{COH} \qquad\qquad \text{HOC} \diagup \quad \diagdown \text{COH} \\
\quad \text{CH}_2 \qquad\qquad\qquad\qquad\qquad\qquad \\
\text{C}_3\text{H}_7\text{.OC.C} \diagdown \quad \diagup \text{C}-------\text{HC}_2-------\text{C} \diagdown \quad \diagup \text{C.CH}_3 \\
\text{CO} \qquad\qquad\qquad\qquad\qquad\qquad \text{COH}
\end{array}
$$

Durch Einwirkung von Formaldehyd auf *Filicinsäure* lässt sich *Albaspidin* herstellen, dem man danach die folgende Konstitution zuschreibt:

$$
\begin{array}{c}
\text{C.(CH}_3\text{)}_2 \qquad\qquad\qquad\qquad\qquad \text{C.(CH}_3\text{)}_2 \\
\text{HOC} \diagup \quad \diagdown \text{COH} \qquad\qquad \text{HOC} \diagup \quad \diagdown \text{COH} \\
\text{C}_3\text{H}_7\text{.OC.C} \diagdown \quad \diagup \text{C}-------\text{CH}_2-------\text{C} \diagdown \quad \diagup \text{C.CO.C}_3\text{H}_7 \\
\text{CO} \qquad\qquad\qquad\qquad\qquad\qquad \text{CO}
\end{array}
$$

Die Substanz ist in Alkohol wenig, in Äther und Benzol leicht löslich.

---

[1]) BOEHM: Arch. exp. Path. 38, 35, 1896. Liebigs Ann. 302, 171, 1898. 307, 249, 1899; 318, 230, 253, 1901; 329, 269, 318, 321, 338, 1904.

[2]) KARRER: C. 1920, III, 378.

FeCl₃ färbt die alkoholische Lösung rot, Schwefelsäure gelb. Bei Behandlung mit Zinkstaub in alkalischer Lösung entsteht Buttersäure und Filicinsäure.

Pharmakologische Bedeutung besitzt nach KRAFT [1]) das *Filmaron*, dem im wesentlichen die anthelmintische Wirkung der Farndroge zugeschrieben wird (KRAFT [1])). Es enthält 4 Phloroglucinbutanongruppen in diphenylmethanartiger Bindung.

$$C_3H_7.OC.C \quad CO \quad C \text{——} CH_2 \text{——} C \quad CO \quad C.CO.C_3H_7$$

(chemische Strukturformel)

Voɪ der *Flavaspidinsäure* $C_{23}H_{28}O_8$ existieren eine α Form mit dem Fp 92° und eine β Form vom Fp 156°. Beide Substanzen sind alkohollöslich.

In relativ grosser, jahreszeitlich stark differierender Menge tritt im Filixrhizom die *Filixsäure* auf, die mit Alkohol gekocht Albaspidin abspaltet. In der Alkalischmelze entstehen Isobuttersäure und Phloroglucin; bei der Reduktion mit Zinkstaub erhält man neben Buttersäure und Phloroglucin noch Methyl-, Di- und Trimethylphloroglucin sowie Filicinsäure. Man schreibt daher der Substanz die folgende Konstitution zu:

(chemische Strukturformel)

Die Filixsäure wurde ausser im Wurmfarn auch in *Dryopteris rigida* und in *Athyrium Filix-femina* gefunden. Sie ist in Alkalien leicht löslich und aus dieser Lösung fällt die zuvor physiologisch unwirksame Substanz in amorpher und jetzt pharmakologisch wirksamer Form aus. Fp 184.5°.

---

[1]) KRAFT: Chem. Zentr. II, 400, 1896; 533, 1903; Arch. Pharm. 242, 489, 1904

*Phloraspin* $C_{23}H_{28}O_8$ ist in Chloroform und Keton löslich, ebenso in absolutem Alkohol. Letztere Lösung färbt sich mit $FeCl_3$ rotbraun. Fp 211°.

Aus *Dryopteris spinulosa* hat POULSSON [1]) eine Reihe weiterer verwandter Stoffe isoliert. Das von ihm gefundene *Polystichin* soll allerdings nach HAUSMAN mit dem von BOEHM aus dem Wurmfarnrhizom extrahierten Aspidin identisch sein, dem BOEHM die folgende Konstitution zuschreibt:

$$C.(CH_3)_2 \qquad\qquad C.CO.C_3H_7$$

$$OC\diagup\diagdown COH \qquad HOC\diagup\diagdown COH$$

$$C_3H_7.CO.C\diagdown\diagup C ---- CH_2 ---- C\diagdown\diagup C.CH_3$$

$$COH \qquad\qquad C.OCH_3$$

Für die übrigen in Farnen gefundenen Stoffe gibt POULSSON die folgenden Formeln an:

Polystichalbin $C_{22}H_{26}O_9$
Polystichinin $C_{18}H_{22}O_6$
Polystichocitrin $C_{15}H_{22}O_9$
Polystichoflavin $C_{24}H_{30}O_{11}$.

An Stelle der Filixsäure enthält das Rhizom von *Dryopteris athamantica* die nahestehende *Pannasäure* (KÜRSTEN [2]), HEFFTER [3])). Hinsichtlich der pharmakologischen Wirkung gibt TSCHIRCH an, dass Aspidinal völlig wirkungslos sei, Filixsäure u. Flavaspidinsäure wirken zwar lähmend, aber nicht anthelmintisch. Der Hauptwirkungsstoff ist das Filmaron. Es scheint aber, als ob der Wirkungsgrad der Droge ausserdem im erheblichen Mass vom Gehalt an ätherischen und besonders fetten Ölen in einer noch wenig durchsichtigen Weise abhängig sei.

Über die physiologische Bedeutung dieser Stoffe ist wie über diejenige der Phenolderivate im Ganzen wenig Sicheres bekannt. Sieht man von einer oberflächlichen Deutung als Schutzstoffe ab, so möchte man ihnen am ehesten eine bedeutsame Rolle bei der biologischen Oxydation beimessen. Es ist durchaus wahrscheinlich, dass die Phenoldehydrasen den Wasserstoff der phenolischen Hydroxydgruppen der Phloroglucinkomponenten sehr leicht beweglich machen.

Anderseits scheinen die Ketongruppen, wie sie in der Filicin-Flavaspidin- und Filixsäure, sowie im Filmaron und Aspidin vorhanden sind, gar wohl zur Aufnahme von Wasserstoff geeignet.

In der Polypodiacee *Matteuccia orientalis*, deren Rhizom die chinesische Droge Kuan-chung liefert, ist von früheren Autoren schon eine Reihe von

---

[1]) POULSSON: Arch. exp. Path. 35, 97, 1894, 41, 246, 1898.
[2]) KÜRSTEN: Arch. Pharm. 229, 258, 1891.
[2]) HEFFTER: Arch. exp. Path. 38, 458, 1896.

Farnsäuren festgestellt worden wie Filmaron, Flavaspidinsäure, Albaspidin, Aspidinol. Diesen Stoffen hat in einer neueren Untersuchung MUNESADA[1]) einen in Methylalkohol kristallisierenden Stoff Matteuccinol vom Fp 167–170 und der Zusammensetzung $C_{14}H_{14}O_4$ hinzugefügt. Verseifungszahl 201.

## § 11. Aetherische Oele.

— Obwohl diese weitverbreiteten Stoffe sicherlich auch den Pteridophyten nicht fehlen, finden sich in der Literatur nur gelegentlich spärliche Hinweise auf ihr Auftreten in Rhizomen (*Dryopteris Filix-mas*) und in Sporen. All diese Angaben tragen indes deutlich den Stempel von Nebenbefunden. Auf die nahe chemische Verwandtschaft der Sporopollenine mit Terpenen ist bereits hingewiesen worden. Bei der Destillation von *Equisetum palustre* erhielt BAUP ein ätherisches Öl von aromatischem Geruch. Fp = 58–65°. Harzartige Stoffe: Auf den Unterseiten der Blätter der Gattungen *Pityrogramma*, *Notolaena*, und *Cheilanthes* stehen köpfchenförmige Drüsenzellen, die kristallini- sche Ausscheidungen erzeugen. Diese letzteren besitzen je nach Art weisse oder gelbe Farbe, worauf die bei Gärtnern gebräuchliche Bezeichnung Gold- und Silberfarne Bezug nimmt. Während man diese Excrete früher für Wachse hielt, hat KLOTZSCH [2]) sie den kampherartigen Stoffen beigezählt (Pseudostearophene). STRASBURGER schrieb ihnen fettartigen Charakter zu, wohingegen WIESNER ihre Nichtzugehörigkeit zu Fetten wahrscheinlich machte. CZAPEK (I, S. 182) sagt, dass sie auch mit Wachsen nichts zu tun hätten. ZOPF [3]) hat dann die Stoffe rein gewonnen, indem er die Farnblätter für nur sehr kurze Zeit in Äther eintauchte. In der ätherischen Lösung fiel dann beim Einengen ein gelber Stoff, bei Bildung grösserer Kristalle ein roter Körper aus, der in kaltem und heissem Wasser, wie auch in Benzin unlöslich ist, sich aber in Aceton, Alkohol, Äther und Chloroform gut löst. Die alkoholische Lösung färbt sich mit $FeCl_3$ braunrot, reagiert neutral und zersetzt sich in Gegenwart von Alkali und Ammoniak ziemlich rasch unter Entfärbung. Der Gehalt der Blätter an solchem Gymnogrammen ist gering: 228 grössere und kleinere Farnblätter ergaben eine Ausbeute von 2,06 g. Elementare Zusammensetzung $C_{18}H_{18}O_3$.

Bei fraktionierter Lösung in Petroläther wurde aus dem Rohextrakt ein zweiter fast farbloser, wachsartiger Stoff (Fp. 63–64°) von neutraler Reaktion isoliert. Er erwies sich in kaltem Alkohol, Äther und Petroläther schwer löslich, dagegen leicht löslich in den heissen Flüssigkeiten. Ob dieser Körper ebenfalls von den Drüsenzellen stammt oder aber aus der Epidermis extrahiert wurde, ist nicht entschieden.

Das silberweisse Sekret von *Pityrogramma calomelanos* (Fp 140–142°) reagiert ebenfalls neutral und löst sich in Alkohol und Äther. $FeCl_3$ färbt Lö-

---

[1]) MUNESADA: Ber. Ohara Inst. für landw. Forschung II, 4, 429, 1924.
[2]) KLOTZSCH: Mon. Ber. d. Berl. Akad. Dezember 1851.
[3]) ZOPF: Ber. d. Deutsch. Bot. Ges. 24, 264, 1906.

sungen des C a l o m e l a n s rotbraun. Die Elementaranalyse ergab die Zusammensetzung $C_{20}H_{22}O_6$.

Über die Bildung harzartiger Stoffe und den Ort ihrer Entstehung wurde neuerdings von WEEVERS [1]) die ältere Theorie TSCHIRCHS (vgl. auch HÖHLKE) wieder aufgenommen, wonach die harzartigen Stoffe ein Produkt der Zellmembran darstellen sollen und in einer sog. resinogenen Schicht, die nach TSCHIRCH ein biologisches Membrankolloid im Sinne HANSTEEN CRANNERS darstelle, entstehen sollen. Nach WEEVERS wäre die Sekretbildung eine umgeleitete Ligninsynthese wie die Gummosis eine Art verunglückte Pektinsynthese darstellte.

Im Rhizom von *Dryopteris Filix-mas* werden die harzartigen Stoffe durch Drüsen ausgeschieden, welche besonders in der Nähe der Leitbündel kranzartig in die Interzellularen hereinragen. Das Wurmfarnharz ist löslich in Äther, Chloroform, Schwefelkohlenstoff, Benzol und in Alkalien. Dass die Harzbildung auf Kosten der Stärke erfolge, ist mehr eine naheliegende Vermutung als eine bewiesene Tatsache. Harzartige Stoffe wurden auch aus dem Destillationsrückstand von *Equisetum palustre* gewonnen.

**§ 12. Gerbstoffe.** — Schütze [2]) erwähnt in seiner Schilderung tropischer Farne das Vorkommen von Gerbstoffschläuchen in Stamm und Blattstielen, deren Inhalt sich mit Eisenchlorid schwarz färbt. Als Bestandteil der Droge aus Wurmfarnrhizom wird die F i l i x g e r b s ä u r e erwähnt, die in Mengen von 4,5–11,7% (KRUSE) auftritt. Sie ist ein rotbraunes, in Wasser wenig lösliches Pulver von der Zusammensetzung $C_{41}H_{36}NO_{18}$ (nach REICH $C_{82}H_{76}N_2O_{38}$). Mit Kalkwasser erhitzt, spaltet sie P y r r o l ab, mit $H_2SO_4$ erhitzt, geht sie in F i l i x r o t ($C_{26}H_{18}O_{12}$) über, das wohl mit der T a n n a s p i d s ä u r e von LUCK identisch ist. Bei der Hydrolyse wird Zucker abgespalten.

Die P r o t o f i l i x s ä u r e, ebenfalls ein Bestandteil des Wurmfarnrhizoms zerfällt unter hydrolytischer Einwirkung in P r o t o c a t e c h u s ä u r e und P h l o r o g l u c i n. Sie dürfte in enger stofflicher Beziehung zu den Phlorogluciden des Wurmfarns stehen.

In *Equisetum palustre* wurden Gerbsäuren höchstens in Spuren gefunden.

**§ 13. Alkaloide.** — Die Giftigkeit von *Equisetum palustre* wurde früher zu Unrecht auf dessen Gehalt an Akonitsäure zurückgeführt. Demgegenüber hat LOHMANN [3]) nicht nur die Unschädlichkeit dieser Säure für

[1]) WEEVERS: Kon. Akad. Wetensch. Amsterdam 37, 181, 1934.

Vrgl. auch TSCHIRCH u. STARK: Die Harze, S. 33; FRANCESCONI: Riv. ital. delle ess. e dei prof. 1829; LEHMANN, C.: Planta I, 1926, S. 343.

[2]) SCHÜTZE: Zur physiologischen Anatomie einiger tropischen Farne besonders der Baumfarne. Diss. Berlin 1905.

[3]) LOHMANN, J.: Arb. d. Deutschen Landw. Ges. Heft 100, 1904.

seine Versuchstiere nachweisen, sondern auch aus dem Schachtelhalm ein Alkaloid E q u i s e t i n isolieren können, das sich als starkes Herzgift erwies (Vgl.auch BAUP).

Bekannter und besser definiert ist das L y c o p o d i n aus *Lycopodium complanatum*. Hierbei handelt es sich um eine zweisäurige, bitter schmeckende Base vom Siedepunkt 114–115°, die in Äther ,Alkohol, Wasser, Chloroform und Benzol löslich ist. Ihre Zusammensetzung entspricht der Formel $C_{32}H_{52}N_2O_3$ (BOEDEKER [1])). Im tropischen *Lycopodium Saururus* entdeckte BARDET [2]) das P i l i j a n i n $C_{15}H_{24}N_2O$, das abführend und zum Erbrechen reizend wirkt. Fp. 64–65°. Es wird für ein Nicotinderivat (vielleicht Oxyamylnicotin) [3]) gehalten.

Aus *Pteridium aquilinum* hat TODA [4]) zwei nicht näher beschriebene Basen neben reichlichen Mengen von Betain und Cholin isolieren können. Der letztere Körper, der als wichtigste Lecithinkompomente eine bedeutsame physiologische Rolle spielt, wurde überdies auch in *Equisetum limosum* und *Phyllitis Scolopendrium* gefunden.

**§ 14. Weitere organische Stoffe in Pteridophyten.** — Es soll versucht werden in diesem Kapitel die spärlichen Angaben über organische Stoffe in Pteridophyten zu sammeln, die nur selten vorkommen oder über die bei Pteridophyten wenig Sicheres bekannt ist.

Der aus den Blättern höherer Pflanzen durch FRANZEN bekannt gewordene H e x y l e n a l d e h y d ist auch in den Blättern von *Pteridium aquilinum* gefunden worden.

Über das Auftreten von Fetten und Ölen wird ebenfalls nur spärlich berichtet. BÄSECKE gibt an, dass in alten Rhizomteilen von *Davallia bullata* mit dem Verschwinden von Stärke sich vikariierend Fett einstellt. TSCHIRCH gibt für das Rhizom des Wurmfarns 6% fettes Öl an, das neben Ölsäure noch Palmitin- u. Cerotinsäure enthält. Im Öl von *Dryopteris spinulosa* treten neben der Ölsäure noch die Linol- u. Linolensäure auf.

BRACONNET gibt für *Equisetum maximum* 0,40 g Fett und Chlorophyll pro 500 g Frischgewicht an. Relativ fettreich dagegen sind vielfach die Sporen, doch haben wir weder über die Natur noch über die Menge der Fette nähere Angaben finden können. Eine Ausnahme macht nur das Fett der Sporen von *Lycopodium clavatum*. Nach LANGER [5]) enthalten die Sporen dieses Bärlapps 49,34% Fett. 80–86% dieses Fetts sind Glyceride der

---

[1]) BOEDEKER: Lieb. Ann. 208, 363, 1881.

[2]) Vgl. hierzu: ARATA und CANZONERI: Gazz. Chim. ital. 22, I, 146, 1892; ADRIAN: Compt. rend. 102, 1322, 1886.

[3]) DOMINGUEZ: Chem. Zentralbl. 1932, I, 3452.

[4]) TODA: Journ. of Bioch. 2.433, 1913.

[5]) LANGER: Arch. Pharm. (3) 27, 241, 625 (1889).

Lycopodiumölsäure, deren Konstitution der folgenden Formel entspricht:

$$\begin{matrix} CH_3 \\ CH_3 \end{matrix} \Big\rangle CH . CH : C \Big\langle \begin{matrix} COOH \\ (CH_2)_9 . CH_3. \end{matrix}$$

Ausserdem [1]) enthält das Fett noch 3,2% Dioxystearinsäure, 1.13% Stearinsäure, 0.85% Palmitinsäure und 2,0% Myristinsäure. In den Sporen von *Lycopodium Selago* fand KEEGAN [2]) 47% Fettsäureglyceride und freie Fettsäuren. In den Sporen von *L. clavatum* wurden überdies noch gefunden: 7.8% Glycerin, 3% Zucker, davon 2.1% Saccharose; entgegen den Angaben von RIEGEL, FRITSCHE und WINKLER scheint Stärke gänzlich zu fehlen, das Auftreten von Äpfel- und Citronensäure bedarf der Bestätigung.

**§ 15. Die Mineralstoffe der Pteridophyten.** — Der Mineralstoffwechsel und Gehalt der Pteridophyten an Aschenstoffen schliesst sich, soweit er überhaupt näher studiert ist, eng an denjenigen der Blütenpflanzen an. Es sollen an dieser Stelle lediglich einige Besonderheiten hervorgehoben werden. Auf die hohen K i e s e l s ä u r e m e n g e n, die in manchen Pteridophytenaschen gefunden werden, ist bereits an anderer Stelle hingewiesen worden. Daneben verdient besondere Erwähnung der hohe A l u m i n i u m g e h a l t der Asche von bestimmten *Lycopodium*-Arten. Wie die folgende Zusammenstellung ausweist, ist diese Besonderheit jedoch kein durchgängiges Familienmerkmal. Vielmehr wechselt der Al-Gehalt bei verschiedenen Arten der Familie ausserordentlich stark.

Für *Lycopodium complanatum* gibt ALDERHOLDT [3]) einen Tonerdegehalt von 51,85–57,36% an, in *Lycopodium clavatum* bestimmte CHURCH 15,24% $Al_2O_3$, während SALMS und ALDERHOLDT auf wesentlich höhere Werte, nämlich 27 bez. 26,5% kamen. In den Sporen derselben *Lycopodium*-art wies LANGER einen $Al_2O_3$-Gehalt von 15.3% nach. Einen exorbitant hohen Aluminiumwert fand COUNCLER bei *Lycopodium chamaecyparissus* mit 39,07%. Des weiteren wurde der Aluminiumgehalt noch in den folgenden Pteridophyten bestimmt: *Lycopodium alpinum* 33,5% (CHURCH), *L. cernuum* 10,09%, *L. Selago* 7,29%, *L. annotinum* 18,1%, *Salvinia natans* 1,86%, *Psilotum triquetrum* und *Marsilia* Spuren.

Auffallend gering ist der Al-Gehalt bei epiphytischen Lycopodien wie z.B. *Lyc. Phlegmaria*, das nur 0,45% $Al_2O_3$ in der Asche aufweist, während *Lyc. Billardieri* gar kein Aluminium enthält.

Über die physiologische Bedeutung des Aluminiums in diesen Pflanzen und über die Abhängigkeit seines Vorkommens von der Bodenunterlage scheinen keine näheren Angaben vorzuliegen.

Von den *Filicinae* schien nach einer Untersuchung LANGERs unter den

---

[1]) RATHJE: Arch. Pharm. 246, 692, (1908).

[2]) KEEGAN: Botan. Zentr. 96, 575, (1904).
Vgl. auch TSCHIRCH: Handb. d. Pharmak.

[3]) Vgl. TSCHIRCH: Handbuch der Pharmakognosie und CZAPEK II, 371.

geprüften Farnen nur eine neuseeländische Cyatheacee einen besonders hohen Aluminiumgehalt (19,65%) aufzuweisen.

Gesamtaschenanalysen liegen von den folgenden Formen vor:

| | $K_2O$ | $Na_2O$ | $CaO$ | $MgO$ | $Mn_3O_4$ | $Fe_2O_3$ | $P_2O_5$ | $SO_3$ | $SiO_2$ |
|---|---|---|---|---|---|---|---|---|---|
| *Lycopodium annotinum* . | 37,29 | 1,49 | 8,54 | 6,35 | 4,00 | 1,35 | 6,52 | 12,56 | 3,52 |
| *Ophioglossum* . . . . . | 64,10 | 3,53 | 14,65 | 4,60 | 0,47 | 0,19 | 3,44 | 5,44 | 3,58 |

Gegenüber der chemischen Beschaffenheit des Untergrundes verhalten sich die einzelnen Arten, selbst aus einer Gattung sehr verschieden; im allgemeinen aber herrscht doch eine ziemlich weitgehende Indifferenz gegenüber der chemischen Bodenbeschaffenheit vor. Ganz fehlt es indessen auch nicht an Bodenspezialisten. Da eine zusammenfassende Darstellung nach dieser Richtung hin fehlt, können hier nur einige wenige, besonders auffällige Formen herausgegriffen werden, und das nur mit Bezugnahme auf die Anforderungen der Pflanzen an die B o d e n r e a k t i on. Diese Frage ist in der älteren Literatur noch unter dem Stichwort K a l k- u n d K i e s e l - s ä u r e p f l a n z e n ohne nähere Angaben über die Grenzen der den Pflanzen erträglichen H-Konzentrationen behandelt worden. Ein Beispiel für die Uneinheitlichkeit der Reaktionsanprüche der Pflanzen innerhalb der engsten systematischen Einheiten mag die Gattung *Dryopteris* sein. Es ist bekannt, dass *Dryopteris rigida* nur auf K a l k gedeiht, während *D. cristata* eine ausgesprochene M o o r p f l a n z e ist und das verwandte *Polystichum lonchitis* auf a l l e n B ö d e n fortkommt. Ähnliche Differenzen sind auch innerhalb der Gattung *Asplenium* bekannt geworden: *A. viride* wächst auf a l l e n B ö d e n, *A. adulterinum* und *A. Adiantumnigrum* sind spezifische S e r p e n t i n f o r m e n, *A. septentrionale* f l i e h t den Kalk, *A. fontanum* s u c h t ihn. U r g e s t e i n s f o r m e n sind u.a. weiterhin *Allosorus crispus, Notholaena Marantae*, während *Phyllitis Scolopendrium* und *Blechnum Spicant* k a l k l i e b e n d e Farne sind. *Pteridium aquilinum* wird im allgemeinen als verlässlicher Säurebodenanzeiger bewertet. DE LITARDIÈRE [1] beschreibt indessen einen Standort des Adlerfarns mit einem ph von 8,33. Von *Cystopteris fragilis* gibt MUSSACH [2] an, dass die Sporenkeimung optimal bei einem ph von 6–6,8 verläuft. Nähere Angaben über amerikanische Farne und deren Abhängigkeit von der Bodenreaktion macht STEAGALL. CRAW [3] beschreibt die Bodenreaktionen der natürlichen Farnassoziationen von Indiana.

Die *Lycopodiinae* scheinen im allgemeinen saure Standorte zu bevorzugen, während *Equisetum arvense* und *E. maximum* schweren lehmreichen

---

[1] DE LITARDIÈRE: Bull. Soc. Bot. France 80, 230–233.
[2] MUSSACH: Beih. botan. Zentralbl. 51, 204–254, 1933.
[3] CRAW: Butler University. Botan. Studies, 2, 151–158, 1932.

Boden aufsuchen. Im ganzen ist also die Zahl der Leitpflanzen unter den Farnen nach unseren heutigen Kenntnissen nicht sehr gross. Es erscheint indessen wahrscheinlich, dass ihre Zahl durch weitere exakte Untersuchungen doch noch erheblich vermehrt werden könnte. Dass es unter den Pteridophyten nicht an Mineralstoffspezialisten fehlt, das erweisen die Aluminium- und Kieselsäurepflanzen, wie auch die Tatsache, dass der K a l i - r e i c h t u m von *Pteridium aquilinum* früher sogar technisch ausgewertet wurde.

**§ 16. Die Assimilation.** — Über die Assimilation von Farnen in Abhängigkeit von Standort und Entwicklung sind wir durch die schöne Arbeit NILS JOHANSSONS [1]) eingehender unterrichtet worden. Unter Verwendung der LUNDEGÅRDHschen Methoden der Assimilationsmessung und in Anlehnung an seine theoretischen Darlegungen hat JOHANSSON die Standortsbedingungen der untersuchten Farne einer eingehenden quantitativen Prüfung unterzogen und die Reaktionsweise seiner Versuchspflanzen gegenüber den ökologischen Faktoren ermittelt. Auf diese Weise erhalten wir nicht nur Einblick in den tatsächlichen Verlauf und die Grösse der Assimilation bestimmter Farne, vielmehr wird auch die Bedeutung der einzelnen ökologischen Faktoren für das Ausmass der Kohlehydratbildung abschätzbar.

Bereits ältere Untersuchungen liessen JOHANSSON vermuten, dass die Schliesszellenbewegungen Verlauf und Grösse der Assimilation am natürlichen Standort massgeblich beeinflussen. Daher wurden die Anteile der Einwirkung der Aussenfaktoren auf die Grösse der Spaltöffnungsfläche getrennt zu bestimmen versucht.

Schon hinsichtlich der Lichtwirkung ergaben sich interessante Differenzen bei Farnen schattiger und sonniger Standorte. Als Versuchsobjekt der ersteren ökologischen Gruppe stand die s.g. *Dryopteris austriaca* zur Verfügung. Hier bewirkte unter sonst günstigen Aussenbedingungen bereits eine Lichtintensität von 1% der natürlichen maximalen Sonnenbestrahlung Öffnung der Stomata, so dass man annehmen darf, dass unter natürlichen Bedingungen das Licht nur eine geringe Rolle für die Öffnung der Spalten spielt. Tatsächlich erfolgte nur an ganz regnerischen Tagen bereits in den frühen Nachmittagsstunden ein Dunkelschluss der Stomata. Viel bedeutender scheint die Luftfeuchtigkeit für den Öffnungszustand der Spalten zu sein: bereits eine solche von 75% wirkte sich als Öffnungsdepression aus. Infolge der dadurch bedingten Herabsetzung der Luftfeuchtigkeit machte sich gewöhnlich auch bei Erhöhung der Temperatur über 15° eine deutliche Schliessungstendenz bemerkbar, im Zusammenhang mit einer sich rasch verschlechternden Wasserbilanz. Die Schattenpflanze ist also insofern an ihren Standort angepasst, als ihre Schliesszellen empfindlich nur auf d i e Faktoren reagieren, deren relative Konstanz durch den Standort selbst

[1]) JOHANSSON N.: Svensk. Bot. Tidskr. 1923 and 1926.

weitgehend gesichert erscheint. Ganz anders verhält sich *Dryopteris spinulosa* als Vertreter des S o n n e n p f l a n z e n t y p s . Wohl spielen auch hier die Lichtverhältnisse am natürlichen Standort keine Rolle für das Zustandekommen der Spaltenöffnung, aber auch gegenüber den übrigen Aussenfaktoren erweist sich der Spaltöffnungszustand dieser Pflanze bedeutend unabhängiger. Zwar setzt auch hier eine ungünstig werdende Wasserbilanz die Öffnungsweite herab, aber die Drosselung ist eine erheblich geringere als beim Schattenfarn. Die Folge hiervon ist, dass beim Sonnenfarn Öffnungsgrad und Öffnungszeit der Spalten grösser sind als bei den Schattenformen. Im Hinblick auf die Loftfieldsche Typisierung müssen die Farne dem Luzernetyp zugeordnet werden: unter günstigen Bedingungen sind die Spalten tagsüber geöffnet, nachts geschlossen. Bei sich verschlechternder Wasserbilanz können die Spalten tagsüber zeitweise oder im Extremfall ganz geschlossen werden; dann erfolgt eine kompensatorische nächtliche Öffnung der Stomata.

Bei den engen Beziehungen zwischen Sättigungsdefizit, Temperatur und Wasserbilanz der Pflanzen konnte nach den oben dargelegten Ermittlungen auch ein erheblicher Einfluss der Temperatur auf die Öffnungsweite der Spalten vermutet werden, was in der Tat durch den Versuch bestätigt wurde: Erhöhung der Temperatur wirkte sich spaltenschliessend aus.

In gleichlaufenden Versuchen wurde nun der Einfluss der W a s s e r b i l a n z auf die $CO_2$-Assimilation ermittelt. Dabei konnte eine völlige Konkordanz zwischen Spaltöffnungsweite und Assimilationseffekt festgestellt werden, so dass JOHANSSON zu dem Schluss kommt, d a s s u n t e r n a t ü r l i c h e n V e r h ä l t n i s s e n d i e S p a l t ö f f n u n g s f l ä c h e i n e r s t e r L i n i e d i e G r ö s s e d e r A s s i m i l a t i o n b e s t i m m t . Die Wirkung der Aussenfaktoren auf die Assimilation ist daher stets eine indirekte, die über eine primäre Einwirkung auf den Öffnungsgrad der Spalten läuft. Daher erweist sich der Lichteinfluss auf die Assimilation auch nicht als konstant, sondern in deutlich erkennbarer Abhängigkeit von andern die Schliesszellenbewegung regulierenden Faktoren.

L i c h t i n t e n s i t ä t u n d A s s i m i l a t i o n b e i F a r n e n :

Auch bei den Farnen treten hinsichtlich der Lichtwirkung auf die Assimilation die beiden Typen der Sonnen- und Schattenpflanzen auf.

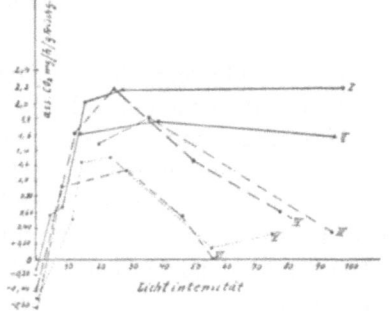

FIG. 2. — *Dryopteris austriaca.* Abhängigkeit der Assimilation von der Lichtintensität.( n. JOHANSSON).

1. Typus der Schattenpflanze: *Dryopteris austriaca:*

Fig. 2 gibt eine anschauliche Vorstellung von der Lichtwirkung in Abhängigkeit anderer Faktoren: Die Kurven I und II sind an trüben Tagen

mit hoher Luftfeuchtigkeit aufgenommen worden: mit zunehmender Lichtintensität steigt die Assimilation zunächst an, bis zwischen 25 und 30% der
natürlichen Strahlungsintensität ein Maximum in der Assimilation erreicht
wird. Von weiterer Steigerung der Lichtintensität bleibt die Assimilation
unberührt, die Kurven laufen der Abscissenachse parallel weiter. Stehen
die Öffnungsbewegungen der Schliesszellen jedoch bereits unter einer Licht-
depression, wie das an hellen Tagen (Kurve III und IV) oder am Nachmit-
tag der Fall ist (Kurven V und VI), so verschiebt sich einerseits das Assimi-
lationsoptimum in die Bezirke geringerer Lichtintensität, und anderseits
fällt die Assimilation bei Erhöhung der Beleuchtung ziemlich rasch ab.

2. Sonnenpflanzen: *Pteridium aquilinum* und *Dryopteris spinulosa*:

Fig. 3 gibt die Beziehungen zwischen Lichtintensität und Assimilation
bei *Pteridium aquilinum* wieder:

Gegenüber dem Schattentypus fällt zunächst das höhere Lichtoptimum

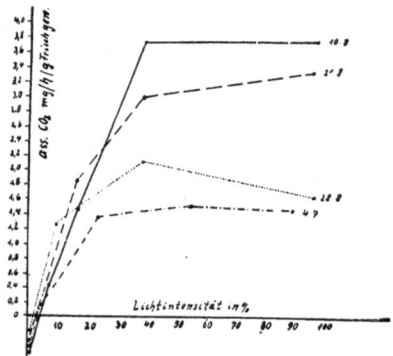 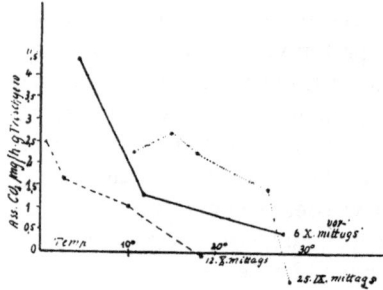

FIG. 3. — *Pteridium aquilinum*. Abhängigkeit der Assimilation von der
Lichtintensität. (n. JOHANSSON).

FIG. 4. — *Dryopteris austriaca*. Ab-
hängigkeit der Assimilation von der
Temp. bei 0,03% $CO_2$ und 20–35%
Licht. (n. JOHANSSON).

auf, das in Abhängigkeit vom Sättigungsdefizit der Luft zwischen 30 und
100% der natürlichen Strahlungsintensität zu liegen kommt. Selbst an völlig klaren Tagen mit relativ geringem Luftfeuchtigkeitsgehalt steigt die As-
similationskurve gelegentlich noch bis zur sehr hohen Lichtintensitäten an;
niemals erfolgt ein starker Assimilationsabfall mit Erhöhung der Lichtintensität, wie er für die Schattenfarne charakteristisch ist.

Die Konkordanz mit den Beziehungen zu Spaltöffnungsbewegung in Abhängigkeit von der Belichtung bei den beiden ökologischen Typen tritt hier
sehr klar in Erscheinung. Licht- und Schattenfarne sind assimilationsmäs-
sig an die Licht- und Feuchtigkeitsverhältnisse ihrer Standorte angepasst
durch eine r e g u l i e r t e  R e a k t i o n s w e i s e  i h r e r  S c h l i e s s-
z e l l e n . Wo durch Überdosierung der Lichtintensität eine Assimilations-
hemmung eintritt, sind in erster Linie Schliesszellenbewegungen verantwortlich zu machen, gegen die sog. plasmatische Faktoren, wie Enzym-

zersetzung, Inaktivierung der Chloroplasten, Stärke- und Zuckeranhäufung in den Plastiden an Einfluss stark zurücktreten.

Assimilation und Temperatur:

Auch in der Temperaturwirkung auf die Assimilation unterscheiden sich Sonnen- und Schattenfarne recht beträchtlich. Bei den Schattenpflanzen liegt das Temperaturoptimum der Assimilation erstaunlich niedrig. (Vgl. Fig. 4).

Unter natürlichen Bedingungen (Lichtintensität 20–35%, $CO_2$-Gehalt 0,03%) wurde die höchste Assimilationsleistung in einiger Abhängigkeit von andern nicht näher definierten Aussenfaktoren unterhalb 15°C gemessen. Bei weiterem Temperaturanstieg fällt dann die Assimilation rasch ab. Besonders steil ist der Abfall der Kurven von etwa 26° ab, wo dann Assimilation und Atmung sich vielfach nur eben noch das Gleichgewicht halten. Der Kompensationspunkt liegt indessen nicht fest, sondern hängt wesentlich von der Öffnungsfläche der Spalten und damit von der absoluten Höhe der Assimilationskurven ab, derart, dass der Kompensationspunkt um so niedriger liegt, je geringer das absolute Ausmass der Assimilation ist. Zweifellos aber ist die Temperaturwirkung auf die Assimilation kein einheitlicher Vorgang. Wie bereits angedeutet, hat JOHANSSON die Atmung der Pflanze mit in die Assimilationsbestimmung eingeschlossen, so dass bereits eine verschiedene Temperaturempfindlichkeit dieser beiden physiologischen Vorgänge zu komplizierten ökologischen Assimilationskurven führen könnte. Dass die Temperatur die rein chemischen Assimilationsprozesse weit über 10° hinaus beschleunigt, wird ohne weiteres ersichtlich, wenn die Schliesszellenbewegung durch erhöhte Kohlensäuretension der Atmosphäre kompensiert wird. Die Fig. 5 gibt das Ergebnis eines solchen Versuchs wieder: tatsächlich steigt hier die Assimilation mit der Temperatur bis etwa 38° in linearer Abhängigkeit von der Temperatur. Der Temperaturquotient liegt bei etwa 1,5.

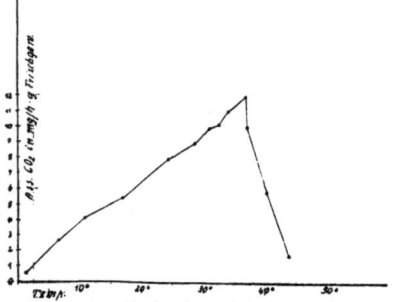

Sonnenfarne: *Pteridium aquilinum*: Hier liegt der Temperaturoptimalpunkt etwas höher (15°) unter natürlichen Bedingungen. Die Assimilation ist absolut gemessen wesentlich höher, und daher liegt auch der Kompensationspunkt selbst unter ungünstigen Verhältnissen noch über 30°. Sogar an

FIG. 5. — *Dryopteris austrica*. Abhängigkeit der Assimilation von der Temperatur bei hohem $CO_2$-Gehalt u. schwachem Licht. (n. JOHANSSON).

sehr heissen Tagen geben daher die Sonnenfarne noch einen Assimilationsüberschuss, wenn auch das Ausmass desselben bei niederer Temperatur steigt.

Dass diese Verschiedenheit der Reaktionsweise der Sonnen- und Schat-

tenfarne auf Temperatureinflüsse massgeblich von der Schliesszellenbewegung bestimmt ist, konnte JOHANSSON mit vergleichenden Versuchen über
den Einfluss der Aussenfaktoren auf Schliesszellenbewegung und Assimila-
tionshöhe wahrscheinlich machen: Unter günstiger Wasserökonomie (vor-
mittags) steigt die Assimilation mit der Temperatur bis auf etwa 32° an.
Der Temperaturquotient ist dabei etwa 2. Unter diesen Verhältnissen trat
auch keine Verengerung der Spalten ein, im Gegensatz zu nachmittaglich
angesetzten Versuchen, in denen Spaltenweite und Assimilation symbath
gehend, sich mit Temperatursteigerung verringerten. Interessant erscheint
noch der Vergleich der Temperaturoptima der Farne und derjenigen anderer Pflanzen in Bezug auf die Assimilation. Für Kartoffel und Tomate gibt
LUNDEGÅRDH ein Optimum von 20°, für die Zuckerrübe ein breites Opti-
mal-Gebiet zwischen 20 und 30° an. Nur Frühlingsblüher, wie *Anemone*

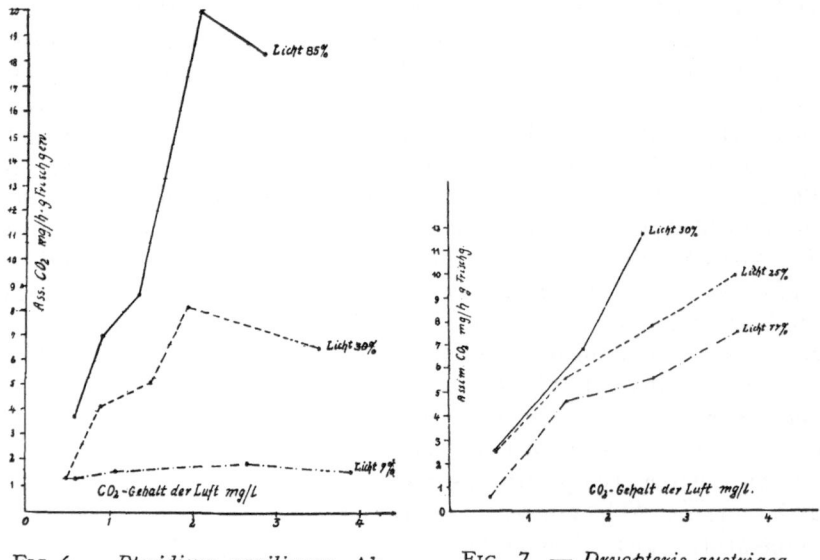

FIG. 6. — *Pteridium aquilinum*. Abhängigkeit der Assimilation vom $CO_2$-
Gehalt der Luft bei drei verschiedenen Lichtintensitäten.
(n. JOHANSSON).

FIG. 7. — *Dryopteris austriaca*.
Abhängigkeit der Assimilation
vom $CO_2$-Gehalt der Luft bei
drei verschiedenen Lichtintensitäten. (n. JOHANSSON).

*nemorosa* weisen ein ähnlich tiefes Temperaturoptimum der Assimilation
wie die Farne auf.

Der Assimilationsquotient wurde in allen untersuchten Fällen gleich 1
gefunden, so dass man annehmen kann, dass der Gaswechsel tatsächlich
das Ausmass der stattgehabten Assimilation richtig widerspiegelt.

Der Einfluss der $CO_2$-Konzentration auf die Assimilation.

TSCHESNOKOV und BAZYRINA [1]) haben in einer kurzen Betrachtung dar-

---

[1]) TSCHESNOKOV u. BAZYRINA: Planta 1930, S. 473.

auf hingewiesen, dass das BLACKMANsche Faktorengesetz ebenso wenig
strenge Gültigkeit für sich in Anspruch nehmen kann, wie die LUNDE-
GÅRDHsche Formulierung der Spannung des Licht- und Kohlensäurefak-
tors, da alle äusseren Faktoren die $CO_2$-Assimilation nur indirekt beeinflussen, während die direkten Faktoren quantitativ nicht zu fassen sind. Hinsichtlich der $CO_2$-Einwirkung auf die Zuckerbildung müsste die Konzentration an den Reaktionsoberflächen bekannt sein, um gesetzmässige Beziehungen zwischen $CO_2$-Tension und Photosynthese ermitteln zu können,
ganz ungeachtet der Beeinflussung anderer physiologischer Prozesse durch
die Kohlensäure. Trotz dieser gewichtigen Vorbehalte ergeben die JO-
HANSSONschen Versuche doch recht bemerkenswerte Ergebnisse. Sehr klar
zum Ausdruck kommt die Abhängigkeit der $CO_2$-Einwirkung von der Lichtintensität, und zwar in doppelter Weise: einerseits als ein die Spaltenweite
regulierender Faktor und anderseits als Energiequelle des endothermen
photosynthetischen Vorgangs. In allen Fällen ist eine Steigerung der Assimilation bei Erhöhung des $CO_2$-Gehalts der Atmosphäre zu bemerken. Bei
Sonnenfarnen (Vgl. Fig. 6) geht die durch vermehrte $CO_2$-Darbietung induzierte Assimilationssteigerung der applizierten Lichtintensität durchaus
über den ganzen Lichtbereich bis zur vollen Sonnenstrahlung symbath. Bei
den Schattenfarnen (Vgl. Fig. 7) hingegen vermindert sich die Assimilationssteigerung oberhalb der für die Spaltenöffnung kritischen Lichtintensität von etwa 30% wieder. Ökologische Bedeutung scheinen diese Zusammenhänge vor allem für Schattenfarne zu besitzen, an deren Standort
erhöhte $CO_2$-Konzentration in Luft und Bodenwasser gegenüber derjenigen
eines sonnigen Standorts nachweisbar waren.

Will man die maximale Assimilationsleistung der Farne verschiedener
ökologischerTypen vergleichen, so kommt man zu recht verschiedenen
Werten je nach der angewandten Bezugsgrösse; sehr klare Verhältnisse
ergeben sich, wenn man auf das Frischgewicht der Pflanzen bezieht, während pro Einheit der Blattfläche sich wenig diskutierbare Resultate ableiten lassen. Vom ökonomischen Standpunkt aus interessierte vor allem die
Bezugnahme der Assimilationsleistung auf die Einheit des Trockengewichts, weil sie eine Art Rentabilitätsquote des investierten biologischen
Kapitals darstellt. Leider liegen derartige Angaben nicht vor, immerhin
ergeben sich einige Zusammenhänge auch aus der von JOHANSSON gegebe-
nen Assimilationsbezugnahme auf das Frischgewicht.

Pro 1g Frischgewicht werden von den untersuchten Farnen pro Stunde
die folgenden Mengen $CO_2$ (in mg) assimiliert:

Dryopteris austriaca . . . . . . . . . . 2,44 mg
Polypodium vulgare . . . . . . . . . . 2,24 ,,
Pteridium aquilinum . . . . . . . . . . 3,94 ,,
Dryopteris spinulosa . . . . . . . . . . 4,72 ,,

Wenn auch bei den Schattenfarnen (1 und 2) das relative Trockengewicht
niedriger als bei den Sonnenfarnen (3 u. 4) ist, so scheinen doch die Sonnen-

farne zur stärkerer Produktion organischer Substanz befähigt als die Schattenpflanzen, sofern eine Einzelbeobachtung eben überhaupt generelle Schlüsse zulässt.

Fasst man das gesamte Tatsachenmaterial über die Zusammenhänge zwischen Assimilationsleistung und Aussenfaktoren des Standorts zusammen, so kann man sagen, d a s s  s o w o h l  S o n n e n -  w i e  S c h a t t e n f a r n e  b e m e r k e n s w e r t e  Z ü g e  d e r  A n p a s s u n g  a n  i h r e n  S t a n d o r t  t r a g e n.

C h l o r o p h y l l b i l d u n g  i m  V e r l a u f  d e r  B l a t t e n t w i c k l u n g.

Einer der wichtigsten inneren Faktoren der Assimilation ist der Chlorophyllgehalt der Blätter. Auch hinsichtlich der Entstehung des grünen Farbstoffs, im Verlauf der pflanzlichen Entwicklung zeigen die einzelnen Farne recht abweichende Bilder:

*Dryopteris austriaca* verfügt schon in den noch eingerollten Blättern über einen ansehnlichen Chlorophyllgehalt, der sich im Verlauf der Blattentwicklung weiter bis auf den dreifachen Wert des anfänglichen Gehalts steigert. Auch die ökologische Antipode, die Sonnenform *Dryopteris spinulosa* weist bereits in den eingerollten Blättern hohen Chlorophyllgehalt auf. Im Gegensatz zur Schattenpflanze vermindert sich derselbe jedoch mit der sich vollziehenden Aufrollung der Blätter ungemein rasch, so dass er innerhalb eines Monats auf die Hälfte des Anfangswerts reduziert ist. Dabei färben sich die Blätter rasch gelb; ob diese Verfärbung nur eine Folge des Chlorophyllrückgangs oder zugleich auch einer Neubildung von gelben Farbstoffen ist, wurde leider nicht ermittelt. Der Verlauf der Chlorophyllkurve lässt vermuten, dass *Dryopteris spinulosa* ursprünglich eine Schattenpflanze war, die erst sekundär auf Grund weitgehender Anpassung sonnige Standorte zu erobern vermochte.

*Pteridium aquilinum* zeichnet sich durch einen besonders niederen Chlorophyllgehalt seiner jungen Blätter aus, der sich indes während der Vegetationsperiode bis auf den neunfachen Betrag des Anfangsgehalts steigert.

D i e  E n t w i c k l u n g  d e r  p h o t o s y n t h e t i s c h e n  A k t i v i t ä t  d e r  B l ä t t e r.

Eingerollte Blätter von *Polypodium*, deren Chlorophyllgehalt bei etwa 4 mg pro 1 g Frischgewicht lag, wurden ohne erkennbare assimilatorische Tätigkeit befunden. Erst im Verlauf der Streckung und Aufrollung der Blätter wurde bei 25% Lichtgenuss der Kompensationspunkt erreicht. Eine bemerkenswerte aktive Assimilationsbilanz ist indes erst im völlig aufgerollten Blatt zu konstatieren. Dann allerdings warten diese Blätter sehr bald mit der vollen Assimilationsleistung auf.

Bekanntlich unterscheidet man nach dieser Richtung hin zwei Typen von Pflanzen:

1. d e n  H a f e r -  o d e r  G e r s t e n t y p u s, der seine volle assimilatorische Potenz erst in einem längeren Entwicklungsverlauf erreicht, was

man mit der Bildung eines Assimilationsenzyms in Zusammenhang brachte (IRVING) [1]).

2. den H e l i a n t h u s t y p, dessen Kotyledonen sofort die volle assimilatorische Fähigkeit entfalten, da in den Reserveorganen angeblich (BRIGGS) [2]) das Assimilationsenzym bereits in hinreichender Menge präformiert sein soll.

Von den Farnen neigt *Pteris* dem Helianthustyp, *Polypodium* mehr dem Hafertypus zu.

### § 17. Tagesperiodische Schwankung der Atmung von Farnblättern.

— Zwischen einem vor-und nachmittäglichen Atmungsmaximum liegt eine zeitlich mehr oder weniger ausgedehnte Atmungsdepression um die Tagesmitte. Dabei handelt es sich offenbar nicht um eine autonome Periodizität wie sie DELEANO [3]) und BENNET–CLARK[4]) an Blättern höherer Pflanzen festgestellt haben, sondern um eine durch Aussenfaktoren induzierte Rhythmik. Unter diesen scheint die Wassersättigung der Pflanze die wichtigste Rolle zu spielen, deren Tages-Kurve in ihrem Verlauf eine auffallende Ähnlichkeit mit der Atmungskurve besitzt. (Vgl. KRASSNOSELSKI–MAXIMOV [5])). Dementsprechend sind mittägliches Wasserdefizit und Atmungsdepression bei den Sonnentypen der Farne am grössten. (Vgl. Fig.8).

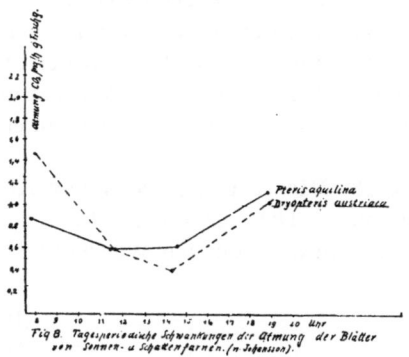

FIG. 8. — Tagesperiodische Schwankungen der Atmung der Blätter von Sonnen- u. Schattenfarnen. (n. JOHANSSON).

Leider besitzen wir über den Zusammenhang zwischen Atmungsintensität und Wassergehalt nur statistische Angaben ohne Kausalanalyse (ILJIN [6])). Aus den Untersuchungen HORNS [7]) und TOLLENAERS [8]) ist zwar bekannt geworden, dass der Saccharosegehalt der Blätter mit abnehmendem Wassergehalt ansteigt, aber wir können in einer Zuckerkonzentrationserhöhung in dem vorliegenden Ausmass keine Ursache einer Atmungsdämpfung

---

[1]) IRVING: Ann. of Bot. 24, 805.

[2]) BRIGGS: Roy. Soc. Proc. Ser. B. Bd. 91, 249, 94, 12.

[3]) DELEANO: Zs. für Bot. 1911, 3, 657.

[4]) BENNET–CLARK: Notes Brt. Trinity Coll. Dublin 1932, 4, 233.

[5]) KRASSNOSELSKI–MAXIMOV: Ber. Deutsch. Bot. Ges. 43, 527.

[6]) ILJIN: Flora 16, 379.

[7]) HORN: Bot. Archiv 3, 137.

[8]) TOLLENAER: Omzettingen van koolhydraten in het blad van *Nicotiana tabacum* L. Dissertation Wageningen 1925.

sehen (Vgl. auch DAVIS und SAWYER [1])) und MILLER [2]). Nachdem KIDD
und WEST [3]) entwicklungsperiodische Schwankungen der Plasmapermeabi-
lität für Rohrzucker wahrscheinlich machen konnten, wären in diesem
Zusammenhang Untersuchungen über die tagesperiodischen Schwankungen von hohem Interesse, denn nach den neueren Erfahrungen über die
Lokalisation des desmolytischen Zuckerabbaus (WERTHEIMER [4])) können
Permeabilitätsfragen für die Geschwindigkeit der Atmungs- und Gärungs-
prozesse von bestimmender Wichtigkeit sein.

Dass auch die mit dem Wasserverlust gekoppelte Erschwerung des Gas-
austausches nicht ohne Einfluss auf die Atmung bleibt, darf nach mancher-
lei Beobachtungen an höheren Pflanzen und an Moosen (MAYER und PLAN-
TEFOL [5])) kaum bezweifelt werden. Die starke Anreicherung der Inter-
zellularluft an $CO_2$ erschwert die Oxydation des Zuckers offenbar durch
Blockierung der aktiven Oberflächen der Oxydationskatalysatoren, so dass
der Atmungsquotient beträchtlich ansteigt, als Ausdruck der einsetzenden
Gärung. Eine umfassende Analyse dieser Zusammenhänge erschiene nicht
nur bei Farnen, sondern allgemein erwünscht.

### § 18. Die Abhängigkeit der Atmung der Farne von der Tempe-
ratur. — Die rein chemische Komponente der Atmung spricht naturgemäss in der üblichen Weise auf Temperaturschwankungen an; die phy-

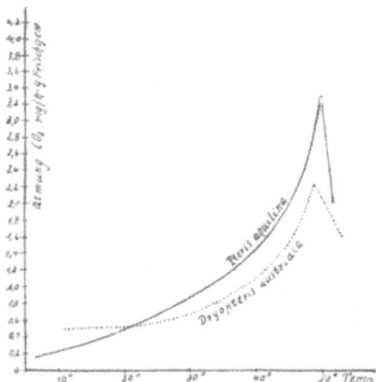

siologische Komponente gibt der Tem-
peratur-Atmungskurve den Charakter einer Optimumkurve. Das Temperaturoptimum liegt bei etwa 50° und
ist sehr scharf ausgeprägt. Hinsichtlich der Temperaturquotienten ergibt
sich ein bemerkenswerter Unterschied
zwischen Sonnen- und Schattenfarnen. Bei den ersteren liegt die Atmung
bei niederen Temperaturen (10°) relativ tief, um mit steigenden Temperaturen rasch anzuwachsen. Der $Q_{10}$-
Wert zwischen 10 und 20° liegt bei
etwa 2 und fällt nur auf etwa 1,8
zwischen 40 und 50°. Im Gegensatz
hierzu zeigen Schattenfarne bei niede-

FIG. 9. — Temperatur-Abhängig-
keit der Atmung von Farnblättern.
(n. JOHANSSON).

ren Temperaturen relativ hohe Atmungsintensität. Die $Q_{10}$-Werte liegen
bis 25° bei 1,1 bis 1,2, steigen zwischen 25 und 35° auf 1,5 und erlangen erst

[1]) DAVIS and SAWYER: J. Agric. Science 6, 406, 1914; 7, 352, 1916.
[2]) MILLER: Ebenda 27, 785, 1924.
[3]) KIDD und WEST: Proc. Roy. Soc. London 1930, 106, S. 93.
[4]) WERTHEIMER: Protsplasma 21, 522, 1934.
[5]) MAYER u. PLANTEFOL: Compt. Rend. Acad. Sc. Paris, 1925 S. 181.

im Temperaturbereich zwischen 40 und 50° den Normalwert. 2. Die Atmung der Sonnenfarne ist also wesentlich temperaturabhängiger als diejenige der Schattenfarne. (Vgl. Fig. 9). Im Bereich zwischen 10 und 47° steigt die Atmung von *Pteridium aquilinum* um den 11 fachen, diejenige von *Dryopteris austriaca* dagegen nur um den 4 fachen Betrag. Die Schattenfarne besitzen also ein physiologisches System, um die Temperaturwirkung auf den chemischen Prozess der Atmung abzubremsen, das indes nur bei niederen und mittleren Temperaturen zur Auswirkung kommt, d.h. in dem normalen Temperaturbereich ihres Standorts. Da die Atmungsversuche mit voll turgeszenten Blättern ausgeführt wurden, kann der Grad der Wassersättigung der Pflanzen keinen Einfluss auf die gemessene Atmungsintensität genommen haben.

Als weiterer Faktor, der die Atmungsintensität der Blätter wesentlich beeinflusst, wurde das B l a t t a l t e r ermittelt.

**§ 19. Abhängigkeit der Atmungsintensität vom Blattalter.** — Unter Bezugnahme auf gleiche Blattflächen und auf die Gewichtseinheit produzieren junge Blätter in gleicher Zeit grössere $CO_2$-Mengen als alte Organe. Die höchste Atmungsintensität wurde an Blättern gemessen, die noch völlig schneckenförmig eingerollt waren. Mit der Streckung und Aufrollung der Blätter ging die Atmungsintensität rasch zurück. Der Gradient ist allerdings nicht unabhängig von äusseren und inneren Faktoren: so wird die Atmungsintensität junger Blätter durch die Temperatur erheblich stärker beeinflusst als bei alten. Zweifellos spielt bei diesen Änderungen der Atmungsintensität der Wassergehalt der Blätter während der Entwicklung eine bedeutsame Rolle.

Bei *Pteridium aquilinum* sinkt der Wassergehalt mit fortschreitender Entwicklung und symbath mit ihm auch die Atmung, bei *Dryopteris spinulosa* dagegen bleibt die Atmungsintensität wie auch der Wassergehalt im Verlauf der Blattentwicklung nahezu konstant.

Ob — wie JOHANSSON annimmt — dem gemessenen Atmungsmaximum der Blätter in Knospenlage ein Atmungsanstieg vorausgeht analog der Keimlingsentwicklung von Samen, erscheint im Hinblick auf die völlig differenten inneren Entwicklungsbedingungen zumindest sehr zweifelhaft.

Auch muss dahingestellt bleiben, in wie weit die hohe Atmungsintensität der Knospenblätter nur ein Ausdruck ihres relativen Plasmareichtums ist, und bis zu welchem Grad auch hierbei Änderungen in der Menge und Aktivität der Atmungsfermente während der Blattentwicklung eine Rolle spielen. Während die Knospenblätter hinsichtlich ihres Energiestoffwechsels noch eine negative Bilanz aufweisen, wird bereits im ersten Stadium der Blattaufrollung und Streckung der Kompensationspunkt erreicht.

**§ 20. Das Atmungsmaterial.** — So ausgiebig die Frage nach dem Atmungsmaterial in der pflanzenphysiologischen Literatur diskutiert ist, fehlt uns doch eine planmässige Untersuchung gerade bei Pteridophyten so gut wie ganz. Bei der auf anderen Gebieten beobachteten Annäherung der Gefässkryptogamen an die Phanerogamen wurde auf eine Bestätigung dieser allgemeinen Regel hinsichtlich der Atmung und ihres chemischen Ablaufs verzichtet. Wir können zur Diskussion dieser Frage daher nur gelegentliche Angaben und Äusserungen heranziehen.

So berichtet BÄSECKE [1]), dass die Zellen der jungen Blätter von *Phyllitis Scolopendrium* vollgepfropft mit Stärke seien; im Verlauf der Blattentwicklung nimmt dann der Stärkegehalt rasch ab, und im voll erwachsenen Blatt soll sich Stärke nur mehr innerhalb der Endodermis gefunden haben. Diese Stärke wird offenbar u.a. auch im Atmungsprozess verbraucht, denn nach 8–10 tägiger Verdunklung waren die Blätter völlig stärkefrei. Dabei wird offensichtlich auch die in den Rhizomen gespeicherte Stärke mobilisiert und dem hungernden Blatt zugeführt, denn Blätter, die während der Verdunklung in Verbindung mit dem Rhizom geblieben waren, verloren ihre Stärke etwas langsamer als abgeschnittene Wedel. Die quantitativen Untersuchungen ergaben im belichteten Normalblatt pro 100 g Frischgewicht einen Stärkegehalt von 14,49 g neben 4,256 g Monosacchariden, während Disaccharide auffallenderweise gänzlich fehlten. Nach $3^1/_2$ tägiger Verdunklung fanden sich neben 3,056 g Zucker nur mehr 8,975 g Stärke. Da indes die während der Versuchszeit ausgeschiedene $CO_2$-Menge nicht gemessen wurde, lässt sich über den Anteil der Kohlehydrate an der Menge der insgesamt veratmeten Stoffe nichts sagen. Hierüber möchte aber die Grösse des Atmungsquotienten Aufschluss geben. Wir haben daher einige wenige ergänzende Versuche hierüber mit einem im Gewächshaus gezogenen Farn (*Blechnum brasiliense*) angestellt. Die Versuche wurden mit einzelnen Fiederblättchen nach der manometrischen Methode ausgeführt: Frischgewicht 0,400 g, Temperatur 26°.

| Zeit | $O_2$-Verbrauch | $CO_2$-Bildung | $CO_2 : O_2$ |
|---|---|---|---|
| 0 —20. Minute | 328    cmm | 185,6 cmm | 0,57 |
| 21.—30.   ,, | 133,5   ,, | 88,8   ,, | 0.59 |
| 31.—40.   ,, | 165    ,, | 93,1   ,, | 0,56 |
| 41.—50.   ,, | 139,6   ,, | 87,2   ,, | 0,63 |
| 51.—60.   ,, | 138    ,, | 90,6   ,, | 0,66 |
| Mittel | 151    ,, | 90,9   ,, | 0,60 |

Die Atmungsintensität innerhalb einer Stunde änderte sich unter den herrschenden Versuchsbedingungen also nur unerheblich. Wohl aber erhöhte sich der Atmungsquotient immerhin um 10% über sein Mittel. Der niedrige Quotientenwert lässt darauf schliessen, dass im ausgewachsenen Blatt neben Kohlehydraten noch ein anderer offenbar energiereicherer Körper verbrannt wird, der — wie aus dem Anstieg des Quotientenwertes zu erkennen ist-physiologisch besonders leicht angreifbar sein muss. Leider

[1]) BÄSECKE: Bot. Zeitg. 66, 25, 1908.

sind wir über die Farne auch in biochemischer Hinsicht so ungewöhnlich dürftig unterrichtet, dass wir mit unseren Deutungen durchaus spekulativ bleiben müssten. Zwar berichtet Bäsecke über eine beobachtete Umwandlung von Kohlehydraten in Fett bei einem Farn (*Davallia bullata*), aber seine Angaben beziehen sich auf das Rhizom der Pflanze. Nur soviel scheint festzustehen, dass es sich bei der von uns gemessenen Atmung nicht um eine Hungererscheinung einer kohlehydratarmen Pflanze und dadurch bedingten Eiweissabbau handeln kann, denn einerseits spricht schon die kurze Versuchszeit gegen eine solche Annahme anderseits aber müsste sich unter derartigen Umständen der Atmungsquotient während des Versuchs vermindert und nicht erhöht haben. Eine direkte Zuckerbestimmung behob endlich alle Zweifel: es ergaben sich pro 100 g Frischgewicht 870 mg löslichen vergärbaren Zucker und 940 mg Stärke. Das ist zwar auffallend viel weniger als Bäsecke [1]) bestimmt hatte, aber man muss bedenken, dass die von letzteren angewandte Reduktionsmethode nicht nur die Kohlehydrate sondern alle reduzierenden Substanzen erfasste, und deren gibt es in Farnen zweifellos nicht wenige (Phloroglucinderivate). Im übrigen aber entspricht der von uns bestimmte Kohlehydratwert demjenigen der Blätter höherer Pflanzen viel eher als der exorbitante Wert, den Bäsecke gefunden hatte, und der über 50% des gesamten Trockengewichts ausmachen würde. Bei den im Versuch herrschenden Bedingungen hätte die Atmung mindestens 18 Stunden durch die blatteigenen Kohlehydrate gespeist werden können. Man muss also annehmen, dass in alten Farnblättern ein sauerstoffarmer Körper verbrannt oder doch oxydiert wurde. Dieser nicht näher zu definierende Stoff wird offenbar erst im Verlauf der Blattentfaltung gebildet, denn die jungen, noch in Knospenlage befindlichen Blätter weisen einen Atmungsquotienten von 1,07 (Mittel von 6 Bestimmungen) unter gleichen Versuchsbedingungen, also ziemlich reine Kohlehydratatmung auf. Dabei ist die Atmungsintensität der jungen Blätter ausserordentlich viel grösser als diejenige der alten: der Sauerstoffverbrauch liegt etwa 6 mal so hoch, die $CO_2$-Produktion ist 3,4 mal so gross wie bei ausgewachsenen Blättern.

**§ 21. Die anaerobe Atmung.** — Auch über die intramolekulare Atmung der Farne liegen in der Literatur keine Angaben vor. Zur Ausfüllung dieser Lücke haben wir einige Versuche mit verschiedenen Farnen (*Blechnum brasiliense* und *Polypodium aureum*) angestellt. Dabei liess sich feststellen, dass auch die Farne in Abwesenheit von Sauerstoff zur A l k o h o l b i l d u n g befähigt sind. Allerdings entspricht die gebildete Alkoholmenge dem aus der alkoholischen Gärungsformel abzuleitenden Wert auch dann nicht, wenn die präformierte (als Carbonat) im Blatt eingelagerte $CO_2$ in Abzug gebracht wird. Neben der Zuckervergärung läuft

---

[1]) Bäsecke: Bot. Zeitung 66, 25, 1908.

also noch ein anderer $CO_2$ liefernder anaerober Prozess ab, der nicht weiter definiert werden kann. Zur Bildung flüchtiger Säuren kommt es hierbei nicht. Der Quotient Alkohol: $CO_2$ lag bei *Polypodium aureum* um 0,4 bei *Blechnum brasiliense* um 0,6.

Wenn auch die Gesamtheit der bei der anaeroben Atmung sich abspielenden Teilprozesse noch nicht bekannt ist, so kann doch nicht bezweifelt werden, dass die Farne wie die höheren Pflanzen über e i n  k o m p l e t t e s  z y m a t i s c h e s  S y s t e m  verfügen. Trotzdem aber vermögen sie — wieder in Übereinstimmung mit den Blättern der meisten höheren Pflanzen- eine Anaerobiose nur begrenzte Zeit zu ertragen. Für die von uns untersuchten Formen waren nach 22 stündiger Anaerobiose bei 37° bereits die ersten Spuren einer beginnenden Schädigung zu erkennen. Wahrscheinlich wird für die Resistenz die $H^.$-Konzentration des Zellsafts von erheblicher Bedeutung sein (Vgl. PAECH [1]). Als energetischer Vorgang aber durfte diese Gärung für die Farne ebenso bedeutungslos sein wie für die höheren Pflanzen.

### § 22. Der Stickstoff-Stoffwechsel.

— Der Umfang unserer Kenntnisse hinsichtlich des N-Umsatzes der Farne steht leider in einem krassen Missverhältnis zur Wichtigkeit dieser physiologischen Frage. So wissen wir über die chemische Zusammensetzung der Eiweisse in den Farnen so gut wie nichts. Das hängt natürlich mit den grossen technischen Schwierigkeiten der Eiweissisolierung aus vegetativen Pflanzenorganen zusammen. Von den höheren Pflanzen kennen wir wenigstens in Einzelfällen die chemische Zusamensetzung der Sameneiweisse. Dagegen sind die Farne als „wenig geeignete Objekte" ausserhalb der Untersuchungen geblieben. Durch die ausführliche Darstellung eines genauer untersuchten Einzelfalles, der die Ableitung allgemein gültiger Schlüsse und Vorstellungen über den N-Umsatz bei Farnen überhaupt gesattet, glauben wir am ehesten noch ein hinreichend geschlossenes Bild dieser Stoffwechselsphäre vermitteln zu können.

Aber auch die Dynamik des N-Stoffwechsels ist bei den Farnen kaum untersucht. Einen ersten Vorstoss nach dieser Richtung hat im Zusammenhang mit der Frage der physiologischen Bedeutung der Säureamide G. SCHWAB [2] gemacht, der mir in freundlicher Weise seine Analysenwerte zur vorliegenden Darstellung zur Verfügung stellte.

Die lebhaftesten Umsetzungen im N-Stoffwechsel vollziehen sich in den Farnen im Zusammenhang mit der rhythmisch verlaufenden Entwicklung naturgemäss während des Austreibens zu Beginn der neuen Vegetationsperiode. Ein mehr oder weniger umfangreiches Rhizom birgt die zum Aufbau der oberirdischen Organe notwendigen Stoffreserven. Hinsichtlich der

---

[1] PAECH: Planta 1935, 24, S. 78.
[2] G. SCHWAB: Planta 1936, 25, S. 579.

Kohlehydrate liegen die Verhältnisse sehr einfach, da neben der Stärke alle andern Reservestoffe als Kohlehydratreserven stark zurücktreten; dagegen findet sich der Stickstoff bereits im ruhenden Rhizom in verschiedener organischer Bindung vor.

Das ruhende Rhizom: An N-haltigen Fraktionen wurden im Rhizom von *Dryopteris Filix-mas* neben dem Total-N die folgenden bestimmt:

| Pro 100 g Frischgewicht: | | | | | |
|---|---|---|---|---|---|
| Eiweiss-N | Löslicher N | Ammoniak-N | Glutamin- N | Asparagin- N | Total-N |
| 233 mg | 231 mg | 10 mg | 109,5 mg | 12,3 mg | 454 mg |

Auch bei den Farnen tritt die Besonderheit der vegetativen Reserveorgane[1]) in dem relativ hohen Gehalt an löslichem N, der hier den Eiweissgehalt sogar noch etwas übertrifft, hervor. Besonders hoch innerhalb des löslichen N ist der Gehalt an Glutamin, das nahezu die Hälfte der Gesamtfraktion ausmacht. Demgegenüber tritt das andere Amid Asparagin ebenso wie der Ammoniak sehr stark zurück. Man wird nicht fehlgehen, wenn man diese Glutaminanhäufung in Zusammenhang mit der Rückwanderung N-haltiger Substanzen aus dem im Herbst vergilbenden Blättern bringt. Dafür sprechen vor allem die im Verlauf der jahresperiodischen Neubildung von oberirdischen Organen vor sich gehenden Änderungen im Amidgehalt des Rhizoms, wovon die

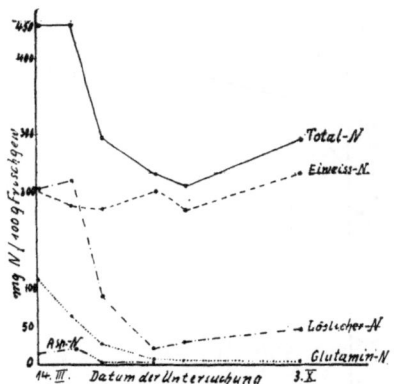

FIG. 10. — N- Stoffwechsel im Rhizom des Wurmfarns im Verlauf der saisonalen Entwicklung.

folgende Kurve einen Überblick zu vermitteln vermag.

Der N-Umsatz im austreibenden Rhizom: Zunächst tritt völlig eindeutig die Tatsache in Erscheinung, dass nur die in der Fraktion Löslicher N zusammengefassten N-haltigen Substanzen die Rolle beweglicher Reservestoffe spielen, ihr Gehalt geht von etwa 240 mg N pro 100 g Frischgewicht zu Beginn des Austreibens bis auf 21 mg im Juni zurück; dann tritt eine leichte Erhöhung auf 45 mg bis Oktober ein. Leider liegen keine Analysenwerte späterer Untersuchungstermine vor. Aus den Frühjahrswerten kann jedoch erschlossen werden, dass die Auffüllung des Rhizons mit löslichem N erst mit der Rückwanderung von N aus absterbenden oberir-

---

[1]) Vgl. GRÜNTUCH: Planta 7, 388, 1929.

dischen Organen erreicht wird. Das Rhizom sendet also den bereits in gelöster Form im Frühjahr vorliegenden N in die neu entstehenden Organe hinaus und zieht ihn aus den im Herbst absterbenden Blättern wieder zurück. Hierbei spielt vor allem das Glutamin eine hervorragende Rolle. Es tritt als bevorzugter Wanderstoff ganz klar in Erscheinung, denn nicht nur absolut, sondern auch relativ erleidet dieses Amid im Rhizom während des Austreiben den stärksten Rückgang (Abfall auf etwa 3% des Anfangswertes). Es wird praktisch vollkommen nach den neugebildeten Organen abgeleitet. Das darf jedoch nicht ohne Einschränkung im Sinne einer Bestätigung der Pfefferschen Auffassung von der Bedeutung der Amide als Wanderstoffe gedeutet werden, worauf wir weiter unten noch zurückkommen werden, denn bezüglich des relativen N- Gehalts ist das hier völlig zurücktretende Asparagin dem Glutamin noch überlegen.

Gegenüber den Amiden und wohl auch den Aminosäuren kommt dem Ammoniak als Wanderstoff (in Form der Ammonsalze) keine Bedeutung zu. Die geringe Menge der im Zellsaft der Farne (s. S. 355) vorhandenen

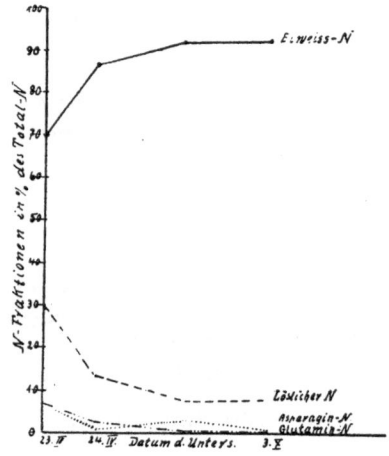 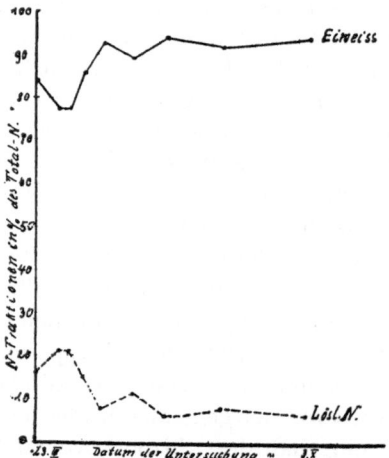

FIG. 11.—*Dryopteris Filix-mas*: Änderungen im Gehalt an N haltigen Stoffen im Verlauf der Entwicklung der Achsenorgane.

FIG. 12. — *Dryopteris Filix-mas*: Schwankungen im Einweiss- u. Löslichen-N im Verlauf der Blattentwicklung.

freien H· schliesst eine weitergehende Ammoniakspeicherung offenbar aus.

Sehr auffällig tritt die Stabilität der Rhizomeiweisse im Verlauf der sommerlichen Neubildung von Organen hervor. Ihr Gehalt nimmt in keiner Phase der Entwicklung auch nur vorübergehend ab. Die Rhizomeiweisse sind also nur zum allergeringsten Teil Reserveeiweisse. Eine kleine Abnahme im Eiweissgehalt der Rhizome während der Organneubildung kann freilich durch die Bezugnahme auf das variable Frischgewicht maskiert sein. Immerhin kann man feststellen, dass die Eiweissabnahme des Rhizoms im Verlauf der Entwicklung relativ nicht grösser ist als die Abnahme des Rhizomfrischgewichtes. Unbeschadet der daraus notwendig werdenden Korrektion kann man sagen, dass der lösliche N die bei weitem wichtigste N-Reservefraktion des Rhizoms darstellt.

Die N-Ableitung aus dem Rhizom spiegelt sich in gegensinnigen Veränderungen des Gehalts von Blattstiel und Blattflächen an N-haltigen Substanzen wider.

B l a t t s t i e l e : (Vgl. Fig. 11). Im Stadium der Aufrollung der Blätter stauen sich die löslichen N- haltigen Stoffe im Blattstiel. Glutamin und Asparagin in fast gleichen Mengen auftretend, machen zusammen rund 50% des gesamten löslichen N aus. Der starken Amidauswanderung aus dem Rhizom entspricht also eine vorübergehende Anhäufung im Blattstiel. Das deutliche Hervortreten des Asparagins möchte als Ergebnis einer bevorzugten Verarbeitung von Glutamin im Rahmen der die Neubildung von Pflanzenmaterial begleitenden Eiweisssynthesen gedeutet werden. Nach den Untersuchungen von G. SCHWAB ist in höheren Pflanzen ein reichliches Auftreten von Gustamin mit einer gesteigerten Aktivität der Glutaminase gekoppelt. Im Blattstiel des e r w a c h s e n e n Blattes nimmt dann der lösliche N und mit ihm der Gehalt an Amiden rasch ab, und es ergeben sich Verhältnisse wie in der Blattlamina, d.h. nahezu 92% des Total- N ist als Eiweiss deponiert. Die geringe Bedeutung der Ammonsalze als Wanderstoffe tritt auch in den Blattstielen in Erscheinung.

B l ä t t e r : (Vgl. Fig. 12). Imponierend tritt bei der Blattentwicklung die Fähigkeit dieser Organe zur Eiweissbildung zu Tage. Vergleicht man zunächst die Werte für Total-N in den Blättern verschiedener Entwicklungsstufen, so fällt von rasch wieder ausgeglichenen Schwankungen abgesehen die relative Konstanz sofort ins Auge (Fig. 13); auf die Frischgewichtseinheit fällt in allen Entwicklungsphasen die gleiche Stickstoffmenge. Die Zufuhr von Stickstoff aus dem Rhizom, zum Teil-besonders im späteren Entwicklungsverlauf-natürlich auch aus dem Substrat, hält also mit der Frischgewichtsvermehrung durchaus Schritt. Man wird auf Grund dieser zu Tage tretenden Korrelation geneigt sein, die N-

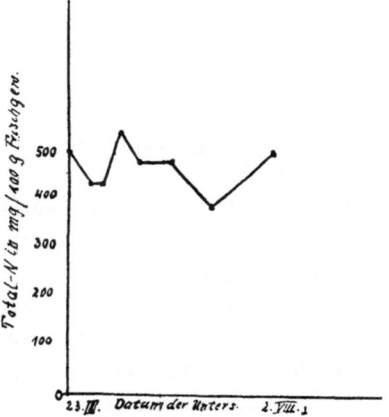

FIG. 13. — *Dryopteris Filix-mas*: Schwankungen im Gehalt an Total-N im Verlauf der Blattentwicklung.

Zufuhr als bestimmenden Faktor der Substanzneubildung aufzufassen. Was nun das Verhältnis Löslicher: Eiweiss-N betrifft, so macht sich mit dem Alter der Blätter ein deutlicher Anstieg des relativen Eiweissgehalts geltend, d.h. der zugeleitete N wird in den Blättern nicht sofort in vollem Umfang

zu Eiweiss synthetisiert. Nach den Untersuchungen von Paech [1]) ist das
Verhältnis Eiweiss: Löslichem N in erster Linie durch den Kohlehydrat-
spiegel der betreffenden Organe bestimmt. Es erscheint auch durchaus wahrscheinlich, dass die Kohlehydratzufuhr aus Reserveorganen und Assimilation mit der stürmischen Neubildung (vom 23. III. bis 8. V. wuchs ein Blatt
von 5 cm Länge auf 85 cm heran) nicht Schritt halten kann, so dass
leicht vorübergehend ein relativer Kohlehydratmangel einzutreten vermag. Mit der fortschreitenden Auffüllung des Kohlehydratdefizits steigt
dann auch der Eiweissgehalt der Blätter bis auf 94% des Total- N an. Die
gelegentliche Stauung an Löslichem N kommt auch in einem zeitweisen
Anstieg der Amidmengen (Fig. 14) im Blatt zum Ausdruck. Das Überwiegen des Asparagins lässt vermuten, dass das Glutamin auch im Blatt
einer rascheren Verarbeitung anheimfällt als das Asparagin, das sich im
Stoffwechsel der Versuchspflanze fast ein wenig zellfremd ausnimmt. Mit

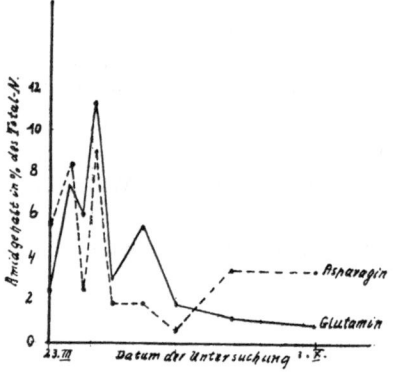

FIG. 14. — *Dryopteris Filix-mas*:
Schwankungen im Amidgehalt der
Blätter im Verlauf der Entwicklung
der Pflanze.

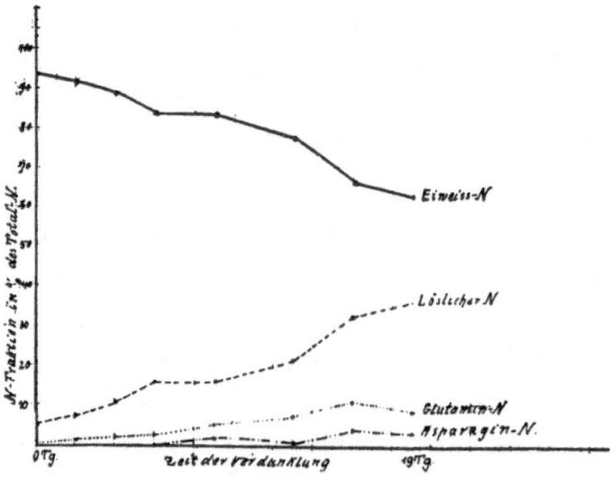

FIG. 15. — *Dryopteris Filix-mas*: Wirkung
länger dauernder Verdunklung auf den N-
Stoffwechsel der Blätter.

Erhöhung des Kohlehydratspiegels gehen dann beide Amide auf geringe
Beträge zurück. Ammoniak häuft sich auf keiner Entwicklungsstufe in
erheblichen Mengen an.

Bedeutung der Amide im N-Haushalt der Farne.

Wir haben oben die alte Streitfrage nach der physiologischen Bedeutung
der Amide [2]), ob Wanderstoffe oder Entgiftungsstoffe offen lassen müssen. Das gehäufte Auftreten der Amide im Rhizom konnte nach beiden
Richtungen gedeutet werden, weil nicht bekannt war, in welcher Form der
Lösliche N dem Rhizom aus den Blättern zugewandert war. Eine Entscheidung der Frage war möglich, wenn man die Ableitung der Löslichen N haltigen Substanzen nach dem Rhizom verhinderte. Durch gleichzeitige Ver-

[1]) Paech: Planta 1935, 24, S. 78.

[2]) Mothes: Planta 1925, I, S. 472.

dunklung abgeschnittener Blätter wurden für den N-stoffwechsel künstlich die Bedingungen vergilbender Blätter geschaffen. Die folgende Tabelle gibt das Ergebnis eines solchen Verdunklungsversuches wieder: (vgl. auch Fig. 15).

| Verdunklungszeit in Tagen . . . . . | 0 | 0 | 4 | 6 | 9 | 13 | 16 | 19 |
|---|---|---|---|---|---|---|---|---|
| % Eiweiss-N. . . . . | 94 | 92,4 | 88,8 | 83,6 | 84,2 | 78,3 | 66,9 | 63,5 |
| % Löslicher N . . . . | 6 | 7,6 | 11,2 | 16,4 | 19,8 | 21,7 | 33,1 | 36,5 |
| % Ammon.-N . . . . | 0,3 | 0,3 | 0,3 | 0,4 | 0,6 | 0,7 | 2,5 | 2,8 |
| % Glutamin-N . . . . | 0,4 | 1,8 | 2,3 | 2,6 | 5,8 | 8,3 | 11,2 | 8,8 |
| % Asparagin-N . . . | 0,03 | 0,6 | 0,9 | 0,3 | 2,4 | 1,2 | 4,3 | 3,1 |
| % Rest-N . . . . . . | 5,3 | 5,0 | 7,7 | 13,1 | 7,0 | 11,5 | 15,1 | 21,8 |

Der im Abbau entstehende Ammoniak wird zunächst fast quantitativ als Glutamin entgiftet. Erst bei stärker fühlbar werdendem Kohlehydratmangel tritt auch Asparagin in Erscheinung, das bezeichnenderweise auch schon von erheblichen Ammoniakmengen als dem Symptom verschärften Kohlehydratmangels begleitet wird.

Im Sinn von RUHLAND und WETZEL ist *Dryopteris Filix-mas* somit als typische Amidpflanze zu bezeichnen; im Gegensatz zu den meisten höheren Amidpflanzen erfolgt jedoch die Ammoniakentgiftung nicht in Form von Asparagin-, sondern von Glutaminbildung.

# Chapter XII

# ÖKOLOGIE DER EXTRATROPISCHEN PTERIDOPHYTEN

von

H. Gams (Innsbruck)

## I. Autökologie

**§ 1. Geschichtliches.** — Obgleich die meisten der unzähligen Untersuchungen über die Morphologie, Entwicklungsgeschichte und Physiologie sämtlicher Pteridophytengruppen von Al. Braun, Mettenius und Hofmeister (1846–71) bis zu Bower, Campbell, Claussen, Goebel [1]) und ihren vielen Schülern auch autökologische Beobachtungen enthalten und auch synökologische Beziehungen von viel mehr Biozönotikern untersucht worden sind als bei den Moosen [2]), gibt es noch keine Gesamtdarstellung der Ökologie aller lebenden Pteridophyten. Am nächsten kommt diesem Ziel Christs Geographie der Farne (s. Kap. XIII § 1), welche jedoch nur die Filicineen mit Ausschluss der Hydropteriden und noch ohne Einbeziehung der experimentellen Ökologie behandelt.

**§ 2. Die Lebensformen als Ausdruck der Gesamtökologie.** — Sämtliche Stämme der Pteridophyten haben, sobald einmal das Übergewicht des Sporophyten über den Gametophyten erlangt war, schon im Devon und Karbon die Baumform oder phanerophytische Lebensform erreicht und es ist kaum zweifelhaft, dass alle lebenden Pteridophyten (vielleicht mit Ausnahme der Psilotales, falls diese wirklich von niedrigen Psilophyten abgeleitet werden können) von immergrünen Phanerophyten mit ursprünglich autotrophen, adnaten Prothallien abzuleiten sind [3]).

---

[1]) Von seiner Dissertation in Bot. Zeitung 1877 bis zur III. Auflage seiner Organographie 1933, insbesondere Morphol. u. biol. Studien (Ann. Jard. bot. Buitenzorg VII–XXXIX, 1888–1928), Pflanzenbiologische Schilderungen (Marburg 1889 –91) und Archegoniatenstudien (Flora LXXVI–CXXII, 1892–1928).

[2]) Gams in Manual of Bryology 1932 p. 323.

[3]) F. O. Bower, The Origin of a Land Flora. London 1908.

H. Mölholm Hansen, En Undersögelse over de Raunkiaerske Livsformers Palaeontologi. Raunkiaer-Festschr. Kopenhagen 1930.

W. Zimmermann, Der Baum in seinem phylogenetischen Werden. Ber. Deutsch. Bot. Ges. XLVIII, 1930.

Am ursprünglichsten erscheinen demnach unter den lebenden Pteridophyten neben einigen epiphytischen Lycopodien mit noch vorwiegend autotrophen Prothallien die Baumfarne und am weitesten abgeleitet die Kryptophyten und Therophyten im Sinne RAUNKIAERS, zu welchen unter den Lycopodialen *Phylloglossum* und *Isoetes*, unter den Filicineen die Ophioglossaceen, *Ceratopteris*, Marsiliaceen, *Salvinia* und *Azolla* gehören. Die Vereiniging der letztgenannten Heterosporen als Hydropterides ist heute als nur ökologische, nicht phylogenetische Klassifikation anerkannt. Die Versuche, die Ophioglossaceen und die sekundär z. T. (*Trichomanes*-Arten) sogar wurzellos gewordenen adnaten Farne direkt von Bryophyten abzuleiten (PRESL 1841, PRANTL 1875 u.a.), haben fehlgeschlagen, denn auch die scheinbar so einfachen Hymenophyllaceen, Vittarieen usw. sind sehr wahrscheinlich nur reduziert, und die Ähnlichkeit mit adnaten Moosen [1] beruht nur auf ökologischer Konvergenz (s. GOEBEL in Flora CXXIV 1930)

Mit Bezug auf die Lebensformen ergeben sich folgende Progressionen:

*a.* Beim Sporophyten von Phanerophyten über Chamaephyten und Hemikryptophyten einerseits zu sekundären Adnaten (Epixylen und Epipetren), andrerseits zu Geophyten (inkl. Therophyten) und sekundären Hydrophyten. Unter den heutigen Lycopodialen und Psilotalen überwiegen die Chamaephyten, unter den Filicineen namentlich der gemässigten und kalten Klimate die Hemikryptophyten, unter den Articulaten (*Equisetum*) die Geophyten. Bei den Lycopodialen (*Phylloglossum*) und *Equisetum*, wo Chamaephyten fast direkt zu Geophyten geworden sind, scheint das hemikryptophytische Stadium wegen der Empfindlichkeit der durch die mikrophyllen Sprosse zu wenig geschützten Prothallien nicht beständig zu sein. Die sommergrünen Farne und Equiseten sind zweifellos durchwegs von immergrünen, die heteroblastischen Equiseten (*Heterophyadica* A.Br.) von homoblastischen (*Homophyadica* A.BR.) abzuleiten, sodass also die übliche Anordnung in *Vernalia, Subvernalia* und *Aestivalia* umzukehren ist.

*b.* Beim Gametophyten geht die Entwicklung vom autotrophen, adnaten Prothallium, das im allgemeinen ganz mit anakrogynen Jungermannialen übereinstimmt und nur bei einigen adnaten Farnen (*Trichomanes, Schizaea* u.a.) und Lycopodien (TREUBS *Phlegmaria*-Typ) sekundär atavistisch zur Fadenform zurückgekehrt ist, einerseits bei den Isosporen zu fakultativen Geophyten (Knöllchenbildung bei *Anogramma*- und *Equisetum*-Arten [2]) und obligatorisch mykotrophen Geophyten (*Lycopodiaceae, Psilotaceae, Ophioglossaceae*), andrerseits bei den Heterosporen (*Selaginella, Isoetes, Marsiliaceae, Salvinia, Azolla*) zum Verzicht auf die selbständige Lebensform, indem die Prothallien die Makro- und Mikrosporen nicht mehr ganz verlassen. Die in mehreren Stämenm unabhängig aufgetretene Hete-

---

[1] GAMS in Manual of Bryol. p. 328. H. MEUSEL, Wuchsformen und Wuchstypen der europäischen Laubmoose. N. Acta Leopold. N. F. III, Halle 1935.

[2] K. LUDWIGS, Untersuchungen zur Biologie des Equiseten. Flora CIII, 1911.

rosporie scheint direkt Vorbedingung für die Aufgabe der Selbständigkeit des Gametophyten, welche den Blütenpflanzen eine so viel grössere ökologische Amplitude als den Archegoniaten und damit ihre Überlegenheit ermöglicht hat. Warum die heterosporen Articulaten schon in der Juraformation erloschen sind, ist noch unbekannt.

FIG. 1. — *Dicksonia squarrosa*-Bestand (mit *Schefflera digitata, Fuchsia excorticata, Freycinetia Banksii, Knightia excelsa, Dysoxylum spectabile* und *Rhopalostylis sapida*) in einem feuchten Tal bei Huia nächst Auckland (New Zealand).
Photo L. CRANWELL.

Bei Berücksichtigung beider Generationen ergibt sich folgende Reihe der Lebensformen:

1. Phanerophyten mit autotrophen Prothallien (Baum- und Kletterfarne, einige südamerikanische *Equiseta*).

2. Chamaephyten und Adnate mit wenigstens teilweise autotrophen Prothallien (viele isospore Farne, *Lycopodium* subgen. *Urostachya* p.p., *Equisetum* subgen. *Sclerocaulon = Hippochaete*).

3. Chamaephyten mit mykotrophen Prothallien (meiste Lycopodien und Psilotales).

4. Hemikryptophyten mit autotrophen, kurzlebigen Prothallien (meiste isospore Farne).

5. Therophyten mit autotrophen, halb geophytischen Prothallien (*Anogramma*).

6. Geophyten und Helophyten mit selbständigen, autotrophen Prothallien (viele Farne, wie *Pteridium, Dryopteris* subgen. *Thelypteris* und *Gymnocarpium; Equisetum* subgen. *Malacocaulon = Euequisetum*).

7) Geophyten mit geophytischen, mykotrophen Prothallien (*Phylloglossum, Ophioglossaceae*).

8. Hydrophyten mit selbständigen, autotrophen Prothallien (*Ceratopteris*).

9. Chamaephyten mit unselbständigen Prothallien (*Selaginella*).

10. Helo- und Hydrophyten, Geo- und Therophyten mit unselbständigen Prothallien (*Isoetes, Marsiliaceae, Salvinia, Azolla*).

**§ 3. Wasserhaushalt.** — Im Wasserhaushalt nehmen die Pteridophyten Zwischenstellungen zwischen den ganz vom Wassergehalt der Umgebung abhängigen (poikolohydren [1])) Thallophyten und Bryophyten und den in verschieden hohem Grad ihren Wasserhaushalt regulierenden (homoiohydren [1])) Blütenpflanzen ein. Die autotrophen Prothallien der meisten Filicineen und Equiseten, die Luftsprosse der meisten Farne und Equiseten, bei manchen Farnen (z.B. vielen Hymenophyllaceen und Wasserfarnen) auch die ganzen Pflanzen vermögen Austrocknung ebensowenig zu ertragen wie die meisten Algen und Moosprotonemata. Nach DOSDALL [2]) welken die Sprosse des sonst so widerstandsfähigen *Equisetum arvense* bei gleichen Wurzelbedingungen früher als z.B. die von *Phaseolus vulgaris* und *Helianthus annuus*.

Der osmotische Wert des Zellsafts der grünen Gewebe schwankt nach WALTERS Messungen bei Schattenfarnen und *Equisetum* zwischen nur 8 und 16 Atm. und bei höherer Konzentration (*Blechnum Spicant* bei 24, *Polypodium vulgare* erst bei 35 Atm.) sterben die meisten Farnblätter ab. Dagegen vertragen die meisten Rhizome, selbst solche sehr hygrophiler Arten (z.B. *Adiantum capillus-Veneris* und *Equisetum*), die Sporokarpien von *Marsilia*, die Sporen der meisten Pteridophyten, die Knollenprothallien vieler Ophioglossaceen und die Prothallienknöllchen von *Anogramma*- und *Equisetum*-Arten längere Lufttrockenheit. Bei *Anogramma (Gymnogramme) leptophylla* und *chaerophylla* perennieren nur die Prothallien, wogegen

FIG. 2. — *Selaginella tamariscina* (oben ausgebreitet, unten eingerollt) mit *Cyclophorus lingua* an sonnigen Felsen im nördlichen Ussuri-Gebiet. — Photo LAZARENKO.

---

[1]) H. WALTER, Die Hydratur der Pflanze. Jena 1931.

[2]) L. DOSDALL, Water requirement and adaptation in *Equisetum*. Plant World XXII, 1919.

die kurzlebigen Sporophyten nur in dauernd dampfgesättigter frostfreier Luft vegetieren [1]).

Die Sprosse mehrerer *Selaginellen*, wie der amerikanischen *S. lepidophylla* [2]), der afrikanischen *S. imbricata* und der ostasiatischen *S. tamariscina* (Fig. 2), und die immergrünen Blätter vieler in vollem Sonnenlicht wachsender Farne (in geringem Grad z.B. viele Arten von *Polypodium, Asplenium, Phyllitis* und *Woodsia*, in höchstem Grad Arten von *Pellaea, Ceterach, Notholaena* (Fig. 4 und 12) und *Cheilanthes*, somit CHRISTs dicht spreuschuppiger „*Cheilanthes*-Typ") vermögen, obgleich auch ihr osmotischer Wert normalerweise nur 13–17 Atm. beträgt (bei *Asplenium Ruta-muraria* fand WALTER bis 19,4, bei *Cheilanthes*-Arten bis 24 Atm.), nach völliger Lufttrockenheit wieder aufzuleben [3]). Die bei der Austrocknung erfolgende Einkrümmung, welche wenig treffend als Xerotropismus, besser als Trockenstellung bezeichnet wird, erfolgt meist nach oben (*Selaginella, Polypodium, Ceterach, Cheilanthes* usw., auch *Cardiomanes reniforme*, s. Fig. 3), seltener (z.B. *Asplenium*-Arten) nach unten. Viele Arten vertragen mehrmonatige (einige *Selaginellen* mehrjährige) vollständige Austrocknung und leben nach Wasseraufnahme durch die Blätter (bei *Cheilanthes, Woodsia, Elaphoglossum* usw. besonders durch Vermittlung der Spreuschuppen, bei *Selaginella* der Ligula) wieder auf, wobei oft das doppelte, in einzelnen

FIG. 3. — *Trichomanes (Cardiomanes) reniforme* mit sich bei Trockenheit oberwärts einrollenden Blättern, in dichter Matte auf Lava, nur durch Zweige von *Metrosideros tomentosa* und Horste von *Astelia Banksii* etwas beschattet. Rangitoto Island (New Zealand).
Photo L. CRANWELL.

---

[1]) GOEBEL, Entwicklungsgeschichte der Prothallien von *Gymnogramme leptophylla*. Bot. Zeitung XXXV, 1877 und in Flora LXXII 1889. GAMS in Beitr. z. geobot. Landesaufnahme d. Schweiz XV 1927 p. 107 u. 356.

[2]) LECLERC DU SABLON, Sur la réviviscence du *Selaginella lepidophylla*. Bull. Soc. Bot. de France XXXV 1888. J. P. WOJENOWIC, Beiträge zur Morphologie, Anatomie und Biologie der *Selaginella lepidophylla*. Diss. Breslau 1890. J. C. TH. UPHOF, Physiological Anatomy of xerophytic Selaginellas. New Phytologist XIX 1920. E. F. ANDREWS, Habits and habitats of the north american resurrection fern. Torreya XX 1920.

[3]) V. BR. WITTROCK, Biologiska ormbunkstudier. Acta Horti Bergiani I 1891. A. BORZI, Xerotropismo nelle felci. N. Giorn. Bot. Ital. XX 1913. F. L. PICKETT in Bull. Torrey Bot. Club L 1923, LIII 1926 and Amer. Fern Journ. XXI 1931.

Fällen das 6–8 fache des Trockengewichts an Wasser aufgenommen wird. Diese Dürreresistenz des Laubes ist bei den Farnen aber seltener als bei den Moosen und wohl durchwegs eine sekundäre Anpassung.

FIG. 4. — Die xerophytische *Notholaena Fendleri* KUNZE zwischen Lavablöcken in der Halbwüste bei Tucson (S. Arizona). — Photo J. C. TH. UPHOF.

**§ 4. Temperatur.** — Als ökologischer Faktor wirkt die Temperatur besonders indirekt über den Wasser- und Stoffhaushalt, doch sind namentlich viele Arealgrenzen thermisch bestimmt. Die meisten Baum- und Kletterfarne sind auf das megatherme Tropengebiet beschränkt, doch erreichen Baumfarne z.B. am Kilimandscharo 2500, auf Celebes 2700 und in Ecuador 3420 m Höhe. Südlich reichen sie bis ins Kapland, Tasmania, Auckland, S.-Brasilien und Juan Fernandez (Fig. 5). Windende Farne fehlen zwar Europa und dem Mittelmeergebiet, doch reicht *Lygodium japonicum* von Japan und Shensi bis Neuseeland und *L. palmatum* von New Hampshire und Pennsylvania bis S.-Chile. Die Mehrzahl der adnaten Pteridophyten, vor allem der Hymenophyllaceen, ist sehr empfindlich inbezug auf thermische und hygrische Kontinentalität, doch reichen einige der „hyperozeanischen" *Hymenophyllum*-Arten sehr weit nach Norden (Faer Öer, W.-Norwegen, Sachalin) und Süden (S.-Georgia, Auckland, Campbell-I., Chatham-I., Kerguelen) und auf den Bergen des tropischen Afrika, Asien und Amerika steigen einige bis über 3600 m, wobei sie gleich den hyperozeanischen

Moosen (s. Man. of Bryol. p. 340) und Flechten an entweder ganz frostfreie oder gut schneegeschnützte Standorte gebunden sind [1]).

Nach absteigender Empfindlichkeit gegen Winterkälte ergibt sich folgende Reihe: Baumfarne, *Woodwardia* und *Trichomanes* — *Hymenophyllum peltatum* und *Asplenium marinum* — *Hymenophyllum tunbridgense* und *Asplenium lanceolatum* — *Gymnogramme leptophylla* und *Osmunda* — *Matteuccia* und *Phyllitis* — *Blechnum Spicant* und *Polystichum*-Arten (Fig. 5). Die grosse Mehrzahl aller Pteridophyten ist an das Waldklima gebunden, doch gehen viele an lokalklimatisch begünstigten Orten mit gutem Schneeschutz weit über die thermischen Waldgrenzen, sowohl Chamaephyten (*Lycopodium andinum* in Bolivia bis 5000 m, *L. Saururus* auf den Kamerun-Bergen bis 3300, auf der Quimzacruz-Kordillere bis gegen 5200 m mit *Polypodium stipitatum*), wie Hemikryptophyten im Schutz von Blöcken und Felsspalten (viele Polypodiaceen auf den Gebirgen Südasiens und Südamerikas bis über 4000 m, besonders Arten von *Cystopteris, Woodsia, Asplenium* und *Polystichum*, so *P. Duthiei* im Himalaya bis ca. 5700 m, auf den Anden auch *Pellaea* und *Jamesonia*), Geophyten (*Cryptogramma Stelleri* im Himalaya bis über 5000 m, *Botrychium Lunaria* auf vielen Gebirgen bis zur Schneegrenze, in Grönland bis 71°40', in Südamerika bis Patagonien), Helophyten und Hydrophyten (*Isoetes lacustre* bis Magerö 71°7').

Höhengrenzen der in Grönland, Europa, Nordafrika und Mittelasien [2]) am höchsten steigenden Pteridophyten:

| | Scoresby-Sound in E-Grönland | Jotunheimen in Norwegen | Alpen | Kaukasus | Marokkanischer Atlas | Tianschan u. Dshung. Alatau | Pamir |
|---|---|---|---|---|---|---|---|
| Botrychium Lunaria . . . . | — | — | 3000 | 3000 | — | 3050 | 3300 |
| Cystopteris fragilis . . . . . | 1255 | 1500 | 3000 | 3100 | 3700 | 3100 | ca. 4500 |
| Athyrium alpestre . . . . . . | — | 1870 | 2250 | 2800 | — | — | — |
| Cryptogramma crispa . . . . | — | 1500 | 2730 | 3000 | — | — | — |
| Polypodium vulgare . . . . . | — | — | 2600 | 2250 | — | 2800 | — |
| Asplenium septentrionale . . . | — | ? | 2820 | 2800 | — | ca. 2100 | — |
| ,,      viride . . . . . . . | — | ? | 2950 | 2800 | 3700 | 3100 | — |
| ,,      Ruta-muraria . . | — | — | 2900 | 2200 | 3700 | 3100 | — |
| Lycopodium Selago . . . . . | 940 | 1940 | 3120 | 2600 | — | ca. 2700 | — |
| Equisetum variegatum . . . . | 1255 | ? | 2545 | 2600 | — | — | — |

[1]) J. E. HOLLOWAY, Studies in the New Zealand Hymenophyllaceae. Trans. N. Zeal. Inst. LIV–LV, 1923–24. H. HANDEL-MAZZETTI, Symbolae Sinicae VI, 1929. Ferner die Arealkarten bei CHRIST (Geogr. d. Farne T. II, III), HOLMBOE (Naturen 1927) u. DEGELIUS (Acta phytogeogr. suec. VII 1935 p. 212).

[2]) Nach N. HARTZ, R. JÖRGENSEN, BRAUN–BLANQUET, GROSSHEIM, MAIRE (Mém. Soc. sc. nat. Maroc XXXIII 1932) u.a., Mittelasien nach O. u. B. FEDTSCHENKO in Acta Horti Petrop. XXXVIII 1924 u. Bull. Jard. Bot. USSR 1927.

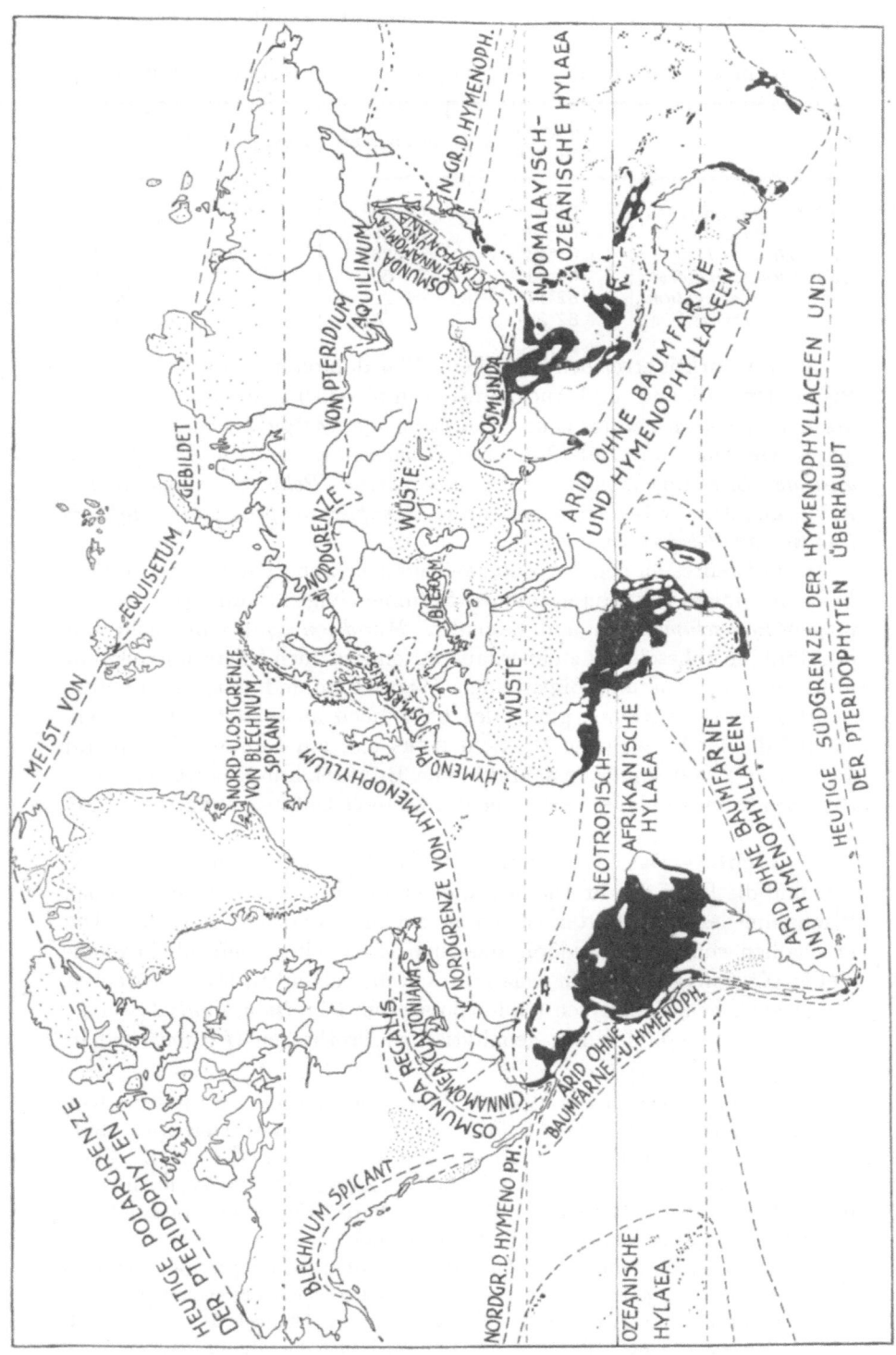

FIG. 5. — Die Pteridophyten-reichsten Hylaea-Gebiete (schwarz), die Pterido-
phyten-ärmsten Gebiete (punktiert) und einige Trocken- und Kältegrenzen.

Die 6 am weitesten in die Arktis vorgedrungenen Pteridophyten sind [1]):

| | Grönland und Ellesmereland | Spitzbergen | Novaja Zemlja | Taimyr |
|---|---|---|---|---|
| *Cystopteris fragilis* . | 81°47′ | ca. 78° | 73°20′ | 68°30′ |
| *Woodsia glabella* . . | 81°47′ | ca. 78° | — | ca. 70° |
| *Lycopodium Selago* . | 81°43′ | ca. 78° | ca. 72° | 74°27′ |
| *Equisetum scirpoides*. | 69°33′ | 79°56′ | 73°20′ | ca. 72° |
| ,,      *variegatum* | 82°17′ | ca. 78° | 74°30′ | 74°27′ |
| ,,      *arvense* . | 82°29′ | ca. 79° | 73°20′ | 74°51′ |

In der Antarktis erreichen heute keine Pteridophyten den 60. Breitegrad. Bis zu den Falkland- und Auckland-Inseln (51–53°) gehen u.a. *Trichomanes*-Arten, *Schizaea australis* und *Gleichenia cryptocarpa*, bis etwa 55° (Fuegia, St. Georgia, Campbell-I.) mehrere *Hymenophyllum*-Arten, *Polypodium australe*, *Blechnum tabulare* und *penna-marina*, *Polystichum mohrioides*. Arktis und Antarktis gemeinsam sind *Hymenophyllum peltatum*, *Cystopteris fragilis* und *Polypodium vulgare*.

Am resistentesten gegen vorübergehend sehr hohe Hitzegrade sind die Rhizome einiger Geophyten und die Stämme einiger Baumfarne. So vermögen *Equisetum*-Arten und besonders *Pteridium aquilinum* nicht nur Waldbrände unbeschadet zu überdauern, sondern auch Überschüttung mit vulkanischer Asche und Bimssteinen. Wieweit sie und einige Baumfarne selbst nach Überströmen glühender Lava noch auszutreiben vermögen, worauf die Beobachtungen W. SCHIMPERS u.a. am Gunung Guntur auf Java, von YOSHII am Komagatake, von BURKILL am Kilauea, von ROBYNS am Rumoka u.a. [2]) deuten, bedarf weiterer Untersuchung.

## § 5. Licht.
— Das Lichtbedürfnis für die Sporenkeimung, das Wachstum und die Form der Prothallien, die Bildung der Gametangien und das Wachstum des Sporophyten ist schon sehr oft untersucht worden [3]). Die Prothallien wohl aller Pteridophyten, auch der in vollem Sonnenlicht wachsenden Xerophyten, sind an gedämpftes Licht gebunden. Dass die Sporen der eigentlichen Geophyten (Ophioglossaceen, *Pteridium*, *Equisetum* u.a.) und der Chamaephyten mit geophytischen Prothallien (*Lycopodium*) in

---

[1]) OSTENFELD, Flora Arctica 1902. H. ROSENDAHL in Medd. om Grönland LVI 1918. CHRISTENSEN in Amer. Fern Journ. I 1921. M. RIKLI, Das Ausklingen der Pteridophytenflora in der Polaris. Ber. Schweiz. Bot. Ges. XLII 1933 u.a.

[2]) J. BURKILL in Proc. Linn. Soc. London CXXXVIII 1926, Y. YOSHII in Bot. Magaz. Tokyo XLVI 544 1932, W. ROBYNS in Verh. Belg. Kolon. Inst. I 1932, ferner die S. 396 genannten Arbeiten über Krakatau.

[3]) K. BITTNER, Über Chlorophyllbildung im Finstern bei Kryptogamen. Österr. Bot. Zeitschr. LV 1905. ISABURO NAGAI, Physiologische Untersuchungen über Farnprothallien. Flora CVI 1914. W. NIENBURG in Ber. Deutsch. Bot. Ges. XLII 1924, A. MUSSACK in Beih. Bot. Cbl. LI 1933. Dort auch die ältere Literatur.

völliger Dunkelheit keimen, haben bereits HOFMEISTER, MILDE u.a. festgestellt, aber dasselbe gilt nach SCHELTING, LAAGE, H. FISCHER, NAGAI u.a. auch z.B. für die Osmundaceen, *Polypodium vulgare* u.a., wogegen z.B. viele Asplenien, *Ceterach, Phyllitis, Aneimia* u.a. Lichtkeimer sind. Die Prothallien der Osmundaceen, von *Cryptogramma* u.a. vermögen, wie schon BORODIN [1]) gezeigt hat, auch im Dunkeln wenigstens Antheridien zu bilden. Den Einfluss verschiedenen Lichts haben BURGERSTEIN und STEPHAN [2]) untersucht (s. auch S. 368–370).

Farnprothallien und junge Sporophyten mit Primärblättern werden sehr regelmässig in Höhlen bis fast an die Chlorophyllgrenze bei einem relativen Lichtgenuss von 1/1000–1/1700 getroffen, ja selbst im völlig dunkeln Höhleninnern um nur kurze Zeit brennende Lampen, so solche von *Cystopteris fragilis* und *Polypodium vulgare*, doch werden bei einem Lichtgenuss von unter 1/300 keine Sori mehr gebildet [3]). Für die Bildung normaler Prothallien mit befruchtungsfähigen Archegonien brauchen aber die meisten Pteridophyten mit Ausnahme derer mit Knollenprothallien und für die normale Ausbildung des Sporophyten alle Pteridophyten Licht in verschiedener Menge. Wenn aber erst Rhizome gebildet und Reservestoffe gespeichert sind, vermögen nach KAROLINE BITTNER wohl sämtliche Filicineen und einige Selaginellen, nicht aber die Lycopodien und Equiseten, auch in völliger Dunkelheit Blätter mit Chlorophyll zu bilden. An Stelle dorsiventraler Prothallien und *Selaginella*-Sprosse werden im Dunkeln isolaterale gebildet. Das Lichtbedürfnis der Tropho- und Sporophylle der meisten Pteridophyten schwankt innerhalb sehr weiter Grenzen. Das Optimum der meisten Wald- und Felsfarne dürfte nach den Messungen LÄMMERMAYRS [4]) u.a. bei einem relativen Lichtgenuss von 1/10–1/50 liegen. Meist in vollem Sonnenlicht wachsen z.B. *Asplenium Ruta-muraria, Ceterach, Cheilanthes, Cyclophorus* (Fig. 2) und andere Xerophyten, einige *Lycopodien* (z.B. *L. alpinum* und *inundatum*) und die meisten *Equiseten* und Heterosporen mit Ausnahme einiger *Selaginellen* und in grösserer Wassertiefe lebender *Isoeten* und *Pilularien*.

**§ 6. Boden (Edaphismus).** — Bei allen Archegoniaten scheinen die ursprünglicheren Sippen saure mineralische und Humusböden vorzuziehen und alkalische Böden zu meiden, sodass die basiphilen (minimialkaline

---

[1]) I. BORODIN in Bull. Acad. St. Petersb. XII 1868.

[2]) A. BURGERSTEIN in Ber. Deutsch. Bot. Ges. XXXVI 1908. J. STEPHAN in Planta V 1928 u. Jahrb. f. wiss. Bot. LXX 1929.

[3]) F. MORTON u. H. GAMS, Höhlenpflanzen. Speläol. Monogr. V, Wien 1925. Dort die älteren Arbeiten. MAHEU et GUÉRIN in Bull. Soc. Bot de Fr. 1935.

[4]) L. LÄMMERMAYR in Jahrb. Gymnas. Leoben IX–X 1907–8, Denkschr. Akad. Wien LXXXVII–XCII 1911–15 u. Mitt. Naturw. Ver. Steiermark LIV 1918.

nach WHERRY) durchwegs von oxyphilen (mediaciden nach WHERRY, aci-
dophilen nach AMANN, APINIS u.a. [1])) abzuleiten sind.

Aus der bereits sehr grossen Zahl von allerdings nur zum kleineren Teil
einwandfreien und in vielen regionalen Vegetationsmonographien zerstreuten Aciditätsmessungen von. Pteridophytensubstraten geht folgendes hervor:

Die grosse Mehrzahl der Pteridophyten, insbesondere auch der tropischen, scheint euryion mit Optimum im sauren Bereich (meist zwischen
pH 5 und 6). Das gilt für die meisten *Isoetes*- und *Equisetum*-Arten, viele
Ophioglossaceen, Osmundaceen und Polypodiaceen, so *Phegopteris Dryopteris* (Amplitude pH 3,4–7,3), *Pteridium aquilinum* (von pH 3,2, welchen
Wert RUTTNER an den Solfataren von Samosir auf Sumatra mass, bis 7,6
nach F. CHODAT [2])), *Polypodium vulgare* (pH 4,5–8,5 nach WHERRY) und
*polypodioides* [3]). Euryion mit Optimum um den Neutralpunkt (subneutro-
phil) oder im alkalischen Bereich (eurybasiphil) sind die meisten Selaginellen (so wohl alle europäischen Arten), *Ceterach officinarum*, *Polystichum
lobatum* und *lonchitis*, *Camptosorus rhizophyllus* [4]), *Equisetum hiemale* und
*maximum*.

Unter den stenoionen Pteridophyten überwiegen besonders in den warmen Ländern, worauf zuerst WHERRY hingewiesen hat, die oxyphilen. Zu
ihnen gehören wohl alle *Lycopodiaceae* (die grösste, fast bis zum Neutralpunkt reichende Amplitude scheinen das besonders weit verbreitete *Lycopodium Selago* und das nahestehende nordamerikanische *L. lucidulum* aufzuweisen), *Schizaeaceae* und *Hymenophyllaceae*, *Blechnum Spicant*, *Cryptogramma crispa*, *Gymnogramme leptophylla* u.a. Stenobasiphil sind dagegen die
Kalk- und Dolomitfelsen bewohnenden *Phyllitis*, *Adiantum*, *Cheilanthes*
usw. sowie wahrscheinlich alle *Salvinia* und *Azolla*. Stenoneutrophil scheinen mehrere Ophioglossaceen, die meisten *Woodsia*- und *Notholaena*-Arten,
*Cystopteris fragilis* [5]) und viele Asplenien, darunter die Serpentinfarne [6]).

---

[1]) E. T. WHERRY in Amer. Fern Journ. X–XI 1920–22 u. XVII 1927. J. J. RO-
BINOVE and C. D. LA RUE in Pap. Michigan Acad. sc. IX 1929. A. APINIS in Acta
Horti bot. Univ. Latv. VIII–X 1935–36.

[2]) F. CHODAT, La concentration en ions hydrogènes du sol. Thèse Genève 1924.
F. RUTTNER in Tropische Binnengewässer, Suppl. XI z. Arch. f. Hydrob. 1932.

[3]) R. M. HARPER in Amer. Fern Journ. IX 1919. L. J. PESSIN in Ecology VI 1,
1925.

[4]) E. J. HILL in Fern Bull. IX 1901. E. T. WHERRY in Journ. Washington Acad.
sc. VI. 1916.

[5]) A. MUSSACK, Untersuchungen über *Cystopteris fragilis*. Beih. Bot. Cbl. LI 1932.

[6]) SADEBECK in Sitzungsber. Ges. f. Bot. Hamburg III 1887. LÄMMERMAYR in
Sitzungsber. Akad. Wien 1926, Mitt. Naturw. Ver. Steiermark LXVII 1930 u.
Pflanzenareale I 8 1928. SUZA und ZLATNIK in Bull. Acad. sc. Bohême 1928. MESSERI
in N. Giorn. Bot. Ital. XLIII 2. 1936.

Viele Farngattungen enthalten sowohl oxyphile wie neutrophile und basiphile Arten, so z.B.:

| | oxyphil | neutrophil | basiphil |
|---|---|---|---|
| Ophioglossum: | arenarium | vulgatum | Engelmannii |
| Dryopteris s. l.: | Oreopteris, cristata. | Braunii, Goldiana | Villarsii, Robertiana |
| | Phegopteris u.a. | | |
| Asplenium: | septentrionale und | Adiantum-nigrum, | Ruta-muraria, viride, |
| | montanum | cuneifolium, fallax | fissum, Seelosii, u. a. |
| Woodsia: | ilvensis? | alpina u.a. | glabella |
| Cryptogramma: | crispa u.a. | | Stelleri |
| Cystopteris: | | fragilis | regia und montana |

Die beiden Serpentin-Asplenien, von denen A. fallax HEUFLER 1856 = adulterinum MILDE 1865 von den beiden Entdeckern wohl mit Recht für einen Bastard des euryionen A. Trichomanes mit dem basiphilen A. viride gehalten worden ist, wogegen A. cuneifolium VIVIANI 1806 = serpentini TAUSCH 1839 von dem subneutrophilen A. Adiantum-nigrum abstammt und nach ZLATNIK Serpentinböden von pH 6,9–7,8 vorzieht, sind nach LÄMMERMAYR u.a. nicht absolut serpentinstet, sondern kommen auch auf Magnesit, Granit und Sandsteinen vor. Die mit ihnen z.B. in Mähren, Steiermark, von dessen Serpentinböden LÄMMERMAYR 20 Farnarten anführt, und in Oberitalien oft vergesellschaftete Notholaena Marantae wächst in den Südalpen hauptsächlich auf neutralem Porphyrschutt. Die Vorliebe dieser Farne für Serpentin dürfte daher in erster Linie auf der neutralen Reaktion der meisten Serpentinböden, in zweiter auf der geringen Konkurrenz durch andere Pflanzen beruhen.

Obligatorische Halophyten scheint es unter den Pteridophyten nicht zu geben, da selbst das oft als solcher angeführte Acrostichum aureum der Mangroven nach RUTTNER u.a. ebenso gut auch in alkalischem Süsswasser (pH 7,3–8,3) gedeiht und auch in der Mangrove nur wenig Salz speichert [1]), die ziemlich zahlreichen in Brackwasser gefundenen Isoetes-Arten durchwegs saures Süsswasser vorziehen und die hyperozeanischen Asplenien (s. S. 387.) wohl nicht wegen einer Vorliebe für Salzluft, sondern nur wegen ihrer Frostempfindlichkeit in Nordeuropa an Strandfelsen gebunden sind.

Ebensowenig scheint es wirklich düngerliebende Pteridophyten zu geben. Die Isoetes-Arten sind streng an ungedüngte Böden und oligotrophe Gewässer gebunden und meiden die eutrophen wohl weniger wegen ihrer Reaktion als wegen des die Konkurrenz höherer Hydrophyten fördernden Nährstoffgehalts. Die meisten Farne und alle Lycopodien werden durch Beweidung und Düngung rasch vertrieben. Am widerstandsfähigsten sind auch hiegegen wie gegen Brände (S. 390) einige Geophyten, wie Botrychium Luna-

---

[1]) H. WALTER u. M. STEINER in Zeitschr. f. Bot. XXX, 1936. Vgl. auch S. 437.

*ria* und *Pteridium aquilinum*, die auch in regelmässig gemähtenund beweideten Wiesen aushalten können. Schwache Beweidung begünstigt auch *Dryopteris Filix-mas* und *spinulosa*, *Athyrium alpestre* und die amerikanischen *Dennstaedtia punctilobula* [1]) und *Hypolepis repens*. Mehrere Asplenien, *Ceterach*, *Cystopteris*, *Cryptogramma crispa* u.a. gehen im Schutz von

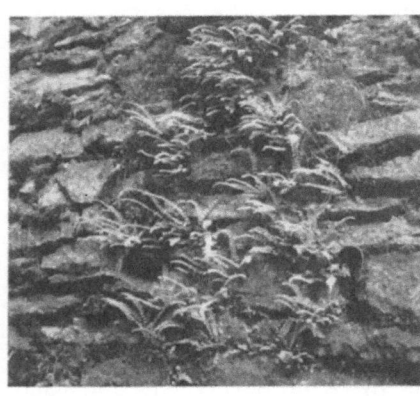

Trockenmauern (Fig. 6) über ihr natürliches Areal hinaus und erscheinen daher in manchen Gegenden als mehr oder weniger „hemerophil", sind jedoch z.B. gegen Rauchgase fast ebenso empfindlich wie die meisten Moose und Flechten.

Wirkliche Kulturlandbewohner sind, abgesehen von den Wasserfarnen der tropischen Reisfelder, nur einige *Equisetum*-Arten, vor allem *E. palustre* (niederdeutsch „Duwock") und *E. arvense*, das als lästiges Ackerunkraut und Begleiter von Wegen und Eisenbahnen dem Menschen von allen Pteridophyten am weitesten sowohl in

FIG. 6. — *Asplenium Trichomanes* an einer Klostermauer in Vianden, Luxemburg. Photo van Steenis.

Wüstengebiete wie in die Arktis gefolgt ist [2]).

**§ 7. Vegetative Vermehrung.** — Vegetative Vermehrung ist besonders bei stärker abgeleiteten Typen sehr verbreitet, sowohl beim Gametophyten (Sprossungen langlebiger Prothallien, z.B. bei *Anogramma*) wie beim Sporophyten, bei diesem sowohl durch unterirdische Rhizome, die bei *Pteridium* und *Equisetum* bis über 4 m lang werden (Fig. 14), und Wurzelbrut (z.B. bei *Ophioglossum*), wie durch oberirdische Sprossteile. Verbreitung durch losgelöste Rhizomstücke spielt namentlich bei helophytischen Equiseten eine Rolle [3]).

Rein vegetative Vermehrung durch Bruchblätter mit austreibenden

---

[1]) H. J. Lutz in Journ. of Agricult. Research XLI 7, 1930.

[2]) E. Korsmo, Unkräuter im Ackerbau der Neuzeit. Berlin (Springer) 1930. Derselbe, Ugressfrö. Oslo (Gyldendal) 1935, Wehsarg in Arb. Deutsch. Landw. Ges. XXXL 1927, Buchli in Beitr. Geob. Landesaufn. d. Schweiz XIX 1936. Über die Bekämpfung des Duwocks: P. Sorauer in Zeitschr. f. Pflanzenkrankh. VIII 1898, C. A. Weber in Arb. d. Deutsch. Landw. Ges. LXXII 1902 u. Deutsch. Landwirtsch. Presse 1927, G. Stockfleth in Ill. Landw. Zeitung XXV 1905.

[3]) R. Sernander, Den skandinaviska vegetationens spridningsbiologi 1901 p. 170. E. Crosetti e P. Fontana, Sulla disseminazione di una crittogame vascolare alpina per mezzo delle correnti d'acqua. Boll. Nat. Siena XXVI 1907.

Achselknospen kommt bei *Isoetes* vor [1]). Verlängerte Blätter mit einwurzelnden Spitzen sind besonders bekannt bei *Camptosorus rhizophyllus* (Fig. 7) und *sibiricus*, *Woodwardia radicans*, *Blechnum Sodiroi*, *Polystichum craspedosorum*, Arten van *Asplenium*, *Leptochilus*, *Fadyenia*, *Trichomanes* u.a. Bei einzelnen *Asplenien* werden die sprossenden Blätter geradezu zu Ausläufern.

Regelmässige Gemmenbildung findet sich bei vielen Tropenfarnen (*Dryopteris prolifera*, *Asplenium viviparum* u.a., bei *Hymenophyllaceae* und *Vittarieae* auch an den Prothallien) und der nordamerikanischen *Cystopteris bulbifera* [2]), bei *Psilotum* [3]), *Lycopodium Selago*, *lucidulum*, *reflexum* u.a., deren Gemmen abgeschleudert werden können [4]).

FIG. 7. — *Camptosorus rhizophyllus*, das „wandernde Blatt" (walking leaf) auf saurem Boden von Quarz-Glimmerschiefer in West-Virginia (Jefferson Co., Keys Ferry). — Photo E. T. WHERRY.

Ausnahmsweise Proliferation ist aber sehr viel weiter verbreitet und steht bei vielen Land- und Wasserfarnen ähnlich wie bei den viviparen Gräsern wohl in ursächlichem Zusammenhang mit den viel untersuchten Hemmungen der geschlechtlichen Fortpflanzung (Apogamie und Aposporie) infolge abnorm erhöhter Chromosomenzahlen (näheres hierüber in Kap. VI). Für die Fernverbreitung sind vielleicht die „Dauerbrutknospen" von Bedeutung, die GOEBEL (Organogr. III) von dem philippinischen *Cheilanthes varians* beschreibt.

## § 8. Sporenausstreuung und Verbreitungsagentien. — Die normale Verbreitungseinheit (Diaspore) der meisten Pteridophyten ist die Spore und das weitaus wichtigste Verbreitungsagens der Wind (Anemochorie). Bei den meisten Lycopodien und Farnen scheinen, abgesehen von der

---

[1]) K. GOEBEL, Über Sprossbildung auf *Isoetes*blättern. Bot. Zeitung XXXVII 1879.

[2]) W. KUPPER in Flora XCVI 2, 1906. W. BALLY ebenda IC 1909.

[3]) SOLMS–LAUBACH in Ann. Jard. bot. Buitenzorg IV 1884.

[4]) HEGELMAIER in Bot. Zeitung XXX 48, 1872. J. E. HOLLOWAY, Studies in the New Zealand.... *Lycopodium* II. Methods of vegetative propagation. Trans. N. Zeal. Inst. XLIX 1917. GOEBEL, Organographie III (2. Aufl. 1918, 3. 1933).

Emporhebung der Sporangien und Sori auf besonders ausgebildeten Sprossen und Blattabschnitten und dem Kohäsionsmechanismus des Leptosporangiaten-Annulus, besondere Ausstreuvorrichtungen zu fehlen [1]). Die meisten sind wohl Ombrohydrochore (Regenschwemmlinge und Regenballisten [2])).

Wie wirksam die Anemochorie ist, zeigt besonders die Wiederbesiedlung von Krakatau [3]), wo schon 3 Jahre nach der Katastrophe von 1883 unter den ersten Ansiedlern 10 Polypodiaceen waren; 1897 waren *Ophioglossum moluccanum* und *Lycopodium cernuum* dazugekommen, 1919 je 1 Cyatheacee, Hymenophyllacee, Schizaeacee, Marattiacee und Equisetacee. Die Gesamtzahl der Pteridophyten stieg von 1919 bis 1929 von 49 auf 63. Ähnliches berichten SCHIMPER und ERNST von den Laven des Gunung Guntur auf Java, ROBYNS vom Rumoka in Kongo u.a. Es handelt sich aber immer um Arten der Nachbargebiete. Eigentliche Kosmopoliten sind unter den Pteridophyten ebenso spärlich wie unter den Moosen (S. 454).

Im Gegensatz zu den Lycopodien und meisten Filicineen werden die Sporen von *Selaginella* [4]) und *Equisetum* aktiv ausgeschleudert. Die Stacheln der Makro- und Mikrosporen von *Selaginella* und die Elateren der *Equisetum*-Sporen begünstigen das Zusammenhängen mehrerer Sporen und damit die Wahrscheinlichkeit des Zusammenwachsen mehrer Prothallien und der Befruchtung.

Als Verbreitungseinheit (Diaspore) dienen in manchen Fallen ganze Sporangien, bei *Isoetes* Bruchblätter mit den Sori, bei den Marsileaceen die Sporokarpien. Bei *Isoetes* und den Wasserfarnen überwiegt naturgemäss die Hydrochorie, doch sind bereits auch mehrere Fälle von Zoochorie bekannt. Die Sporen dreier amphibischer *Isoetes* werden in Südafrika durch Regenwürmer, vielleicht auch durch eine Schnecke verbreitet [5]). Ob das häufige Zusammenvorkommen der submersen *Isoetes* mit Salmoniden [6]) und Cypriniden (z.B. Brachsen) nur auf ähnlichen Ansprüchen an den Gewässertypus beruht oder mit Verbreitung durch Fische zusammenhängt,

---

[1]) J. HIGGINS and C. E. WATERS, Ejection of Fern Spores. Asa Gray Bull. V 1897.

[2]) P. MÜLLER, Über Samenverbreitung durch den Regen. Ber. Schweiz. Bot. Ges. XLV 1936.

[3]) TREUB in Ann. Jard. bot. Buitenzorg VII 1888, PENZIG ebenda XVIII 1902, DOCTERS VAN LEEUWEN ebenda XXXI–XXXII 1921–22, A. ERNST in Vierteljahrsschr. Naturf. Ges. Zürich LII 1907 u. LXXIX 1934. S. auch S. 423.

[4]) GOEBEL in Flora LXXXVIII 1901, NEGER ebenda CIII 1911.

[5]) A. V. DUTHIE, The method of spore dispersal of three South African Species of *Isoetes*. Ann. of Botany XLIII 1929.

[6]) H. MARCAILHOU–D'AYMÉRIC, Coexistence des *Isoetes* et des truites dans la plupart des lacs de l'Ariège, des Pyrénées-Orientales et de l'Andorre. C. R. Congr. soc. sav. 1899, Paris 1900.

bleibt weiter zu untersuchen, ebenso, ob auch adnate Farne durch Tiere verbreitet werden [1]).

Die Makrosporen von *Marsilia* und *Pilularia* bilden einen Schwimmring aus Gallerte, der wohl nicht nur das Schwimmen an der Wasseroberfläche, sondern auch Transport durch Insekten und Wasservögel ermöglicht [2]). In hnlicner Weise dürften auch die Glochidien der Massulae von *Azolla* nicht äur das Zusammenhaften von Makro- und Mikrosporen, sondern auch epizoische Verbreitung begünstigen. Das europäische Areal von *Salvinia natans* zeigt einen deutlichen Zusammenhang mit den wichtigsten Vogelzugsstrassen [3]), und so dürfte das zerrissene und unbeständige Areal der meisten Wasserfarne mit Epizoochorie in Verbindung stehen.

### § 9. Allgemeine Keimungs- und Wachstumsbedingungen. — Die

meisten Pteridophytensporen behalten ihre Keimfähigkeit sehr lange, viele Farnsporen 10–20 Jahre, manche jedoch, wie die chlorophyllreichen Sporen von *Osmunda*, nur wenige Tage. Die meisten Filicineensporen sind auf nassem Torf sowohl in gedämpftem Licht, wie (namentlich die der Geophyten, s. § 5) auch im Dunkeln leicht zur Keimung zu bringen, diejenigen der meisten Lycopodien und Equiseten sehr viel schwerer [4]).

Die Sporen der stenooxyphilen Arten (s. § 6) keimen am besten in sehr verdünnten sauren Nährlösungen, die der subneutrophilen *Cystopteris fragilis* (s. S. 392) am besten in 1–2% Nährlösung von pH 6,5, die von *Equisetum arvense* in Lösungen mit einem osmotischen Druck von 3,5 Atm. [5]).

Auch die Prothallien der xerophytischen Farne wachsen nur in dauernd feuchter Luft und nicht nur die der Wasserfarne können auch direkt auf Nährlösung schwimmend kultiviert werden. Die Benetzbarkeit der Sporen ist verschieden, z.B. bei den meisten *Lycopodien* sehr gering, bei *L. Selago* aber gross.

---

[1]) G. E. MATTEI, Probabili relázioni biologiche fra Imenofillee ed animalcoli. Malphighia XXXI 1928.

[2]) A. SCHULTZ, Beiträge zur Kenntnis der Entwicklung der Macrogametophyten von *Pilularia globulifera* L. und *Marsilia quadrifolia* L. Planta XXV 1936.

[3]) K. WEIN, Die Verbreitung der *Salvinia natans* im südwestlichen Europa in ihren Beziehungen zum Vogelzug. Beih. LXI zu FEDDES Repert. 1930.

[4]) W. HOFMEISTER, Vergleich. Unters. 1851, in Flora XXV 1852 u. Pringsh. Jahrb. III 1863. H. BRUCHMANN, Über die Prothallien und Keimpflanzen mehrerer europäischer Lycopodien, Gotha 1898 u. in Flora CI 1910. N. LAGERHEIM in Arkiv f. Bot. VI 1906 u. Svensk Bot. Tidskr. IJ 3 1908. WUIST in Bot. Gazette IL 1910. A. LAAGE in Beih. Bot. Col. XXI 1907. H. FISCHER ebenda XXVII 1911. G. KLEBS in Sitz. ber. Akad. Heidelberg 1916–17. W. NIENBURG in Ber. Deutsch. Bot. Ges. XLII 1924. H. LASSER in Flora CXVII 1924. W. KARPOWICZ in Bull. Acad. Polon. 1927. Die Marburger Dissertationen von W. DÖPP 1927, E. GERHARDT 1927, H. WALDMANN 1928, W. STREY 1931, W. SCHMELZEISEN 1932, K. H. GÖBEL 1935 u.a.

[5]) R. GISTL in Ber. Deutsch. Bot. Ges. XLVII 1929.

Nicht nur die perennierenden, mykotrophen Prothallien der Lycopodialen und Ophioglossaceen, sondern auch die autotrophen der Osmundaceen, von *Anogramma* u.a. können jahrelang vegetieren und sich durch
Sprossung vermehren, ohne dass es zur Befruchtung und Embryobildung
kommen muss. Für die Befruchtung ist immer Wasser, meist Regenwasser
oder Tau, erforderlich, Bei den Isosporen erfolgt sie regelmässig auf dem
Gametophyten, bei den Heterosporen mit zunehmender Selbständigkeit,
bei einigen Selaginellen, wo das stark reduzierte Mikroprothallium früh
abstirbt, manchmal schon auf dem Sporophyten [1]).

Die Ansicht LÜSTNERS (Diss. Wiesbaden 1898), dass ein Zusammenhang
zwischen Netzskulptur der Sporen und unterirdischer Keimung bestehe,
konnte BRUCHMANN nicht bestätigen. Nach seinen Befunden keimen die
Sporen mehrerer Lycopodien (z.B. *L. annotinum* und *clavatum*) erst nach
einigen Jahren [2]), die Prothallien werden erst nach 12–15 Jahren ge-
schlechtsreif und bleiben etwa 20 Jahre am Leben. Auch die Prothallien
und Protokorme der Ophioglossaceen vegetieren rein saprophytisch
mindesten 8–10 Jahre. Damit stimmen die Beobachtungen finnischer Forscher, dass nach Brandrodung Lycopodien erst nach 30–35 Jahren wieder
erscheinen, die Beobachtungen über die Wiederbesiedlung vulkanischer
Böden (§ 8) und von Gletschern verlassener Moränenböden, wo sowohl hemikryptophytische wie geophytische Polypodiaceen schon nach wenigen
Jahren (so *Phegopteris Dryopteris* und *Phegopteris* am Jostedalsbre nach
spätestens 7 Jahren [3]) wieder erscheinen, wogegen nach meinen Beobachtungen an mehreren Alpengletschern *Botrychium Lunaria* erst nach mehr
als 30 und *Lycopodium Selago* kaum vor 60 Jahren auftaucht, woran neben
der langen unterirdischen Entwicklung auch die langsame Versauerung
der anfänglich meist neutral reagierenden Moränenböden schuld sein mag.

Über das Alter, welches die einzelnen Pteridophyten erreichen können,
liegen wenig verlässliche Beobachtungen vor. Baumfarne, wie *Todea
barbara*, erreichen in Kultur ein Alter von mehreren 100 Jahren. *Lycopodien* werden sicher mehrere Jahrzehnte, die meisten Hemikryptophyten
und Geophyten wohl nur einige Jahre alt. *Ceratopteris thalictroides* und
*Salvinia natans* sind meist einjährig, können aber auch mehrjährig kultiviert werden.

## II. Synökologie s. lat. (= Biocönotik)

**§ 10. Sociabilität und Dispersion.** — Die Häufungsweise der Individuen einer Art (Sociabilität) und die Gleichmässigkeit ihrer Verteilung

---

[1]) F. LYON, A study of the sporangia and gametophytes of *Selaginella apus* and
*rupestris*. Bot. Gazette XXXII 1901.

[2]) BRUCHMANN in Flora CI 1910.

[3]) K. FAEGRI in Bergens Mus. Aarbok VII 1933.

(Dispersion) hängen in erster Linie von den Verbreitungsmitteln (§ 7 und 8) ab und sind daher ebenso sehr autökologische wie synökologische Eigenschaften. Die grosse Mehrzahl der reinen Anemochoren und im besonderen die meisten mykotrophen Geophyten zeigen sehr geringe Geselligkeit und starke Überdispersion, sodass z.B. die meisten Ophioglossaceen nur in vereinzelten Individuen oder kleinen Gruppen wachsen. Bei den Adnaten, kriechenden Hemikryptophyten und Rhizomgeophyten kommt es dagegen durch vegetative Ausbreitung zunächst zur Bildung geschlossener Herden, gelegentlich von Hexenringen [1]), dann von grossen, mehr oder weniger geschlossenen Vereinen mit deutlicher Unterdispersion, so insbesondere bei vielen Waldfarnen und Equiseten, ähnlich auch bei wurzelnden und schwimmenden Wasserfarnen. In der Regel werden nur die Gesellschaften dieser mehr geselligen Pteridophyten als eigene Vereine bewertet.

**§ 11. Wettbewerb (Competitio), Symbiose und Parasitismus. —** In Wettbewerb treten die Pteridophytenindividuen nicht nur mit solchen der gleichen Art und anderer Pteridophyten, sondern sehr oft auch mit andern Pflanzen der gleichen Lebensformenklasse, so Baumfarne mit Gymnospermen, Palmen und Dikotylenbäumen, Lycopodien mit Zwergsträuchern, adnate Hymenophyllaceen und Vittarieen mit Moosen und Flechten, Equiseten mit Juncaceen und Cyperaceen, Ophioglossaceen mit Orchideen, *Isoetes* mit *Litorella*, Salviniaceen mit Lemnaceen usw. Vielfach führt dieser Wettbewerb zu mehr oder weniger gesetzmässigen Vereinigungen.

Die besonders innigen Vergesellschaftungen mit Symbionten und Parasiten sind schon in Kap. III–V behandelt. Zur Ergänzung seien noch einige Fälle von Schädigung durch andere Pflanzen angeführt: Die chasmophytischen Farne haben nicht nur unter der Konkurrenz von Blütenpflanzen zu leiden, sondern werden ebenso wie die adnaten Farne nicht selten von adnaten Moosen und besonders Flechten (*Diploschistes*- und *Parmelia*-Arten) überwuchert. Auf *Dryopteris Filix-mas* und *Matteuccia* habe ich wiederholt *Cuscuta europaea* beobachtet. Die kanarisch-westmediterrane *Orobanche trichocalyx* soll nach DESPREAUX [2]) auf *Pteridium aquilinum* schmarotzen.

**§ 12. Die Pteridophytenvereine. —** Vom Devon bis ins Mesophytikum wurde die gesamte Landvegetation der Erde von grossenteils phanerophytischen Pteridophyten, vom Karbon an auch von Pteridospermen und vom Perm an auch von Gymnospermen beherrscht, doch treten spätestens im Verlauf der permokarbonischen Vereisungen der Südhemi-

---

[1]) W. N. CLUTE, Fairy Rings formed by *Osmunda*. Fern Bull. IX 1901.
[2]) BARKER, WEBB et BERTHELOT, Phytogr. Canar. III 185, 1836–50.

sphäre (*Glossopteris*-Flora) auch schon adnate und hemikryptophytische Formen auf. Heute beherrschen Pteridophyten nur noch ausnahmsweise ganze Biozönosen (besonders adnate, chasmophytische und helophytische), sondern treten meist in untergeordneten Vereinen (societies) in von Blütenpflanzen beherrschten Soziationen bzw. Konsoziationen auf. Für ihre Klassifikation gilt das im Man. of Bryol. p. 328–336 ausgeführte. In der folgenden Übersicht sind nur Sporophytenmerkmale berücksichtigt.

1. Oberirdische Überdauerungsknospen vorhanden . . . . . . . . . . . . . . . . . . . . . . .2.
1+. Oberirdische Überdauerungsknospen (ausgenommen in einigen Fällen blattbürtige Brutknospen) fehlen . . . . . . . . . . . . . . . . . . . . . . . . . . . . . . . . . . . . .4.
2. Mit aufrechtem, oberirdischem Holzstamm (Phanerophyten): *Baumfarne* (§ 12).
2+. Ohne aufrechten Holzstamm. . . . . . . . . . . . . . . . . . . . . . . . . . . . . . . . . . . .3.
3. Oberirdische Achsen windend oder der ganze Stamm einer festen Unterlage angedrückt:                              *Lianen- und Adnatenvereine* (§ 13).
3+. Oberirdische Achsen aufsteigend oder auf meist loser Unterlage kriechend:                                        *Chamaephytenvereine* (§ 14).
4. Überdauerungsknospen in der Bodenfläche oder zwischen Gesteinsstücken:                                           *Hemikryptophytenvereine* (§ 15).
4+. Überdauerungsorgane im Boden oder in Wasser (Kryptophyten) . . . . . . . .5.
5. Überdauerungsorgane (Knollen oder Rhizome) ganz im Boden eingesenkt:                                             *Geophytenvereine* (§ 16).
5+. Überdauerungsorgane mindestens periodisch unter Wasser oder schwimmend:                                         *Helo- und Hydrophytenvereine* (§ 17).

**§ 13. Baumfarnvereine.** — Die meisten Dicksoniaceen und Cyatheaceen der immerfeuchten Regen- und Nebelwälder der südhemisphärischen Gebirge bilden oft aus mehreren Arten (s. das Titelbild, Fig. 1, 8 und S. 439) bestehende Vereine, die wohl zu einer einzigen Vereinigung (federatio) *Alsophilion* zusammengefasst werden können und meist mit anderen, oft dominierenden Phanerophytien, Epixylien (s. § 13) u.a. zu mehrschichtigen Soziationen verbunden sind, für die auf die Darstellungen HOLTTUMS, CHRISTS (Geogr. d. Farne) und die dort angegebene Literatur verwiesen sei. Mehrere Arten sind sehr widerstandsfähig gegen Feuer; einzelne vertragen auch leichten Frost. Einzelne Arten werfen das Laub ab (*Alsophila comosa* in Sikkim). Resistentere Arten reichen bis S-Florida [1]), S-Japan [2]) (*Cyathea boninsinensis, Alsophila acaulis*), Argentinien (*Dicksonia Sellowiana*), Kapland (*Hemitelia capensis*) und zu den Chatham-Inseln.

Mehrere südhemisphärische *Blechnum*-Arten (besonders *Bl. capense* und *tabulare*) bilden mit ihren bis meterhohen Stämmen recht euryözische und weit verbreitete Vereine, die wohl als *Blechnion subantarcticum* zusammen-

---

[1]) UNDERWOOD in Proc. Ind. Acad. Sc. 1891, ABAXON in Ann. Rep. Smiths. Inst. 1911, SMALL in Journ. E. Mitchell Sc. Soc. XXXV 1920, DOBBIE in Amer. Fern Journ. XIX 1929.

[2]) G. MASAMUNE, Floristic and geobotanical studies in the island of Yakusima. Mem. Taihoku Univ. 1934.

gefasst werden können. Das *Blechnetum socialis* der Anden von Bolivia und Ecuador erreicht über der Waldgrenze 3000–4000 m Höhe.

Eine weitere phanerophytische Vereinigung (*Odontosorion*) bilden die unbegrenzt fortwachsenden Stachelklimmer der vorwiegend westindischen Gattung *Odontosoria* (bramble ferns) [1].

Ob sich auch unter den ähnlich undurchdringliche Dickichte bildenden, aber zumeist hemikryptophytischen und geophytischen *Gleichenieta*, die CHRIST sowohl mit den Odontosorieta wie mit den Pteridieta verbindet Phanerophytien befinden, ist mir nicht bekannt (s. S. 443).

FIG. 8. — Farnwald in der Waiho-Schlucht, Wertland (New Zealand): links *Dicksonia* sp., cf. *lanata* var. *gregalis*, in der Mitte *Lycopodium volubile*, rechts *Blechnum procerum* und *Hemitelia Smithii*, vorn *Leptopteris superba*.

Photo TEICHELMANN.

**§ 14. Lianen- und Adnatenvereine.** — Die meisten echten Farnlianen sind rein tropisch, so *Stenochlaena palustris* (*Acrostichum scandens*), mehrere *Dennstaedtia*-und *Lygodium*-Arten, *Blechnum* (*Salpichlaena*) *volubile, Lathyropteris madagascariensis* u.a. Bis in extratropische Gebiete reichen *Lygodium palmatum* [2]) und *scandens* und das neuseeländische *Lycopodium volubile* (Fig. 8).

Die Adnaten, die meist zu unbestimmt als Epiphyten bezeichnet wer-

---

[1]) J. MASSART in Bull. Soc. r. Bot. Belg. XXXIV 1905.
[2]) SAUNDERS in Fern Bull. VIII–IX 1900–01, BLODGETT in Torreya I 1901.

den, obgleich viele von ihnen ebenso gut epipetrisch wachsen, gehören ebenfalls zum weitaus grössten Teil den Tropen und feuchten Subtropen an, so die höchst spezialisierten Oleandren, Platycerien, und Drynarien (s. S. 429–433 und die klassischen Werke SCHIMPERS, GOEBELS und CHRISTS), ebenso die ökologisch den Hymenophyllaceen ähnlichen Vittarioideen [1]). Es scheint auch nur in den Tropen nereidische Farne, d.h. wasserbewohnende Felshafter, zu geben: einige südamerikanische und malayische Asplenien, *Polypodium pteropus* [2]), *Blechnum Francii* und das westafrikanische *Elaphoglossum palustre*. Die weitaus meisten Adnaten sind immergrün. Eine Ausnahme bildet z.B. *Davallia canariensis* [3]).

In extratropische Gebiete reichen, abgesehen von einigen Gelegenheitsepiphyten und Gelegenheitsepipetren (z.B. Selaginellen) nur einige vorwiegend adnate Gattungen von Polypodiaceen und Hymenophyllaceen, von denen die meisten sowohl epixyl (epiphytisch) wie epipetrisch wachsen, in den trockenern und kälteren Gebieten meist nur epipetrisch, so *Cyclophorus lingua* (Fig. 2) und *Camptosorus sibiricus* in Nordasien, *C. rhizophyllus* (Fig. 7) in Nordamerika.

Von den vielen vorwiegend epiphytisch wachsenden Polypodien, die der Lebensform nach grossenteils an der Grenze von Chamaephyten, hemikryptophytischen Chasmophyten und eigentlichen Adnaten stehen, reichen nur ganz wenige in extratropische Gebiete, so *Polypodium polypodioides* [4]) in N.-Amerika bis Pennsylvania, *P. lineare* in Ostasien bis Japan und Ussurigebiet und das dank seiner Vielgestaltigkeit und Euryözie fast kosmopolitisch gewordene *P. vulgare*, der weitaus verbreitetste Farnepiphyt der Nordhemisphäre [5]).

Es wächst oft chasmophytisch mit Asplenien, *Cystopteris* usw. [6]), selbst geophytisch in sauren Wald- und Heideböden, bildet aber auch einen eigenen, wirklich adnaten Verein *Polypodietum vulgaris*, der sowohl an Bäumen wie auf Fels mit Isobryalen- und Stictaceen-Vereinen in Kon-

---

[1]) GOEBEL in Flora LXXXII 1896 u. N. F. XVII 1924, E. BRITTON and A. TAYLOR in Torrey Bot. Club VIII 3 1902.

[2]) GIESENHAGEN in Flora LXXVI 1892. A. ERNST in Ann. Jard. Buitenzorg VII 2 1908.

[3]) C. BOLLE, Standorte der Farne auf den Kanar. Inseln. Zeitschr. Ges. f. Erdkunde Berlin 1863–66.

[4]) SMALL in Torreya III 1903, R. HARPER in Amer. Fern Journ. IX 1919, L. PESSIN in Ecology VI 1925.

[5]) V. WITTROCK in Acta Horti Bergiani II 1894, P. FLICHE in Bull. Soc. Bot. de France XLIX 1902, A. SHADOVSKY in Trud. Bot. Mus. Petersb. X 1913, D. JOHNSON in Bot. Gazette LXXII 1921, R. BENEDICT in Amer. Fern Journ. XII 1922, B. PETTERSSON in Mem. Soc. F. Fl. Fenn. VI 1930 u.a.

[6]) E. WETTER, Ökologie des Felsflora kalkarmer Gesteine. Jahrb. St. Gall. Naturw. Ges. LV 1918. F. FIRBAS in Beih. Bot. Zentralbl. XL 1924. E. OBERDORFER in Mitt. Bad. Landesver. f. Naturk. N. F. III 1934.

kurrenz tritt und nur mit ähnlichen Vereinen amerikanischer und ostasia-
tischer Polypodien vereinigt werden kann.

Das Areal der wohl ausnahmslos adnaten Hymenophyllaceen [1]) umschliesst völlig das der ebenfalls adnaten Vittarioideen und das der Baumfarne, geht aber in allen Richtungen darüber hinaus (Fig. 5). Während sie
in den tropischen und südhemisphärischen Regen- und Nebelwäldern grossenteils epixyl wachsen, besonders auch auf Baumfarnen, und zusammen
mit Vittarioideen und kleinen Polypodien eine Art epiphytischen Trocken-
torfs („urru" auf Costarica) [2]) bilden, leben sie im extratropischen Gebiet
fast ausschliesslich epipetrisch an kalkfreiem, dauernd bergfeuchtem Gestein

Fig. 9. — *Hymenophyllum tunbridgense* an senkrechten, schattigen Sandsteinfelsen
im Zickzackschlüff bei Grundhof, Luxemburg. — Photo van Steenis.

an vor direktem Sonnenlicht und Frost geschützten Orten. Besonders
weit reichen *Trichomanes speciosum* (bis Alabama, Rio de Janeiro, Irland,
Japan, noch im Pliozän bis Bulgarien), *parvulum* (bis Japan und Wladi-
wostok), *caespitosum* und *flabellatum* (S.-Afrika, Maskarenen, Chatham-
u. Falkland-Inseln), *Hymenophyllum tunbridgense* (bis Schottland, Mittel-
deutschland, Westitalien und Kroatien, das ähnliche *H. Fomini* im Pliozän
des Kaukasusgebiets), *peltatum* (bis zu den Faer Öern und S. Georgia) und

[1]) Giesenhagen, Die Hymenophyllaceen. Marburg 1890. Holloway, Studies
in the New Zealand *Hymenophyllaceae*. Trans. N. Zealand Inst. LIV–LV, 1923–24.

[2]) C. Werckle in Soc. Nac. Agricult. Costaric. S. José 1909. S. auch S. 436.

*multifidum* (bis zu den Auckland-, Campbell- und Antipodeninseln, s. Fig. 5).

Sowohl *H. tunbridgense* (Fig. 9) wie *H. peltatum* (= *Wilsonii*) [1]) sind so regelmässig vergesellschaftet mit dem ähnlich, aber noch weiter verbreiteten Lebermoos *Diplophyllum albicans*, dass sie trotz ihrer Geselligkeit mehr als Charakterarten der ozeanischen Fazies des *Diplophylletum albicantis* denn als Dominanten eines eigenen Hymenophylletums erscheinen.

**§ 15. Chamaephytenvereine.** — Abgesehen von den kurzstämmigen Marattialen, Osmundaceen und Polypodiaceen, welche von den phanerophytischen zu den hemikryptophytischen überleiten, und den immergrünen Equiseten, welche von ihren phanerophytischen Vorfahren zu Geophyten und Helophyten führen, gehören hierher nur die Lycopodien, Psilotalen und Selaginellen.

Fig. 10. — *Lycopodium annotinum* mit *Medeola virginica* und *Aralia nudicaulis* in einem Tal des Allegany State Park, Cattaranga County, W. New York.
Photo Laurence E. Hicks.

Von eigenen *Lycopodium*-Vereinen kann im extratropischen Gebiet kaum gesprochen werden, obgleich mehrere Arten, wie die weit verbreiteten *L. annotinum* (Fig. 10) und *clavatum*, oft mehrere m² bedeckende Herden bilden. Die weitaus meisten sind mehr oder weniger konstante

---

[1]) L. v. Heufler in Verh. Zool. Bot. Ges. Wien XX 1870; F. Graves in Ann. of Scott. Nat. Hist. 1899; G. Rouy in Rev. Bot. syst. I; E. Klein, *H. tunbr.*, Luxemburg 1916; L. Rodway in Proc. R. Soc. Tasmania 1913/14; Morton u. Gams in Speläol. Monogr. V. 1925 p. 188.

und charakteristische Bestandteile von Ericaceen-Vereinen, so *L. anno-tinum* und *complanatum* von *Myrtilleta*, *L. clavatum* auch von anderen *Vaccinieta* und *Calluneta*, *L. alpinum* von *Empetro-Vaccinieta* und *Loise-leurieta*. An bestimmte Baumarten sind die *Lycopodien* ebensowenig ge-bunden wie die Waldfarne, wohl aber sind die von ihnen besiedelten Vac-cinieta usw. örtlich an bestimmte Nadelwaldkonsoziationen gebunden, sodass sie z.B. als lokale Charakterarten der Fichten-Konsoziation er-scheinen. können *L. Selago* und *lucidulum* treten dank ihrer besonders gros-sen ökologischen Amplitude in sehr verschiedenen Chamaephytien und Hemikryptophytien der Heidewälder, Zwergstrauch- und Grasheiden bis zu eigentlichen Schneeböden auf [1]). *L. inundatum* ist oft mit *Andromeda polifolia*, *Oxycoccus* und *Drosera* vergesellschaftet, meist auf nacktem Torf in sauren Grasmooren (*Rhynchosporion* oder *Scheuchzerion*).

Fig. 11. — *Psilotum triquetrum* mit tief in Lava-Klüfte eingesenkten Rhizomen auf der Insel Rangitoto (New Zealand). — Photo L. Cranwell.

Die Psilotalen (*Psilotum* und *Tmesipteris*) wachsen gleich vielen *Ly-copodien* in ihrem tropischen und australisch-neuseeländischen Haupt-areal epiphytisch (besonders auf *Cyathea* und *Dicksonia*), daneben aber auch an und zwischen Felsen und im extratropischen Gebiet — *Psilotum triquetrum* bis Korea und S.-Japan — vorwiegend an Gestein, wobei sie

[1]) J. W. Trail in Ann. Scott. Nat. Hist. 1910.

mit ihren tief eingesenkten Prothallien [1]) und Rhizomen eine Zwischenstellung zwischen Chamaephyten, Adnaten und Geophyten einnehmen
(Fig. 11).

Die Selaginellen zeigen viel weniger als die Lycopodien Beziehungen zu
Heiden und Heidewäldern als zu Chasmophytien und selbst Epipetrien.

Nur die am weitesten nordwärts reichende und am höchsten steigende
*S. Selaginoides* (= *spinulosa*) wächst regelmässig in basiphilen Zwerg-
strauch- und Grasheiden (Dryadeta, Cariceta usw.) bis zu Helophytien und
Schneeböden, meist in vollem Sonnenlicht. Die europäischen *S. helvetica*
und *denticulata*, die ostasiatische *S. tamariscina*, die nordamerikanische *S.
rupestris* u.a. wachsen vorzugsweise an beschatteten, mässigfeuchten Felsen von neutraler bis alkalischer Reaktion und treten dabei weniger mit
andern Gefässpflanzen (am häufigsten mit Asplenien und Crassulaceen) als
mit Moosen (*Distichium, Ditrichum, Fissidens* u.a.) und Flechten (besonders *Cladonia*) in Konkurrenz. Das *Selaginelletum helveticae*, das ich
aus dem Wallis beschrieben habe [2]), geht in sekundären Fragmenten einerseits auf Felsflächen, andrerseits auf Alluvionen über.

**§ 16. Hemikryptophytia.** — Hiezu die mehr oder weniger streng an
Felsspalten gebundenen Chasmophytia, die ohne scharfe Grenze einerseits
in die Farnwiesen, andrerseits in Geophytien übergehen. Unter den Chasmophytien sind viele Arten von *Asplenium* über weite Gebiete so konstant
und charakteristisch, dass ich die alpinen Chasmophytia 1927 [2]) als *Asplenion* zusammengefasst habe, BRAUN–BLANQUET [3]) als *Asplenietales
rupestres*. Die von ihm und H. MEIER gegebene Gliederung, welche auch
chamaephytische Chasmophytia umfasst, ist jedoch wenig natürlich.

Nach den Ansprüchen an die Bodenreaktion, Wärme und Feuchtigkeit
ergeben sich folgende ökologische Reihen:

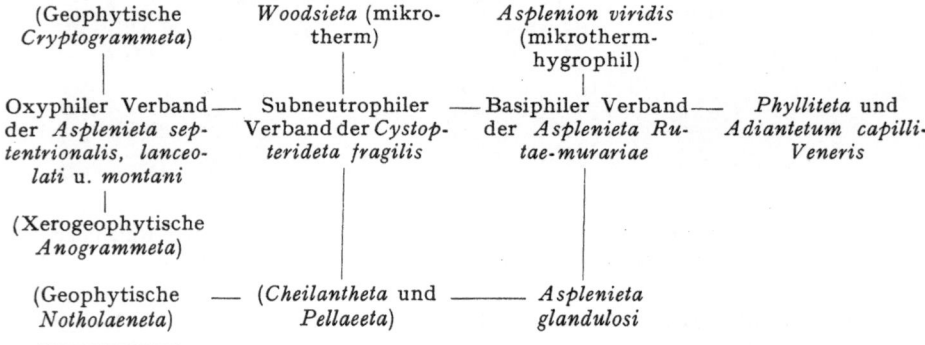

| (Geophytische *Cryptogrammeta*) | *Woodsieta* (mikro-therm) | *Asplenion viridis* (mikrotherm-hygrophil) | |
|---|---|---|---|
| Oxyphiler Verband— der *Asplenieta sep-tentrionalis, lanceo-lati* u. *montani* | Subneutrophiler Verband der *Cystop-terideta fragilis* | —Basiphiler Verband— der *Asplenieta Ru-tae-murariae* | *Phylliteta* und *Adiantetum capilli-Veneris* |
| (Xerogeophytische *Anogrammeta*) | | | |
| (Geophytische *Notholaeneta*) | — (*Cheilantheta* und *Pellaeeta*) — | *Asplenieta glandulosi* | |

[1]) A. A. LAWSON in Trans. R. Soc. Edinb. LI/LII 1917, J. E. HOLLOWAY in
Trans. New Zeal. Inst. LI 1918 u. LIII 1921.

[2]) GAMS in Beitr. z. geobot. Landesaufn. d. Schweiz XV 1927 p. 554/5.

[3]) H. MEIER et J. BRAUN–BLANQUET, Classe Asplenietales rupestres. Prodome
des Groupements végétaux II 1934.

Die 3 Hauptverbände umfassen eine grössere Zahl von Unionen, so der oxyphile die eurasiatischen *Asplenieta septentrionalis* mit dem *Saxifragetum mixtae* der Pyrenäen, dem *Phyteumetum serrati* Korsikas, dem *Primuletum hirsutae* der Westalpen usw., die atlantisch-westmediterranen *Asplenieta lanceolati* mit *Cheilanthes pteridioides*, *Antirrhinum asarina* usw. und das nordamerikanische *Asplenietum montani*.

Der subneutrophile Verband der *Cystopterideta fragilis* umfasst mehrere trotz ihrer z.T. fast kosmopolitischen Verbreitung bisher verkannte Unionen mit mehreren Nebenreihen, so vor allem die Union der *Cystopteris fragilis* (besonders oft mit *Asplenium Trichomanes* und Crassulaceen, in Amerika auch mit *Pellaea*-Arten), der mikrothermen *Woodsia alpina* und *ilvensis*, die ich entgegen mehreren Literaturangaben in den Alpen wie in Skandinavien nur auf schwach saurem Gestein gesehen habe, des stärker xerothermen *Ceterach officinarum* [1]), der Serpentin-Asplenien (s. § 6), der mit diesen oft vergesellschafteten, aber auch einen eigenen, mehr geophytischen Verein bildenden *Notholaena Marantae* und der übrigen xeromorphen *Notholaena*-, *Cheilanthes*- und *Pellaea*-Arten [2]) (Fig. 4 und 12).

FIG. 12. — Standorte der extrem xerophytischen Farne *Pellaea atropurpurea* und *glabella*, *Cheilanthes Feei* und *Notholaena dealbata* in der sonst farnfreien Prairie von Minneapolis (Kansas) an Sandsteinblöcken.
Photo J. H. SCHAFFNER.

Der basiphile Verband der *Asplenieta Rutae-murariae* umfasst eine grosse Zahl von namentlich in den Mittelmeerländern sehr artenreichen, nach Norden rasch verarmenden Vereinen, wie die *Potentilleta caulescentis* und *Clusianae*, die vorwiegend westmediterranen *Asplenieta fontani*, an die sich die stärker xerothermen *A. glandulosi* [3]) anschliessen, das *Asplenietum Seelosii* der Dolomiten [4]) und die illyrisch-ostalpinen *Asplenieta fissi* [5]). Dem mehr mikrothermen und hygrophilen Verband *Asplenion viridis*, der viel weiter nordwärts

---

[1]) L. DESSALLE et A. REYNIER in Bull. Soc. Sc. d. Basses-Alpes XIX (1919) 1920.

[2]) F. L. PICKETT in Bull. Torrey Bot. Club L 1923, LIII 1926 u. Amer. Fern Journ. XXI 1931.

[3]) R. MOLINIER, Etudes phytosociologiques et écologiques en Provence occidentale. Ann. Mus. d'Hist. nat. Marseille XXVII (1934) 1935.

[4]) L. DIELS, Einige Bemerkungen zur Ökologie des *Asplenium Seelosii* Leyb. Verh. Bot. Ver. Prov. Brandenb. LVI 1914.

[5]) HORVAT, Vegetationsstudien in den Kroatischen Alpen II. Bull. Acad. Yougosl. XXV 1931; MARZELL in Jahrb. d. Ver. z. Schutz d. Alpenpfl. VI 1934.

reicht und höher steigt, gehören u.a. *Cystopteris montana* und *regia* und *Woodsia glabella* an. Sehr viel frostempfindlicher sind dagegen die *Phylliteta*[1]) und das vorwiegend schattige, triefende Tuffstellen (stillicidia) bekleidende, aber auch recht dürreresistente Zwergformen ausbildende *Adiantetum capilli-Veneris*[2]).

Die *Asplenieta septentrionalis, Trichomanis, Rutae-murariae* usw. besiedeln in durchwegs stark verarmten Fragmenten oft Trockenmauern[3]) (Fig. 6).

Ähnlich wie die europäischen *Asplenium-, Phyllitis-* und *Adiantum-*Arten besonders tief in Höhlen eindringen (s. § 5), ist dies auch in amerikanischen Höhlen der Fall[4]), so auf Masatierra[5]): *Blechnum auriculatum, Dryopteris inaequalifolia, Histiopteris incisa, Hypolepis rugulosa* u.a.

An die mehr oder weniger streng chasmophytischen *Asplenieta* usw. schliessen sich zunächst solche Farnvereine an, die sowohl chasmophytisch, wie als Bewohner von Wald- und Heideböden wachsen: das weitverbreitete *Asplenium Adiantum-nigrum* mit seiner Waldsteppen bewohnenden Serpentinrasse *cuneifolium, Phyllitis Scolopendrium*, das amerikanischostasiatische *Adiantum pedatum* und vor allem die zumeist immergrünen *Polystichum*-Arten, die wohl in mehrere Verbände zusammenzufassen sind. Auf der Südhemisphäre besonders weit verbreitet ist das *Polystichetum capensis*. Besonders viele Arten hat Südostasien, wo mehrere Arten bis zur Schneegrenze steigen (*P. Duthiei* im Himalaya bis ca. 5700 m). Nur wenige (*P. lonchitis, lobatum, angulare = setiferum* und das oft, aber nicht immer sommergrüne *Braunii*) reichen nach Sibirien und Europa, wo sie immer noch fragmentarische, sehr hoch steigende Vereine bilden (*P. lonchitis* auf den Alpen bis 2610, am Tianschan bis 2750 m), ähnlich *P. lonchitis, scopulorum* und *acrostichoides* in N.-Amerika.

An diese immergrünen Vereine schliessen sich als Ausstrahlung eines anderen immergrünen Verbandes der Südhemisphäre das frostempfindliche, oxyphile, meist mit Myrtilleta und Piceeta verbundene *Blechnetum Spicantis*[6]) und die grosse Federation der sommergrünen Waldfarne (Farnwiesen, CHRISTS „Formation der bodenständigen strunklosen Farne") oder das *Athyrion* an. Seine Vereine bilden ökologische Reihen, die von den vor-

---

[1]) F. MORTON in Österr. Bot. Zeitschr. LXIV 1914 und LXXV 1925. V. VOUK ibidem LXV/LXVI 1915/16. G. LUSINA in Ann. di Bot. XV 1920.

[2]) E. HOFMANN u. MORTON in Bot. Archiv XXIV 1929.

[3]) W. WEST, Mural Ecology. Journ. of Bot. XLIX 1911.

[4]) S. PRICE, Some Ferns of the Cave Region of Stone County, Miss. Fern Bull. XII 1904. R. HARPER, The Fern grottoes of Citrus County, Florida. Amer. Fern Journ. VI 1916.

[5]) C. SKOTTSBERG, Notes on the Vegetation in the Cumberland Bay Caves, Masatierra, Juan Fernandez Isles. Ecology XVI 1935.

[6]) A. PALMGREN in Mem. Soc. pro Fauna et Flora Fenn. VII 1932, J. HOLMBOE in Avh. Norsk. Vid. Akad. IX 1937.

genannten immergrünen Hemikryptophytien über sommergrüne Hemikryptophytien zu Geophytien führen und regelmässig mit Vereinen an-
giospermer Hochstauden (Aconiteta, Adenostyleta u.a.) und Hochgräser
(Calamagrosteta) in Wettbewerb treten und mit diesen zusammen Bestandteile von Laub- und Nadelwaldkonsoziationen („Farn-Typen" der Finnen[1])
bilden, ohne doch enger an bestimmte Waldbäume gebunden zu sein.

Die grösste ökologische Amplitude und die weiteste Verbreitung im ganzen holarktischen Waldgebiet haben die in den Wäldern vom „Farntyp"
oft vergesellschafteten *Athyrium Filix-femina, Dryopteris Filix-mas* und
*spinulosa* s. lat., kleinere die mikrothermen *Dryopteris fragrans* und *Matteuccia Struthiopteris*, die nordamerikanisch-ostasiatischen *Onoclea sensibilis*
und *Osmunda Claytoniana*, die nordamerikanischen *Dryopteris noveboracensis* und *marginalis, Athyrium thelypteroides, Dennstaedtia punctilobula*
u.a., die kaukasischen *Dr. oreades, Raddeana* u.a. Die *Matteuccia-, Onoclea-*
und *Osmunda*-Arten bilden stärker hygrophile bis helophytische Vereine;
*Dryopteris spinulosa, fragrans* und *barbigera* und *Athyrium alpestre* subark-
tisch-subalpine Vereine, die besonders in Krummholz-Konsoziationen auftreten und im Schutz von Blöcken und Schneegruben[2] weit über die Waldgrenze steigen.

Als stenooxyphile Vereine von geringerer Verbreitung seien noch die der
*Dryopteris Oreopteris* (oft mit *Nardeta* und *Pteridieta*) und der ähnlich wie
*Blechnum Spicant* in der Holarktis ganz isoliert dastehenden, sowohl auf
saurem Sand wie in Sphagneta wachsenden *Schizaea pusilla* (curly grass der
nordamerikanischen Pine barrens [3]) genannt.

**§ 17. Geophytenvereine.** — Die Geophytia umfassen unmittelbar an
die Chasmophytia anschliessende Schuttvereine (Lithophytia) und die
Vereine der Wald-, Heide- und Wiesenböden besiedelnden autotrophen
Rhizomgeophyten und mykotrophen Knollengeophyten. Schuttstrecker
im Sinne SCHROETERS [4] sind die in § 15 genannte *Notholaena Marantae,
Dryopteris Villarsii* und *Cystopteris regia* der Kalkalpen und vor allem die
*Cryptogramma* (= *Allosorus*)-Arten der Hochgebirge: die oxyphile, meist
sehr artenarme Geröllvereine bildende *Cr. crispa* mit ihren Verwandten,
wie der in N.-Amerika z.B. mit *Cystopteris fragilis, Pellaea Bridgesii, densa*
und *Breweri* vergesellschafteten *Cr. acrostichoides* (Fig. 13) und die basi-

---

[1]) A. K. CAJANDER in Acta forest. fenn. I 1909 u. XX 1921; Y. ILVESSALO
ibidem XXXIV 1929; V. KUJALA in Comm. Inst. quaest. forest. Finl. X 1926 u.
XXII 1936.

[2]) „*Allosoreto-Athyrion alpestris*" bei R. NORDHAGEN in Bergens Mus. Aarbok 1936.

[3]) C. SAUNDERS in Linnean Fern Bull. IV 1896; E. BRITTON ibidem IV 1896,
Bull. Torrey Bot. Club XXVIII 1901, Plant World IV 1901; R. BENEDICT in
Amer. Fern Journ. III 1913.

[4]) C. SCHRÖTER, Das Pflanzenleben der Alpen, Zürich 1908, 2 Aufl. 1926 p. 709.

phile, im Himalaya bis 5500 m steigende *Cr. Stelleri*. Diese leitet mit ihren im Gegensatz zu den vorigen nicht aufsteigenden, sondern horizontal kriechenden Rhizomen zu den Schuttwanderen [1]) über, deren Hauptvertreter die ebenfalls basiphile *Dryopteris (Gymnocarpium) Robertiana* ist. In den Kalkalpen ist sie meist mit *Rumex scutatus, Moehringia muscosa, Valeriana montana* und *Petasites niveus* vergesellschaftet [2]).

FIG. 13. — *Cryptogramma acrostichoides* im kalifornischen Gebirge, ca 3000 m.
Photo E. SMITH.

An diese Rohboden-Geophytien schliessen sich die autotrophen Geophytien der Humusböden an, zunächst die der nächstverwandten *Dryopteris Linnaeana = Phegopteris Dryopteris* und ihrer asiatischen Verwandten (*Dr. oyamensis* und *remota*). Erstere Art ist keineswegs, wie der zunächst nur diese Art bezeichnende Name *Dryopteris* (Eichenfarn) besagt, an Eichenwälder gebunden, sondern wächst, meist mit Hylocomieta und Myrtilleta, in sehr verschiedenen Laub- und Nadelwäldern, in denen ein besonderer „Dryopteris-Typ" unterschieden wird, sowie zwischen Blöcken bis über die Waldgrenze [3]).

Noch weniger an bestimmte Wälder gebunden ist die stenoxyphile, häufig auch auf sauren Rohböden und trotz ihrem Namen nur ausnahmsweise auch unter Buchen wachsende *Dryopteris Phegopteris* (= *Phegopteris polypodioides*) und die mit ihr in N.-Amerika oft vergesellschaftete *Dr. (Phegopteris) hexagonoptera*.

Ähnlich euryion wie *Dr. Linnaeana* (s. 392), aber zufolge der meist tiefer (5–20 cm tief) liegenden, bis über 4 m langen, sehr derben Rhizome (Fig. 14) viel widerstandsfähiger ist das kosmopolitische *Pteridium aquilinum* (= *Pteris aquilina*), das keineswegs nur trockene bis mässig feuchte, nicht oder wenig beschattete Humusböden, sondern auch sehr verschiedenartige mineralische Böden besiedelt und namentlich auf brandgerodeten Flächen sehr grosse Ausdehnung erlangt, ebensogut in den tropischen Alang-Alang-Wiesen (*Imperatetum cylindricae*) und *Gleichenia*-Heiden

[1]) Siehe Note 4 Seite 409.

[2]) H. JENNY, Vegetationsbedingungen und Pflanzengesellschaften auf Felsschutt. Beih. Bot. Cbl. XLVI 1930.

[3]) A. PRESCOTT in Fern Bull. XIX 1911; A. K. CAJANDER in Acta forest. fenn. XX 1921; V. KUJALA in Comm. Inst. forest. Finl. X 1926; C. MALMSTRÖM u. K. LUNDBLAD in Skogsförsöksanst. Exkurs. led. XI/XII 1926/27; R. CHING in Contrib. Biol. Labor. Sc. Soc. China IX 1933.

(in Ostasien u.a. mit *Osmunda japonica* und *Botrychium lanuginosum*), wie in den verschiedensten Laub- und Nadelwäldern (besonders Querceta, Castaneeta, Pineta und Junipereta), Ericaceen- und Ginsterheiden [1]).

Zur Bekämpfung des in manchen Weidegebieten (W.-Europa, Australien, Neuseeland u.a.) zu einem lästigen Weideunkraut gewordenen *Pteridium* sind besondere Mähmaschinen gebaut worden [2]).

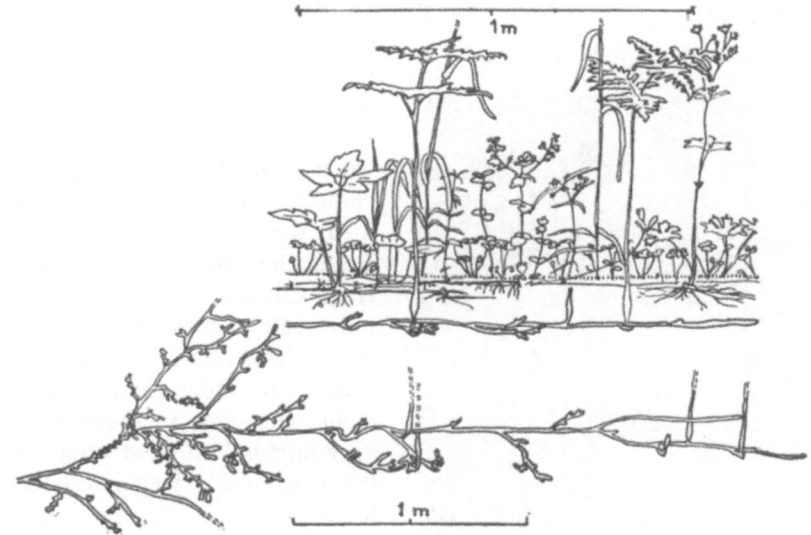

FIG. 14. — *Pteridium aquilinum* in einem finnischen Hainwald mit *Milium*, *Aegopodium, Geranium silvaticum, Oxalis* usw. — Nach V. KUJALA 1926.

Viele Analogien mit diesen Polypodiaceenvereinen haben die *Equiseteta*. Das hygrophil-oxyphile *Equisetum silvaticum* wächst oft mit Dryopterideta in versumpfenden Wäldern. *E. hiemale* und *maximum* wachsen besonders in Laubwäldern mit *Polysticheta* und *Cariceta*; *E. pratense* [3]) und *arvense* ähnlich wie *Pteridium* und oft mit ihm zusammen in verschiedenen Wäldern und Heiden, sowie apophytisch als Ruderalpflanzen. Die übrigen sind vorwiegend Helophyten, doch vermögen einige, wie *E. palustre* und *ramosissimum*, auch oberflächlich sehr trockene Standorte zu besiedeln.

---

[1]) J. VALLOT in Rev. gén. de Bot. 1896; GILLOT et DURAFOUR in·Bull. Soc. Nat. Ain 1904; L. A. BOODLE in Journ. Linn. Soc. XXXV 1904; T. W. WOODHEAD, Diss. Zürich 1906; M. BÜSGEN in Zeitschr. f. Forst- u. Jagdwesen 1915; F. CHODAT, Thèse Genève 1924; V. KUJALA in Comm. Inst. forest. Finl. X 1926; TH. BIELER in C. R. Acad. Paris CLXXXV 1927; C. E. ANGST in Publ. Puget Sound Biol. Stat. V 1927; H. LUTZ in Yale School Forest. Bull. XXXVIII 1934; H. CONARD in Amer. Midl. Natur. XVI 1935 u.a.

[2]) N. Brit. Agriculturist LXXXVIII 1936, Scott. Farmer XLIV 1936, Nature 24. X 1936.

[3]) A. MAILLEFER in Bull. Soc. Vaud. sc. nat. LVIII 1934.

Eine Zwischenstellung zwischen Hemikryptophyten, Geophyten und Therophyten nimmt *Anogramma* (*Gymnogramme*) *leptophylla* ein (s. GOE-BELS S. 386 genannte Dissertation). In den winterfeuchten Subtropen wächst sie gesellig auf im Winter feuchten, im Sommer dürren Wald- und Fels-böden, an ihrer Nordgrenze in den Süd- und Zentralalpen nur in dauernd frostfreien Felsklüften mit *Ceterach*, *Asplenien* und wintergrünen Moosen (*Timmiella*, *Targionia*, *Sphaerocarpus* u.a. [1])).

FIG. 15. — *Botrychium onondagense* UND. in der Nadelstreu eines *Thuja*-Bergwalds bei Quebec in Kanada. — Photo MOUSLEY.

Die Ophioglossaceen bilden im Gegensatz zu den autotrophen Rhi-zomgeophyten meist keine selb-ständigen Vereine, sind aber auch wie die ebenfalls mykotrophen Or-chideen, mit denen sie oft zusam-men wachsen, von der übrigen Vegetation sehr wenig abhängig. Nach CHRIST „wimmeln sie derart universell und zerstreut über die ganze Erde hin, dass kaum ein Punkt zu finden sein dürfte, wo nicht eine oder mehrere Formen vorkommen"; aber dennoch sind die meisten sehr wenig gesellig. So habe ich das von allen am weitesten verbreitete *Botrychium Lunaria* in den verschiedensten Wiesengesell-schaften von der *Festuca vallesiaca*-Steppe bis zu regelmässig gemähten und beweideten *Agrostis*- und *Tri-setum*-Wiesen (dort die bis 30 cm hohe f. *robustum* SCHUR, meist mit *Orchis*-Arten zusammen) und bis in die hochalpinen Elyneta (f. *nanum* CHRIST, oft mit *Nigritella*) gesehen.

Wesentlich geselliger scheint nur das von seinem nordamerikanischen Hauptareal bis Brasilien und Eurasien ausstrahlende *B. virginianum* zu sein, das in nordamerikanischen Laubwäldern recht hohe Frequenz erreicht, so in einem von CAIN [2]) in Indiana untersuchten Urwald mit 76% die höchste der dortigen Farne. Auch im Ostbaltikum wächst es gesellig in Laubwäldern, besonders auf Schlagflächen [3]).

Die meisten übrigen Botrychien wachsen ähnlich wie *B. Lunaria* ganz

---

[1]) MILDE in Bot. Zeitung XX 1862, GAMS in Bull. Soc. Murith. XXXIX 1916, Speläol. Monogr. V 1925 u. Beitr. Geobot. Landesaufn. XV 1927.

[2]) ST. A. CAIN in Amer. Midland Nat. XV 5 1934.

[3]) N. MALTA, Zur Verbreitung der Gattungen *Ophioglossum* und *Botrychium* in Lettland. Acta Horti Bot. Univ. Latv. III 1928.

vereinzelt in Wäldern (Fig. 15) und häufiger in voll besonnten Mager-
wiesen, oft an recht ungewöhnlichen Standorten [1]).

Die „universelle und dabei ganz lokale, stets auf weite Zwischenräume getrennte Verbreitung bei einer unerschöpflichen, minimalen Variation"
der *Ophioglossum*-Arten nennt CHRIST „das grösste Rätsel in der Geographie der Farne". Das von allen Arten auf der Nordhemisphäre am weitesten verbreitete *O. vulgatum* ist entschieden neutrophil [2]) und wächst daher
am häufigsten in neutralen Sumpfwiesen (in Mitteleuropa besonders in
Molinio-Phragmiteta mit *Colchicum* und Orchideen), doch habe ich es auch
in trockenen Kastanienwäldern und im Überschwemmungsgürtel von Voralp-
penseen gefunden, SKUJA [3]) auf einem Brachacker. Die Bodenansprüche
der nordamerikanischen Arten behandeln BENEDICT und WHERRY [4]). Das
von KOMAROV [5]) von Thermen am Vulkan Uzon auf Kamtschatka beschriebene *O. thermale* wächst auf warmem, tonigem Kalkboden.

**§ 18. Helo- und Hydrophytenvereine.** — Sowohl die vorwiegend
chamaephytischen Lycopodialen (s. § 14), wie teils phanero-bis hemikryptophytische (*Gleichenia, Schizaea, Woodwardia, Dryopteris* u.a.) und geophytische Farngattungen (*Thelypteris*) enthalten mehr oder weniger streng an
das Sumpfleben angepasste Helophyten. Während die *Osmunda*-Arten
(Fig. 5 und 16), *Schizaea pusilla, Woodwardia (Lorinseria) areolata* u.a. in
erster Linie Waldpflanzen sind, die aber doch oft in Mooren, namentlich
Sphagnum-Mooren wachsen, ist *Woodwardia (Anchistea) virginica* bereits
eine eigentliche Sumpfpflanze, die z.B. in den nassen Mooren von Bailey
Pont in Connecticut mit ihren Rhizomen ausgedehnte, teils untergetauch-
te, teils flutende Matten bildet [6]).

Der am weitesten verbreitete Sumpffarn ist *Thelypteris palustris* (= *As-
pidium* oder *Dryopteris Thelypteris*), die sowohl auf der Nord-, wie auf der
Südhemisphäre vorzugsweise nasse Schwingmoore mit schwach saurer
Reaktion bewohnt, regelmässig mit *Sphagna, Carices* und *Eriophora*, in
N.-Amerika besonders auch mit *Woodwardia virginica, Scirpus america-*

---

[1]) J. H. SCHAFFNER in Ohio Natur. X 1909, E. J. WINSLOW in Am. Fern Journ.
IV 1914, D. L. DUTTON in Fern Bull. XVIII 1920.

[2]) W. N. CLUTE in Fern Bull. IX 1901, R. C. BENEDICT in Am. Fern Journ. IV
1914, BENEDICT, SHELPS and WHEELER ibidem V 1915, E. T. WHERRY ibidem XI
1921, G. CHANVEAUD in Bull. Soc. Bot. de France LV 1908, A. GIUGNI–POLONIA
in Verh. Schweiz. Naturf. Ges. 1914, W. KOCH in Jahrb. St. Gall. Naturw. Ges.
LXI (1925) 1926.

[3]) Sieh Seite 412 Note 3.

[4]) Sieh Note 1 Seite 392.

[5]) W. L. KOMAROV in Feddes Repert. XIII 1914 u. Flora Kamtsch. I 1927.

[6]) S. POWER, Floating Islands. Bull. Geogr. Soc. Philadelphia XII 1914.

*nus, Hypericum*-Arten und *Eupatorium perfoliatum* [1]). Ähnlich verhalten sich *Dryopteris cristata* und die tropische *Dr. gongylodes*.

FIG. 16. — *Osmunda cinnamomea* in Sphagnum-Moor, Cranberry Island, Buckeye Lake, Licking County, Ohio. Im Vordergrund *Oxycoccus macrocarpus*, zu beiden Seiten Büsche von *Toxicodendron vernix*. — Photo JOHN H. SCHAFFNER.

FIG. 17. — *Equisetum praealtum* Raf. = *hiemale* var. *affine* an einem Bachufer. Columbus, Ohio. Nächst *E. arvense* die in den oestlichen Verein. Staaten verbreitet- ste Art. — Photo JOHN H. SCHAFFNER.

Die helophytischen *Equiseta* bilden 3 ökologische Gruppen: die geschlossene Sumpfwiesen (Cariceta, Junceta usw.) bewohnenden *E. palustre, ramosissimum* usw., die vorzugsweise Mineralböden an fliessenden Gewässern besiedelnden und bis ins Hochgebirge und in die Arktis vorgedrungenen *E. variegatum* und *scirpoides* (ähnlich verhalten sich auch das vorzugsweise waldbewohnende und in Auenwäldern besonders häufige *E. hiemale* und seine Verwandten in N.-Amerika und Asien, Fig. 17) und das weit in stehende und fliessende Gewässer vordringende *E. limosum* (= *E. heleocharis* und

[1]) M. L. FERNALD, A Study of *Thelypteris palustris*. Contrib. Gray Herb. Harvard Univ. 1929. H. CONARD in Amer. Midland Naturalist XVI 1935.

*fluviatile*), das allein oder mit *Phragmites, Schoenoplectus-* und *Carex*-Arten ausgedehnte, torfbildende Bestände bildet. Einige vorwiegend an periodisch überschwemmten Orten wachsende *Equiseta* (*E. litorale, Moorei, trachyodon* u.a.) sind wohl hybridogen [1]).

Die Isoetaceen und Marsileaceen bilden ökologische Reihen, welche von nur vorübergehend überschwemmten Geophyten über Amphiphyten zu dauernd submersen Arten oligotropher Gewässer führen. Die *Isoetes*-Arten teilte bereits A. Braun in Terrestria (*I. hystrix, Duriei* u.a.), Amphibia (viele Arten besonders in den westlichen Mittelmeerländern, S.-Afrika und N.-Amerika, so *I. Dodgei* und *Engelmannii*, Fig. 18) und Aquatica (*I. lacustre, echinosporum* und Verwandte). Von den terrestrischen und amphibischen Vereinen, die sich nicht scharf trennen lassen, beschrieben Braun–Blanquet und Moor [2]) aus N.-Afrika die von *I. hystrix* mit *Juncus capitatus* und *Cicendia* und von *I. adspersum* mit *Marsilia pubescens* und *Nitella capitellata*; aus S.-Frankreich die von *I. setaceum* mit *Marsilia pubescens, Pilularia minuta* und *Juncus pygmaeus* und von *I. Duriei* mit *Juncus capitatus* und *Aira capillaris*. Den meisten Vereinen des westmediterranen „*Isoetion*" gemeinsam sind u.a. *Riccia*-Arten, *Juncus bufonius, Lythrum hyssopifolium* und *Mentha pulegium*, welche zusammen mit *Marsilia strigosa* auch in den östlichen Mittelmeerländern und bis in die Kaspi-Niederung ganz ähnliche Vereine bilden [3]). Allen diesen Vereinen und auch denen anderer Marsileaceen und der submersen *Isoeten* gemeinsam ist, dass sie weiches Wasser vorziehen und wohl schwachen Chlorid-, aber keinen höheren Karbonatgehalt vertragen. Die südhemisphärischen Arten wachsen u.a. mit *Ophioglossum-, Eriocaulon-* und *Xyris*-Arten

FIG. 18. — *Isoetes Engelmannii* A. Br., North Carolina, Haywood Co. Photo H. C. Blomquist.

[1]) O. Holmberg in Bot. Notiser 1920, G. Samuelsson in Vierteljahrsschr. Nat. Ges. Zürich LXVII 1922, A. Becherer in Candollea IV 1929, J. Kümmerle in Mag. Bot. Lap. XXX 1931.

[2]) J. Braun–Blanquet, Un joyau floristique et phytosociologique „l'Isoetion" méditerranéen. Comm. S. I. G. M. A. XLII 1936. M. Moor, Zur Soziologie der Isoëtalia. Beitr. z. geobot. Londesaufn. d. Schweiz XX 1936 u. Prodr. d. Pflges. 4, 1937.

[3]) H. Glück, Biol. u. morph. Unters. III 1911 u. in Beih. Bot. Cbl. XXXIX 1923. B. Keller, F. Leisle u.a. in Rastitelnost Kasp. Nismennosti I 1936.

zusammen, so *I. Welwitschii* im Hochland von Benguella. Ähnlich verhält sich auch *Phylloglossum Drummondii*. Die Sporokarpien vieler *Marsilia*-Arten (bei *M. hirsuta* auch Rhizomknöllchen) und die Rhizomknollen vieler Isoeten vertragen langdauernde Austrocknung; so konnte DURIEU Knollen von *I. Duriei, hystrix* und *setaceum* nach 2–6 und WITTROCK solche des mexikanischen *I. Pringlei* noch nach 11 Jahren zum Austreiben bringen [1]). Diese ephemeren, amphibischen Geophyten gehören daher zu den am weitesten in die Wüsten Afrikas, Asiens und Australiens vordringenden Pteridophyten.

Im Gegensatz zu diesen dürreresistenten *Isoeto-Marsileeta* zeigen die *Pilularia*-Arten und die dauernd submersen *Isoeten* eine vorwiegend ozeanische Verbreitung [2]). Ihre Vereine können als *Isoeto-Litorellion* zusammengefasst werden, da die europäischen *I. lacustre* und *echinosporum* ebenso regelmässig mit *Litorella lacustris*, wie die nordamerikanischen Arten mit *L. americana* und das südamerikanische *I. Savatieri* mit *L. australis* vergesellschaftet sind. Alle wachsen am häufigsten auf sandigem Boden in saurem Wasser von pH 4,5–6,8 (ebenso die sie regelmässig begleitenden *Lobelia Dortmanna* und *Elatine*-Arten), doch ertragen sowohl *I. lacustre* und *echinosporum* wie *I. macrosporum* auch neutrale bis alkalische Reaktion [3]) und gehen bis ins Brackwasser, ähnlich auch die nordamerikanischen *I. riparium* und *saccharatum*, welches z.B. am Wicomico River in Maryland mit *Tillaea aquatica* und *Sagittaria subulata* zusammen wächst [4]).

Während die europäische *Pilularia globulifera*, die ebenfalls dem *Isoeto-Litorellion* angehört, weniger hoch steigt und nordwärts reicht als die bis zum Nordkap reichenden *Isoetes*-Arten, erreicht *P. Mandoni* auf den bolivianischen Anden 5000 m Höhe. *Isoetes hypsophilum* hat HANDEL-MAZZETTI aus einem westchinesischen Gebirgssee im 3600 m. Höhe beschrieben (Symb. Sin. VI 1929).

In schroffem Gegensatz zu den *Isoeten* und genannten Marsileaceen stehen die vorwiegend freischwimmenden (pleustischen) Arten von *Ceratopteris*, *Salvinia* und *Azolla*, die ganz an sich stark erwärmende, nährstoffreiche, alkalische Gewässer gebunden sind und nur ausnahmsweise auch Landformen bilden. Die Tropenkosmopoliten *Ceratopteris thalictroides* [5]), *Salvinia natans* [6]), *Azolla pinnata* u.a. wachsen ähnlich wie die Lemnaceen, *Pistia*,

---

[1]) V. BR. WITTROCK in Acta Horti Berg. I 1891, C. R. BARNES in Bot. Gazette 1895, H. GLÜCK in Flora CXV 1922.

[2]) A. DONAT in Pflanzenareale I8 1928 u. III 8 1933, S. THUNMARK in Acta phytogeogr. suec. II 1931, G. SAMUELSSON ibidem VI 1934, E. OBERDORFER in Ber. Naturf. Ges. Freiburg i. Br. XXXIV 1934, K. JÖNS in Schr. Naturwiss. Ver. Schlesw. Holst. XX 2, 1934.

[3]) J. IVERSEN in Bot. Tidsskr. XL 1929, N. C. FASSETT in Trans. Wisconsin Acad. XXV 1930, L. R. WILSON in Ecol. Mon. V 1935.

[4]) G. H. SHULL in Bot. Gazette XXXVI 1903.

[5]) Y. YABE and K. YASUI in Bot. Mag. Tokyo XXVII 1913.

[6]) K. YASUI in Bot. Mag. Tokyo XXIV 1910 and Ann. of Bot. XXV 1911.

*Eichhornia*, und *Trapa* und oft mit diesen (auch einigen *Marsilia* Arten) somit als Glieder des *Lemnion*, sehr gesellig, oft auch als Unkräuter von Raisfeldern. Die meist unbeständigen Vorkommnisse von *Salvinia* und *Azolla* ausserhalb der Tropen und Subtropen sind meist durch Ornithochorie zu erklären (§ 8). Da *Azolla filiculoides* bis ins mittlere Pleistozän auch am Niederrhein und Don wuchs [1]), so braucht ihr neuerliches Auftreten in Europa nicht nur auf Einschleppung aus Aquarien zu beruhen.

## III. Angewandte Ökologie

**§ 19. Humus- und Torfbildung, Leitfossilien.** — Als Humus- und Torfbildner stehen die heutigen Pteridophyten hinter den Laubmoosen und Cyperaceen zurück. Die meisten Waldfarne, insbesondere die des *Athyrion*, geben einen nur schwach sauren, relativ fruchtbaren Mullboden, in dem nur ausnahmsweise stärker saurer Trockentorf gebildet wird. Mikroskopisch sind solche Böden leicht an den Treppentracheiden und glatten *Athyrium*-Sporen zu erkennen.

Die einzigen Pteridophytentorfe, die heute so verbreitet sind und solche Mächtigkeit erreichen, dass sie in Moorprofilen ausgeschieden werden, sind von *Equisetum limosum* (leicht kenntlich an den glänzendschwarzen Rhizomen mit derben Epidermen aus isodiametrischen Zellen und Tracheiden mit breiten Spiralbändern) und *Thelypteris palustris* (kenntlich an den Treppentracheiden und stachligen Sporen) gebildet. Der *Woodwardia*-Torf (S. 413) scheint noch nicht genauer untersucht.

Die Sporen, die sich wohl von allen Pteridophyten dank ihrem hohen Gehalt an Sporenin [2]) sehr gut erhalten und zufolge der charakteristischen Skulpturen in den meisten Fällen bis zum Genus, in vielen (*Lycopodium, Isoetes, Thelypteris* u.a.) bis zur Art bestimmbar sind, finden in der besonders von LAGERHEIM [3]) begründeten quantitativen Mikrofossilanalyse steigende Verwendung, sowohl in paläozoischen, mesozoischen und tertiären (s. HIRMER), wie in quartären Ablagerungen. Als fazielle Leitfossilien benützt worden sind insbesondere die namentlich in eiszeitlichen Sedimenten häufigen Mikrosporen von *Selaginella selaginoides* [4]), in interglazialen Seeablagerungen Sporen von *Salvinia* und *Azolla* (s. § 17), in spät- und post-

[1]) P. A. NIKITIN in Trudy Komm. Tschetv. Per. III 1933, F. FLORSCHÜTZ in Verh. Geol.-Mijnb. Gen. v. Nederl. IV 1919, Nederl. Kruidk. Arch. 1928 u. Geol. en Mijnbouw 1935.

[2]) F. ZETZSCHE und Mitarbeiter in Ann. d. Chemie 1926 u. Helvet. Chim. Acta XIV–XV 1931–32; O. KÄLIN, Diss. Bern 1933; F. KIRCHHEIMER in Centralbl. f. Mineral. 1932, Bot. Archiv XXXV 1933 u. Ber. Schweiz. Bot. Ges. XLII–XLIV 1933–34.

[3]) N. LAGERHEIM in Geol. Fören. Förh. Stockholm XXIV 211, 1902; ferner die Pollenatlanten von G. ERDTMAN in Arkiv f. Bot. 1923 u. H. MEINKE in Bot. Archiv 1927.

[4]) W. BEIJERINCK in Proc. Akad. Amsterdam XXXVI 1 1933 u. Nederl. Kruidk. Arch. XLIII 1933; F. FIRBAS in Biblioth. bot. CXII 1935.

glazialen auch solche von *Isoetes* [1]). Bei seinen Mikrofossilanalysen aus feuerländischen Mooren hat AUER [2]) die Sporen von *Hymenophyllum, Asplenium, Blechnum, Cystopteris* und *Lycopodium* getrennt gezählt und aus den mit ihrer Hilfe konstruierten Diagrammen Schlüsse über die Verschiebung der Regenwaldgrenzen gezogen.

**§ 20. Nutzpflanzen.** — Die Pteridophyten haben zwar viel geringere wirtschaftliche Bedeutung als die Coniferen, Palmen, Gramineen u.a., finden aber doch vielerlei Verwendung: Als Speise dient Stärkemehl aus Stämmen und Rhizomen von Baumfarnen (neuerdings auf Hawaii auch industriell verwertet [3]) und aus den Rhizomen mehrerer Waldfarne, so in den verschiedensten Ländern (z.B. Neuseeland, Japan und Kanaren) von *Pteridium aquilinum* mit der tropischen var. *esculentum*, dann die Rhizomknollen von *Equisetum arvense*, die noch im 18. Jahrhundert auf Island und in Norwegen als jardneter, stukneper usw. allgemein gegessen und vermahlen wurden [4]), die Sporokarpien der australischen *Marsilia nardu* und *Drummondii*, die als nardu oder addo ein wichtiges Volksnahrungsmittel bilden. Die fleischigen Blätter von *Ceratopteris* werden als Gemüse gekocht (früher auch die von *Ophioglossum*, z.B. in Südschweden als herrekål), *Azolla* ähnlich wie *Eichhornia* als Schweinefutter und Gründünger gesammelt [5]).

Die süssschmeckenden Rhizome von *Polypodium vulgare* werden nicht nur wie *P. glycyrrhiza* gekaut, sondern es werden aus ihnen wie auch aus denen von *Matteuccia*, die in Skandinavien auch als Ziegenfutter dienen, bierähnliche Getränke hergestellt (siril-drikke in Norwegen). Zur Likörbereitung (Capillaire, Syrupus ad longam vitam )sind namentlich in Frankreich *Asplenium Trichomanes, Ruta-muraria* u.a. verwendet worden [6]). Die Cumarin-reichen *Polypodium phymatodes* (Malaya) und *pustulatum* (Australien) werden zum Parfümieren gebraucht.

Mehrere Polypodiaceen finden seit dem Altertum allgemeine Verwendung als Anthelminthica (*Rhizoma filicis* von *Dryopteris Filix-mas*, ähnlich von der amerikanischen *Dr. marginalis* und der südafrikanischen *Dr. athamantica* = uncomocomo), gegen Leber-, Milz- und Harnleiden (besonders *Phyllitis Scolopendrium*, z.B. in Südtirol auch *Asplenium septentrionale* = Harngras, ähnlich auch *Ceterach, Botrychium* und *Lycopodium*). Zum Blutstillen werden die Spreuschuppen von *Alsophila*- und *Cibotium*-Arten und besonders mehrere *Equisetum*-Arten gebraucht, von denen Aufgüsse auch neuerdings gegen Rheumatismen und Tuberkulose empfohlen werden.

---

[1]) E. OBERDORFER in Ber. Naturf. Ges. Freiburg i. Br. XXXI 1931.

[2]) V. AUER in Acta Geogr. V 2, Helsingfors 1933.

[3]) A. MARQUES in Agron. colon. VI 1922.

[4]) J. HOLMBOE in Avh. Norske Vid. Akad. 1929.

[5]) A. CHEVALIER in Rev. bot. appl. et agronic. colon. 1926, s. auch Amer. Bot. X 1906 p. 71.

[6]) E. ROLLAND, Flore populaire 1896; P. SÉBILLOT, Le Folklore de France III 1906; ROLAND DE BONAPARTE, Usage et folk-lore des fougères. La Nature XLVII 1919.

Mehrere Farne (u.a. *Polypodium* und *Botrychium*) und Lycopodien werden sowohl als Aphrodisiaca (auch beim Vieh) wie als Abortiva verwendet [1] [2].

Das Sporenmehl von *Lycopodium* (seltener auch von *Equisetum*) wird seit altersher zu Pudern und Feuerwerk benützt. Technische Verwendung finden mehrere Baumfarne zu Bauzwecken (so *Dicksonia Sellowiana* in Brasilien zu Zäunen), zu Zigarrenkisten und Hüten (besonders auf den Philippinen [3]), die Liane *Stenochlaena palustris* zu Schiffstauen, die wollähnlichen Spreuschuppen mehrerer mittelamerikanischer und malayischer Baumfarne, sowie des ostasiatischen *Cibotium barometz* und der makaronesischen *Culcita macrocarpa* (feto abrum der Azoren) als Polstermaterial.

Besonders die immergrünen Equiseten dienen als Scheuermaterial für Metall (Zinnkraut), die Blätter vieler Waldfarne (besonders *Pteridium*) als Streu, auch zum Abhalten von Ungeziefer, *Lycopodium clavatum* und *annotinum* zum Milchseihen, zu Kränzen u.a. Zu dekorativen Zwecken finden viele Farne Ostasiens, der Mittelmeerländer (z.B. *Adiantum capillus-Veneris* und *Pteris cretica*) und Amerikas (z.B. *Polystichum acrostichoides*, Christmas fern) Verwendung, ähnlich auch mehrere Selaginellen. Das Aufstellen oder Aufhängen der Auferstehungsfarne (resurrection ferns) *Selaginella tamariscina* in Ostasien und *lepidophylla* in Amerika hängt wie die ähnliche Verwendung von Anastatica mit magischen Vorstellungen zusammen.

**§ 21. Zauberpflanzen.** — HILDEGARD von Bingen im 12., BRUNFELS und BOCK im 16. Jahrhundert bezeichnen die Waldfarne als die vornehmsten aller Zauberpflanzen: „Kein kraut ist da meer hexen werck und teuffels gespenst mit getriben würt" (BRUNFELS). Bis ins 17. Jahrhundert war der Glaube in Europa allgemein verbreitet, dass in der Johannisnacht gesammelter „Farnsamen" Wunder wirke. Die Synode von Ferrara erliess 1612 ein Verbot „ne quis ea nocte filices filicumque semina colligat". Ähnliche Vorstellungen wurden auch an die Rhizomknollen von *Equisetum* geknüpft [2].

Zu den besonders heilsamen und zauberkräftigen „Rutae" werden seit alter Zeit *Asplenium*-, *Adiantum*- und *Botrychium*-Arten gezählt. *Asplenium Trichomanes* und *Ruta-muraria* galten u.a. ähnlich wie *Polytrichum* als „Widerton", d.h. als Mittel zu Gegenzauber. Zu Liebeszauber, zur Förderung wie auch zur Verhinderung von Milchproduktion u.a. werden *Lycopodium*-Arten und ganz besonders *Botrychium Lunaria* als eine der noch heute z.B. in mehreren Alpenländern volkstümlichsten Zauberpflanzen verwendet [4].

---

[1] S. Note 5 S. 418 und die folgenden:

[2] H. MARZELL, Unsere Heilpflanzen. Freiburg 1922, näheres in den pharmakognostischen Werken von TSCHIRCH u.a.

[3] R. C. BENEDICT in Amer. Fern Journ. V 1915; C. B. ROBINSON in Philipp. Journ. Sc. Bot. VI 1911.

[4] H. MARZELL, Die Mondraute (*Botrychium lunaria*) als Kraut des Mondes. Schweiz. Archiv f. Volkskunde XXXI 1931.

CHAPTER XIII

# THE ECOLOGY OF TROPICAL PTERIDOPHYTES

by

R. E. HOLTTUM (Singapore)

**§ 1. Introductory and Historical.** — By far the greatest bulk of literature on pteridophytes is systematic or morphological. Scattered through this literature is a good deal of information relevant from an ecological standpoint, but it is very scattered, and much of it may doubtless have been overlooked. The task of examining it all is an impossible one. The ecological literature is in a similar case, as references to the ecology of pteridophytes are usually contained in papers of a more or less general nature. The number of papers restricted to a consideration of the ecology of pteridophytes is extremely small. On the other hand, there is much relevant information to be found in papers which do not mention pteridophytes as such; as an example may be cited comparative studies of conditions of light, moisture and temperature in the various strata of the forest. The literature therefore is a vast one; not more than a small fraction of it can be cited, and it is feared that a good deal of it has been overlooked.

The only considerable summary of information on the ecology of pteridophytes is in CHRIST's work *Die Geographie der Farne* (Jena 1910). In that book is a list of all the more important literature up to 1910; nearly all of it is systematic. Among the many papers cited, the only one of more than passing interest from an ecological standpoint is COPELAND's *Comparative Ecology of the San Ramon Polypodiaceae* (Philippine Journal of Science, 2C, 1907). I know of no other paper of any length devoted entirely to the ecology of tropical pteridophytes.

Of more recent general ecological works which are important, the following may be cited [1]:

SHREVE, F., A Montane Rain Forest. Carnegie Institution of Washington, 1914.

McLEAN, R. C. Studies in the Ecology of Tropical Rain-forest. Journ. Ecol. 7: 5–54, 121–172. 1919.

BROWN, W. H. Vegetation of Philippine Mountains. Manila 1919.

---

[1] Cf. also SCHIMPER's Pflanzengeographie, 3rd. edition by F. C. VON FABER (Jena 1935), pp. 327–588.

DAVIS, T. A. W. and P. W. RICHARDS. The vegetation of Moraballi Creek, British Guiana. Journ. Ecol. 21 : 350–384 (1933) and 22 : 106–155 (1934).

ALLEE, W. C. Measurement of Environmental Factors in the Tropical Rain forest of Panama. Ecology, 7: 273–302. 1926.

OLIVER, W. R. B. New Zealand Epiphytes. Journ. Ecol. 18: 1–50. 1930.

COCKAYNE, L. The Vegetation of New Zealand. Die Vegetation der Erde, XIV. 2nd edn. 1928.

The following works on ferns, chiefly of a systematic nature, have ecological information:

CHRISTENSEN, C. and R. E. HOLTTUM. The Ferns of Mount Kinabalu. Gardens Bulletin, S. S. 7, part 3. 1934.

HOLTTUM, R. E. On *Stenochlaena*, *Lomariopsis* and *Teratophyllum* in the Malayan Region. Gardens Bulletin, S. S. 5: 245–313. 1932.

The subject of the ecology of tropical pteridophytes is of far too wide a scope to be dealt with adequately in a short chapter of this nature. Ferns and their allies have assumed almost all the forms of growth and of adaptation found among flowering plants [1]), except that at the present day they have few representatives of tree habit, and they have not produced species which are able to endure quite such extremes of desert and arctic conditions as certain phanerogams. A full account of the ecology of ferns would therefore be almost a treatise on ecology as a whole.

There are certain generalisations which may be made as a preliminary to a consideration of the subject in detail. In the first place, pteridophytes are rarely dominant in any vegetation. There are exceptions, naturally; but though such exceptions may be important from a vegetation standpoint, they are concerned with a very small proportion of species of ferns. Ferns are in fact largely dependent on other plants to provide them with the conditions of shelter and support which they need; and they are rarely so abundant as to have a profound effect on the other units of the vegetation, which would not be greatly modified if the ferns were all removed. The few cases in which ferns are dominant are not permanent or climax vegetation units, but of a more or less temporary successional nature.

The second generalisation which one may make is that the existence of fern plants is ultimately dependent on the occurrence, at least seasonally, of conditions in which the gametophyte generation can live and develop, and the youngest stages of sporophyte growth become established. We are accustomed to see and to marvel at the greatly varied form and adaptation of the sporophytes, which are the ferns as we know them, but indeed there must be nearly as much variety of adaptation among the gametophytes. It is true that if a prothallus of *Platycerium* grew upon the forest floor, the resulting sporophyte, if produced, would find itself in uncongenial surroundings, and would not develop very far; but it is also true that the *Platycerium* prothallus must be able to develop in relatively exposed

[1]) See GAMS, this Manual pp. 382–384.

positions on a tree trunk, in which prothalli of many ferns would be unable to exist. How much specialisation exists in the gametophyte generation is little known, and there is scope for a great deal of experimental investigation of the subject. Most of the investigations so far carried out have been morphological rather than physiological or ecological.

The third generalisation is that in each main vegetation type existing in various parts of the world, though the species of ferns and Lycopods may vary from continent to continent, their growth forms and the parts they play in the vegetation complex are very much the same. Thus SHREVE's description of the rain forest of Jamaica gives a picture of fern growth closely paralleled by that of BROWN in the Philippines. There are naturally the extreme types of specialisation in form and structure, which are different in the different groups of ferns in different parts of the world; but the main types remain the same, though represented by different species or genera.

In the present chapter it is impossible to do more than describe briefly the principal roles played by pteridophytes in the various types of vegetation, and to mention in particular some of the more remarkable cases of ecological specialisation. As by far the greater proportion of fern species exist in the moister parts of the tropics, it is appropriate to devote this chapter to tropical regions alone. In this connection it may be remarked that the present writer's personal experience is almost entirely concerned with the vegetation of the Malayan region. He cannot write from first hand experience of the American and African tropics, and he cannot hope to consult all the relevant literature on the ferns of those regions. As above remarked, the available literature indicates that ecologically the pteridophytes are essentially similar throughout the tropics, and therefore it is hoped that the present account will not be of a seriously unbalanced nature.

**§ 2. Spore Dispersal.** [1]) — This subject is one that borders on the general distribution of ferns. It is well established that spores may be distributed freely by wind, and the factor which limits the ability of a fern to spread and establish itself on new ground is not the ability to travel over long distances, but the ability to survive during the process of transport. PICKETT [2]) showed that spores of *Camptosorus rhizophyllus* survived throughout the winter, and this must be the case also with most of the ferns of seasonal climates. Fern spores carried by winds in the tropics must be able to withstand full exposure to sun and relatively dry air during the day for considerable periods if they are to travel far. It is very probable that some cannot withstand these conditions, as for example the extremely thin-walled spores of *Lomagramma*. Few observations of this nature seem to have been made.

**§ 3. The Gametophyte Generation.** — In conditions of shade of the forest, a considerable proportion of fern spores must be able to germinate;

---

[1]) See also GAMS, in this Manual, p. 395.     [2]) Amer. Journ. Bot. 1: 477–498. 1914.

perhaps all except those which require conditions of extreme moisture. It is the ferns which grow in exposed places — the more xerophytic epiphytes, and the terrestrial sun-ferns — which require specialised adaptations to more or less adverse conditions. Probably no prothallus will grow under conditions of full exposure to the sun. Even where the adult sporophytes are fully exposed, it is always found that they begin life in some more or less sheltered spot, where the sun shines only for a short time during the day or where protection is afforded by a rock crevice or a hole in the ground. In climates with a wet and dry season, the growth of the prothalli will take place in the former, and the young plant must have become sufficiently established at the onset of the dry weather. Even in Malaya, where there is no long dry season, there is a large mortality of young sporophytes precariously starting life in exposed positions. TREUB [1]) noted that on the bare pumice of Krakatau, after the eruption of 1883, the first covering was of *Cyanophyceae*, and that the prothalli of ferns grew on the gelatinous substratum so formed. This condition appears to be a general one in Malaya on newly exposed earth surfaces (such as steep banks) which are unsuitable for the growth of seedlings of phanerogams. But even with the assistance of the Cyanophyceae, prothalli growing on banks fully exposed to the open air, though sheltered from the sun, must be able to withstand a considerable amount of drying. PICKETT [2]) made some observations on the prothalli of *Camptosorus rhizophyllus*, which grows on dry limestone ledges and on rocks in stream beds, and *Asplenium platyneuron*, a fern of open woods on high ridges and dry hillsides. He found that under experimental conditions the prothalli had marked drought resistance, and that they would survive prolonged drought of the degree of severity found in nature, so long as they were not exposed to direct sunlight. I have not seen comparable observations made on tropical ferns. [3]).

The climbing ferns which start life in the very moist low levels of the tropical forest are an interesting case. *Lomagramma* prothalli for example are invariably found on constantly moist rocks beside small streams in deep shade. The young plants creep over the rocks with long slender rhizomes, increasing in size up to a limit; the adult form of the plant is developed only when a rhizome meets a tree and climbs up it. The large fronds produced on the high-climbing rhizome are able to exist under moderately dry air conditions, and it is evidently the requirements of the prothalli which limit the positions in which these ferns can grow. If the prothalli could grow on the bark of a tree, the species would be much more abundant There must be a great deal of specialisation of prothalli to growth on rock or soil surfaces, on humus, among bryophytes, or on bark; or to various conditions of moisture and acidity.

---

[1]) Ann. Buitenz. 7: 213–223. 1888.

[2]) L.c.

[3]) See also GAMS, in this Manual, p. 385.

The *Ophioglossaceae* present a special case, on account of the saprophytic nature of the prothalli, and the obligate presence of mycorrhiza;
*Lycopodium*, *Psilotum* and *Tmesipteris* also come into this category.
CAMPBELL [1]) made some interesting observation on the prothalli of tropical
species of *Ophioglossum*. He found those of *O. pendulum* abundantly
present (to the extent of several hundred) between the old leaf bases of a
plant of *Asplenium Nidus*, 30–40 cm below the crown of living leaves. Such
prothalli must have existed for many years. In Singapore they grow
commonly among the old leaf-bases of *Platycerium*, and it is usually only
when the *Platycerium* plant is in an old and enfeebled condition that the
sporophytes of *O. pendulum* begin to appear. CAMPBELL found that infection by the mycorrhizal fungus took place from the substratum at a very
early stage. The prothalli of the terrestrial *Ophioglossaceae* are wholly subterranean (with few exceptions), and the young sporophytes are also
provided with an endophytic fungus. CAMPBELL [2]) also records the presence of an endophytic fungus in the green prothalli of four species of *Gleichenia*, and of other primitive ferns.

The prothalli of *Lycopodium* may be either entirely subterranean as in
*L. clavatum* and most terrestrial species; or they may have an exposed
green upper part, as in *L. cernuum*; in the epiphytic species the prothallus
is slender and branched, the branches extending through the humus substratum of dead bark etc in which the whole is embedded. All have endophytic fungi, though *L. cernuum* and others are able to carry on photosynthesis.

## § 4. Lowland tropical Rain-forest. — I. *General.* — A number of
descriptions of this type of forest have been written from various parts of
the world, and all give a similar picture as regards the general conditions
under which pteridophytes live, though the species of trees and other plants
vary from place to place. There are naturally sub-types of evergreen forest,
some more open and some more shady, and within these the conditions
available for pteridophytes vary. It is impossible here to do more than
indicate the genaral nature of the pteridophytes and the positions they
occupy in the vegetation.

As regards physical conditions of humidity, temperature and light,
several series of observations have been published. All of these indicate
that there is a gradual change from the ground level to the crowns of the
tallest trees. The humidity of the air is at a maximum near the ground, and
the light intensity, daily range of temperature, and air movements are at a
minimum. In primitive high forest of the Malayan lowlands, HAINES [3])

---

[1]) Ann. Buitenz. 21: 138. 1907.
[2]) Amer. Naturalist, 42: 154, 1908.
[3]) Bull. Rubber Res. Inst. of Malaya No. 4. 1931.

found that the temperature at the soil surface did not vary more than 1 °C during three months, the temperature being about 25°C. At a height of only one metre above ground level the temperature range was 7° C, the mean remaining about the same. The range of temperature at a depth of 5 cm in soil exposed to the full sun reached a maximum of 45°C, with a greatest daily range of 20°C; this indicates the degree of protection afforded by the forest.

BROWN has described similar conditions in Dipterocarp forest in the Philippines. DAVIS and RICHARDS have described forest in the lowlands of British Guiana; they also found great constancy of conditions on the forest floor, the air very still, humidity never falling below a rather high value, and little change of temperature. Light has been measured among others by MOREAU [1]) in African forest, ALLEE in Panama, and CARTER [2]) in British Guiana; the intensity at ground level is of the order of .1 to 1 per cent of the full light outside the forest.

In such forest, which affords the maximum range of conditions for plant growth, pteridophytes grow on the ground, on rocks, as low-level epiphytes on the bases of tree trunks, as climbers reaching moderate or great heights on the trees, and as high-level epiphytes in the lighter and more fluctuating conditions of the crowns of the larger trees. The degree of specialisation of form and structure also goes in approximately the above order, the most specialised types being the high-level epiphytes.

Il. *Terrestrial ferns* are many and varied in form. Their abundance usually depends on the amount of light, and in typical high forest they are scattered, never forming close patches except sometimes in lighter places (though here they may often be crowded out by other plants) and near streams. Most terrestrial forest ferns have a short stock, erect or creeping, and rather large fronds. A considerable proportion of them show some dimorphism; this usually takes the form of a somewhat contracted lamina in the fertile fronds, which stand erect, on longer stipes than the broader sterile fronds which bend away in a rosette around them. If there is a fairly pronounced dry season, the fertile fronds are often produced then and not at other times. This form of dimorphism undoubtedly helps the distribution of spores, raising the sporangia well above the surrounding leaves and exposing them to drier air and any slight wind that may exist. Tree ferns usually show no such dimorphism, but their fronds are raised to a higher level in the vegetation. They are not usually abundant except near streams or in the lighter parts of the forest. They are more frequent on mountains, where a moister air is combined with a greater light intensity. Terrestrial forest ferns are always rather thin in texture (except those that grow in the more open type of forest).

---

[1]) Journ. Linn. Soc. Zool. 39: 285. 1935.

[2]) See MOREAU's paper.

SHREVE measured transpiration rates of terrestrial shade ferns and found them greater than those of typical herbaceous plants of the forest floor. He also estimated that *Diplazium celtidifolium* transpired through the epidermal surface at least as much as through the stomata. McLEAN found that the xylem supply of shade ferns is sufficient for only the limited transpiration occurring in the forest and quite inadequate to supply requirements of transpiration of the leaves if placed in dry air. Thus ferns of this type are specialised to allow of considerable degree of transpiration in an atmosphere of high humidity. Some of them have a limited degree of adaptability, and may respond to drier conditions by reduction of leaf size and increased thickness of lamina.

FIG. 1. — *Dryopteris dissecta* with *Cyathodium cavernarum* on damp rocks inside birds'-nest cave near Niah (Sarawak). This association received only light of low intensity; association furthest from mouth of cave. — Photo P. M. SYNGE.

The largest ferns of the terrestrial type are the species of *Angiopteris*, which may reach enormous dimensions. They are usually only found in rather moist situations, often near streams. When exposed to dry weather, the leaves collapse owing to lack of turgor of the swollen base of the stipe; they have fairly considerable power of recovery after wilting. Whether the collapse of the fronds is of any advantage is doubtful. The smallest plants of the terrestrial class are some species of *Trichomanes*, and *Selaginellas*, which are usually confined to this habitat, though a few species are adapted to more exposed positions. *Selaginellas* never occupy a dominant place in any vegetation, though the climbing species may form small thickets.

III. *Rock ferns*. There are certain species of ferns which differ in no essential respect from the terrestrial forest species just mentioned except that they grow onlyon rocks. They may have a creeping rhizome which is firmly attached to rocks on steam-banks, the roots often reaching down into crevices or to moss-covered places; or they may have short erect stocks usually growing in crevices of rock. In the Malay Peninsula, rocky places in the forest, except in the case of limestone hills, are rare and almost confined to the banks of streams, and rock-ferns are therefore not abundant. A few species seem to be confined to limestone, but most are not specialised as regards the kind of rock; the requirement seems to be adequate aeration for the roots. Such ferns, if of any size, must have a fairly moist atmosphere, and the usual streamside position supplies this; alternatively, they must have the power to rest in dry seasons. Limestone rock ferns of regions with even a moderate dry season may lose their leaves when terrestrial ferns do not. The species of *Adiantum* appear to be mostly rock-ferns; of these, *A. tenerum* and others have articulate leaflets which are deciduous.

IV. *Climbers*. Climbing species of ferns may be divided into the more hygrophytic types, which only climb a short distance and are confined to the lowest stratum of the forest, and those which climb up the trunks of the tall forest trees. The low climbers, being rooted in the earth, are not greatly different in type from terrestrial ferns, but commonly have smaller leaves which without the climbing stem would be little raised above the ground level. The high climbers are a very specialised class, and among them are some remarkable species, notably of the genera *Lomariopsis*, *Teratophyllum* and *Lomagramma*. These ferns have often two types of leaves, one borne by the parts of the plant in the lowest layers of the forest, and the other on the high-climbing parts; the former I have termed *bathyphylls* [1], a name more suitable than juvenile leaves or "water-leaves", which assume more than we know about their nature. The upper leaves may be termed *acrophylls*. The bathyphylls of *Teratophyllum* are very various in form, and are characteristic for the various species; in fact, it is sometimes easier to discriminate species from their bathyphylls than from their acrophylls, which, climbing into the upper air, have assumed a rather uniform simply pinnate type. The uniformity of this type led to the indiscriminate lumping of many species together, and to a failure to distinguish the remarkable differences between the genera concerned. The bathyphylls are usually

---

[1] Gardens Bulletin S.S. 5: 246. 1932.

more compound than the acrophylls, thinner in texture, and are borne by slender rhizomes creeping on rocks, on the ground or on the lowest parts of tree trunks. It has been assumed by KARSTEN [1]) and CHRIST that one of their functions is to act as water-absorbing organs. They have certainly some power of water absorption, but it is fairly certain that the plant depends far more on its very well-developed root system than on its bathyphylls for water supply. The bathyphylls of climbing stems lie close to the supporting tree trunk, the rachis horizontal, but any leaflets below the rachis usually stand obliquely away from the trunk, thus facing more towards the light. Stomata are borne by the parts not in close contact with the trunk. In the case of *Teratophyllum aculeatum*, the youngest stages have only the lamina on the lower side of the rachis developed. The fertile fronds of these ferns are acrostichoid, and again are very much alike. They are borne on the higher parts of the climbing stem, where there is drier air and the spores are sure of dispersal. But they are not at all common, and it seems likely that they are only produced when a rather pronounced dry season affects the humidity of the air in the forest; or it may be that the fall of a neighbouring tree exposes the upper parts of the fern suddenly to unusually dry conditions. These climbing ferns make the best of

FIG. 2. — *Teratophyllum aculeatum* var. *montanum*:. bathyphylls on tree trunk in forest. Malay Peninsula, alt. 1200 m.
The photograph shows how the downwardly directed pinnae grow outwards away from the trunk, to catch the light; the upper pinnae are closely pressed to the trunk. In young stages of this species, only the downward-growing pinnae are developed. — Photo R. E. HOLTTUM.

---

[1]) Ann. Buitenz. 12: 143. 1895.

br th worlds; they have their roots in the moist bottom of the forest, and develop there a frond system suited to such conditions; and by their strong climbing stems they climb to a lighter place, on the trunks of trees which may otherwise be without tenants, and there produce suitable larger and more resistent sterile and fertile fronds. None of them climb to the full height of the crown of the trees, except *Stenochlaena*, which never starts life in the very shady tall forest, but in more open forest or scrub.

V. *Low-level epiphytes*. These are essentially delicate ferns, found only in moist or shady conditions. In most ordinary high forest the conditions do not seem to be sufficiently moist for the full development of this group. It is only when one comes near a stream, so that very high constant humidity occurs, that one finds them at all abundantly. They are best developed in the mountain forest (see below). With this proviso, some further observations on them may be given here. Some of them are found both on rocks or trees, but most seem to be specialised to either the one or the other substratum. *Hymenophyllaceae* are largely found in this class; they often form pure patches on the trunks of trees, or the species may be mixed, and bryophytes also occur along with them. Some of them have the power of curling up when dried and remaining so for a time without harm, but others cannot withstand this. There is little published information on the behaviour as regards resistence to drying of tropical *Hymenophyllaceae*. It is clear that they are on the whole very exacting, and it is only in the most humid places that they are abundant, where atmospheric saturation must be nearly complete for considerable periods, or where spray or dew maintain their surface moist during a large part of the time. Other low-level epiphytes are less hygrophytic. *Antrophyum*, with its spongy mass of very hairy roots and succulent leaves, comes into this group; also *Asplenium tenerum*, with a spongy mass of outward-growing roots and a miniature nest-habit. Such root-masses absorb a large amount of water and tide the plants over rainless periods.

FIG. 3. — *Antrophyum callaefolium*, growing abundantly on andesite rocks, G. Kanaga, Java. Common on rocks where such occur, in the absence of rocks it will also grow on tree trunks.
Photo VAN STEENIS.

VI. *High-level epiphytes*. These epiphytes grow on the branches of the crowns of the tallest trees in the forest. They have to withstand rather dry conditions during the day, and may have to pass many days without rain, since few localities are without a short dry season. Similar habitat conditions are found on isolated trees in the open, provided that the crowns are sufficiently shady; the ferns of this group are therefore most commonly seen and most accessible on such trees, upon which they often grow near the ground. Palms with persistent leaf-bases afford peculiarly favourable positions for such ferns. They become weeds in an *Elaeis* plantation. Some of them may even grow on rocks, in exposed or lightly shaded places. They have developed a variety of methods of protection against drought, and include some of the most remarkable of all ferns. Some members of the group are not very resistent and they accordingly occupy the more shady positions. Among these less xerophytic plants are the epiphytic species of *Lycopodium*, which hang pendulous below the large tree branches, often rooting in a group of other epiphytes, in the less exposed parts of the crown.

FIG. 4. — *Asplenium africanum*: Nigeria, Shasha rainforest, one of the most frequent epiphytes. — Photo P. RICHARDS.

FIG. 5. — Very young *Platycerium Stemaria* and young *Asplenium africanum* on Cola sp., Shasha Forest Reserve, Nigeria. — In the interior of the forest this species is found only at great heights above the ground, in the tops of tall trees: in more open situations however (where the light intensity is higher) it may be found close to the ground level, as is a general occurrence among "high level epiphytes". Photo RICHARDS.

Most high-level epiphytes start their life in association with bryophytes, which are often the pioneers of epiphytic vegetation, affording shelter and moisture for fern prothalli and for orchid seedlings. The hardier ferns which are first established may in turn afford shelter for others. For example, the enormous spongy mass of roots of *Asplenium Nidus* affords shelter for other species such as *Ophioglossum pendulum*, other species of *Asplenium* etc. and the water trickling down from it after heavy rain will be utilised by other epiphytes which are not actually growing upon it.

Thus is developed a vegetation upon the substratum provided by the branches of the dominant trees. The epiphytes are opportunists which have made use of this precarious site. unusable by other plants; they are adapted to it, and to no other. They may be subdivided into various groups according to their methods of combating the enemy of drought.

a. *Nest-epiphytes. Asplenium Nidus* is the chief of these. In its capacious nest are caught fallen leaves; through a layer of these grow each successive crop of Asplenium fronds, holding the accumulating mass securely between

FIG. 6. — *Asplenium Nidus*, showing nest-form and mass of roots and humus below. Rainforest nr. Tjibodas, G. Gedeh, Java. — Photo P. SYNGE.

their bases, where the leaves gradually rot, the roots of the fern penetrating among them. In this way gigantic masses of roots and humus are developed over long periods of years, continuing until their weight breaks the branch, or the tree itself falls.

b. *Bracket-epiphytes* are similar in effect, but instead of a nest or cup the humus collectors are short sessile bracket leaves or in some cases the enlarged bases of ordinary leaves. These grow over the creeping rhizome of the fern (itself protected by a thick layer of scales), collect humus and protect the roots. The *Drynaria* group are the chief representatives of this type. *Platycerium* is a different type with overlapping erect bracket leaves, and (usually) pendulous leaves bearing the fertile areas. In both the *Drynaria* and *Platycerium* groups the bracket leaves are persistent, but the others are deciduous and may be shed in the dry season (where such occurs).

c. *Simple coriaceous leaves*. These leaves are usually small in size, often simple, thick and coriaceous in texture, with some water storage capacity and restricted transpiration, often covered with hairs or scales. The species of *Cyclophorus* (see GIESENHAGEN) [1]), and *Elaphoglossum*, are the most striking of the group. *Cyclophorus* may have trichome hydathodes capable of absorbing water; excretory hydathodes are also present in many species.

FIG. 7. — *Platycerium grande*. Botanic Garden, Brisbane, Queensland. The erect bracket-fronds curl back towards the tree as they die, holding the dead leaves they have collected. — Photo E. H. WILSON.

*Drymoglossum piloselloides* represents a very reduced fleshy type, which often grows right to the ends of the branches of a tree, sometimes even covering the leaves where these are long-lived, clinging very closely with its numerous root hairs. These ferns can withstand considerable exposure to drought, sometimes becoming flaccid and shrivelled in dry weather. The rhizomes are often wide-creeping and are almost always well protected with scales; the fronds are often articulate, and are shed when old. They are often associated with orchids, and together the epiphytes may make a very dense covering on the branches of the tree, the leaves protecting the roots from extreme exposure. The roots often grow in a close thin mat with lichens, algae and bryophytes.

d. *Succulent rhizome*. Almost all the larger epiphytic ferns have a rather stout fleshy rhizome, which doubtless serves to some extent for water storage, but there are a few ferns in which this is especially developed. The most remarkable cases are those in which the fleshy rhizome offers shelter to ants; they are comparable to phanerogams such as *Hydnophytum*. *Polypodium sinuosum* and allied species, and *Lecanopteris*, are the chief examples; KARSTEN [2]) mentions (and figures) also a very curious species called *P. imbricatum*, with a segmented thin flattened rhizome, up to 12 cm wide and only 8 mm thick. He does not say that it is ant-inhabited. The *P. sinuosum* group have scaly rhizomes and are lowland plants; *Lecanopteris* have naked rhizomes and are found only on mountains; but all are functionally similar. DOCTERS VAN LEEUWEN [3]) records

---

[1]) Der Farngattung *Niphobolus*. 1901.
[2]) L.c.
[3]) Ber. der deutsch. bot. Ges. 47: 90. 1929.

that the sporangia of these ferns have many oil drops in the thin wall-cells, and that for this reason they are sought by ants, which distribute the spores to places suitable for germination. It is very doubtful whether the ants protect the plants in any way, and it is certain that the plants can grow in the absence of ants; but probably the ants do them some service in bringing food to the roots. The more elaborate species of the *P. sinuosum* group have large crustose rhizomes, covering a branch of a tree, but they never attain the enormous size of *Lecanopteris*. Old trees in mountain forest sometimes have a large proportion of the crown branches covered with a growth of this fern, the horny black rhizomes, lacking leaves, having a fungus-like aspect. Only the growing tips of the plant have fronds, but the older parts, often dead, continue to act as ant-nests. The ants also bring other small seeds (of orchids, *Dischidia* etc.) which likewise have oil in their coats; these germinate in the crevices of the rhizome and there find shelter for their roots. In this way the fern may be the beginning of a mass of epiphytes.

CHRIST refers to two interesting ferns from Tropical America, *Polypodium Brunei* and *P. bifrons*. Both these bear tuberous growths on the rhizome, and it is probable that they serve as water-storage organs. The tubers are hollowed in the upper surface, and afford some protection for roots. HOOKER also mentions that the tubers of *P. bifrons* are ant-inhabited.

*e. Less specialised types.* There are many larger-leaved epiphytes, usually not occupying the more exposed positions, and often taking advantage of the shelter afforded by nest-ferns and other pioneers, which are less elaborately adapted to epiphytic life than those already mentioned. They usually have rather fleshy rhizomes, but their leaves are often fairly thin in texture. Pinnae, or whole leaves, are often deciduous.

*f. Nutrition of high-level epiphytes.* This is a problem which needs further study. The character of the bark of the tree may have an important influence on the ability of the pioneer epiphytes to settle, and in some cases may determine the presence or absence of epiphytes. Many epiphytes are humus collectors, and so develop what amounts to a soil for their roots. Others make use of the smaller-scale humus collection of lichens and bryophytes. Mineral elements and nitrogen must be provided to a small extent by bark decomposition, but still it seems surprising that there is a sufficient supply. A recent publication of LAUSBERG [1]) may perhaps throw some light on the subject. Dealing with certain herbaceous plants, he finds that the leaves excrete on their surfaces a relatively large amount of mineral salts. If tree-leaves behave in the same way, the nutrition of epiphytes like *Drymoglossum* is largely explained. The excreted salts would be washed down the branches and trunks by rain, and will in part be available for any epiphytes which may be present.

---

[1]) Jahrb. Wiss. Bot. 81: 769. 1935.

The water supply of these epiphytes is the crucial factor in their existence. They are only found abundantly in the moist tropics, and in continuously moist warm temperate climates such as that of New Zealand. Though they are well protected against water loss, and can tolerate considerable loss of water without injury, there are limits to their endurance; they cannot survive a very prolonged dry season, and they must have a considerable amount of rain during the growing season. Even in the most equable climate, rainfall is intermittent, and between rains the regular heavy nightly dew must be an important factor in the water supply of such epiphytes.

VII. *Riverside ferns.* River-sides afford conditions of high humidity

FIG. 8. — *Dipteris Lobbiana* on rocks in streambed, Johore, 100 m, The narrow coriaceous leaf-segments are typical of many ferns of this habitat, which is subject to frequent flooding with swiftly-flowing water. — Photo R. E. HOLTTUM.

combined with more light than exists in the depths of the forest. They are therefore especially suited to epiphytes of all kinds, and the wealth of these surpasses by far that of ordinary forest. Especially is this the case where the river is not too wide; where the trees on the banks lean over and almost meet overhead, and where primitive high forest on either side protects these from drying winds. "High-level" epiphytes are found in the more exposed positions, and often quite low down, and they may occur on rocks also. Deeper under the shade and nearer the water, on the lower parts of the tree trunks, are masses of *Hymenophyllaceae* and other hygrophytes. Small trees on the bank are often laden on all their branches with pendulous bryophytes, among which are small ferns and sometimes epiphytic *Selaginella*s.

The river banks have also their characteristic fern-flora, and in this there is often a distinct zonation. Where the banks are rocky, the rocks, which are subject to inundation from time to time by floods, are sometimes covered with ferns of the *Dipteris Lobbiana* type, with creeping rhizome on the rock surface and erect fronds with narrow leaflets. There are also

Fig. 9. — *Trichomanes papillatum* on each bank of forest stream, Malay Peninsula, 1500 m. — Photo R. E. Holttum.

other smaller ferns (such as *Lindsaya* spp. in Malaya). Where the bank is not rocky, different types of ferns occur, often forming a close cover. Higher up the bank are other ferns which will not withstand frequent submersion, but which require the moist and relatively light conditions of this situation; not all of these are specialised to river banks, but many are so. Where the forest has been cleared, and only secondary growth, or a mere fringe of trees, remains, the wealth of species is much less. The chief reason is doubtless the wind, which prevents the attainment of the necessary degree of humidity. In Malaya, there is a remarkable species which is confined to the banks of tidal streams, beyond the reach of salt water, but where there is a regular rise and fall of water level. *Tectaria*

FIG. 10. — *Polypodium taeniatum* and *Hymenophyllum holochilum* on small tree beside stream, Penang, 700 m. — Photo R. E. HOLTTUM

*semibipinnata* grows in the mud, so that at low tide it is fully exposed, but at high tide often almost completely submerged.

**§ 5. Mangrove.** — The larger trees of well-grown mangrove often bear the usual high-level epiphytes. The more hygrophilous low-level epiphytes are usually only found in areas of mangrove furthest from the sea; here they are often very abundant, much as on riverside trees. A few ferns, notably *Humata parvula*, seem almost confined to this habitat.

FIG. 11. — *Acrostichum aureum*, which normally grows in salt water, in chemically controlled not-salt water near Depok (W. Java) with *Susum anthelminthicum* (right). This fern often persists in areas which were formerly tidal, but have been artificially cut off from contact with the sea. — Photo van Steenis.

The most characteristic fern of the mangrove is of course *Acrostichum aureum* (and allied species). These ferns grow in the drier parts of the mangrove area, though they will stand immersion by the tides. They grow most vigorously in open places, and after clearing they may form such dense thickets that the establishment of the tree species becomes difficult. Like other mangrove plants, these ferns transpire freely. Cope-

LAND records that the stomata occupy 35% of the lower surface. Young plants of these species are often found in sheltered places on rocky sea shores, but in such positions they rarely attain any size.

§ 6. **Mountain plants in the moist tropics.** — I. *Mid-mountain forest.* — BROWN divides mountain forest in the Philippines into mid-mountain and mossy forest, and this is a good general distinction, but naturally the topography of ridge and valley make considerable local differences of conditions; and the vegetation near the summit of a small mountain will be very different from the vegetation at the same altitude on a larger mountain. Also windward and leeward slopes may have quite

FIG. 12. — Virgin forest on the East side of Mt. Gedeh in West-Java, about 2000 m above sea-level. *Dryopteris* spec. div. along a forest-path.
Photo DOCTERS VAN LEEUWEN.

different conditions of moisture. Mid-mountain forest is lower in stature than lowland forest and therefore tends to be lighter at the ground level. It is also usually moister on account of more or less regular mist and cloud. The pteridophyte vegetation is very similar to that of the lowland forest, but usually much richer, especially as regards terrestrial ferns and low-level epiphytes. Tree-ferns are also frequent in the undergrowth, but the most conspicuous growth of these (usually confined to a few robust species) is in clearings and beside streams, where light is stronger. The flora generally is different in species from the lowland forest. The high-level epiphytes are sometimes enormously developed on the old trees, forming masses of humus larger than the branches on which they rest; ferns have a conspicuous place among such epiphytes, but are accompanied by a great variety of angiosperms.

FIG. 13. — Belgian Congo: Ituri, Lubero, 2000 m. Typical growth of tree ferns in a rather open place in valley forest (*Cyathea Manniana* etc.).
Photo H. M. KING LEOPOLD III.

II. *Mossy Forest* [1]). In mossy forest the conditions are extremely moist, owing to regular cloud and mist. There are certainly several types of mossy forest, according to conditions of exposure, temperature, degree of clouding etc. In some cases hepatics predominate (often in large masses) and in others mosses form almost a pure covering on the trees. The mosses only occur abundantly in the most humid localities. Mossy forest of whatever type always consists of short trees, often close together. It is usually most developed on ridge tops; not in valley bottoms, where conditions for tree growth are more favourable and higher forest occurs. BROWN found that a hair hygrometer recorded 100% humidity all the time in the mossy forest, but he moted that the hairs became wet with mist, and in the drier periods this was slow in evaporating. He concluded that the humidity never fell much below the maximum. The light is not so strong as might be expected, on account of the clouding. SHREVE in Jamaica dealt with a forest almost

[1]) See RICHARDS, Journ. Ecol. 24: 340–356. 1936.

approximating to mossy forest, but stated that complete saturation was a transitory state even in the most humid situations.

The mossy forest typically consists of a single layer of tree-crowns, with sparse undergrowth. The trunks of the trees are bryophyte-covered to a

FIG. 14. — Old secondary forest on the S. side of Mt. Guntur (Goentoer) in West-Java, about 1500 m above sea-level. *Cyathea* species, one of the more robust tree-ferns, which can stand exposed conditions. — Photo DOCTERS VAN LEEUWEN.

greater or less extent, and also bear numerous hygrophytic epiphytes including *Hymenophyllaceae*, *Grammitis*, and the small species of *Polypodium*. The exposed parts of the crowns may bear very xerophytic epiphytes. The terrestrial vegetation includes a number of ferns, and also sometimes the terrestrial Lycopods, which however are more abundant in the open thickets on ridges. The ferns are typically of a rather xerophytic type. In Malaysia the species of *Plagiogyria* are characteristic; these have erect fertile fronds, but their most remarkable feature is the slimy covering of the young expanding fronds and the white aerophores which penetrate it. A similar condition is found in some other ferns [1]), of several genera, most of which inhabit similar situations. The mucilage is secreted

[1]) GOEBEL. Ann Buitenz. 36: 84. 1926.

by glands on the surface or on scales. It protects the very young fronds
from desiccation, but in many cases it would seem to be quite superfluous

FIG. 15. — *Lycopodium pinifolium* in moss forest on G. Gegerbentang (1750 m)
he young sterile branches at first horizontal, later pendulous. — Photo VAN STEENIS.

for this purpose, and is dispensed with by most ferns. It may be of value in
sudden dry weather.

SHREVE's remarks on the conditions of "montane rain-forest" apply
to the moister mountain forest generally. "The prolonged occurrence

of rain, fog and high humidity, at relatively low temperatures, places the vegetation of a montane rain-forest under conditions which are so unfavourable as to be comparable with the conditions of many extremely arid regions. The collective physiological activities of the rain-forest are continuous but slow; those of the arid regions are rapid, but confined to very brief periods". "The desert plant loses far more water per unit area than does the plant of the rain forest" (each under the normal condition in which it grows).

The reason for the dwarfing of the trees, and for the peculiar character of some of the terrestrial plants (such as *Plagiogyria*) is probably the accumulation of a raw acid humus similar to that of "hochmoor" in colder climates. VON FABER [1]), speaking of the craterplants of Java, noted that the epiphytes grew in a very acid substratum composed chiefly of moss, in which little nitrification took place. He found this quite different from the humus of *Asplenium Nidus* in the mid-mountain forest at Tjibodas. *Sphagnum* is not usually well developed in mossy forest, as the light is inadequate, but it is often abundant in more open places on mountain ridges.

## § 7. Sun-ferns of tropical lowlands. — These ferns are fairly sharply distinct from the forest ferns, though they vary in the degree of exposure they will tolerate. They may be divided into the two main groups of long-creeping and short-creeping or tufted ferns. The former often form almost pure patches of vegetation, and may be important from a successional standpoint. The latter are only locally dominant, and usually occur along with other plants, though they may also play an important part in the establishment of new vegetation. The most interesting observations on the colonisation of new ground by sun-ferns were those of TREUB [2]) on Krakatau three years after the great eruption. He recorded that ferns composed almost the whole of the vegetation of the island (other than the strand-flora) at that time, but gives little information as to their relative abundance. One of the species found by TREUB was *Pityrogramma calomelanos*, an American species which has become well established in Asia and is one of the first plants to occupy bare ground.

The establishment of sun-pteridophytes can best be seen on newly cut earth banks, or on abandoned clearings. WARDLAW [3]) described the colonisation of an abandoned St. Lucia banana plantation, the soil of which had become impoverished. Under these conditions *Cyathea* sp. and *Cecropia* grew abundantly, the former equalling in height the stunted banana plants after two years. WARDLAW also remarked that on slightly weathered rock surfaces, landslides, cuttings etc. *Gleichenia* spp., *Lygodium* spp. and *Lycopodium cernuum* cover the ground. All these are surface-

[1]) Die Kraterpflanzen Javas. Buitenzorg. 1927.
[2]) L.c.
[3]) Journ. Ecol. 19: 60–64. 1931.

rooting plants, and if conditions are moist enough for the prothalli to develop, they soon become established.

In Malaya a continuous growth of *Gleichenia* spp. is also found in places where a hard compacted soil (often resulting from abandoned cultivation) offers little opportunity for the establishment of angiosperm seedlings. The prothalli develop in the more sheltered positions, on banks or in hollows, and once established the ferns spread rapidly. *Pteridium* cannot colonise such hard surfaces as *Gleichenia*, as it has a subterranean rhizome, whereas that of *Gleichenia* is superficial. But *Pteridium*, when established, can withstand burning, which kills *Gleichenia* entirely.

FIG. 16. — *Gleichenia linearis* (two varieties) on edge of forest near Singapore. A typical thicket-forming tropical sun-fern. — Photo R. E. HOLTTUM.

*Pteridium* tends to become established as a thicket-fern in rather poor sandy soils and where the vegetation has been burnt; it is especially characteristic of areas which are subject to periodical clearing for cultivation, and in such cases *Cibotium barometz* may be associated with it; *Cibotium* has a thick prostrate caudex which is not killed by light burning.

*Lycopodium cernuum* does not grow luxuriantly in the full sun, but may accompany Gleichenia in thickets. It has found its optimum condition in some old Hevea plantations in which the soil, having been clean-weeded for years, has become very hard. When weeding ceases, the *Lycopodium* may become established, and under the light shade of the *Hevea* trees it often grows more luxuriantly than *Gleichenia*. *L. cernuum* spreads very rapidly by means of runners which root and produce new erect shoots; in fact, almost any prostrate branch may root, and so renew its vigour. The species is clearly specialised like the *Gleichenia*s to a compacted soil surface, the result of heavy rain on cleared ground. If that surface is broken up by cultivation, seedlings of many plants are easily established; the soil may not be seriously impoverished, but its physical condition is un-

FIG. 17. — *Blechnum orientale* and *Scleria* in the solfatares near Samosir (Sumatra, Lake Toba). In the foreground *Eleocharis dulcis*. *Blechnum orientale* is a tropical sun-fern tolerating a very wide range of conditions. — Photo K. HERMANN.

suitable for most plants. On the most exposed steep cuttings of roadsides it may take years for a covering of ferns to become established. *Gleichenia* rhizomes will start to grow over the bare surface from the neighbouring vegetation, during the wetter weather, to be dried off by a few weeks of bright sun.

A *Gleichenia* thicket may persist for a considerable time, but in the ordinary course of events it is gradually invaded by trees and shrubs; some of the hardier of these may have become established at the same time as the fern, but develop much more slowly. The young trees in time produce a low secondary forest, in which the *Gleichenia* persists (often associated with *Nepenthes*) in a less vigorous condition, its fronds often growing much longer than in the open, and climbing some distance up the trees. In wetter places, *Stenochlaena palustris* is a thicket-forming fern, which in time forms a dense mass of lianes on the trees which are established in its protection. These ferns clearly form an important preliminary stage in the development of new vegetation on unfavourable sites.

The *Lygodium*s, where they occur, are a distinctive feature of the secondary scrub vegetation. Sporangia are only borne on the parts of the fronds fully exposed to sun and air. Young plants may often be seen also in moderately shady forest, but in such places they usually remain dwarfed. *Lygodium*s resemble in habit some of the climbing orchids which need shelter from the shade of shrubs for their roots, but only flower on the branches which project freely into the air above the bushes.

Where good high forest has been cleared, quite different conditions prevail from those of the abandoned cultivated areas already mentioned. The soil, though rapidly deteriorating on the sudden exposure to sun and rain, is rich and of good texture. If the clearing is not planted with a crop which effectively covers the ground, it at once becomes invaded by secondary forest trees of many kinds, which become rapidly established. *Gleichenia* and other ferns often gain a footing, but in the dense growth they do not usually form a prominent feature. Sometimes *Nephrolepis* becomes established and may form dense thickets, its slender runners forming a very rapid means of propagation.

There are certain ferns which are characteristic of swampy ground, and are abundant where such is kept free from tree growth. Some of these have creeping rhizomes and tend to form thickets; others are short-stemmed species, invading the vegetation where they have opportunity.

Water-ferns are few and nearly all are sun-plants. *Ceratopteris* grows freely in shallow water, but never seems to be a dominant plant. *Azolla* and *Salvinia* are locally abundant as floating plants, and *Marsilia* with its floating leaves is found in rice-fields and other shallow water. *Polypodium pteropus* is recorded by ERNST [1]) as a facultative water-fern. It usually grows on rocks beside streams or in stream beds, where it is occasionally submerged.

**§ 8. Mountain fern-thickets.** — These are very similar in aspect to the fern thickets of the lowlands, but are formed chiefly of different species (though some, such as *Gleichenia* spp. and *Pteridium*, have a great altitudinal range). They are always developed in clearings, the species varying with the altitude and degree of exposure. SHREVE describes thickets in Jamaica which are very similar in aspect to those found in Malaya; the American *Odontosoria*s are matched by the thorny *Hypolepis* and *Dennstaedtia*s of the Malayan region. *Lycopodium casuarinoides* and

---

[1]) Ann. Buitenz. 22: 103. 1908.

Fig. 18. — *Polydium Féei* var. *vulcanicum* on an old lava-stream in the crater of Mt. Gedeh in West-Java, about 2600 m above sea-level. Poorly developed subalpine vegetation, *Anaphalis javanica, Gahnia javanica* Z. et M., *Gaultheria fragantissima*, and the lichen: *Stereocaulon graminosum.* — After Docters van Leeuwen.

other species are locally important constituents of such thickets. The thickets are usually gradually invaded by tree growth (often from the edges) but on exposed ridges and summits they may persist for long periods and are perhaps to be regarded as a kind of sub-climax vegetation type. Among ferns of this group perhaps *Matonia pectinata* is the most interesting. It is confined to exposed ridges and summits of mountains in the Malayan region, growing sometimes in quite dense thickets in sunny places, but also in open ridge forest, where it may attain a large size. Sometimes it grows in association with *Sphagnum*. GATES [1]) has described a *Sphagnum* bog in the tropics. Such bogs are not infrequent on Malayan mountains. Ferns of the section *Eugleichenia* are often associated with them, also other ferns of similar habit (*Hypolepis* etc.).

Tree-ferns of the more resistant types often form almost pure patches of vegetation in valley clearings, but they give way in time to forest. Their trunks, covered with a thick mass of roots, afford shelter for epiphytes.

The ferns on the upper slopes of the volcanoes of Java are of considerable interest. *Polypodium Féei* var. *vulcanicum* may form a close pure terrestrial growth almost up to the crater's edge and is one of the few members of the *Vaccinium* association which forms the woody vegetation approaching most nearly to the crater. In form, it is a typical "high-level" epiphyte with coriaceous leaves. The other ferns which come nearest to the crater are typical mountain ridge ferns.

**§ 9. Evergreen rain-forests of the subtropics.** — We may take New Zealand as a typical example of this type of vegetation, as it has received more study than most other regions. The general aspect of the vegetation is often very similar to that of Malaya, except that the species are different and their number, especially of the epiphytes, is less. This vegetation has been well described by COCKAYNE, OLIVER and others. COCKAYNE writes of ferns in lowland forest: "Ferns, as a whole, are of particular physiognomic importance. The most conspicuous are the tree ferns, especially *Cyathea dealbata*, *C. medullaris* (sometimes 15 m high and the fronds 6 m long), *Hemitelia Smithii*, *Dicksonia squarrosa* (slender and not particularly tall) and *Dicksonia fibrosa*. Of the smaller ferns, *Blechnum discolor* is specially physiognomic, occurring as it does in far-reaching colonies". The terrestrial vegetation of *Metrosideros robusta* forest of the North Island sometimes consists of a dense growth of *Hymenophyllaceae*, including *Trichomanes reniforme* and *Hymenophyllum flabellatum*. In *Metrosideros lucida* forest of the South Island *Hymenophyllaceae* also abound. At Franz Josef glacier, the end of which is 213 m above sea-level, this forest comes to within a few metres of the ice; the forest here contains *Hemitelia Smithii* of low stature, several *Hymenophyllaceae*, *Hypolepis tenuifolia*, *Histiopte-*

---

[1]) Journ. Ecol. 3: 24–30. 1915.

*ris incisa, Blechnum procerum, B. lanceolatum, Asplenium bulbiferum, A. flaccidum, Polystichum vestitum, Polypodium diversifolium, P. Billardieri, Lycopodium volubile.* The approach of such a wealth of pteridophytes to a glacier is a remarkable phenomenon.

Ferns play an important part in the developmental stages of vegetation, fern-thickets of *Pteridium, Gleichenia* etc. grading into shrub thickets. The *Pteridium* is evergreen as in the moist tropics, and forms a permanent dense thicket difficult for other species to enter. *Blechnum procerum* may cover steep slopes, especially in river gorges. The sun-ferns sometimes develop a different habit when they occur in *Leptospermum* forest. The *Pteridium* becomes a scrambling liane, other species are less xerophytic in habit, and some are sterile. In *Sphagnum* bogs *Gleichenia* and *Lycopodium* spp. occur, the former being important members of transition stages to scrub.

The epiphytes are of essentially the same character as the tropical ones, but the more specialised types are absent. All true epiphytes are found in rain forest; few occur where the rainfall is less than 60 cm. They are variously adapted to conditions of greater or less humidity and light, but as elsewhere there is no rigid zonation. Moisture rather than shade is the essential factor for *Hymenophyllaceae*, though they will tolerate considerable shade. Insufficient light prohibits the growth of the sun-loving epiphytes. Two of the Hymenophylla only occur on tree-fern trunks, which are also favoured by some of the other ferns, and by *Tmesipteris*. Some of the *Hymenophyllaceae* are more resistent to drought than others and climb higher on the trees. The root-hairs of a number of the epiphytes contain fungus hyphae; whether these have any function in the nutrition of the ferns is unknown. The only species comparable with the more resistant tropical epiphytes are *Cyclophorus serpens* and *Polypodium diversifolium;* these may grow on rocks as well as on trees. Two species of *Lycopodium* hang pendulous below the branches of trees in the same way as the tropical epiphytic species. Podocarp forest, with a dense canopy of foliage, contains the greatest variety of epiphytes.

**§ 10. Pteridophytes of less favourable climates.** — In less favourable climates than those of the moist tropics and subtropics, fern growth is a less conspicuous element in the vegetation, and the species occurring must be able to withstand, to a greater or less degree, and for a greater or less period, conditions of drought or cold, or both. There is no sharp division between one climate and another, and it is difficult to divide the various classes of pteridophyte floras.

Other regions with similar climates possess a fern flora of similar aspect to that of New Zealand. Tasmania, the Canary Islands, parts of Japan, and of the coastal regions of South America, which have equable oceanic climates, have a considerable wealth of ferns, but the number of epiphytic species is less than in the tropics.

·The Monsoon regions border on the continuously moist tropics, and there is a gradual change from one to the other, the dry season lengthening and increasing in intensity. In such climates there is abundance of rain in the wet season, and it is a question of the ferns being able to exist during the drought; this they usually do by adopting a deciduous habit. The leaves of terrestrial ferns may shrivel, unless they are in sheltered localities or near streams, in which case this may be unnecessary. Epiphytes may lose their leaves, a process facilitated in many cases by articulation of pinnae or stipes. In some cases, fleshy leaves of epiphytes are not shed, but curl up, and in this wilted condition are able to continue to exist. Drynarias lose their vegetative leaves and retain the dead bracket leaves, which protect the rhizome and roots. In some cases this shedding of leaves and resting is obligatory; in others it depends on external conditions. GOEBEL [1]) remarked that the Drynarias of Java retained their leaves throughout the dry season if the plants were in sheltered positions, and were also ever-green in hothouses in Europe. *D. Fortunei*, however, loses its leaves sea-sonally even in the uniform hothouse climate. Parallel cases might be cited in phanerogams; e.g., a bare stage is obligatory in *Bombax mala-baricum* but not in *Tectona grandis*, though both are normally deciduous in seasonal climates. In some cases one species replaces another as a uniform climate is replaced by a seasonal one. Thus, in the seasonal climate of the north of the Malay Peninsula, *Platycerium grande* takes the place of *P. coronarium*, which is abundant in the south. *P. grande* will grow in the uniform climate of Singapore, but will not reproduce itself; *P. coro-narium* cannot withstand a prolonged dry season. A deciduous habit is resorted to by different ferns in conditions of different severity; and within the same region different habitats vary in the conditions of exposure and dryness. A limestone rock-fern may need to drop its fronds in a short dry season during which most terrestrial ferns can continue to vegetate; another rock-fern may wilt but retain its leaves in a living condition. In the mountains of the monsoon regions, rain is more abundant than in the low country, and the dry season less severe; conditions approximate more to those of the mountains of the continuously moist regions.

If the wet season is shorter or has less rain, the conditions for pteri-dophyte development are less favourable. Epiphytes are the first group to disappear. Terrestrial ferns become chiefly confined to woodlands, where such exist, except for a limited number of xerophytic species. In open country, and on high mountains beyond the limit of tree growth, a re-markable series of xerophytes occur; these are perhaps best developed on the Andes, the species of Jamesonia being particularly notable. These ferns have short-creeping rhizomes, often rooting deeply, the stipes slender and wiry, the lamina reduced, often much dissected, the small segments often

---

[1]) Ann. Buitenz. 39: 142. 1928.

concave beneath, the surfaces waxy, scaly or hairy, sometimes very hea-
vily protected. CHRIST gives a list of such plants, including *Jamesonia,
Doryopteris* and *Aneimia* from the Andes, *Pellaea* from Mexico and South
Africa, Mohria from the Kalahari and East Africa, *Actiniopteris* from
Abyssinia and India, *Platyzoma* from Australia, *Adiantum capillus-veneris*
from Mesapotamia and central Asia, *Onychium melanolepis* from Sinai,
*Ceterach officinale* from Morocco, *Cheilanthes farinosa* from Somaliland.
These ferns are the nearest approach to desert plants which occur among
pteridophytes. *Adiantum capillus-veneris* (as well as some of the others
mentioned) is very widely distributed, and may grow in exposed rocky
places in wetter climates; it is found for example in Penang. According to
CHRIST the other principal fern-type of the high Andes is *Elaphoglossum,*
the fronds of which are simple and entire. The woodland forms are broad
and often almost glabrous; the species of the exposed uplands are reduced
in size, thick in texture, and very heavily protected with a covering of
scales and hairs. Some species of *Cyclophorus* in China are similar in form
and habitat.

A few species of *Lycopodium* are also adapted to dry conditions on the
Andes; they have fleshy or coriaceous leaves, often curling up in drought.
Some remarkable species of *Selaginella* have a similar habit. *S. lepidophylla*
is perhaps the most notable of these; it occurs in Mexico, and the plants
roll themselves into the form of close valls during dry weather, persisting in
this condition for long periods, and uncurling during the infrequent rainy
periods.

Chapter XIV

# GEOGRAPHIE

von

Hubert Winkler (Breslau)

**§ 1. Literatur.** — 1. J. D'Urville, De la distribution des Fougères sur surface du globe terrestre. Ann. Sc. Nat. 6, Paris 1825. — 2. J. G. Baker, On the geographical distribution of Ferns. Transact. Linn. Soc. 26/1, London 1868. — 3. K. M. Lyell, A geographical handbook of all the known Ferns. London 1870. — 4. W. J. Hooker and J. G. Baker, Synopsis Filicum, ed. 2, London 1874. — 5. J. Palacky, O rozšiřeni kapradi na světě (Über die Verbreitung der Farnartigen). Sitzbungsber. k. böhm. Ges. d. Wiss. 1885. Prag. 1886. — 6. A. Engler u. K. Prantl, Die Natürl. Pflanzenfam. 1/4. Leipzig 1902. — 7. Gillot et Durafour, Repartition topogr. de Pteris aquilina. Bull. Soc. Natur. de l'Ain, 1904. — 8. C. Christensen, Index Filicum. Hafniae 1906 u. Suppl. — 9. H. Christ, Die Geographie der Farne. Jena 1910. — 10. N. E. Pfeiffer, Monograph of the Isoetaceae. Annals Missouri Bot. Garden, 9, St. Louis Mo. 1922. — 11. E. Irmscher, Pflanzenverbreitung und Entwicklung der Kontinente. Mitt. a. d. Inst. f. Allgem. Bot. in Hamburg, 5, 1922; II. Teil ebenda, 8, 1929. — 12. W. Studt, Die heutige und frühere Verbreitung der Koniferen und die Geschichte ihrer Arealgestaltung. Ebenda, 6, 1926. — 13. F. O. Bower, The Ferns, II u. III. Cambridge 1926 u. 1928. — 14. J. H. Schaffner, Geogr. Distribution of the species of Equisetum in relation to their phylogeny. Amer. Fern Journ. 20, 1930. — 15. A. Donat, Einige Isoetiden, II; in E. Hannig u. H. Winkler, Die Pflanzenareale, 3. Reihe, Heft. 8, 1933. — 16. M. Rikli, Das Ausklingen der Pteridophytenflora in der Polaris und deren pflanzengeographische Beziehungen zu ihren Nachbargebieten sowie zur Alpenflora. Ber. d. Schweizerischen Botan. Ges. 42, 1933.

**§ 2. Geschichtliches.** — Die Wissenschaft der Pteridophyten-Geographie hat erst eine kurze Geschichte. J. D'Urville (1) gab 1825 nach der

---

Anmerkung: Herrn Carl Christensen, der im 18. Kapitel dieses Werkes die Klassifikation bearbeitet hat, bin ich für die Kontrolle der giltigen Namen und für zahlreiche Hinweise zu grossem Dank verpflichtet. Das der Tabelle zugrunde gelegte System entspricht dem Christensenschen. Die Namen der bei der 15. Fam. (*Polypodiaceae*) noch angegebenen Unterfamilien sind in jenem Kapitel des Handbuchs nachzusehen. W.

damals üblichen statistischen Art eine Übersicht über den Anteil der Farne an der Gefässpflanzenflora einzelner Gebiete. Wenn G. BAKER seine Abhandlung (2) mit der Feststellung beginnt, dass wir von keiner grossen natürlichen Pflanzenordnung die Verbreitung so gut kennten, wie von den Farnen, so nimmt er damit ihr Ergebnis vorweg. Er will sagen, dass von keiner grossen Gruppe die herbarmässig niedergelegten Grundlagen für die Darstellung der Verbreitung so vollständig und — nicht zuletzt dank W. HOOKERs und seiner eigenen Bemühungen — so kritisch durchgearbeitet seien wie von den Farnen. In der Einleitung legt er kurz die klimatischen Ursachen der Verteilung der Farne dar, die durch ihr Fehlen oder ihre Zahl und Üppigkeit „with the precision of an hygrometer" die trocknen und feuchten Gebiete der Erde anzeigen. Die 1. kurze Tabelle stellt diese Beziehungen — unter Vergleich mit der Temperatur — zusammen. Es werden darin 10 „Distrikte" unterschieden, die, obwohl man sie hauptsächlich als Vegetationsbereiche ansehen muss, vielfach mit den von CHRIST später umrissenen Farn-,,Florenreichen" zusammenfallen. Den Hauptteil der Abhandlung bildet die 2., etwas unübersichtliche Tabelle mit der vollständigen Aufzählung der von BAKER anerkannten Arten und ihrer Verbreitungsgebiete. Zum Schluss werden die einzelnen Distrikte noch kurz floristisch charakterisiert, mit Betonung des statistischen Gesichtspunktes. Endemismus und Disjunktionen (auch schon einzelne Grossdisjunktionen) werden kurz berührt. So müssen wir diese Abhandlung BAKERs als die eigentliche Begründung der Wissenschaft von der Geographie der Farne ansehen. Das zwei Jahre später veröffentliche Handbuch LYELLs (3) bringt einen Fortschritt durch seine übersichtliche und ins einzelne gehende Aufzählung der Areale, mit Hervorhebung der Endemen, aber keine Vertiefung der allgemeinen Gesichtspunkte. — PALACKYs kurze (tschechisch geschriebene) Arbeit (5) umfasst zwar als erste die gesamten Pteridophyten, zeigt in ganz grossen Zügen auch die vorweltliche Verbreitung, bietet aber weder tatsächlich noch theoretisch Neues.

Trotz dieser Vorgänger konnte CHRIST 1910 in seiner „Geographie der Farne" (9) noch schreiben, dass bis dahin keine „einlässliche Darstellung der Farnverbreitung" gegeben worden sei. CHRIST hat das Verdienst, durch Vergleich der Farn- und Blütenpflanzen ausdrücklich mit der Vorstellung aufgeräumt zu haben, dass die Farnpflanzen (und die Sporenpflanzen überhaupt) vermöge der unbegrenzten Verbreitbarkeit ihrer Wanderorgane weite und unbestimmte Areale, also eine diffusere, weniger charakteristische Verbreitung als die Blütenpflanzen hätten. Für die volle Erkenntnis von der Bedeutung des Endemismus und der Disjunktion bei den Farnen und für die Festlegung der Farn-Florenreiche ist das Werk des schweizerischen Pteridologen bis heute die Hauptgrundlage geblieben.

**§ 3. Die Bedeutung der Umwelt für die Farnverbreitung.** — Ihr Wohngebiet wird den Pteridophyten in starkem Masse von klimatischen

Bedingungen vorgeschrieben: sie sind im allgemeinen mesotherme Hygrophyten. Erst bei einem Regenfall von mehr als 200 cm im Jahre entfalten sie ihre ganze Fülle; alle Erdstriche, die nicht mindestens 60 cm Niederschlag empfangen, sind arm an Pteridophyten, da die Ausbildung xeromorpher Sippen nur in beschränktem Masse erfolgt ist; die meisten sind Schattenpflanzen, so dass das Areal der Pteridophyten nahezu mit dem Waldareal zusammenfällt. „Der baumlose hohe Norden, das Steppengebiet Zentral- und Vorderasiens, die Prärie Nordamerikas, der Wüstengürtel vom Indus bis Marokko, die Pampas, die Geröllfelder Patagoniens und die peruanisch-chilenische Küste, Zentral- und Westaustralien, Südwestafrika sind fast ohne Farne (und Pteridophyten überhaupt), und innerhalb der Waldgebiete sind grosse Hochländer und trockene Gebiete nur von kleineren, xerophytisch angepassten Arten bewohnt, während nur in den sie durchfurchenden Schluchten im Schutz von Galleriewäldern ein üppigeres Farnleben möglich ist" (9). Küstenländer, Halbinseln, wie Florida, Inseln, wie Wight und Irland, die Canaren und Azoren, Madagaskar, die Maskarenen und Seychellen, die Philippinen und Borneo, Neu-Guinea, Neu-Caledonien, Tasmanien, Neu-Seeland, Juan Fernandez, Hawaii, sind bevorzugte Sammelpunkte der Pteridophyten.

Ihre Hauptmasse, besonders die der Farne und von diesen wieder der stattlichen, vor allem baumförmigen, ist auf die warmen Erdgebiete beschränkt. Von den eusporangiaten Farnen gehen nur die kleinen *Ophioglossaceae*, von den leptosporangiaten hauptsächlich *Polypodiaceae*, ferner *Osmunda*-Arten, wenige kleine *Schizaeaceae* und winzige *Hymenophyllaceae* bis in die kalten nördlichen Gebiete, während in die für Farnwuchs äusserst geeignete gemässigte Südhemisphäre viele tropische Gattungen, selbst *Cyatheaceae*, vordringen. Die unter gleichmässigeren Bedingungen lebenden Wasserfarne (*Marsiliaceae, Salviniaceae*) sind stark verbreitet in den nördlichen aussertropischen Zonen, und untergetaucht lebende Arten der Gattung *Isoetes* findet man fast auf der ganzen Erde, am häufigsten ausserhalb des Aquatorialgürtels.

Alpine und arktische Farne gibt es nur wenige, ebenso streng desertische, weil alle Extreme in Temperatur und Trockenheit den Farnen widerstreben. Hochnordisch ist die Gattung *Woodsia*. Zu den Alpinen gehören vor allem die kleinen Pflänzchen der Gattung *Polystichum*, die in China und im Himalaya bis in die Schneeregion bei mehr als 5000 m Höhe vordringt; in den Anden, von 3400 m an aufwärts, zahlreiche — ebenfalls winzige — Vertreter von *Elaphoglossum, Jamesonia* u.a.

Ähnlich den *Filicales* verhält sich die grosse Gattung *Selaginella*. Sie ist fast ganz auf die schattigsten und feuchtesten Stellen des tropischen und subtropischen Regenwaldes beschränkt; aussertropische Verbreitung haben nur wenige Arten, von denen *S. selaginoides* arktisch-alpin ist. Wohl am weitesten nach Süden — bis Tasmanien — ist *S. Preissii* vorgedrungen. Die bekannten, im trocknen Zustande eingerollten Rosetten von *S. lepi-*

*dophylla* und Verwandten, *S. rupestris, borealis, mongholica* u.a. mit mehr oder weniger ausgesprochenem Xerophytencharakter bewohnen trockene Gebiete der Neuen und Alten Welt. Von *Lycopodium* finden die zahlreichen epiphytischen Arten natürlich nur in den Regenwäldern, sei es der heissen oder der südlichen temperierten Zone, günstige Existenzbedingungen; die am Boden wachsenden haben aber oft ein auffallend geringes Wärmebedürfnis; sie kommen daher vorzugsweise ausserhalb des Tropengürtels oder innerhalb auf den hohen Gebirgen — in den Anden manche an der Schneegrenze — vor. *L. Selago* geht im den Alpen bis zu Höhen von 3120 m. Vielleicht spielt bei dieser Beschränkung aber auch die Verbreitung der Pilze eine Rolle, mit denen die Lycopodien in Symbiose leben.

Von den Gattungen der *Psilotales* ist die monotypische *Tmesipteris* in Australien, auf seinen Inseln und Neu-Seeland verbreitet und geht von da — bis zu den Philippinen — in die Tropenzone über. Die Gattung *Psilotum* ist mit 2 Arten in den Tropen und Subtropen beider Hemisphären zerstreut, nach Norden bis Japan und Florida, nach Süden bis Neu-Seeland. Die Vertreter beider Gattungen sind Humusbewohner.

Die *Equisetales* neigen offenbar von allen Pteridophyten am meisten nach der oligothermen Seite hin und haben auch das geringste Bedürfnis nach Luftfeuchtigkeit; viele lieben aber feuchten Boden oder leben sogar im Wasser. Die meisten Arten der einzigen Gattung *Equisetum* sind ausserhalb der Tropen beheimatet. *Equisetum arvense, variegatum, silvaticum* und *scirpoides* dürften von allen Pteridophyten am weitesten nach Norden gehen, bis Grönland, Island und Spitzbergen und bis ins arktische Amerika und Sibirien. In den Alpen steigen einzelne Arten bis 3000 m auf. (Vergl. S. 385—390).

### § 4. Die Merkmale der Areale. A. Die Grösse und zonale Verteilung.

a. Kosmopolitische und zonenumfassende Sippen. — Die Grösse der Pteridophytenareale ist ebenso verschieden wie die der Angiospermen. Kosmopolitische Arten gibt es längst nicht so viele, wie man bei der leichten Verbreitbarkeit der Sporen voraussetzen könnte; kaum 2½ Dutzend Farngattungen, die in der Tabelle (S. 460) etwa unter den Arealtypen 24–28 angeführt sind, bewohnen den Erdball in annähernder Gesamtausdehnung. Einige der weitest verbreiteten Arten sind folgende: *Pteridium aquilinum*, wohl zugleich der geselligste Farn, kommt in allen Klimaten vor, vom Äquator bis zum Polarkreis und südwärts bis Neu-Seeland. Noch weiter nach Norden, bis Grönland, dringt *Cystopteris fragilis* vor, der andrerseits bis ins tiefe Südamerika und in die südl. gemässigte Zone der Alten Welt geht. Annähernd gleich kommt ihm *Asplenium Trichomanes*. Neben *Pteridium* ist *Osmunda regalis* „der universelle Farn schlechthin" (s. S. 389). Fast kosmopolitisch — mit Ausschluss trockner Gebiete — sind *Lycopodium Selago* und *clavatum*.

Selbst in der Tropenzone, der Hauptverbreitungszone der Pteridophyten, gibt es nur wenige um den ganzen Erdball herumgehende Arten ( P a n t r o p i s t e n ). Am allgemeinsten und gleichmässigsten sind verbreitet: *Gleichenia linearis; Pteris vittata, quadriaurita, biaurita; Histiopteris incisa; Asplenium praemorsum; Nephrolepis cordifolia, acuta, exaltata; Acrostichum aureum* (in der Mangrove); *Dryopteris gongylodes, dentata; Didymochlaena lunulata; Doryopteris concolor.* Pantropische Gattungen gibt es in erheblich grösserer Zahl, wie die Arealtypen 1 und 12–28 der Tabelle ausweisen.

Einige *Cryptogramma*-Arten sind z i r k u m p o l a r e Glazialpflanzen. *Woodsia ilvensis, alpina* und *glabella; Dryopteris dilatata, cristata* und *spinulosa* bewohnen das subarktische Europa und Nordamerika und als Relikte die Gebirge des gemässigten Teiles der Nordhalbkugel. Von den 200 Farnarten des nordamerikanisch-eurasischen Waldgebietes haben sich etwa 20 % gleichmässig in diesem ganzen Gebiet auszubreiten vermocht, wie *Matteuccia, Athyrium Filix-femina, Ophioglossum vulgatum;* dazu *Lycopodium annotinum, alpinum* und *complanatum,* das bis in die alpine Region der Tropenzone vorstösst; schliesslich *Equisetum pratense, palustre, hiemale, variegatum, scirpoides.* Nur sechs, noch dazu mono- oder oligotypische Gattungen, gehören allein der nördlichen gemässigten Zone an. (Arealtypus 64, 65).

Vorwiegend s ü d l i c h - g e m ä s s i g t, aber in der Alten und Neuen Welt bis in die Tropen vordringend, ist *Polystichum capense, Blechnum capense* und *tabulare.* Allein den t i e f e n S ü d e n (Arealtypus 66–68) bewohnen nur sieben — wiederum monotypische — Gattungen, *Todea, Thyrsopteris, Serphyllopsis, Cardiomanes, Leptolepia, Loxsoma* und die Lycopodiacee *Phylloglossum,* keine davon aber alle Landmassen zugleich. Aus grösseren Gattungen sind nur *Polystichum mohrioides* und *Blechnum penna marina* zirkumpolarantarktisch.

B i p o l a r e Sippen gibt es unter den Pteridophyten natürlich nur sehr wenige, da die Zahl ihrer aussertropischen Vertreter an sich schon recht gering ist. ,,Arktische Spuren in der Subantarktis'' nennt CHRIST einige kleine *Botrychium*-Arten (*Botrychium Lunaria*). Von den *Filicales* ist am auffälligsten die leptosporangiate Farngattung *Pleurosorus* (Arealtypus 63) mit drei schwach getrennten Arten, von denen die eine in den Gebirgen Südspaniens, die anderen teils in Neu-Seeland und Australien, teils im südlichen Chile vorkommen.

b. E n d e m e n. — Der Endemismus spielt bei den Pteridophyten eine früher ungeahnte Rolle, vornehmlich bei den heterosporen, die ihr Wohngebiet in der Regel nur ausdehnen können, wenn Sporen beiderlei Geschlechts die gleiche Verbreitung erfahren. Bei den etwa 700 *Selaginella*-Arten ist eine Verbreitung über das Mediterrangebiet, von den canarischen Inseln bis Syrien (*S. denticulata*) oder über Costarica-Guiana-Nordargentinien (*S. radiata*), über Japan-China-Ostindien-Java (*S. involvens*), über Sumatra–Java–Borneo–Neuguinea (*S. intermedia*) schon weit zu nennen; sie

kommt kaum zwanzigmal vor. Am weitesten verbreitet sind die systematisch etwas isoliert stehende *S. helvetica*, nämlich in den Gebirgen
Mittel- und Südeuropas, bis zum Kaukasus, ferner in Persien, dem Amur-
gebiet, der Mandschurei und in Japan; und *S. proniflora* (=*S. Belangeri*),
von Ostindien über die Sundainseln, Philippinen und Neu-Guinea bis
Queensland. Die überwiegende Zahl der *Selaginella*-Arten hat nach unserer
heutigen Kenntnis eine sehr enge Verbreitung; zahlreich sind die Insel-
endemismen, doch wird auch von den die Kontinente bewohnenden
Arten allermeist eine enge oder sehr enge Beschränkung angegeben. Wie
weit in der Gattung *Selaginella* der Endemismus den durchschnittlichen
Pteridophytenendemismus überwiegt, zeigen folgende Tatsachen: Madagaskar weist rund 46 % Endemen aller Pteridophyten auf, aber 70 %
in der Gattung *Selaginella*, und für Neu-Guinea ist das Verhältnis derselben
Sippen 60 % und 80 %.

Ähnlich liegen die Arealverhältnisse bei der viel artenärmeren *Isoetes*,
was die verbreitungshemmende Wirkung der Heterosporie um so mehr
betont, als diese Gattung vorzüglich Wasser- und Sumpfpflanzen enthält.

Von den isosporen Pteridophyten weist die monotypische Psilotaceen-
gattung *Tmesipteris* — besonders neben dem pantropischen *Psilotum* —
ein verhältnismässig beschränktes Areal auf: Australien und die benach-
barten Inseln, Neu-Caledonien, einige polynesische Inseln, die Philippinen.
*Equisetum* ist mit stark beschränkten Arten im Himalaya, in Japan, in
Peru und Mexico vertreten. Die Schilderung des Endemismus der *Filicales*,
wie sie CHRIST bereits gibt, trifft in ihren Grundzügen auch nach unserer
heutigen Kenntnis noch zu. In dieser grossen Klasse kann man alte und
junge Typen mit Sicherheit bezeichnen und so zwischen dem konserva-
tiven und progressiven Endemismus gut unterscheiden. Beide weisen, wie
bei den Blütenpflanzen, zahlreiche Fälle engster, oft punktartiger Verbreitung auf. ,,So gut wie die Canaren unter den *Statice* ihre Nobiles haben, die
auf einem einzigen Felsen vorkommen, so sehr sind die edlen *Polystichum
Webbianum, falcinellum, drepanum* auf einzelne Punkte in Madeira und
*Matonia sarmentosa* auf einige Höhlenstandorte bei Sarawak auf Borneo
isoliert'' (9). Alte Endemismen sind noch *Archangiopteris Henryi* und
*Neocheiropteris palmatopedata* in Yünnan; die Gattung *Dipteris*, von deren
halbem Dutzend Arten nur eine (*D. conjugata*) über das ganze indisch-
malayische Gebiet verbreitet ist; *Loxsoma Cunninghami* in Neu-Seeland
und die nahe verwandten drei *Loxsomopsis*-Arten, je in Costarica, Ecuador
und Bolivien; *Thyrsopteris elegans* auf Juan Fernandez. *Botrychium* kommt
in ganz Afrika mit einer auf die Grasregion des Kamerunberges beschränkten Art vor. — Als Musterbeispiele für Neuendemismus nennt CHRIST die
zahlreichen *Polystichum*-, *Polypodium*- und *Athyrium*-Formen Chinas,
*Aneimia* und *Doryopteris* in Brasilien, *Elaphoglossum* in den Anden. Bei
den artenreichen Cyatheaceen-Gattungen *Alsophila*, *Hemitelia*, *Cyathea*
ist strenger Endemismus die Regel· manche sind auf einzelne Vulkane

beschränkt. Da zur Beurteilung dieser Verhältnisse die Artenzahl von Bedeutung ist, wird sie in Spalte *e* der Tabelle berücksichtigt.

**B. Die meridionale Verteilung.** — Der Begriff „Disjunktion" in seiner älteren, allgemeinen Fassung hat wohl Spekulationen über die Verbreitungsmöglichkeiten der zerstreut lebenden Sippen anregen, aber auf die Entstehungsgeschichte der Pflanzenareale überhaupt kaum ein Licht werfen können; erst seine Fassung im IRMSCHER'schen Sinne der „Grossdisjunktion" hat hier weitergeführt. Da IRMSCHERS Schemata zugleich eine übersichtliche Vorstellung von der Verbreitung geben, so sei der Versuch gemacht, diese Schemata auf die Hauptmenge der Pteridophytengattungen anzuwenden, wobei die oben erörterte Bedeutung der Tropenzone für diese Pflanzengruppe besonders hervorgehoben wird. Die folgende Tabelle ergibt auch leicht eine Übersicht über die Verbreitung der Familien; auf die der Gattung untergeordneten Sippen konnte bei der Beschränkung des Raumes nicht eingegangen werden.

### Übersicht über die Verbreitung der Pteridophyten.

| *a.* Nr. des Arealtypus | *b.* Arealtypus | *c.* Nr. der Gattung | *d.* Gattungsname | *e.* Zahl der Arten | *f.* Systematische Stellung (Filicales, Familien 1.–21.) | *c.* Nr. der Gattung |
|---|---|---|---|---|---|---|
| **Ia** | **I. Rein tropisch** *a.* In d. ganzen Tropenzone (pantropisch) | | | | | |
| 1. | 1 2 3 4 | 1. | Didymochlaena | 1 | 15,12 | 1. |
| | | 2. | Ithycaulon | 10 | 15,1 | 2. |
| | | 3. | Lomariopsis | 20 | 15,12 | 3. |
| **Ib** | *b.* Mit einer Lücke | | | | | |
| 2. | 1 2 . 4 | 4. | Loxoscaphe | 8 | 15,10 | 4. |
| 3. | 1 . 3 4 | 5. | Cyclopeltis | 6 | 15,12 | 5. |
| | | 6. | Syngramma | 20 | 15,6 | 6. |
| 4. | . 2 3 4 | 7. | Drymoglossum | 6 | 15,14 | 7. |
| | | 8. | Hymenolepis | 13 | 15,14 | 8. |
| | | 9. | Monogramma | 2 | 15,7 | 9. |
| **Ic** | *c.* Mit zwei Lücken | | | | | |
| 5. | 1 2 . . | 10. | Adiantopsis | 16 | 15,6 | 10. |
| | | 11. | Anisosorus | 2 | 15,5 | 11. |
| | | 12. | Microstaphyla | 3 | 15,15 | 12. |
| | | 13. | Trachypteris | 2 | 15,6 | 13. |

| Gr. | No. | Name | n | 1 | 2 | 3 | 4 | 5 | 6 | 7 | 8 | 9 | 10 | 11 | 12 | 13 | 14 | 15 | 16 | 17 | 18 | 19 | 20 | 21 | No. |
|---|---|---|---|---|---|---|---|---|---|---|---|---|---|---|---|---|---|---|---|---|---|---|---|---|---|
| **6.** 1 . 3 . | 14. | Hemionitis | 8 | | | | | | | | | | | | | | | 15,6 | | | | | | | 14. |
| | 15. | Helminthostachys | 1 | 1 | | | | | | | | | | | | | | | | | | | | | 15. |
| | 16. | Phanerosorus | 2 | | | | | | | | 8 | | | | | | | | | | | | | | 16. |
| | 17. | Cystodium | 1 | | | | | | | | | | | | | 13 | | | | | | | | | 17. |
| | 18. | Acrophorus | 3 | | | | | | | | | | | | | | | 15,12 | | | | | | | 18. |
| | 19. | Acrosorus | 5 | | | | | | | | | | | | | | | 15,14 | | | | | | | 19. |
| | 20. | Aglaomorpha | 11 | | | | | | | | | | | | | | | 15,14 | | | | | | | 20. |
| | 21. | Christiopteris | 3 | | | | | | | | | | | | | | | 15,14 | | | | | | | 21. |
| | 22. | Craspedodictyon | 6 | | | | | | | | | | | | | | | 15,6 | | | | | | | 22. |
| | 23. | Diplaziopsis | 1 | | | | | | | | | | | | | | | 15,10 | | | | | | | 23. |
| | 24. | Diplora | 4 | | | | | | | | | | | | | | | 15,10 | | | | | | | 24. |
| | 25. | Davallodes | 12 | | | | | | | | | | | | | | | 15,3 | | | | | | | 25. |
| | 26. | Grammatopteridium | 2 | | | | | | | | | | | | | | | 15,14 | | | | | | | 26. |
| | 27. | Hemigramma | 6 | | | | | | | | | | | | | | | 15,12 | | | | | | | 27. |
| **7.** . . 3 4 | 28. | Leptochilus | 10 | | | | | | | | | | | | | | | 15,14 | | | | | | | 28. |
| | 29. | Lomagramma | 14 | | | | | | | | | | | | | | | 15,12 ? | | | | | | | 29. |
| | 30. | Merinthosorus | 1 | | | | | | | | | | | | | | | 15,14 | | | | | | | 30. |
| | 31. | Nematopteris | 2 | | | | | | | | | | | | | | | 15,14 | | | | | | | 31. |
| | 32. | Platytaenia | 1 | | | | | | | | | | | | | | | 15,2 ? | | | | | | | 32. |
| | 33. | Prosaptia | 15 | | | | | | | | | | | | | | | 15,14 | | | | | | | 33. |
| | 34. | Schizostege | 3 | | | | | | | | | | | | | | | 15,5 | | | | | | | 34. |
| | 35. | Scleroglossum | 7 | | | | | | | | | | | | | | | 15,14 | | | | | | | 35. |
| | 36. | Stenolepia | 1 | | | | | | | | | | | | | | | 15,10 | | | | | | | 36. |
| | 37. | Sphaerostephanos | 5 | | | | | | | | | | | | | | | 15,12 | | | | | | | 37. |
| | 38. | Stenosemia | 3 | | | | | | | | | | | | | | | 15,12 | | | | | | | 38. |
| | 39. | Scyphularia | 7 | | | | | | | | | | | | | | | 15,3 | | | | | | | 39. |
| | 40. | Taenitis | 3 | | | | | | | | | | | | | | | 15,2 ? | | | | | | | 40. |
| | 41. | Tapeinidium | 12 | | | | | | | | | | | | | | | 15,2 | | | | | | | 41. |
| | 42. | Vaginularia | 6 | | | | | | | | | | | | | | | 15,7 | | | | | | | 42. |

**Id** — *d.* Mit drei Lücken

| Gr. | No. | Name | n | 1 | 2 | 3 | 4 | 5 | 6 | 7 | 8 | 9 | 10 | 11 | 12 | 13 | 14 | 15 | 16 | 17 | 18 | 19 | 20 | 21 | No. |
|---|---|---|---|---|---|---|---|---|---|---|---|---|---|---|---|---|---|---|---|---|---|---|---|---|---|---|
| | 43. | Danaea | 30 | | | 3 | | | | | | | | | | | | | | | | | | | 43. |
| | 44. | Regnellidium | 1 | | | | | | 6 | | | | | | | | | | | | | | | | 44. |
| | 45. | Loxomopsis | 3 | | | | | | | | | | 10 | | | | | | | | | | | | 45. |
| | 46. | Hymenophyllopsis | 2 | | | | | | | | | | | 11 | | | | | | | | | | | 46. |
| | 47. | Metaxya | 1 | | | | | | | | | | | | | | 14 | | | | | | | | 47. |
| | 48. | Adenoderris | 2 | | | | | | | | | | | | | | | 15,12 | | | | | | | 48. |
| | 49. | Amphiblestra | 1 | | | | | | | | | | | | | | | 15, 5 | | | | | | | 49. |
| | 50. | Ananthocorus | 1 | | | | | | | | | | | | | | | 15, 7 | | | | | | | 50. |
| | 51. | Anaxetum | 1 | | | | | | | | | | | | | | | 15,14 | | | | | | | 51. |
| | 52. | Anetium | 1 | | | | | | | | | | | | | | | 15,7 | | | | | | | 52. |
| | 53. | Anopteris | 1 | | | | | | | | | | | | | | | 15,5 | | | | | | | 53. |
| | 54. | Atalopteris | 3 | | | | | | | | | | | | | | | 15,12 | | | | | | | 54. |
| | 55. | Camptodium | 2 | | | | | | | | | | | | | | | 15,12 | | | | | | | 55. |
| | 56. | Campyloneurum | 25 | | | | | | | | | | | | | | | 15,14 | | | | | | | 56. |
| | 57. | Cochlidium | 10 | | | | | | | | | | | | | | | 15,14 | | | | | | | 57. |
| | 58. | Craspedaria | 10 | | | | | | | | | | | | | | | 15,14 | | | | | | | 58. |
| | 59. | Cyclodium | 2 | | | | | | | | | | | | | | | 15,12 | | | | | | | 59. |
| | 60. | Dictyoxiphium | 1 | | | | | | | | | | | | | | | 15,2 | | | | | | | 60. |
| | 61. | Enterosora | 1 | | | | | | | | | | | | | | | 15,14 | | | | | | | 61. |
| **8.** 1 . . . | 62. | Eschhatogramme | 4 | | | | | | | | | | | | | | | 15,14 | | | | | | | 62. |
| | 63. | Fadyenia | 1 | | | | | | | | | | | | | | | 15,12 | | | | | | | 63. |
| | 64. | Goniopteris | 70 | | | | | | | | | | | | | | | 15,12 | | | | | | | 64. |
| | 65. | Hecistopteris | 1 | | | | | | | | | | | | | | | 15,7 | | | | | | | 65. |
| | 66. | Holodictyum | 2 | | | | | | | | | | | | | | | 15,10 | | | | | | | 66. |
| | 67. | Hypoderris | 4 | | | | | | | | | | | | | | | 15,12 | | | | | | | 67. |
| | 68. | Jamesonia | 18 | | | | | | | | | | | | | | | 15,6 | | | | | | | 68. |
| | 69. | Llavea | 1 | | | | | | | | | | | | | | | 15,6 | | | | | | | 69. |
| | 70. | Marginariopsis | 1 | | | | | | | | | | | | | | | 15,14 | | | | | | | 70. |
| | 71. | Maxonia | 1 | | | | | | | | | | | | | | | 15,12 | | | | | | | 71. |
| | 72. | Neurocallis | 1 | | | | | | | | | | | | | | | 15,5 | | | | | | | 72. |
| | 73. | Odontosoria | 12 | | | | | | | | | | | | | | | 15,2 | | | | | | | 73. |
| | 74. | Olfersia | 1 | | | | | | | | | | | | | | | 15,12 | | | | | | | 74. |
| | 75. | Ormoloma | 2 | | | | | | | | | | | | | | | 15,1 | | | | | | | 75. |
| | 76. | Paltonium | 1 | | | | | | | | | | | | | | | 15,14 | | | | | | | 76. |
| | 77. | Phlebodium | 2 | | | | | | | | | | | | | | | 15,14 | | | | | | | 77. |
| | 78. | Plecosorus | 2 | | | | | | | | | | | | | | | 15,12 | | | | | | | 78. |
| | 79. | Polybotrya | 24 | | | | | | | | | | | | | | | 15,12 | | | | | | | 79. |
| | 80. | Pterozonium | 4 | | | | | | | | | | | | | | | 15,6 | | | | | | | 80. |
| | 81. | Rhipidopteris | 4 | | | | | | | | | | | | | | | 15,15 | | | | | | | 81. |

| | | | 1. | 2. | 3. | 4. | 5. | 6. | 7. | 8. | 9. | 10. | 11. | 12. | 13. | 14. | 15. | 16. | 17. | 18. | 19. | 20. | 21. | |
|---|---|---|---|---|---|---|---|---|---|---|---|---|---|---|---|---|---|---|---|---|---|---|---|---|
| | | 82. Saccoloma | 1 | | | | | | | | | | | | | | 15,1 | | | | | | | 82. |
| | | 83. Saffordia | 1 | | | | | | | | | | | | | | 15,6 | | | | | | | 83. |
| | | 84. Soromanes | 1 | | | | | | | | | | | | | | 15,12 | | | | | | | 84. |
| | | 85. Stigmatopteris | 20 | | | | | | | | | | | | | | 15,12 | | | | | | | 85. |
| | | 86. Trismeria | 2 | | | | | | | | | | | | | | 15,6 | | | | | | | 86. |
| 9. | . . . .<br>. 2 . .<br>. . . . | 87. Ochropteris | 1 | | | | | | | | | | | | | | 15,5 | | | | | | | 87. |
| | | 88. Psammiosorus | 1 | | | | | | | | | | | | | | 15,3 | | | | | | | 88. |
| | | 89. Archangiopteris | 4 | 2 | | | | | | | | | | | | | | | | | | | | 89. |
| | | 90. Macroglossum | 2 | 2 | | | | | | | | | | | | | | | | | | | | 90. |
| | | 91. Christensenia | 2 | | 3 | | | | | | | | | | | | | | | | | | | 91. |
| | | 92. Matonia | 2 | | | | | | | 8 | | | | | | | | | | | | | | 92. |
| | | 93. Arthromeris | 10 | | | | | | | | | | | | | | 15,14 | | | | | | | 93. |
| | | 94. Cerosora | 1 | | | | | | | | | | | | | | 15,6 | | | | | | | 94. |
| | | 95. Cheilanthopsis | 1 | | | | | | | | | | | | | | 15,6 | | | | | | | 95. |
| | | 96. Diacalpe | 1 | | | | | | | | | | | | | | 15,12 | | | | | | | 96. |
| | | 97. Dictyocline | 1 | | | | | | | | | | | | | | 15,12 | | | | | | | 97. |
| | | 98. Egenolfia | 10 | | | | | | | | | | | | | | 15,12 | | | | | | | 98. |
| | | 99. Heterogonium | 4 | | | | | | | | | | | | | | 15,12 | | | | | | | 99. |
| 10. | . . . .<br>. . 3 .<br>. . . . | 100. Lecanopteris | 9 | | | | | | | | | | | | | | 15,14 | | | | | | | 100. |
| | | 101. Lithostegia | 1 | | | | | | | | | | | | | | 15,12 | | | | | | | 101. |
| | | 102. Luerssenia | 1 | | | | | | | | | | | | | | 15,12 | | | | | | | 102. |
| | | 103. Neocheiropteris | 2 | | | | | | | | | | | | | | 15,14 | | | | | | | 103. |
| | | 104. Oreogrammitis | 1 | | | | | | | | | | | | | | 15,14 | | | | | | | 104. |
| | | 105. Parasorus | 1 | | | | | | | | | | | | | | 15,3 | | | | | | | 105. |
| | | 106. Peranema | 1 | | | | | | | | | | | | | | 15,12 | | | | | | | 106. |
| | | 107. Photinopteris | 1 | | | | | | | | | | | | | | 15,14 | | | | | | | 107. |
| | | 108. Psomiocarpa | 1 | | | | | | | | | | | | | | 15,12 | | | | | | | 108. |
| | | 109. Pycnoloma | 3 | | | | | | | | | | | | | | 15,14 | | | | | | | 109. |
| | | 110. Quercifilix | 1 | | | | | | | | | | | | | | 15,12 | | | | | | | 110. |
| | | 111. Tectaridium | 1 | | | | | | | | | | | | | | 15,12 | | | | | | | 111. |
| | | 112. Teratophyllum | 9 | | | | | | | | | | | | | | 15,12 | | | | | | | 112. |
| | | 113. Platyzoma | 1 | | | | | | 7 | | | | | | | | | | | | | | | 113. |
| | | 114. Stromatopteris | 1 | | | | | | 7 | | | | | | | | | | | | | | | 114. |
| | | 115. Aspleniopsis | 1 | | | | | | | | | | | | | | 15,6 | | | | | | | 115. |
| | | 116. Cionidium | 1 | | | | | | | | | | | | | | 15,12 | | | | | | | 116. |
| 11. | . . . .<br>. . . 4<br>. . . . | 117. Dendroconche | 1 | | | | | | | | | | | | | | 15,14 | | | | | | | 117. |
| | | 118. Diellia | 6 | | | | | | | | | | | | | | 15,2 ? | | | | | | | 118. |
| | | 119. Hemipteris | 1 | | | | | | | | | | | | | | 15,5 | | | | | | | 119. |
| | | 120. Neurosoria | 1 | | | | | | | | | | | | | | 15,6 | | | | | | | 120. |
| | | 121. Oenotrichia | 3 | | | | | | | | | | | | | | 15,1 | | | | | | | 121. |
| | | 122. Sadleria | 7 | | | | | | | | | | | | | | 15,9 | | | | | | | 122. |
| | | 123. Thysanosoria | 1 | | | | | | | | | | | | | | 15,12 | | | | | | | 123. |
| IIa | II. *Tropisch u.*<br>*aussertrop.*<br>a. P a n t r o p.<br>u. i. der n ö r d l<br>Zone | | | | | | | | | | | | | | | | | | | | | | | |
| | | 124. Antrophyum | 50 | | | | | | | | | | | | | | 15,7 | | | | | | | 124. |
| | | 125. Bolbitis | 80 | | | | | | | | | | | | | | 15,12 | | | | | | | 125. |
| | | 126. Coniogramme | 20 | | | | | | | | | | | | | | 15,6 | | | | | | | 126. |
| 12. | . . 3 .<br>1 2 3 4<br>. . . . | 127. Ctenitis | 150 | | | | | | | | | | | | | | 15,12 | | | | | | | 127. |
| | | 128. Loxogramme | 35 | | | | | | | | | | | | | | 15,14 · | | | | | | | 128. |
| | | 129. Microlepia | 45 | | | | | | | | | | | | | | 15,1 | | | | | | | 129. |
| | | 130. Nephrolepis | 30 | | | | | | | | | | | | | | 15,3 | | | | | | | 130. |
| | | 131. Stenoloma | 13 | | | | | | | | | | | | | | 15,2 | | | | | | | 131. |
| | | 132. Tectaria | 200 | | | | | | | | | | | | | | 15,12 | | | | | | | 132. |
| 12a. | . 2 3 .<br>1 2 3 4<br>. . . . | 133. Onychium | 6 | | | | | | | | | | | | | | 15,6 | | | | | | | 133. |
| IIb | b. P a n t r o p.<br>u. i. d. s ü d l.<br>Zone | | | | | | | | | | | | | | | | | | | | | | | |
| 13. | . . . .<br>1 2 3 4<br>1 2 . . | 134. Doryopteris | 35 | | | | | | | | | | | | | | 15,6 | | | | | | | 134. |

| Gr. | Verbreitung | Nr. | Gattung | Zahl | 1 | 2 | 3 | 4 | 5 | 6 | 7 | 8 | 9 | 10 | 11 | 12 | 13 | 14 | 15 | 16 | 17 | 18 | 19 | 20 | 21 |
|---|---|---|---|---|---|---|---|---|---|---|---|---|---|---|---|---|---|---|---|---|---|---|---|---|---|
| 14. | . . . . / 1 2 3 4 / 1 . . 4 | 135. | Dicksonia | 17 | | | | | | | | | | | | | | 13 | | | | | | | 135. |
| 15. | . . . . / 1 2 3 4 / . 2 . 4 | 136. | Marattia | 50 | | | | 3 | | | | | | | | | | | | | | | | | 136. |
| | | 137. | Hemitelia | 100 | | | | | | | | | | | | | | 14 | | | | | | | 137. |
| IIc | c. Pantrop. u. i. der nördl. u. südl. Zone | | | | | | | | | | | | | | | | | | | | | | | | |
| 16. | . . 3 . / 1 2 3 4 / . 2 . . | 138. | Oleandra | 35 | | | | | | | | | | | | | | | 15,4 | | | | | | 138. |
| 17. | . . 3 . / 1 2 3 4 / . . . 4 | 139. | Alsophila | 300 | | | | | | | | | | | | | | 14 | | | | | | | 139. |
| | | 140. | Ceratopteris | 2 | | | | | | | | | | | | | | | 15,6 | | | | | | 140. |
| | | 141. | Diplazium | 375 | | | | | | | | | | | | | | | 15,10 | | | | | | 141. |
| | | 142. | Lindsaya | 150 | | | | | | | | | | | | | | | 15,2 | | | | | | 142. |
| | | 143. | Platycerium | 8 | | | | | | | | | | | | | | | 15,14 | | | | | | 143. |
| 18. | 1 . . . / 1 2 3 4 / . 2 . . | 144. | Pityrogramma | 20 | | | | | | | | | | | | | | | 15,6 | | | | | | 144. |
| 19. | . . 3 . / 1 2 3 4 / . 2 . 4 | 145. | Dicranopteris | 5 | | | | | | | | 7 | | | | | | | | | | | | | 145. |
| | | 146. | Cyathea | 300 | | | | | | | | | | | | | | 14 | | | | | | | 146. |
| | | 147. | Acrostichum | 3 | | | | | | | | | | | | | | | 15,5 | | | | | | 147. |
| | | 148. | Cyclophorus | 100 | | | | | | | | | | | | | | | 15,14 | | | | | | 148. |
| | | 149. | Schizoloma | 25 | | | | | | | | | | | | | | | 15,2 | | | | | | 149. |
| | | 150. | Vittaria | 80 | | | | | | | | | | | | | | | 15,7 | | | | | | 150. |
| | | 151. | Psilotum | 2 | | | | | | | | | | | | | | | | | | | | 20 | 151. |
| 20. | 1 . . . / 1 2 3 4 / 1 2 . 4 | 152. | Schizaea | 30 | | | | | | 5 | | | | | | | | | | | | | | | 152. |
| | | 153. | Dennstaedtia | 70 | | | | | | | | | | | | | | | 15,1 | | | | | | 153. |
| 21. | . . 3 . / 1 2 3 4 / 1 2 . 4 | 154. | Sticherus | 100 | | | | | | | | 7 | | | | | | | | | | | | | 154. |
| | | 155. | Ctenopteris | 200 | | | | | | | | | | | | | | | 15,14 | | | | | | 155. |
| | | 156. | Grammitis | 100 | | | | | | | | | | | | | | | 15,14 | | | | | | 156. |
| | | 157. | Histiopteris | 3 | | | | | | | | | | | | | | | 15,5 | | | | | | 157. |
| | | 158. | Hypolepis | 40 | | | | | | | | | | | | | | | 15,1 | | | | | | 158. |
| 22. | . 2 3 . / 1 2 3 4 / 1 2 . . | 159. | Elaphoglossum | 400 | | | | | | | | | | | | | | | 15,15 | | | | | | 159. |
| 23. | 1 . 3 . / 1 2 3 4 / . . . 4 | 160. | Lygodium | 40 | | | | | | 5 | | | | | | | | | | | | | | | 160. |
| 24. | 1 . 3 . / 1 2 3 4 / 1 2 . 4 | 161. | Pellaea | 80 | | | | | | | | | | | | | | | 15,6 | | | | | | 161. |
| | | 162. | Azolla | 6 | | | | | | | | | | | | | | | | 16 | | | | | 162. |
| 25. | . 2 3 . / 1 2 3 4 / . 2 . . | 163. | Leptogramma | 10 | | | | | | | | | | | | | | | 15,12 | | | | | | 163. |
| 26. | . 2 3 . / 1 2 3 4 / . 2 . . | 164. | Dryopteris | 150 | | | | | | | | | | | | | | | 15,12 | | | | | | 164. |
| 27. | 1 2 3 . / 1 2 3 4 / . 2 . 4 | 165. | Marsilia | 65 | | | | | | | 6 | | | | | | | | | | | | | | 165. |
| | | 166. | Pteridium | 1 | | | | | | | | | | | | | | | 15,5 | | | | | | 166. |
| | | 167. | Botrychium | 30 | 1 | | | | | | | | | | | | | | | | | | | | 167. |
| | | 168. | Ophioglossum | 50 | 1 | | | | | | | | | | | | | | | | | | | | 168. |
| | | 169. | Hymenophyllum | 300 | | | | | | | | | | 9 | | | | | | | | | | | 169. |
| | | 170. | Trichomanes | 330 | | | | | | | | | | 9 | | | | | | | | | | | 170. |
| | | 171. | Adiantum | 200 | | | | | | | | | | | | | | | 15,6 | | | | | | 171. |
| | | 172. | Asplenium | 650 | | | | | | | | | | | | | | | 15,10 | | | | | | 172. |
| | | 173. | Athyrium | 180 | | | | | | | | | | | | | | | 15,10 | | | | | | 173. |

| | | No. | Genus | 1 | 2 | 3 | 4 | 5 | 6 | 7 | 8 | 9 | 10 | 11 | 12 | 13 | 14 | 15 | 16 | 17 | 18 | 19 | 20 | 21 | |
|---|---|---|---|---|---|---|---|---|---|---|---|---|---|---|---|---|---|---|---|---|---|---|---|---|---|
| **28.** | 1 2 3 .<br>1 2 3 4<br>1 2 . 4<br>(kosmopolitisch) | 174. | Blechnum | 180 | | | | | | | | | | | | | | 15,9 | | | | | | | 174. |
| | | 175. | Cheilanthes | 130 | | | | | | | | | | | | | | 15,6 | | | | | | | 175. |
| | | 176. | Cyclosorus | 200 | | | | | | | | | | | | | | 15,12 | | | | | | | 176. |
| | | 177. | Cystopteris | 18 | | | | | | | | | | | | | | 15,10 | | | | | | | 177. |
| | | 178. | Notholaena | 60 | | | | | | | | | | | | | | 15,6 | | | | | | | 178. |
| | | 179. | Polypodium | 50 | | | | | | | | | | | | | | 15,14 | | | | | | | 179. |
| | | 180. | Polystichum | 225 | | | | | | | | | | | | | | 15,12 | | | | | | | 180. |
| | | 181. | Pteris | 250 | | | | | | | | | | | | | | 15,5 | | | | | | | 181. |
| | | 182. | Thelypteris | 500 | | | | | | | | | | | | | | 15,12 | | | | | | | 182. |
| | | 183. | Equisetum | 24 | | | | | | | | | | | | | | | | 17 | | | | | 183. |
| | | 184. | Lycopodium | 180 | | | | | | | | | | | | | | | | | 18 | | | | 184. |
| | | 185. | Selaginella | 700 | | | | | | | | | | | | | | | | | | 19 | | | 185. |
| | | 186. | Isoetes | 65 | | | | | | | | | | | | | | | | | | | | 21 | 186. |

**IId**    *d. In den Tropen* e i n e  Lücke

| | | No. | Genus | 1 | 2 | 3 | 4 | 5 | 6 | 7 | 8 | 9 | 10 | 11 | 12 | 13 | 14 | 15 | 16 | 17 | 18 | 19 | 20 | 21 | |
|---|---|---|---|---|---|---|---|---|---|---|---|---|---|---|---|---|---|---|---|---|---|---|---|---|---|
| **29.** | 1 2 3 .<br>1 2 3 .<br>1 2 . . | 187. | Osmunda | 12 | | | 4 | | | | | | | | | | | | | | | | | | 187. |
| | | 188. | Woodsia | 38 | | | | | | | | | | | | | | 15,11 | | | | | | | 188. |
| **30.** | 1 . . .<br>1 2 3 .<br>1 2 . . | 189. | Aneimia | 90 | | | | 5 | | | | | | | | | | | | | | | | | 189. |
| **31.** | . 2 3 .<br>1 2 3 .<br>1 . . 4 | 190. | Salvinia | 10 | | | | | | | | | | | | | | | 16 | | | | | | 190. |
| **32.** | . 2 . .<br>1 2 3 .<br>1 2 . 4 | 191. | Anogramma | 7 | | | | | | | | | | | | | | 15,6 | | | | | | | 191. |
| **33.** | 1 2 3 .<br>1 . 3 .<br>1 . . . | 192. | Phyllitis | 8 | | | | | | | | | | | | | | 15,10 | | | | | | | 192. |
| **34.** | . 2 . .<br>1 . 3 4<br>. . . 4 | 193. | Culcita | 10 | | | | | | | | | | | | 13 | | | | | | | | | 193. |
| **35.** | . . 3 .<br>1 . 3 4<br>. . . . | 194. | Plagiogyria | 30 | | | | | | | | | | | 12 | | | | | | | | | | 194. |
| | | 195. | Cibotium | 12 | | | | | | | | | | | | 13 | | | | | | | | | 195. |
| | | 196. | Gymnopteris | 9 | | | | | | | | | | | | | | 15,6 | | | | | | | 196. |
| **36.** | 1 . 3 4<br>. . . 4 | 197. | Paesia | 12 | | | | | | | | | | | | | | 15,5 | | | | | | | 197. |
| **37.** | 1 2 3 .<br>1 2 . . | 198. | Pleopeltis | 90 | | | | | | | | | | | | | | 15,14 | | | | | | | 198. |
| **38.** | . 2 3 .<br>. 2 3 4<br>. 2 . 4 | 199. | Davallia | 35 | | | | | | | | | | | | | | 15,3 | | | | | | | 199. |
| **39.** | . . 3 .<br>. 2 3 4<br>. 2 . 4 | 200. | Phymatodes<br>(incl. Selliguea) | 100 | | | | | | | | | | | | | | 15,14 | | | | | | | 200. |
| **40.** | . . 3 .<br>. 2 3 4<br>. . . 4 | 201. | Angiopteris | 100 | 2 | | | | | | | | | | | | | | | | | | | | 201. |
| **41** | . . 3 .<br>. 2 3 4<br>. 2 . . | 202. | Lepisorus | 25 | | | | | | | | | | | | | | 15,14 | | | | | | | 202. |
| **42.** | . 2 3 4<br>1 . . 4 | 203. | Arthropteris | 15 | | | | | | | | | | | | | | 15,3 | | | | | | | 203. |
| **43.** | . 2 3 4<br>. 2 . 4 | 204. | Gleichenia | 6 | | | | | | 7 | | | | | | | | | | | | | | | 204. |

| Nr. | Verbreitung | Nr. | Name | Zahl | 5 | 6 | 14 | 15 | 20 | Nr. |
|---|---|---|---|---|---|---|---|---|---|---|
| **44.** | . . 3 .<br>. 2 3 4<br>. . . . | 205.<br>206.<br>207.<br>208. | Cyrtomium<br>Drynaria<br>Humata<br>Microsorium<br>(incl. Colysis) | 12<br>20<br>40<br>40 | | | | 15,12<br>15,14<br>15,3<br>15,14 | | 205.<br>206.<br>207.<br>208. |
| **45.** | . 2 3 4<br>. 2 . . | 209. | Stenochlaena | 4 | | | | 15,15 | | 209. |
| **IIe** | ε. In den Tropen zwei Lücken | | | | | | | | | |
| **46.** | i 2 . .<br>1 2 . . | 210. | Anapeltis | 12 | | | | 15,14 | | 210. |
| **47.** | i 2 . .<br>. 2 . . | 211. | Lonchitis | 8 | | | | 15,5 | | 211. |
| **48.** | 1 2 3 .<br>1 . 3 . | 212. | Woodwardia | 10 | | | | 15,9 | | 212. |
| **49.** | . 2 3 .<br>. 2 3 .<br>. . . . | 213. | Hypodematium | 3 | | | | 15,12 | | 213. |
| **50.** | . 2 . .<br>. 2 3 .<br>. 2 . . | 214.<br>215. | Actiniopteris<br>Ceterach | 1<br>6 | | | | 15,5<br>15,10 | | 214.<br>215. |
| **51.** | . . 3 .<br>. . 3 4<br>. . . . | 216.<br>217.<br>218. | Dipteris<br>Cheiropleuria<br>Leucostegia | 8<br>1<br>20 | | | | 15,13<br>15,14<br>15,3 | | 216.<br>217.<br>218. |
| **52.** | . . . .<br>. . 3 4<br>. . . 4 | 219.<br>220. | Doodia<br>Tmesipteris | 8<br>3 | | | | 15,9 | 20 | 219.<br>220. |
| **IIf** | f. In den Tropen drei Lücken | | | | | | | | | |
| **53.** | 1 . . .<br>1 . . .<br>1 2 . . | 221. | Marginaria | 25 | | | | 15,14 | | 221. |
| **54.** | 1 2 . .<br>1 . . .<br>.. . . 4 | 222. | Pilularia | 6 | | 6 | | | | 222. |
| **55.** | i . . .<br>1 2 . . | 223. | Gymnogramme | 60 | | | | 15,6 | | 223. |
| **56.** | 1 . . .<br>1 . . . .<br>. . . . | 224.<br>225. | Bommeria<br>Phanerophlebia | 4<br>9 | | | | 15,6<br>15,12 | | 224.<br>225. |
| **57.** | i . . .<br>1 . . . | 226. | Lophosoria | 1 | | | 14 | | | 226. |
| **58.** | . 2 . .<br>. 2 . . | 227. | Mohria | 3 | 5 | | | | | 227. |
| **59.** | 1 2 3 .<br>. . 3 .<br>. . . . | 228. | Matteuccia | 3 | | | | 15,8 | | 228. |
| **60.** | . . 3 .<br>. . 3 .<br>. . . . | 229.<br>230.<br>231.<br>232. | Brainea<br>Drymotaenium<br>Lemmaphyllum<br>Monachosorum | 1<br>2<br>5<br>1 | | | | 15,9<br>15,14<br>15,14<br>15,12 | | 229.<br>230.<br>231.<br>232. |

| | | No. | | | 1. | 2. | 3. | 4. | 5. | 6. | 7. | 8. | 9. | 10. | 11. | 12. | 13. | 14. | 15. | 16. | 17. | 18. | 19. | 20. | 21. | |
|---|---|---|---|---|---|---|---|---|---|---|---|---|---|---|---|---|---|---|---|---|---|---|---|---|---|---|
| **61.** | . . . 4<br>. . . 4 | 233. | Leptopteris | 6 | | 4 | | | | | | | | | | | | | | | | | | | | 233. |
| **III** | *III. Rein ausser-tropisch* | | | | | | | | | | | | | | | | | | | | | | | | | |
| **62.** | 1 2 3 .<br>. . . .<br>1 2 . . | 234. | Cryptogramma | 7 | | | | | | | | | | | | | | | 15,6 | | | | | | | 234. |
| **63.** | . 2 . .<br>. . . .<br>1 . . 4 | 235. | Pleurosorus | 3 | | | | | | | | | | | | | | | 15,10 | | | | | | | 235. |
| **64.** | 1 . 3 .<br>. . . . | 236. | Camptosorus | 2 | | | | | | | | | | | | | | | 15,9 | | | | | | | 236. |
| | . . . . | 237. | Onoclea | 1 | | | | | | | | | | | | | | | 15,8 | | | | | | | 237. |
| **65.** | . . 3 .<br>. . . .<br>. . . . | 238. | Monachosorella | 2 | | | | | | | | | | | | | | | 15,12 | | | | | | | 238. |
| | | 239. | Pleurosoriopsis | 1 | | | | | | | | | | | | | | | 15,10 | | | | | | | 239. |
| | | 240. | Saxiglossum | 1 | | | | | | | | | | | | | | | 15,14 | | | | | | | 240. |
| | | 241. | Sinopteris | 2 | | | | | | | | | | | | | | | 15,6 | | | | | | | 241. |
| **66.** | . . . .<br>. 2 . 4 | 242. | Todea | 1 | | 4 | | | | | | | | | | | | | | | | | | | | 242. |
| **67.** | . . . .<br>. . . .<br>1 . . . | 243. | Serphyllopsis | 1 | | | | | | | | | | | 11 | | | | | | | | | | | 243. |
| | | 244. | Thyrsopteris | 1 | | | | | | | | | | | | | 13 | | | | | | | | | 244. |
| **68.** | . . . .<br>. . . .<br>. . . 4 | 245. | Cardiomanes | 1 | | | | | | | | | | | 11 | | | | | | | | | | | 245. |
| | | 246. | Loxsoma | 1 | | | | | | | | | | 10 | | | | | | | | | | | | 246. |
| | | 247. | Leptolepia | 1 | | | | | | | | | | | | | | | 15,1 | | | | | | | 247. |
| | | 248. | Phylloglossum | 1 | | | | | | | | | | | | | | | | | | 18 | | | | 248. |

**§ 5. Die Florengebiete,** die CHRIST für die *Filicales* unterscheidet, lassen sich auf die Gesamtheit der Pteridophyten übertragen. Sie sollen deshalb in enger Anlehnung an ihn mit den nötig erscheinenden Erweiterungen gekennzeichnet werden.

1. Die Flora des kaltgemässigten nördlichen Waldgebietes beider Halbkugeln mit dem arktisch-alpinen Element setzt sich zusammen aus besonderen Arten, die alle im Süden Verwandte haben, und aus solchen des südlicheren temperierten und sogar tropischen Waldgebietes selbst. Über die Zahl der Farnarten, die das ganze zirkumpolare temperierte Florengebiet durchziehen, vergl. § 4, A, a (S. 455). In der Alten Welt ist dessen Farnflora ungemein gleichförmig, aber im feuchten Osten und besonders Westen gehäuft, während in Nordamerika die meisten Farne dem feuchten Waldland im Osten angehören, hinab bis zum Golf von Mexico und an die Grenze der trocknen Hochländer. Im pazifischen Westen dehnt sich dieses Florenreich nur über den höheren Norden aus, von Oregon aufwärts; die kalifornische Küste rechnet bereits zur mexikanischen Flora.

Die floristische Schilderung der Untergebiete, wie sie CHRIST so anschaulich gibt, muss hier unterbleiben. Erwähnt sei über CHRIST hinaus, dass die Gattung *Equisetum* und auch *Lycopodium* in dem nördlichen Waldgebiet eine verhältnismässig nicht geringe Rolle spielt und von den etwa 65 *Isoetes-*

Arten mehr als 20 dort auftreten, vorzüglich im neuweltlichen Anteil.

2. Die Mediterranflora mit dem atlantischen Westrand Europas und dem Kaukasus. Erstere, die noch in Syrien gut vertreten ist, trägt — entsprechend dem sommertrockenen Klima — deutlich xerophilen Charakter, und gerade auch die endemischen Mittelmeerfarne gehören der Gruppe der kleinen Felsen-Asplenien an. In den höheren Gebirgen erscheinen die Arten der nördlichen Waldregion. Die vom Klima unabhängigere Gattung *Isoetes* bildet auch in dieser Flora einen recht merklichen Einschlag. Der vom Golfstrom beherrschte atlantische Westrand Europas hat seine eigene Farnflora, in die sich allerdings Mittelmeertypen einmischen. Die atlantische leitet sich in letzter Linie von dem altafrikanischen Gebiete ab. Besonders hervorzuheben ist Irland, dessen Farnflora durch ihren südlichen Charakter in so hoher nördlicher Breite einzig auf der Erde dasteht. Der Kaukasus lässt an seinem West- und Ostfuss mediterrane und desertische Einflüsse erkennen; seine Wald- und Alpenregion aber ist noch ganz mitteleuropäisch, mit einigen Endemismen und einer fern-östlichen Spur. Unsere Waldfarne sind namentlich am feuchten Ostrande des Pontus mächtig vertreten.

3. Die warmtemperierte chinesisch-japanische Flora reicht vom Tsin ling shan bis zum Hochplateau von Tibet, von dem zahlreiche Täler und Schluchten in das innere China hinablaufen. Die Südgrenze dieses reichen Florengebietes schiebt sich weiter nach Süden vor, als CHRIST es angenommen hatte, bis Nordburma und Tonkin, wenn hier die Durchsetzung mit malayischer Flora auch schon stärkere Geltung bekommt. Wie für die Blütenpflanzen, so ist dieses Florenreich auch für die Farne „ein wahres Reich der Mitte" (9), ein Zentrum allerersten Ranges. Überhaupt gehen wohl nirgends auf der Erde die geographischen Eigentümlichkeiten der Blütenpflanzen und der Farne so genau parallel wie hier. Die hervorstechendste besteht darin, dass in beiden Gruppen der Reichtum von Arten erstaunlich, die punktförmige Zerstreuung und Vereinzelung der Arten und selbst der Individuen zum grossen Teil aber verblüffend ist. Gleichlaufend damit geht ein Auftreten vielgliederiger Formenreihen und Formenkreise, die sich im Laufe der seit langem ungestörten Entwicklung des Landes, unterstützt von seiner edaphischen und klimatischen Mannigfaltigkeit, herausgebildet haben. Hier kann sich also ein alter Endemismus (*Archangiopteris, Neocheiropteris, Dipteris* u.a.) mit einem reichen progressiven (*Polystichum, Polypodium* s. lat. *Cheilanthes, Dryopteris* u.a.) zusammenfinden. CHRIST schätzt, dass die Annahme von 800 Farnarten in China nicht zu hoch gegriffen sei. Von diesem Reichtum gehen nach allen Seiten Ausstrahlungen, reichlich nach Westen, dem Fusse des Himalaya entlang, auch nach Osten in das benachbarte Japan, weniger bis zu den Philippinen, Borneo und Java. Die erheblich ärmere Farnflora Japans hat ihren eigenen Zusammenhang mit

dem malayischen Archipel über die Liukiu-Inseln und Formosa; auf Sachalin und Korea, wohl auch auf Jeso, mischt sich die Farnflora des nördlichen Waldgebietes mit den japanischen Typen. Der Farnendemismus Japans ist wesentlich ein Neuendemismus, der sich an die chinesische Flora anlehnt. Die Halbinsel Korea hat vorwiegend eine japanische Farnflora, nur im Gebirge und im Norden ein nordasiatisches Gepräge.

4. Die indisch-malayische Farnflora umfasst das ganze Monsungebiet der altweltlichen Tropen bis nach Hawaii und den östlichen Gruppen der polynesischen Inselflur und ist trotz dieses gewaltigen Umfanges eine geschlossene Einheit. Ihrem Reichtum, der sich nach den Rändern zu allerdings abschwächt, ist nur die Fülle der tropisch-amerikanischen Farnflora zu vergleichen. Allein das Zentrum des Gebietes, die insulare Malaya samt der Halbinsel Malakka, weist heute sicher weit über 1500 Arten auf. Vorderindien entbehrt, sogar in seinen klimatisch bevorzugten Teilen, wo auch ein schwacher Endemismus waltet, vieler bezeichnender malayischer Leitfarne, während ein namhafter xerophiler Anteil meist afrikanischer oder aus Amerika über Afrika gewanderter Arten eindringt. Die trockene Nordhälfte Ceylons teilt noch den verarmten und afrikanisch-xerophytischen Charakter des Dekkan-Plateaus. In dem Hochwald der Insel wiederholt sich die Farnflora der feuchteren südlichen Gebirge Vorderindiens, aber mit starker Bereicherung an echt malayischen Formen.

Den Himalaya-Abhang, besonders Sikkim, Assam und die zum chinesischen Plateau aufsteigenden oberen Flussläufe von Obertonkin, Ober-Annam, Yünnan, soweit sie mit tropischem Regenwalde erfüllt sind, bezeichnet CHRIST als die Nordgrenze der indisch-malayischen Farnflora. Hier mischen sich in den eisernen malayischen Bestand merklich chinesische Elemente, und ein deutlicher Endemismus macht sich geltend, der im oberen Regenwalde von Yünnan durch sehr alte Typen gekennzeichnet ist (vergl. S 456). Nach Nordwesten zu, bei Simla, ist die Farnflora recht verarmt und natürlich xerophil betont.

Der südlichste Rand Chinas, also Honkong und die nahen Taimoschan-Berge des Festlandes, gehört vorwiegend zur indo-malayischen Flora. Formosa, dessen grössere nördliche Hälfte schon ausserhalb der Tropen liegt, hat in seinen waldigen Gebirgen eine reiche, in ziemlich gleichen Teilen aus malayischen und chinesischen Elementen gemischete Farnflora. Der Endemismus dieser Kontinentalinsel ist neu und ebenfalls jenen beiden Floren angelehnt. In Tonkin, Süd-Annam und Siam ist der malayische Einfluss herrschend.

Die ganze Fülle der malayischen Flora findet sich auf der Halbinsel Malakka, an deren Südspitze die archaistische Farnkolonie des Mt. Ophir hervorragende Endemen beherbergt; vieles davon ist Borneo gemeinsam, jener grössten der Sundainseln, von der allein weit über 800 Arten und

Varietäten von Farnen, z.T. ausgezeichnete Endemismen, bekannt sind, ferner wohl ein Dutzend *Lycopodium*-Arten und etwa 20 *Selaginella*, davon beinahe die Hälfte endemisch. In dieser fast aller xerophilen Elemente entbehrenden hygrothermen Flora bergen selbst die höheren Gebirgsstufen nur wenige boreale Formen. Das australe Element zeigt sich in Spuren. Sumatra schliesst sich auch in seiner Farnflora der Malayischen Halbinsel eng an, weist fast nur weit verbreitete Arten und wenige Endemen auf. Java ist in seinem immerfeuchten Westteil viel reicher, vor allem auch an Baumfarnen; sein Endemismus ist malayischer Neuendemismus. In dem trocknen Ostjava treten die Farne sehr zurück. Östliche Typen sind im ganzen auf Java noch selten.

Anders ist es schon auf Celebes. Der Grundstock der Farnflora ist auch hier eine rein malayische Auslese, doch bestehen hervorragende Beziehungen zu Polynesien und Melanesien, zu dem eine ganze Flur von Inseln hinüberleitet. Über die schmale Brücke von Celebes führt dieser Wanderzug weiter nach den Philippinen, wo sich der Artenreichtum der malayischen Farnflora mit mehr als 600 Arten wiederum zu einem Höhepunkt steigert. *Selaginella* ist mit etwa 30 Arten vertreten, von denen nur der siebente Teil über die Inselgruppe hinausgeht. Besonders ausgezeichnet ist die Farnflora der Philippinen durch die grosse Einheit des ganzen Archipels, in dem sich auch die leitenden, z.T. höchst originellen endemischen Arten über die ganze Inselgruppe hin erstrecken. Abgeleitete Formengruppen oder -kreise um eine prägnante Art sind, wie bei gewissen Blütenpflanzen, z.B. *Elatostema*, auf den Philippinen eine häufige Erscheinung. Der Austausch ihrer Flora mit Borneo ist gering, ebenso mit Formosa. Die Bergregion besitzt einen reicheren Anteil kontinentaler Arten aus der chinesisch-japanischen Flora. Australische Einstrahlungen sind wohl stärker als auf den Sundainseln.

Für Neu-Guinea, das auch heute noch längst nicht eingehend erforscht ist, gab BRAUSE bereits 1921 fast 1000 Farnarten an, von denen rund 600 — hauptsächlich Bergarten — als endemisch bezeichnet wurden. Eine der bemerkenswertesten ist die monotypische Gattung *Papuapteris*, eine hochalpine Pflanze, die noch die Jamesonien der Anden hinter sich lässt. 250 Arten kommen zugleich in der Malaya vor; überhaupt hat die Farnflora von Neu-Guinea ein entschieden malayisches Gesicht. Ferner bestehen starke Beziehungen zu den Philippinen, Polynesien, Indien und Ceylon, Neu-Caledonien und Australien, schwächere zu China-Japan und Formosa. Die Gattung *Selaginella* wies 1921 bereits 58 Arten (48 endemisch) auf, *Lycopodium* 22 (7 endemisch). An Neu-Guinea schliesst sich Nordost-Australien eng an.

Aus Neu-Caledonien strahlt eine Menge von Arten nach den Neu-Hebriden, Viti und Samoa aus; die Insel ist der Brennpunkt des polynesisch-melanesischen Bezirkes der malayischen Farnflora, mit einem „höchst kräftigen und dabei (wegen der tiefen und reichlichen Blattfiederung) bizarren Endemismus". Diese Variante der malayischen Farnflora ist durch eine Anzahl uralter xerophytisch veranlagter Typen charackterisiert. Bedeutend ist schon die Einstrahlung neuseeländisch-australischer Formen. *Schizaea fistulosa* könnte man fast zirkumantarktisch nennen. — Von den etwa 130 Farnarten der Neu-Hebriden sind die allermeisten neucaledonisch und ostmalayisch; daneben ist ein schwacher Neoendemismus zu erkennen. — Viti, der grösste und zugleich zentralste der polynesischen Archipele, hat — besonders auf der feuchten Passatseite — eine weniger originelle Farnflora als Neu-Caledonien, und viele ihrer eigentümlichen Arten teilt die Viti-Gruppe noch mit Samoa. Dem extremen äquatorialen Regenwaldklima Upolus entspricht die energische Stammentwicklung auch bei sonst stammlosen Gattungen. Trotz der sehr artenreichen Farnflora der Samoa-Inseln sind diese doch kein Florenzentrum, sondern ein Sammelpunkt allgemein in dieser Inselflur verbreiteter ostmalayischer Arten. Der nicht sehr bedeutende Endemismus ist neogen. Tahiti, durch ungeheure, nur höchst lückenhaft von Atollen durchsetzte Zwischenräume von der Malaya getrennt, weist dennoch eine durchaus malayische Farnflora, mit einigen eigenartigen Endemismen auf.

Wenn irgend einer Inselgruppe, so gebührt der von allen Festländern und grösseren Inseln weit abliegenden Hawaiischen die Bezeichnung ozeanisch. Da ausserdem die vulkanischen Gebirge 3000 m übersteigen, ist ihre Farnwelt ausserordentlich reich. Übermächtig ist der Endemismus alter Typen, die aber keine Monotypen sind, sondern sich durch einen neueren Entfaltungsprozess in Formengruppen auflösen, so dass fast 3/4 aller Gefässkryptogamen von Hawaii Endemen sind. Die dazwischen noch auftauchenden fremden Arten sind fast durchweg pantropisch oder doch tropenvag und scheinen — auch die amerikanischen — den Weg hierher über die alte Welt gefunden zu haben.

5. Die a u s t r a l i s c h - n e u s e e l ä n d i s c h e Flora. — Während die Nordostspitze Australiens noch voll der malayischen Flora angehört, und in den Trockengebieten seines Inneren und des Südwestens die Pteridophyten bedeutungslos sind, ist das temperierte Regenwaldgebiet Ostaustraliens, Tasmanien und Neu-Seeland mit Auckland- und Campbell-Insel zu einer Floreneinheit zusammenzufassen, in der überraschenderweise trotz der südlichen Lage nicht nur eine der üppigsten und massigsten, sondern auch systematisch interessantesten Farnfloren auftritt, ein „Farnmikrokosmus", wie CHRIST Neu-Seeland nennt. Die Ursache liegt in der überaus feuchten Atmosphäre und der für solche Breiten ungewöhnlichen Ausgeglichenheit des Klimas. Xerophyten treten hier natürlich ganz zurück.

Dieses Florenreich, in Sonderheit Neu-Seeland, weist reichlichen Altendemismus auf; nicht wenige seiner Formen streifen in die westliche Halbkugel oder doch bis Süd-Afrika hinüber: zwei Hinweise darauf, dass die australisch-neuseeländische Farnflora (wie die entsprechende Phanerogamenflora) das „Relikt einer grösseren Südflora ist, die zur Tertiärzeit oder früher sich ausbildete, ausstrahlte und jetzt noch die ozeanischen Punkte besetzt hält, die ihr klimatisch ihre Existenz sichern". (9).

6. Die tropisch-afrikanische Farnflora der Waldgebiete ist arm und ebenso eintönig-gleichartig, wie das Land orographisch und klimatisch homogen ist; „localisierte Seltenheiten scheint es im schwarzen Kontinent nicht zu geben" (9). Trotz eines spärlichen Altendemismus und ziemlich gut entwickelten, allerdings auf wenige Gattungen beschränkten Neoendemismus bietet das tropische Afrika nicht das Bild eines Zentrums, sondern einer Kolonie, die von der Malaya und Südamerika aus spät und spärlich besiedelt worden ist. Eine spezifische Einwirkung in umgekehrter Richtung ist kaum zu spüren. Selbst die Hochgebirge des tropischen Afrika ermangeln in ihrer nach oben hin an tropischen Elementen immer mehr verarmenden Farnflora der Fülle und Originalität, bis auf die bedeutende Beimischung aus der süd- und randafrikanischen xerophilen Gebirgsflora.

7. Die afrikanische Süd- und Randflora ist an Pteridophyten nicht minder reich und originell als an Blütenpflanzen. Und wie sich die südafrikanische Phanerogamenflora von der tropisch-afrikanischen aufs schärfste unterscheidet, so auch die Farnflora. „Nichts kontrastiert energischer mit den frondosen, breit angelegten Gestalten der Waldflora, als die xerotherm eingestellten Felsen- und Buschheidefarne, als die halb, ja völlig desertisch ausgestatteten kleineren, kompakten Arten mit polierten Spindeln und zottiger Bekleidung, Bewohner eines Klimas und einer Unterlage, die ebenso trocken sind wie das tropische Tiefland feucht, ebenso gebirgig wie das Kongobecken tiefländisch" (9). Und wie jene phanerogame Xerophytenflora Südafrikas zugleich die Randflora des Kontinentes darstellt, die überall da, wo das Randgebirge heute noch vorhanden ist, zutage tritt und erst im Mittelmeergebiet ausklingt, so auch die Farne der Süd- und Randflora, von denen einzelne noch längs der atlantischen Küste bis Island hinauf und weiter zu verfolgen sind.

Auch auf den Kranz der Inseln, die im Osten und Westen das afrikanische Festland umgürten, zeigen sich deutliche Reste der altafrikanischen xerophytischen Randflora. Im Osten erweist sich ihre Expansivkraft so stark, dass sie mit einzelnen bezeichnenden Arten bis nach Arabien, Vorderindien und sogar China und Java übergreift; auch im Norden erstreckt sie sich weit nach dem gemässigten, ja kühlen Europa. Der Gesamtcharakter dieser Inselfloren hat freilich seine eigene Ausprägung, meist als deutlicher Mischcharakter. Sehr genau ist uns heute durch die Bearbeitung C. CHRISTENSENs die Mascarenen-Region mit ihrem Kernstück Madagaskar bekannt. Von den rund 500 Pteridophytenarten dieser Insel sind 232 (46, 3 %) endemisch, davon etwa 40 ohne nähere Beziehungen zu Nachbarfloren. Für den ganzen Archipel (Madagaskar, Comoren, Mascarenen, Seychellen) erhöht sich die Zahl der Endemen noch um annähernd 60.

Demgegenüber finden sich etwa ein Dutzend kosmopolitische und etwa 20 pantropische Arten. Das afrikanische Element steuert etwas mehr als 100 Arten bei (davon beinahe 30 südafrikanisch); fast genau soviele das „östliche" Element (Afrika und dem trop. Asien, besonders der Malaya, viele auch der Polynesia gemeinsam); etwa 25 das „westliche" (afrikanisch-amerikanische) Element und etwa 10 das „südliche" Element, d.h. Arten der südlichen temperierten Zone. Die vier atlantischen Archipele im Nordwesten Afrikas (die Canaren, Madeira, die Azoren und Kapverden) beherbergen in ihren höheren Gebirgen mehrere mitteleuropäische und in den tieferen Lagen zahlreiche Mittelmeerfarne. Einige Arten gehören der afrikanischen Randflora an. Die Kapverden haben noch tropische Formen, bis zu echten Epiphyten. Die grosse Zahl endemischer Farne entspricht der hohen Selbständigkeit der atlantischen Flora überhaupt.

Die kleineren ozeanischen Inseln im Westen der Südhälfte von Afrika „haben von jeher durch ihre, wie von allen Winden zusammengewehte Flora die Aufmerksamkeit auf sich gezogen" (9). Die Farnflora von Ascension ist paläotropisch, überwiegend afrikanisch. St. Helena nennt CHRIST „ein Wunder von Endemismus" (darunter wenige Altendemen). Sonst ist die Farnflora der Insel deutlich altafrikanisch, mit sehr heterogenen ozeanischen Einflüssen, auch einigen amerikanischen Beimengungen. Das von BAKER behauptete Überwiegen des amerikanischen Elementes auf Tristan da Cunha lehnt CHRIST ab; nach ihm ist die Farnflora auch hier im ganzen afrikanisch mit sparsamem antarktischen und noch schwächerem trop.-amerikanischen Einfluss.

Die Gruppe St. Paul—Neu-Amsterdam, die zwar dem australischen Festlande näher als dem afrikanischen liegt, hat von den Mascarenen doch noch geringeren Abstand, woraus sich der Mangel an klaren Beziehungen zur australisch-neuseeländischen Farnflora und der (indirekte) Einfluss der südafrikanischen erklärt. Ausserdem sind antarktische, malayische und trop.-subtropische amerikanische Einschläge bemerkbar.

8. Die mexikanisch-kalifornische Farnflora umfasst das äusserst xerophytische Hochland von Mexiko mit seinen Gebirgen, ohne die tropischen Galleriewälder, und klingt nach Nordwesten durch die Chaparales von Neu-Mexiko und Arizona in Kalifornien aus. Hauptsächlich auf dem inneren mexikanischen Plateau entfaltet sich bis auf die dürrsten Standorte eine der eigenartigsten Floren der Welt, in der die Farne fast durchweg nur mit stark xerophilen, vielfach neuendemischen Arten vertreten sind. Zwergiger Wuchs, drahtartig starre Ausbildung, Beschuppung, Haarbekleidung oder Wachsbelag, lederig-dickes Blattgefüge treten als Verdunstungsschutz mit oder für einander auf; einzelne Arten leben hier noch epiphytisch. Diese mexikanische Flora herrscht auch in Arizona, Neu-Mexiko und auf der südkalifornischen Halbinsel; „aber auch Kalifornien bis etwa zum 43°. nördlicher Breite, bis an das Waldgebiet Oregons hinauf, ist dieser xerophilen Flora untertan" (9).

Die meisten Waldfarne Nordost-Amerikas fehlen hier. In entgegengesetzter Richtung aber dringen zahlreiche Xerophytenformen der mexikanisch-kalifornischen Farnflora bis ins Innere von Nordamerika vor.

9. Der tropisch-amerikanischen Farnflora ist ein Reichtum der Gestaltung eigen, der dem malayischen kaum nachsteht, auch an Stärke des Endemismus nicht. Wenn auch die Neotropen solcher Kraftgestalten wie der Drynarien und Platycerien fast oder ganz entbehren, so weist die Gattung *Pteris* in Südamerika Riesenformen auf, wie sie in der Malaya bei dieser Gattung nur selten zu finden sind. Die neotropische Farnwelt hat „einen Charakter der Frondosität, eine Tendenz zu möglichst grossen, breiten und dabei reichgeteilten Blattspreiten, welche von jeher die Reisenden begeistert hat" (9).

Am weitesten nördlich erstreckt sich die geschlossene tropisch-amerikanische Waldflora in Mexiko, dessen offenes Binnenland von zahlreichen Schluchten mit tropischem Galleriewald durchsetzt ist. In diesen ist eine Menge von Farnen vertreten, die allgemein durch das tropische Amerika bis Brasilien und zu den Antillen gehen. Keines der tropischen Gebiete Mexikos sondert sich farnfloristisch heraus; der atlantische und der pazifische Abhang des Hochlandes sind wenig verschieden. Etwa der vierte bis dritte Teil der Arten mögen endemisch sein. Im Süden geht diese Flora durch das waldige Chiapas nach Guatemala über, das prägnanter Endemen fast ganz entbehrt, und erlischt im wesentlichen in der tiefen Depression Nicaraguas.

Dagegen ist die Landenge von Costarica mit ihren mehr als 3000 m hohen Vulkanen durch die feuchten Luftströmungen zweier Meere hochbegünstigt. Das fördert so mächtigen Wuchs der Farne, wie er sonst in der Neuen Welt nicht wieder erreicht wird. Die Arten des Regenwaldes, die schon einen starken Einschlag aus Südamerika bekommen, sind bis in die Höhen hinauf massgebend; nur in einer mittleren Lage bei San José und Karthago treten Xerophyten auf. Zur andinen Flora steht eine erhebliche Zahl z.T. nivaler Formen der hohen Gebirgsgipfel in Beziehung. Einigermassen ist auch noch antillischer Einfluss spürbar. Doch ist Costarica auch.die Stätte eines „glänzenden, allgegenwärtigen" Endemismus zum Teil mit hervorragenden Archaeotypen. „Es dürfte kaum ein Land im tropischen Amerika gefunden werden, in welchem sich so viele Florenbestandteile von Grenzgebieten neben so viel endemischen Farnen zusammendrängen" (9).

Westindien, durch die kleinen Antillen und Trinidad mit dem Festlande in Verbindung, bietet einen wahren Inbegriff der tropisch-amerikanischen Farnflora dar. CHRIST spricht von der „relativen Vollständigkeit an tropisch-amerikanischen Arten, die Westindien besitzt". Alle grossen und viele kleinen Inseln dieses Archipels sind gebirgig; die Gebirgssippen sind neben manchen andinen Arten vielfach Endemen. Überhaupt ist der Farnendemismus Westindiens nicht auffallend gering, wie GRISEBACH meinte. Die Farnflora der ganzen Inselgruppe ist recht homogen, ein Zug, der stark an die Philippinen erinnert (vergl. S. 466). Das Zentrum der antillischen Farnflora ist nach neueren Nachweisungen URBANS und CHRISTENSENS nicht Jamaika, wie CHRIST angenommen hatte, sondern ohne Zweifel Hispaniola mit etwa 650 Arten. Die Farnflora von Jamaika und Südost-Kuba ist freilich nur wenig ärmer und auch sehr ähnlich, aber die Gebirge Hispaniolas haben einen höheren Prozentsatz andiner Arten. Weniger Besonderheiten bietet Portoriko.

Auf der Halbinsel Florida, die nicht in die Tropen hineinreicht, kommt doch eine reichliche Auswahl allgemein verbreiteter tropisch-amerikanischer Farne vor, die alle auch in Westindien vorhanden sind. Doch wachsen sie nicht auf der Ostseite der Halbinsel, die durch eine Inselflur mit Westindien verbunden ist, sondern in den Hammocks der Westküste, den zwischen Nadelwald eingestreuten Laubwaldkomplexen. Wenige Arten tropischer Herkunft gehen bis Alabama, Texas und in die mittleren Staaten der Union hinauf.

Die tropische Farnflora des ungegliederten südamerikanischen Kontinents ist — wie die des allein vergleichbaren Afrika — im ganzen ziemlich homogen, von Guayana bis Rio Grande do Sul, Paraguay und dem nördlichen Argentinien. Eine Oase mit reicher Eigengestaltung der Farnflore — wie seiner Flora überhaupt — stellt das orographisch und klimatisch ausgezeichnete Gebiet des Roraima- und Duida-Gebirges dar; es sei nur eine so eigentümliche Form wie *Hymenophyllopsis* erwähnt. Dem Farnwuchs wenig günstig ist die grosse Hyläa. Ihren periodischen Überschwemmungen sind einige Gruppen der Gattung *Trichomanes*, *Polypodium*-Arten u.a. durch Kletterwuchs angepasst. Selbst in dem bereits subandinen Bolivien ist die Trivialität nur wenig gemildert: kaum ein Sechstel des Artbestandes, noch dazu allgemein-neotropischen Gattungen angehörig, ist endemisch. Im immergrünen Walde der gebirgigen Küstenprovinzen Südbrasiliens kommt eine bedeutende Zahl charakteristischer Arten hinzu, zum Teil Ausstrahlungen der andinen Flora oder bereits Vertreter der südlichen gemässigten Fazies. In Paraguay erfährt die tropische Farnflora wegen der südlichen Lage und der geringen Erhebung des Landes eine „fast normal zu nennende Abschwächung" (9), die in Nordargentinien ganz allmählich zum Erlöschen führt.

10. Die südbrasilianische Camposflora zeigt nochmal eine wunderbare xerotherme Anpassung. In die Senkungen, besonders mit Waldbestand, ist eine immer noch stattliche Zahl von Regenwaldfarnen vorgedrungen und in die Hochgebirge ein gewisser andiner Bestandteil. Zur eigentlichen Camposflora gehört vor allem die extrem xerophile Gattung *Aneimia*, die hier eines ihrer Zentren hat, ebenso wie *Doryopteris*.

11. Die Andenflora. Während durchaus nicht alle Hochgebirge der Tropen eine eigene Farnflora hervorgebracht haben, hat die der

Anden, die nordwärts bis Costarica und Mexiko, seitlich bis Südbrasilien und Westindien vorgedrungen ist und im tiefen Süden in die antarktische Flora übergeht, eine besondere Prägung. Stellenweise geht zwar die amerikanische Waldflora bis in die subalpine Stufe hinein, und es fehlen nicht einige kleine nordamerikanische *Asplenium*-Arten, mit *A. viride* und *Trichomanes* verwandt. Die eigentlichen Andenfarne aber sind äusserst eigenartig, aufs schärfste angepasst an das hochalpine, ja nivale Klima. Eine der charakteristischsten Gattungen ist *Elaphoglossum* mit einfachen, zungenförmigen, in der Hochregion grundständigen, durch alle Mittel xerophil ausgestalteten, meist starr lederigen Blättern. Von ihrem andinen Hauptgebiet strahlt sie nicht nur über die Hochländer Amerikas (ausgenommen Kalifornien und die Vereinigten Staaten), sondern über Afrika bis tief nach Polynesien in identischen oder homologen Arten aus: „ein energisch sich ausdehnendes, weil trefflich ausgerüstetes Geschlecht" (9). *Jamesonia, Gymnogramme, Cheilanthes, Aneimia, Notholaena, Pellaea,* seien noch genannt. Dass auch die Farnflora eines so langen, durch alle Breiten der Südhalbkugel hingestreckten Hochgebirges nicht gleichartig ist, bedarf kaum einer Erwähnung.

12. Die s ü d c h i l e n i s c h e Farnflora entfaltet sich in dem sehr regenreichen Gebiete südlich des 37. Breitengrades zu ausserordentlicher Üppigkeit. Echte Baumfarne werden allerdings durch *Blechnum* mit meterhohen Stämmen ersetzt. Einige Arten sind hier endemisch, andere, denen sich *Gleichenia* hinzugesellen, gehen bis in den tiefen antarktischen Süden hinab. Durch einige *Polystichum-* und *Asplenium*-Arten hängt die südchilenische Flora mit der neuseeländisch-australischen zusammen. Nirgends in Südamerika ist die Gattung *Hymenophyllum* so vielfältig; von ihren Arten ist etwa die Hälfte endemisch. Trotz ihrer um mehrere Breitengrade nördlicheren Lage einer trockenen Festlandküste gegenüber schliessen sich die Juan Fernandez-Inseln dem feuchten valdivischen Florengebiet an. Doch kommen Baumfarne vor, hauptsächlich die altendemische verzweigte *Dicksonia Berteroana* und die uralte, auf Masafuera beschränkte *Thyrsopteris elegans*, die einzige Art ihrer Gattung, mit zuweilen mannesstarkem Stamm.

13. Die a n t a r k t i s c h e Farnflora lässt sich bei der weiten Zerstreuung der winzigen, von Pflanzen bewohnbaren Landreste des ehemaligen antarktischen Kontinents mit der des Hohen Nordens natürlich nicht vergleichen; kaum 10 Arten spricht CHRIST wirklichen antarktischen Charakter zu, vor allem *Blechnum tabulare* und *B. penna marina*; *Polystichum mohrioides*; *Polypodium australe*; *Schizaea australis* und *fistulosa,* die aber fast alle nach Norden die Grenze der Antarktis weit überschreiten.

Der M e n s c h hat durch Zurückdrängung des Waldes in weitem Masse e i n s c h r ä n k e n d auf die Verbreitung der Pteridophyten, besonders der Farne, gewirkt. A n t h r o p o c h o r e Farne gibt es nur sehr wenige und in sehr beschränktem Umfange: *Ceratopteris* ist ein Unkraut der Reisfelder; in Costarica findet sich auf den Ziegeldächern eine merkwürdige Pflanzengesellschaft zusammen, zu der auch einige *Polypodium*-Arten gehören (9, S. 333). In Europa hat das calciphile *Asplenium Ruta-muraria*, ursprünglich Felsenpflanze, in Mauerritzen eine weite künstliche Verbreitung gefunden. *Equisetum arvense* und *palustre* sind als Unkräuter fast Kosmopoliten geworden. (Vergl. S. 394).

# Chapter XV

# GEOGRAPHIE UND ZEITLICHE VERBREITUNG DER FOSSILEN PTERIDOPHYTEN

## von

## Max Hirmer (München)

Bei der Beurteilung der geographischen Verhältnisse der vorzeitlichen Pflanzenwelt ist es eine selbstverständliche Notwendigkeit, gleichzeitig die stratigraphische d.h. zeitliche Aufeinanderfolge ins Auge zu fassen. Nicht selten haben ja Pflanzen oder Pflanzengruppen zu gewissen Zeiten der Erdgeschichte eine mehr oder minder weite Verbreitung über die Erdoberfläche besessen, um in Laufe der Erdepochen allmählich auf ein mehr oder minder beschränktes Areal zurück gedrängt zu werden.

Was in Hinblick auf die geographischen Verhältnisse der fossilen Pflanzenwelt derzeit das brennendste Problem ist, ist der Umstand, dass wir bis ins Alt-Tertiär hinein verfolgen können, dass eine Pflanzenwelt von subtropischem bis warm-temperierten Charakter sich bis in hohe nördliche Breiten ausgedehnt hat, und dass weiterhin fossile Pflanzen auch aus Gebieten bekannt sind, die heutigentags der Antarktis angehören. Wenn auch an dieser Stelle darauf näher einzugehen unmöglich ist, so muss doch auch hier betont werden, dass eine grosse Menge Tatsachen in betreff der geographischen Verbreitung der fossilen Pflanzenwelt, die noch vor wenigen Jahren unverständlich erschienen, heute begreiflich werden, sobald man die Wegener'sche K o n t i n e n t a l - V e r s c h i e b u n g s - T h e o r i e ins Auge fasst.

Hier sei in diesem Zusammenhang nur beispielsweise hingewiesen auf die vollkommene Gleichförmigkeit der Pflanzenwelt von Beginn des Auftretens der ersten Landpflanzen im Obersilur bis zum Erlöschen der Psilophyten-Flora zu Ende des Mitteldevons und weiterhin auf die ebenso grosse Gleichförmigkeit der oberdevonischen und unterkarbonischen Floren, ganz gleichgültig, ob man die Florenräume Europas und Nordamerikas sowie Sibiriens und Ostasiens oder Südamerikas, Südafrikas und Australiens ins Auge fasst. Die enorme Gleichförmigkeit, sie sich abermals gegen Trias-Ende und zu Beginn der Jura-Zeit wieder in der Geschichte der vergangenen Pflanzenwelt bemerkbar macht, sei gleichfalls hier hervorgehoben. Desgleichen die bis in die Arktis reichende Ausdehung subtropischer bis warmtemperierter Pflanzen noch bis zum Ende des älteren Tertiärs.

Nicht zuletzt spricht ein gewichtiges Wort zugunsten der genannten Theorie die Tatsache der Gleichförmigkeit der Floren der heute südhemisphärischen Gebiete von Südamerika, Afrika und Australien mit der Flora Indiens vom Ende der Karbonzeit und durch das ganze Perm und das Mesozoikum hindurch.

Wenn bis jetzt die ehedem auf der Erde und vor allem zu gewissen Zeitepochen, vorhandene und — gemessen an den heutigen Verhältnissen — fast unverständliche

Gleichförmigkeit der Pflanzenwelt im Bereich weitester Erdgebiete hervorgehoben wurde, so muss doch andererseits auch darauf hingewiesen werden, dass sich z.B. schon im jüngeren Palaeozoikum, etwa von Beginn, vor allem aber von Ende des Oberkarbon ab zweifellos eine Anzahl Florenprovinzen herausmodelliert haben, die allerdings später, vor allem gegen Trias-Ende zufolge der schon oben erwähnten Gleichförmigkeit wieder sehr strak verwischt worden sind. Es muss auch weiterhin noch darauf hingewiesen werden, dass, wenn auch in der Oberkreide und im Alt-Tertiär — wie weiter oben schon bemerkt — eine subtropische bis warm gemässigte Pflanzenwelt sich weit nach Norden hin, ja selbst bis ins Gebiet der hohen Arktis hinein aufdecken lässt, dennoch m e r i d i o n a l  bereits eine Differenzierung vor sich gegangen zu sein scheint.

Soviel über  a l l g e m e i n e  Probleme der historischen Pflanzengeographie.

Für die im folgenden des näheren darzustellenden pflanzengeographischen Verhältnisse der fossilen  P t e r i d o p h y t e n  sei als besonders wichtig nochmals zusammengestellt:

I. Die erdweite Gleichmässigkeit der Pflanzenwelt vom Pteridophyten-Beginn (Obersilur) bis zum Ende des Unterkarbons.

II. Dass im Oberkarbon und vor allem kurz vor Beginn der Permzeit sich zweifellos eine Anzahl, in ihren Elementen deutlich voneinander verschiedene Florenräume heraus gebildet haben, nämlich

1. Der  e u r a m e r i s c h e  F l o r e n r a u m, umfassend Nordamerika und Europa.

2. Der  A n g a r a - F l o r e n r a u m,  beginnend im Westen mit den Gebieten der Dwina und Petschora über den Ural hinweg bis in die Gebiete des Amur und Ussuri, mit einer südlichen Grenze im Gebiet des Russischen Tarbagatai und südlich der grossen Kohlenbecken von Kusnetzk und Minussinsk.

3. Der  C a t h a y s i a - F l o r e n r a u m,  umfassend die Gebiete von Shansi und Korea in Ostasien, seine typische Pflanzenwelt im obersten Oberkarbon und im Unterperm entwickelnd. Diesem Florengebiet dürfte auch noch Sumatra zugehören, wovon eine Flora aus der Zeit des stephanischen Oberkarbons bekannt ist.

4. Der  G o n d w a n a - F l o r e n r a u m,  umfassend die Gebiete des mittleren und südlichen Südamerika nebst der Antarktis, ferner Süd- und Mittelafrika, sodann Indien südlich des Himalaya-Bogens und Australien nebst Tasmanien und Neuseeland.

III. Wenn die Florenelemente des Gondwana-Raumes zu Ende der Karbon- und während der Permzeit im wesentlichen auf die gerade oben genannten Gebiete beschränkt waren und sich zu dieser Zeit lediglich noch Beziehungen zur Flora des Angara-Raumes aufdecken lassen, so ist kein Zweifel, dass vom Beginn der Trias an ein Hereinströmen der Gondwana-Elemente in das euramerische Gebiet statt hatte, wie umgekehrt auch ein

Wandern von Pflanzen aus dem euramerischen Florenraum nach den Gondwana-Gebieten unverkennbar ist.

Was die Belegung der hier skizzierten Grundlinien der pflanzengeographischen Verhältnisse der ausgestorbenen Pflanzenwelt betrifft, so ist klar, dass die hier gemachten Angaben unmöglich allein an Hand der pflanzengeopgraphischen Verhältnisse der Pteridophyten herausgeschält werden können und in ihr volles Licht erst kämen, wenn neben der Darstellung der Verhältnisse bei den Pteridophyten auch die primitiveren Gymnospermen-Gruppen herangezogen werden könnten.

Die folgende Uebersicht beschränkt sich im allgemeinen auf eine systematische Aufzählung der Pteridophyten-Gattungen unter Angabe ihrer zeitlichen und geographischen Verbreitung [1]).

Es ist dabei selbstverständlich, dass die fossile Pflanzenwelt des euramerischen Pflanzenraums wesentlich besser bekannt ist als die der übrigen Erdteile. Um aber ein klares Bild davon zu geben, welche Pflanzen in den Angara-, Cathaysia-, und Gondwana-Florenräumen massgebend gewesen sind, sind einige diese Räume betreffende listenförmige Zusammenstellungen der systematischen Uebersicht angeschlossen.

Von der sehr grossen in Frage kommenden L i t e r a t u r sei in ernster Linie auf die glänzende Darstellung von A. C. SEWARD: „Plant Life through the ages.'' verwiesen. Was neuere Veröffentlichungen über n i c h t-euramerische Florenräume betrifft, seien genannt die Arbeiten von B. SAHNI für Indien, von A. B. WALKOM für Australien, von A. L. DU TOIT für Afrika, M. ZALESSKY für den Angara-Raum, von T. G. HALLE und S. KAWASAKI für das Cathaysia-Gebiet, von W. J. JONGMANS und W. GOTHAN für Sumatra. Zusammenfassende Darstellungen finden sich in einigen der Abschnitte „Paläobotanik'' (v. M. HIRMER) in „Fortschritte der Botanik''. Zusammenfassende Darstellungen über den euramerischen Florenraum sind für das Palaeozoikum in erster Linie W. GOTHAN zu danken. An neuesten Untersuchungen über das Mesozoikum der Arktis seien noch die Arbeiten von A. C. SEWARD und M. T. HARRIS aufgeführt.

---

[1]) Die hinter dem Gattungsnamen in Klammern beigefügte Zahl bezeichnet — wo möglich — die Zahl der bis jetzt beschriebenen Arten.

## PSILOPHYTALES

Erste Landpflanzen; älteste Pteridophytengruppe. Erdweit verbreitet sowohl über die nordhemisphärischen Gebiete: Nordamerika (südlichstes Ost-Kanada und westliche Vereinigte Staaten: Wyoming) u. Europa: Schottland, Nord- u. Mittel-Norwegen, Nordfrankreich, Belgien und Rheinlande (Hunsrück, Gegend von Siegburg bei Bonn und Elberfeld) und Böhmen als auch im Bereich der heute südhemisphärischen Gebiete des alten Gondwana-Florenraumes: Viktoria in Australien, Südafrika und Südamerika (Falklands-Inseln). Beginnend im Obersilur und erlöschend im oberen Mitteldevon. Nach allem was bis jetzt bekannt ist, waren die Psilophytales bis zum Beginn des Mitteldevon die einzigen Landpflanzen, sind aber bereits als Heide-artige Massenvegetation aufgetreten. Erhalten z.T. in ausgezeichneten strukturbietenden Resten (aus dem Old-Red-Sandstein von Aberdeen in Schotland) teils in inkrustierten ,,Abdrücken'' unter Erhaltung organischen Materiales.

Bekannt sind folgende:

*Rhynia* (3), *Hornea* (1), *Taeniocrada* (1), *Yarravia* (2), *Hedeia* (1), *Bucheria* (1), *Zosterophyllum* (3), *Hicklingia* (1), *Loganella* (1); *Sporogonites* (3) *Pseudosporochnus* (1) *Psilophyton* (ca. 4), *Arthrostigma* (1), *Dutoitia* (1), *Dawsonites* (1); *Asteroxylon* (2), *Thursophyton* u. *Hostimella* ferner als Übergangsform zu den Lycopodiales: *Haplostigma* (1) und als Übergangsformen zu den Filicales: *Pectinophyton* u. *Protopteridium* (vgl. S. 555)

## LYCOPODIALES

Eine im Florenbild des jüngeren Palaeozoikums, vor allem in Karbon, zufolge der vielen baumförmigen Arten tonangebende Pteridophyten-Gruppe. Von der Mitte des Perm ab allergrösstenteils nur noch durch krautige Formen vertreten und im Florenbild zurücktretend, wenn auch durch alle Zeiten als Gruppe im ganzen nachweisbar. Im Oberdevon und vor allem im Unterkarbon nicht nur in den Gebieten der Nordhemisphäre sondern auch im Bereich des Gondwana-Florenraumes hervorragend entwickelt.

### A. Lepidophyta:

*Lepidodendron* reich an Arten im euramerischen Karbon, einschliesslich dem der Arktis, ferner eine Anzahl von Arten in den anderen paläozoischen Florengebieten: Sumatra, Cathaysia-Raum (Shansi und Korea), Angara-Raum und Gondwana-Raum, hier teils in Formen nahestehend denen des euramerischen Raumes, teils in eigenen. — *Lepidophloios* vorwiegend karbonisch und euramerisch, wenn auch dem Gondwana-Gebiet nicht fehlend, so im Oberkarbon von Brasilien. — *Sigillaria*. Unterkarbon bis oberes Rotliegendes, hauptsächlich im euramerischen Florenraum und hier sehr artenreich, jedoch auch mit einigen Arten, insbesondere des Subsigillaria-Kreises, in den anderen Florenräumen. — *Bothrodendron* offenbar rein karbonisch und euramerisch; was unter diesem Namen aus dem Bereich der Gondwana-Gebiete beschrieben ist, dürfte bestenfalls damit verwandt, sicher nicht identisch sein. — *Porodendron*. Oberdevon bis Unterkarbon, Arktis, Europa und Angara-Raum (Kirgisen-Steppe). *Pleuromeia* (ca. 3) Untertrias von Europa, Ostasien und vielleicht auch von Südafrika.

### B. Palaeozoische Lycopodiales von mehr minder baumförmiger Gestalt [1]):

*Haplostigma* Oberdevon bis Unterkarbon des Gondwana-Raums. — *Archaeosigillaria*. Oberdevon u. Unterkarbon v. Nordamerika bzw. Europa; hieher wohl auch die „*Protolepidodendron*"-Arten des Oberdevons v. Neu-Südwales. — *Cyclostigma* Oberdevon u. Unterkarbon von Europa u. Nordamerika sowie von Australien. — *Heleniella* u. *Helenia* Oberdevon des Donetz, letztere auch Perm d. Ural. — *Eleuterophyllum* und *Lepidodendropsis* Unterkarbon, ersteres in Europa, letzteres ebenda und im östl. Nordamerika. — *Protasolanus* u. *Thaumasiodendron*, beide aus dem Unterkarbon von Mitteleuropa. — *Asolanus, Omphalophloios* u. *Ulodendron* aus dem Oberkarbon von Europa u. Nordamerika, letztere Gattung auch aus dem Unterkarbon v. Neu-Südwales angegeben. — *Pinacodendron* u. *Phialophloios* beide im Oberkarbon von Mitteleuropa, letzteres Monotyp des Saargebietes. — *Maroesia* Oberkarbon (Stefan) v. Sumatra. — Gattungen des Angara-Florenraumes sind: *Angarodendron, Caenodendron, Caragandites, Demetria, Eichwaldia, Lophiodendron, Micheevia, Paikhovia, Rimnocladon, Signacularia, Viatcheslavia*, die meisten permisch, einige unterkarbonisch oder vom Unterkarbon bis ins Perm reichend. — *Cyclodendron* Perm v. Südafrika. —

### C. Krautige Lycopodiales:

Unter diesen die ältesten Vertreter der Lycopodiales; sodann finden sie sich ferner im Karbon neben den in jener Zeit so zahlreichen baumförmigen Gestalten der Gruppe und stellen von der Trias ab bis zur Gegenwart wieder allein das Kontingent der Lycopodiales.

---

[1]) Nicht festgestellt, ob ligulat oder eligulat.

*Baragwanathia* Obersilur v. Australien (Victoria); *Drepanophycus* Unterdevon des Rheinlandes (Siegburg); *Protolepidodendron* (3) Mitteldevon v. Mitteleuropa; *Lycopodiopsis* Perm von Brasilien; *Leptophloëm* Oberdevon u. Unterkarbon von Nordamerika u. Australien; *Poecilitostachys* Buntsandstein v. Frankreich; *Lycostrobus* Rhät. v. Schonen. Dies alles Formen unsicherer Stellung innerhalb der Gruppe.

*Lycopodites* vom Oberdevon ab; *Selaginellites* vom Oberkarbon ab; *Nathorstia* und *Isoetites* von der Unterkreide ab bekannt.

### D. Samentragende Lycopodiales:
*Lepidocarpon* Unter- und Oberkarbon v. Mitteleuropa; *Miadesmia* Oberkarbon von England.

## ARTICULATALES

Eine im Gegensatz zu den neuzeitlichen Verhältnissen im Mesozoikum uns insbesondere im Palaozoikum reich und mannigfaltig entwickelte Gruppe. Wenn auch die Mehrzahl der Gattungen und Formen des Paläo- und Mesozoikums aus dem euramerischen Florengebiet bekannt ist, so spielen gewisse Articulatales auch in den heute grösstenteils der Südhemisphäre angehörenden Gondwana-Gebieten keine geringe Rolle, und zwar — im Gegensatz zu den Lycopodiales, die im Gondwanaraum reichlich nur während des Oberdevon u. Unterkarbon entwickelt waren — vorwiegend im jüngeren Palaeozoikum und noch durch das ganze Mesozoikum hindurch.

### A. Protoarticulatineae:
*Calamophyton* Mitteldevon v. Elberfeld, *Hyenia* ebenda u. Norwegen.

### B. Pseudoborniineae:
*Pseudobornia* Oberdevon d. Bären-Insel.

### C. Sphenophyllineae und Cheirostrobineae:
*Sphenophyllum* Bgt. formenreiche Gattung, nicht nur vom Oberdevon bis zum Rotliegenden im euramerischen Florenraum, sondern auch im obersten Karbon und Perm der Angara-, Cathaysia- und Gondwana-Florenräume. — *Cheirostrobus* Monotyp aus dem Unterkarbon von Schottland.

### D. Equisetineae:
**1. Asterocalamitaceae:** *Asterocalamites*. Unterkarbon der ganzen Nordhemisphäre wie auch des Gondwana-Florenraumes. — *Autophyllites'*

*Pothocitopsis* u. *Sphenasterophyllites* unbedeutende Gattungen des Karbons v. Mitteleuropa.

**2. Calamitaceae:** *Mesocalamites* u. *Protocalamites*, beides Übergangsformen zw. Asterocalamites u. Calamites, bisher nur aus dem Unterkarbon u. unterem Oberkarbon Europas bekannt. *Calamites*[1]) hervorragend wichtige Gattung des Karbons und Rotliegenden sowie mit Resten im Zeckstein, in der mannigfaltigsten Weise erhalten; vorwiegend euramerisch, selten im Cathayisa- u. Gondwana-Florenraum; im Angara-Raum vorwiegend durch *Paracalamites* vertreten.

**3. Equisetaceae:** *Equisetites* vom Karbon bis zur Gegenwart, überall; *Phyllotheca* artenreich; Charakterpflanzen des jüngeren Paläozoikums (insbesondere des Perms) der Angara- und Gondwana-Florenräume; im Gondwana-Gebiet sowie in Europa u. Asien auch im Mesozoikum.

**4. Anhang: Articulatales unsicherer Stellung:**
*Neocalamites* Jungpaläozoikum des Cathaysia- Angara- u. Gondwana-Florenraumes, im Mesozoikum noch im Gondwana-Gebiet sowie in Europa u. Nordamerika. — *Schizoneura* Im Jungpaläozoikum vorwiegend im Gondwana-, spärlicher im Cathaysia-Gebiet; im Mesozoikum gleichfalls im Gondwana-Gebiet sowie in Nordamerika, Europa u. Asien. — *Lobatannularia* Perm der Angara- und vor allem des Cathaysia-Florenraumes.

## CLADOXYLALES

Nur aus dem älteren Paläezoikum (Mitteldevon bis Unterkarbon) von Europa bekannte Pflanzengruppe mit den Gattungen *Cladoxylon* u. *Voelkelia*.

## FILICALES

**A: Eusporangiatae:**
**1. Aneurophytineae:** Die hieher gehörigen Gewächse, die zunächst unter die Samen-tragenden Pteridospermen eingereiht wurden, stellen einen Vorläufer-artigen, aber offenbar bereits heterosporen Farn-Typus von baumförmiger Gestalt dar. *Aneurophyton* (1) aus dem oberen Mittel-Devon von Elberfeld. — *Eospermatopteris* (1) aus dem Oberdevon von New York.

---

[1]) Bezüglich der Hilfsgattungen zu den Stämmen (*Athropitys, Calamodendron, Athrodendron*), zu den Blatt-tragenden Zweigen (*Asterophyllites, Annularia*, u.a.), zu den Blüten (*Calamostachys, Palaeostachya, Metacalamostachys, Cingularia* u.a.) sowie der Wurzeln (*Astromyelon* u.a.) vgl. Abschnitt A r t i c u l a t a l e s (S. 514–517).

**2. Coenopteridineae:** rein p a l a e o zoisch (Ober-Devon bis Unter-Perm); die überwiegende Zahl der — meist in strukturbietenden Resten bekannten — Gattungen und Arten ist in M i t t e l-E u r o p a gefunden; nur wenige stammen von N o r d-A m e r i k a, der A r k t i s, von S i-b i r i e n und von A u s t r a l i e n. Im Einzelnen gilt Folgendes:

**a. Zygopteroideae:**

**1) Stauropteridaceae:** *Stauropteris* (2) Unterkarbon v. Schottland, Mittl. Ober-Karbon v. Mittel-Europa.

**2) Etapteridaceae:** *Dineuron* (2) Unterkarbon v. Schottld. u. Mittel-Frankreich; — *Metaclepsydropsis* (2) Unterkarbon v. Schottld. u. Thüringen; — *Diplolabis* (1) Unterkarbon v. Schottld., Schlesien u. Mittel-Frankr.; — *Zygopteris* s. str. (einschl. *Botrychioxylon*) Mittl. Oberkarbon v. Nord-Engld. u. Rottlieg. v. Sachsen; — *Asteropteris* Ober-Devon v. New York; — *Etapteris* (6) Unterkarbon-Rotlieg. v. Mittel-Europa; — *Coryne-pteris* einschl. *Alloiopteris* (ca. 20) Mittl. Oberkarbon v. Mittel-Europa; — *Gymnoneuropteris* (1) Unterkarbon v. Kärnten; — *Cephalotheca* (3) Oberdevon d. Bären-Insel.

**3) Clepsydraceae:** *Clepsydropsis* (1) Unterkarbon v. Thüringen; — *Asterochlaenopsis* Unterkarbon d. Kirgisen-Steppe; — *Protoclepsydropsis* (1) Unterkarbon v. Schottld.; — *Austroclepsis* (1) Oberdevon bezw. Unterkarbon v. Australien; — *Ankyropteris* (7) Mittl. Oberkarbon bis Rotlieg. v. Mittel-Europa; — *Asterochlaena* (2) Rotlieg. v. Chemnitz; — *Menopteris* (1) ebendaher.

**b. Botryopteroideae:**

**1) Botryopteridaceae:** *Botryopteris* (5) Unterkarbon bis Rotliegend. v. Mittel-Europa; — *Grammatopteris* (2) Rotlieg. v. Mittel-Frankr. u. Sachsen.

**2) Tubicaulidaceae:** *Tubicaulis* (3) Mittl. Oberkarbon bis Rotlieg. v. Mittel-Europa.

**c. Anachoropteroideae:**

**Anachoropteridaceae:** *Anachoropteris* (1) Mittl. Oberkarb. v. Europa; — *Gyropteris* (1) Unterkarbon v. Niederschlesien; — *Zeilleria* (ca. 6) Unterkarbon bis Mittl. Oberkarb. v. Europa.

**3. Marattiineae:** Vom Karbon bis zur Jetzt-Zeit reichende Farn-Gruppe mit einem Höhepunkt der Entwickelung gegen Ende der Karbon-Zeit und im Rotliegenden und einem weiteren Höhepunkt während der oberen Trias und des Beginns der Jura-Zeit. Wenn auch während des Paläozoikums der Euramerische Florenraum und während des Mesozoikums die heute der Nordhemisphäre angehörigen Gebiete (Europa, das Nördlichere Asien sowie Nord-Amerika das Gross der Marattiineen hervorgebracht hat, so muss doch hervorgehoben werden, dass Marattiineën weder im Paläo- noch im Mesozoikum den Gondwana-Gebieten gefehlt haben. Im Einzelnen gilt Folgendes:

**a. Pecopteroideae:** bei den paläozoischen Formen sind sowohl eine grosse Menge (vielfach struktur-bietend erhaltener) Stammreste (*Psaronius, Megaphyton, Caulopteris, Stipidopteris*) als eine Menge von verschiedenen Gattungen (*Asterotheca, Scolecopteris, Diplasterotheca, Acitheca, Crossotheca* z. T., *Tetrameridium, Ptychocarpus, Danaeites* u.a.) in Form fruktifizierender Wedel-Reste bekannt. Nur wenige im Mesozoikum.

S t a m m - R e s t e : *Megaphyton* (ca. 10) Oberkarbon (Westfal) bis Rotliegendes v. Europa u. Nord-Amerika; entsprechende Strukturbiedende Reste (distiche *Psaronius*) (ca. 9) nur aus dem Rotliegenden v. Europa bekannt. — *Caulopteris* (nebst *Stipidopteris*) (ca. 25) Oberkarbon (Westfal) bis Rotliegendes v. Europa u. Nord-Amerika; entsprechende strukturbietende Reste (quirlige u. schraubige *Psaronius* (ca. 25) aus dem Oberkarbon (Westfal u. Stefan) von Europa sowie dem Permo-Karbon v. Brasilien; ferner Perm des Angaraflorengebietes und Untertrias v. Baden.

W e d e l-R e s t e : *Asterotheca* (ca. 25) vielleicht schon ab Unterkarbon, sicher ab Oberkarbon (Westfal) bis Lias; im Paläozoikum in Europa u. Nordamerika, sowie in Ostasien (Cathaysia-Flora) und Sumatra; ferner im nördl. Südafrika (Süd-Rhodesia). — *Scolecopteris* (ca. 4) Oberkarbon (Westfal) bis Rotlieg. v. Europa u. Nordamerika. — *Diplasterotheca* Rotlieg. v. Mittel-Frankr. — *Acitheca* (1—3) Oberkarbon (Stefan) bis Rotlieg. Europa u. Sumatra. — *Ptychocarpus* (ca. 5) Oberkarbon (ob. Westfal) bis Rotliegendes von Nordamerika, Europa, Ostasien, sowie von Sumatra u. Süd-Rhodesia — *Crossotheca* (z. Teil) Oberkarbon-Rotlieg. von Europa u. Nordamerika — *Tetrameridium* (1) Oberkarbon (Westfal) v. Oberschlesien. — *Danaeites* (1) Oberkarbon (Westfal) d. Saargebietes. — *Parapecopteris* (1) Oberkarbon (Stefan) v. Mittel- u. Süd-Frankreich. — *Bernouillia* (3) u. *Merianopteris* (1) Keuper v. Mittel-Europa. —

**b. Danaeoideae:** nur an Hand der (wenig durchgefiederten) Wedel und erst seit dem Mesozoikum bekannt.

*Marattiopsis* (ca. 7) Rhät bis Unterkreide von Europa u. Ostasien sowie Grönland. — *Danaeopsis* (3) Obertrias von Europa u. Nordamerika.

**Anhang:** die unter *Weichselia* gehenden Wedelreste sind erdweit vom Wealden bis zur Mittel-Kreide verbreitet; zugehörige Stammreste: *Paradoxopteris*. Hieher auch die Gattungen: *Nathorstia* weitverbreitet im Mesozoikum u. *Rienitsia* Trias v. Australien.

**4. Ophioglossineae:** kaum fossil bekannt; mit Sicherheit jedenfalls erst von E o z ä n an: *Ophioglossites eozaenicum*, Eozän v. Verona, dem rez. *O. lusitanicum* nahestehend. (Der gelegentlich immer wieder in der

Literatur wiederkehrende Versuch die Angehörigen der karbonischen (über wiegend unterkarb.) Gattung *Rhacopteris* in Beziehung mit den Ophioglossineen zu bringen, ist ganz abwegig).

### Filicales Leptosporangiatae:
#### 1. Osmundaceae:

In Wedel- und Sporangien-Resten bereits seit dem Karbon bekannt, in Stammresten seit dem Perm nachweisbar. Permische Stammreste, besonders von dem primitiveren protestelischen Bau-Typ in vergleichsweise besonders grosser Gattungs- und Arten-Fülle aus dem Angara-Florenraum bekannt. Wedelreste im mittleren Mesozoikum (Oberste Trias und Alt-Jura) erdweit verbreitet.

*a.* S t a m m r e s t e : *Anomorrhoea* (1), *Bathypteris* (1), *Chasmatopteris* (1), *Jegosigopteris* (1), *Petscheropteris* (1), *Thamnopteris* (4) und *Zalesskya* (4) sämtliche aus dem Perm des Angara-Florenraumes, die meisten vom Ural, einige aus dem Petschora-bezw. dem Kusnezk-Becken. Die siphonostelischen bezw. diktyostelischen *Osmundites* (ca. 7) seit dem Jura bekannt, aus Europa u. Nordamerika sowie den Gondwana-Gebieten (Neuseeland und Paraguay).

*b.* W e d e l r e s t e : S p o r a n g i e n allein (*Todeopsis* u. *Sturiella*) schon aus dem Unterkarbon von Zentr.-Frankr. bekannt. W e d e l r e s t e selbst erst seit dem Oberkarbon nachgewiesen: *Discopteris* (5) Oberkarbon (Westfal u. Stefan) v. Europa; — *Kidstonia* (1) Oberkarb. v. Kleinasien. Mesozoisch: *Todites* nebst den entsprechenden sterilen Resten· *Cladophebis* in sehr zahlreichen Arten erdweit vom Keuper aufwärts. — *Speirocarpus* (4) Mittl. Keuper v. Nordamerika u. Europa. *Osmundites* vom Lias aufwärts u. weit verbreitet, im Alttertiär auch noch im Gebiet der heutigen Arktis.

#### 2. Schizaeaceae:

In Vorläufern schon seit der Mitte des Oberkarbon, in ganz typischen Formen seit dem Rhät nachgewiesen.

*Senftenbergia* (3) Oberkarbon (Westfal) v. Europa; — *Klukia* (5) Rhät bis Wealden v. Europa bis Ostasien; — *Norimbergia* (1) Lias v. Franken; — *Schizaeopsis* (1) Unterkreide v. Virginien und *Schizaeopteris* Oberkreide v. Japan; — *Lygodium* Unterkreide bis Miozaen v. Europa, Nordamerika u. Ostasien. Möglicherweise zugehörig auch die im Wealden erdweit, auch über die Gondwanagebiete verbreitete *Ruffordia* sowie *Pelletiera*. Zugehörige stammreste vielleicht: *Tempskya* (5) aus der Unterkreide v. Nordamerika u. Europa.

#### 3. Marsiliaceae:

Fossil n i c h t mit Sicherheit nachgewiesen.

#### 4. Gleicheniaceae:

Der rezenten Gattung zweifellos nahestehende Formen sind bereits aus dem Mittl. Keuper und dem Lias bekannt; e c h t e Gleichenien (*Gleiche-nites* s. str.) kennt man seit der Unterkreide und zwar von der Sekt. *Eu-Gleichenia* aus d. Unterkreide von Mitteleuropa sowie von Grönland einerseits u. Patagonien andererseits; Sekt. *Mertensia* (4) ebenfalls aus der Unterkreide von Mitteleuropa u. Nordamerika sowie der Arktis (Spitzbergen u. Grönland) und v. Argentinien. Die nur foss. Sekt. *Didymosorus* (6) von der Unterkreide bis zum Eozän, in Mitteleuropa u. Grönland.

Zum Verwandschafts-Kreis der Gleicheniaceen gehören auch noch neben der mesozoischen *Mertensites* (3) des Mittl. Keuper v. Virginia u. Lunz eine Anzahl paläozoischer Gattungen; so: *Oligocarpia* (6) Oberkarbon (Westfal.) v. Mitteleuropa; *Monocarpia* Oberkarbon (Stefan) v. Sumatra; *Shansitheca* Stefan bezw. Perm v. Ostasien (Shansi) u. Süd-Rhodesia; ferner *Boweria, Sturia* u. *Dendraena*, sämtl. drei aus dem Oberkarbon v. Mitteleuropa.

### 5. Matoniaceae:

Im Mesozoikum erdweit verbreitet und zu dieser Zeit auch die grösste Formen-Mannigfaltigkeit erreichend. Vom oberen Jura ab Areal-Reduktion. *Phlebopteris* (4) Mittl. Keuper bis Unterkreide, Grönland u. östl. U. S.A., Europa, Nord-Afrika, Ost-Asien und Australien; — *Selenocarpus* (1) nur Lias v. Franken; — *Matonidium* (2) Ober-Jura bis Unterkreide v. Europa u. Indien; — *Matoniella* (1) Unt. Oberkreide v. Mähren.

### 6. Hymenophyllaceae:

Vorkommen in der Vergangenheit noch etwas problematisch, was bei der extrem angepassten Familie auch nicht Wunder nehmen kann. Die hieher gerechneten fossilen Formen können ebensogut dem Dicksoniaceen-Kreis angegliedert werden, bezw. dazu Vorläufer darstellen.

*Hymenophyllites* Oberkarbon (Westfal) v. Europa und wieder im Rhät-Lias von Polen sowie von Argentinien.

### 7. Loxsomaceae:

Vielleicht gehört hieher *Stachypteris* aus dem Mittl. Jura v. England.

### 8. Dicksoniaceae und Cyatheaceae:

Sowohl in Stamm- als in fruktifizierenden Wedelresten seit der Trias bekannt, von den Gebieten der Nordhemisphäre und denen des Gondwana-Florenraumes.

S t a m m r e s t e: *Protopteris* (8) Wealden bis Tertiär v. Grönland u-Europa; — *Rhizodendron* und *Oncopteris* Kreide v. Mittel-Europa.

W e d e l r e s t e: *Coniopteris* Trias u. Jura, erdweit verbreitet, Arktis, Europa bis Ostasien und Gondwana-Gebiete; ebenso die jurassische *Eboracia*; *Gonatosorus* u. *Alsophilites*, beide aus dem Jura v. Polen.

**9. Dipteridaceae:** eine ähnlich wie die *Matoniaceae* im Mesozoikum weitest verbreitete und wesentlich formenreichere Familie als dies die heute form- und arealbeschränkten Nachzügler ahnen lassen. Im Bereich des Gondwana-Florenraumes ist die Familie mit einer Anzahl von Arten dreier Gattungen vertreten.

**a. Camptopteroideae:** *Camptopteris* (2) Mittl. Keuper bis Rhät v. Mittel-Europa; – *Dictyophyllum* (ca. 16) artenreichste Gattung, Obertrias bis Unterkreide, Europa bis Ostasien, Australien, Neuseeld. u. Südamerika (Chile); – *Thaumatopteris* (2) Rhät u. Lias v. Mittel-Europa; – *Clathropteris* Obertrias bis Unterkreide v. Nordamerika, Europa, Nordafrika, Asien bis Ostasien, und Südamerika (Chile); — *Oishia* (1) Rhät-Lias von Ostasien.

**b. Dipteroideae:** *Hausmannia* u. *Protorhipis* sehr artenreich; in Vorläufern vielleicht schon vom Rotliegenden ab; sicher vom Rhät ab, im ganzen nordhemisphärischen Gebiet und in Australien; – *Podoloma* Eozän v. Südengland, der rez. Dipteris coniugata nahestehend.

**c. Goeppertelloideae:** *Goeppertella*, Lias von Europa u. Asien (Tonkin).

### 10. Polypodiaceae:

Diese heute sowohl in Hinblick auf Vielgestaltigkeit wie auf Gattungs- und Artenfülle grösst-entwickelte Farn-Familie scheint erst vergleichsweise s p ä t in Erscheinung getreten zu sein. Die ältesten sind *Davallia* aus dem Lias v. Polen; weiterhin *Asplenium* aus der Unterkreide v. Grönland; ferner die Pterideen *Adiantum* (ebendaher und auch aus dem Jura v. Ostasien bekannt) sowie *Onychiopsis* (der rez. *Cryptogramma* nahestehend) aus dem Wealden von Europa und Ostasien, von Südafrika u. Südamerika; – *Polypodium* wird gleichfalls schon aus dem Jura v. Nordamerika (Oregon) angegeben. Aus dem Tertiär kennt man ferner noch ausser den schon von älteren Zeiten her bekannten oben genannten noch die Gattungen *Onoclea*, *Aspidium*, *Nephrodium*, *Scolopendrium*, *Cheilanthes* u. *Drynaria*.

### 11. Salviniaceae:

Fossil nur in der Gattung *Salvinia* bekannt, in einer Anzahl Arten seit der Ober-Kreide.

### Filices incertae sedis:

Die hierher gehörigen Formen, die wohl (mindestens zum grössten Teil) den eusporangieten Filices angehören oder doch zwischen diesen und den leptosporangiaten Filices vermitteln, sind zu nicht geringem Teil wichtige Leitpflanzen, sowohl in stratigraphischer als z. T. auch pflanzengeografischer Hinsicht. Genannt seien:

*Archaeopteris* (11) Oberdevon. Oestl. Nordamerika, Bären-Insel und Europa bis zum Donetzbecken. – *Rhacopteris*: Untergr. *Anisopteris* (ca. 14) Unterkarbon von Europa und des Gondwana-Florenraumes (hier mit *Lepidodendron* tonabgebend), nur mit wenigen Arten ins Oberkarbon hereinreichend; *Eurhacopteris* (1) Oberkarbon v. Mitteleuropa. – *Hymenotheca* (4)

Oberkarbon. *Renaultia* (11), *Dactylotheca* (8), *Myriotheca* (2), *Cyclotheca* (1), *Sphyropteris* (6), *Urnatopteris* (2) u.a. sämtliche dem Oberkarbon des euramerischen Florenraumes angehörend, z. T. (*Dactylotheca*) bis ins Rotliegende reichend. – *Anomopteris* (1) und *Neuropteridium* (5) beide aus dem Buntsandstein v. Europa. Inwieweit *Neuropteridium validum* und ähnliche der Gondwanagebiete wirklich zu *Neuropteridium* im Sinn der europäischen Arten gehören, steht dahin. – *Chiropteris* (3), Obertrias bis Kreide von Europa u. Amerika sowie der Gondwana-Gebiete.

| Palaeozoische Pteridophyten im Bereich des Angara-Florenraumes | Unter-Karbon | | | Jüngeres Palaeozoikum (Perm) | | | | | | | | |
|---|---|---|---|---|---|---|---|---|---|---|---|---|
| | Ural | Kirgisen Steppe | Kusnetzk-Becken | Petschora-Becken | Ural | Russ. Tarbagatai | Kusnetzk-Becken | Minussinsk-Becken | Untere Tunguska | Angara-Fluss | Buku-Muren Steppe | Amur Geb. u. Ussuri |
| | 1 | 2 | 3 | 1 | 2 | 3 | 4 | 5 | 6 | 7 | 8 | 9 |
| **Lycopodiales** (Anordnung alphabetisch) | | | | | | | | | | | | |
| *Angarodendron Obrutschewi* | | | | | | | | + | | + | | |
| *Caenodendron primaevum* | | + | | | | | | | | | | |
| *Caragandites rugosus* | | + | | | | | | | | | | |
| *Demetria asiatica* | | | + | | | | | | | | | |
| *Eichwaldia biarmica* | | | | | + | | | | | | | |
| *Helenia inopinata* | | | | | + | | | | | | | |
| *Knorria anceps* | | | | | + | | | | | | | |
| „    *mamillaris* | | | | | + | | | | | | | |
| *Lepidodendron kirgisicum* | | + | | | | | | | | | | |
| „    *ostrogianum* | | | + | | | | | | | | | |
| „    *Schmalhauseni* | | | | | | | + | | | | | |
| „    *Veltheimianum* | | + | | | | | | | | | | |
| „    *spec.* | | | | | | | | + | | | | |
| *Lepidostrobus (? Bothrostrobus) spec.* | | + | | | | | | | | | | |
| *Lophiodendron superum* | | | | | | | | + | | | | |
| „    *tyrganense* | | | + | | | | | | | | | |
| *Micheevia pulchella* | + | | | | | | | | | | | |
| „    *rimnensis* | + | | | | | | | | | | | |
| „    *uralica* | + | | | | | | | | | | | |
| *Paikhovia Tschernowi* | | | | | + | | | | | | | |
| *Porodendron tenerrimum* | | + | | | | | | | | | | |
| *Rimnocladon minutum* | + | | | | | | | | | | | |
| *Signacularia Noinskii* | | | | | + | | | | | | | |
| *Tomistachys thyrsiculus* | | | | | | | | + | | | | |
| *Viatcheslavia vorcutensis* | | | | | + | | | | | | | |
| **Articulatales.** | | | | | | | | | | | | |
| *Sphenophyllum denticulatum* | | | | | | | + | | | | | |
| „    *Stuckenbergi* | | | | | + | | | | | | | |
| *Asterocalamites scrobiculatus* | | + | + | | | | | | | | | |
| *Calamites gigas* | | | | | + | | | | | | | |

| Palaeozoische Pteridophyten im Bereich des Angara-Florenraumes | Unter-Karbon | | | Jungeres Palaeozoikum (Perm) | | | | | | | | |
|---|---|---|---|---|---|---|---|---|---|---|---|---|
| | Ural | Kirgisen Steppe | Kusnetzk-Becken | Petschora-Becken | Ural | Russ. Tarbagatai | Kusnetzk-Becken | Minussinsk-Becken | Untere Tunguska | Angara-Fluss | Buku-Muren Steppe | Amur Geb. u. Ussuri |
| | 1 | 2 | 3 | 1 | 2 | 3 | 4 | 5 | 6 | 7 | 8 | 9 |
| *Paracalamites decoratus* | | | | cf. | + | | | | | | | |
| „ *izylensis* | | | | | | | + | | | | | |
| „ *Kutagorae* | | | | cf. | + | | | | | | | |
| „ *laticostatus* | | | | + | | | | | | | | |
| „ *robustus* | | | | | | | + | | | | | |
| „ *sibiricus* | | | | | | | + | | | | | |
| „ *similis* | | | | | + | | | | | | | |
| „ *striatus* | | | | | + | | + | | | | | |
| *Athropitys linearis* | | | | | + | | | | | | | |
| *Annularia asteriscus* | | | | | | | + | | | | | |
| cf. *Annularia spec.* | | | | | | | + | | | | | |
| *Lobatannularia Schtschurowskii* | | | | | | | + | | + | | | |
| *Phyllotheca bardensis* | | | | | + | | | | | | | |
| „ *deliquescens* | | | | cf. | | | + | + | + | | | |
| „ *equisetoides* | | | | | | | | | + | | | |
| „ *macrostachya* | | | | | + | | | | | | | |
| „ *pauciflora* | | | | | | | | | + | | | |
| „ *peremensis* | | | | | + | | | | | | | |
| „ *Sokolowskii* | | | | | | | + | | | | | |
| „ *stellifera* | | | | | | | + | | + | | | |
| „ spec. div. | | | | | | | | | + | + | + | |
| *Equisetites Czekanowskii* | | | | | | | | | + | | | |
| Anhang: | | | | | | | | | | | | |
| *Taibia tyrganensis* | | | | | | | + | | | | | |
| | | | | | | | | | | | | |
| **Filicales S. Str.** | | | | | | | | | | | | |
| *Asterochlaenopsis kirgisica* | | + | | | | | | | | | | |
| *Psaronius sibiricus* | | | | | | | + | | | | | |
| *Tschernowia synensis* | | | | + | | | | | | | | |
| *Anomorrhoea Fischeri* | | | | | + | | | | | | | |
| *Bathypteris rhomboidalis* | | | | | + | | | | | | | |
| *Chasmatopteris principalis* | | | | | + | | | | | | | |
| *Jegosigopteris Jaworskii* | | | | | | | + | | | | | |
| *Petscheropteris splendida* | | | | + | | | | | | | | |
| *Thamnopteris Gwynne-Vaughani* | | | | + | | | | | | | | |
| „ *kazanensis* | | | | | + | | | | | | | |
| „ *Kidstoni* | | | | | + | | | | | | | |
| „ *Schlechtendahli* | | | | | + | | | | | | | |
| *Zalesskya diploxylon* | | | | | + | | | | | | | |
| „ *fistulosa* | | | | | + | | | | | | | |
| „ *gracilis* | | | | | + | | | | | | | |
| „ *uralica* | | | | | + | | | | | | | |
| *Cladophlebis adnata* | | | | + | | | + | | | | | |
| „ *tychtensis* | | | | | | | + | | | | | |
| „Pecopteris” *synica* | | | | + | | | + | | | | | + |

## Palaeozoische Pteridophyten im Bereich des Cathaysia-Florenraumes in Ost-Asien und auf Sumatra

| Sumatra | Kt (Korea) | J | K | Yu u (Shansi) | Yu o | SU u | SU o | SO u | SO o | Taxon |
|---|---|---|---|---|---|---|---|---|---|---|
| | | | | | | | | | | **Lycopodiales:** |
| — | — | × | — | — | — | — | — | — | — | Lepidodendron aculeatum |
| — | — | — | — | 1 | — | — | — | — | — | „ Gaudryi |
| × | — | — | — | — | — | — | — | — | — | „ mesostigma |
| × | — | — | — | — | — | — | — | — | — | „ molle |
| — | × | × | — | — | — | — | — | — | 1 | „ oculis felis |
| × | — | — | — | — | — | — | — | — | — | „ Posthumi |
| — | — | × | — | — | — | — | — | 2 | ? | „ spec. |
| — | — | — | — | — | — | — | — | — | 1 | Lepidostrobus spec. |
| — | — | × | — | — | — | — | — | — | — | Lepidostrobophyllum latisquamum |
| — | — | × | — | — | — | — | — | — | — | „ longitriangulare |
| — | — | × | — | — | — | — | — | — | — | „ spec. |
| — | — | × | — | — | — | — | — | — | — | Sigillaria acutangula |
| — | — | × | — | — | — | — | — | — | — | „ cf. semipulvinata |
| — | — | × | — | — | — | — | — | — | — | „ yajidoensis |
| — | — | × | — | — | — | — | — | — | — | „ spec. A |
| — | — | × | — | — | — | — | — | — | — | „ spec. B |
| — | — | × | — | — | — | 1 | — | — | — | ? Sigillariostrobus spec. |
| × | — | — | — | — | — | — | — | — | — | Maroesia rhomboidea |
| × | — | — | — | — | — | — | — | — | — | Stigmaria asiatica |
| × | — | × | — | 1 | 1 | 1 | — | — | — | „ ficoides |
| — | — | × | — | — | — | — | — | — | — | „ rugulosa |
| × | — | — | — | — | — | — | — | — | — | ? Lycopodites spec. |
| | | | | | | | | | | **Articulatales:** |
| — | — | — | — | — | — | 2 | — | — | — | Sphenophyllum costae |
| cf. | — | × | — | — | — | 2 | — | — | — | „ emarginatum |
| — | — | — | — | — | — | — | 1 | — | — | „ fimbriatum |
| — | — | × | — | — | — | — | — | — | — | „ macrophyllum |
| — | — | × | — | — | — | — | — | — | — | „ macrotruncatum |
| × | — | × | — | 2 | — | 3 | — | — | — | „ oblongifolium |
| — | — | × | — | — | — | — | — | — | — | „ orientale |
| — | — | × | — | — | — | — | — | — | — | „ pseudocostae |
| — | — | × | — | — | — | 2 | — | — | — | „ rotundatum |
| — | — | — | × | — | — | — | — | 2 | 2 | „ sino-coreanum |
| — | — | — | × | — | — | — | — | — | — | „ speciosum |
| × | — | × | — | — | — | 2 | 2 | — | — | „ Thonii |
| — | — | × | — | — | — | ? | 1 | 2 | 1 | „ „ , var. minor |
| × | — | × | — | ? | — | — | — | — | 2 | „ verticillatum |
| — | — | — | × | — | — | — | — | — | — | „ spec. |
| — | — | — | — | — | — | 1 | 1 | — | — | Bowmanites laxus |
| × | — | × | — | — | — | 1 | — | — | 3 | „ spec. |
| — | — | × | — | — | — | — | — | — | — | Calamites cf. Cisti |
| × | — | × | — | — | ? | 1 | — | — | — | „ Suckowi |
| × | — | — | — | — | — | — | — | — | — | „ jubatus |
| — | — | — | — | — | — | — | — | — | 1 | Asterophyllites longifolius |
| × | — | — | — | — | — | — | — | — | — | „ (Calamocladus) spec. |
| — | — | — | — | — | — | — | — | 2 | 1 | Annularia crassiuscula |
| — | — | — | — | — | — | — | 3 | ? | — | „ gracilescens |
| — | — | — | × | — | — | — | — | — | — | „ maxima |
| — | — | × | × | — | — | ? | 2 | — | — | „ mucronata |
| — | — | × | — | — | — | — | — | — | — | „ orientalis |
| — | — | — | × | — | — | — | — | — | — | „ papilioniformis |
| — | — | — | — | 1 | — | — | — | — | — | „ pseudostellata |
| — | — | — | × | — | — | — | — | — | — | „ Shirakii |
| × | — | × | — | — | 3 | ? | — | — | — | „ stellata |
| × | — | — | — | — | — | — | — | — | — | „ spec. |

| Sumatra | Korea | | | Shansi | | | | | | Palaeozoische Pteridophyten im Bereich des Cathaysia-Florenraumes in Ost-Asien und auf Sumatra |
| | Kt | J | K | Yu | | SU | | SO | | |
| | | | | u | o | u | o | u | o | |
|---|---|---|---|---|---|---|---|---|---|---|
| — | — | — | ? | — | — | — | — | 3 | 3 | Lobatannularia ensifolia |
| — | — | — | × | — | — | — | — | — | 1 | „ heianensis |
| — | — | × | — | — | — | — | — | — | — | „ inaequifolia |
| — | — | — | × | — | — | — | — | 1 | — | „ lingulata |
| — | — | × | — | — | — | 3 | 1 | — | — | „ sinensis |
| — | — | × | — | — | — | — | 2 | — | — | Macrostachya huttoniaeformis |
| × | — | — | — | — | — | — | — | — | — | Palaeostachya incrassata |
| × | — | — | — | — | — | — | — | — | — | Calamiten-Blüten incert. gen. |
| — | — | — | × | — | — | — | — | — | — | Neocalamites Meriani |
| — | — | — | × | — | — | — | — | — | — | Schizoneura striata |
| — | — | — | × | — | — | — | — | — | — | ? Schizoneura polyfolia |
| — | — | — | × | — | — | — | — | — | — | Phyllotheca cf. australis |

**Filicales:**

| Sumatra | Kt | J | K | Yu-u | Yu-o | SU-u | SU-o | SO-u | SO-o | |
|---|---|---|---|---|---|---|---|---|---|---|
| × | — | — | — | — | — | — | — | — | — | Asterotheca arborescens |
| × | — | × | — | — | — | — | — | — | — | „ Candolleana |
| × | — | — | — | — | — | — | — | — | — | „ cf. Cisti |
| × | — | — | — | — | — | — | — | — | — | „ cf. Daubrei |
| × | — | × | — | — | — | — | — | 2 | — | „ hemitelioides |
| — | — | — | — | — | — | — | — | — | 2 | „ Norinii |
| — | — | — | × | — | — | — | — | 3 | 3 | „ orientalis |
| × | — | × | — | — | — | — | — | — | — | „ oriopteridia |
| × | — | — | — | — | — | — | — | — | — | „ spec. A |
| × | — | — | — | — | — | — | — | — | — | „ spec. B |
| × | — | × | — | — | — | — | — | — | — | Acitheca polymorpha |
| × | — | — | — | — | — | — | — | — | — | Scolecopteris Verbecki |
| × | — | — | — | — | — | — | — | — | — | Ptycocarpus unitus |
| × | — | — | — | — | — | — | — | — | — | Monocarpia Posthumi |
| cf. | — | × | — | — | — | 2 | 3 | — | — | Oligocarpia Gothani |
| — | — | — | — | — | — | — | 1 | — | ? | Shansitheca Kidstoni |
| — | — | × | — | — | — | 2 | — | — | — | „ palaeosilvana |
| — | — | — | × | — | — | — | — | — | — | Discopteris kogendoensis |
| — | — | × | — | — | — | 3 | — | — | — | Cladophlebis Nystroemi |
| — | — | — | × | — | — | — | — | — | — | Shirakia bilobifolia |

**Abkürzungen in der Floren-Liste:**
Kt: K o t e n-Schichten von Korea
J : J i d o-Schichten von Korea
K : K o b o s a n-Schichten von Korea
Yu: Y u e h e m e n k o u-Schichten von Shansi (u: untere, o: obere.)
SU: U n t e r e S h i h h o t s e-Schichten von Shansi (u: untere, o: obere)
SO: O b e r e S h i h h o t s e-Schichten von Shansi (u: untere, o: obere)
  Die Zahlen in den Spalten bezeichnen den Häufigkeitsgrad (3 = maximal).
  Das Vorkommen von Djambi auf S u m a t r a gehört dem obersten Oberkarbon (Stefan) an.
E s  e n t s p r e c h e n :

| Gebiet von Shansi (China): | Gebiet von Korea: | dem geologischen Alter nach: |
|---|---|---|
| Obere S h i h h o t s e-Schichten | K o b o s a n-Schichten | wohl O b e r e s U n t e r-P e r m, keinesfalls jünger als Mittel-Perm. |
| Untere S c h i h h o t s e-Schichten | J i d o-Schichten | U n t e r e s U n t e r-P e r m. |
| Y u e h e m e n k o u -Schichten | K o t e n-Schichten | entweder noch u n t e r. s t e s U n t e r-P e r m- oder S t e f a n i s c h e s O b e r k a r b o n. |

| Pteridophyten der paläozoischen Schichten des gesamten Gondwana-Florenraumes | Ober-Silur bis Mittel-Devon | | | | Ober-Devon bis Unter-Karbon | | | | Ober-Karbon bis Perm | | | |
|---|---|---|---|---|---|---|---|---|---|---|---|---|
| | Austr. | Indien | S. Afr. | S. Amer. | Austr. | Indien | S. Afr. | S. Amer. | Austr. | Indien | S. Afr. | S. Amer. |
| **Psilophytales:** | | | | | | | | | | | | |
| *Thursophyton Milleri* | + | | | | | | | | | | | |
| *Hostimella spec.* | + | | | | | | | | | | | |
| *Sporogonites Chapmanni* | + | | | | | | | | | | | |
| „    *Chapmanni* fo. *minor* | + | | | | | | | | | | | |
| *Hedeia corymbosa* | + | | | | | | | | | | | |
| *Yarravia oblonga* | + | | | | | | | | | | | |
| „    *subsphaerica* | + | | | | | | | | | | | |
| *Zosterophyllum australicum* | + | | | | | | | | | | | |
| *Dutoitia pulchra* | | | + | | | | | | | | | |
| Psilophytales von *Rhynia*-Typus | | | | + | | | | | | | | |
| **Lycopodiales:** | | | | | | | | | | | | |
| *Baragwanathia longifolia* | + | | | | | | | | | | | |
| *Haplostigma irregulare* | | | + | cf. | | | | | | | | |
| *Cyclodendron Leslii* | | | | | | | | | | | + | |
| „    *Mathieui* | | | | | | | | | | | + | |
| *Cyclostigma australe* | | | | | + | | | | | | | |
| „    *spec.* | | | | | + | | | | | | | |
| „*Protolepidodendron" lineare* | | | | | + | | | | | | | |
| „    *yalwalense* | | | | | + | | | | | | | |
| *Leptophloem australe* | | | | | + | | | | | | | |
| *Lepidodendron australe* | | | | | + | | + | | | | | |
| „    *Clarkei* | | | | | + | | | | | | | |
| „    cf. *dichotomum* | | | | | + | | | | | | | |
| „    cf. *nothum* (= ? *L. australe*) | | | | | | | | + | | | | |
| „    *obovatum* | | | | | | | | (+) | | | | (+) |
| „    *Osbornei* | | | | | + | | | | | | | |
| „    *pedroanum* | | | | | | | | + | | | + | + |
| „    *rimosum* | | | | | | | | (+) | | | | (+) |
| „    cf. *scutatum* | | | | | + | | | — | | | | |
| „    *Veltheimianum* | | | | | + | | cf. | | | | | |
| „    *vereenigingense* | | | | | | | | | | | + | |
| „    *Volkmannianum* | | | | | + | | cf. | | | | | |
| *Lepidostrobus spec.* | | | | | | | | (+) | | | | (+) |
| *Lepidophyllum spec.* | | | | | | | | (+) | | | | (+) |
| *Knorria spec.* | | | | | + | | | (+) | | | | (+) |
| *Lepidophloios laricinus* | | | | | | | | | | | | (+) |
| *Sigillaria australis* | | | | | | | | | | | | + |
| „    *Brardi* | | | | | | | | | | | + | + |
| „    *muralis* | | | | | | | | | | | | + |
| „    *spec.* | | | + | | | | | | | + | + | + |
| „*Bothrodendron" spec.* | | | | | | | | | | + | | |
| *Ulodendron minus* | | | | | + | | | — | | | | |

| Pteridophyten der paläozoischen Schichten des gesamten Gondwana-Florenraumes | Ober-Silur bis Mittel-Devon | | | | Ober-Devon bis Unter-Karbon | | | | Ober-Karbon bis Perm | | | |
|---|---|---|---|---|---|---|---|---|---|---|---|---|
| | Austr. | Indien | S. Afr. | S. Amer. | Austr. | Indien | S. Afr. | S. Amer. | Austr. | Indien | S. Afr. | S. Amer. |
| *Ulodendron spec.* | | | | | + | — | — | — | | | | |
| *Lycopodiopsis Derbyi* | | | | | | | | | | | | + |
| *Stigmaria ficoides* | | | | | + | — | — | — | | | + | + |
| „ *spec.* | | | | | + | — | — | (+) | | | | (+) |
| Megasporen von *Lepidophyten* | | | | | | | | | | | | + |
| **Articulatales:** | | | | | | | | | | | | |
| *Sphenophyllum oblongifolium* | | | | | | | | | | | + | + |
| „ *Thonii* | | | | | | | | | | | + | — |
| „ *Thonii var. minor* | | | | | | | | | | | + | — |
| „ *speciosum* | | | | | | | | | + | + | + | — |
| „ *spec.* | | | | | ? | | | | | | + | — |
| *Asterocalamites scrobiculatus* | | | | | + | — | — | + | | | | |
| *Calamites Suckowi* | | | | | — | — | — | (+) | | | | (+) |
| „ *varians* | | | | | + | — | — | — | + | | | |
| cf. *Calamites* spec. | | | | + | | | | | | | | |
| *Calamostachys spec.* | | | | | — | — | — | (+) | | | | (+) |
| *Annularia australis* | | | | | | | | | | | | + |
| *Neocalamites Carreri* | | | | | | | | | | | | + |
| *Equisetites calamitinoides* | | | | | | | | | | | | + |
| „ (? *Schizoneura*) *morenianus* | | | | | | | | | | | | + |
| *Phyllotheca australis* | | | | | | | | | + | — | + | + |
| „ *deliquescens* | | | | | | | | | + | + | — | + |
| „ *Etheridgei* | | | | | | | | | + | — | — | — |
| „ *Griesbachii* | | | | | | | | | — | + | — | — |
| „ *indica* | | | | | | | | | — | + | — | — |
| „ *robusta* | | | | | | | | | + | + | — | — |
| „ *uluguruana* | | | | | | | | | — | — | + | — |
| „ *Whaitsi* | | | | | | | | | — | — | + | — |
| „ *Zeilleri* | | | | | | | | | — | — | + | — |
| „ *spec.* | | | | | | | | | + | — | + | — |
| *Schizoneura africana* | | | | | | | | | — | — | + | — |
| „ *gondwanensis* | | | | | | | | | + | + | + | ? |
| „ *Wardi* | | | | | | | | | — | + | — | — |
| „ *spec.* | | | | | | | | | + | — | + | + |
| *Actinopteris bengalensis* | | | | | | | | | — | + | — | — |
| **Filicales:** | | | | | | | | | | | | |
| 1. **Coenopteridaceae.** | | | | | | | | | | | | |
| *Austroclepsis australis* | | | | | + | — | — | — | | | | |
| *Botrychioxylon spec.* | | | | | + | — | — | — | | | | |
| 2. **Marattiaceae.** | | | | | | | | | | | | |
| *Psaronius brasiliensis* | | | | | | | | | — | — | — | + |
| *Tietea singularis* | | | | | | | | | — | — | — | + |

| Pteridophyten der palaeozoischen Schichten des gesamten Gondwana-Florenraumes | Ober-Silur bis Mittel-Devon | | | | Ober-Devon bis Unter-Karbon | | | | Ober-Karbon bis Perm | | | |
|---|---|---|---|---|---|---|---|---|---|---|---|---|
| | Austr. | Indien | S, Afr. | A. Amer. | Austr. | Indien | S. Afr. | S. Amer. | Austr. | Indien | S. Afr. | S. Amer. |
| *Asterotheca arborescens* | | | | | | | | | — | + | — | — |
|     „      *spec. A* | | | | | | | | | — | + | — | — |
|     „      *spec. B* | | | | | | | | | — | + | — | — |
| *Ptychocarpus unitus* | | | | | | | | | — | + | — | — |
|     „      *unitus var. emarginata* | | | | | | | | | — | + | — | — |
| 3. Osmundaceae. | | | | | | | | | | | | |
| *Cladophlebis ? australis* | | | | | | | | | + | — | — | — |
|     „      *Roylei* | | | | | | | | | + | + | — | — |
|     „      *spec.* | | | | | | | | | — | + | — | — |
| 4. cf. Gleicheniaceae | | | | | | | | | | | | |
| *Shansitheca* cf. *Kidstoni* | | | | | | | | | — | + | — | — |
| 5. Incertae sedis. | | | | | | | | | | | | |
| *Archaeopteris Howittii* | | | | | + | — | — | — | | | | |
| *Rhacopteris inaequilatera* | | | | | — | + | — | — | | | | |
|     „      *intermedia* | | | | | + | — | — | — | | | | |
|     „      cf. *Machanecki* | | | | | + | — | — | + | | | | |
|     „      *ovata* | | | | | + | — | — | — | | | | |
|     „      *? Roemeri* | | | | | + | — | — | — | | | | |
|     „      *septentrionalis* | | | | | + | — | — | — | | | | |
|     „      *Szajnochai* | | | | | — | — | — | + | | | | |
|     „      *Weissiana* | | | | | — | — | — | + | | | | |
|     „      *Wilkinsoni* | | | | | + | — | — | — | | | | |
|     „      *spec.* | | | | | + | — | — | — | | | | |
| *Neuropteridium plantianum* | | | | | | | | | — | — | — | + |
|     „      *validum* | | | | | | | | | — | + | + | + |
|     „      *tasmanense* | | | | | | | | | + | — | — | — |

(+) bedeutet, dass nicht feststeht, ob das betreffende Südamerikanischer Vorkommen dem Unter- oder Oberkarbon bzw. Perm zuzurechnen ist.

Die Vorkommen-Angaben (der Erdteile) sind ausgefüllt nur in denjenigen formationsgruppen, in welchen die betreffenden Pflanzen überhaupt vorkommen können.

| Pteridophyten der mesozoischen Schichten des gesamten Gondwana-Florenraumes | Unter-Trias | | | | Ober-Trias bis Unter-Kreide | | | | Ober-Kreide | | | |
|---|---|---|---|---|---|---|---|---|---|---|---|---|
| | Austr. | Indien | S. Afr. | S. Amer. | Austr. | Indien | S. Afr. | S. Amer. | Austr. | Indien | S. Afr. | S. Amer. |
| **Lycopodiales:** | | | | | | | | | | | | |
| cf. *Pleuromeia spec.* | — | — | + | — | | | | | | | | |
| *Lycopodites gracilis* | | | | | | + | | | | | | |
| „ *victoriae* | | | | | + | | | | | | | |
| „ spec. A | | | | | | | | | + | | | |
| „ spec. B | | | | | | | | | + | | | |
| **Articulatales:** | | | | | | | | | | | | |
| *Neocalamites Carreri* | — | — | + | — | + | — | + | — | | | | |
| „ *hoerensis* | | | | | + | | | | | | | |
| *Equisetites approximatus* | | | | | | | | + | | | | |
| „ *Nicoli* | | | | | + | | | | | | | |
| „ cf. *platyodon* | | | | | | | + | | | | | |
| „ *rajmahalensis* | cf | + | — | | | | | | | | | |
| „ *rotiferum* | | | | | + | | | | | | | |
| „ *wonthaggiensis* | | | | | + | | | | | | | |
| „ spec. div. | + | — | + | | + | — | + | | + | — | — | — |
| *Phyllotheca australis* | + | | | | + | | | | | | | |
| „ *concinna* | + | | | | | | | | | | | |
| „ *Hookeri* | + | | | | | | | | | | | |
| „ *minuta* | | | | | + | | | | | | | |
| „ *spec.* | + | | | | | | | | | | | |
| *Schizoneura* cf. *africana* | | | | | + | | | | | | | |
| „ *gondwanensis* | cf. | + | + | | — | + | + | | | | | |
| „ spec. α (*Seward*) | — | — | + | | + | — | + | + | | | | |
| „ spec. β (*Seward*) | | | | | | — | + | | | | | |
| „ *spec.* | | | | | + | | | | | | | |
| **Filicales:** | | | | | | | | | | | | |
| 1. Marattiaceae. | | | | | | | | | | | | |
| *Asterotheca denmeadi* | | | | | + | | | | | | | |
| „ *Fuchsi* | | | | | | | + | | | | | |
| ? „ *Hillae* | | | | | + | | | | | | | |
| *Marattiopsis macrocarpa* | | | | | | + | | | | | | |
| „ *Muensteri* | | | | | | | + | ? | | | | |
| *Danaeopsis cacheutensis* | | | | | | | | + | | | | |
| *Nathorstia alata* | | | | | | | | + | | | | |
| „ *willcoxi* | | | | | | | | | + | — | — | — |
| *Rienitsia spathulata* | + | — | — | | | | | | | | | |
| *Weichselia reticulata* | | | | | | + | | | | | | |
| 2. Osmundaceae. | | | | | | | | | | | | |
| *Osmundites Dunlopi* | | | | | + | | | | | | | |
| „ *Gibbiana* | | | | | + | | | | | | | |
| „ *Kolbei* | | | | | | | + | | | | | |

| Pteridophyten der mesozoischen Schichten des gesamten Gondwana-Florenraumes | Unter-Trias | | | | Ober-Trias bis Unter-Kreide | | | | Ober-Kreide | | | |
|---|---|---|---|---|---|---|---|---|---|---|---|---|
| | Austr. | Indien | S. Afr. | S. Amer. | Austr. | Indien | S. Afr. | S. Amer. | Austr. | Indien | S. Afr. | S. Amer. |
| *? Osmundites*-Stämme | + | | | | | | | | | | | |
| *Todites Goeppertiana* | | | | | | | + | | | | | |
| „  *Roesserti* | | | | | | | + | | | | | |
| „  *Williamsoni* | | | | | + | | + | | | | | |
| „  *spec.* | + | | | | | | | | | | | |
| *Cladophlebis* cf. *Albertsi* | | | | | + | | | | | | | |
| „  *antarctica* | | | | | + | | + | | | | | |
| „  *australis* | + | | | | + | | + | | + | | | |
| „  *Browniana* | | | | | | + | cf. | | | | | |
| „  *concinna* | | | | | | + | | | | | | |
| „  *denticulata* | | | | | + | + | + | + | | | | |
| „  *indica* | | | | | | + | | | | | | |
| „  *Johnstoni* | | | | | + | | | | | | | |
| „  *nebbensis* | | | | | | + | | | | | | |
| „  *oblonga* | | | | | | | + | | | | | |
| „  *Roylei* | cf. | | | | + | | | | | | | |
| „  *tasmanica* | | | | | + | | | | | | | |
| „  *spec.* (cf. *sublobata*) | | | | | | | + | | | | | |
| „  *spec.* | + | | | | | | | | | | | |
| 3. S c h i z a e a c e a e. | | | | | | | | | | | | |
| cf. *Klukia exilis* | | | | | | | + | | | | | |
| *Ruffordia acrodentata* | | | | | | | | | + | | | |
| „  *Goepperti* | | | | | + | | + | | | | | |
| „  *Mortoni* | | | | | | | | | + | | | |
| 4. G l e i c h e n i a c e a e. | | | | | | | | | | | | |
| *Gleichenites (Microphyllopteris) acutus* | | | | | + | | | cf. | | | | |
| „  *gleichenoides* | | | | | + | + | | | + | | | |
| „  cf. *micromerus* | | | | | | | + | | | | | |
| cf. *Gleichenites (Microphyllopteris) pectinata* | | | | | + | | | | | | | |
| *Gleichenites rewahensis* | | | | | | + | | | | | | |
| „  *sanmartini* | | | | | | | + | | | | | |
| *Gleichenia spec.* | | | | | | | + | | | | | |
| 5. M a t o n i a c e a e. | | | | | | | | | | | | |
| *Phlebopteris polypodioides* | | | | | + | | | | | | | |
| *Matonidium indicum* | | | | | | + | | | | | | |
| 6. H y m e n o p h y l l a c e a e. | | | | | | | | | | | | |
| *Hymenophyllites Mendozaensis* | | | | | | | + | | | | | |
| „  *spec.* | | | | | | | + | | | | | |
| 7. D i c k s o n i a c e a e | | | | | | | | | | | | |
| *Coniopteris arguta* | | | | | | | + | | | | | |
| „  *delicatula* | | | | | + | | | | | | | |
| „  *hymenophylloides* | | | | | + | + | | + | | | | |
| „  „  var. *australis* | | | | | + | | | | | | | |
| „  cf. *lobata* | + | | | | | | + | | | | | |

| Pteridopyten der mesozoischen Schichten des gesamten Gondwana-Florenraumes | Unter-Trias | | | | Ober-Trias bis Unter-Kreide | | | | Ober-Kreide | | | |
|---|---|---|---|---|---|---|---|---|---|---|---|---|
| | Austr. | Indien | S. Afr. | S. Amer. | Austr. | Indien | S. Afr. | S. Amer. | Austr. | Indien | S. Afr. | S. Amer. |
| *Coniopteris* cf. *nephrocarpa* | | | | | — | — | — | + | | | | |
| *Eboracia lobifolia* | | | | | + | + | — | cf. | | | | |
| 8. D i p t e r i d a c e a e | | | | | | | | | | | | |
| *Dictyophyllum acutilobum* | | | | | + | — | — | — | | | | |
| „ *Carlsoni* | | | | | — | — | — | + | | | | |
| „ *Davidii* | | | | | + | — | — | — | | | | |
| „ *? obtusilobum* | | | | | + | — | — | — | | | | |
| „ *rugosum* | | | | | + | — | — | — | | | | |
| *? Dictyophyllum spec.* | | | | | + | — | — | — | | | | |
| *Clathropteris polyphylla* | | | | | — | — | — | + | | | | |
| *Protorhipis Buchii* | | | | | + | + | — | — | | | | |
| *Hausmannia dichotoma* | | | | | — | + | — | — | | | | |
| „ cf. *Pelletieri* | | | | | | | | | + | — | — | — |
| „ *Wilkinsi* | | | | | | | | | + | — | — | — |
| 9. P o l y p o d i a c e a e. | | | | | | | | | | | | |
| *Onychiopsis Mantelli* | | | | | — | + | — | — | | | | |
| cf. *Onychiopsis psilotoides* | | | | | — | — | — | + | | | | |
| 10. I n c e r t a e s e d i s. | | | | | | | | | | | | |
| *Chiropteris copiapensis* | | | | | — | — | + | + | | | | |
| „ *cuneata* | | | | | — | — | + | — | | | | |
| „ *Etheridgei* | | | | | + | — | — | — | | | | |
| „ *lacerata* | | | | | + | — | — | — | | | | |
| „ *tasmanica* | | | | | + | — | — | — | | | | |
| „ *Zeilleri* | | | | | — | — | + | — | | | | |
| *? Chiropteris spec.* | | | | | + | — | — | — | | | | |
| *Neuropteridium moombracense* | | | | | + | — | — | — | | | | |

In sämtlichen Tabellen betreffend den *Gondwana*-Florenraum bedeutet *Austr.* neben Australien auch noch gegebenenfalls Tasmanien und Neuseeland.

Die Vorkommen-Angaben (der Erdteile) sind ausgefüllt nur in denjenigen formationsgruppen, in welchen die betreffenden Pflanzen überhaupt vorkommen können.

Chapter XVI

PSILOPHYTINAE

von

Richard Kräusel (Frankfurt a. M.)

**Geschichte, Allgemeines.** — Die Gruppe hat ihren Namen nach der Gattung *Psilophyton* erhalten, die Dawson bereits vor etwa 80 Jahren im Unterdevon Kanadas entdeckte, und von der er später einen Wiederherstellungsversuch veröffentlichte. Aber lange Zeit fanden seine Angaben wenig Beachtung. Man konnte ihm an anderen Stellen Irrtümer nachweisen und bezweifelte so auch sein *Psilophyton*. Heute wissen wir, dass seine Rekonstruktion im Wesentlichen richtig war. Es brauchte lange Zeit zu dieser Einsicht. Während auf dem Gebiete der Karbonflora und auch für die des Oberdevons grosse Fortschritte zu verzeichnen waren, blieben die Pflanzen der älteren Devonzeit eine terra incognita oder wurden, wo sie Bearbeiter fanden, völlig falsch, z.B. als Algen gedeutet. Den ersten wesentlichen Fortschritt brachte die Untersuchung des norwegischen Devons durch Nathorst und vor allem Halle (1916), neuerdings durch Høeg fortgesetzt. Dann erschienen die grossartigen Untersuchungen von Kidston & Lang (1917—1924). Begünstigt durch die einzigartige Erhaltung der Pflanzen in dem verkieselten Torfbett von R h y n i e in S c h o t t l a n d, konnten sie Morphologie und Innenbau weitgehend aufklären und zum ersten Male wieder Rekonstruktionen wagen. Seitdem ist der Umfang des Schrifttumes gewaltig angeschwollen. Neue Arbeiten von Lang und Cookson schlossen sich an. Die längst bekannte Devonflora B ö h m e n s wurde von Kräusel & Weyland revidiert, die weiterhin in einer Reihe von Arbeiten neue und ältere Funde im deutschen Devon bekannt machten, ohne bislang zu einem Abschluss gekommen zu sein. Darüber hinaus sind die Psilophyten (und sonstige Devonpflanzen) heute zu einem beliebten Gegenstand der Betrachtung geworden. Unter Verzicht auf Einzelheiten sei auf einen Bericht von 1936 verwiesen [1]), von dem aus leicht eine Über-

---

[1]) Kräusel, R. Neue Untersuchungen zur paläozoischen Flora: Rheinische Devonfloren. — Ber. D. Bot. Ges. **54**, 1936, 307.

Halle, T. G. On *Drepanophycus*, *Protolepidodendron* and *Protopteridium*, with notes on the Palaeozoic flora of Yunnan. — Palaeont. Sinic. **A 1,** Fasc. 4, 1937.

Lang, W. H. On the plant-remains from the Downtonian of England and Wales. — Phil. Transact. Roy Soc. London **B 227,** 1937, 245.

Principi, P. La vita vegetale nei primi periodi della storia della terra. — Atti Soc. Sci. Genova **1,** 1936, 189.

sicht des gesamten, die Psilophyten betreffenden Schrifftums gewonnen werden kann.

Die Psilophyten beschränken sich durchweg auf Schichten, die n i c h t j ü n g e r  a l s  m i t t e l d e v o n i s c h sind. In Australien kommen sie zusammen mit den nach der paläontologischen Lehrmeinung angeblich auf das Silur beschränkten Graptolithen vor. Besonders kennzeichnend sind sie aber für das Unterdevon. Man kennt sie aus Europa und Asien, Nord- und Südamerika, Südafrika und Australien. Offenbar waren sie damals w e l t w e i t  v e r b r e i t e t und für diese Zeit ebenso kennzeichnend, wie später etwa die Gymnospermen oder seit der Kreide die Angiospermen.

Die allgemeine Bedeutung der Psilophyten beruht also einmal darauf, dass sie die ä l t e s t e n  u n s  b e k a n n t e n  G e f ä s s p f l a n z e n darstellen, zum andern aber auf ihrem  e i n f a c h e n  B a u. Anfangs ge- machte Versuche, diesen als abgeleitete Reduktions-oder Anpassungser- scheinung zu deuten, können angesichts der Zahl der bisher bekannten Formen als widerlegt gelten. So einfach gebaute Pflanzen wie *Hornea* nähern sich dem Sporophyten gewisser Moose, andererseits umfassen die Psilophyten Vorläuferformen, die morphologisch zu den verschiedenen Gruppen der höheren Pteridophyten hinführen, für deren Ableitung sie den Schlüssel geben. Damit stellen sie ein s t a m m e s g e s c h i c h t l i c h e s M a t e r i a l  a l l e r e r s t e n  R a n g e s dar, worüber im Abschnitt „Phylogenie" mehr zu finden ist.

### Psilophytales

Kleinwüchsige Pflanzen ohne echte Blätter und Wurzeln, in Rhizom und Luft- sprosse gegliedert. Sporangien, soweit bekannt, mit mehrschichtiger Wandung, mit oder ohne Kolumella, ohne besonderen Öffnungsmechanismus, isospor, end- ständig an den Luftsprossen oder an kürzeren Verzweigungssystemen, Sprosse un- verzweigt bis gabelig oder ± monopodial bis sympodial verzweigt, blattlos, nackt oder mit Drüsen oder Dornen besetzt, Rhizom verschieden gestaltet, knollenförmig bis kriechend verzweigt, Achsen, soweit bekannt, protostelisch, der Xylemteil rund bis sternförmig zerklüftet, mit Schrauben- und Ringtracheiden, vielleicht auch schon Tüpfelzellen vorhanden, Phloem das Xylem umgebend, aus gestreckten, dünnwandigen Zellen aufgebaut, Epidermis mit Spaltöffnungen, diese wie bei den höheren Pflanzen gebaut.

### Fam. I. Rhyniaceae

Rhizom und Luftsprosse gabelig verzweigt, nackt, Sporangien endständig, mit mehrschichtiger Wandung, ganz von den durch Tetradenteilung gebildeten Sporen erfüllt.
*Rhynia* KIDST. & LANG, 2 Arten im mittleren Old Red Sandstein von Schottland.

### Fam. II. Horneaceae

Rhizom knollenförmig, gelappt, Luftsprosse gabelig verzweigt, nackt, Sporangien endständig, mit Kolumella.
*Hornea* KIDST. & LANG, 2 Arten, mittleren Old Red von Schottland.
*Sporogonites* HALLE, Formgattung für isolierte endständige Sporangien vom Typus I und II, Unter-Mitteldevon.

Manual of Pteridology                                         32

*Yarravia* LANG & COOKS., ähnlich dem vorigen, Sporangien aber zu mehreren im unteren Teil synangienartig vereinigt, Obersilur(?), Australien.

*Cooksonia* LANG, desgleichen, Sporangienträger aber gegabelt, Obersilur, England.

## Fam. III. Psilophytaceae

Rhizom kriechend, wie die Luftsprosse gabelig verzweigt, diese nackt oder mit Drüsen besetzt, ihre Enden ± eingerollt bezw. die Sporangien tragend, diese so endständig, meist gruppenförmig vereinigt, Protostele klein, rund, Epidermis mit Spaltöffnungen.

*Psilophyton* DAWSON, mehrere Arten im Unterdevon, weit verbreitet, Alte und Neue Welt.

*Dawsonites* HALLE, fruchtende Sprosse unklarer Herkunft, aber vom Bau der *Psilophyton*-Fruktifikation.

*Taeniocrada* WHITE, Sporangien am Ende kurz verzweigter Seitensprosse, mehrere Arten in Unter-Mitteldevon, *T. decheniana* massenhaft im „Haliseritenschiefer" des rheinischen Unterdevons.

*Hicklingia* KIDST. & LANG, Luftsprosse nackt, büschelförmig beieinander stehend, am Ende keulenförmige Sporangien tragend, Unterdevon.

## Fam. IV. Zosterophyllaceae

Luftsprosse bei einigen ± büschelförmig angeordnet, wenig verzweigt, Sporangien an kurzen Seitenästen sitzend, ± ausgeprägt zu Ständen vereinigt.

*Zosterophyllum* DAWS. Sporangien zu kleinen ährenähnlichen Ständen vereinigt, ± nierenförmig, sich mit breitem Querriss öffnend, 3 Arten im Obersilur(?), Australien, Unterdevon, Schottland, Rheinland.

*Barinophyton* WHITE, *Pectinophyton* HØEG, *Bucheria* DORF, verzweigte, fruchtende Sprosse mit ± einseitig orientierten Sporangien.

## Fam. V. Sciadophytaceae

Luftsprosse thallusartig-sternförmig vereinigt (vergl. *Thallomia* HEARD aus dem Silur?), selten verzweigt, von einer sehr schmalen Stele durchzogen, oben köpfchenartig verdickt, die Vermehrungsorgane (Sporangien?) tragend.

*Sciadophyton* STEINM., Unterdevon des Rheinlandes, Kanada.

*Dutoitia* HØEG, Südafrika (Stellung sehr zweifelhaft).

## Fam. VI. Asteroxylaceae

Luftsprosse im unteren Teil mehr monopodial, im oberen mehr gabelig verzweigt, die unteren dicht mit schuppenartigen Blättern besetzt, diese nach oben immer kleiner werdend, obere Sprosse schliesslich nackt („*Thursophyton*"- und „*Hostimella*"-Stadium), Stele sternförmig, ohne oder mit Innengewebe (Mark?), mit Abzweigungen zu den Blättern, aber nicht in diese eintretend, Sporangien klein, birnförmig, am Ende der obersten (ob aller?) nackten Spross-Systeme.

*Asteroxylon* KIDST. & LANG, 2 Arten, Mitteldevon, Schottland, Rheinland. Hierher wahrscheinlich viele der als *Thursophyton* NATH. (beblättert) und *Hostimella* POT. & BERNH. (nackt) bezeichneten, sehr häufigen, aber nur im Abdruck erhaltenen Stücke.

## Fam. VII. Pseudosporochnaceae

Luftspross hochwüchsiger als bei den übrigen, Anatomie unbekannt, Hauptachse bis zur halben Höhe unverzweigt, dann sich büschelförmig in wenige Äste teilend diese gabelig verzweigt, seitlich mit büschelförmig sich gabelig aufteilenden feineren Sprossen, deren letzte Enden die länglich-schmalen Sporangien tragen. *Pseudosporochnus* (STUR) POT. & BERN., Mitteldevon d. Alten und Neuen Welt.

Weiter werden noch die wichtigsten „Vorläufer-Formen" zusammengestellt, deren Bau zu den höheren Pteridophyten überleitet, aber doch $\pm$ psilophyten-artige Primitivzüge aufweist. Sie lassen sich an *Asteroxylon* bezw. *Pseudosporochnus* anschliessen.

### Fam. VIII. Drepanophycaceae

*Drepanophycus* GOEPP. (*Arthrostigma* DAWS.), Sprosse bis zu mehreren cm dick, mit sehr dünner Stele, unregelmässig mit dorn-oder stachelartigen Blättern besetzt, diese mit Leitstrang, z.T. nierenförmige Sporangien tragend, häufig im Unterdevon d. Alten und Neuen Welt.

*Baragwanathia* LANG & COOKS. dem vorigen ähnlich , Blätter aber lang, schlaff, Obersilur( ?), Australien.

*Haplostigma* SEWARD, nur vegetative Sprosse bekannt, Südafrika.

Hieran schliessen sich Formen wie *Protolepidodendron* KREJČI mit Gabelblättern, Unterbis Mitteldevon, *Barrandeina* STUR mit verzweigten Blättern und *Duisbergia* KR. & WEYL., Mitteldevon, letztere mit komplizierter gebauter Stele (ähnlich *Schizopodium* HARRIS, Australien). Sie alle zeigen schon $\pm$ deutliche Anklänge an die echten *Lycopodiales*.

### Fam. IX. Protopteridiaceae

*Protopteridium* KREJČI, Mitteldevon (*Milleria* LANG), Sprosse noch unregel-mässig räumlich verzweigt (z. T. *Hostimella*-Typus), Stelenbau und Sporangien-anordnung aber an Farne und Pteridospermen erinnernd.

Hieran schliessen sich Formen wie *Aneurophyton* KR. & WEYL. (*Eospermatopteris* GOLDRING) und *Cladoxylon scoparium* KR. & WEYL. aus dem Mitteldevon, die gewöhnlich schon zu den Farnen gerechnet werden.

Zum Schluss seien die Gattungen zusammengestellt, für die Wiederherstellungs-versuche vorliegen: *Psilophyton* (DAWSON), *Hornea, Rhynia, Asteroxylon* (KIDSTON & LANG, KRÄUSEL & WEYLAND), *Taeniocrada, Zosterophyllum, Sciadophyton, Pseudosporochnus, Drepanophycus, Protolepidodendron, Barrandeina, Duisbergia, Protopteridium, Aneurophyton, Cladoxylon scoparium* (KRÄUSEL & WEYLAND). Sie sind zum grösseren Teil in der auf S. 496 genannten Schrift zusammengestellt.

CHAPTER XVII

LYCOPODIINAE

by

J. WALTON (Glasgow) and A. H. G. ALSTON (London)

**Introductory.** — Without giving a detailed account of the earlier systems of classification in which divisions of this group were included a few examples may be given to illustrate how the present system of classification has evolved. DILLENIUS in his Historia Muscorum (1741) distinguished *Selago, Lycopodium, Selaginoides,* and *Lycopodioides* as distinct genera. *Selaginoides* included *Selaginella Selaginoides,* while *Lycopodioides* included other species of *Selaginella. Calamaria* was the name given to what is now termed *Isoëtes.* Homospory or heterospory and the presence or absence of a ligule were not recognised as of important diagnostic value until near the end of the following century. LINNAEUS in the Species Plantarum (1753) included the Lycopods among the Mosses while *Isoëtes* was placed with the Filices. SPRENGEL in the Systema Vegetabilium (1827) recognised the existence of dimorphic sporangia but did not regard it as of diagnostic value. He included *Psilotum* and *Lycopodium* in the *Lycopodeae* but his *Lycopodium* included *Selaginella* as well as *Lycopodium.* ENDLICHER (1836) classified the Lycopods as follows:

Ordo XXV Isoëteae (*Isoëtes, Stigmaria* and *Lepidodendron Veltheimianum*).
Ordo XXXVI Lycopodiaceae (*Psilotum, Lycopodium, Lycopodites, Selaginites. Lycopodium* including *Selago, Lepidotis,* and *Stachygynandrum*).
Ordo XXXVII Lepidodendreae (*Lepidodendron, Ulodendron, Lepidostrobus, Cardiocarpon*).

That there were two kinds of spores in certain species of *Selaginella* was known to DILLENIUS, but the exact nature of the megaspores was long disputed and some botanists thought that they were gemmae. The nature of the spores of *Lycopodium* was shown by LINDSAY in 1792, who germinated those of *L. cernuum*; they were previously thought to be pollen grains. BEAUVOIS created his genera *Stachygynandrum, Diplostachium,* and *Selaginella* for those species which he knew to have megaspores — "semences" as he called them. ROBERT BROWN in 1820 used the presence of megaspores for his main sectional division, and BRONGNIART (Hist. Veg. Foss. 1837) recognised heterospory in *Stachygynandrum* (= *Selaginella* Spring) and *Isoëtes*. He considered however that secondary wood in *Sigillaria* was an index of gymnospermous affinity, a mistake that WILLIAMSON corrected. Recently it has been shown that some *Selaginellae* of the *rupestris* group have megaspores only. The number of megaspores is very variable; sporangia with a single spore are found in *S. monospora*, while DUERDEN records as many as 42 in cones of *S. Willdenovii*. The fossil *Selaginellites Suissei* had 16 to 20 megaspores in each sporangium. The arrangement of the sporangia has been little studied and may be useful for separating groups of species.

STUR in 1875 detected the presence of a ligular put on the leaf cushion of *Lepidodendron* and compared it with the pit which contains the ligule in *Selaginella* and *Isoëtes*. SOLMS–LAUBACH demonstrated the presence of the ligule in a structurally preserved *Lepidodendron* in 1892. SPRING (1842–50) in his classic monograph dealt very fully with the *Lycopodiaceae*. He recognised four genera, 1. *Lycopodium*, 2. *Selaginella*, 3. *Tmesipteris*, and 4. *Psilotum*. He subdivided the genera into sections and groups of species and drew attention to the dorsal rhizophores of the *Selaginellae* with articulate stems. SPRING's work was followed by BAKER's Fern Allies (1887) which separates the *Selaginellaceae* from the *Lycopodiaceae* but leaves *Psilotum* and *Tmesipteris* in the former. In PRITZEL's classification (in Engler and Prantl, 1900) the *Lycopodiales* are divided into the two groups *Eligulatae* (incl. *Lycopodiaceae* and *Psilotaceae*) and *Ligulatae* (incl. *Selaginellaceae, Lepidodendraceae, Bothrodendraceae, Sigillariaceae*, and *Isoëtaceae*). LOTSY in 1909 divided those Archegoniates in which the sporophyte is the conspicuous part of the life-history into two divisions based on the numbers of flagella on the spermatozoid. As the nature of the spermatozoid in the majority of the groups is unknown, for they are fossils, this classification has little to recommend its adoption. It also has the effect of placing *Isoëtes* in closer proximity to the Ferns than to the *Lepidodendraceae* and *Selaginellaceae* with which it agrees closely in several important morphological characters. The *Psilotaceae* are now placed in the *Psilotales*.

The following system of classification is based very largely on that given by HIRMER (Handbuch d. Paläobotanik, 1927) for the fossil representatives of the group.

### Lycopodiales

Microphyllous pteridophytes with usually frequently branched axes, upright or creeping. Sporangia borne singly on the upper side of the fertile leaf, near to its base. Sterile leaves and fertile leaves not differing greatly in structure. Sporangium with wall consisting of several layers (Eusporangiate type). Gametophytes small, tuberous.

Div. I *Eligulatae* (a) *Homosporeae.*

## Fam. I. Lycopodiaceae

Shoots generally of radial construction. Leaves arranged in spirals or whorls, rarely dimorphic. Fertile leaves usually aggregated in definite cones. Sporangia homosporous. Gametophytes tuberous, green or colourless, with mycorrhiza. Sperm biciliate.

Gen. I **Lycopodium**. — A large genus containing both terrestrial species char-acteristic of acid soils in the temperate regions, and tropical epiphytes. Sporangia compressed, unilocular, dehiscing by a single slit, solitary in the axils of more or less modified leaves. The classification given here mainly follows that of HERTER, though *Urostachys*, which he raised to generic rank in his later papers, is kept as a subgenus. Subgenus I. *UROSTACHYS*. Branching regularly dichotomous. Stems rooting at the base only.

This subgenus, which is sometimes separated as a genus, contains the great major-ity of the species of *Lycopodium*. It may be divided into three sections, as follows:

Sect. 1. *Selago*. Erect terrestrial species. Leaves and sporophylls similar.

Sect. 2. *Subselago*. Pendent epiphytes. Leaves and sporophylls similar. Leaves gradually diminishing in size towards the apex of the stem.

Sect. 3. *Phlegmaria*. Strobili strongly differentiated from the vegetative part.

Subgenus II. *CLAVATOSTACHYS*. Branching apparently monopodial, owing to unequal dichotomy. Strobili terminal, peduncled. Leaves multifarious. Strobili 1–20, creeping plants.

A subgenus containing the cosmopolitan and variable *L. clavatum* and several other temperate or mountain species.

Subgenus III. *COMPLANATOSTACHYS*. Branching apparently monopodial. Strobili terminal, usually peduncled. Leaves distichous and dimorphous. Strobili terminal 1–20, rarely more, creeping or climbing plants.

The species of this subgenus show a parallelism with *Selaginella* in their dim-orphous leaves: this is especially marked in *L. volubile*, a climbing species. In *L. casuarinoides* the apices of the leaves are hyaline and deciduous.

Subgenus IV. *CERNUOSTACHYS*. Branching apparently monopodial. Plants with tree-like branches springing from a creeping stem, which roots at intervals. Strobili 50–100, not peduncled. Valves of the sporangium unequal.

A small and very distinct group containing only the common tropical *L. cernuum* and a few allied species.

Subgenus V. *INUNDATOSTACHYS*. Branching apparently monopodial. Sporophylls similar to the leaves; strobili terminal.

The European *L. inundatum* and the north American *L. alopecuroides* belong to this subgenus.

Subgenus VI. *LATERALISTACHYS*. Branching apparently monopodial; strobili lateral.

There are only two species in this subgenus. *L. laterale* and *L. diffusum*; they are natives of Australia and New Zealand.

Gen. II. **Phylloglossum**. — Small plant with a very short axis arising from an annually produced tuber. The leaves terete, subulate, arranged in a tuft round the base of the axis. Roots arise above the tuber but below the leaves. Axis terminates in a strobilus of the Urostachys type. Sporangia homosporous, unilocular, dehiscing by a longitudinal slit. Sporophylls broadly ovate, imbricated.

A single species, *P. Drummondii* is found in Australia and New Zealand. This genus is very similar to *Lycopodium*, especially to *L. cernuum*, in its mode of development, but different from most species in its vegetative characters. It has been suggested by BOWER that *Phylloglossum* should be regarded as a permanently embryonic form of Lycopod. Its chief interest lies in the discussion of the morphol-ogy of the tuber. Miss SAMPSON considered that it was a modified branch, while WORSDELL compared it with the rhizophore of *Selaginella*. OSBORN has shown that accidentally detached leaves will grow and form adventitious tubers. Tubers are not confined to this genus of Lycopods but have been recorded in *Lycopodium tuberosum*. HOLLOWAY considered that the protocormous tubers of certain other species of

*Lycopodium* were xerophytic adaptations rather than persisting rudiments of a primitive organ.

Gen. III. **Spencerites** (Carboniferous). — Sporophyll peltate with a leaf-like umbo. Eligulate. Spores winged. Only the cone known.

Note: — While it is true that the fossil plants *Pinacodendron, Lepododendropsis,* and *Cyclostigma* are heterosporous, it is still possible that they may prove to be *Ligulatae.* The structure of the base of the leaf in those genera is not known. The majority of their leaves were fertile and the ligule if it were present in the usual position on the abaxial side of the sporangium would not be seen on the stem after the abscission of the leaves.

Div. II. *Ligulatae* (a) *Heterosporeae.*
a. Herbaceous types.

## Fam. II. Selaginellaceae

Genus **Selaginella.** — Shoot usually creeping and dorsiventral with leaves in four ranks. Ligule present at base of each leaf and sporophyll. In most species three roots are present at the base of the stem axis in the sporeling. The rest of the root system is produced from rhizophores, special branches of the shoot system. The roots and rhizophores fork dichotomously. Heterosporous, gametes biciliate.

I. Subgenus *EUSELAGINELLA.* Leaves and sporophylls uniform.

1. Group of *S. Selaginoides* .Sporophylls spirally arranged.

The most primitive group of species with uniform leaves. There are only two species, *S. Selaginoides,* which is widespread in the north temperate zone, *S. deflexa* a closely allied species from Hawaii. *S. Selaginoides* is the only species in which secondary xylem elements have been recorded. The stele is solid at the base of the shoot, with a central pith in the upper part. The root system arises from a swollen knot at the base of the hypocotyl which may be compared with the rhizophoric part of *Isoetes.*

2. Group of *S. pygmaea.* Strobili tetragonous. Leaves decussately arranged, at least in the lower part of the stem.

The two species, *S. pygmaea* and *S. gracillima* are annuals. Miss Duthie has shown that anisophylly may be brought about in *S. pygmaea* by one-sided illumination, and it has been suggested that isophylly is secondary rather than primitive in this species. *S. gracillima* has a radial stele with, solid core of xylem.

3. Group of *S. uliginosa.* Strobili tetragonous. Leaves decussate.

A single species, *S. uliginosa* with, according to Miss Stele, a well-developed solenostelic rhizome and erect aerial shoots; the erect shoots have four steles.

4. Group of *S. rupestris.* Strobili tetragonous. Leaves linear, spirally arranged.

A group of xerophytic species with many narrow, needle-shaped, spirally arranged leaves, and a single banded stele.

II. subgenus *STACHYGYNANDRUM.* Leaves dimorphous, sporophy s uniform.

This group contains the vast majority of the species of *Selaginella,* namely those with uniform sporophylls and some or all of the leaves dimorphous.

Series 1. D e c u m b e n t e s. Prostrate species with their main stems rooting throughout. Connected with *Euselaginella* by *S. sanguinolenta* which may have either uniform or dimorphic leaves. Goebel and Suessenguth have shown that dimorphic leaves are induced by wet conditions in this species. Some of the species of this group have hairs on the surface of the leaves, e.g. *S. hispida, S. Lindenii, S. alligans* climbs by means of rhizophores which develop disc-like pads similar to those of the tendrils of *Parthenocissus tricuspidata. S. serpens* is well known for the periodic changes in the colour of its leaves.

Series 2. A s c e n d e n t e s. Suberect species, often with a single banded stele; usually with conspicuous rhizophores in the axils of the lower branches.

*S. trachyphylla* is remarkable for the rough papillose surface of the leaves.

Series 3. S a r m e n t o s a e. Suberect scandent or rarely prostrate, polystelic species, with distant, entire leaves.

*S. Willdenovii* and *S. caesia* are remarkable for the metallic sheen of their leaves. This is due to the presence of particles in the cutin of the epidermis which reflect certain rays of light. *S. picta* is normally variegated.

Series 5. C a u l e s c e n t e s. A large group of erect species. The simple erect part of the stem is usually homophyllous and the branched frond-like part heterophyllous. Some species *S. Braunii*, *S. Vogelii*, and *S. biformis* have unicellular hairs on their stems and branches. *S. haematodes*, *S. umbrosa*, and *S. erythropus* are remarkable for their bright red stems. The winged megaspores of *S. caulescens* are specially noteworthy. Banded steles are usual.

Series 6. C i r c i n a t a e. A small group of xerophilous species derived either from the C a u l e s c e n t e s or from the D e c u m b e n t e s through some such species as *S. siamensis*. *S. lepidophylla* and *S. pilifera* are sold under the name "Resurrection plant". The stems curl up when dry and expand when moistened. *S. pallescens* is known to have a banded stele.

Series 7. A r t i c u l a t a e. A group of tropical American species with articulate stems. The articulations are often very conspicuous in dried specimens and (according to HARVEY–GIBSON) are due to hypertrophy of the middle layers of the cortex. According to HIERONYMUS they enable the plant to propogate itself vegetatively, by the breaking up of the stems. An important character seems to be that the rhizophores are dorsal and extra-axillary in this group while they are ventral and axillary in the other groups. Two steles are usual, but 3–5 are sometimes found.

III. Subgenus *HOMOSTACHYS*. Leaves dimorphous. A small group of species with creeping stems and loose strobili. The sporophylls are equal in *S. helvetica*, and dimorphous in the other species *S. pallidissima*, *S. Savatieri*, and *S. Rothertii*. They are probably allied to *S. denticulata* which has a variety with dimorphous sporophylls that is found in wet places.

IV. Subgenus *HETEROSTACHYS*. Sporophylls dimorphous. A large group of heterophyllous species characterised by their strongly dimorphous sporophylls. The sporophylls are of two kinds, the large sporophylls on the upper surface of the stems in the same plane as the small, median leaves, and the small sporophylls on the lower surface on the same plane as the large leaves. Vegetative reproduction has been reported by Miss BANCROFT in two Indian species, and tubers in *S. abyssinica* by CHIOVENDA.

*Selaginellites* Zeiller (Extinct) (e.g. *S. Suissei* Zeill., *S. primaevus* Gold.).

Leaves heterophyllous as in *Selaginella*. Ligule not detected. Heterosporous.

## Fam. III. Isoëtaceae

Axes short, erect, with crowded, acicular leaves all of which may be fertile. Base of axis swollen with a meristem which produces roots in regular succession. Roots dichotomously forked.

*Isoëtes*. Axis very short. Rhizophoric part of axis lobed, root producing meristem in grooves between lobes. All the leaves may be fertile. Heterosporous, gametes polyciliate. The most recent classification of the species is that of PFEIFFER (Ann. Miss. Bot. Gard. IX, pp. 29–232), which is based on the sculpturing of the megaspores.

*Nathorstiana* (Cretaceous). Axis very short. Lower part of leaf bearing axis sterile, upper part fertile. Spores unknown. Rhizophoric region obscurely lobed. Roots in vertical series.

*Pleuromeia* (Triassic). Axis short, leaf-bearing part of axis elongated, the older part with widely separated leaf scars. Sterile leaves long. Fertile leaves short, almost orbicular, grouped at distal end of axis. Rhizophoric base of stem lobed. Roots distributed uniformly over lobes. Heterosporous, ligulate.

*Note*. Some authors have considered that the evidence is in favour of the sporangium being on the abaxial surface of the sporophyll. This evidence is not reliable. Shrivelled or compressed leaves of *Isoëtes* appear to have the spore mass projecting on the abaxial surface. The apparent projection of the spore mass in *Pleuromeia* can easily be accounted for by supposing that the softer tissues have been compressed by the surrounding sediment.

### Fam. IV. Miadesmiaceae (Carboniferous)

Axes slender, branched. Small leaves with large ligules. Heterosporous. The sporangium invested with an integument covered with multicellular hairs. *Miadesmia.* Only genus.

b. *Arborescent types.*

### Fam. V. Lepidodendraceae (Palaeozoic)

Trees with main trunk bearing a crown of branches. Branching often apparently dichotomous. Leaves vary in size on the same plant, larger leaves on the trunk and main branches, smaller leaves on the ultimate branches. Leaves deciduous with a persistent base. Ligule present in a pit on the leaf base. The surface of the trunk and branches is covered with the contiguous diamond-shaped leaf bases. The leaf scar shows the smaller scars of vascular bundle and two parichnos strands. Cones borne terminally on the ultimate branches or on special lateral peduncles on the trunk or main branches. Microsporangia and megasporangia in the same cone or in different cones. The basal part of the axis (*Stigmaria*) rhizophoric. There are four main rhizophoric branches each of which may branch equally. The roots are distributed regularly over the rhizophore and its branches. The rhizophoric meristems are situated at the extremities of the root bearing branches (cf *Pleuromeia*).

*Lepidodendron* (incl. *Lepidophyllum, Lepidostrobus, Stigmaria* (in partem), *Knorria, Aspidiopsis, Aspidiaria* etc.).

Leaf cushion rhombic with leaf scar in the middle or upper part of it.

*Lepidophloios* (incl. *Halonia*). Leaf cushion drooping so that the leaf scar appears to lie below the centre of the leaf cushion. Cones borne on the ends of peduncles on the trunk and main branches.

### Fam. VI. Lepidocarpaceae (Carboniferous)

Cones bearing megasporophylls. Each megasporophyll with a single megaspore in the sporangium. Prothallus developed in the megaspore before the sporophyll or spore become detached. Sporangium enclosed in an integument. Micropyle in the form of a slit-like opening over the upper surface of the megasporangium. — *Lepidocarpon.*

### Fam. VII. Bothrodendraceae (Carboniferous)

General habit like *Lepidodendron.* Leaves attached to the young branches by diamond shaped leaf cushions. As the branches matured the leaves became detached and the leaf bases became gradually stretched flat and obliterated. As a result the leaf scars appear widely separated on the almost smooth surface of the branch. Leaf scar shows the scars of leaf trace and usually two parichnos strands. The ligular pit may be seen on the branch surface at some distance from the leaf-scar. Two vertical rows of branch scars are present on the trunks and main branches which probably represented the points of attachment of cone peduncles. Heterosporous. *Bothrodendron.*

Note. In the surface characters of the shoots and the behaviour of the leaf cushions *Bothrodendron* is intermediate between *Lepidodendron* and some species of *Sigillaria.*

### Fam. VIII. Sigillariaceae (Palaeozoic)

Trees with unbranched or rarely branched axes. Leaves long (up to 1 metre in length) attached to the later formed part of the trunk or branches, separating when older by an absciss layer. The persistent leaf bases with various arrangements. Leaf scar shows scar of leaf trace and two parichnos strands. Cones borne on special peduncles. Cone peduncles grouped in zones on the trunk and branches. Rhizophoric base similar to *Lepidodendron.*

*Sigillaria.*

Group I. *Eusigillaria.* Leaf cushions always clearly marked and distinct.

Sub-group *Favularia.* Leaf scars distinct on six-sided leaf cushion and arranged in numerous vertical series.

Sub-group *Rhytidolepis.* Horizontal boundaries becoming less pronounced or disappearing entirely. The leaf bases joining to form vertical ribs on the stems and branches.

Group II. *Subsigillaria* (incl. *Clathraria* and *Leiodermaria*). Leaf cushions never united to form ribs and in some species so indistinct that the leaf scars appear to be situated on a smooth stem surface.

### Lycopodiales Incertae Sedis.

1. Herbaceous types. Ligules or ligular pits not demonstrated (Some heterosporous) *Lycopodites,* spp.

2. *a.* Arborescent types with leaf cushions. Ligules or ligular pits not demonstrated. Palaeozoic.

*Archaeosigillaria*
*Omphalophloios*
*Ulodendron*
*Phialophloios*
*Thaumasiodendron*
*Leptophloem*
*Arctodendron*
*Porodendron* (probably = *Bothrodendron*)
Etc.

2. *b.* Arborescent types without clearly marked leaf cushions or ligular pits.
*Cyclostigma*        (Heterosporous) Palaeozoic
*Pinacodendron*         ,,              ,,
*Asolanus*                              ,,
*Protasolanus*                          ,,
*Lycostrobus* (Heterosporous. Cone only, Mesozoic).

Note. *Cyclostigma* includes *Cyclostigma Kiltorkense* Haughton sp. and does not include such forms as *C. Kidstoni* and *C. Wijkianum* of some authors which are definitely known to be ligulate, and must be included in *Bothrodendron.* The presence or abscence of parichnos scars on the leaf scar is not a safe criterion. There are beautifully preserved specimens of *Lepidophloios Scoticus* in which the parichnos marks are not visible.

# Chapter XVIII

## PSILOTINAE

von

### Max Hirmer (München)

Die Gruppe, von der fossile Formen bis jetzt noch nicht bekannt geworden sind, umfasst die beiden lebenden Gattungen: *Psilotum* Sw. und *Tmesipteris* Bernh. Jede der beiden Gattungen umschliesst nur eine geringe Anzahl Arten. Die Gruppe ist, worauf später noch einzugehen sein wird, in ihrer phylogenetischen Stellung die am meisten umstrittene, insoferne sie in Beziehung einerseits zu den *Lycopodiales*, andererseits zu gewissen *Articulatales*, insbesondere, zu *Sphenophyllum*, sowie natürlich auch zu den *Psilophytales* gebracht worden ist. Erst durch die neueren und neuesten Entdeckungen an den devonischen *Psilophytales* konnte — ohne dass damit gesagt werden soll, dass die Gruppen stammesgeschlichtlich irgendwie n ä h e r miteinander in Beziehung stehen — eine gewissen Klarheit in die Ausdeutung der morphologischen Verhältnisse der *Psilotales* gebracht werden, und sicher steht jedenfalls fest, dass es sich beiden *Psilotales* um eine derjenigen Pteridophyten-Gruppen handelt, die wie die Articulatales und Lycopodiales in den Komplex der m i k r o p h y l l e n Pteridophyten gehört.

**1. G a m e t o p h y t.** Die erst vor weniger als 20 Jahren entdeckten Gametophyten von *Psilotum* und *Tmesipteris* haben miteinander gemeinsam, dass sie — ebenso wie das übrigens für die Jugendzustände des Sporophyten zutrifft — unterirdisch lebende, chlorophyllfreie Holosaprophyten sind, ausgerüstet mit einer endotrophen, letzten Endes auch exotrophen Mykorrhiza, welch letztere durch Infektion der Rhizoiden erreicht wird. Die Gametophyten beider Gattungen stimmen in ihrer G e s a m t e r s c h e i n u n g weitgehend überein, sind radiär zylindrisch und häufig mehr oder minder gabelig verzweigt; sie haben sehr langsam wachsende Apikal-Meristeme an den jeweiligen Gabelteilstücken ihres Gesamtkörpers, sind ringsum mit langen, aber einzelligen Rhizoiden besetzt und tragen die Geschlechtsorgane, sowohl die Antheridien als die Archegonien, über ihren Gesamtkörper verteilt.

Hervorzuheben ist, dass der Gametophyt von *Psilotum* offenbar b i polar gabelteilig ist, im Gegensatz zu dem offensichtlich u n i polar gestalteten Gametophyten von *Tmesipteris*.

Die Tatsache der B i polarität des Gametophyten von *Psilotum* ist insoferne von hervorragendem Interesse als der Gametophyt von *Psilotum* somit die primitivste Grund-Gestaltung aufweist, die überhaupt möglich ist.

Die offenbar aus e i n e r Epidermiszelle hervorgehenden [1]) A n t h e r i d i e n sind gross und von kugeliger Gestalt und springen insbesondere bei *Psilotum* erheblich über die Oberfläche des Gametophyten hervor; es ist aber zu betonen, dass die j u n g e n Antheridien-Anlagen noch kaum über die Oberfläche hervortreten. Die in grosser Zahl im Antheridium gebildeten S p e r m a t o z o i d e n sind mehr minder keulenförmig und schraubig-gewunden und polyziliat. Letztere Eigentümlichkeit, die zweifellos nur die stammesgeschichtlich fortgeschrittenen Pteridophyten charakterisiert, ist insofere von besonderem Interesse als sie sich hier bei Gametophyten findet, deren Gesamtorganisation offensichtlich noch eine noch primitive ist. Die A r c h e g o n i e n, deren Oeffnung offenbar durch Abbrechen des oberen Halsteiles erfolgt, sind bei den beiden Gattungen insoferne etwas verschieden gebaut als bei *Psilotum* die Halsreihe aus 6, bei *Tmesipteris* aus 4 Etagen von Zellen besteht. Zahlenverhältnisse der Halskanalzellen sowie Existenz der Bauchkanalzelle sind nicht mit Sicherheit ermittelt.

**2. S p o r o p h y t.** Der E m b r y o ist exoskop, d.h. der Sprossscheitel der bei b e i d e n Gattungen nur u n i polaren Sporophyten-Anlagen ist dem Halsteil des Archegoniums zugekehrt. Der Fussteil erreicht in älteren Embryonalzuständen gegen den Gametophyten hin insoferne eine Oberflächen-Vergrösserung, als er mehr minder weitgehend lappig zerteilt sein kann. Was aus der Embryonalanlage hervorgeht, ist, wie schon eingangs betont, ein zunächst blatt- und chorophylloser Holosaprophyt von u n i polarer Gesamtfassung. Es sei hier schon darauf hingewiesen, dass die Gesamtgestaltung des jungen Sporophyten weitehendst an die Gestaltung der primitiveren *Psilophytales* wie *Rhynia* u.a. erinnert.

Die s p ä t e r e n Z u s t ä n d e des Sporophyten zeigen bei stets unipolarer Grundgestaltung eine Differenzierung in ausgesprochen rhizomartige, chlorophyllfreie und mit Mykorhiza versehene sowie ringsum mit Rhizoiden besetzte Rhizome, denen jede echte Wurzelbildung fehlt und anderseits in beblätterte und letzten Endes Sporophylle tragende, mehr oder minder gabelteilige Lichtsprosse. Den Uebergang zwischen Rhizom und eigentlichen Lichtspross vermittelt eine oft über eine Anzahl Centimeter ausgedehnte Uebergangszone; diese ist in ihrem unteren Teil noch völlig blattlos um an ihrem oberen Teil mehr minder weitgehend mit Schuppenblättchen besetzt zu sein [2]).

An den eigentlichen L i c h t s p r o s s e n ist die B e b l ä t t e r u n g bei den beiden Gattungen verschieden. Bei *Psilotum* sind die Laubblätter lediglich auf kleine leitbündellose Schuppen reduziert; bei *Tmesipteris* finden sich wohl-entwickelte Blätter von lederiger Textur und von Anfang an vertikaler Ausrichtung der Spreite. In die Blätter von *Tmesipteris* tritt ein Leitbündel ein, das aber nicht zu ihrer Spitze, sondern nur etwas bis zur Mitte weit durchzieht.

Die W u c h s f o r m d e r L i c h t s p r o s s e ist eine verschiedene; sie ist bei *Psilotum* in der Regel aufrecht und ziemlich reichlich gabelteilig, bei *Tmesipteris* dessen Lichtsprosse nur wenig gabelteilig sind, sind sämtliche Formen von *T. tannensis* halb aufrechte oder hängende Epiphyten, während *T. Vieillardi* aufrechten Wuchs hat.

Die xeromorphe B l a t t s t r u k t u r bei *Tmesipteris* bezw. die Reduktion der Schuppen bei *Psilotum* erscheint verständlich in Hinblick auf die Tatsache, dass abgesehen von *T. Vieillardi* sämtliche Arten Epiphyten sind.

A n a t o m i e: Die offensichtlich primitivste Fassung der S t e l e liegt bei den Lichtsprossen von *Psilotum* vor, die stellat-protostelisch sind bei wechselnder Zahl (3–10) der Sternvorsprünge im Querschnitt. Protoxylem exarch, gelegentlich Spuren von Sekundärxylem. Kern der Protostele nicht mehr aus Tracheiden sondern aus Sklerenchym gebildet. Bei *Tmesipteris* besteht die Stele aus einer Anzahl um das zentrale Mark geordneter Bündel, die in ihrem Inneren in der Regel selbst noch etwas Parenchym enthalten, bei *T. Viellardi* sind die Verhältnisse insoferne

---

[1]) Dies scheint wenigstens sicher für *Tmesipteris* festzustehen.

[2]) Diese Übergangszone springt auffälliger ins Auge bei *Tmesipteris*, wo am eigentlichen Lichtspross-Teil echte Laubblätter gebildet werden während bei *Psilotum* sich auch am oberen Lichtspross-Teil nur Schuppenblätter bilden.

noch primitiver als bei *T. tannensis* als markständig Xylemstränge vorhanden sind, — ein wohl letzter Rest einer ursprünglich — wie jetzt noch bei *Psilotum* — protostelischen Gesamtfassung des Xylems. Das Phloëm bei *Tmesipteris* ist kein einheitlicher Zylindermantel sondern umgibt mehr minder vollständig jedes der einzelnen Stelenbündel.

Die im Querschnitt aus nur wenigen Tracheiden bestehenden B l a t t s p u r - s t r ä n g e im Lichtspross von *Tmesipteris* (bei *Psilotum* sind wie oben bemerkt — überhaupt keine Blattspuren entwickelt) sind entweder solid oder haben etwas Parenchym in der Mitte.

Hervorzuheben ist, dass die Parenchymstellen zwischen den einzelnen Xylem-bündeln der Stelenperipherie von *Tmesipteris* keinerlei Beziehung zu den Blatt-spurabgängen aufweisen also keineswegs Blattlücken darstellen. In der Ueber-gangsregion von Rhizom zu Lichtsprossen reichen die Blattspurstränge lediglich bis an die Ansatzstelle der auf Schuppen reduzierten Blätter, — eine Erscheinung die in der gleichen Ausführung bei der Psilophytale *Asteroxylon mackiei* festgestellt ist.

V e g e t a t i v e  V e r m e h r u n g kann bei *Psilotum triquetrum* zufolge Brut-körperbildung aus den Rhizoiden der Rhizome stattfinden. Es handelt sich um Zellkörper, die zunächst und auch noch nach Ablösung von der Mutterpflanze chlo-rophyllfrei sind und holosaprophytisch leben, die aber nach allmählicher Erstar-kung chlorophyllhaltige Lichtsprosse entwickeln können.

Die S p o r o p h y l l e sind in ihrer Deutung sehr umstritten. Tatsache ist, dass bei *Psilotum* ein dreiteiliges Sporangium zwischen den Teilen des in 2 Stummel kollateral geteilten Schuppenblattes sitzt, und dass bei *Tmesipteris* ein zweiteiliges Sporangium zwischen den kollateralen Teilstücken des in etwa ein Drittel seiner Länge geteilten Sporophylls angebracht ist. Es handelt sich somit im Prinzip bei b e i d e n  G a t t u n g e n um e i n  u n d  d i e s e l b e Erscheinung, die nur dadurch äusserlich stärker verschieden erscheint, dass bei *Psilotum* die Sporophylle wie die vegetativen Blätter auf keine Stummel reduziert sind im Gegensatz zu *Tmesipteris*, wo vegetative Blätter und Sporophylle bei vertikaler Ausrichtung breitflächig ent-wickelt sind. Es ist aber zu betonen, dass bei *Psilotum* u n d *Tmesipteris* die vege-tativen Blätter einfach, die Sporophylle dagegen kollateral geteilt sind. Vergleichs-weise belanglos ist, dass die Sporangien bei *Psilotum* in der Regel 3-teilig, bei *Tmesipteris* nur 2-teilig sind, wobei, — beiläufig bemerkt, — es im Prinzip nicht viel anders ist, wenn man die Sporangien, statt sie als geteilt aufzufassen, als Spo-rangiengruppen von 2 bezw. 3 weitgehendst miteinander verwachsenen Sporangien betrachtet.

Das Lageverhältnis und die morphologische Beziehung des Sporangiums bezw. Synangiums zu dem sterilen laubigen bezw. schuppigen Teil des Sporophylls wird wohl am besten erfasst, wenn mann annimmt, dass das G e s a m t sporophyll der Psilotaceen als ein serial geteiltes aufzufassen sei, wobei der a b axiale noch einmal kollateral teilende Abschnitt laubig bezw. schuppig ausgebildet und wobei der a d axial stehenden Abschnitt fertil und zum Sporangiophor wird, an dessen Distal-ende das erwähnte 2 bezw. 3 geteilte Sporangium bezw. das 2 oder 3 teilige Synan-gium getragen wird. Der Trägerstiel des Sporangiophors, dessen Länge stets exakt der ungeteilten Strecke des kollateral geteilten abaxialen Sporophyll-Abschnittes entspricht, ist so weitgehend mit dem letzteren verwachsen, dass es den Anschein hat, als sässe das Sporangium bezw. das Synangium zwischen den kollateralen Teil-hälften des abaxial serialen Teilstückes.

Es ist offensichtlich, dass eine derartige Auffassung die Sporophyll-Gestaltung der Psilotaceen in weitgehendem Masse mit der vieler *Articulatales*, insbesondere von *Sphenophyllum*, in Beziehung bringt, wobei ausdrücklich betont sein soll, dass es sich dabei zunächst lediglich um eine rein formale morphologische Konvergenz im *morphogenetischen* Sinn handeln kann, nicht aber um Dinge, die als Ausdruck einer *phylogenetischen* und somit engeren Beziehung der genannten Gruppen gedeutet werden brauchen.

Die S p o r a n g i e n w a n d ist mehrschichtig, wobei der Inhalt der inneren Schichten im Laufe der Entwicklung des Sporangiums weitgehendst für die Aus-bildung der Sporen mit verwandt wird. Das Exothezium (äusserste Sporangium-wand) weist keine besondere Wandverdickung auf, wohl aber sind die Seitenwände verholzt. Tapetenzellen sind nicht entwickelt. Dagegen ist das sporogene Gewebe selbst durchsetzt von einer Anzahl steriler Zellen die den Sporenmutterzellen homolog sind, physiologisch jedoch die Funktion der Tapetenzellen erfüllen.

**3. Zur Stammesgeschichte.** Es ist ganz offenbar, dass die *Psilotales* in vieler Hinsicht morphologisch ausserordentlich interessante Verhältnisse bieten und dass das Verständnis dieser Verhältnisse erst gegeben wird unter Heranziehung fossiler Formen gleichgültig, ob man diese nun stammesgeschichtlich damit enger verwandt auffasst oder nicht.

Im folgenden sei nochmals zusammenfassend herausgestellt, was die Gruppe der Psilotales an besonders interessanten Strukturen aufweist.

1. Prinzipielle Gleichheit zwischen Gametophyt und dem blattlosen Jugendzustand des Sporophyten; dass der eine der Haplophase angehört und selbstredend Geschlechtsorgane trägt, während der andere der Diplophase angehört, tut nichts zur Sache. Das wesentliche ist, dass beide, Gametophyt und junger Sporophyt, in einer Gestaltung vorliegen, die bei den einfachsten uns bekannten Pflanzen: den Psilophytalen *Rhynia* u.a. durch die ganze Phase des Sporophyten und wohl sicher auch des uns derzeit noch nicht bekannten Gemetophyten vorgelegen haben.

2. Bipolare Gestaltung des Gametophyten von Psilotum: Es ist ganz offensichtlich, dass wie schon oben hervorgehoben, Bipolarität den primitivsten Gestaltungstypus verwirklicht.

3. Sporophyllgestaltung. Es ist schon darauf hingewiesen worden, dass nach der hier vorgetragenen Auffassung über die der Sporophyllgestaltung zugrunde liegende seriale und kollaterale Teilung weitgehende formale Beziehungen aufgedeckt werden zu den Verhältnissen, wie sich bei vielen Articulatales, insbesondere bei *Sphenophyllum* und Verwandten finden.

Es sei hier aber noch darauf hingewiesen, dass es mangels exakter, unter dieser Fragestellung durchgeführter entwicklungsgeschichtlicher Untersuchungen zur Zeit nicht zu entscheiden, wohl aber sehr gut vorstellbar ist, dass der Sporophyll-Gestaltung aller mikrophyllen Typen ein prinzipiell gleicher Gestaltungsvorgang, eben der der serialen Durchteilung der Gesamt-Sporophyll-Anlage zugrunde liegt; es ist nicht von der Hand zu weisen, dass es sehr wahrscheinlich ist, dass sich die Sporophyll-Gestaltung aller Lycopodiales auf einen ähnlichen Vorgang zurückführen lässt wie sich ja auch — um einen Blick in den Bereich der Gymnospermen zu werfen, — die Gestaltung der Megasporophylle der Coniferen auf einen prinzipiell gleichen Vorgang: den der primären Serialteilung hat zurückführen lassen.

Wiewohl es sich bei derartigen Gestaltungsvorgängen gegebenenfalls nur um formale Konvergenzen im Sinn rein morphogenetischer Vorgänge handeln kann, ist doch anderseits möglich dass die hier angedeuteten der Sporophyllgestaltung der mikrophyllen Typen voraussichtlich zugrunde liegende seriale Teilung, eben weil sie voraussichtlich allen mikrophyllen Typen gemeinsam ist, als Ausdruck ihrer phylogenetischen engeren Beziehungen untereinander gedeutet werden kann.

Bezüglich der *Literatur* sei Verwiesen auf die Darstellungen in F. O. Bower (Primitive Land Plants) und K. Goebel (Organographie) sowie auf die schönen Einzeluntersuchungen von A. A. Lawson, J. E. Holloway und B. Sahni.

CHAPTER XIX

ARTICULATAE

von

MAX HIRMER (München)

Während derzeit nur noch die einzige Gattung *Equisetum* mit kaum mehr
als 30 Arten lebt, hat die Gruppe der Articulatales zur Zeit des jüngeren
Palaeozoikums eine vergleichsweise sehr erhebliche Formenfülle und Mannigfaltigkeit entwickelt. Eine grosse Anzahl von z. T. artenreichen Gattungen ist aus dieser Zeit auf uns gekommen, teils an Hand strukturbietender Reste, teils an Hand von inkohltem Material oder im Abdruck.
Ueber die Zeit des eigentlichen Höhepunktes der Kraft-Entfaltung der
Articulatales hinaus haben sich weiterhin noch bis ins Mseozoikum eine
Anzahl Gattungen erhalten; so sind besonders aus der jüngeren Trias noch
stattliche Formen bekannt. Wenn heute nur krautige Formen existieren,
während noch aus dem älteren Mesozoikum, vor allen aber dem jüngeren
Paläozoikum viele viel stattlichere und z. T. sogar baumförmige Articulatales bekannt sind, so darf daraus nicht generall auf einen allmählichen
Schwund der konstitutionellen Kraft der Gruppe geschlossen werden. Sind
doch krautige, unseren kleineren Schachtelhalmen recht ähnliche Formen,
schon aus dem Karbon bekannt, und war — wie das übrigens auch für die
*Lycopodiales* gilt — die Existenz der grossen baumförmigen Typen an die
im Karbon u. Rotliegenden herrschenden Klimaverhältnisse angepasst und
allergrösstenteils zufolge des zur sind diese Permzeit einsetzenden Klima-
Umschwunges umgekommen.

Für die S p e z i e l l e  S y s t e m a t i k der Gruppe gilt Folgendes.
Bekannt sind:

I) **Protoarticulatineae**
   1. C a l a m o p h y t a c e a e :
   *Calamophyton* KR. U. WLD. ⎫
   2. H y e n i a c e a e :          ⎬ (Mitteldevon).
   *Hyenia* NATH.                   ⎭

II) **Pseudoborniineae**
   P s e u d o b o r n i a c e a e :
   *Pseudobornia* NATH. (Oberdevon)

**III) Sphenophyllineae**
    1. C h e i r o s t r o b a c e a e :
       *Cheirostrobus* Scott (Unterkarbon)
    2. S p h e n o p h y l l a c e a e :
       *Sphenophyllum* Bgt. (Oberdevon-Alttrias; hauptsächlich Karbon u. Perm)

**IV) Equisetineae**
    1. A s t e r o c a l a m i t a c e a e :
       *Asterocalamites* Schimp. (Oberdevon-Unterkarbon)
       *Pothocitopsis* Nath. (Unterkarbon)
       *Autophyllites* Grd. Eury (Oberkarbon)
       *Sphenasterophyllites* Sterzel. (Oberkarbon)
    2. C a l a m i t a c e a e :
       *Protocalamites* Scott (Unterkarbon)
       *Mesocalamites* Hirmer (Unter- zu Oberkarbon-Grenze)
       *Calamites* Suckow (Oberkarbon u. Rotliegendes)

        *Arthropitys* Goepp.    ⎫
        *Arthrodendron* Will u. Scott  ⎬ Strukturbietende Stamm- u. Zweigreste
        *Calamodendron* Bgt.   ⎭
        *Astromyelon* Will.   ⎫
        *Myriophylloides* H. u. C.  ⎬ Wurzelreste
        *Pinnularia* L. u. H.   ⎭
        *Asterophyllites* Bgt.   ⎫
        *Annularia* Stbg.   ⎪
        *Lobatannularia* Kawasaki  ⎬ Blattreste u. beblätterte Zweige
        *Calamariophyllum* Hirmer  ⎭
        *Calamostachys* Schimp.   ⎫
        *Palaeostachya* Weiss   ⎪
        *Metacalamostachys* Hirmer  ⎬ Blüten
        *Cingularia* Weiss  ⎪
        *Macrostachya* Schimper  ⎪
        *Huttonia* Stbg.   ⎭
    3. E q u i s e t a c e a e :
       *Equisetites* Stbg. (Karbon-Altkreide).
       *Equisetum* L. (Jung-Kreide bis Jetzt)
       *Phyllotheca* Bgt. (Oberstes Karbon bis Unterkreide)
    4. A n h a n g :
       *Neocalalamites* Halle (Trias)
       *Schizoneura* Schimper (Oberstes Karbon-Jura)

    **I. Protoarticulatineae:** Von den echten Articulaten dadurch unterschieden, dass weder in der Stammverzweigung noch in der Blattanordnung derjenige Exaktheits-Grad erreicht ist, der die echten Articulatales charakterisiert. Sowohl bei *Calamophyton* Kr. u. Wld. als bei *Hyenia* Nath. ist der oberirdische Stamm unregelmässig gabelig verzweigt; an den unteren Stammteilen Blätter noch nicht wirtelig sondern noch unregelmässig gestellt; erst in den höheren Verzweigungsgraden echte Wirtelstellung der Blättchen; diese mehrfach gabelteilig. Sporophylle schildförmig, bei *Calamophyton* mit je einem, bei *Hyenia* mit je mehrerer Sporangien an jeder Schildhälfte, Bei *Hyenia* entspringen die Lichtsprosse starken Rhizomen, in Quirlen zusammenstehend, aber nur die zenithwärts gerichteten austreibend.

    **II. Pseudoborniineae:** Gesamtwuchs ähnlich dem von *Hyenia;* Blätter regelmässig in 4-gliedrigen superponierten Quirlen an den Aesten 2ten, d.i. letzten Grades; bei den Blüten steter Wechsel von fertilen u. sterilen Elementen.

    **III. Sphenophyllineae:** *Sphenophyllum*, Bgt. Artenreiche Gattung. Pflanzen mit sehr langem, aber dünnem, maximal kaum über 1 cm starkem, reich verzweigtem Spross mit deutlicher Gliederung in Internodien und Knoten; an letzteren die Blätter (stets in superponierten Quirlen) und die Abgänge der Aeste und Wurzeln. Spross wie bei *Pseudobornia* und den *Protoarticulaten* noch massiv gebaut, jedoch-im Gegensatz zu den genannten Formen-mit vergleichsweise starker Xylem-Bildung. In der Rinde starke und mehrfache Periderm-Bildung. Primärxylem der Stele im Querschnitt dreieckig, ganz aus weitlumigen Zentripetalmetaxylem-Tracheiden. Protoxylem an den drei Kanten, davon die Blattspurstränge abgehend; Sekundärxylem-Bildung frühzeitig einsetzend; zunächst an den Breitseiten der Xylemsäule, später auch vor den Kanten, vor letzteren die Elemenete kleinlumiger. Wie im Metaxylem vielreihig gestellte Tüpfel an den Radialwänden. Fehlen normaler „Mark"strahlen; diese jedoch vertreten durch Züge kleinlumigeren Parenchyms, die vertikal und horizontal zu den Sekundärxylemtracheiden verlaufen. Im Phloëm Siebröhren und Parenchym. Spätere Phellogene aus dem Sekundär-Phloëm gebildet, Wurzeln di- bis tetrarch mit reichlich Sekundärxylem — und Periderm-Bildung.

Wie die Anatomie ist auch die äussere Erscheinung des Sprosses von der Dreizahl beherrscht: Die Blättchen zu je zwei, drei oder vier mal dreien in superponierten Quirlen. Die in ihrer Grundform keilförmigen Blättchen sind bei den ältesten Formen (des Oberdevon und Unterkarbon) stark zerschlitzt, bei den jüngsten Formen (vom oberen Oberkarbon ab) ungeteilt und meist grösser als bei den älteren Formen. Ein Teil der Arten heterophyll, indem feiner zerteilte Blättchen an den Sprossen niedrigen, unzerteilte an den Sprossen höheren Verzweigungungsgrades zu finden sind. Daher die ursprünglich irrtümliche Auffassung, dass die Sphenophyllen, die in Wirklichkeit offenbar Lianen waren, Wasserpflanzen gewesen seien. Bei manchen homophyllen Arten sind die (hier immer nur 6) Blättchen jeden Quirls nahezu in einer (der tragenden Sprossachse anliegenden) Ebene ausgebreitet, eine Art Blattmosaik bildend.

Blüten von verwickeltem Bau, indem die Sporophyllenlagen, in der Regel serial geteilt und die serialen Teilabschnitte ihrerseits oft nochmals serial und in der überwiegenden Mehrzahl der Fälle auch noch kollateral geteilt sind.

Derartiges wird am besten an Hand des als Vorläufer-Typ der Sphenophyllen anzusprechenden nur in der Blüte, nicht in vegetativen Teilen bekannten *Cheirostrobus* SCOTT dargetan. Hier ist die Sporophyll-Anlage zunächst serial und dann dreifach kollateral durchgeteilt. Dabei sind die abaxialen Abschnitte steril, die adaxialen fertil. Die einzelnen fertilen Abschnitte (Sporangiophore genannt) sind ihrerseits schildförmige Bildungen mit je 4 Sporangien. Die Mannigfaltigkeit, die für die Blüten der *Sphenophyllum-A*rten selbst festgestellt ist, entsteht nun dadurch dass einerseits die einzelnen Sporangiophore in ihrer Fertilität reduziert sein können, indem ihre Schilder nicht 4, sondern nur etwa 2 oder 1 Sporangium tragen und dass anderseits die Zerteilung des ad- und des abaxialen Abschnittes bei den einzelnen Arten bald eine grössere, bald eine geringere ist und dass letzten Endes sogar der ad- und abaxiale Abschnitt fertil sein können.

Das Wesentliche für das Verständnis der *Sphenophyllum*-Blüten besteht einerseits in der Tatsache der offensichtlich primären serialen Durchteilung der Sporophyll-Anlage und andererseits darin, dass sowohl zufolge der starken Reduktion der Fertilität der einzelnen Sporangiophore als der damit allenfalls verknüpften Kürzung der Sporophyllstiele Bildungen zustande kommen können, die eine frappante Aehnlichkeit mit den Sporophyllen der *Lycopodiales* besitzen.

**IV. Equisetineae.** Kerngruppe der Articulaten (vom Devon-Ende bis zur Jetztzeit), mit einer grösseren Anzahl baumförmiger Vertreter im Paläozoikum. Strukturell gekennzeichnet durch den Besitz eines mehr oder minder weiten, zentralen Hohlzylinders im Stamm, den nur ein dürftiger Mantel von Markparenchym einhüllt, und an dessen Rand eine grössere Anzahl kollateral gebauter Primärbündel ziehen. Je nach der Art der Stellung der Blätter d.h. ob diese in superponierten oder alternierenden Quirlen gestellt sind, ziehen die Leitbündel im Stamm gerade hoch oder sind untereinander maschenartig verbunden. Paläozoische Formen oft mit beträchtlicher Holzbildung; indes ist zu betonen, dass neben solchen Formen zweifellos eine ganze Anzahl mit nur beschränkter Sekundärxylem-, aber mit wohl beträchtlicher Rindenparenchym-Entwicklung existierten.

**1. Asterocalamitaceae.** Formen mit mehr minder reich gabelteiligen, in superponierten Quirlen stehenden Blättern. Blüten denen von *Equisetum* ähnlich mit ungeteilten schildförmigen Sporophyllen, gelegentlich mit Unterbrechung der Sporophyllserien durch einen Quirl steriler Blätter. Xylem (bei *Asterocalamites*) mit endarchen Primärbündeln; Primärxylem noch undeutlich in Faszikular- und Interfaszikular-Xylem gesondert, daher noch wie siphonostelisch wirkend.

**2. Calamitaceae.** Vorläufer-Gattung: *Mesocalamites* HIRMER. Noch mit Unregelmässigkeiten in der Alternanz der Blattquirle. (Ueberleitung von *Asterocalamitaceae* zu *Calamitaceae*.)

*Calamites* SUCKOW mit meist grösster Regelmässigkeit in der Organstellung: Blätter in alternierenden Quirlen.

Verschiedene Weisen des Gesamtaufbaues und der Verzweigung der Stämme, verschiedene Typen von Laubblatt-tragenden Sprossen und eine Anzahl verschiedener Blütentypen lassen die artenreiche Gattung mannigfaltig in ihrer äusseren Erscheinung sein. Bedauerlicherweise sind in vielen Fällen die einzelnen (noch unter eigenen „Gattungs''-namen gehenden) Organkategorien nicht in Bezug auf ihre Zusammengehörigkeit untereinander geklärt, was mit der Fragilität der Pflanzen zusammenhängt. Immerhin ist das Bild im ganzen klar.

Der prinzipielle Gesamtaufbau ist dem von *Equisetum* vergleichbar, sowohl in der Gliederung der Stämme in Knoten (daran Blätter, Aeste und Wurzeln) und in Internodien als in der Wachstumsweise: wie bei *Equisetum* haben sich von im Boden weit verzweigten kriechenden vieljährigen Rhizomen die oberirdischen Sprosse erhoben.

Hinsichtlich der Beblätterung ist zu unterscheiden zwischen den schmalen unter einander nicht verwachsenen Blättern, die an den Knoten der eigentlichen Stämme stehen und den wesentlich grösseren und breiteren Blättern, die an eigenen, vorwiegend der Assimilation dienenden Spross-Systemen getragen sind.

Je nach der Art der Stellung und Fassung dieser B l ä t t e r unterscheidet man eine Anzahl von Formen, die, da man ihre Zugehörigkeit zunächst noch nicht kannte, oder auch heute teilweise noch nicht kennt, mit eigenen Hilfsgattungs-Namen belegt sind: *Asterophyllites* BGT. ein bis mehrfach verzweigte Sprosssysteme wobei die verhältnismässig langen und schmalen Blätter der einzelnen Quirle so nach oben gebogen sind, dass sie in ihrer Gesamtheit wie ein Becher um den sie tragenden Spross herum stehen. *Annularia* STBG. mehrfach verzweigte Sprosssysteme, wobei die Blätter jeden Quirls in einer einzigen Ebene ausgebreitet sind. Es ist äusserst wahrscheinlich, dass letzten Endes die Blättchen aller Blattquirle in eben derselben Ebene ausgebreitet waren, in welcher die Gesamtverzweigung der mehrfach, stets aber nur in einer einzigen Ebene zwei-zeilig verzweigten Beblätterungssysteme erfolgt ist. *Lobatannularia* KAWASAKI. Bei diesen ähnlich wie bei *Annularia* gebauten Verzweigungssystemen ist aus der ganz in einer Ebene erfolgenden Verzweigung und Blattorientierung die letzte Konsequenz gezogen, insoferne als die dem Licht abgekehrten Blättchen jeden Quirls wesentlich kürzer als die ihm zugekehrten sind, sodass ein ausgesprochenes Blattmosaik entsteht.

B l ü t e n vielgestaltig: neben einer Anzahl nur mangelhaft bekannter Formen (*Macrostachya* SCHIMPER und *Huttonia* STBG.) sind andere strukturell sehr wohl aufgeklärt. Für alle gilt, dass ein steter Wechsel zwischen Sporangien-tragenden, in der Regel schildförmigen Elementen (Sporangiophoren) und sterilen Hüllblättchen (Brakteen) vorhanden ist. Dass dieser stete Wechsel wie bei *Cheirostrobus* und den Spenophyllen als zufolge serialer Spaltung der Sporophyll-Einheit zustande gekommen auszudeuten sei, liegt nahe anzunehmen.

Bei *Calamostachys* SCHIMPER stehen fertile und sterile Elemente in gleichmässigen Vertikalabständen übereinander, die Glieder der fertilen Quirle sämtliche superponiert, die der sterilen sind in grösserer Zahl vorhanden und alternieren untereinander; bei *Palaeostachya* WEISS stehen die fertilen Elemente jeweils in der Achsel der nächst unteren sterilen. Bei *Metacalamostachys* HIRMER liegen die Dinge umgekehrt, sodass — unter Koppelung steriler und fertiler Elemente — die fertilen u n t e r d e n sterilen, also abaxial liegen. Das gleiche gilt für die häufigere Art (*C. typica*) der Gattung *Cingularia* WEISS [1]), einer Form, die sich von den übrigen

---

[1]) bei *C. cantrilli* KIDSTON sind ad- u n d abaxialer Abschnitt fertil.

eben aufgeführten Blüten dadurch unterscheidet, dass die fertilen Abschnitte nicht schildförmig, sondern flächig ausgebildet sind und an ihrem äusseren Unterrand 4 zu je zweien hintereinander liegende Sporangien tragen. Dass diese letzteren Formen eine Ausdeutung der Gesamtblütenstruktur im Sinne der *Sphenophyllum*-Blüten sehr nahe legen, liegt auf der Hand.

Tragung der Blüten verschieden; entweder grosse, mehrfach verzweigte Blütenstände mit kleinen Einzelblüten oder vergleichsweise grosse Blüten einzeln oder zu wenigen am alten Holz der Stämme entspringend.

Betont sei, dass weder die Art der Beblätterung noch die der Blüten für bestimmte der unten zu schildernden drei Grundformen der Stämme spezifisch sind, dass vielmehr die einzelnen Stammgrundformen zum Teil diesen, zum Teil jenen Typ von Beblätterung und Blüten getragen haben.

Mannigfaltig wie Beblätterung und Blüten bei den Calamiten ist die Art der Ausbildung ihrer Stämme.

Die s t r u k t u r b i e t e n d erhaltenen Stammreste werden auf Grund ihrer feineren Anatomie in 3 Hilfsgattungen gruppiert: *Arthropitys* GOEPPERT, *Arthrodendron* WILL. u. SCOTT, *Calamodendron* BGT. Der Leitbündelverlauf ist zufolge der Tatsache, dass die Bündel der aufeinander folgenden Internodien alternieren und die Bündel an den Knotenlinie miteinander in Verbindung treten, maschig. Es hat sich aber zeigen lassen, dass der Bündelverlauf der Calamiten eigentlich ein eustelischer ist und die Maschenbrücke am Knoten nur eine sekundäre und in der Regel schwächere Bildung ist.

Die Holzbildung am Stamm ist bald eine sehr beträchtliche gewesen, *C. gigas* u.a., zum Teil müssen die Stämme Zeit ihres Lebens von mehr parenchymatischer Beschaffenheit gewesen sein, mit weiter Markhöhle und nur geringer Holzbildung, wie z.B. *C. undulatus* und im Extrem *Arthropitys Jongmansi*. Uebrigens herrscht bei denjenigen Formen, bei welchen eine stärkere Sekundärxylem-Entwicklung statt hat, insoferne eine Verschiedenheit als manche (*A. Hirmeri*) eine Sekundärxylem-Bildung aus nahezu n u r Tracheiden aufweisen, andere (*A. Felixi*) Sekundärxylem unter Beimischung s e h r reichlichen Parenchyms zeigen, wieder andere (z.B. *A. communis*) diesbezüglich die Mitte halten.

Es muss noch darauf hingewiesen werden, dass die p h y s i o l o g i s c h e A n a t o m i e der Calamitenstämme uns noch Rätsel aufgibt. Wir haben gute Gründe anzunehmen, dass die weiten Atemhöhlen im Sprosszentrum der Durchlüftung des Stammes gedient haben; fest steht jedenfalls, dass bei einsetzendem sekundären Dickenwachstum ihr Durchmesser konkordant zufolge eines sehr beträchtlichen Dilatationsvermögens der primären Markstrahlen immer mehr und mehr vergrössert worden ist.

Wir wissen auf Grund strukturbietender Reste auch, dass die primäre Rinde zum mindesten gewisser Calamiten eine Menge weiter, lakunöser Räume enthalten hat, und wir wissen auch, dass mittels der an den Knotenlinien befindlichen Infra- und Supranodalkanäle eine Art Querverbindung zwischen der Markhöhle einerseits und diesen lakunösen Räumen der Aussenrinde durch das ganze Sekundärxylem und das anschliessende Phloëm hindurch stattgefunden hat; aber wir haben keine Anhaltspunkte dafür, dass die aussergewöhnlich glatte Stammrinde der Calamiten ihrerseits irgendwelche Stellen besessen habe, die den Lentizellen unserer Bäume entsprächen und mit Hilfe deren und mit dem zentrale Hohlraum der Markhöhle in einer direkten Verbindung mit der die Pflanzen umgebende Aussenluft gestanden wäre.

Es versteht sich von selbst, dass die Calamiten an ihren B a s a l p a r t i e n ganz ähnlich verzweigt gewesen sind wie wir das für die heute lebenden Schachtelhalme kennen (vgl. S. 520), dass also auch aus der Basis der Stämme immer wieder von neuem Stammanlagen sich entwickeln konnten, falls der eine oder der andere der schon gebildeten Stämme durch irgend einen Umstand zum Absterben und zum Zusammenbrechen kam und wir müssen auch auf Grund einer ganzen Anzahl von Funden annehmen, dass immer eine Anzahl einem einzigen Individuum angehörende Stämme sich nebeneinander erhoben haben und dies sicher sogar auch bei Formen mit reicher Astbildung [1]) statt gehabt hat.

---

[1]) Dass derartiges — mutatis mutandis — auch heute möglich ist, wissen wir von dem für manche Coniferen (*Sequoia sempervirens*, *Thuja gigantea* u.a.) Bekannten

W u r z e l n an den Knoten der unterirdischen Sprossteile entweder ringsum gleichmässig um die Knotenperipherie verteilt, und zwar einzeln oder zu grossen Büscheln zusammenstehend.

Nach ihrer ä u s s e r e n E r s c h e i n u n g s f o r m lassen sich die Calamiten-Stämme folgendermassen gruppieren:

A. *Gruppe Eucalamites* WEISS. Hieher werden gezogen diejenigen Sprossreste, bei welchen an allen oder doch fast allen Knoten der Stämme Blatt-tragende Aeste getragen wurden. Je nach der Gliederzahl der Astquirle unterscheidet man die Formen der:

*a) Carinatus-Gruppe:* Aeste an jedem Knoten nur zu zweien oder in Einzahl.

*b) Cruciatus-Gruppe:* Aeste zu mehr als zwei in alternierenden Quirlen an allen Knoten oder so, dass wie bei manchen Arten (z. B. *C. Brongniarti*) einer grösseren Periode Ast-tragender Knoten einer ohne Astabgänge folgt; vegetative Aeste und Blüten untereinander am selben Knoten abwechselnd.

Es ist nicht ausgeschlossen, dass die hieher gezogenen Sprossreste z.T. nur die ± oberen Teile von Calamiten darstellen, deren Stämme zunächst nicht Blatt- und Blütenäste trugen, somit zunächts im wesentlichen unverzweigt (also vom Typ *Stylocalamites*) waren; andererseits finden sich innerhalb der Gruppe *Eucalamites* in der derzeitigen Fassung auch zweifellos Formen, die eucalamitoide Verzweigung schon von der Basis an besassen, wie z.B. der in allen Teilen bestbekannte *C. carinatus*.

B. *Gruppe Calamitina* WEISS. Stämme dadurch ausgezeichnet, dass Asttragung nur an in bestimmten Intervallen aufeinander folgenden Knoten stattfindet, sodass zwischen den die Aeste (in grösserer Anzahl quirlig) tragenden Knoten immer eine ganze Anzahl von Knoten sich befindet, an denen keinerlei Astabgänge statthaben (umgekehrte Periodizität wie bei *Eucalamites Brongniarti*). Die Gestaltung und vor allem die relative Länge der zwischen den Ast-Knoten befindlichen Internodien ist dabei eine nach den Arten verschiedenen. Bezüglich dieser Gruppe im ganzen ist zu bemerken, dass es nicht sicher feststeht, ob nicht etwa die hier zugerechneten Stammpartien nur die oberen Teile grosser in ihren unteren Teilen unverzweigter Calamiten-Stammteile vom Typ *Stylocalamites* (vgl. unten) gewesen sind, oder ob die alle nicht sehr kräftigen Stämme von unten her den für *Calamitina* geschilderten charakteristischen Verzweigungsmodus besessen haben.

C. *Gruppe Stylocalamites* WEISS. In diese Gruppe rechnet man alle diejenigen Formen von *Calamites*, an deren Stammresten eine Verzweigung in der Regel nicht zu beobachten ist; unter ihnen kennt man kleinere Formen, wie den von uns genau studierten *C. Schulzi*, dessen Stämme bei nur wenig Meter Höhe und kaum einem Dezimeter Durchmesser, in ihrem oberen Teil an den Knoten mit verhältnismässig langen Blättern besetzt waren und die grossen Blüten in Menge getragen haben. Andererseits kennt man aber auch ziemlich grosse Formen, wie den sehr häufigen *C. Suckowi*, sowie den möglicherweise die Oberpartie von *C. Suckowi* darstellenden *C. Cisti* und letzten Endes muss als eine besonders grosse Form *C. gigas* genannt werden, deren Stamm, bei starker Holzbildung, bis über 1 m Durchmesser erreicht hat. Ob diese letzteren grossen Formen, gleichgültig ob sie nun starke Holzbildung wie *C. gigas* hatten oder mehr krautiger Konsistenz wie *C. Suckowi* und *C. Cisti* gewesen sind, in ihrei ganzen zweifellos sehr beträchtlichen Länge völlig unverzweigt gewesen sind, ähnlich wie das für den kleineren, oben erwähnten *C. Schulzi* feststeht, ist zweifelhaft. Astabgänge sind jedenfalls an den zum Teil sehr zahlreichen Resten dieser Stämme kaum bekannt. Andererseits ist schwer vorzustellen, wie diese Stämme ohne alle Blatt-tragenden Zweige die nötige Menge von Assimilaten zusammen gebracht haben konnten; vor allem wäre es da.nicht verständlich, wo, wie bei *C. gigas* zufolge starker Sekundärxylem-Bildung Durchmesser von, wie bemerkt, über 1 Meter erreicht, also grosse Massen organischen Materials gebildet worden sind und andernteils doch sogar angenommen werden muss, dass der das Assimilationsparenchym enthaltende p r i m ä r e Rindenmantel zufolge der starken Ausdehnung der Sekundärxylems mehr minder bald gesprengt und abgeworfen worden sein dürfte. Die dem Stamm selbst ansitzenden, offenbar bei allen Arten

---

her, wo trotz grosser Kronenentwicklung eine ganze Anzahl einem einzigen Individuum angehörende Stämme dicht gedrängt dem Boden entspriessen.

sehr schmalen Stammblätter dürften in gar keinem Falle für die Assimilationsarbeit viel bedeutet haben,

Es ist schon betont worden, dass mit dem Ende des Paläozoikums und dem Umschwung der Klimaverhältnisse, mit anderen Worten: mit dem Heraufkommen einer aussergewöhnlich trockenen, letzten Endes über weiten Gebieten unserer Erde zur Wüstenbildung führenden Epoche, das Ende der grossen baumförmigen Articulaten-Gestalten gegeben war; wenn auch kleinere Calamiten-Formen noch bis in den Zechstein herein gereicht zu haben scheinen, so bedeutet eben doch das Ende dieser Zeit auch endgültig das Ende der Calamiten als solchen. Für die Trias geblieben sind nur die:

**3. Equisetaceae.** *Equisetites.* Die bereits im Karbon an Hand kleiner krautiger Formen nachgewiesene Gattung gleicht im Prinzip der jetzt lebenden Gattung *Equisetum*; nur dass sie in z.T. wesentlich beträchtlicheren Dimensionen aus dem Buntsandstein (*E. Mougeoti* BGT.), und aus dem unteren und mittleren Keuper bis zum Lias bekannt ist. Von den diesbezüglichen Formen hat das sehr häufige *E. arenaceus* JÄGER einen Stamm von noch bis 25 cm Durchmesser erreicht bei bis zu 130 Blättern im Knoten und mehr oder minder reichlicher Verzweigung. Seine rein *Equisetum*-artigen Blüten dürften in Gruppen zu mehreren an kleinen ihrerseits massenhaft aus der Stammspitze hervorbrechenden Aesten getragen worden sein.

Im älteren Mesozoikum finden sich noch die schon im Perm beginnenden Gattungen *Phyllotheca* BGT., *Neocalamites* HALLE, und *Schizoneura* SCHIMPER; *Phyllotheca* ist dadurch interessant, dass die Sporophylle in einer Anzahl von Quirlen, aber immer wieder in Abwechslung mit einem Quirl steriler Blätter am Sprosse stehen.

**Equisetum** L. Einzige lebende Gattung mit ca. 32 Arten. Artikulation im Spross-Aufbau besonders betont, insoferne als zufolge der scheidigen Verwachsung der Blätter die unmittelbar über den Knoten anschliessenden Internodienpartien in ihrer Ausbildung etwas anders gefasst sind als die restlichen über die Blattscheiden hervorragenden Teile der Internodien: an dem umscheideten Teil ist die Epidermis weniger stark mechanisch versteift, der Spross daher jeweils unmittelbar über dem Knoten besonders brüchig.

S p r o s s - A u f b a u mittels tetraedrischer Scheitelzelle, primäre Gestaltung somit eine zunächst schraubenförmige, trotzdem die gesamte Organabgliederung präzis wirtelig ist. [1]

B l a t t. Differenzierung derart, dass bei den vegetativen Blättern nur die jeweils oberen Teile der Gesamt-Blattanlage zur Blattbildung verwendet werden, während der untere Teil zur Berindung des Spross-Internodiums dient. Sporophylle dagegen aus der Gesamt-Blattanlage hervorgehend. [2]

[1] Es wäre aber irrtümlich zu schliessen, dass demzufolge das Moment der quirligen Organbildung, wie sie für alle e c h t e n Articulaten charakteristisch ist, eine abgeleitete Erscheinung sei; ist es doch vielmehr sehr viel wahrscheinlicher, dass es eine abgeleitete Erscheinung ist, wenn der Sprossscheitel des heutigen *Equisetum* mittels einer Scheitelzelle und nicht mit einem vielzelligen Meristems arbeitet; ist letzteres doch zweifellos der ursprünglichere Typ von Scheitel-Fassung und ist es auch schwer anzunehmen, dass die sehr breiten Scheitel, wie sie noch zum Teil die Sprosse der mesozoischen *Equisetites*-Arten und vor allem vieler paläozoischer *Calamites* besessen haben, mittels einer Scheitelzelle an Stelle der alle primitiveren Pteridophyten charakterisierenden vielzelligen Scheitelmeristems gewachsen seien.

[2] Bedenkt man, dass für einen gut Teil der Articulaten seriale Spaltung der Sporophylle als so gut wie sicher angenommen werden darf, und bedenkt man weiterhin, dass die entwicklungsgeschichtlichen Bilder des *Equisetum*-Sporophylls derartige sind, dass in ihnen eine ähnliche Spaltung in einen distalen und einen proximalen Teil statt haben könnte, wie er sich aus der Anlage der sterilen Blätter

S p o r o p h y l l rein schildförmig [1]), an der Schildfläche — und zwar primär randständig — die 8–10 nachträglich gegen die Unterseite der Schildfläche verschobenen Sporangien entspringend.

---

tatsächlich vollzieht, so scheint es wohl möglich, dass bei *Equisetum* noch generelle Dinge nachklingen, die in einem anderen Verwandtschaftskreis der Articulaten, insbesondere dem von *Sphenophyllum* und Verwandten, sich weiter gehend ausgewirkt haben und, in Form einer ganz offensichtlich erkennbaren s e r i a l e n O r g a n s p a l t u n g in Erscheinung tretend, die eigenartige Sporophyll- und Blütenfassung jener Formen bedingt haben. (vgl. hiezu das S. 513 Gesagte).

[1]) Es sei noch hervorgehoben, dass die Sporophylle von *Equisetum* zwar — ihrer Entstehung nach — zweifellos dorsiventrale Organe sind, dass aber über diese primäre Dorsiventralität hinweg ihre Gestaltung eine offensichtlich radiäre, eben allseits rein schildförmige, ist.

Derartige schildförmige Bildungen sind bis in die neueste Zeit zweifellos falsch gedeutet worden, indem — wie das auch noch z.B. in GOEBEL's Organographie dargelegt wird — sie sich aus zwei Teilen: einem oberen, welcher der Blattfläche steriler Blätter entspräche und einen unteren, der einen Auswuchs von deren Unterseite darstelle, zusammensetzen sollen. Es ist aber mit allem Nachdruck zu betonen, dass derartige Deutungen auf Grund der paläontologischen Befunde an schildförmigen Bildungen bei Formen der verschiedensten — und zwar mikro- wie makrophyller -Verwandtschaftkreise völlig abwegig sind. Wir wissen jetzt sehr wohl, dass die letzten Ausgliederungen an den wedelartigen Pteridospermen-Sporophyllen letzten Endes schildförmig sein können und dass an solchen schildförmigen Bildungen die synangial gefassten Sporangien sich randständig befinden (so seien diesbezüglich für Mikrosporophylle *Crossotheca*, *Aulacotheca*, *Whittleseya* u.a., für Megasporophylle *Lepidopteris* genannt); und es muss betont werden, dass derartige schildförmige Bildungen sich zurück führen lassen auf glockenförmig eingetiefte, in ihrem sporogenen Komplex in eine Anzahl Teile gegliederter, in ihrer Grundform jedoch noch von dem eiförmigen *Rhynia*-Sporangium ableitbarer Sporangientypen, wie sie für die Psilophyten an Hand von *Yarravia* nachgewiessen sind. Es spielt dabei keine Rolle, ob es sich dabei handelt um endständige Bildungen an Systemen, die noch nicht in Spross und Blatt gegliedert sind, wie das ja für die *Psilophytales* von *Rhynia*-Verwandtschaft zutrifft, oder ob es sich handelt um die letzten und endständigen Ausgliederungen grosser vielfach verzweigter Wedelbildungen, wie sie bei den *Pteridospermae* vorliegen, oder ob es sich handelt um die kleinen, in sich weiter nicht verzweigten Ausgliederungen mikrophyller Pflanzen-Typen wie *Equisetum* u. Verwandten; und es muss noch betont werden, dass es bei der Beurteilung dieser letzteren Formen auch belanglos ist, ob es sich, wie bei *Equisetum* u. anderen um ein echtes ganzes Sporophyll, oder wie bei den Sporangiophoren von *Sphenophyllum* und *Cheirostrobus* und vielleicht auch von *Calamostachys* und *Palaeostachya* um die serialen und kollateralen Teilabschnitte von Sporophyllen handelt. Das wesentliche ist, dass es abwegig wäre, derartige Bildungen um jeden Preis von einem dorsiventralen Blatt abzuleiten; handelt es sich doch vielmehr bei ihnen um eine zwar in der Mehrzahl der Fälle seitenständig ausgegliederte Organbildung, die zwar theoretisch dorsiventral, praktisch aber stets von vornherein radiär gestaltet ist. Das was bei der formalen Betrachtung solcher Phänomene noch viel zu sehr übersehen wird, was aber zu ihrem wirklichen Verständnis nicht genug hervorgehoben werden kann, ist die Tatsache,

S p o r a n g i e n typisch eusporangiat, d.h. mit bis zur Reife mehrschichtiger Wand. Präzises Stomium nicht vorhanden; lediglich durch besondere Zellanordnung nahe der späteren Aufrisstelle angedeutet. S p o r e n mit Perispor, dieses in 2 sich kreuzende Bändchen aufgeteilt, deren Hälften bei Austrocknung sich ausstrecken und das Aneinanderhaken der Sporen ermöglichend das Schwebevermögen in der Luft vergrössern. G e s c h l e c h t s v e r t e i l u n g: Primär halb männlich, halb weiblich; dies aber nur zu folge p h ä n o- nicht genotypischer Geschlechtsbestimmung; [1]) denn es ist letzten Endes so, dass die zu weiblichen Prothallien prädestinierten Sporen bei mangelhafter Ernährung (zu geringer Belichtung und Dichtsaat) zu männlichen herabgedrückt werden können, und dass anderseits ältere weibliche Prothallien, soferne keine Befruchtung der an ihnen gebildeten Archegonien stattgefunden hat, zur Bildung von Antheridien übergehen, also auch auf den männlichen Zustand herab gedrückt werden können.

P r o t h a l l i e n (Abb. 10) als flächenhafte Bildung beginnend, später in mehrzellschichtigen Zustand übergehend, ausgesprochen dorsiventral. An der Basis mehr walzenförmig, gegen das Oberende zu von annährend fächerförmiger Gestalt; an der dem Licht abgekehrten Unterseite werden Rhizoiden, andere dem Licht zugewandten Assimilationslappen gebildet. Am Oberende des mehrzellschichtigen Prothalliumkörpers ein Randmeristem, das nach der Unterseite weiterhin Rhizoiden, nach der Oberseite hin abwechselnd Assimilationslappen und Archegonien bildet. In Fällen, wo weibliche Prothallien letzten Endes vermännlicht werden, kann der ganze Meristem-Rand in die Bildung von Antheridien übergehen. Spermatozoiden polyziliat.

Der E m b r y o von *Equisetum* kann als ein besonders klares Beispiel für den Typus des Pteridophyten-Embryos angeführt werden, nämlich eines Embryos, an welchem nicht, wie das bei den primär stets allorrhizen Samenpflanzen der Fall ist, ein Sprosspol und ein diesem diametral entgegengesetzter Wurzelpol ausgebildet sind, sondern eines Embryos, an welchem die erste Wurzel ebenso wie alle übrigen eine seitenständige Entstehung hat. Im übrigen zeigt der Embryo eine Gliederung in eine zum Sprosspol werdende, dem Archegonienhals zugekehrte (exoskope)

---

dass Symmetrien im Pflanzenkörper sich überschneiden und überdecken können; nicht zuletzt kann als ein Beweis dafür nochmals auf die oben schon erwähnte Tatsache zurückgegriffen werden, dass zwar die primäre Organbildung bei *Equisetum* zufolge der Tätigkeit der Sprosscheitelzelle eine schraubige ist, dass aber trotzdem der Gesamtaufbau des Sprosses letzten Endes ein rein quirliger ist; dass also, — um das Allgemeine hervorzuheben — die zunächst schraubige Symmetrie und damit sukzedane Organentstehung überdeckt wird von quirliger und simultaner Organdifferenzierung und Ausgliederung; und Entsprechendes gilt für die gesamten Schildbildungen: ihrer Anlage nach können sie dorsiventral sein, aber ihre allenfallsige Dorsiventralität wird frühest aufgehoben und überdeckt durch rein radiäre Organgestaltung.

[1]) Eine frühzeitige während der Sporenentwicklung einsetzende phaenotypische Geschlechtsbestimmung hat offenbar auch bei verschiedenen karbonischen heterosporen *Calamostachys*-Arten stattgefunden: Entweder (*C. Casheana*) bildet von 4 Sporangien eines Sporophylls eines nur Megasporen oder (*C. Binneyana*) im selben Sporangium finden sich Tetraden mit 4 Megasporen, oder 3 Mega- und 1 Mikrospore, oder 2 Mega- und 2 Mikrosporen, oder 1 Mega- und 3 Mikrosporen. Derartiges spricht eindeutig für phaenotypische Geschlechtsbestimmung, da genotypische nur während der Reduktionsteilung stattfinden kann und somit im Fall von *C. Casheana* nicht den Sporeninhalt eines g a n z e n Sporangiums erfassen könnte, anderseits im Fall von *C. Binneyana* bei genotypischer Geschlechtsbestimmung Mega- und Mikrosporen innerhalb einer Tetrade immer im Verhältnis 2 : 2 nicht aber 3 : 1 oder 1 : 3 vorhanden sein müssten.

Hälfte und eine dem Archegoniumhals abgekehrte (endoskope) Hälfte, welch-letztere zunächst als haustorialer Embryo-Fuss ausgebildet ist, an der seitenständig die Anlage der ersten Wurzel erfolgt und die schliesslich resorbiert wird. (ABB. 11).

Die *Equisetum*-Pflanzen im ganzen sind reichlich verzweigte Gewächse. Es sind perennierende Pflanzen, die sich immer von neuem aus dem im Boden kriechenden Rhizom verzweigen. Der aus dem Embryo zunächst entstehende erste Spross bildet insgesamt nur wenige Internodien und an seinen Knoten befinden sich in der Regel nur aus 3 Blättern gebildete Scheiden. Verhältnismässig früh, im übrigen aber in ganz analoger Weise wie bei der Verzweigung der Sprosse überhaupt, gliedert sich an seinem untersten Knoten ein Seitensprosse ab, der bereits wesentlich kräftiger gebaut ist und vielgliedrige Blattscheiden besitzt; an ihm finden weitere Verzweigungen statt; bereits einer der Sprosse dritten Verzweigungsgrades kann schliesslich indem er statt zenithaufwärts zu wachsen, mehr minder horizontal oder sogar steil abwärts wächst und in den Boden eindringt, die Anlage des ersten unterirdischen Rhizomsprosses darstellen, der reichlich Wurzeln erzeugt und von dem aus nun zunächst die weitere Abgliederung der im Laufe der kommenden Jahre entstehenden zenithwärts gehenden Sprosse statt hat. Auch aus den Basalpartien dieser letzteren können weitere Ausgliederungen zenithwärts gerichteter Lichtsprosse statthaben.

Die Verzweigung der Sprosse — gleichgültig ob an ober- oder unterirdischen Systemen — erfolgt lediglich an den Knoten, jedoch nicht in der Achsel der scheidig verwachsenen Blätter sondern in Alternanz damit.

Bei manchen Arten entstehen an den unterirdischen Sprossen ein bis wenig gliedrige Sprosse, deren Internodien zu eiförmigen Bildungen anschwellen können und als Reservestoffbehälter dienen, eine Erscheinung die bereits von manchen mesozoischen Equisetites-Arten her bekannt ist, nur dass entsprechend der zum Teil sehr beträchtlichen Grösse dieser Formen ihre Knollen die Dimensionen unserer Kartoffelknollen erreichen konnten.


Die feinere Systematik der Gattung beruht auf der Gestaltung der Spaltöffnungs-Nebenzellen, sowie der Anordnungsweise der Spaltöffnungen, ferner auf der Gesamtausbildung der fertilen Sprosse, auf der Form der Blüte und endlich auf der Endodermis-Ausbildung und Verteilung im Spross.

Es wird unterschieden:

**Sektion I: Eu-Equisetum** SAD. Spaltöffnungs-Nebenzellen in gleicher Höhe wie die übrigen Epidermiszellen; Blüte meist stumpf.

**Subsektion 1 : Heterophyadica** A. BR. fertile und sterile Sprosse verschieden, die fertilen ohne Chlorophyll und ohne Spaltöffnungen.
Gruppe Ametabola (Vernalia) fertile Stengel nach dem Abblühen absterbend.
Gruppe Metabola (Subvernalia) fertile Sprosse nach dem Abblühen ergrünend und wie die sterilen verzweigt.

**Subsektion 2 : Homophyadica** A. BR. (**Aestivalia**) fertile Sprosse von Anfang an mit Chlorophyll und den sterilen Sprossen ähnlich;. Blüten schwarz.
Gruppe des *E. palustre* Sprosse mit einheitlicher Aussen-Endodermis.
Gruppe des *E. Heleocharis* jedes Spross-Leitbündel mit eigener Endodermis.

**Sektion II: Hippochaete** MILDE Spaltöffnungs-Nebenzellen unterhalb des Niveaus der übrigen Epidermis, daher Spalten mit Vorhof; allermeist äussere und innere Gesamt-Endodermis; Blüten spitz; sämtliche von homophyadischem Spross-Typ.
Gruppe Pleiosticha MILDE Spaltöffnungen in je 2 oder mehreren Längsreihen, grösste Arten bis 5 m u. mehrere cm im Durchmesser.
Gruppe Ambigua MILDE Spaltöffnungen an der gleichen Pflanze bald ein-bald mehrreihig.
Gruppe Monosticha MILDE Spaltöffnungen nur ein-reihig.

Zur Literatur. In betreff der älteren Werke zu *Equisetum* sei verwiesen auf R. SADEBECK's Darstellung in ENGLER-PRANTL's Natürl. Pflanzenfamilien. Eine neueste kurzgefasste Bearbeitung der lebenden u. fossilen Formen ist in der 4. Aufl. von R. WETTSTEIN, Handbuch d. Systemat. Bot. (bearbeitet von F. v. WETTSTEIN u. M. HIRMER) zu finden. Für die Kenntnis der f o s s i l e n *A r t i - c u l a t a l e s* sei in erster Linie verwiesen auf die Arbeiten von WILLIAMSON, WILLIAMSON u. SCOTT, HIRMER sowie KNOELL (A n a t o m i e), JONGMANS sowie KIDSTON u. JONGMANS (S y s t e m a t i k, b e s. d. C a l a m i t e n), SCOTT, BOWER sowie HIRMER (G e s a m t - A u f b a u u. B l ü t e n b a u); KRÄUSEL u. WEYLAND (ä l t e s t e F o r m e n) u.a. Z u s a m m e n f a s s e n d e D a r - s t e l l u n g e n in SCOTT's Studies, BOWER's Primitive Land Plants und HIRMER's Handbuch d. Paläobotanik.

CHAPTER XX

FILICINAE

by

CARL CHRISTENSEN (Copenhagen)

**History.** — In the basic work of modern taxonomy, the S p e c i-
e s  P l a n t a r u m  1753 of LINNÆUS, the Pteridophyta formed the first
order, *Filices*, of the 24th class *Cryptogamia*. LINNÆUS described 12 genera
of ferns with 192 species besides 4 genera of "fern-allies" (*Equisetum, Marsilia, Pilularia, Isoetes*), while *Lycopodium* was placed in the second order:
*Musci*. LINNÆUS based his genera exclusively upon the characters of the
sori (superficial or marginal, round, linear or confluent, etc.) thus using
the same characters of genera which almost all pteridologists regarded
as the most important ones for a century and a half. The sporangia
and spores were the only organs of reproduction known for a century and
were believed to play the same rôle as the flowers of the Phanerogams and
their characters were naturally considered as being of principal taxonomic
value. During the latter half of the 18th century several botanists (e.g.
ADANSON and J. E. SMITH) found that the sori of some ferns were covered
with a protecting membrane, the indusium, which was absent in others,
and this difference, not used by LINNÆUS, was found important enough,
not only to split up the Linnæan genera but also to arrange all genera in
two series, *Indusiatae* and *Exindusiatae*, which treatment was followed by
almost all the pteridologists of the 19th century. J. E. SMITH, who described
a number of genera (1793), divided the known genera into two series, *An-
nulatae* and *Exannulatae* (sporangia with or without a ring), but the first
botanist who made a serious attempt to arrange all known ferns in a
convenient sequence was the Swede OLOF SWARTZ in his S y n o p s i s
F i l i c u m (1806). He divided the ferns into three orders: *Gyratae* (= *An-
nulatae* SMITH) with two suborders: *Nudae* and *Indusiatae*, *Spurie
Gyratae* (our *Schizaeaceae, Osmundaceae, Gleicheniaceae* and *Angiopteris*),
and *Agyratae* (*Marattiaceae, Ophioglossaceae*). He described 38 genera of
ferns with 710 species besides numerous dubiae. In 1810 a new enumeration
of the known species was published by C. F. WILLDENOW in the fifth

volume of his edition of LINNÆUS' *Species Plantarum*, but his classification
was rather retrogressive. During the following three decades the genera
were grouped together by R. BROWN, G. F. KAULFUSS, C. L. BLUME, G.
KUNZE and others, into families which are still maintained under their
original names.

All the genera described before 1836 were based on soral characters only,
but in that year C. B. PRESL published his revolutionary work T e n t a-
m e n  P t e r i d o g r a p h i a e. In this and some later works he created a
large number of new genera based on the  v e n a t i o n  and partly on the
number of vascular bundles in the leaf-stalk. His ideas were, however, only
accepted by few pteridologists such as the English gardener JOHN SMITH,
and A. L. A. FÉE (G e n e r a  F i l i c u m  1852), while most writers
retained the old Swartzian classification, WILLIAM HOOKER, for example,
in his large and important work  S p e c i e s  F i l i c u m  (5 vols. 1844–64)
with descriptions of all known species of *Polypodiaceae*. Though HOOKER
in  G e n e r a  F i l i c u m  (1839–42) adopted more of PRESL's and J.
SMITH's new genera, he reduced them to subgenera, totally neglecting
others, thus reverting to the large Swartzian genera. In his compendious
work,  S y n o p s i s  F i l i c u m  (1865–68, new edition 1874), for the
greater part worked out after HOOKER's death by J. G. BAKER and containing descriptions of 2235 species, the conception of genera is the same,
and as this conservative work was the only handbook of ferns during three
decades we find the peculiarity that practically all of the hundreds of
new species described in these years were referred to the same few genera
adopted by SWARTZ a century ago. It must be admitted that the progress
of fern-taxonomy since the days of PRESL was slow and new ideas were few
and rarely accepted. FÉE for example claimed to have found an available
generic character in the number of cells of the annulus, and J. SMITH
(H i s t o r i a  F i l i c u m  1875) divided the *Polypodiaceae* into two
divisions: *Eremobrya* (fronds articulated with the rhizome) and *Des-
mobrya* (fronds adherent) and he succeeded in delimiting several natural
genera or groups combining more characters. His work was overlooked for
many years but it is now generally admitted that his ideas were sound and
that he was really a pioneer of modern pteridology. The keenest pteridologist of the 19th century was no doubt GEORG METTENIUS. He was a pupil
of HOFMEISTER and ALEX. BRAUN and used their comparative method in
his fern-studies, basing his results not only on studies of the external
morphology but also on microscopical researches bearing chiefly on the
development of the sori. He understood the importance of the vegetative
characters such as the venation and he proposed (1856) the classification
of the kinds of venation (venatio Drynariae, v. Goniophlebii, etc., etc.) still
used. His wide-ranging and penetrating studies were interrupted by his
early death (1866) but it is probable that his results, if completed and
published, would have inaugurated a new period of fern-taxonomy,

the development of which the Synopsis Filicum brought to a stand-still.

The theory of evolution, widely accepted after 1880 as a basis for systematic works, had for a long time no effect upon the classification of the ferns. As late as 1897 H. CHRIST in D i e  F a r n k r ä u t e r  d e r  E r d e made the first successful attack against the dominating influence of HOOKER and BAKER. Basing the genera on the structure of the whole plant his classification was a step in the right direction but can still not be said to be a natural one. Far more in accordance with the modern view of phylogeny as the basic principle of systematics was the wholly new classification of the *Polypodiaceae* by L. DIELS in the fourth volume of ENGLER u. PRANTL: D i e  n a t ü r l i c h e n  P f l a n z e n f a m i l i e n (1899), which was followed by CARL CHRISTENSEN in I n d e x  F i l i c u m (1905–6), an alphabetical enumeration of all known ferns (149 genera with about 6000 species). These two works have since been the standard- works of fern-taxonomy and have been followed by nearly all later writers in spite of their unnatural sequence of the families and misconception of some genera and their right systematic position. During the last thirty years much work has been done in order to clear up the phylogeny and mutual relationships of the living ferns partly by comparing them with extinct species, partly by a thorough study of the development of all organs including the gametophyte which never has been considered before by the taxonomists. Among the scientists who have contributed largely in this way to our knowledge K. GOEBEL and F. O. BOWER are the most prominent, the former in his A r c h e g o n i a t e n s t u d i e n  and  O r g a n o-g r a p h i e  d e r  P·f l a n z e n, the latter in numerous papers, the results of which he collected in the important work T h e  F e r n s (3 vols. 1923–28). The studies of these and other botanists have proved that soral characters, although of high systematical value, are alone insufficient to clear up the phylogeny of living ferns; it is now evident that these characters are, in many cases, evolutionary states arrived at within otherwise very different groups; on the other side the structure of the dermal appendages (hairs and trichoms) and other organs, the stelar structure etc. appear to be of high value as genotypical characters. Neither GOEBEL nor BOWER ventured to propose a new classification based upon their numerous important results of their studies which as a whole agree with the conclusions arrived at by taxonomists. Among these E. B. COPELAND alone has tried to arrange the oriental genera of *Polypodiaceae* into a phyletic sequence (1929), but omitting as he does the African and American genera and the other families his essay is incomplete. It contains however many new and fertile points of view and must be seriously considered by anyone, who tries to work out a new phyletic classification of ferns based on the results of modern research. Such an attempt is made in the following scheme which in addition is based on an intimate knowledge of four-fifths of all known ferns, now numbering round 9000 species. It is not conclusive, too many points remaining obscure and it encounters two difficulties which still cannot be overcome. Many important features observed in the anatomical and morphological structure of the sporophyte or gametophyte of single or a few related species cannot be ascribed with certainty to all species of the same genus and one must therefore always remember that hardly one character ascribed to a genus is to be found in all its species. The linear sequence of families and genera moreover gives no clear idea of the highly interwoven phylogeny of recent ferns and it is perhaps impossible to demonstrate to others the reticulated phylum as it appears to the trained taxonomist.

## Some important taxonomic works

(For other works may be referred to CARL CHRISTENSEN: Index Filicum with supplements where a nearly complete list is to be found)

### A. General

J. E. SMITH, Tentamen botanicum de Filicum generibus.... Mém. Ac. Turin v. V. 1793.

O. SWARTZ, Synopsis Filicum. Kiliæ. 1806.

C. F. WILLDENOW, Linné Species Plantarum, ed. quarta, v. V. 1810.

G. F. KAULFUSS, Enumeratio Filicum. Lipsiæ. 1824.

M. DESVAUX, Prodrome de la famille des fougères. — Mém. Soc. Linn. Paris, v. VI. 1827.

H. SCHOTT, Genera Filicum. Vindobonae. 1834.

C. B. PRESL, Tentamen Pteridographiae. — Pragae 1836. — Supplementum. 1845.

————, Epimeliae botanicae. Pragae 1849.

W. J. HOOKER, Genera Filicum. London 1838–42. 120 pl.

————, Species Filicum v. I–V. London 1844–64. 304 pl.

A. L. A. FÉE, Genera Filicum. Paris et Strasbourg. 1850–52. 32 pl.

G. METTENIUS, Filices horti botanici Lipsiensis. Leipzig 1856. 30 pl.

T. MOORE, Index Filicum. Synopsis of the genera. London 1857. 84 pl.

W. J. HOOKER and J. G. BAKER, Synopsis Filicum. London 1865–68. Sec. edition 1874.

J. SMITH, Historia Filicum. London 1875.

M. KUHN, Die Gruppe der Chaetopterides. Berlin 1882.

H. CHRIST, Die Farnkräuter der Erde. Jena 1897.

L. DIELS and others, Pteridophyta in Engler u. Prantl: Die nat. Pflanzenfam. 1899–1900.

CARL CHRISTENSEN, Index Filicum. Hafniae 1905–06. Supplementa I–III. 1913. 1917, 1934.

F. O. BOWER, The Ferns. v. I–III. — Cambridge 1923–28.

E. B. COPELAND, The oriental genera of *Polypodiaceae*. — Univ. Calif. Publ. Bot. v. XVI. 1929.

### B. Illustrated works (general)

C. SCHKUHR, Vier und zwanzigste Klasse.... oder Kryptogamische Gewächse. v. I. Wittenberg 1804–09. 219 pl.

W. J. HOOKER et R. K. GREVILLE, Icones Filicum. London 1827–31. 240 pl.

G. KUNZE, Die Farrnkräuter .... Leipzig 1840–51. 140 pl.

W. J. HOOKER, 1. and 2. century of ferns. London 1854, 1861.

————, Filices exoticae. London. 100 pl.

J. G. BAKER, in Hook. Ic. Plant. pl. 1601–1700. 1886–87.

## C. Local

J. Milde, Filices Europae et Atlantidis.... Lipsiæ 1867.
C. Luerssen, Die Farnpflanzen. Rabenhorst: Krypt. Flora 2. Aufl. v. III. 1884–89.
M. Kuhn, Filices africanae. Lipsiæ 1868.
T. R. Sim, The ferns of South Africa. Capetown 1892. 159 pl. — 2. edition. Cambridge 1915. 186 pl.
Carl Christensen, The Pteridophyta of Madagascar. — Dansk Bot. Ark. v. VII. 1932. 80 pl.
C. L. Blume, Flora Javae. Filices. Bruxelles 1828–29, 95 pl.
R. H. Beddome, Ferns of Southern India. Madras 1863–64. 271 pl.
———, Ferns of British India, with supplement. Madras 1865–70 and 1876, 390 pl.
———, Handbook to the ferns of British India, Calcutta 1883. Supplement 1892.
C. B. Clarke, A review of the Ferns of Northern India. — Trans. Linn. Soc. London. II. s. Bot. v. I. 1880. 36 pl.
C R. W. K. v. Alderwerelt v. Rosenburgh, Malayan Ferns. Batavia 1909. Supplement 1917.
M. Ogata, Icones Filicum Japoniae. Tokyo 1928–36, 350 pl. (unfinished).
R. C. Ching, Icones Filicum Sinicarum. Peiping 1930–35, 150 pl. (unfinished).
F. M. Bailey, Lithograms of the Ferns of Queensland. Brisbane 1892. 191 pl.
W. J. Robinson, A taxonomic study of the Pteridophyta of the Hawaiian Islands. Bull. Torr. Bot. Cl. v. XL. 1913.
E. B. Copeland, Ferns of Fiji. Bishop Mus. Bull. no. 59.
———, Pteridophytes of the Society Islands. Ibid. no. 93.
A. L. A. Fée, Histoire des Fougères et des Lycopodiacées des Antilles. Paris 1866. 34 pl.
———, Cryptogames vasculaires du Brésil. Paris 1869–73. 108 pl.
D. C. Eaton, The Ferns of North America, 1877–80. 81 pl.
J. G. Baker and others, Pteridophyta in Martius: Fl. bras. v. I pars 2. 1859–84.
A. Sodiro, Cryptogamae vasculares quitenses. Quiti 1893.
G. S. Jenman, West Indian and Guiana Ferns. Trinidad 1898–1908.
L. M. Underwood and W. R. Maxon in North American Flora. v. XVI. 1909 (unfinished).
W. R. Maxon, Pteridophyta, in Sci. Survey of Porto Rico and the Virgin Islands v. V, pt. 3. 1926.

## D. Some important monographs

A. L. A. Fée, Histoire des Acrostichées. Strasbourg 1845. 64 pl.
———, Histoire des Vittariées, and Histoire des Antrophyées (3. et 4. mémoire). 1852 5 pl.
de Vries et P. Hartig, Monographie des Marattiacées. Leiden 1853. 8 pl.
G. Mettenius, Über einige Farngattungen I–VI. 1857–59.
R. van den Bosch, Synopsis Hymenophyllacearum. Ned. kr. Arch. v. IV. 1859. — Supplements ibid. v. V. 1861–63.
A. Braun, Über die Marsiliaceen-Gattungen *Marsilia* und *Pilularia*. — Monatsber. Berl. Akad. 1863 und 1870.
J. Milde, Monographia generis Osmundae. Vindobonæ 1868.
———, Botrychiorum monographia. Verh. zool. bot. Ges. Wien v. XIX. 1869.
K. Prantl, Untersuchungen zur Morphologie der Gefässkryptogamen. I. Die Hymenophyllaceen. 1875. II. Die Schizaeaceen. 1881.
———, Systematische Übersicht der Ophioglosseen. Ber. Deutsch. bot. Ges. v. I. 1883.
H. Christ, Monographie des Genus *Elaphoglossum*, 1899.
K. Giesenhagen, Die Farngattung *Niphobolus*. Jena 1901.
Carl Christensen, A Monograph of the genus *Dryopteris*. Part I–II (American sp.). Vid. Selsk. Skr. s. VII v. IV. 1907, v. X. 1913, s. VIII v. VI. 1920.
———, Taxonomic Fern-studies. I. (Drymoglossinae et Cochlidiinae). Dansk Bot. Arkiv. v. V. no. 3. 1929.
E. B. Copeland, *Leptochilus* and genera confounded with it. Phil. Journ. Sci. v. XXXVII. 1928. 30 pl.
———, The fern genus *Plagiogyria*. Ibid. v. XXXVIII. 1928, 15 pl.

**Series I. Filices Eusporangiatae.** — Sporangium originating from a group of cells, its wall consisting of several layers of cells. Homosporous; no perispore. Antheridia sunk in the tissue of the prothallium. A suspensor observed in several species.

Order I. **O p h i o g l o s s a l e s**

Characters of the only family:

### Ophioglossaceae

S p o r o p h y t e: Terrestrial (rarely epiphytic) herbs with short, fleshy, usually naked (never scaly) rhizome and fleshy roots. Leaves one to several, not circinnate in bud and without stipular base, the fertile ones consisting of a basal common stalk which bears at the apex a sterile, simple or dissected blade and a stalked fertile spike or panicle, the s p o r o p h y l l. Sporangia marginal in two rows, rather large, sessile or with a very short, thick stalk, without annulus. Spores very numerous (up to 15000 in *Ophioglossum*, 1500–2000 in *Botrychium*), tetraedral. — G a m e t o p h y t e: Prothallium a hypogean, tuber-like body, saprophytic (usually without chlorophyll) and with mycorrhiza. — 3 genera with about 30 species.

A. Sterile blade and sporophyll more or less compound, the veins free. Sporangia free.

*Helminthostachys* Kaulf. Sporophyll a slender but compound spike. Sporangia opening by a vertical slit. Sterile blade subdigitate-palmately divided. — 1 sp. in tropical Asia-Polynesia.

*Botrychium* Sw. Sporophyll a compound panicle. Sporangia subsessile, opening by a transverse slit. Sterile blade simply pinnate or decompound. — Ca. 30 sp. in the Arctic and Northern temperate zones, few in the Tropics and Antarctic.

B. Sterile blade simple, linear to ovate and reniform, rarely (in subgenus *Cheiroglossa*) palmately divided, with reticulated veins, Sporophyll a simple spike, the large sporangia sunk and coalescent, opening by a transverse slit.

*Ophioglossum* L. Small terrestrial herbs spread throughover the globe, or larger tropical epiphytes (subgenera *Ophioderma* and *Cheiroglossa*). Ca. 50 sp.

Order II. **M a r a t t i a l e s**

S p o r o p h y t e: Terrestrial plants with an erect, massive or horizontal rhizome. Leaves circinnate in bud, articulated to the rhizome by fleshy swellings and with stipule-like enlargements at the base of the stalk, pinnately or, in *Christensenia*, digitately divided; mucilage-canals and tannin-cells in the tissue. Sporangia borne superficially on the veins below, forming linear or circular sori, either free or coalescent in "synangia", sessile and without annulus or, in *Angiopteris*, with a very rudimentary one, all the sporangia of the same sorus developing simultaneously, opening by an internal, vertical slit. Spores tetraedral or bilateral, very numerous (1400–7000). — G a m e t o p h y t e: Prothallium green, flat, variously shaped, long-living, a central part forming the massive cushion which produces the sexual organs on the under side. Endophytic fungus present.

The few genera of this order were all hitherto referred to a single family, *Marattiaceae*, but it seems more appropriate to divide them into two (if not four) families:

### Family 1. **Angiopteridaceae**

Sporangia free, closely placed in two rows, forming linear sori, surrounded by fringed scales ("false indusium") and with a rudimentary annulus at the top. Leaves uniform, pinnately divided with free veins. — 3 genera.

*Angiopteris* Hoffm. Mighty ferns with bipinnate leaves. "Recurrent false veins" often present between the true ones. Sori near the edge, consisting of 10–30 sporangia. — Ca. 100 closely allied sp. (partly geographical races?) from Madagascar through Trop. Asia to Japan and eastern Polynesia.

*Macroglossum* Campbell. (2 Malayan sp.) and *Archangiopteris* Christ et Gies. (4 (or 1) sp. incl. *Pseudomarattia* Hayata. China–Tonkin–Formosa) have simply pinnate leaves without recurrent veins and a high number of sporangia (40–60 and up to 160).

## Family 2. **Marattiaceae**

Sporangia coalescent in synangia. Recurrent veins and rudimentary annulus none. — 7 genera with ca. 80 species.

*Marattioideae.* Leaves uniform, pinnately divided with usually free veins. Synangia linear, supramedial or near the edge, sporangia oval, opening by an internal slit, with or without indusial scale. One genus:

*Marattia* Sw. Rhizome erect, radial. Leaves bi-tripinnate; synangia near the margin, linear-oval with 16–40 loculi. Ca. 50 sp. in the Tropics of both hemispheres.

*Christensenioideae.* Rhizome creeping. Leaves uniform, digitate with reticulate veins. Synangia circular with a depression in the centre, scattered over the under surface; loculi 10–20, opening by an internal slit; indusial scales none. One genus: *Christensenia* Maxon. — 2 Malayan sp.

*Danaeoideae.* Rhizome oblique or horizontal. Leaves subdimorphic, simply pinnate (rarely entire) with free veins and nodose swellings on stipes and rachis. Synangia extended from midrib to margin, very close and covering the whole under surface of the fertile pinnae, each sporangium opening by a terminal pore, without indusial scales. One genus:

*Danaea* Sm.–Ca. 30 sp. in Trop. America.

**Series II. Filices Leptosporangiatae.** — Sporangium originating from a single superficial cell, its wall consisting of a single layer of cells. Homosporous or heterosporous. Antheridia raised over the surface of the prothallium. No suspensor.

## Order III. **F i l i c a l e s**

Leaf circinnate in bud. All, *Marsileaceae* excepted, homosporous.

## Family 1. **Osmundaceae**

S p o r o p h y t e : Terrestrial ferns with massive, hard, erect rhizome without scales. Leaves uniform or dimorphous, pinnately divided, not jointed to the rhizome, with stipular expansions covered with glandular hairs at stipe-base, the axes often with woolly hairs; veins free. Sporangia either marginal on much reduced segments or superficial on the under side of unaltered pinnules, not forming real sori and without indusium, all developing simultaneously ("Simplices" of Bower), globose or pyriform, short stalked and with a rudimentary annulus formed by few thick-walled cells near the distal end, dehiscent by a slit running from these cells across the top to the ventral side. Spores rather numerous (up to 500), tetraedral, green, without perispore, ephemeral. — G a m e t o p h y t e : Prothallium green, cordate, flat with a percurrent thickened midrib, large and fleshy, long living. — 3 genera with ca. 20 species.

A. Sporangia marginal on much reduced pinnules.

*Osmunda* L. Leaves wholly or partially dimorphous, in dense crowns, arranged in two circles, the inner fertile and erect, the outer sterile and spreading. Sporangia wholly replacing the leaf-tissue of all or some pinnæ. — 12 species, mostly in swamps, in the temperate and tropical regions.

B. Sporangia superficial on the veins of unaltered pinnules.

*Todea* Willd. Leaf thick with stomata, bipinnate. Caudex up to 1 m high and thick. — 1 sp. (*T. barbara*) in Australia, New Zealand and South Africa.

*Leptopteris* Presl. Leaf very thin, like that of a filmy fern, without stomata, finely dissected. Caudex sometimes subarborescent. — 6 sp. from New Zealand–Polynesia–Melanesia.

## Family 2. **Schizaeaceae**

S p o r o p h y t e : Mostly terrestrial and xerophilous ferns with a horizontal or erect rhizome clothed with simple hairs or, in *Mohria*, with scales. Leaves erect or twining, dichotomously or pinnately divided, of diverse form, the veins free and mostly dichotomously branched, rarely reticulate. Fertile pinnae, except in *Mohria*, much reduced (sporangiophores). Sporangia obovoid or pyriform, large and sessile or, in *Lygodium*, short stalked, marginal in origin but soon forced into a superficial position by different outgrowths from the fertile segment, which protect them as "false indusia", in two rows (solitary in *Mohria*) on the vein, not forming real sori but each a "monangial sorus" or "monangium", provided with a complete transverse annulus at the apical end and a definite stomium, opening by a vertical slit. Spores tetraedral or bilateral, without perispore, rather numerous (128–256). — G a m e t o p h y t e : Prothallium green, flat with a lateral cushion or, in *Schizaea*, filamentous with some swollen, pale-green cells with mycorrhiza, the simplest and most primitive of all prothallia known. — 4 genera with ca. 160 species.

A. Prothallium filamentous, spores bilateral; no scales.

*Schizaea* Sм. Rhizome oblique or short-creeping, of an advanced protostelic structure. Leaves polystichous, simple or forked, unicostate, grass- or rush-like, or flat and repeatedly dichotomous with uni- or pluricostate divisions. Sporangiophores terminal on the costae, either spuriously digitate in a penicillate tuft with the sporangia apparently in four rows (subgenus *Actinostachys*) or pinnate with the sporangia in two distinct rows (subg. *Euschizaea* and *Lophidium*), partly protected by the indusium-like recurved margins. — Ca. 30 sp. in the tropical and southern temperate regions, one (*S. pusilla*) in Atlantic N. America.

B. Prothallia flat, green, spores tetraedral.

*Lygodium* Sw. Rhizome creeping, solenostellic, without scales. Leaves monostichous, twining, of indefinite growth, dichotomously palmately or pinnately divided, veins free, rarely reticulate. Sporangia in two rows on marginal sporangiophores or specialized lobes of otherwise not altered pinnules, each borne near the end of the vein and enveloped in a pocket formed by outgrowths from the lobe. — Ca. 40 sp. in tropical and subtropical regions, one (*L. palmatum*) in Atlantic N. America.

*Anemia* Sw. Rhizome short creeping or oblique, dictyostelic with polystichous leaves, or, in subgenus *Anemiaebotrys*, creeping and solenostelic with distichous leaves, without scales but usually densely coated with silky hairs. Leaves pinnately divided, sometimes wholly dimorphous but usually only the basal pair of pinnæ fertile with extremely reduced leaf-tissue, erect or sometimes spreading. Sporangia closely placed in two rows on the ultimate fertile segments which are either semiterete without spurious indusium or narrowly foliose with recurved, indusium-like margins. — Ca. 90 sp. nearly all in tropical America, a few in Africa and Madagascar, one in India.

*Mohria* Sw. Rhizome short, dictyostelic, scaly. Leaves polystichous, pinnately divided, in habit like *Cheilanthes*. A single sporangium borne near the end of each vein of the unaltered fertile leaf, more or less protected by the revolute margins. — 3 sp. in Africa.

### Family 3. **Marsileaceae**

S p o r o p h y t e : Aquatic or semiaquatic herbs with long creeping, solenostelic, branched, hairy rhizomes. Leaves in two rows, circinnate in bud, linear or 2-4 foliolate, clover-like, bearing ventrally at the base or lower part of the stipes one to several, sessile or stalked s p o r o c a r p s, which are interpreted as highly specialized pinnae. The sporocarp is a hairy, firm, globose or ovoid-oblong body of different structure (see the genera) containing the sori, rupturing by one or four slits. Sori four to several, each consisting of a peripherical receptacle which bears numerous, in origin marginal sporangia enveloped by a soft tissue from the inner side ("indusium"). S p o r a n g i a o f t w o k i n d s : M e g a s p o r a n g i a below, each containing one very large (up to 0.5 mm long) megaspore coated with a thick layer of mucilage, and M i c r o s p o r a n g i a with numerous minute microspores. Some elongated cells in the sporangia of *Pilularia* are interpreted as a distal vestigial annulus. Spores of both kinds tetraedral. — G a m e t o p h y t e : The male prothallium consists of two vegetative cells formed by divisions in the basal part of the microspore, the rest of which develops two antheridia each with 16 sperm-cells; spermatozoids with a large number of cilia. The female prothallium is formed in a hemispherical, thin-walled projection of the megaspore and consists chiefly of a single archegonium, while the cavity of the spore is filled with nourishment for the embryo; neck of archegonium with only two rows of cells. Embryonal development essentially as in other leptosporangiate ferns. — 3 genera with about

70 species. In spite of the widely different habit of the sporophyte this family seems to have its nearest relatives in the *Schizaeaceae*, and I follow, therefore, BOWER in placing it here and not most other writers in associating it with the similarly heterosporous but otherwise abundantly different family, the *Salviniaceae*.

*Pilularia* L. Leaves linear, grass-like, each bearing one globose sporocarp, which inside the thick wall is divided into four loculi, each containing a sorus and splitting by a longitudinal rupture and evacuating the sporangia in a mass of mucilage. — 6 species in Europe, America and Australia.

*Marsilea* L. Leaves 4-foliolate, each bearing one to several ovoid-oblong sporo-carps which rupture longitudinally at maturity into two halves, each containing several sori which are emitted all together borne at intervals on a wormlike band of mucilaginous tissue. — Ca. 65 species in temperate and tropical regions, many in Australia.

*Regnellidium* LINDMAN. Leaves bifoliolate. — 1 sp. in Brazil.

### Family 4. **Gleicheniaceae**

S p o r o p h y t e: Terrestrial and mostly xerophilous ferns with long creeping and usually branched, protostelic (rarely solenostelic) rhizome clothed with scales or hairs. Leaves solitary, not jointed to the rhizome, uniform, of a straggling habit, apparently dichotomously but in reality pinnately divided; after the development of one pair of pinnae the stipes ends in a dormant, paleaceous or hairy bud and the same may be repeated by the axes of the primary and subsequent pinnae in some-what different ways within the systematic groups; veins free and usually forked, the axes often chaffy or hairy. Sporangia as a rule few in exindusiate sori in a single row at each side of the midrib, superficial (sometimes sunk in cavities), placed round a receptacle and all developing simultaneously, broadly pear-shaped, sub-sessile, provided with a complete, transverse annulus and opening by a vertical slit; spores tetraedral or bilateral, without perispore, numerous (some hundreds) or few (16–32) in *Platyzoma*. — G a m e t o p h y t e: Prothallium green, flat with a midrib and marginal proliferations; neck of archegonium straight, antheridia very large with several hundreds of spermatozoids. — About 120 species, all but 1–2 usually referred to one genus, *Gleichenia*, which is an aggregate of 3 (or 5) natural genera. Many species form dense thickets in dryer places (e.g. savannahs) in tropical and subtropical regions.

*Dicranopteris* BERNHARDI. Rhizome and axes with hairs only. Secondary and tertiary axes naked, i.e. not pectinato-pinnate; segments long and plane with dorsal sori consisting of 6–12 sporangia. A small genus with two sections: *Heteropterygium* (DIELS): Axis protostelic, spores tetraedral (type: *D. linearis*, one of the common-est ferns of the Old World) and *Acropterygium* (DIELS): Axis solenostelic, spores bilateral (type: the American *D. pectinata*).

*Sticherus* PRESL. Rhizome protostelic, with scales. Secondary and subsequent axes pextinato-pinnate (*Eu-Sticherus*, syn. *Holopterygium* DIELS, type *S. laevigatus*), or the secondary ones only developed and bipinnate (*Hicriopterus* PRESL, syn. *Diplopterygium* DIELS, type *S. glaucus*). Segments and sori as in *Dicranopteris* but sporangia few (2–6); spores bilateral. — Ca. 100 sp.

*Gleichenia* SMITH. Like *Sticherus* but segments very small, roundish, concave with terminal sori, which in some species are sunk in deep cavities. — Ca. 10 sp. in the southern hemisphere.

*Platyzoma* R. BR. and *Stromatopteris* METT. are monotypic genera (Australia and New Caledonia respectively) with more clustered, simply pinnate leaves.

### Family 5. **Matoniaceae**

S p o r o p h y t e: Terrestrial ferns with complicated solenostelic, creeping rhizome clothed with hairs. Leaves solitary, uniform, dichotomous, of different habit in the two genera, coriaceous, the veins mostly free outside the soral area. Sporangia few together in superficial sori, simultaneous in origin, covered by a coriaceous, peltate indusium developed from the centre of the convex receptacle, sessile or short-stalked, provided with an irregular, more or less complete, sinuous annulus without definite stomium, opening laterally. Spores tetraedral, without perispore, typically 64. — G a m e t o p h y t e little known; prothallium green. — An archaic family of two genera with four species, probably nearest related to the *Dipteridoideae* amongst living ferns.

*Matonia* B. BR. Leaf erect, pedate ("a very perfect example of a catadromic helicoid dichopodium", BOWER) with pinnatifid divisions. Sori uniseriate, costular, seated on a plexus of veins. — 2 Malayan sp., forming thickets like *Gleichenia*.

*Phanerosorus* COPELAND. Leaf pendent, repeatedly dichotomous but usually of a pinnate habit because of the abortion of some terminal buds, with entire, forked pinnae. Sori medial, on a single, mostly free vein. — 2 sp. in Borneo and New Guinea.

### Family 6. **Hymenophyllaceae**

S p o r o p h y t e : Minute to medium-sized, terrestrial or epiphytic herbs with protostelic, never scaly rhizome, which is mostly wide-creeping and branched with distichous, dorsiventral leaves, or, in some species of *Trichomanes*, erect, radial with the leaves arranged spirally; root-system weakly developed, roots sometimes wholly absent but substituted by much hairy, leafless branches from the rhizome. Leaves uniform, rarely (in *Trichomanes* sect. *Feea*) dimorphous, often with simple or branched hairs but without scales, with few exceptions (e.g. *Cardiomanes*) very thin and translucent, formed of one layer of cells without intercellular spaces and stomata, entire and then usually very small (down to 5 mm long) or variously mostly pinnately divided with 1-nerved ultimate segments. Sori terminal on a vein and marginal, immersed or commonly more or less free, sometimes subpetiolulate, consisting of a two-valved or cylindrical to campanulate, at best two-lipped indusium, which encloses a long, cylindrical receptacle with numerous sporangia, these originating in basipetal succession (*Gradatae*, BOWER), of various sizes, short-stalked, globose or compressed, provided with an oblique annulus without definite stomium-cells, opening laterally by a long slit. Spores tetraedral, without perispore, few to many (32 to 420 in one sporangium, BOWER). — G a m e t o p h y t e : Prothallium of *Trichomanes* filamentous, branched, resembling a green Alga, with the archegonia produced at the end of short branches (archegoniophores), of *Hymenophyllum* flattened, strap-like, never cordate, with antheridia and archegonia borne together on a lateral lobe. — 4 genera with over 600 species.

*Trichomanes* L. Indusium cylindrical to campanulate with entire or two-lipped mouth, receptacle often very long and exserted as a long bristle. Prothallium filamentous. — Ca. 330 sp. of widely different habit, belonging to about ten, mostly very distinct subgenera (or better genera), nearly all tropical, one (*T. speciosum*) in western Europe.

*Cardiomanes* PR. Like *Trichomanes* in sori but the entire, reniform frond coriaceous, several cells in thickness. — 1 sp. in N. Zealand.

*Serpyllopsis* V. D. B. A single terminal trichomanoid sorus, habit of *Hymenophyllum*. — 1 sp. in temp. S. America.

*Hymenophyllum* SMITH. Indusium more or less deeply divided into two entire or dentate valves, receptacle short, rarely a little exserted. Prothallium flattish. Ca. 300 sp. mostly tropical (two in W. Europa), belonging to two subgenera: *Sphaerocionium* PR. (*Euhymenophyllum* auct.): Segments and valves entire, and *Leptocionium* PR.: Segments and generally also valves dentate.

### Family 7. **Loxsomaceae**

S p o r o p h y t e : Terrestrial ferns of davallioid habit with creeping solenostelic rhizome coated with hairs. Leaves pinnately divided, uniform, coriaceous, not jointed to the rhizome, the veins free. Sori marginal in origin, terminal on the veins, consisting of a cup-like indusium resembling that of *Trichomanes* enclosing a columnar receptacle, which is elongated beyond the mouth of the indusium and bears the basipetal sporangia with hairs (paraphyses) between. Sporangia pyriform, subsessile, obliquely ascending, provided with a complete annulus, different in the two genera. Spores tetraedral, without perispore, 64 in each sporangium. — G a m e t o p h y t e : Prothallium of the common cordate type with bristle-like hairs below.

A small family of doubtful systematic position, in soral characters mostly resembling *Trichomanes* but in general habit and anatomy the *Dennstaedtioideae*. — 2 genera with 4 species.

*Loxsoma* R. Br. Sporangia with only some of the cells of the annulus thickened, without definite stomium-cells, opening by a median longitudinal slit. — 1 sp. in New Zealand.

*Loxsomopsis* Christ. Sporangia with all cells of the annulus thickened and with a well differentiated stomium, opening by a lateral slit. — 3 sp. in Andes (Costa Rica — Bolivia).

### Family 8. **Hymenophyllopsidaceae**

S p o r o p h y t e: Terrestrial ferns of *Hymenophyllum*-like habit and venation with erect, stout, solenostelic rhizome, c l o t h e d  w i t h  s c a l e s. Leaves radial, pinnately divided, thin (3 layers of cells), without stomata. Sori terminal on the veins, provided with a leafy, green, lobed indusium, which is fixed either to the vein at the very base only or connate to the fertile lobe at one (the outer) or both sides. Sporangia few, large, subsessile, in shape and structure closely resembling those of *Alsophila* with a similar broad, oblique annulus. Spores globose. — G a m e t o p y t e unknown. — 1 genus:

*Hymenophyllopsis* Goebel, Flora, N. F. XXIV p. 3. (1929). — 2 sp. in Guiana (Mt. Roraima and Mt. Duida) and Venezuela.

Quite unique as to several characters, in habit and thin leaves resembling the *Hymenophyllaceae*, in scale-characters and structure of the sporangia the *Cyatheaceae*, still abundantly different from these families as well as from all *Polypodiaceae*.

### Family 9. **Plagiogyriaceae**

S p o r o p h y t e: Terrestrial ferns with a radial, dictyostelic, massive and hard, scaleless stem which may bear slender, solenostelic runners. Leaves in dense rosette, dimorphous, simply pinnate, resembling those of *Blechnum* sect. *Lomaria*, the stipes at base fleshy and swollen, trigonous and bearing two rows of tuberculiform pneumatophores, when young covered with mucilaginous hairs, the veins forked. Sori exindusiate but protected by the revolute margins of the narrow fertile pinnæ, superficial on the branches of the forked veins, developing in a mixed succession, long-stalked, provided with an  o b l i q u e  annulus and a definite stomium, opening by a lateral slit. Spores tetraedral-globose, few (48), without perispore. — G a m e t o p h y t e of the ordinary leptosporangiate type. — One genus:

*Plagiogyria* (Kunze) Mettenius. — About 30 species, mostly from Asia (Malaya to Japan), a few from America (Mexico and the Greater Antilles to Bolivia).

While older pteridologists referred the species of this genus to *Lomaria* it was later placed in the *Pteridoideae*. In spite of having some important characters in common with the *Polypodiaceae* (habit, mixed sori, few spores) it differs as to others widely from all groups of that family, and as it shows in some essential features affinity to the more primitive types of ferns, especially the *Osmundaceae* (anatomy, stipe-base, lack of scales, annulus) it seems preferable with Bower to place this isolated genus in a family of its own.

### Family 10. **Dicksoniaceae**

S p o r o p h y t e: Mostly tree-ferns with an upright, often lofty, radial, complicated dictyostelic, rarely decumbent trunk, with hairs but without scales. Leaves in a crown, very large, their bases as a rule hidden among a dense mass of hairs, pinnately divided, often partially dimorphous, coriaceous. Sori marginal and terminal on the veins, protected by a cup-shaped or two-valved indusium. Sporangia basipetal, of somewhat different structure, opening by a lateral slit. Spores tetraedral, without perispore, few (48–64). — Gametophyte? — 3 genera belonging to two subfamilies:

T h y r s o p t e r o i d e a e. Leaf decompound, partially dimorphous, the basal pinnæ being fertile with the ultimate divisions wholly reduced to the sori. Indusium at first subglobose, later a broad entire cup. Sporangia with a short, thick stalk and a complete, twisted annulus consisting of about 50 cells, half of which are thickened, the rest forming a lateral stomium. One genus:

*Thyrsopteris* KUNZE with one sp., *T. elegans*, in Juan Fernandez. A small tree, 1–1.5 m high, spreading by runners.

D i c k s o n i o i d e a e. Indusia two-valved. Sporangia long-stalked, the annulus more or less oblique with a definite stomium. — 2 or 4 genera:

*Dicksonia* L'HÉRITIER. Lofty tree-ferns with partially dimorphous leaves, the fertile pinnae being more or less contracted. Outer valve of the indusium formed by a slightly altered leaf-tooth. — 17 sp. in the southern hemisphere.

*Cibotium* KAULFUSS. Mostly arborescent ferns with a rather short (1–5 m) trunk or a short decumbent caudex and uniform leaves. Outer valve of the indusium similar to the inner. — 13 sp. in East Asia, Hawaiian Is. and Central America.

The third genus *Culcita* PRESL (*Balantium* auct., non KAULFUSS), usually placed in this subfamily, agrees with the former genera in characters of the sori (indusium two-lipped, sori gradate, annulus oblique) but as to others (no epigaeous stem, oblique ultimate segments and whole habit) it approaches *Dennstaedtia* and COPELAND may be right in associating it with that genus. — 9 sp. in trop. and subtrop. regions, one in the Atlantic Isl. and S. Spain (*C. macrocarpa*, syn. *Balantium culcita*).

*Cystodium* J. SM., a monotypic Malayan genus may tentatively be placed here as a probable derivative of *Dicksonia* with bipinnate, uniform fronds and a more reduced inner indusium-valve and slightly oblique annulus.

### Family 11. **Cyatheaceae**

S p o r o p h y t e: Mostly tree-ferns with an often lofty, radial, dictyostelic trunk (except in Protocyatheaceae) clothed with hairs or scales at the top and often with characteristic scars of fallen leaves and adventitious roots, or, these remaining, covered with a thick layer of roots. Leaves spirally arranged in a crown, rarely partially dimorphous, pinnately divided, the stipes often spiny and densely scaly, sometimes with "aphlebia" (abortive pinnae) at base; veins with few exceptions free, simple or forked. Sori superficial and dorsal on the veins, round, exindusiate or with a more or less complete, inferior, cup-shaped indusium. Sporangia provided with a complete, o b l i q u e annulus, interrupted by a definite stomium, opening by a lateral slit, often mixed with hairs (paraphyses). Spores tetraedral, without perispore, 64. G a m e t o p h y t e: Prothallium of the usual cordate type, when old with bristles on both sides and small scales. Antheridium with 2 cap-cells.

A large family of ca. 700 very uniform species, mostly found in tropical mountain-forests and mostly very local in distribution. BOWER has recently divided it into two families, the first of which shows several primitive characters, resembling the *Gleicheniaceae*.

P r o t o c y a t h e a c e a e. Caudex short on creeping, solenostelic, clothed with hairs but lacking scales. Sori exindusiate, sporangia large, short-stalked, all simultaneous in origin. — 2 otherwise widely different monotypic (?) American genera: *Lophosoria* PR. and *Metaxya* PR.

C y a t h e a c e a e. Caudex erect, mostly arborescent, dictyostelic, clothed with scales. Sporangia small, long-stalked, basipetal in origin. The 3 genera usually adopted are weakly characterized and several writers have recently referred all species to one: *Cyathea*.

*Alsophila* R. BROWN. Indusium none. — Ca. 300 sp.

*Hemitelia* R. BROWN. Indusium a basal, interior scale. Veins in some American species anastomosing. — Ca. 100 sp.

*Cyathea* SMITH. Indusium cup-shaped, at first wholly embracing the sorus. — Ca. 300 sp.

### Family 12. **Polypodiaceae**

S p o r o p h y t e extremely diverse in habit, size and structure, briefly characterized under the subfamilies. Sporangia usually forming sori of different shape and position and with or without indusium, as a rule mixed in origin, long-stalked and provided by a vertical, incomplete annulus opening by a transverse slit; spores few, usually 32 or 64, with or without perispore. — G a m e t o p h y t e: Prothallium typically green, flat, cordate; antheridia with one cap-cell except in *Woodsia*.

By far the largest family, comprising about 170 genera with ca. 7000 species. It is divided .below into 15 subfamilies. Most of these are well characterized and were perhaps better dealt with as families. They are not very closely related to each other but probably separate branches from an ancient common stock, and the sequence does not indicate therefore an evolutionary progress. The subfamilies with marginal sori have retained most primitive characters and are therefore placed first, but others, especially the *Polypodioideae*, include forms which are no doubt recent representatives of very old ferns. Further we find within several subfamilies forms which totally lack scales and because of this primitive character are placed under the heading C h a e t o p t e r i d e s before the forms with scales, L e p i- d o p t e r i d e s. Uniting all Chaetopterides of the different subfamilies into one subfamily would be too unnatural; it is much more probable that the existence of such primitive forms within otherwise very different groups proves that these are derived separately from ancient ferns and independently have developed forms with flattish scales. The same may be said on the a c r o s t i c h o i d forms with the sporangia spread over the leaf, which are to be found in some subfamilies or smaller groups, but this character is presumably developed relatively late. The subfamily *Acrosticheae* of older pteridologists including all acrostichoid ferns does not exist as a natural one.

## Subfamily I. **Dennstaedtioideae**

Sori round, terminal on the veins, borne on punctiform or transversely oblong but never fused receptacles, marginal in origin but in *Microlepia* withdrawn from the margin, protected by a double and indistincly two-lipped or single indusium; sporangia basipetal or mixed in origin with long, thin, three-rowed stalks, often mixed with filiform paraphyses; spores without perispore. — Terrestrial ferns with creeping, solenostelic rhizome clothed with hairs, (rarely erect, dictyostelic and with scales); leaves not jointed to the rhizome, uniform, 1–3-pinnate with free veins, as a rule of herbaceous texture. – 8 genera.

· To this subfamily those ferns belong which show more primitive characters than the other *Polypodiaceae*, and it connects this latter with *Dicksoniaceae*. The genus *Culcita*, mentioned under the latter family, might as well be placed here, and an unbroken series of genera leads, on the other side, directly to the lindsayoid and pteridoid ferns. It is therefore impossible to draw a definite boundary between it and the following subfamilies with marginal sori. As confined here it includes only the genera with definite, never fused sori.

I. *Dennstaedtieae* Indusium ·double, consisting of an inner or lower flap and an outer or upper one that appears as a more or less modified marginal lobe from the fertile segment.

A. C h a e t o p t e r i d e s. Rhizome wide-creeping, solenostelic, with hairs. Both indusia connate.

*Dennstaedtia* BERNHARDI. Sori marginal, the entirely connate indusia forming a ·cyathiform, often reflexed pocket. Pinnules usually unequal-sided. — Ca. 70 tropical sp., one in N. America.

*Microlepia* PRESL. Sori rather far from the margin, the connate indusia cupuli- form. Pinnules as a rule equilateral. — 45 sp., only one American.

*Leptolepia* METT. and *Oenotrichia* COPELAND are small Malayan-Polynesian genera near *Microlepia*, but the inner indusium fixed at the base only.

B. L e p i d o p t e r i d e s. Rhizome creeping or suberect with some scales. Indusia not connate.

*Saccoloma* KAULFUSS. Rhizome creeping, leaves simply pinnate. Upper indusia ·of all sori fused and forming a membranous, sinuous and continuous margin to the pinna. — 1 American sp.

*Ormoloma* MAXON. Similar but upper indusia not fused and membranous. — 2 American sp.

*Ithycaulon* COPELAND. Rhizome short, dictyostelic, leaves decompound. — Ca. 10 tropical sp.

*Orthiopteris* COPELAND. A monotypic genus (Fiji) of doubtful position, with erect rhizome but without scales and with the sori of *Davallia*.

II. *Hypolepideae* Indusium single, the inner wholly aborted, the outer a more or less modified flap which is recurved and covers the sorus, sometimes absent. Rhizome creeping, with hairs, solenostelic.

*Hypolepis* BERNHARDI. — Ca. 40 sp. of dennstaedtioid or dryopteroid habit. Intermediate between *Dennstaedtia* and the chaetopteroid *Pteridoideae*.

## Subfamily II. **Lindsayoideae**

Sori marginal or intramarginal and terminal (except in *Taenitis*), either single and round or coenosori borne on the fused vein-tips, mixed and indusiate (except in *Taenitis*); indusium usually double, consisting of the more or less modified or unaltered margin of the fertile segment, to which the lower indusium may be connate at three sides or fixed at the base only; stalk of sporangium three-rowed; spores without perispore. — Terrestrial or epiphytic ferns with creeping, proto- or solenostelic rhizome, which (in *Lindsaya*) is of a special "lindsayoid" type and clothed with bristles or narrow, lanceolate, castaneous scales consisting of 2–4 rows of cells, or both kinds with intermediate forms intermixed; leaves not jointed to the rhizome, pinnately divided, rarely simple and dimorphic. — 9 genera.

*Tapeinidium* (PR.) C. CHR. Sori as in *Microlepia* but sometimes two fused. Fronds generally coriaceous, 1–3-pinnate with free veins. — Ca. 10 Malayan-Polynesian sp.

*Stenoloma* FÉE. (*Sphenomeris* MAXON). Sori single or often two fused, the inner indusium fixed at base only or partially connate at the sides. Fronds decompound of definite growth, mostly of herbaceous texture, with three veins. — 18 sp.

*Odontosoria* (PR.) FÉE. Sori as in *Stenoloma*. Fronds decompound of indefinite growth, often spiny. — 11 American sp.

*Lindsaya* DRYANDER. Sori single or more often two to many or all fused into coenosori, borne on a vascular commissure connecting the vein-tips, the inner indusium fixed at base only. Fronds uniform, once or twice pinnate, rarely simple, the pinnae trapeziform or semilunate with the lower margin intact and sterile, of some species cuneiform or equal-sided; veins free or sparsely reticulate. — Ca. 150 tropical sp. — A small group of Malayan sp. (*Isoloma* J. SM.) with equilateral, ovate-cordate to lanceolate pinnae were formerly included in *Lindsaya*, now usually but unnaturally referred to *Schizoloma*.

*Schizoloma* GAUDICHAUD. Sori as in *Lindsaya*, but fronds dimorphous with reticulate venation. — 1 Malayan sp. (*S. cordatum*).

To this tribe may be referred preliminarily the following genera:

*Taenitis* WILLD. Sporangia aggregated in long, linear, exindusiate coenosori medial between the midrib and margin or sometimes near the latter. Fronds simple or impari-pinnate with reticulated veins. — 3 Malayan sp. — In sori and habit very different from the genuine lindsayoid genera but without doubt closely related to *Schizoloma*.

*Platytaenia* KUHN, a monotypic Malayan genus, is an acrostichoid derivative of *Taenitis* with dimorphic fronds.

*Diellia* BRACKENRIDGE, a small genus endemic in the Hawaiian Islands, resembles *Lindsaya* in habit and sori, but more probably it is of asplenioid origin.

*Dictyoxiphium* HOOKER, a monotypic American genus with simple fronds, with marginal, continuous coenosori without distinct upper indusium, venation reticulate.

## Subfamily III. **Davallioideae**

Sori and indusia essentially as in the former subfamily but never fused; stalk of sporangium one-rowed (always?), spores without perispore. — Mostly epiphytic ferns with creeping, dictyostelic rhizome clothed with peltately fixed and usually broad scales; leaves not jointed to the rhizome, pinnately divided, rarely simple, as a rule deltoid, the veins free. — 7 genera.

*Davallia* SMITH. Sori intramarginal on the tips of the ultimate segments, the lower indusium semicylindrical or cupuliform, free at the outer edge only. Fronds uniform or nearly so, usually compound, of rigid texture. — Ca. 35 sp. in the Old World's Tropics, one (*D. canariensis*) extending to S. W. Europe. Related small Malayan genera are: *Scyphularia* FÉE, *Parasorus* V. A. V. R. and *Trogostolon* COPELAND.

*Humata* CAVANILLES. Lower indusium fixed at base only. Small ferns of rigid texture with simple or 1–2-pinnate fronds, often dimorphous. — Ca 40 Malayan-Polynesian sp., one (*H. repens*) extending to Madagascar.

*Leucostegia* PRESL. Sori as in *Humata* but fronds usually decompound, uniform and of thin texture. — 20 Asiatic-Polynesian sp.

*Davallodes* COPELAND. Sori of some species as in *Davallia*, of others as in *Leucostegia*, but all with characteristic rhizome-scales which from a broad, peltate base are suddenly contracted into a long bristle. — 12 Malayan sp.

The following genera are, though looking very different, now usually associated with the davallioid ferns:

*Nephrolepis* SCHOTT. Sori terminal, round, in an intramarginal row, furnished with a reniform, lunate or peltate indusium. Rhizoma erect or scandent, paleaceous. Leaves uniform, simply pinnate with articulated pinnae and free veins. — Ca. 30 tropical sp., some much cultivated.

*Arthropteris* J. SMITH. Sori terminal, superficial, uniserial, exindusiate or with a reniform indusium as in *Dryopteris*. Rhizome wide-creeping, paleaceous, Leaves simply pinnate with a joint on the stipes, the veins free. — Ca. 15 tropical sp.

*Psammiosorus* C. CHR. Like *Arthropteris* but sori dorsal on the sparsely anastomosing veins and scattered, exindusiate. — 1. sp. in Madagascar.

### Subfamily IV. **Oleandroideae**

Sori dorsal, superficial, round, in a wavy line usually near the midrib, with reniform indusia. Rhizome creeping or erect, suffrutescent, with imbricated scales, the leaves solitary or in distant whorls, simple with the stipes jointed as in *Arthropteris*, usually uniform, the veins free. — 1 genus of doubtful but probably davallioid relationship:

*Oleandra* CAVANILLES. — Ca. 35 tropical sp.

### Subfamily V. **Pteridoideae**

Sori marginal or intramarginal, as a rule continuous coenosori borne on a vascular commissure connecting the vein-ends, protected by the reflexed, membranous leaf-margin, and, in the two first-named genera, also with a thin inner indusium; stalk of sporangium three-rowed, the spores without perispore. — Terrestrial ferns with erect or creeping rhizome, which is as a rule solenostelic but often with perforations, with hairs or scales; leaves uniform or subdimorphic, not jointed to the rhizome, pinnately (rarely digitately) divided, the veins free or anastomosing without free included veinlets. — 12 genera.

I. C h a e t o p t e r i d e s. Rhizome with hairs which in the two last-named genera are mixed with narrow scales.

*Pteridium* GLEDITSCH. Receptacle marginal, a lower indusium present; sporangia at first gradate, later mixed; veins free. — 1 cosmopolitan sp. (*P. aquilinum*) with many geographical subspecies.

*Paesia* ST. HILAIRE. A small tropical genus of somewhat simpler structure but hardly different from *Pteridium*.

*Lonchitis* L. Lower indusium absent but replaced by hairs; sporangia mixed; rhizome stout, hairy, veins reticulate; sori as a rule confined to the sinuses of the pinnae. — 8 African, 1 American sp.

*Anisosorus* TREVISAN. Like *Lonchitis* but rhizome with narrow scales and veins free. — 1 American, 1 African sp.

*Histiopteris* J. SMITH. Rhizome creeping with hairs and scales, veins sparsely reticulate. Lower indusium absent, sporangia at first basipetal. — 8 sp.

II. L e p i d o p t e r i d e s. Rhizome with scales. Sori intramarginal without lower indusium; sporangia mixed.

*Pteris* L. Veins free (§ *Eupteris*) or forming costal areoles (§ *Campteria*) or reticulate (§ *Litobrochia*). — Ca. 250 tropical and subtropical sp. *Anopteris* DIELS, *Actiniopteris* LINK, *Ochropteris* J. SM., *Hemipteris* ROSENSTOCK, *Schizostege* HILLE-BRAND and *Amphiblestra* PRESL are mostly monotypic genera with the sori of *Pteris*, differing in morphological characters.

Acrostichoid genera, probably derivated from the Pteridoideae

Rhizome stout, erect or climbing, of a complicated dictyostelic structure, with scales. Leaves simply impari-pinnate. Indusium none.

*Neurocallis* FÉE. Sporangia borne partly on an intramarginal, linear receptacle, partly on and between the reticulated veins in the outer half of the not much contracted fertile pinnae. — 1 W. Indian sp.

*Acrostichum* L. Sporangia spread over the surface without receptacle. Rhizome stout, erect, leaves uniform but fertile and sterile fronds sharply separated or the upper pinnae only being fertile; veins densely reticulate. — 1 (or 3 to several) sp. (*A. aureum*) in brackish water or lakes in the tropics.

*Stenochlaena* J. SMITH (sensu propria). Sporangia borne partly on a medial linear receptacle, partly on and between the branched, diplodesmic vein-ends of the linear, much contracted fertile pinnae. Rhizome climbing; pinnae articulated, veins forming costular areoles. — 3 sp. in Malaya and Africa.

## Subfamily VI. **Gymnogrammeoideae**

Sori superficial, without real indusia but often protected by pseudo-indusia formed by modified margined lobes or by the continuous, reflexed, more or less modified leaf- margin, either confined to the distal ends of the veins and then round or more often confluent into a continuous or interrupted intramarginal line but (except in *Onychium*) without a connecting commissure between the vein ends, or oblong to linear following the course of the veins, sometimes to the midrib; a subacrostichoid state is arrived at by some genera; sporangia mixed, mostly large and short-stalked, the spores (except in *Cheilanthopsis*) globose-tetraedral without perispore. — Ferns of diverse habit, mostly terrestrial with solenostelic (rarely dictyostelic) rhizome, the leaves not jointed. — 27 genera of uncertain mutual relationship and probably representing more evolutionary lines, but most of them no doubt related to the *Pteridoideae*, with which they were associated by most authors. They are here divided into five tribes.

A. *Cryptogrammeae*. Sori borne on the apical part of the free veins, usually oblong, protected by the reflexed, continuous leaf-margin. Rhizome creeping or ascending, scaly, the leaves usually more or less dimorphous or, in *Llavea*, partly dimorphous, 2–4-pinnate.

a. With vascular commissure connecting the vein-ends; leaves uniform or sub-dimorphous. :

*Onychium* KAULFUSS. — 6. sp. in East-Asia.

b. Without commissure:

*Neurosoria* METT. (1 sp. in N. Australia). *Cryptogramma* R. BR. (7 sp.). *Llavea* LAGASCA (1 Mexican sp.).

B. *Ceratopterideae*. Sporangia solitary on the veins, spherical, subsessile with a few- to many-celled, wholly or partly indurated annulus, when young protected by the reflexed leaf-margin. — Aquatic ferns with dimorphous, pinnate, often viviparous leaves, the sterile floating, the fertile erect and more divided, the veins reticulate, the full-grown axis dictyostelic.

*Ceratopteris* BRONGNIART. — 2 or 4 sp.

This isolated genus is usually placed in a family of its own (*Parkeriaceae*) but BOWER is probably right in regarding it as a relative of the *Cryptogrammeae* adapted to aquatic life.

C. *Gymnogrammeae*. Sori linear, following the course of the veins, becoming reticulate in species with reticulate venation, sporangia scattered, large. Rhizome mostly erect or short-creeping.

I. C h a e t o p t e r i d e s. Rhizome with hairs only. Leaves uniform.

*Pterozonium* FÉE. Leaves entire, cordate or suborbicular with free veins; sori forming a medial band, with paraphyses. — 4 sp. in N. South America.

*Jamesonia* HOOK. et GREV. Leaf simply pinnate, of indefinite growth, with small, roundish, often imbricated and tomentose pinnae with free veins; paraphyses none. — 18 sp. in the high Andes and Central-Brazil. Hardly different from the following genus.

*Gymnogramme* Desvaux (*Psilogramme* Kuhn). Leaves of definite or indefinite growth, 1-pinnate-decompound, mostly of thin texture and with flexuous rachises, often very hairy with articulated hairs, the veins free; paraphyses none. — 60 American sp.

*Aspleniopsis* Mett. Leaf closely resembling some species of *Asplenium* with simply pinnate leaves and unequal-sided pinnae, but the exindusiate, densely paraphysate sori and solenostelic, hairy rhizome agree better with the other genera of this tribe. — 1 sp. in New Caledonia.

*Syngramma* J. Smith. Leaves simple, lanceolate or ovate, or pinnate, the veins united near the margin by a continuous, intramarginal vein or forming one or more series of intramarginal areoles; sori paraphysate. — 20 Malayan-Polynesian sp.

*Craspedodictyum* Copeland. Differs from *Syngramma* in the palmately divided leaves and lack of paraphyses. — 5 Malayan-Polynesian sp.

II. L e p i d o p t e r i d e s. Rhizome with scales, in some genera also with silky hairs.

*Coniogramme* Fée. Leaves rather large, uniform, 1–2-impari-pinnate with broad, entire pinnae, generally of thin texture and thinly pubescent, the veins free or sometimes areolate; sori linear with small paraphyses, not reaching the distinct terminal hydathodes. — Ca. 20, mostly East-Asiatic sp.

*Anogramma* Link. Small ferns with very small and nearly naked rhizome and caespitose, subdimorphous, thin, subglabrous, 2–3-pinnate leaves with free veins; paraphyses none. — Sporophyte ephemerous (annual?) but the lobed prothallia perennial with tuber-like archegoniophores which eventually may propagate vegetatively. — 7 sp., 1 (*A. leptophylla*) also in Western and Southern Europe.

*Pityrogramma* Link (*Ceropteris* Link). Leaves uniform, caespitose, 1–3-pinnate, covered by white or yellow powder beneath, rarely sparsely hairy, the veins free; sori often confluent, without paraphyses. — Ca. 20 sp., mostly American, some much cultivated with numerous hybrids (silver- and gold-ferns).

*Trismeria* Fée. Like the former genus but leaves subdimorphous with simple or trifoliolate pinnae with subentire subdivisions. — 2 American sp.

*Gymnopteris* Bernhardi. Leaves uniform, usually caespitose, 1–2-pinnate, more or less densely covered with hairs, the veins normally free, paraphyses none. — 9 American and Asiatic, hardly congeneric sp.

*Bommeria* Fournier. Like the former genus but leaf palmately divided, broadly pentagonous-deltoid, the veins free or areolate. — 4 sp. in Mexico and S. W. U.S.A.

*Hemionitis* L. Leaves caespitose, subdimorphous, the sterile short-stalked in rosette, the fertile on long stalks, cordate, entire or palmately lobed, soft-hairy with articulated hairs, the veins and sori regularly areolate. — 7 American, 1 Asiatic sp.

*Trachypteris* André. Leaves dimorphous, the sterile entire, in rosette, densely covered with lanceolate scales beneath, the fertile on long stalk, pinnate, subacrostichoid; veins areolate. Xerophytes. — 1 or 2 sp. in Galapagos and S. America, 1 in Madagascar.

D. *Adianteae*. Sori close to the margin, the sporangia borne along and sometimes also between the ends of the veins on the underside of the sharply reflexed, membranous or coriaceous, continuous margin, or separate, oblong or orbicular, often in deep sinuses placed lobes, which are veined and conceal and protect the sori. — Terrestial ferns of elegant but very diverse habit, the rhizome erect or creeping, scaly, the leaves uniform, usually with ebenous, polished stipes and rachises, rarely entire, generally 1–5-pinnate, or pedate, sometimes hairy or glaucous, the ultimate pinnules either unilateral, trapeziform or cuneate-flabellate, often articulated, the veins free, very rarely areolate (§ *Hewardia*).

*Adiantum* L. — Ca. 200 sp., only one (*A. capillus-Veneris*) in W. and S. Europe.

E. *Cheilantheae*. Sori borne on the distal region of the veins, sometimes extending rather far toward the midrib, round or more often confluent into an intramarginal line, in most genera protected by the modified, veinless, continuous reflexed margin or lobes of the segment. — Terrestrial, often rock-loving, xerophytic ferns with erect or creeping, in some species dictyostelic, scaly rhizome, often black, polished stipes and palmately or pinnately divided, usually uniform leaves.

The 8 genera of this tribe run confusingly together and the classification of the species is quite uncertain.

*Adiantopsis* FÉE. Sori borne on the tips of the veins on a punctiform receptacle, protected by modified, orbicular or oblong, reflexed lobules. Leaves herbaceous with black stipes, 1–4-pinnate or digitate-pinnate, nearly glabrous, the veins free. — 14 sp. in America, 2 in Madagascar.

*Cheilanthes* SWARTZ. Sori on the thickened vein-ends, generally cofluent, protected by the reflexed, more or less modified, usually interrupted margin. Mostly xerophytic ferns with erect or creeping, densely scaly, sometimes dictyostellc rhizome and 1–4-pinnate leaves, often with very small, roundish ultimate pinnules, and densely hairy or scaly or, in the section *Aleuritopteris*, farinose beneath with white or yellow powder. Veins free. — Ca. 130 sp.

*Notholaena* R. BROWN. Differs from *Cheilanthes* chiefly by lacking the modified, reflexed margin. — Ca. 60 sp., 1 (*N. Marantae*) in S. Europe.

*Sinopteris* C. CHR. et CHING. Like some species of *Cheilanthes* § *Aleuritopteris* with white powder but sorus consisting of a single (rarely two) very large, sessile sporangium with a very broad complete annulus of 32 cells. — 2 Chinese sp. A prototype of *Cheilanthes* and quite unique by the monangial sori.

*Pellaea* LINK. Sori as in *Cheilanthes* but the reflexed margin continuous as in *Pteris*. Mostly coriaceous, glabrous, xerophytic ferns with fronds of diverse habit and division, the stipes as a rule black and polished, the hidden veins free or, in a few species, areolate. — Ca. 80 sp.

*Doryopteris* J. SMITH. Like *Pellaea* but leaves usually more or less dimorphous, simple or palmato-pinnatifid and the veins areolate. — Ca. 35 sp.

*Saffordia* MAXON. Like *Doryopteris* is habit and venation but lamina covered with scales beneath and the subacrostichoid sori confluent, forming an intramarginal band without marginal pseudo-indusium. — 1 sp. in Peru.

? *Cheilanthopsis* HIERONYMUS. A monotypic Chinese genus of doubtful systematic position, connecting the sori of *Adiantopsis* with the habit, articulated hairs and bilateral spores of *Woodsia*.

### Subfamily VII. **Vittarioideae**

Sori exindusiate, superficial or immersed coenosori; sporangia with many-celled annulus, their stalk single-celled at base but consisting of several cells below the capsule, usually mixed with club-shaped or filiform paraphyses; spores bi-or triplanate, without perispore. — Epiphytic ferns with creeping, protostelic (the first 3 genera) or dorsiventrally dictyostelic rhizome without sclerenchyma and clothed with characteristic clathrate scales. Leaves non-articulate usually close together, uniform, simple or rarely forked, with spicular cells in the epidermis, veins free or fused. Prothallia anomalous: irregularly lobed and propagating vegetatively by gemmae. — 7 (8) genera.

A very natural subfamily, apparently quite isolated but possibly developed from some gymnogrammeoid ancestors and specialized to epiphytic life.

A. Veins free, rhizome protostelic.

*Hecistopteris* J. SMITH. Fronds very small, moss-like, cuneiform and dichotomously cleft; sori superficial. — 1 American sp.

*Monogramma* SCHKUHR. Fronds simple, grass-like with a midrib without lateral veins. Coenosori borne on the midrib and immersed, paraphyses club-shaped. Very small ferns, in structure the simplest of all *Polypodiaceae*. — 1 sp. in the S. E. African Isles and 1–2 in Malaya.

*Vaginularia* FÉE. Like *Monogramma* but sori borne at both sides of the midrib on short lateral veins, paraphyses filiform. — 6 Malayan-Polynesian sp.

B. Venation areolate; rhizome usually dorsiventrally dictyostelic.

*Vittaria* SMITH. Fronds entire, grasslike and sometimes 1 m long rarely oblanceolate or dichotomously forked, with a percurrent or evanescent midrib and distant, oblique, undivided veins, which are connected at the end by an intramarginal vein, on which the usually deeply immersed coenosori are borne. — Ca. 80 tropical sp.

*Ananthocorus* UNDERWOOD et MAXON. Like *Vittaria* but veins forming several rows of areoles and the intramarginal coenosori nearly superficial. — 1 American sp.

*Antrophyum* KAULFUSS. Fronds usually tufted, entire, from linear to broadly oblanceolate with a percurrent or soon evanescent midrib and regularly reticulate venation with two to several rows of areoles. Coenosori superficial or immersed, borne on some or all veins and often reticulate; paraphyses wanting or club-shaped or filiform. — Ca. 40 sp. The American sp. (ca. 10) have been segregated as a genus (*Polytaenium* DESV.), characterized by percurrent midrib and lack of paraphyses.

*Anetium* SPLITGERBER. Like *Anthrophyum* but subacrostichoid, the sporangia also borne on the mesophyll. — 1 American sp.

### Subfamily VIII. **Onocleoideae**

Sori superficial, dorsal on the veins, globose, borne on cylindrical or convex receptacles, furnished with a membranous, inferior and shell-shaped or globose indusium (wanting in one species) and besides protected by the firm, brown, incurved, continuous or interrupted leaf-margin; sporangia basipetal in origin, large with many-celled annulus, the spores large, bilateral, with or without perispore. — Terrestrial ferns with dictyostelic, paleaceous rhizome, the leaves not jointed, pinnate, dimorphous, the fertile ones much contracted, the stalk with two vascular bundles. — A small subfamily of two genera, forming a link between *Cyatheaceae* and the blechnoid ferns.

*Matteuccia* TODARO. (*Struthiopteris* WILLD.). Rhizome short with underground runners; leaves bipinnatifid with free veins. Sori of *M. Struthiopteris* medial and biserial and all those of the same group protected by a marginal lobe, of the other species (§ *Pentarhizidium* HAYATA) uniserial and intramarginal, protected by the continuous, dark-brown leaf-margin; indusium fugacious, shell-shaped, absent in *M. intermedia;* perispore present. — 3 sp., *M. Struthiopteris* (Ostrich-fern) in the whole N. temperate zone, 2 in East Asia.

*Onoclea* L. Rhizome creeping, the sterile leaf pinnate with copiously anastomosing veins without free included veinlets, the fertile bipinnate with globose pinnules, each with a single sorus wholly enveloped by the globose indusium; no perispore. — 1 sp. (*O. sensibilis*) in Eastern N. America and East Asia.

### Subfamily IX. **Blechnoideae**

Sori short or long coenosori borne on commissures parallel to the midrib of the pinna, in a single line or, in *Doodia*, in 1–3 rows at each side, protected by a marginal (the incurved, modified margin) or superficial, intramarginal to subcostal, exterior and introrse, membranous indusium; sporangia with many-celled annulus and three-rowed stalk, spores large, bilateral, with or without perispore. — Terrestrial, sometimes subarborescent ferns with erect or creeping, dictyostelic often stoloniferous, paleaceous rhizome, the leaves not jointed, as a rule simply pinnatifid to bipinnate. — 5 genera.

BOWER interpreted the indusium of *Eu-Blechnum* and other genera as homologous with the incurved, modified margin of *Lomaria* and subsequently he regarded all genera as being exindusiate. This interpretation can not be accepted; the leafy part of the fertile pinnae outside the sori (the "flange") is wholly of the same nature as the same part of the sterile ones.

*Blechnum* L. Coenosori long, continuous (rarely interrupted); no perispore. Veins (of sterile pinnae) normally free. — Ca. 180 sp., only one (*B. spicant*) in the northern temperate region. — Usually divided into three subgenera (or genera):

*Lomaria* WILLD. (*Struthiopteris* SCOPOLI). Leaves dimorphous, the fertile much contracted, linear, indusium marginal or intramarginal.

*Blechnum* L. (*Eu-Blechnum*). Leaves uniform, coenosori subcostal or medial.

*Salpichlaena* J. SMITH. Leaves uniform or subdimorphous, twining, bipinnate, the veins united by an intramarginal vein. — American liana.

*Sadleria* KAULFUSS. Sori as in *Eu-Blechnum*, perispore present; all veins forming costal areoles (venatio Doodiae). Small tree-ferns with coriaceous, uniform, bipinnate leaves. — 7 sp. in Hawaii.

*Woodwardia* SMITH. Coenosori interrupted, in a subcostal row, sunk in the mesophyll, no perispore. Leaves mostly large, uniform or, in subgenus *Lorinseria*, dimorphous; venatio Doodiae or 2–4 rows of areoles. — 10 sp. in N. America and East Asia, one (*W. radicans*) in the Atlantic Isl. and S. W. Europe.

*Doodia* R. BROWN. Like *Woodwardia* but sori superficial, in 1–4 rows, and perispore present; leaves small. — 8 sp. in Polynesia and Australia.

*Brainea* J. SMITH. Sori exindusiate, single or confluent into coenosori as in *Woodwardia* or even acrostichoid, perispore present. — 1 East Asiatic sp. (*B. insignis*), a small tree of rather doubtful systematic position, apparently a link between the *Blechnoideae* and *Cyatheaceae*, approaching the latter in anatomical structure, gradate sori and complete annulus, by HAYATA therefore placed in a subfamily of its own.

## Subfamily X. **Asplenioideae**

Sori superficial, usually oblong to linear, more or less oblique to the midrib, borne either on one or on both sides of the fertile vein, as a rule indusiate; indusium membranous, either single, oval to linear, opening inwards or outwards, or double with the two parts placed back to back on the same vein and free or, in *Athyrium*, fused at the upper end; sporangia small with thin, one-rowed stalks, spores bilateral with perispore (unless in *Cystopteris*). — Mostly terrestrial ferns with erect, creeping or climbing, paleaceous and normally dictyostelic rhizome; leaves unless in some species of *Asplenium* not jointed to the rhizome, of very diverse habit, simple to decompoundly pinnate, uniform, the stipes with 1–2 vascular bundles, often more of less paleaceous, rarely pubescent.

A large subfamily of 8–10 genera, the systematic position of which is rather puzzling. The large, central genus *Asplenium*, one of the most distinct of fern-genera, seems to form a special branch with many twigs from the stock of lepto-sporangiate ferns, originating from unknown ancestors and now in the state of rapid development. Certain of its groups show however distinct resemblance to other asplenioid genera, thus to *Phyllitis* and relatives, which approach the *Blechnoideae* and rather unnaturally were referred to that subfamily by BOWER, and through the problematical genus *Diplazium* to *Athyrium*, which evidently is related to *Dryopteris* (*Eudryopteris*) and *Cystopteris*, which latter might be placed as well in the *Asplenioideae* (or even united with *Athyrium*) as in the *Dryopteridoideae*. These facts seem to prove that the subfamily must be placed between the *Blechnoideae* and *Dryopteridoideae*.

A. *Asplenieae*. Sporangia inserted direct on the veins, usually on one side only, indusium single, rarely double, straight. Rhizome-scales firm, clathrate with dark cell-walls.

*Phyllitis* HILL. (*Scolopendrium* ADANSON). Sori geminate in opposite pairs, those of two neighbour veins parallel and close, at last confluent, their indusia opening towards each other. Leaves caespitose, simple or slightly lobed, the veins free or (in the American subgenera *Antigramma* and *Schaffneria*) areolate. — 8 sp., *P. Scolopendrium* and 2 others in S. and W. Europe.

*Camptosorus* LINK. Sori similar but less distinctly geminate, the leaves small, simple, prolonged into a long rooting tail ("walking fern"), the veins often united near the midrib. — 1 sp. in N. America, 1 in N. E. Asia.

*Diplora* BAKER (incl. *Triphlebia* BAKER). Sori as in *Phyllitis*, rhizome climbing, leaves simple or pinnate with free veins. — 4 Malayan-Melanesian sp.

*Asplenium* L. Sori and indusia oval to linear, rarely geminate or double, in the subgenus *Loxoscaphe* roundish and seated near the tip of the narrow ultimate lobe which together with the oval indusium forms a subdavallioid sorus. Terrestrial (often rock-loving) or epiphytic ferns of very diverse habit and rhizome, leaves simple to decompound, mostly simply pinnate but in several species extremely variable in the same individual and often propagating vegetatively by buds on the rachis or veins or by the rooting prolonged apex of leaf or pinnæ or by stolons, the veins free or (§ *Thamnopteris*) united along the margin or, in some small groups (*Asplenidictyum*, *Holodictyum*), areolate without free included veinlets. Many species, even of different groups, hybridize in nature. — Ca. 650 sp.

Small e x i n d u s i a t e genera derivated from *Asplenium* are *Ceterach* GARSAULT (leaves covered with scales beneath, 3–4 sp.), *Pleurosorus* FÉE (3 sp.) and *Pleurosoriopsis* FOMIN (1 sp.), both with articulated hairs.

B. *Athyrieae*. Sporangia borne on receptacles with a vascular strand branched off from the vein. Indusia variable (see genera). Rhizome-scales usually large and soft with thin cell-walls.

*Diplazium* Sw. At least some of the costular sori with double, not fused indusium,

the rest usually simple and straight. Mostly large, terrestrial ferns with simple to
1–3-pinnate leaves, the veins free or in some smaller groups (*Anisogonium* Pr.,
*Diplaziopsis* C. Chr.) areolate. — Ca. 375 tropical sp. — A rather problematical
genus; while some of its groups (e.g. *Monomelangium* Hayata) hardly may be
separated from *Asplenium* the great bulk of species ought perhaps to be united
with *Athyrium* as did Milde and Copeland.

*Athyrium* Roth. Sori round to linear, the indusia of typical species hooked, i.e.
double as in *Diplazium*, but the anterior one is at the distal end crossing the vein
and continuous with the posterior, usually shorter one, of others hippocrepiform
or even reniform and indiguishable from these of *Dryopteris*, or simple, linear and
straight as in *Asplenium*, or reduced to a small inferior scale as in *Cystopteris* or
even totally absent (*Cornopteris* Nakai). Terrestrial, usually erect ferns with
flaccid, herbaceous, 1–3-pinnate leaves with free veins, without hairs but often
rather paleaceous. — Ca. 180 sp., mostly East-Asiatic, few in the tropics.

*Cystopteris* Bernhardi. Sori round with an inferior, vaulted, acuminate indusium. Veins free. — 18 sp. of different habit, most of which probably should be
transferred to different groups of *Athyrium*, leaving *C. fragilis* (the most
widely distributed of all ferns) and its immediate relatives in *Cystopteris*. Confined
thus the genus is distinct enough but of doubtful systematic position. In most
structural characters it agrees as well with *Athyrium* as with *Dryopteris* (*Eudryopteris*), differing from both in the absence of perispore, but the two bundles of the
stipes bring it nearer to the former genus.

*Stenolepia* v. A. v. R. Sori as in *Cystopteris*, leaves with articulated hairs. —
1 Asiatic sp.

## Subfamily XI. **Woodsioideae**

Sori superficial, dorsal near the vein-ends, round, borne on a circular receptacle
with tracheids, indusiate; indusium surrounding the receptacle at base, either
splitted up into narrow segments (*Euwoodsia*) or calyciform and at first wholly
embracing the sorus, later on rupturing (§ *Physematium*); sporangia basipetal, few, with three-rowed stalk, spores bilateral with perispore. — Small terrestrial ferns with paleaceous dictyostelic, erect rhizome, the leaves pinnate-
bipinnate with free veins, often chaffy or with articulated hairs, the stipes with
two vascular bundles, never articulated to the rhizome but in *Euwoodsia* with a
joint. — 1 genus:

*Woodsia* R. Brown. — 38 sp. mostly in the northern and arctic regions. — In
some characters of the sporophyte (basal, circular indusia, basipetal sporangia)
and the gametophyte (prothallium with articulated hairs and two-celled antheridium cap) the genus approaches *Cyathea;* on the other side it is connected with the
*Dryopteridoideae* by the following genera.

Genera   between   the   Woodsioideae   and   Dryopteridoideae

Isolated and mostly monotypic genera of remote mutual relationship, resembling
*Dryopteris* in important characters (sori mixed, stalk with more vascular bundles)
but indusia basal and circular as in *Woodsia* or a basal scale as in *Cystopteris*.
A. Indusium cyathiform.

*Diacalpe* Blume. Indusium wholly embracing the sorus as in *Cyathea*, coriaceous.
Leaf 2–3-pinnate with free veins. — 1 Asiatic sp.

*Peranema* Don. Sori s t a l k e d, wholly covered by the globose indusium which
has a small opening at the distal side at base and finally splits into two halves,
otherwise like *Diacalpe*. — 1 Asiatic sp.

*Hypoderris* R. Brown. Sori as in *Woodsia*, scattered over the surface, either
compital on the reticulated veins or on free included veinlets. Rhizome creeping,
leaves entire or pinnatifid with the habit, venation and scales of *Tectaria*. — 4
American sp.

B. Indusium a basal, inferior or lateral scale. Veins free.

*Acrophorus* Blume. Indusium inferior, membranous, early falling. — 2 Asiatic sp.

*Lithostegia* Ching. Indusium lateral, coriaceous, dark brown, sori terminal. Leaf
finely dissected, habit of *Leucostegia*. — 1 Asiatic sp.

## Subfamily XII. **Dryopteridoideae**

Sori superficial, dorsal or sometimes terminal on the veins, round, rarely oblong to linear, indusiate or not, in some derived genera acrostichoid; indusium in origin basal and inferior but soon fixed centrally on the more or less elevated receptacle, either with a sinus at the proximal side and thus becoming reniform, or fully entire and then orbicular and peltate. Sporangia mixed, with three-rowed stalks, spores bilateral with perispore. — Mostly terrestrial ferns with erect or creeping dictyostelic, paleaceous rhizome; leaves not jointed to the rhizome, of very diverse habit, venation and pubescence, uniform except in the acrostichoid genera.

A large subfamily of about 25 genera with ca. 1500 species, representing at least two phyletic lines which have arrived at or perhaps preserved the same soral condition but as to anatomical structure and dermal appendages have followed different lines, some of which have proceeded to an acrostichoid state. While the first tribe (*Dryopterideae*) shows a pronounced resemblance as well as to the *Woodsioideae* through the isolated genera named above and farther to *Cyatheaceae*, as to *Cystopteris* and *Athyrium*, the second (*Thelypterideae*) seems to be without near relatives outside the subfamily, *Arthropteris* only to be considered. The majority of species were recently placed in one vast genus (*Dryopteris*), which is divided below into several genera.

A. *Dryopterideae*. Rhizome and leaves paleaceous with often very numerous large and broad to hairlike scales, the ribs very rarely with simple, grey hairs. Leaf-stalk with 4–7 or more vascular bundles. Veins free or variously united (venatio Goniophlebii, Pleocnemiae, Sageniae, Drynariae), the free ones ending in hydathodes within the margin. Sporangia never setose.

*Dryopteris* ADANSON (sensu propria-*Eudryopteris* auctt.). Sori generally large with large reniform indusia. Rhizome usually oblique or erect and densely paleaceous with broad, soft and often fimbriate scales. Leaves tufted, lanceolate and bipinnatifid (rarely simply pinnate: *Pycnopteris*) or deltoid, 2–3-pinnate and catadromous, never viviparous by buds on rachis, mostly of thick, somewhat fleshy texture and light-green colour, without intestiniform hairs. Veins free, as a rule forked, the midribs decurrent. — Ca. 150 sp., most in the N. temperate region (e. g. the type-species *D. Filix-mas*), many in East Asia and Africa, few in tropical Asia and America.

This genus runs gradually into *Polystichum*. One of the intermediate groups is *Polystichopsis* (*Rumohra* CHING part.) with decompound, anadromous (polystichoid) leaves, some species with reniform, some with peltate indusia. Probably derived from *Dryopteris* are the small American genus *Plecosorus* FÉE with exindusiate sori partly covered by the reflexed leaf-margin and *Papuapteris* C. CHR. (New Guinea) with the habit of *Jamesonia*. More dubious is *Hypodematium* KUNZE (Africa-Asia).

*Polystichum* ROTH. Essentially like *Dryopteris* but indusia orbicular, peltate, the leaves usually coriaceous and often viviparous with distinctly unequal-sided anadromous pinnae and pinnules and aristate marginal teeth. — Ca. 225 sp.

Polystichoid derivatives: *Adenoderris* J. SM., *Cyclopeltis* J. SM. (veins free), *Cyrtomium* PR. (veins goniophlebioid), *Phanerophlebia* PR. (veins free or subgoniophlebioid), all small genera with peltate indusia and simply pinnate, broad pinnae. *Cyclodium* PR. is similar but rhizome creeping and venation meniscioid. It leads directly to two other small American genera with dimorphous fronds and a c r o s t i c h o i d fructification: *Olfersia* RADDI (veins free) and *Soromanes* FÉE (veins united). More resembling *Polystichopsis* with similar decompound but dimorphous leaves and creeping or climbing rhizome are *Maxonia* C. CHR. with apparently peltate indusia and the fully acrostichoid American genus *Polybotrya* H. B. K. (sensu propria). 24 sp.

*Didymochlaena* DESVAUX. Like *Dryopteris* in general structure and scales but sori terminal, oblong with the indusium attached longitudinally. Leaves bipinnate with trapeziform, articulated pinnules very like those of *Lindsaya*. — 1 polymorphous sp.

*Stigmatopteris* C. CHR. (*Dryopteris* part. auctt.). Sori exindusiate, often oblong. Leaves uniform, mostly of thin texture with pellucid internal glands, rachis often ꭡroad-winged, veins free or irregularly united. — 20 American sp. — Approaching

*Tectaria* in habit and through some species with peltate indusia (*Peltochlaena* FÉE) related to the polystichoid genera.

*Ctenitis* C. CHR. (*Dryopteris* part. auctt.). Sori mostly small with small, reniform indusia or exindusiate. Rhizome erect, with a dense mass of mostly narrow, soft scales. Leaves uniform, herbaceous, ovate-oblong and bipinnatifid or deltoid and bipinnate-decompound with catadromous division, more or less paleaceous and costæ invariably rusty-tomentose above with intestiniform (constricted between the cells) hairs; veins free, simple or forked, the midribs not decurrent. — Ca. 150 tropical sp. This genus runs gradually into *Tectaria*. Intermediate small genera are the West-Indian *Camptodium* FÉE (habit of *Tectaria* but veins free and indusia reniform) and the Malayan *Heterogonium* PR. (habit of the bipinnatifid species of *Ctenitis* but venation pleocnemioid and sori oblong, exindusiate). — A c r o s t i-c h o i d  d e r i v a t i v e s with dimorphous fronds: *Atalopteris* MAXON et C. CHR. (3 W. Ind. sp.) and *Psomiocarpa* PRESL. (1 Philippine sp.).

*Tectaria* CAVANILLES (*Aspidium* auctt.). Sori often small, punctiform, in distinct rows or irregularly scattered and then either seated where the veins meet (compital) or terminal on free included veinlets; indusia peltate (§ *Eutectaria*), reniform (§ *Sagenia*) or wanting (§ *Pleocnemia, Arcypteris*). Rhizome erect or creeping, with the stipe-bases usually clothed with lanceolate, dark, thick scales. Leaves uniform or nearly so, simple, lanceolate or ovate-cordate and then often palmately lobed, or 1–3-pinnate and then generally deltoid, mostly herbaceous and with the pubescence of *Ctenitis*, rarely glabrous or with simple hairs; stipes and rachis often winged and black, sometimes gemmiferous. Veins anastomosing either forming costal areoles only (*Pleocnemia*) or more or less irregularly reticulate with or without free included veinlets (venatio Sageniae or Drynariae). — Ca. 200 tropical sp.

A composite genus including more natural groups of not very close mutual relationship but probably related to different free- veined dryopteroid genera. The majority of species is no doubt intimately related to *Ctenitis* but some groups, e.g. *Dictyocline* MOORE, probably to *Cyclosorus*.

T e c t a r i o i d  d e r i v a t i v e s: *Cionidium* MOORE with stalked marginal sori (N. Caledonia), *Fadyenia* HOOK. et GREV. (W. Ind.) and *Luerssenia* KUHN (Sumatra), monotypic genera with simple, dimorphous leaves and indusia like those of *Didymochlaena*. *Tectaridium* COPELAND is extremely dimorphous (Ins.-Phil.) A c r o s t i c h o i d derivatives are the Asiatic genera *Stenosemia* PRESL, *Hemigramma* CHRIST, *Quercifilix* COPELAND and perhaps also *Bolbitis* (see below).

B. *Thelypterideae*. Rhizome and leaves as a rule sparsely paleaceous, rachis and veins, at least above, with simple or branched, never intestiniform hairs; leaf-stalk with 1–2 vascular bundles, veins free and generally simple or united in pairs (venatio Goniopteridis), reaching the margin. Sori mostly small, round, rarely elongated, with reniform indusia or exindusiate; sporangia often setose.

*Thelypteris* SCHMIDEL (*Lastrea* Bory, *Dryopteris* part. auctt.). Rhizome erect or creeping, the small scales as a rule pubescent. Leaves of nearly all species lanceolate and bipinnatifid, narrowed below, rarely deltoid and catadromously compound, naked or with few scales but always more or less pubescent with simple, grey hairs, at least on the costae above, the underside often with large, sessile, red or yellow, globose glands; pneumatophores often present at the base of the pinnæ. Veins free, usually simple, the basal ones not running to the bottom of the sinus between the segments. — Ca. 500 sp., some few in northern regions, e.g. the type-species *T. palustris* (*Dryopteris Thelypteris*) and the type of the old genus *Phegopteris* (*D. Phegopteris*). The other northern species of *Phegopteris* (*D. Linnaeana* and *D. Robertiana*) may properly be segregated and referred to a small genus *Gymnocarpium* NEWMAN. Other distinct groups are the American *Glaphyropteris* PR. with pneumatophores at the base of the segments and very numerous veins and *Steiropteris* C. CHR., approaching *Cyclosorus* and with a "keel" below the sinuses. More doubtful are *Parapolystichum* KEYSS. with decompound leaves resembling those of the compound species of *Thelypteris* but subanadromous and often with cylindrical glandular hairs beneath and *Pteridrys* C. CHR. et CHING (East-Asia).

*Monachosorum* KUNZE and *Monachosorella* HAYATA, two small East-Asiatic genera with exindusiate sori, may be regarded as offshoots of *Thelypteris*.

*Cyclosorus* LINK (*Dryopteris* part. auctt.). Like *Thelypteris* in sori and pubescence (rarely glandular) but venation different. Veins simple and free, the basal ones running to the bottom of the sinus or more often united into an excurrent vein or two to several being united in pairs (venation goniopteroid or meniscioid). Leaves generally less cut, at best bipinnatifid, sometimes gemmiferous. Sori sometimes seated near the ends of the fused veins and confluent with their neighbours, thus forming lunate, exindusiate sori (Asiatic species of the old genus *Meniscium*), sometimes linear and exindusiate (*Leptogramma*, J. SM., *Stegnogramma* BLUME). The indusium of some species (*Sphaerostephanos* J. SM. = *Mesochlaena* R. BR.) resembles that of *Didymochlaena*. Several Malayan species show a tendency to dimorphism and acrostichoid condition. — Ca. 200 tropical sp.

*Goniopteris* PRESL. (*Dryopteris* part. auctt.). Essentially like *Cyclosorus* in sori and venation but rhizome-scales and as a rule also rachis and costae with branched (stellate) hairs. Leaves dark green, simple to bipinnatifid, often gemmiferous. — 70 American sp.

The American species of *Meniscium* SCHREBER, agreeing with the Asiatic ones in sori, are evidently closely related to *Goniopteris*, differing chiefly by the lack of stellate hairs.

### Acrostichoid geneia of probably dryopteroid origin

The following genera agree all in having dimorphous leaves and contracted fertile, acrostichoid fronds. While the two first-named by most writers are regarded as offshoots of the *Dryopteridoideae* the others are usually believed to be derivatives of the *Blechnoideae* or *Asplenioideae*. HOLTTUM has however shown that all are alike in anatomical structure, which corresponds closely to that of most dryopteroid ferns, and it is therefore unnatural to refer them to different subfamilies and better to place them in an appendix to the *Dryopteridoideae*. — Spores with perispore unless in *Lomagramma*.

*Egenolfia* SCHOTT (*Polybotrya* sp. auctt.). Rhizome short-creeping with tufted, pinnate, often proliferous leaves, pinnæ not articulated, veins free. — 9. Asiatic sp.

*Bolbitis* SCHOTT (*Leptochilus* sp. auctt.). Very like *Egenolfia* but veins variously anastomosing, from simply pleocnemioid to densely reticulate with or without free included veinlets. Leaves mostly pinnate-bipinnatifid, rarely simple and then hardly distinguishable from *Leptochilus*, of some American species (*Anapausia* PRESL) of pronounced tectarioid habit. — Ca. 80 sp.

*Lomariopsis* FÉE (*Stenochlaena* sp. auctt.). Rhizome climbing, flattened with large scales. Leaves simply pinnate with the pinnæ (the terminal one excepted) articulate and free veins. — Ca. 20 sp.

*Thysanosoria* GEPP is a peculiar derivative of *Lomariopsis* with retained individual round sori seated on the vein-ends. — 1 Papuan sp.

*Teratophyllum* METTENIUS (*Stenochlaena* sp. auctt.). Rhizome climbing, cylindrical with small scales and often spiny. Leaves trimorphous: besides the normal sterile and fertile fronds ("acrophylls") resembling those of *Lomariopsis* the younger parts of the rhizome produce short-stalked, as a rule finely dissected "bathyphylls" (HOLTTUM). — 8 Malayan sp.

*Lomagramma* J. SMITH. Like *Lomariopsis* but veins areolate without free included veinlets; bathyphylls and acrophylls. Spores without perispore. — Ca. 15 Malayan-Polynesian sp.

### Subfamily XIII. **Dipteridoideae**

Sporangia few together in punctiform, superficial, exindusiate p l e o s o r i without elevated receptacle, mixed with paraphyses, in origin transitional between the simple and mixed sori, provided with a vertical, nearly complete annulus without well defined stomium. Spores tetraedral without perispore. — Terrestrial ferns with solenostelic, creeping rhizome clothed with brown bristles. Leaves solitary, not jointed to the rhizome, dichotomously forked with dichotomously branched principal veins and finer reticulate venation (venatio Anaxeti). — 1 genus.

*Dipteris* REINWARDT. — 8 Malayan-Polynesian sp., forming thickets. An isolated genus, showing several primitive characters and some affinity to *Matonia*. BOWER made it a family of its own but it is no doubt also related to the so-called dipteroid ferns placed in the following subfamily (*Cheiropleuria* etc.). It is therefore preferable to place it here as an archaic group, perhaps resembling one of those from which the recent *Polypodioideae* originate.

## Subfamily XIV. **Polypodioideae**

Sporangia borne on the under side of the lamina, superficial or immersed, in some smaller genera covering the whole surface (a c r o s t i c h o i d) but usually forming definite sori, which always are e x i n d u s i a t e and as a rule globose or oblong (p o l y p o d i o i d) and then borne either terminally or subterminally on free veins, or compital on anastomosing veins (p l e o s o r i) or on a plexus of veins (d i c t y o s o r i), or fused into c o e n o s o r i which are either g y m n o-g r a m m e o i d (parallel to the "main-veins") or d r y m o g l o s s o i d (parallel to the midrib). — Spores without perispore, mostly bilateral. Leaves generally jointed to the rhizome (exceptions: *Loxogramme* and some small free-veined species), of varied habit, structure and venation, this sometimes d i p l o d e s m i c i.e. a secondary vascular system penetrates the receptacles below and parallel to the normal veins.                                                                                 .

A large subfamily of over 1200 species, all but 4–5 tropical and subtropical and mostly of small or medium size, often epiphytic.

The older pteridologists referred the great majority of these species to one genus, *Polypodium*, while others and most modern writers divide them into a consider-able number of genera, but a final natural classification has yet not been arrived at, partly because several of the defined genera run together, and partly because more no doubt important characters found in the anatomical structure of the vegetative and spore-producing organs of some few species cannot be used now as characters of genera including scores of species, most of which are not examined in this respect, and we are therefore compelled at the present to base the genera chiefly on external morphological characters. It is evident that the subfamily includes ferns which at best are very remotely related and derivated from ancestors of not very close relationship. The first four genera named below are thus doubtless archaic ferns related as well to *Matonia* and *Dipteris* as to other genera of this subfamily (*Phymatodes* and others) which may be regarded as derivated from similar ancestors and now being in the state of rapid development. Other genera, especially *Ctenopteris*, are, on the other side, apparently of recent origin without close relatives among other ferns of earlier times or the present epoch. The number of existing genera depends largely on the individual ideas of systematic treatment; it may be small (about 10) or large (50–60). In the following only the better defined genera are briefly charac-terized. The small genera of doubtful value with gymnogrammeoid, drymoglossoid or acrostichoid species are barely named under the genera with polypodioid sori, to which they are intimately related.

I. C h a e t o p t e r i d e s. Rhizome covered with hairs or bristles. Principal veins dichotomously branched, the smaller ones densely reticulate (venatio Anaxeti); diplodesmic. Fronds dimorphous, no definite sori, paraphyses filiform.

*Cheiropleuria* PRESL. Rhizome protostelic. Fronds glabrous, simple or the sterile forked, the fertile linear, acrostichoid. — 1 Asiatic sp..

*Platycerium* DESVAUX. Rhizome with a simple dictyostele. Fronds clothed with stellate hairs, of two kinds: humus-collecting nest-leaves appressed to the tree-trunk and erect or pendent sporophylls which are usually once to several times forked and bear the sporangia either in thick patches on the underside or on a specialized segment. — 10 sp., most Malayan.

A connecting link between this and the following series is *Christiopteris* COPE-LAND. The dictyostelic rhizome is clothed with s c a l e s, which from a flat, peltate base are drawn out into long bristles. Fronds dimorphous, palmate or ter-quinate, the fertile acrostichoid; paraphyses branched hairs. 3–4 isolated sp.

II. L e p i d o p t e r i d e s. Rhizome of different structure, clothed with s c a l e s, rarely naked, the scales of many kinds prohibiting good systematic characters of species or groups of related species. Principal veins pinnate (except in *Neocheiropteris*).

A. *Pleopeltideae*. Venation variously reticulate (venatio Anaxeti, Marginariae and Phlebodii), often diplodesmic. Sori polypodioid, as a rule pleosori or dictyosori, rarely borne on free veinlets, or long coenosori, sometimes acrostichoid. With few exceptions ferns of the Old World.

a. Sporangia mixed with and when young covered with peltate scales. Hairs none.

*Neocheiropteris* CHRIST. Fronds uniform, pedately lobed; sori close to the principal ribs, pleosori or short coenosori. — 2 sp. in Yunnan and Tibet, related to *Christiopteris*.

*Hymenolepis* KAULFUSS. Fronds partially dimorphous, entire, the apical portion of the fertile ones much contracted, spike-like, with two long coenosori and finally covered with sporangia, diplodesmic. — 1 sp. in Africa (*H. spicata*) and about 10 in Malaya-Polynesia.

*Lepisorus* J. SMITH. Fronds generally uniform, entire with the polypodioid, sometimes confluent sori in a single row at either side of the midrib. Scales of rhizome and lamina (mostly few and near the midrib) often clathrate as are those of the sori. — Ca. 25 Asiatic (mostly Chinese) and African sp.

D r y m o g l o s s o i d  derivatives: *Lemmaphyllum* PRESL, fronds dimorphous and *Drymotaenium* MAKINO, fronds uniform.

*Pleopeltis* HUMB. et BONPL. Fronds uniform, entire or rarely pinnatifid, clothed beneath with peltate, orbicular or lanceolate, appressed, usually fimbriate, non-clathrate scales; sori uniserial, polypodioid, rarely confluent (*P. astrolepis*). — A small American genus with one sp. (*P. lanceolata*) also in Africa and India.

D r y m o g l o s s o i d  derivative: *Marginariopsis* C. CHR. 1 sp. in Andes.

b. Sporangia mixed with hairs or paraphyses none.

1. No specialized humus-collecting "scale-leaves".

x. Fronds hairless or rarely with simple hairs.

*Paltonium* PRESL. An American, monotypic genus of doubtful relationship, closely resembling *Hymenolepis* in general habit but without scales in the long, intramarginal coenosori, which are confined to the somewhat contracted distal portion of the lamina. The uniform fronds quite glabrous and naked.

*Loxogramme* PRESL. Fronds uniform or more or less dimorphous, usually close together and  n o t  a r t i c u l a t e d  t o  t h e  r h i z o m e, entire, generally of thick, fleshy texture with hidden veins, without distinct main-veins; sori long, oblique, gymnogrammeoid coenosori. — About 35, mostly Asiatic sp. — A distinct genus of very doubtful relationship, perhaps related to *Grammitis*, but in vegetative characters not unlike *Hymenolepis* and *Paltonium*.

*Phymatodes* PRESL. Fronds simple or pinnatifid, rarely pinnate, uniform or more or less dimorphous, generally of coriaceous texture with hidden venation but as a rule with prominent main-veins, naked and glabrous. In the smaller species the polypodioid (rarely confluent) pleosori are uniserial at either side of the midrib and the venation less complicated, in the larger the often immersed pleosori or dictyosori are placed in one or two regular rows between the main-veins and the venation densely reticulate (venatio Anaxeti) and often diplodesmic. — A chiefly Asiatic-Polynesian genus of about 100 sp. of very different size and habit, epiphytic. It includes several distinct smaller groups, e.g. the American *Anaxetum* SCHOTT.

Derived from *Phymatodes*:

*Selliguea* BORY. Sori  g y m n o g r a m m o i d,  fronds uniform.

*Pycnoloma* C. CHR. and *Grammatopteridium* v. A. v. R. Sori  d r y m o g l o s s o i d, fronds decidedly dimorphous with linear fertile sporophylls.

A small group of Malayan species (*Myrmecophila*) is of special biological interest. The fronds and sori are like those of *Phymatodes* but the rhizome is inflated and hollow, inhabited by ants, and covered with peculiar, imbricated, peltate scales. The group leads directly to the following genus.

*Lecanopteris* BLUME. Rhizome like that of *Myrmecophila* but very soon quite naked. Fronds pinnatifid with the sori immersed in twisted or reflexed marginal outgrowths from the pinnae. — A few Malayan sp.

*Anapeltis* J. SMITH. Not very different from *Phymatodes*, but the uniform, entire fronds usually of thin texture and the venation simpler (venatio Marginariae). Sori in a row at either side of the midrib, rarely subgymnogrammeoid (*Microgramma*). — About a dozen American sp., one ( (*A. lycopodioides*) also in Africa.

*Microsorium* LINK. Fronds simple or rarely pinnatifid, uniform, usually of thin texture with distinct reticulate venation but without strong, prominent main-veins. Sori punctiform, often very small and numerous, irregularly scattered. — About 40 Asiatic sp.

Derived from *Microsorium:*

*Colysis* PRESL. Sori g y m n o g r a m m e o i d. Fronds often pinnatifid or digitate and subdimorphous.

*Leptochilus* KAULFUSS. Fronds decidedly dimorphous, the fertile linear, a c r o s t i c h o i d or d r y m o g l o s s o i d.

A specialized group of biological interest is the small Papuan genus *Dendroconche* COPELAND. Fronds orbicular, humus-collecting, appressed to the tree-trunk; becoming fertile they shoot out from the top a narrow fertile blade.

*Arthromeris* J. SMITH. Differs from *Phymatodes* chiefly by impari-pinnate leaves with the pinnae articulated to the rachis. — Ca. 10 sp. in East Asia (Himalaya-Formosa).

*Phlebodium* R. BROWN. A small American genus of doubtful systematic position but probably a derivative of *Goniophlebium* with venatio Phlebodii and pleosori.

As an appendix to the above series of genera may be mentioned some isolated, systematically very interesting Malayan species, most of which very well might constitute monotypic genera, intermediate between *Phymatodes* and *Drynaria*. They are all coriaceous ferns with pinnatifid or pinnate fronds with venatio Anaxeti and creeping, usually fleshy rhizome. In some the fronds are uniform, subsessile with broad, humus-collecting bases: *Drynariopsis* COPELAND (*Polypodium heracleum*) with very small, scattered, punctiform sori, and *Pseudodrynaria* C. CHR. (*Pol. coronans*) with uniserial, often confluent sori between the main-veins. In others the fronds are partially dimorphous, the upper pinnae only fertile and much contracted, and with humus-collecting bases, but different in soral characters: *Aglaomorpha* SCHOTT has round, distinct pleosori, *Dryostachyum* J. SMITH confluent coenosori, which form dense square patches between the main-veins, and *Merinthosorus* COPELAND has long uninterrupted coenosori at either side of the midrib of the pinnæ. Humus-collecting leaf-bases are lacking in *Holostachyum* COPELAND with sori as in *Dryostachyum* but fronds wholly dimorphous, and *Photinopteris* J. SMITH with partially dimorphous fronds and long coenosori as in *Merinthosorus*.

xx. Fronds clothed with a tomentum of stellate hairs.

*Cyclophorus* DESVAUX (or *Pyrrhosia* MIRBEL). Fronds simple, rarely forked or dimorphous. — Ca. 100 sp. in the Old World, 2 in Andes.

D r y m o g l o s s o i d derivatives: *Drymoglossum* PRESL. Fronds decidedly dimorphous. — Some few sp. in Malaya, one in Madagascar. — *Saxiglossum* CHING. Fronds uniform. — 1 sp. in China.

2. Specialized scale-leaves present.

*Drynaria* BORY. Fronds of two kinds: brown, chartaceous, sessile, slightly lobed, humus-collecting "scale-leaves", and stalked, green, pinnatifid or pinnate, spore-producing fronds with venatio Anaxeti (or Drynariae) and pleosori. — Ca. 20 sp. in the Old World.

B. *Polypodieae.* Veins free or regularly anastomosing (venatio Goniophlebii or Cyrtophlebii). Sori as a rule polypodioid and terminal or dorsal on free veins or free included veinlets, rarely drymoglossoid.

*Polypodium* L. (incl. *Goniophlebium*). Rhizome creeping, fleshy, often glaucous and densely scaly. Fronds pinnatifid or pinnate, rarely simple, conform, non-lepidote. Veins free in some few species but generally goniophlebioid. Sori terminal, uniserial or less frequently in 2–3 rows at each side of the midrib, sometimes immersed. — A natural genus of ca. 50, mostly epiphytic sp. in tropical and subtropical America and Asia-Polynesia, represented in the northern temperate region by a few, free-veined sp. (*P. vulgare* and related forms). Through some Asiatic species with a somewhat more complicated venation the genus runs into *Phymatodes* .

*Marginaria* BORY (*Lepicystis* J. SM.). Differs from *Polypodium* in the fronds being clothed (often densely) with peltate scales. Veins free or goniophlebioid. Scales of the rhizome usually with a black, central stripe. — Ca. 25 American sp. Closely related to *Pleopeltis.*

D r y m o g l o s s o i d derivative: *Eschatogramme* TREV. — 4 American sp.

*Craspedaria* LINK (*Lopholepis* J. SMITH). Rhizome wide-creeping, slender. Fronds usually dimorphous and thinly clothed with thin, lanceolate scales, the veins goniophlebioid. Sori uniserial, the sporangia mixed with hair-like scales. — A few American sp. (*C. vacciniifolia, C. piloselloides,* etc.), presumably related to *Anapeltis.*

*Campyloneurum* PRESL. Rhizome short-or long-creeping. Fronds uniform, of thick texture, simple or pinnate, with venatio Cyrtophlebii. Sori polypodioid, superficial and terminal, mostly in 2–4 rows at each side of the midrib. — Ca. 25 sp. in tropical America.

*Ctenopteris* BLUME. Under this name may be united the numerous small tropical epiphytes hitherto referred to *Polypodium* subgenus *Eupolypodium* but all very far from *P. vulgare.* Rhizome as a rule short and small with cæspitose leaves, which not always are articulated, sometimes creeping or even trailing. Fronds pinnately divided (from lobed to tripinnate), rarely simple, often hairy, the veins free, forked or pinnate, with terminal sori. — About 200 sp.

Taken in this broad sense the genus is an aggregate of several, more or less distinct natural groups or genera, e.g. *Calymmodon* PRESL and *Acrosorus* COPE-LAND. A group of Malayan-Polynesian species (*Cryptosorus* FÉE) with the sori deeply immersed in cavities in the thick lamina and with ciliate scales leads directly to the following genus and should perhaps be united with it.

*Prosaptia* PRESL. Sori apparently davallioid, in deep cavities opening toward the margin. — Ca. 10 sp. in Asia-Polynesia.

*Enterosora* BAKER. A peculiar, monotypic American genus with subentire leaves and the sublinear sori deeply imbedded in the tissue.

*Grammitis* SWARTZ. Fronds simple, at best lobed, linear, usually small and cæspitose, the sori oblong when young, often immersed. — A rather artificial genus of ca. 100 sp., not easily separated from *Ctenopteris.*

Derived genera with coenosori:

*Cochlidium* KAULFUSS. Coenosori close to the midrib, superficial or sunk in a single central groove. — 10 American sp.

*Scleroglossum* v. A. v. R. Coenosori in two parallel, oblique, often marginal grooves, about as in *Vittaria.* — A few Malayan-Polynesian sp.

*Nematopteris* v. A. v. R. Like the former but leaves rush-like with an widened apical fertile portion. — 2 Malayan sp.

## Subfamily XV. **Elaphoglossoideae**

Rhizome suberect or creeping, usually densely paleaceous. Leaves uniform or more often more or less dimorphous, with few exceptions simple and articulated to phyllopodia and with free veins, fertile fronds acrostichoid; spores bilateral. — 4 genera.

Apparently a natural subfamily but of doubtful systematic position and perhaps polyphyletic. It seems to have reached a more advanced evolutionary condition than most other subfamilies.

A. All fronds simple, articulated upon phyllopodia.

*Elaphoglossum* SCHOTT. All fronds simple, often very chaffy, the scales of rhizome and lamina very varied and of high systematic value. Veins free and forked or casually anastomosing or sometimes joined at the end by a submarginal vein. — Ca. 350 sp. in tropical America, ca. 60 in the old World, where Madagascar is especially rich in endemic sp.

> Derivated genus: *Hymenodium* FÉE. Veins regularly reticulate without free included veinlets. — 1 American sp. (*H. crinitum*).

B. Sterile fronds dissected, continuous with the rhizome.

*Microstaphyla* PRESL. Sterile fronds pinnately divided with forked pinnae. — 2 sp. in Andes, 1 in St. Helena.

*Rhipidopteris* SCHOTT. Sterile fronds flabellately divided, the fertile entire, suborbicular. — 4 American sp.

## Order IV. S a l v i n i a l e s

Sporangia basipetal in globose, indusiate sori, some of which consisting of microsporangia with numerous microspores, some of megasporangia with one large megaspore, both kinds without annulus. Prothallia much reduced. — Aquatic, floating plants with a horizontal, branched stem with a single vascular bundle, the leaves not circinnate in bud.

The phylogeny of the two families is obscure. In spite of a certain resemblance to *Hymenophyllaceae* and *Cyatheaceae* in indusial characters they have hardly anything in common with these families, and they are abundantly different, structurally and genetically, from the similarly heterosporous *Marsileaceae*, with which they usually but unnaturally are united into the order *Hydropterides*. The two genera are further usually placed in one family, but in view of the great difference between them such a treatment seems to be unnatural.

## Family 1. **Salviniaceae**

Characters of the only genus:

*Salvinia* GUETTARD. The hairy stem bears no roots but only leaves in alternating whorls of 3, 2 of which are green, floating, entire and usually cordate with a midrib and anastomosing veins, the upper side papillose or grey-hairy or with stalked, branched, hyaline hairs, the under side often densely matted with brown hairs, the third leaf a submersed "water-leaf", divided into numerous root-like filaments which draw nourishment from the water. Sori borne on the water-leaves, either both kinds clustered near the leaf-base or in opposite pairs along the fertile filaments (pinnæ), the smaller ones with megasporangia in few, distant pairs in the basal part and beyond them several close pairs with microsporangia, all surrounded by a globose, thin indusium fixed at the base of the cylindrical receptacle. Microsporangia numerous, long-stalked, imbedded in a periplasmodium originating from the tapetum, each containing 64 microspores. Megasporangia few, short-stalked, with 8 spore-mothercells but producing only one large megaspore with perispore. — G a m e t o p h y t e. The microspores germinate in the sporangium, penetrate its wall and form a prothallium chiefly consisting of two small antheridia, a lower with 4 and an upper with 2 sperm-cells. The germinating megaspore forms a disc-shaped prothallium with few archegonia with a short neck. The embryo produces no root but only a three-lobed leaf; the next two leaves are alternating, the following in whorls. — 10 sp.

## Family 2. **Azollaceae**

Characters of the only genus.

*Azolla* LAMARCK. Stem bearing roots and leaves. Leaves in two alternating rows, divided into two lobes, an upper with assimilating tissue and stomata and below with a hollow filled with mucilage in which threads of an *Anabaena* are always to be found, and a lower, thin, submersed one. Sori borne on cylindrical receptacles on the submersed lobe, wholly surrounded by the indusium, placed in pairs of one large globose microsporangium-sorus with numerous, long-stalked microsporangia and one acorn-shaped sorus with one large megasporangium. The 64 microspores are imbedded in roundish bodies, "massulae", formed by the tapetum and furnished with anchor-shaped "glochidia" on the surface. The only megaspore is furnished with a thick, warted epispore originating from a part of the periplasmodium, the rest of which forms 3 massulæ serving as a floating apparatus. — G a m e t o-p h y t e: The microspores germinate within the free, floating massulæ, which are easily fixed to the warted megaspore by aid of the glochidia and produces a prothallium like a outgrowth from the surface of the massula, consisting of 8 mother cells for the spermatozoids and a few other cells. The megaspore produces a small green pothallium with one archegonium or, if this not being fertilized, eventually more. — 6 sp.

CHAPTER XXI

FOSSILE FILICINAE

von

MAX HIRMER (München)

Filicales der früheren Erdepochen sind sowohl in strukturbietenden als in im Abdruck und inkohlt erhaltenen Resten in beträchtlicher Zahl und nicht selten in ausgezeichneter Erhaltung auf uns gekommen. Während Vorläufer bereits aus dem Oberen Mitteldevon bekannt sind, sind die ersten systematisch sicheren Filicales vom Oberdevon ab bekannt. Es unterliegt keinem Zweifel, dass die *Filicales Eusporangiatae* im Paläozoikum weitaus die herrschenden unter den damals lebenden Farnen gewesen sind, und dass die *Filicales Leptosporangiatae* im Paläozoikum, und da selbst auch nur erst in dessen jüngsten Phasen, noch in untergeordnetem Masse vertreten gewesen sind. Was dieser Art an Farnen im Paläozoikum bereits vorhanden war, kann grösstenteils nur als Vorläufer unserer heute lebenden leptosporangiaten Farn-Familien angesprochen werden. Dagegen sind im Mesozoikum, vor allem von der Ober-Trias und dem Unteren Jura ab nahezu sämtliche derzeit lebenden Familien der leptosporangiaten Farne schon in markanten und eindeutigen Vertretern entwickelt, wobei ein Teil dieser Familien, insbesondere die *Matoniaceae* und *Dipteridaceae* — gemessen an ihren heutigen kleinen Arealräumen und der heute kleinen Zahl ihrer Arten — in nahe zu erdweiter Verbreitung und überdies in beträchtlicher Gattungs- und Artenfülle entwickelt waren. Daneben ist hervorzuheben, dass bis zur Kreidezeit *Schizaeaceae* und *Gleicheniaceae* ein gleichfalls wesentlich weiter gespanntes, bis in die heutigen Polargegenden hineinreichendes Areal umfassten, wenn auch ihr Vorhandensein zweifellos wesentlich weniger das Florenbild bestimmte als dies für die oben erwähnten Matoniaceen und Dipteridaceen gilt.

Vom Oberdevon und besonders vom Unterkarbon ab müssen tonangebend die *Coenopteridineae* gewesen sein, neben einer Anzahl von Formen, die wir heut innerhalb der Filicales nur als *Incertae sedis* (da eben Gruppen eigener Art und mit den familien-mässig festgelegten Filicales noch nicht vergleichbar) betrachten müssen. Erst vom jüngeren Karbon ab nehmen dann Formen beträchtlich an Zahl zu, die unseren jetzt lebenden *Marattiaceae* zweifellos mehr minder nahe stehen. *Ophioglossineae* sind erst seit dem Alt-Tertiär, und auch da nur spärlichst nachgewiesen.

Einzelheiten der z e i t l i c h e n   u n d   g e o g r a p h i s c h e n   V e r b r e i t u n g   der einzelnen Gruppen finden sich in dem Abschnitt Geographie und Zeitliche Verbreitung der Fossilen Pteridophyten, Seite 480–486. Hier seien im Folgenden lediglich die s y s t e m a t i s c h   u n d   m o r p h o l o g i s c h   w i c h t i g e n   T a t s a c h e n   zusammen gestellt, während bezüglich des Umfanges, d.h. dessen, was die einzelnen Gruppen oder Familien an Gattungen enthalten, abermals auf den oben genannten Abschnitt des *Verf.* in diesem Handbuch verwiesen sei.

### A : Filices Eusporangiatae :

**1) Aneurophytineae :** Diese offenbar heterospore (zunächst als Samen-bildendend betrachtete) Gruppe umfasst die baumförmigen *Aneurophyton* und *Eospermatopteris*. Morphologisch-systematisch liegt ihre Bedeutung darin, dass ihre Wedel noch im Wesentlichen nur spindelig gestaltet sind und noch der laubigen Fiederbildung entbehren [1]). Sodann darin, dass sie, obwohl älteste derzeit bekannte Filicalesähnliche Typen, zweifellos zum mindesten schon heterospor sind und so als direkte *Vorläufer der Pteridospermales* zu betrachten sind.

**2) Coenopteridineae :** Die interessantesten der hieher gehörigen Formen sind die der **Stauropteridaceae** und der **Etapteridaceae ;** beide dadurch auffällig und von allen übrigen Farnen abweichend, dass die Verzweigung ihrer übrigens meist erst spindeligen und der laubigen Fiedernbildung entbehrenden Wedel n i c h t   in einer einzigen Fläche entwickelt sind. Dies rührt daher, dass bei den *Etapteridaceae* die Spindeln der Seitenfiedern 1. Ordnung an ihrer Basis quer zur Ebene der Gesamtwedelverzweigung gabelig geteilt sind, und das sich bei *Stauropteris* dieser Modus der Spindelgestaltung auch bei den Wedelfiedern 2., 3., und 4. Grades wiederholt, mit dem Effekt, dass der Wedel von *Stauropteris* praktisch ein ganzer Busch war und keinerlei Ähnlichkeit mit dem uns geläufigen Bild eines Farnwedels aufwies. Während bei *Stauropteris* und einem Teil der *Etapteridaceae* die Differenzierung von Spross und Wedel noch wenig fortgeschritten scheint und die Wedel wohl noch über sich erhebenden Ästen des im ganzen an der Erdoberfläche kriechenden Spross-Systems vergleichbar sind, zeigt ein anderer Teil der Etapteridaceen (*Asteropteris* und *Etapteris* deutliche und exakte Differenzierung in einen allseitig mit Wedeln besetzten und wohl aufrechten, bei *Etapteris* vielleicht lianenartig kletternden Stamm. Stamm-Stele, wo bekannt, meist einfach protostelisch; bei einigen (*Metaclepsydropsis* und *Botrychioxylon*) mit Sekundär-Xylem; bei *Asteropteris* stellat mit massivem Xylem. Blattstiel-Bündel im Querschnitt stets mehr minder H-förmig.

Die übrigen, unter **Clepsydraceae** sowie **Botryopteroideae** und **Anachoropteroideae** aufgeführten Gattungen stehen mit den vorhergehenden nur in mehr minder lockerem Konnex. Die Primitivität ihres Stelenbaues und die Art der Leitbündel-Gestaltung ihrer Wedel sprechen für mehr minder nähere Verwandschaft. Alle haben bereits Wedel, die insgesamt nur in einer einzigen Ebene verzweigt sind.

Sporangien von einer Anzahl Gattungen bekannt, teils einzeln und endständig an den Spindeln letzter Ordnung (*Stauropteris*) teils in Gruppen und gleichfalls endständig bezw. randständig an den Spindeln letzter Ordnung. Öffnung entweder mittels Apikal-Porus (*Stauropteris*) oder mittels mehrzellreihigem Annulus; in letzterem Fall dieser entweder an der Sporangien-Aussenseite und allmählich in das übrige Sporangium-Wandgewebe übergehend (*Diplolabis, Botryopteris*) oder in Form von zwei einander opponierten breiten Bändern (*Etapteris* einschl. *Corynepteris*).

**3) Marattiineae :** Die Gruppe lässt sich formal unterteilen nach der mehr oder

---

[1]) Vgl. das über ähnliche Formen: *Protopteridium* u.a. Seite 555 Gesagte.

minder weitgehenden Durchfiederung des oft sehr grossen Wedels. Starke Wedel-Durchfiederung charakterisiert die *Pecopteroideae*, während weniger starke Durchfiederung, und daher grössere Angleichung an die heute lebenden Marrattiaceen die *Danaeoideae* charakterisiert.

**a) Pecopteroideae:** Wedeldurchfiederung vier- bis fünffach; Fiederchen letzter Ordnung im Vergleich zu den oft über ein bis zwei Quadrat-Meter grossen Wedeln sehr klein; in der Regel ausgesprochen pecopteridisch. Synangien mit Ausnahme derer von *Danaeites* stets aus nur einigen (vier bis sechs) Sporangien gebildet und kreisrund; diese entweder nur an der Basis miteinander verwachsen (*Asterotheca, Scolecopteris, Acitheca, Tetrameridium*), oder bis gegen die Spitze zu zu einem einheitlichen Synangium (ähnlich dem der rez. *Kaulfussia*) verwachsen (*Ptychocarpus*); fast immer flächenständig, nur selten randständig (wie bei *Asterotheca truncata* sowie den *Crossotheca*-Arten, soweit diese zu den Filicales gehören, wie *Cr. pinnatifida* u. Ähnlichen). Synangium von *Danaeites* länglich weckenförmig, aus ca 16 in zwei Reihen angeordneten, ziemlich freien Sporangien bestehend.

Stämme in Anzahl bekannt, teils an Hand strukturbietender Reste, teils im Abdruck und inkohlt. Nach der Wedel-Stellung sind zu unterscheiden Formen mit alternierend zweizeiliger Wedelstellung (*Psaronii distichi* bezw. (wenn im Abdruck) *Megaphyton*, sowie Formen mit allseitiger Wedeltragung; hier Wedel entweder in zwei bis sieben-zähligen alternierenden Quirlen (*Psaronii polystichi verticillati*) oder mit schraubiger Organstellung (*Psaronii polystichi spirales*); die nicht strukturbietend erhaltenen Reste der beiden letztgenannten Stamm-Typen gehen unter *Caulopteris* und *Stipidopteris*. Stämme vielfach baumförmig und ähnlich denen der rezenten Dicksoniaceen und Cyatheaceen, an der Basis mit mächtigem Wurzelmantel; im Innern mit einer der Organstellung entsprechenden Zahl und Anordnung von konzentrisch ineinander gelagerten hufeisenförmigen Stelenplatten, welche bei manchen Arten sämtlich in die Blattspuren übergehen, bei anderen Arten dagegen z.T. rein stamm-eigen sind.

**b) Danaeoideae:** Wedel gross, jedoch nur einfach bis doppelt gefiedert. Synangien stets aus zahlreichen in zwei Parallelreihen gestellten, unter sich aber stets weitgehend freien Sporangien bestehend, bei *Danaeopsis* die ganze Breite die fertilen Fiedern-Hälften bedeckend, bei *Marattiopsis* nur in geringerer Breite dem Fiedernrand folgend. Kreisrunde Synangien hat *Nathorstia*.

**Anhang:** *Paradoxopteris* mit vergleichsweise starkem und langem Stamm, im Bau prinzipiell dem der rezenzen Angiopteris sowie der Psaronii polystichi spirales ähnlich, jedoch wesentlich komplexer. Der vielleicht zugehörige Wedel (*Weichselia*) sehr gross und gabelteilig, die an nackten Fiedern-Spindeln getragenen Synangien ihrerseits noch zu kugeligen Gruppen zusammengefasst.

**4) Ophioglossineae:** Fossil kaum bekannt; im übrigen von den lebenden nicht wesentlich verschieden.

## B: Filices Leptosporangiatae:

**1) Osmundaceae:** Wedel der palaeozoischen und wohl auch der mesozoischen Gattungen grösstenteils stärker durchgefiedert als bei den rezenten Vertretern. Sporangien teils in Gruppen zu vielen (*Discopteris*, hier rand- oder flächenständig) teils über die ganze Fiederchen-Unterseite verstreut (*Todites* und *Speirocarpus*), teils einzeln an der Fiederchen-Unterseite (*Kidstonia*). Stammreste teils protostelisch siphonostelisch (permische Vertreter), teils dictyostelisch, dabei entweder mit primitiverer Protodictyostelie oder mit moderner und bis amphiphloëischer Dictyostelie.

**2) Schizaeaceae:** Wedel der älteren Formen, wie die der oberkarbonischen *Senftenbergia* und der rhätischen bezw. liasischen *Klukia* u. *Norimbergia* denen der rezenten Vertreter unähnlich, mit rein fiedrigem Bau sowie meist mehrfach durchgefiedert und (Ausnahme *Norimbergia*) gross. Von der Kreide ab werden die Formen denen der lebenden Gattungen ähnlich. Sporangium bei der oberkarbonischen *Senftenbergia* noch mit mehrzell-breitem Annulus. — Die wohl auch zur Familie gehörige *Tempskya* hat einen Scheinstamm, der zufolge Verflechtung der Wurzelmäntel des in zahlreiche Gabelarme verzweigten Stämmchens zustande kommt.

**3) Marsiliaceae:** Fossil nicht mit Sicherheit nachgewiesen.

**4) Gleicheniaceae:** Vorläufer-Formen dürften die oberkarbonischen bezw. unter-permischen *Oligocarpia, Monocarpia, Shansitheca, Sturia, Dendraena* und *Boweria*, ferner die mesozoische *Mertensites* sein; deren Wedel noch nicht von typischem Gleicheniaceen-Bau. Echte, der rezenten Gattung *Gleichenia* nächststehende Formen erst seit der Unterkreide bekannt, und zwar in den auch rezent vorhandenen Sektionen *Mertensia* und *Eugleichenia*, sowie in der nur fossil bekannten Sektion *Didymosorus*; bei dieser auf jeder Fieder letzter Ordnung nur zwei Sori über dem jeweils untersten Seiten-Nerven beiderseits des Mittelnerven.

**5) Matoniaceae :** Wedel sämtlicher fossiler Gattungen (*Phlebopteris, Selenocarpus, Matonidium* und *Matoniella*) im Prinzip wie der der lebenden *Matonia*-Arten gebaut d.h. primär monopodial mit jedoch exotroph sympodialer Verzweigung der beiden Basalfiedern des Wedels, wodurch der Wedel fussförmig zerteilt erscheint. Bei *Phlebopteris* und *Selenocarpus* noch keine Indusien-Bildung; bei letzterer Gattung die Sori nierenförmig statt wie bei den anderen Gattungen kreisrund; bei *Matonidium* beginnende Herausmodellierung des Indusium superum.

**6–8) Hymenophyllaceae, Loxomaceae, Dicksoniaceae und Cyatheaceae :** Soweit überhaupt nachgewiesen (vgl. das Seite 484 Gesagte) unterscheiden sich die fossilen Formen nicht nennenswert von den heute lebenden.

**9) Dipteridaceae :** Wedel unmittelbar an der Spreiten-Basis in zwei gleiche Primärhälften geteilt; darauf folgend mehr minder zahlreiche weitere Gabelteilungen. Diese entweder gleichmässig: *Hausmannia* s. str. (vgl. auch die rezente Dipteris lobbiana), oder ungleichmässig, d.h. bei starker Exotrophie sympodialgabelig; Effekt dass an der ersten Wedelgabelstelle zwei Hauptgabelarme entspringen, an welchen in mehr minder dichtgedrängter Abwickelung die einzelnen Hauptnerven der Wedelabschnitte entspringen. Es ist aber selbstverständlich, dass die beiden Haupt-Gabelarme ihrerseits zusammengesetzt sind aus den Basalteilen der sich weiter gabelnden Primär-Gabelarme. In den einfacheren Fällen verlaufen die Hauptgabelarme in sich ungedreht, und die Fläche der beiden Wedel-Hauptgabelteile liegt in einer Ebene. Dabei können die beiden Hauptgabelarme lang sein: *Protorhipis* (Wedel-Hauptteile zu mehr minder einheitlicher Fläche verwachsen, ähnlich auch die meisten der rezenten Dipteris-Arten), ferner *Oishia* (Sympodien hier ausnahmsweise endotroph); oder die Hauptgabelarme sind kurz: *Thaumatopteris* und *Clathropteris*; bei letzerer Gattung sind die Basalpartieen der gesamten Verzweigungen einschl. der ersten zu einer einheitlichen Fläche verwachsen, der Wedel daher handförmig geteilt. In den komplizierteren Fällen sind die Hauptgabelarme in sich tordiert; und zwar entweder nur an der Basis und hier um 180°: *Dictyophyllum* oder sie sind über ihre ganze Erstreckung hin tordiert, sodass die Fläche der beiden Hauptgabelarme in Form einer wendeltreppenartig aufgerollten Tüte erscheint: *Camptopteris*. — Sämtliche freien Wedelabschnitte sind bei allen Gattungen ungeteilt oder einfach gefiedert, lediglich bei den neuestens zu Familie gezogenen *Goeppertella* (ehedem *Woodwardites*) sind sie doppelt; der Wedelbau dürfte hier dem von *Thaumatopteris* entsprechen. — Sporangien teils in mehr minder grossen Sori oder über die Wedelfläche verstreut. Sporenzahl im Sporangium verschieden, 64 bis 512.

**10) Polypodiaceae :** Die fossilen Reste dieser fossil nur spärlich bekannten Familie fügen dem von den Lebenden her bekannten Bild nichts Wesentliches bei.

**11) Salviniaceae :** Fossile Vertreter nicht wesentlich anders als die lebenden.

**Anhang : Filicales incertae sedis.** Die hiehergehörigen zum grösseren Teil dem Palaeozoikum, zum kleineren Teil auch noch dem Mesozoikum entstammenden, Seite 485 in ihren wichtigeren Vertretern aufgeführten Formen dürften zum guten Teil eusporangiat gewesen sein, während einige von ihnen (*Dactylotheca, Renaultia* u.a.) den Typ des leptosporangiaten Farns vorbereitet zu haben scheinen; Sporangium bei diesen jedoch noch ohne Annulus.

Betreff *Literatur* vgl. F.O. BOWER, The Ferns Vol. I u. II. Cambridge 1923/26 sowie F. O. BOWER, Primitive Land Plants, London 1935. M. HIRMER, *Handbuch d Paläobotanik*, Bd. I, Seite 484–692 (dieser Teil mit 240 Abb.) Berlin u. München 1927; ferner R. WETTSTEIN, *Handb. d. Systematischen Botanik*, 4. Aufl. bearbeitet von F. V. WETTSTEIN u. M. HIRMER, Leipzig und Wien 1935. sowie M. HIRMER, Paläobotanik in *Fortschritte der Botanik*, Bde I–VI, Berlin 1932–37. Spezial-Literatur in den Werken angegeben.

CHAPTER XXII

# PTERIDOPHYTA INCERTAE SEDIS

von

MAX HIRMER (München)

Fossile Pflanzen von zwar s i c h e r e r  P t e r i d o p h y t e n-A r t,
aber von unsicherer Stellung i n n e r h a l b der Gruppe der Pteridophyten sind im Laufe der Zeit in grösserer Anzahl bekannt geworden, vor allem
aus den die ältesten Landpflanzen führenden Schichten.

Dabei kennt man einesteils Formen, welche, obwohl die Einzelheiten
ihrer Organisation genügend gut aufgeklärt ist, keiner der bekannten Pteridophyten-Gruppen (*Psilophytales, Psilotales, Lycopodiales, Articulatales*
und *Filicales*) eingereiht werden können, eben weil sie vermutlich Reprä-
sentanten anderer als der genannten Pteridophyten-Gruppen darstellen;
andernteils kennt man seit neuester Zeit eine Anzahl Formen, die offenbar
vermittelnd zwischen einzelnen der oben genannten fünf Pteridophyten-
Gruppen stehen. Es sind meist Formen, die mehr oder minder eng an die
*Psilophytales* anschliessen, darüber hinaus aber überleiten zu den mikrophyllen *Lycopodiales* oder *Articulatales* einerseits oder zu den makrophyllen
*Filicales* im engeren Sinn andererseits.

I m  e i n z e l n e n sei über folgende Formen hier berichtet.

## 1. Vermittelnde Formen.

Z w i s c h e n  *P s i l o p h y t a l e s*  u n d  e c h t e n  *F i l i c a l e s* vermitteln
die aus dem Mitteldevon Böhmens und Norwegens u.a.o. beschriebenen Formen
*Pseudosporochnus* POT. u. BRND., *Protopteridium* KREJČI, *Pectinophyton* HØEG.
Wiewohl — was besonders für *Pseudosporochnus* gilt — die Formen zweifellos an
echte *Psilophytales* anklingen, ist doch klar, dass in der starken und auf Fiedrigkeit
abzielenden Verzweigung der fruktifizierenden Achsensysteme die Bildung typischer Wedel vorbereitet wird, wie sie dann, präziser gefasst, in den mittel — bzw.
oberdevonischen *Aneurophyton* KR. u. WLD. und *Eospermatopteris* GOLDRING
vorliegen. Auch *Aphlebiopteris* GOTHAN und F. ZIMMERMANN (Oberdevon von Niederschlesien) ist eine Form, bei welcher die Wedelbildung allmählich Gestalt annimmt, wobei es sich ganz offensichtlich um einen Typus handelt, in welchem auch
noch m i k r o phylle Elemente (Aphlebien-artige, schraubig gestellte Anhängsel)
an einem schon im ganzen wedelartig, verzweigten, also m a k r o phyllen Spross-
system zu finden sind. Insgesamt liegen alle diese genannten Formen auf derselben
Entwicklungslinie, die schon von KIDSTON und LANG an Hand des ja klassisch gewordenen *Asteroxylon Mackiei* aufgezeigt worden ist.

Uebergangsformen zwischen *Psilophytales* und *Ly-
copodiales* sind gleichfalls bekannt, so *Haplostigma* SEW. aus den mittel-
bis oberdevonischen Witteberg-Series von Südafrika und andere; der unterdevo-
nische *Drepanophycus* KR. u. WLD., das mitteldevonische *Protolepidodendron*
KREJČI sowie *Cyclodendron* KRÄUSEL aus dem Unterdevon Afrikas dürften gleichfalls noch als Uebergangsformen zu betrachten sein, wennschon die Anklänge an
die Lycopodialens sehr eindeutig und weitstgehende sind.

Zwischen *Lycopodiales* und *Articulatales* dürfte *Boegen-
dorffia* GOTHAN u. F. ZIMMERM. stehen. Bekannt sind dreifach verzweigte Achsen;
Verzweigung jeweils in superponierten 2-gliedrigen Quirlen; Gliederung mit
zunehmendem Verzweigungsgrad deutlicher werdend (vgl. die Protoarticulaten
*Hyenia* KR. u. WLD. und *Calamophyton* KR. u. WLD.); an den Knoten einfach gestellte Quirlblättchen. Am Ende der Achsen letzter Ordnung Blüten-artige Zapfen
mit Sporangien-artigen Bildungen in der Achsel der offenbar quirlig gestellten ziemlich breiten Blättchen; während in der Art der Tragung der als Sporangien zu deu-
tenden Körper Beziehungen zu den *Lycopodiales* vorzuliegen scheinen, weist
die Gliedrigkeit der Stengel zweifellos auf Beziehungen zu den *Articulatales* hin.

Ausführlichere Beschreibung und kritischer Kommentar zur Morphologie dieser
Formen nebst Literatur-Nachweis dazu findet sich in den vom HIRMER bearbei-
teten Abschnitten „Palaeobotanik" in den Bänden 1–3 und 5 der Fortschritte der
Botanik.

## 2. Formen, die offenbar eigene Gruppen innerhalb der Pteridophyten darstellen.

*Cladoxylales.* Eine nur aus dem Mitteldevon von Elberfeld und dem Kulm von
Thüringen, Glatz und Schottland bekannte, vorwiegend durch die Untersuchungen
von SOLMS-LAUBACH, SCOTT, PAUL BERTRAND, sowie KRÄUSEL und WEYLAND er-
forschte Gruppe mit polystelem Stamm, wobei sich die Polystelie zweifellos auf eine
sehr ausgedehnte stellate Protostelie zurückführen lässt. Bei dem mitteldevonischen *Cladoxylon scoparium* ist der Stamm in seinem oberen Teil in eine Anzahl unregelmässig gabelteiliger Aeste aufgelöst, die seitenständige, fächerförmig verzweigte
endständig Sporangien-tragende Sprosssysteme letzter Ordnung abgliedern oder
mit mehr minder weit zerschlitzten Aphlebien besetzt sind. Bei den kulmischen
Arten (*Cladoxylon* UNGER mit vier Arten und *Voelkelia* SOLMS-LAUBACH) scheint
eine exaktere Differenzierung in Spross und Wedel durchgeführt zu sein.

Bei den kulmischen *Cladoxylon mirabile* u. *Clad. taeniatum* geben die Haupt-
sprosse (*Cladoxylon* im engeren Sinn) einerseits stärkere, schraubig oder quirlig gestellte Seitenachsen (*Hierogramma* u. *Arctopodium* genannt) ab, andererseits, und
zwar ohne räumliche Bezugnahme auf die *Hierogramma*- und *Arctopodium*-Achsen,
noch wedelartige, doppeltgefiederte Bildungen von *Clepsydropsis*-Struktur. Es ist
aber zu bemerken, dass auch an den ihrerseits entweder gabelig oder fiederig verzweigten *Hierogramma* bezw. *Arctopodium*-Achsen letzten Endes noch wieder
gleichfalls die *Clepsydropsis*-Bildungen entspringen. Es ist somit klar, dass die
*Hierogramma*- und *Arctopodium*-Achsen als intermediäre Bildungen zwischen
Seitenachsen der eigentlichen *Cladoxylon*-Hauptspross-Körpers und Wedelrhachi-
den anzusprechen sind. Es handelt sich also auch hier wieder um eines der Beispiele
der allmählichen Herauskristallisierung des makrophyllen Wedeltyps im stren-
geren Sinn aus sich verzweigenden Sprossachsen-Systemen.

Für alle kulmischen Arten gilt, dass bei Abgliederung der Seitenachsen eine
Anzahl nebeneinander liegender Arme der polystelen *Cladoxylon*-Spross-Stele die
Bündelversorgung mittels Abgliederung ihrer Aussenteile ergeben.

*Barrandeina* STUR aus dem Mitteldevon Böhmens. Stämme sympodial-gabelig
verzweigt, bis ca 4 cm im Durchmesser mit (vielleicht quirlig gestellten) mehrfach geteilten, fächerförmigen Blättern. An diesen Hauptachsen gelegentlich wesentlich schwächere, einmal gabelteilige Sprosse. Daran Organe (Sporophylle?),
die, ein bis zwei Mal gabelgeteilt u. spindelig, eine kleine Anzahl länglich eiförmiger
Sporangien unbekannten Inhalts tragen. Achsenstruktur unbekannt, vielleicht
polystel.

*Duisbergia* KR. u. WLD. aus dem rheinischen Mitteldevon. Dicke an der Basis
keulenförmiger verbreitete Stämme von *Cladoxylon*-ähnlicher Anatomie mit ringsum offenbar in schraubiger Anordnung dicht gedrängt stehenden Blattarti-

gen Bildungen von mehr oder minder stark zerschlitzter gabelnerviger Spreite; abwechslend mit den Blatt-artigen Bildungen kurz gestielte, umgekehrt keulen- bis birnenförmige Körper, offenbar Sporangien. Möglich, dass die Art Beziehungen zu der vorher geschilderten *Barrandeina* hat, wie auch zu der gleichfalls rätselhaften *Broeggeria* NATHORST des norwegischen Mitteldevons und gewissen als *Psygmophyllum* bezeichnete älteren paläozoischen Formen.

*Noeggerathia.* STERNB. Oberkarbon von Mitteleuropa. Pflanzen mit wedelartigen, einfach gefiedert erscheinenden Bildungen, die von manchen Autoren z.B. NEMEJC als echte Wedel angesprochen werden, von GOTHAN u.a. aber als Sprosse gedeutet werden, bei welchen die ursprünglich schraubig angeordneten, breit keilförmigen Blättchen sekundär zweizeilig eingerückt sind. Fruktifikation zapfenartig mit zweizeilig gestellten Elementen. Die basalen brakteenartig, die oberen fertil und zwar mit einer grösseren Anzahl (bei *N. foliosa* 17 und in 3 Reihen, bei *N. vicinalis* in sogar 5–7 Querreihen gestellten) Sporangien. Bei *N. foliosa* sind Mikro- und Megasporangien nachgewisen, beide von annähernd gleicher Grösse; die ersteren jedoch mit einer offenbar grösseren Zahl von Mikrosporen von 90–130 μ, die letzteren mit einer offenbar kleineren Zahl von Megasporen von 800–1000 μ. Die Verteilung anscheinend so, dass im unteren Teil die Mega-, im oberen die Mikrosporophylle waren.

In den Verwandtschaftskreis von *Noeggerathia* dürfte wohl auch *Tingia* aus dem jüngeren Paläozoikum Ostasiens gehören.

Die einschlägige Literatur ist grösstenteils in den Arbeiten von KRÄUSEL u. WEYLAND (Paläontographica Abt. B. Bde. 87 u. 80) sowei von P. BERTRAND (ebenda Bd. 81), sowie in M. HIRMER, Handbuch der Palaeobotanik u. in Fortschritte der Botanik, Bd. VI (Paläobot.) zu finden.

CHAPTER XXIII

PHYLOGENIE

von

W. ZIMMERMANN (Tübingen)

**§ 1. Allgemeine Übersicht. Literatur.** — Wie in der „Phylogenie der Moose" ist unser Ziel, den Wandlungsvorgang, den man Phylogenie nennt, zu schildern.

Wir müssen unser Arbeitsziel, die alleinige Schilderung eines V o r g a n g s, ausdrücklich umreissen, weil bekanntlich in den „vergleichenden" Wissenschaften (in der „vergleichenden" Morphologie und in der „Systematik") die Schilderung der phylogenetischen Wandlung oft fast unauffindbar vermengt ist mit den mehr technischen Fragen der Abgrenzung und Klassifikation, mit subjektiven Rangordnungen u. dgl.

Die Phylogenie als ein Wandlungsprozess beruht — wie wohl jetzt allgemein anerkannt ist — auf einer Veränderung des Erbgutes. Diese Veränderung geht Schritt für Schritt vor sich, einerseits durch „Mutationen" (so nennen wir heute jede Änderung von Erbfaktoren) und andererseits in beschränktem Umfang durch Neugruppierung der Erbfaktoren im Gefolge des Kernphasenwechsels (Bastardierung, Polyploidie u.dgl.). Phylogenie und Ontogenie sind also in der Natur in einer kontinuierlichen Kette („Hologenie", ZIMMERMANN 1934c) verbunden. Eine naturwissenschaftlich-genetische Betrachtungsweise muss diese „hologenetische" Verkettung stets im Auge behalten, wenn sie nicht vom Wege einer sachlichen Schilderung abweichen will. Wir werden im folgenden gelegentlich darauf hinzuweisen haben, dass Unklarheiten über diese Zusammenhänge, z.B. die Vorstellung einer direkten „phylogenetischen" Verwandlung fertiger Pflanzen bzw. ihrer Teile ineinander, zu verhängnisvollen Missverständnissen geführt haben.

Eine phylogenetische Wandlungsreihe ist daher eine von uns Menschen aus der „Hologenie" herausgegriffene Bilderreihe von embryonalen oder „fertigen" Organismen bzw. deren Teilen. Sie zeigt, wie als Ergebnis der Mutationen im Laufe der Jahrmillionen die einzelnen Merkmale, z.B. die Umrisse der Blätter, die Anordnung der Sporangien u.dgl. anders geworden sind als bei den Ahnen, dass sie sich phylogenetisch „gewandelt" haben.

Es ist eine Eigentümlichkeit der Mutationen, dass sich die einzelnen Merkmale bis zu einem hohen Grad selbständig wandeln. So kann die U m w a n d l u n g v o n M e r k m a l e n besonders bei den Pteridophyten mit ihren reichen Fossilfunden isolierter Organe viel sicherer aufgezeigt werden als die einzelnen Sippenzusammenhänge im Stammbaum. Daher stellen wir die „M e r k m a l s p h y l o g e n i e" der „Sippenphylogenie" voran. (Vgl. ZIMMERMANN 1931 und 1934a). Dagegen können wir in der vorliegenden Übersichtsdarstellung natürlich nicht die einzelnen, sich jeweils auf ein Gen beschränkenden „Elementarprozesse" dieser Wandlung (vgl. ZIMMERMANN 1935a) herausarbeiten. Wir müssen Merkmalskomplexe, wie sie in den Organen (Stamm, Blatt, Sporangium usw.) gegeben sind, in ihrer phylogenetischen Wandlung zusammenhängend verfolgen. Doch müssen wir uns auch hier auf einen Überblick über die Hauptwandlungslinien beschränken. Zur ausführlichen Darstellung aller bekannten Ergebnisse und erörterten Probleme bräuchte man ein vielbändiges Werk.

Wenn wir auf die angedeuteten Grundsätze einer streng realistischen Naturerkenntnis achten, liegt schon heute das Bild der Pteridophyten-Phylogenie in den Wandlungen der wichtigsten Organe und Merkmale als fast lückenlos geschlossene und gut gesicherte Bildreihe von einer einfach organisierten Urform zu den mannigfaltigen heutigen Typen vor. Denn diese Wandlungsvorgänge gehören, namentlich dank den phylogenetischen Untersuchungen von F. O. BOWER, D. H. SCOTT, A. G. TANSLEY, E. C. JEFFREY, A. C. SEWARD, E. A. N. ARBER, A. J. EAMES, R. KIDSTON, W. H. LANG, D. T. GWYNNE-VAUGHAN, H. POTONIÉ, J. C. SCHOUTE, O. LIGNIER, CH. u. P. BERTRAND, B. SAHNI, T. G. HALLE — um nur einige bahnbrechende Phylogenetiker zu nennen — zu den grossartigsten und bestbekannten biologischen Erscheinungen. Ich habe versucht, in einer wegen meines Ausdrucks „Telom" „Telomtheorie" genannten Auffassung diesen gewaltigen Wandlungsprozess als eine Umbildung innerlich sowie äusserlich wenig differenzierter thallophytischer Triebe („Telome") einheitlich zu verstehen. Denn das wichtigste Erkenntnisziel jeder Morphologie, auch der phylogenetischen, ist die Einheit des Ausgangsgebildes, ohne die Mannigfaltigkeit der Umbildung zu vergewaltigen.

Überdies sind uns die Pteridophyten auch aus dem Grunde besonders interessant, weil ihre ausgestorbenen Formen zweifellos die Wurzel der Samenpflanzen mitumfassen. Wir werden darum gelegentlich kurze Ausblicke auf die Samenpflanzen-Phylogenie einflechten.

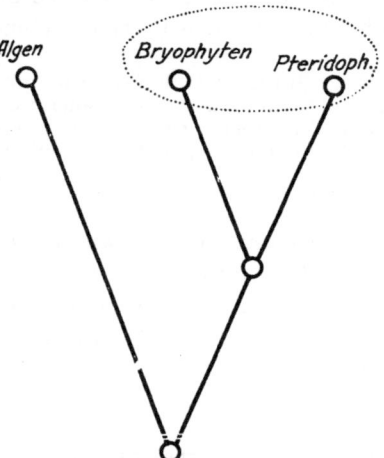

FIG. 1. Allgemein anerkanntes Grundschema der Pteridophyten-Verwandtschaft.

Da die Verwandtschaftsstellung der Pteridophyten innerhalb des gesamten Archegoniaten-Stammes klar ist (vgl. Fig. 1), können wir 3 Hauptabschnitte dieser phylogenetischen Wandlung unterscheiden:

I. Hauptabschnitt (§ 2): Die Wandlungsvorgänge, die zu der letzten mit den anderen Archegoniaten (vor allem den Moosen) gemeinsamen Ahnenform, zum „Ur-Archegoniaten" geführt haben.

Für unsere Phylogenetik der Pteridophyten besitzt auf diesem Wandlungsabschnitt besonderes Interesse die Entstehung des für die Pteridophyten so charakteristischen Generationswechsels und die Entstehung der spezifischen Archegoniaten-Fortpflanzungsorgane.

II. Hauptabschnitt (§ 3 und § 4): Die Wandlungsvorgänge, die vom „Ur-Archegoniaten" zum letzten, allen heutigen Pteridophyten gemeinsamen Ahn, zum „Ur-Pteridophyten" geführt haben.

Unsere Schlüsse sind insofern für diesen Wandlungs-Schritt recht sicher funda-
mentiert, als wir uns auf Grund der Fossilfunde ein gutes Bild vom Aussehen dieses
„Ur-Pteridophyten" machen können. Abgesehen von nebengeordneten Einzelheiten, in denen Übereinstimmung noch fehlt (vgl. unten S. 586), gelten heute
allgemein bei phylogenetisch orientierten Botanikern die frühdevonischen
*Rhynia*-Arten als Vertreter einer den „Ur-Pteridophyten" sehr nahe stehenden
Familie [1]).

Für unsere Übersicht können wir den Aufbau der Ur-Pteridophyten identifizieren mit dem Rhyniaceen-Typ (Fig. 4). Unsere Frage lässt sich damit präzisieren:
W e l c h e   M e r k m a l e   h a b e n   s i c h   a u f   d e m   W e g e   v o m   U r -
A r c h e g o n i a t e n   z u m   U r - P t e r i d o p h y t e n   ( R h y n i a c e e n - T y p )
g e w a n d e l t ? Diese Merkmalswandlung können wir darstellen, indem wir für
den Ur-Pteridophyten

*a*) das Bild des Generationswechsels und des Gametophyten (§ 3), sowie
*b*) das Bild des Sporophyten (§ 4) schildern.


I I I.   H a u p t a b s c h n i t t   (§ 5 — § 7): Die Wandlungsvorgänge,
die vom „Ur-Pteridophyten" zu den verschiedenen, fossil nachweisbaren
und heute lebenden Pteridophyten-Abteilungen geführt haben.

Dieses Stoffgebiet gliedern wir, indem wir

*a*) die Veränderungen des Generationswechsels und des Gametophyten (§ 5),
*b*) die Veränderungen des Sporophyten in seinen vegetativen Teilen (§ 6), sowie
*c*) die Veränderungen des Sporophyten in seinen Fortpflanzungsorganen schildern. (§ 7).


Als Abschluss werfen wir in einem
I V.   H a u p t a b s c h n i t t   (§ 8) einen Blick auf die sich aus diesem
phylogenetischen Ablauf ergebenden   s i p p e n p h y l e t i s c h e n   Z u -
s a m m e n h ä n g e   der Pteridophytengruppen untereinander und mit
ihren Verwandten.


ARBER, E. A. N., On the Past of the Ferns. Ann. of Bot. 1906, **20**, 215.
ARBER, E. A. N., Devonian Floras. Cambridge, 1921.
BASSLER, H., A Sporangiophoric Lepidophyt. Bot. Gaz. 1919, **68**, 73.
BENSON, M., The Grouping of Vascular Plants. New. Phytol. 1921, **20**, 82.
BERTRAND, C. E., Recherches sur les Tmésipteridées. Lille, 1885.
BERTRAND, P., Études sur la Fronde des Zygoptéridées. Diss. Lille, 1909.
BERTRAND, P., L'étude anatomique des fougères anciennes. Progr. Rei Bot. 1911, **4**,
     182.
BERTRAND, P., Importance des Phénomènes de Coalescence. C. R. Congr. Ass.
     Fr. Adv. Sc. Liége, 1924. S. 410.

---

[1]) Nicht-phylogenetische Bedenken, wie sie z.B. neuerdings TROLL (1934)
wieder geäussert hat, ob man *Rhynia* „primitiv" nennen dürfe, können ausser
Betracht bleiben, da wir ja hier die Phylogenie und keine „idealistisch-morpholo-
gischen" Einstufungen betrachten. Dasselbe gilt von den speziellen Einwänden,
die TROLL (Vergleichende Morphologie 1935, S. 47) gegen meine Auffassungen er-
hebt, da Troll durchweg mit rezenten spezialisierten Typen als Ausgangsformen
rechnet und dem Irrtum unterlegen ist, als rechnete ich auch damit.

BERTRAND, P., Observation sur la classification des vrais *Peceopteris*. C.R. Ac. Sc. 1934. 438.

BERTRAND, P., Sur la notion d'espèce à l'époque paléozoique. Proc. 6. intern. bot. Congr. Amsterdam, 1935.

BERTRAND, P., Contribution à l'étude des Cladoxylées de Saalfeld. Palaeontogr., 1935, **80**B, 101.

BOWER, F. O., Studies in the Morphology of Spore-Producing Members. I–V Phil. Trans. Roy. Soc. London, 1894, **185**, 473 bis 1903, **196**, 191.

BOWER, F. O., The Origin of a Land Flora. London, 1908.

BOWER, F. O., Medullation in the Pteridophyta. Ann. Bot. 1911, **25**, 555.

BOWER, F. O., Studies in the Phylogeny of the Filicales I–VIII Ann. Bot. 1920, **24**, 423; 1912, **26**, 269; 1913, **27**, 443; 1914, **28**, 363; 1915, **29**, 495; 1917, **31**, 1; 1918, **32**, 1; 1923, 37, 349.

BOWER, F. O., Hooker Lecture 1917, Journ. Linn. Soc. 1917/20, **44**, 107.

BOWER, F. O., The Primitive Spindle as a Fundamental Feature in the Embryology of Plants. Proc. R. Soc. Edinb., 1922, **43**, 1.

BOWER, F. O., The Ferns (Filicales). Cambridge, 1923–28, **1–3**.

BOWER, F. O., Size and Form in Plants, London, 1930.

BOWER, F. O., Primitive Land-Plants, London, 1935.

BREMEKAMP, C. E. B., Wie soll die Bezeichnung Baum verwendet werden? Ber. D. Bot. Ges. Jg., 1935, **53**, 1935.

BROWNE, I., The Phylogeny and Inter-relation-ships of the Pteridophyta. New Phytol. 1908, **7**, 93.

BROWNE, I., The Inter-Relationships of the Phyla. New Phytol, 1909, **8**, 51.

BROWNE, I., Structure of the Rhizome of *Equisetum giganteum*. Bot. Gaz., 1925, **80**, 48.

BROWNE, I., A New Theory of the Morphology of the Calamarian Cone. Ann. Bot., 1927, **41**, 301.

BROWNE, I., The Noeggerathiae and Tingiae. New Phytol., 1933, **32**, 344

BROWNE, I., Some Views on the Morphology and Phylogeny of the Leafy Vascular Sporophyte. Bot. Review, 1935, **1**, 383 und 427.

BUGNON, P., L'origine phylogénique des Plantes Vasculaires usw. Bull. Soc. Linn. Norm., 1921, **7**, Ser. Bd. **4**, 196.

BUGNON, P., Origine, évolution et valeur des concepts de protoxylème et de métaxylème. Bull. Soc. Linn. Norm., 1924, **7**, Sér. Bd. **7**, 123.

BUGNON, P., Les bases anatomiques de la théorie de la congrescence congénitale. Bull. Soc. Bot. France, 1928, **75**, 25.

CAMPBELL, D. H., Mosses and Ferns. New York 3. Aufl., 1918.

CAMPBELL, D. H., The Eusporangiate Ferns and the Stelar Theory. Amer. Journ. Bot, 1921, **83**, 303.

CAMPBELL, D. H., The relationships of the Anthocerotacae. Flora, 1925, **32**, 62.

CAMPBELL, D. H., Recent contributions to plant evolution. Amer. Nat. 1932, **66**, 481.

CHAUVEAUD, G., La Constitution des Plantes Vasculaires. Paris, 1921.

CHURCH, A. H., Thalassiophyta and the subaerial transmigration. Oxford, 1919.

CONRADI, A., Das System der Farne. Bot. Arch., 1926, **14**, 74.

COOKSON, I. C., On Plant Remains from Silurian usw. Phil. Trans. Roy. Soc. London, 1935, **B**, **225**.

DOMIN, K., Pteridophyta. Nova Encycl. české Akad. Prag, 1929.

DOMIN, K., Phylogenetic Evolution of the Phyllome. Amer. Journ. Bot., 1931, **18**, 237.

EAMES, A. J., On the Occurrence of Centripetal Xylem in *Equisetum*. Ann. Bot., 1909, **23**, 527.

EAMES, A. J., Morphology of Vascular Plants; Lower Groups. New York and London, 1936.

ELIAS, M. K., On a Seed-bearing *Annularia* usw. Univ. Kansas Sc. Bull, 1931, **22**, 115.

FLORIN, R., Zur Kenntnis der paläozoischen Pflanzengattung *Dolerophyllum* Sap. Sv. Bot. Tidskr., 1925, **19**, 171.

FLORIN, R., Zur Kenntnis der paläozoischen Pflanzengattungen *Lesleya* usw., Arkiv för Bot., 1933, **25 A**, No. 19.

Manual of Pteridology

GOEBEL, K. v., Gesetzmässigkeiten im Blattaufbau. Bot. Abh. 1922, **1**, Heft 1.

GOEBEL, K. v., Archegoniatenstudien XIX; Ähnlichkeiten und Parallelbildungen bei Farnen. Flora, 1930, **24**, 367.

GOEBEL, K. v., Organographie der Pflanzen. 3. Aufl. Jena, 1930, **2**.

GOTHAN, W., Lehrbuch der Paläobotanik (1. Aufl. v. H. Potonié) 2. Aufl. Berlin, 1921.

GOTHAN, W., Die Steinkohlenflora der westlichen paralischen Carbonreviere Deutschlands. Arb. Inst. Paläobot. usw., 1929, **1**, 1. und 1931, **1**, 49.

GOTHAN, W. und ZIMMERMANN, F., Die Oberdevonflora von Liebichau und Bögendorf. Arb. Inst. Paläobot. usw., 1932, **2**, 103.

GWYNNE-VAUGHAN, D. T., Remarks upon the Nature of the Stele of *Equisetum*. Ann. Bot., 1901, **15**, 774.

GWYNNE-VAUGHAN, D. T., Some Remarks on the Anatomy of the Osmundaceae. Ann. Bot., 1911, **25**, 525.

HALLE, T. G., Zur Kenntnis der mesoz. Equisetales Schwedens. K. Sv. Vet. Ak. Handl., 1908, **43**, I.

HALLE, T. G., Lower Devonian Plants from Röragen. K. Sv. Vet. Ak. Handl., 1916, **57**, 1.

HALLE, T. G., Palaeozoic Plants from Central Shansi. Pal. Sinica, 1927 A., **2**, 1.

HALLE, T. G., On Leaf-mosaic and Anisophylly in Palaeozoic Equisetales. Sv. Bot. Tidskr., 1928, **22**, 230.

HALLE, T. G., Some Seed-Bearing Pteridosperms from the Permian of China. K. Sv. Vet. Ak. Handl., 1929, **6**, no. 8.

HALLE, T. G., On the Habit of *Gigantopteris*. Förh. Geol. Fören. Stockholm, 1929, **51**, 236.

HALLE, T. G., The Structure of Certain Fossil Spore-bearing Organs believed to belong to Pteridosperms. K. Sv. Vet. Handl., 1933, **12**, no. 6.

HALLE, T. G., Notes on the Devonian Genus *Sporogonites*. Sv. Bot. Tidskr. 30, 613, 1936.

HALLE, T. G., On *Drepanophycus* usw. Palaeontologia Sinica, Ser. A, 1936, **1**, H. 4.

HARRIS, T. M., The Fossil Flora of Scoreby Sound East Greenland. Meddelelser om Gronland, 1931–1935, **85**, Nr. 2, 3 u. 5; **112**, Nr. 1.

HIRMER, M., Zur Kenntnis der Organstellung u. Zahlenverhältnisse i.d. Gattung *Calamostachys*. Flora, 1925, **118/19**, 227.

HIRMER, M., Handbuch der Paläobotanik. München u. Berlin, 1927, **1**.

HIRMER, M., Bemerkungen zur Theorie der serialen Spaltung der Blätter. Ber. D. Bot. Ges., 1933, **51**, 127.

HIRMER, M., Rekonstruktion von *Pleuromeia*. Palaeontogr., 1933, **78 B**, 47.

HIRMER, M., Grundsätzliches zur Rekonstruktion des Lepidophyten-Embryos. Palaeontogr., 1934, **78 B**, 143.

HOLDEN, H. S., Variations in Megaspore Number in *Bothrodendron mundum*. New Phytol., 1932, **31**, 265.

JANCHEN, E., Neuere Vorstellungen über die Phylogenie der Pteridophyten. Mitt. Naturw. Ver. Univ. Wien, 1911, **9**, 33.

JEFFREY, E. C., The Morphology of the Central Cylinder in Vascular Plants. Rep. Brit. Ass. Adv. Sci., Toronto, 1897, 869.

JEFFREY, E. C., The Development, Structure and Affinities of the Genus *Equisetum*. Mem. Boston Nat. Hist., 1899, **5**, no. 5.

JEFFREY, E. C., The Anatomy and Development of the Stem etc. Ann. Bot. 1901, **15**, 779.

JEFFREY, E. C., The Structure and Development of the Stem etc. Phil. Trans. R. Soc. London B., 1902, **195**, 119.

JEFFREY, E. C., The Pteropsida. Bot. Gaz., 1910, **50**, 401.

JONGMANS, W. J., Fossilium catalogus II Plantae, Berlin, 1914, 1915, 1922, 1924, 1930.

JONGMANS, W. J., On the Fructification of *Sphenopteris Hoeninghausi* usw. Geol. Bur. Nederland. Mijngeb. Jaarverslag over 1930, 77.

KIDSTON, R., Fossil Plants of the Carboniferous Rocks of Great Britain I–VI. Mem. Geol. Surv. Gr. Brit., 1923–25, **2**.

KIDSTON, R., On a New Species of *Tempskya* from Russia. Verhandl. Russ. Mineralog. Ges. 1911, **48**, 1.

KIDSTON, R., and GWYNNE-VAUGHAN, D. T., On the Fossil Osmundaceae I–V, Transact. R. Soc. Edinb. 1907, 1909, 1910 und 1915, **45–47**, 50.

KIDSTON, R. u. LANG, W. H., On Old Red Sandstone Plants etc. Transact. Roy. Soc. Edinb., Part I, 1917, **51**, 761, Part II–V, 1921, **52**, 603.

KRÄUSEL, R., Wesen und phylogenetische Bedeutung der ältesten Gefässpflanzen. Ber. D. Bot. Ges., Jg., 1932, **50**, 6.

KRÄUSEL, R., Neue Untersuchungen zur palaeozoischen Flora. Ber. D. Bot. Ges., 1936, **54**, 307.

KRÄUSEL, R. u. WEYLAND, H., Beiträge zur Kenntnis der Devonflora I–III. Senckenbergiana, 1923, **5**, Heft 5/6; Abh. d. Senckenb. Naturf. Ges. 1926, **40**, Heft 2 und 1929, **41**, Heft 7.

KRÄUSEL, R. u. WEYLAND, H., Die Flora des deutschen Unterdevons. Abh. Preuss. Geol. Landesanst., **131**, 1930.

KRÄUSEL, R. u. WEYLAND, H., Pflanzenreste aus dem Devon I–V. Senckenbergiana 1930, **12** und 1932, **14**.

KRÄUSEL, R. u. WEYLAND, H., Die Flora des Böhmischen Mitteldevons. Palaeontogr., 1933, **78 B**, 1.

KRAUS, F., Ein Beitrag zur Kenntnis der Anatomie und Physiologie der Pteridophyten-Spaltöffnungen. Jahresber. Fb. Gymnasium Graz, 1914.

KURSSANOW, L., Eine Notiz zur Frage über Phylogenie des Archegoniums. Bull. natural. Moskau, 1909, 39–43.

LANG, W. H., Alternation of Generations in the Archegoniates. Ann. of Bot., 1898, **12**, 583.

LANG, W. H., Contribution to the Study of the Old Red Sandstone Flora of Scotland I–V. Trans. Roy. Soc. Edinb., 1925, **54**, 293 u. 785.

LANG, W. H., On the Interpretation of the Vascular Anatomy of the Ophioglossaceae. Mem. Proc. Manchester Liter. and Phil. Soc., 1912, **56**, Part II, no. 12.

LANG, W. H., Studies in the Morphology and Anatomy of the Ophioglossaceae I. Ann. Bot., 1914, **27**, 203.

LANG, W. H., A Theory of Alternation of Generations. New Phytol, 1909, **8**, 1.

LANG, W. H., On the Spines, Sporangia, and Spores of *Psilophyton princeps.* Phil. Trans. Roy. Soc. London, Ser. B., 1931, **219**, 421.

LANG, W. H. and COOKSON, I. C., On a Flora, including Vascular Land Plants, associated with *Monograptus* usw. Phil. Trans. R. Soc. London, 1935 **B, 224**.

LIGNIER, O., Sur l'origine des Sphénophyllales. Bull. Soc. Bot. France, 1908, **55**, 278.

LIGNIER, O., Equisetales et Sphenophyllales usw. Bull. Soc. Linn. Norm., 1903, 93.

LIGNIER, O., Essai sur l'évolution morphologique du règne végétale. Bull. Soc. Linn. Norm. 1910 sér. 3, **6**, 35.

LIGNIER, O., Le *Stauropteris oldhamia* Binney et les Coenoptéridées à la lumière de la theorie du mériphyte. Bull. Soc. Bot. France XII, 1912, **59**, Mem. 24.

LIPPS, TH., Neuere Untersuchungen über die Gattung *Weichselia*. Arb. Inst. Paläobot. usw. 1932, **2**, 241.

LOTSY, J. P., Vorträge über Botanische Stammesgeschichte. Cormophyta Zoidogamia, 1909, **2**, 1.

MÄGDEFRAU, K., Die Stammesgeschichte der Lycopodiales. Biol. Zentralbl., 1932, **52**, 280.

MÄGDEFRAU, K., Zur Morphologie und phylogenetischen Bedeutung der fossilen. Pflanzengattung *Pleuromeia*. B. B. Z. 1931, **48**, II, 119.

MEKEL, J. C., Die Entwicklung des Stammes von *Matteuccia Struthiopteris*. Rec. Trav. Bot. Néerl., 1933, **30**, 627.

MEYER, F. J., Die diaplektischen Leitbündel der Lycopodien. Engl. Bot. Jahrb., 1926, **60**, 317.

MOLL, J. W., Phytography as a Fine Art. Leyden, 1934.

NATHORST, A. G., Die oberdevonische Flora des Ellesmerelandes. Rep. 2. Norweg. Arct. Exped. Fram, 1898–1902, 1904, no. 1.

NEMEJC, F., The Morphology and the Systematic Relations of the Carboniferous Noeggerathiae. Preslia. 1931, **10**, 111.

PORSCH, O., Der Spaltöffnungsapparat im Lichte der Phylogenie. Jena, 1905.

POSTHUMUS, O., On Some Principles of Stelar Morphology. Rec. trav. bot. néerl., 1924, **21**, 111.

POSTHUMUS, O., Über die sog. neue Keiltheorie. Ber. D. Bot. Ges. 1931, **49**, 309.

POTONIÉ, H., Über einige Karbonfarne II. Jahrb. Preuss. Geol. Landesanst. f. 1890. Herausgeg. 1892.

POTONIÉ, H., Abbildungen und Beschreibungen fossiler Pflanzenreste. Herausgeg. v. d. Kgl. Preuss. Geol. Landesanst. Lief. I–IX, 1903–1913.

POTONIÉ, H., Grundlinien der Pflanzen-Morphologie. 2. Aufl. Jena, 1912, s.a. Gothan.

RUDOLPH, K., Psaronien und Marattiaceen. Denkschr. Kais. Akad. d. Wiss. Wien, Math. Nat. Kl., 1905/06, **78**, 105.

RUDOLPH, K., Die Entwicklung der Stammbildung bei den fossilen Pflanzen. Lotos, 1921, **69**, 15.

RUSSOW, E., Betrachtungen über das Leitbündel usw. Dorpat, 1875.

SAHNI, B., Observations on the Evolution of Branching in the Filicales. New Phytol., 1917, **16**, no. 1.

SAHNI, B., On the Theoretical Significance of the Psilotaceae. Journ. Indian Bot. Soc. 1923, **3**, 185.

SAHNI, B., On *Tmesipteris Vieillardi*, Dangeard. Philos. Trans. Roy. Soc. London, 1924, Ser. B. **213**, 143.

SAHNI, B., On *Clepsydropsis australis*, etc. Philos. Trans. Roy. Soc. Lond., 1928, Ser. B, **217**, 1.

SAHNI, B., On *Asterochlaenopsis*. Philos. Trans. Roy. Soc. London, 1930, **218**, Ser. B., 447.

SAHNI, B., On a Palaeozoic Tree-Fern, *Grammatopteris Baldaufi* etc. Ann. Bot., 1932, **46**, 863.

SAHNI, B., On the Genera *Clepsydropsis* and *Cladoxylon*. New Phyt., 1932, **31**, 270.

SAHNI, B., On the Structure of *Zygopteris Primaria* (Cotta) etc. Philos. Transact. Roy. Soc. London, 1932, **222**, Ser. B, 29.

SANTSCHI, A., Contribution à l'étude anatomique du système vasculaire du cône d'*Equisetum*. Thèse Lausanne, 1935.

SCHAFFNER, J. H., Geographic Distribution of the Species of *Equisetum*. Amer. Fern. Journ., 1930, **20**, 89 und dort zitierte frühere Arbeiten.

SCHOUTE, J. C., Die Stelärtheorie, Groningen, 1902.

SCHOUTE, J. C., On the Foliar Origin of the Internal Stelar Structure of the Marattiaceae. Rec. trav. bot. néerl. 1926, **23**, 269.

SCHÜEPP, O., Meristeme. Aus Linsbauer Handb. d. Pflanzenanatomie, 1926, **4**.

SCOTT, D. H., On a Palaeozoic Fern, the *Zygopteris Grayi* of Williamson. Ann. Bot. 1912, **26**, 39.

SCOTT, D. H., Studies in Fossil Botany. 1920 u. 1923.

SEWARD, A. C., Fossil Plants I–III. Cambridge, 1898–1917.

SEWARD, A. C., Plant Life through the Ages. Cambridge, 1931.

SEWARD, A. C., Fossil plants from . . . . S. Afr. Quarterl. Journ. Geol. Soc., 1932, Ser. B, 221.

SINHA, B. N., The Origin and Evolution of the Archegonium. Journ. Ind. Bot. Soc. 1928.

SINNOTT, E. W., On Mesarch Structure in *Lycopodium*. Bot. Gaz. 1909, **48**, 138.

SYKES, M. G., Notes on the Morphology of the Sporangium-bearing Organs of the Lycopodiaceae. New Phyt., 1908, **7**, 41.

SZE, H. C., On the Occurrence of a new Species of *Palaeoweichselia* in Kansu. Memoir Nat. Res. Inst. Geol. Acad. Sinica, 1933, no. 13, 59.

TANSLEY, A. G., The Evolution of the Filicinean Vascular System. New Phytol. 1906–08, **6, 7**.

TANSLEY, A. G., The Evolution of Plants. New Phytol., 1920, **19**, 1.

THOMAS, H., The Early Evolution of the Angiosperms. Ann. Bot., 1931, **45**, 647.

THOMAS, H., On some Pteridospermous Plants.... South Africa. Phil. Transact. Roy. Soc. London, Ser. B, 1933, **222**, 193.

THOMAS, H., The Nature and Origin of the Stigma. New Phytol., 1934, **33**, 173.

THOMAS, H., On the Concepts of Plant Morphology. New Phytol. 1935, **34**, 113.

THOMSON, R. B., The Megasporophyll of *Saxegothaea* and *Microcachrys*. Bot. Gaz. 1907, **47**.

TIEGHEM, PH. VAN, Traité de Botanique, 1907, **47**, Paris, 1891, 2. Aufl.

TROLL, W., Grundsätzliches zum Stigmarienproblem. Flora, 1934, N. F., **29**, 94.

TROLL, W., Wurzel. Naturw. Handwörterb. 2. Aufl., 1935, **10**, 682.

UITTIEN, H., Über den Zusammenhang zwischen Blattnervatur und Sprossver-zweigung. Rec. trav. bot. néerl. 1929, **25**, 390.

WEISS, P. L. S., A Re-Examination of the Stigmarian Problem. Proceed. Linn. Soc. London, 1931/32, Part. V, 151.

WETTSTEIN, R., Referat über Zimmermann Phylogenie d. Pfl. Zeitschr. f. Bot. 1930, **24**, 200.

WETTSTEIN, R., Handbuch der Systematischen Botanik. 4. Aufl. 1933, **1**.

WILKOEWITZ, K., Über die Serologie und Morphologie des Farnastes. Bot. Arch., 1929, **23**, 445.

ZEILLER, R., Eléments de Paléobotanique. Paris, 1900.

ZERNDT, J., Triletes giganteus usw. Bull. Ac. Pol. Sc. Bér. B. 1930, S. 1.

ZIMMERMANN, F., Zur Kenntnis von *Eleutherophyllum mirabile*. Arb. Inst. Pa-läobot. u. Petrogr. d. Brennsteine, 1930, **2**, 83.

ZIMMERMANN, W., Die Spaltöffnungen der Psilophyta usw. Zeitschr. f. Bot. 1926, **19**, 129.

ZIMMERMANN, W., Die Georeaktionen der Pflanze. Ergebn. d. Biol. 1927, **2**, 116.

ZIMMERMANN, W., Die Phylogenie der Pflanzen. Jena, 1930.

ZIMMERMANN, W., Der Baum in seinem phylogenetischen Werden. Ber. D. Bot. Ges. 1930a, **48**, (34).

ZIMMERMANN, W., Arbeitsweise der botanischen Phylogenetik usw. Handb. d. biol. Arbeitsmeth., 1931, IX, **3**, 941.

ZIMMERMANN, W., Phylogenie der Bryophyten (aus Verdoorn, Manual of Bryology) Haag, 1932, 433.

ZIMMERMANN, W., Fortpflanzung der Gewächse. Zwischenstufen zwischen Arche-goniaten und Samenpflanzen. Handwörterb. d. Naturwissensch. 2. Aufl. 1934, **4**, 353.

ZIMMERMANN, W., Research on Phylogeny of Species and of Single Characters. Americ. Natural., 1934a, **68**, 381.

ZIMMERMANN, W., Grundfragen der Deszendenzlehre. Moderne Naturwissenschaft. (öff. Vortr. Univ. Tübingen) 1934b, 6 Heft 8.

ZIMMERMANN, W., Genetische Untersuchungen an *Pulsatilla* I–III. Flora 1934c, **129**, 158.

ZIMMERMANN, W., Die phylogenetische Herkunft der gegenständigen u. wirteligen Blattstellung. Jahrb. wissensch. Bot. 1935, **81**, 239.

ZIMMERMANN, W., Untersuchungen zur Gesamtphylogenie der Angiospermen III u. IV. Jahrb. wissensch. Bot., 1935a, **82**, 234.

## § 2. Die phylogenetische Entstehung einer den Archegoniaten gemeinsamen Urform.

— Ich kann hier auf eine eingehende Diskussion verzichten, da ich die gleiche Frage für die Phylogenie der Moose ausführlicher behandelt habe (ZIMMERMANN 1932, S. 436 ff). Wir begnügen uns also mit einer kurzen Schilderung des Ur-Archegoniaten. Der Ur-Archegoniat (vgl. Fig. 2A) besass folgende Eigenschaften:

1) Einen „homologen", besser „homomorphen", Generationswechsel: D.h. die beiden miteinander alternierenden Generationen waren äusserlich gleich wie bei vielen heute noch existierenden marinen Algen (beispielsweise bei

*Dictyota dichotoma*). Sie unterschieden sich lediglich durch die Beschaffenheit der Keimzellbehälter (Sporangien beim Sporophyten, Gametangien
beim Gametophyten), durch die Ausgestaltung der zugehörigen Keimzellen und durch die Chromosomenzahl (Sporophyt diploid, Gametophyt
haploid). Die letztgenannte Eigenschaft sichert eine strenge Alternanz der
Generationen.

2) *Eine thallophytische Gestalt:* dichotom sich gabelnde Triebe ("Telome")
mit einem mechanisch aussteifenden Zentralstrang. Diese Eigenschaften
sind ebenfalls bei heutigen Tangen noch vertreten und kennzeichnen eine
geringe Arbeitsteilung der Organe und Gewebe sowie eine mechanische
Beanspruchung durch Zug (Leben in der Brandungszone!).

3) *Für die Archegoniaten spezifische Keimzellbehälter und Keimzellen:*
Archegonien mit Eiern, Antheridien mit Spermatozoen, [1]) sowie Spo-
rangien mit Sporen (Fig. 5). Die Sporangien waren gleichfalls Telome, d.h.
Thallus-Endtriebe, deren Inhalt unter Reduktionsteilung in Sporen zerfiel.

Die vorgetragene Auffassung entspricht der „Transformationstheorie", (= „homologous theory"). Sie wird neuerdings mehrfach u.a. von EAMES (1936) vertreten.
Ihr entgegengesetzt ist die „Interkalations-theorie" (= „antithetic theory"). Nach
dieser mir unwahrscheinlichen Annahme war der Sporophyt beim Ur-Archegoniaten noch sehr klein, so dass der Generationswechsel anfangs „antithetisch" (=
heteromorph) war und der Sporophyt erst allmählich die heutige Grösse erhalten
hat. Die Interkalationstheorie ist von R. WETTSTEIN und F. O. BOWER vertreten
und von mir früher (z.B. ZIMMERMANN 1932) ausführlicher erörtert worden.

Dieser Ur-Archegoniat hatte sich — vermutlich im Spät-Kambrium oder
Silur — aus küstenbewohnenden Grünalgen, die ja für diese Zeit schon
nachgewiesen sind, herausentwickelt. Selbstverständlich war es ein weiter
Wandlungsweg von solchen Algen bis zu den heutigen Archegoniaten. Der
Verwandtschaftszusammenhang ist daher sehr lose.

Vom Ur-Archegoniaten aus ging die weitere phylogenetische Entwicklung divergent zu den Moosen und zu den Pteridophyten. Bei den M o o -
s e n war es vor allem der G a m e t o p h y t, der sich in stärkerem Masse
ausdifferenzierte, während der auf dem Gametophyten parasitierende
Sporophyt reduziert wurde. In der Entwicklungsreihe, die zum U r -
P t e r i d o p h y t e n führte, wurde dagegen umgekehrt gerade der
S p o r o p h y t reicher differenziert. D.h. der Sporophyt wurde zur
„Haupt-Generation", wogegen der Gametophyt in seinen Ausmassen und
in der Differenzierung von Organen und Geweben zurücktrat.

### § 3. Generationswechsel und Gametophyt des Ur-Pteridophyten.
— Die Reduktion des Gametophyten muss bereits beim Ur-Pteridophyten,
der vermutlich im Ober-Silur gelebt hat, sehr weit fortgeschritten gewesen sein. D.h. der Gametophyt war schon wesentlich kleiner als der

---

[1]) Vgl. ZIMMERMANN, 1932, S. 455.

zugehörige Sporophyt, er war zum „Prothallium" geworden. Im übrigen behielt er die gabelig thallose Gestalt der Ur-Archegoniaten, die ja auch den meisten heutigen Pteridophyten-Prothallien gemeinsam eigen ist. Für die deutliche Reduktion des Gametophyten bei den ältesten Pteridophyten spricht die zunächst sehr auffällige Tatsache, dass man bei fossilen Pteridophyten (einschl. Psilophyten) entweder überhaupt keine Prothallien beobachtet hat, oder doch nur eine viel schwächer als der Sporophyt entwickelte Geschlechtsgeneration. Auch heute erreicht ja kein Prothallium der Pteridophyten eine kormophytische Struktur. Wir haben daher nicht den mindesten Anlass zur Annahme, dass die Pteridophyten-Gametophyten irgendwann einmal ein uns unbekanntes kormophytisches Stadium erreicht hätten.

Auch die vorherrschend gabelige Verzweigung der heutigen Prothallien (vgl. Fig. 2) ist für uns ein Anhaltspunkt, dass der Ur-Pteridophyt noch ein gabeliges Prothallium hatte. Ferner fehlt jeder phylogenetische Anhaltspunkt, dass der Gametophyt des Ur-Pteridophyten die zweifellos primitive Autotrophie aufgegeben hätte.

Das gelegentlich in der Literatur zu findende Argument: „weil heute manche sog., primitive' Pteridophyten, z.B. heutige Lycopodien Gametophyten mit Mykorrhiza besitzen, muss das bei den Ahnen auch so gewesen sein", ist wegen der selbständigen Abwandlung jedes einzelnen Merkmals nicht stichhaltig. Es ist im Grunde ein rein systematisches, aber kein phylogenetisches Argument.

Vermutlich ist die Reduktion und sonstige Umbildung des Gametophyten eine Folge des Uebergangs zum Landleben. Der Gametophyt bedarf vor „Erfindung" des Pollenschlauches des flüssigen Wassers, da seine Spermatozoen nur schwimmend ein Ei erreichen können. Er blieb daher zunächst an flüssiges Wasser gebunden. D.h. der Gametophyt musste klein und in Erdennähe bleiben und musste auch im allgemeinen feuchte Standorte aufsuchen, wo ihm Wasser für den Befruchtungsakt in ausreichendem Masse zur Verfügung stand. [1]

Umgekehrt konnte beim Ur-Pteridopyhten der Sporophyt grösser werden. Ja, es war für das Ausstreuen der Sporen sogar vorteilhaft [1]), dass der Sporangienträger möglichst weit in die Luft ragte. Selbstverständlich bedeutete ein möglichst hoher Sporophyt im Wettkampf um Licht und Luft eine gesteigerte Assimilationsmöglichkeit. Aus dem erstgenannten Grund wurde der Sporophyt bei den Moosen zu einem dem Gametophyten parasitisch aufsitzenden Gebilde. Oder bei den Pteridophyten wurde er zu einem selbständigen und den Gametophyten an Grösse weit übertreffenden Organismus.

---

[1]) Selbstverständlich verstehen sich alle diese Aussagen nur im Sinne der „reinen Oekologie", die — ohne Behauptung von „Zweckursachen" und andere anthropomorphe Konstruktionen — über die natürlichen Lebensbedingungen der Organismen etwas aussagt.

**§ 4. Sporophyt des Ur-Pteridophyten.** — Im Gegensatz zum schlecht überlieferten Gametophyten haben wir ein gutes Bild vom Sporophyten des Ur-Pteridophyten in den ausgezeichnet erhaltenen

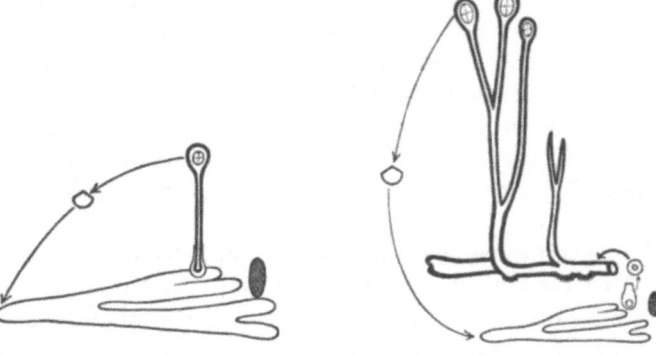

*B* Typus der Bryophyten.   *C* Typus der Pteridophyten.

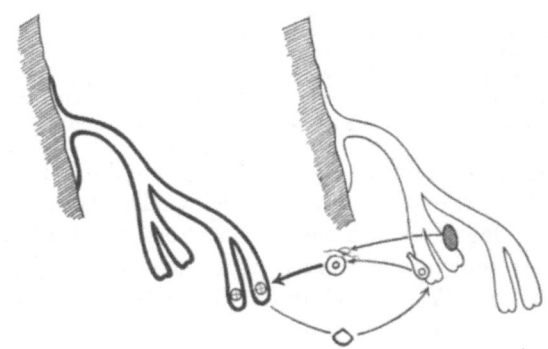

*A* Hypothetische Ausgangsform.

FIG. 2. Abwandlung des Generationswechsels nach der „homomorphen" Auffassung. —
*A* Hypothetische Ausgangsform mit gleichgestaltetem Gametophyt und Sporophyt (= Dictyota-Typ).
*B* Typus der Bryophyten (Sporophyt abhängig vom Gametophyten und reduziert).
*C*. Typus des Pteridophyten (Gametophyt reduziert).

Gametophyt jeweils dünn konturiert, mit Archegonien (flaschenförmig) und Antheridien (eiförmig schraffiert);
Sporophyt dick konturiert, mit Tetrasporen (Sporangien mit Punktkreuz). (Nach ZIMMERMANN 1930, S. 86).

altdevonischen Psilophyten. Sie zeigen ein so übereinstimmendes Bild, dass wir in den gemeinsamen Zügen unbedingt die typischen Merkmale des Ur-Pteridophyten erkennen können. Ein Beispiel ist die Gattung

*Rhynia*. Unter Hinweis auf ihre Schilderung (S. 497) heben wir nur die Hauptpunkte hervor.

H a b i t u s. Die allgemeine Tracht entsprach noch völlig dem thallosen

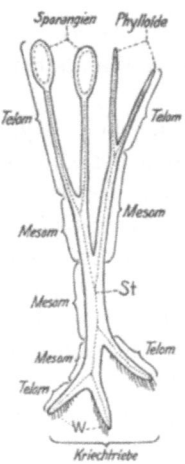

FIG. 3. Tracht der Urpteridophyten (*Rhyniaceae*), (etwas schematisiert, in Anlehnung an TROLL 1934, S. 684). Typische Ausbildung der Telome (oberirdisch als Sporangien und Phylloide, unterirdisch als Kriechtriebe) und der Mesome.
*St* = Stele.
*W* = Wurzelhaare.

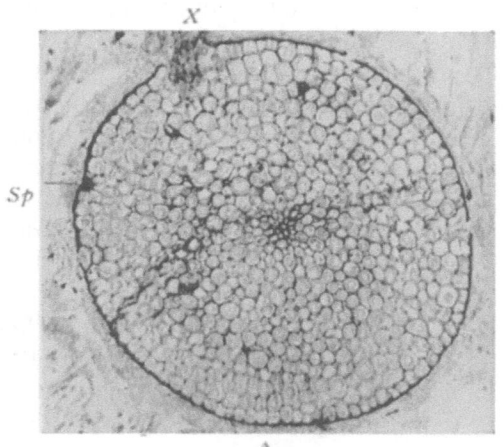

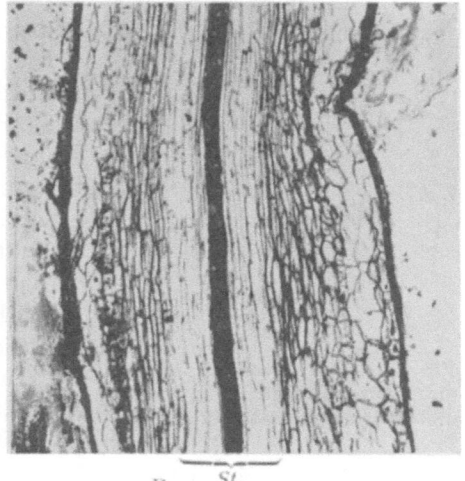

FIG. 4. *Rhynia Gwynne-Vaughani*.
Telom: A) quer, B) längs. — (M.-Devon.)
*St.* = Protostele (wenigzelliger Holzteil, Siebteil gegen die Rinde kaum abgegrenzt).
*Sp.* = Spaltöffnung, wenig eingesenkt.
*X* = angemoderter Teil der Epidermis.
Vergr. 35 mal.     (Aus ZIMMERMANN 1930).

Aufbau, wir wir ihn für die Ur-Archegoniaten annehmen. D.h., wie die Fig. 3 wohl ohne nähere Erläuterungen zur Genüge wiedergibt, ist die ganze Pflanze noch aus wenig differenzierten Telomen und Mesomen [1]) aufgebaut. Als einzigen organisatorischen Fortschritt können wir die Differenzierung der vegetativen Telome bzw. Mesome in Kriechtriebe und aufrechte, assimilierende Lufttriebe bezeichnen. Beide Organen behalten jedoch noch ausgesprochenen

---

[1]) Vgl. unten, S. 570.

„Telomcharakter", ohne Arbeitsteilung der einzelnen Triebe, etwa in Stamm und Blatt.

A n a t o m i e. An Geweben differenzierte sich in *Rhynia* aus: Eine E p i d e r m i s mit Spaltöffnungen, die dem Gametophyt fehlen, eine R i n d e und ein Z e n t r a l z y l i n d e r, bei dem eine äussere Schale als prospektives Phloem aus langgestreckten, aber in der Wandstruktur noch nicht siebröhren-artigen Zellen bestand, (Fig. 4).

Der Kern des Zentralzylinders war das ziemlich einheitliche Xylem, das aus Spiraltracheiden bestand. Man nennt einen derartigen Zentralzylinder oder Stele eine P r o t o s t e l e. (Fig. 15, S. 593). Die Auffassung JEFFREYS, dass in ihr der Ausgangspunkt für die Leitbündelanordnung der übrigen Landpflanzen existiert, hat sich durchgesetzt.

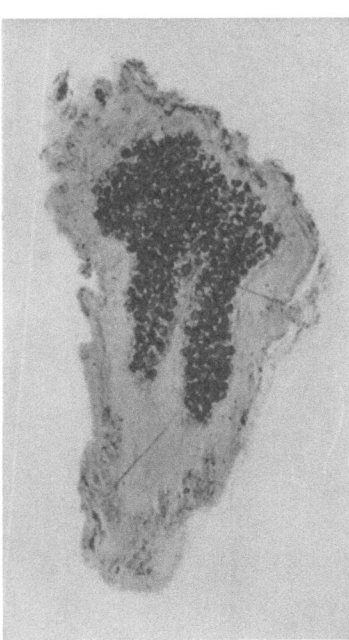

FIG. 5. *Hornea Lignieri*, Sporangium mit Columella. — Vergr. 30 × (Aus ZIMMERMANN 1930).

Die Tracheiden des Zentrums sind bei einigen Psilophyten zarter und werden ontogenetisch zuerst angelegt. Man nennt sie daher Protoxylem. (Fig. 15).

Die S p o r a n g i e n sind die fertilen terminalen Auszweigungen („Telome"). Sie sind aufgebaut aus einer mehrschichtigen („eusporangiaten") Wandung und, als Kern, aus einem sporogenen Gewebe, das typische Luftsporen durch Reduktionsteilung bildet. Die Wandung springt vielfach apikal auf. Manchmal (z. B. bei der Gattung *Hornea* Fig. 5) setzt sich der Zentralzylinder als Ernährungsstrang („Columella") ins Innere der Sporenmasse fort (Vgl. auch unten S. 663).

T e l o m e  u n d  M e s o m e [1]). Die Verwendung dieser beiden Ausdrücke erfordert sowohl in sachlicher wie in begrifflich-nomenklatorischer Hinsicht einige Erläuterungen. Sachlich kennzeichnen diese Ausdrücke die Auffassung, dass die Ur-Pteridophyten entsprechend der „T e l o m t h e o r i e" und Fig. 3 aus lauter Mesomen und Telomen, d.h. aus lauter gleichwertigen, also gabelig verketteten und übereinstimmend einfach organisierten Trieben zusammengesetzt waren, dass sie also noch keinen in Sprosse, Blätter, Sporophylle, Wurzeln usw. gegliederten Aufbau besassen. Wie das die Fig. 3 schildert, nenne ich diese einfachen Endauszweigungen der Triebe wie früher „Telome" und füge nun als neue Bezeichnung den Ausdruck „Mesome" für die prinzipiell ähnlich gebauten „Internodien" zwischen je zwei Gabelstellen ein. Im Laufe der Ontogenie war natürlich jedes „Mesom" zunächst ein „Telom".

---

[1]) Als zusammenfassenden Begriff für Telome und Mesome könnte man den POTONIÉ'schen Begriff „Monosome" verwenden.

Sachlich ist die „Telomtheorie" in ihren Grundzügen heute wohl ziemlich allgemein angenommen worden. Umstritten ist nur die Ausdehnung der Theorie auf einige Spezialgebiete wie die Entstehung des „Lycoblattes" (vgl. unten, S. 586). Auch die Bezeichnung „Telom" bürgert sich mehr und mehr ein. Bedenken treffen auf diesem nomenklatorisch-begrifflichen Gebiet, soweit ich sehe, wiederum nur die Verwendungs g r e n z e des Begriffes „Telom". So haben z.B. WETTSTEIN 1930, S. 201 und BOWER 1935, S. 617 bezweifelt, ob man diesen Begriff auch noch bei komplizierteren Bildungen wie vielen Sprossen und Blättern der Angiospermen verwenden könne. Selbstverständlich existieren, wie ich von Anfang an betont habe (ZIMMERMANN, 1930 S. 66), solche Verwendungsgrenzen. Man kann etwa bei einem ganzrandigen Netznervenblatt einzelne Telome (und Mesome) nicht mehr unterscheiden. Solche Verwendungsgrenzen des Begriffes verhindern aber weder seinen nutzbringenden Gebrauch für primitivere Organe noch die phylogenetische Ableitung der komplizierten Gebilde von den primitiven und noch „typisch" gestalteten „Telomen" und „Mesomen".

Verwendungsgrenzen solcher Begriffe finden sich ja in ganz ähnlicher Weise für andere stark gewandelte Organe, wie die Prothallien. Wollte man wegen Verwendungsgrenzen derartige Begriffe ganz ausschlagen, dürfte auch niemand von einem „Prothallium" mit Archegonien und Antheridien reden, geschweige denn die Gametophyt-Generation der Angiospermen davon ableiten. Kurz, die Begriffe und die Bezeichnungen „Telom" und „Mesom" sind ebenso wie „Archegonien", „Antheridien", usw. in erster Linie für primitive Organisationen der Gefässpflanzen geschaffen.

## § 5. Wandlung des Generationswechsels und des Gametophyten zu ihrer heutigen Gestalt.
— In drei ökologisch verschiedenen Entwicklungslinien hat sich die Gestalt der Prothallien selbständig abgewandelt.

1) F i l i c i n e e n - T y p. Der Gametophyt blieb autotroph und ähnelt oft weitgehend einem wenig entwickelten thallosen Lebermoos. Dieser Typ findet sich bei den *leptosporangiaten Farnen*, den *Osmundales* und *Marattiales* sowie bei den *Equisetales*. Die Hauptwandlung besteht wohl in einer mehr oder weniger weit gehenden Rückbildung und in einer meist deutlich flächenförmigen Entwicklung des zarten Thallus.

Gelegentlich ging die Reduktion bis zu fadenförmigen Gebilden (*Trichomanes*) weiter. Durch die Lage der Meristeme (z.B. bei den *Equisetales* unter der „Lappenkrone"), durch die Behaarung, Stellung der Gametangien und dgl. ergeben sich im Habitus erhebliche Unterschiede, auf die wir im Einzelnen aber nicht eingehen können.

2) L y c o p o d i u m - T y p. Er ist vertreten bei der Gattung *Lycopodium* samt *Phylloglossum*, bei den *Psilotales* und *Ophioglossales*. Das Charakteristikum dieses Typs liegt in der mykotrophen Lebensweise. Im übrigen haben wir bei den *Psilotales* noch einen typisch dichotom sich verzweigenden, ziemlich kräftigen Thallus und beim *Lycopodium cernuum*-Unter-Typ apikal noch eine assimilierende „Lappenkrone". Im allgemeinen ist aber auch hier das thallose Prothallium offensichtlich stark rückgebildet zu einem meist knolligen Gebilde.

3) H e t e r o s p o r e n t y p. Er ist — wie schon sein Name andeutet — durch verschiedene (grosse ♀ und kleine ♂) Sporen und Rückbildung des Gametophyten charakterisiert. Die Grössendifferenzierung geht also Hand in Hand mit einer Geschlechtsdifferenzierung der Sporen. Besonderes Interesse gewinnt der Heterosporen-Typ als Durchgangstyp zur Samenbil-

dung. Dementsprechend vermittelt die Heterosporie (herrschend im Karbon) auch zeitlich zwischen einer Periode vorherrschender Isosporie (Altdevon) und einer Periode vorherrschender Samenreproduktion (Jetztzeit) (vgl. ZIMMERMANN 1932 und 1934a).

Heterosporie hat sich selbständig in 4 Hauptabteilungen der Pteridophyten herausgebildet:

a) L y c o p s i d e n: bei den fossil überlieferten Lepidophyten und bei *Pleuromeia*, sowie bei den bis heute lebenden Selaginellen und *Isoetes*.

b) A r t i c u l a t e n: bei manchen fossil überlieferten Calamiten und Sphenophyllen.

c) N o e g g e r a t h i e e n: soweit bekannt bei allen Vertretern (vgl. NEMEJC).

d) P t e r o p s i d e n: bei unsicher bekannten fossilen Formen, z.B. vielleicht bei *Aneurophyton* und *Eospermatopteris* sowie bei den heutigen Hydropterideen.

Die Grössendifferenz zwischen ♀ und ♂ Sporen war vielleicht nicht gleich mit einem Schlage da. Wenigstens hatten die ersten deutlich heterosporen Pteridophyten (z.B. oberdevonische Lycopsiden wie *Bothrodendron antrimense*) nach J. M. CLARK noch keine scharf gesonderten Makro-und Mikrosporen, sondern auch vermittelnde „intermediäre'' Sporen. Dagegen zeigt das Oberkarbon schon ausserordentlich grosse Makrosporen; Sporen von *Triletes giganteus* (Lycopside) messen nach ZERNDT 6,4 mm.

Mit der Grössenzunahme ging die Zahl der Makrosporen eines Sporangiums zurück. Bei karbonischen Lycopsiden (z.B. Lepidostroben) war meist noch eine grössere Zahl von Makrosporen (4—20) vorhanden. Heute finden wir z.B. bei *Selaginella* nur 4 und bei den Hydropterides nur 1 Makrospore in Makrosporangium (über interessante Ausnahmen vgl. HOLDEN).

Die Prothallienreduktion besteht ökologisch darin, dass die Sporen mehr oder minder ausreichend Reservestoffe von der Mutterpflanze mitbekommen. Sie verzichten daher weitgehend auf eigenes Wachstum und Assimilation. Ihre ganze Entwicklung spielt sich im grossen und ganzen als Zellteilung innerhalb der Sporenmembran ab. Ja, bei den Hydropterideen-Makrosporen bleibt sogar der basale Teil ungeteilt. Ein schwacher Chlorophyllgehalt bei Prothallien heutiger Hydropterideen, sowie Rhizoidhöcker der ♀ *Selaginella*-Prothallien sind rudimentäre Anzeichen einer früheren Autotrophie. Allerdings haben die Rhizoidhöcker eine neue Funktion übernommen: das Auffangen der Mikrosporen, das z.B. bei Lepidophyten von Höckern der Sporenmembran bewirkt wurde.

Die Keimzellbehälter, die Archegonien und Antheridien werden bei der Prothallien-Reduktion gleichfalls, sowohl in ihrer Anzahl wie in ihrer Ausgestaltung reduziert. Am stärksten ist natürlich die Reduktion der Mikrosporen und ihrer Antheridien. Heutige *Isoetes*- Mikrosporen beispielsweise bilden sich nur noch in ganz wenige (ca 9) Zellen um, d.h. in der Hauptsache in eine wenigzellige Wandschicht und in 4 Spermatozoen. Wie bei anderen derartig weitgehenden phylogenetischen Umbildungen ist hier eine ins Einzelne gehende Homologisierung, die vegetatives Prothallium-Gewebe und Antheridien unterscheidet, unmöglich.

Die Heterosporie ist — wie in nicht-phylogenetischem Sinne schon HOFMEISTERS klassische Untersuchungen erkannt haben — eine Zwischenstufe auf dem Wege zur Samenbildung. Eine solche Entwicklungsreihe entspricht aber auch unseren phylogenetischen Vorstellungen (vgl. oben). Zwei Kormophyten-Abteilungen (die *Lepidospermen*, S. 505 und die typischen *Samenpflanzen*, einschliesslich der *Pteridospermen*) haben diesen neuen Schritt über die Heterosporie hinaus zur Fortpflanzung durch Samen getan. Die Möglichkeit, dass auch in der 3. Hauptabteilung, bei den *Articulaten* solche Samenbildung aufgetreten ist, besteht natürlich. Doch bedürfen die diesbezüglichen Angaben z.B. von ELIAS (1931) noch einer eingehenden Überprüfung, um als gesichert zu gelten.

Der entscheidende Wandlungsschritt zu dieser Stufe der Samenbildung liegt beim Makroprothallium, das seine ganze Entwicklung nun auf dem Sporophyten, innerhalb des Sporangiums durchläuft, und das damit das Sporangium zum Samen macht. Die Samenbildung macht also das ♀ Prothallium noch mehr vom Sporophyten abhängig: es erhält vom Sporophyten nicht nur die gesamte Nahrung sondern auch den Lebensraum und wird dementsprechend noch weitergehend reduziert. Bei den höchst differenzierten Samen, bei den Angiospermensamen, unterbleibt ja bekanntlich die zelluläre Unterteilung des Prothalliums vor der sexuellen Kernverschmelzung völlig. Dass die heterosporen *Hydropterides* eine Annäherung durch den nicht zellulär geteilten basalen Teil der Makrosporen zeigen, erwähnten wir schon.

## § 6. Phylogenie des Sporophyten (vegetative Organe).

Der Kernpunkt dieses Wandlungsabschnittes ist die phylogenetische Herausbildung des vegetativen Sprosses und der Wurzel. Daneben haben sich auch einige nicht einwandfrei in diese Typen der Kormophytenorgane einzuordnende Gebilde herausentwickelt (bzw. erhalten!), wie die sog. Wurzelträger der Selaginellen.

Die Herausbildung von Sprossachsen, Blättern und Wurzeln der Pteridophyten ist ein besonders deutlicher Beleg der „Telomtheorie".

## A. Phylogenie der Embryonalstadien und das gegenseitige Verhältnis von Spross und Wurzel.

Wir müssen unsere Betrachtung einleiten mit der Embryonalentwicklung heutiger Kormophyten, (Fig. 6), da die unmittelbaren Anhaltspunkte für die Embryonalgestalt der Ahnen äusserst spärlich sind.

Bei allen heutigen Kormophyten liegt die 1. Teilungswand der befruchteten Eizelle, die künftige Sporophytachse, die von BOWER 1922 erkannte „primitive spindle", mit den beiden Polen, dem Sprosspol und dem „Gegenpol", (meist durch Bewurzelung gezeichnet) fest. Die „primitive spindle" steht nämlich senkrecht auf dieser ersten Teilungsebene.

Im Gefolge der 2., darauf senkrechten Teilungsebene ergibt sich ein für alle Kormophyten-Embryonen charakteristisches doppelt „gegabeltes" Gebilde (Fig. 6, rechte Reihe) mit 4 (Quadranten:) Organbezirken. 2 Organbezirke des Sprosspols liefern das 1. Blatt und den übrigen Teil des Sprosses. Die 2 entgegengesetzten Organbezirke des „Gegenpols" liefern die übrigen Organe eines Embryos (Wurzel, Haustorium und Suspensor).

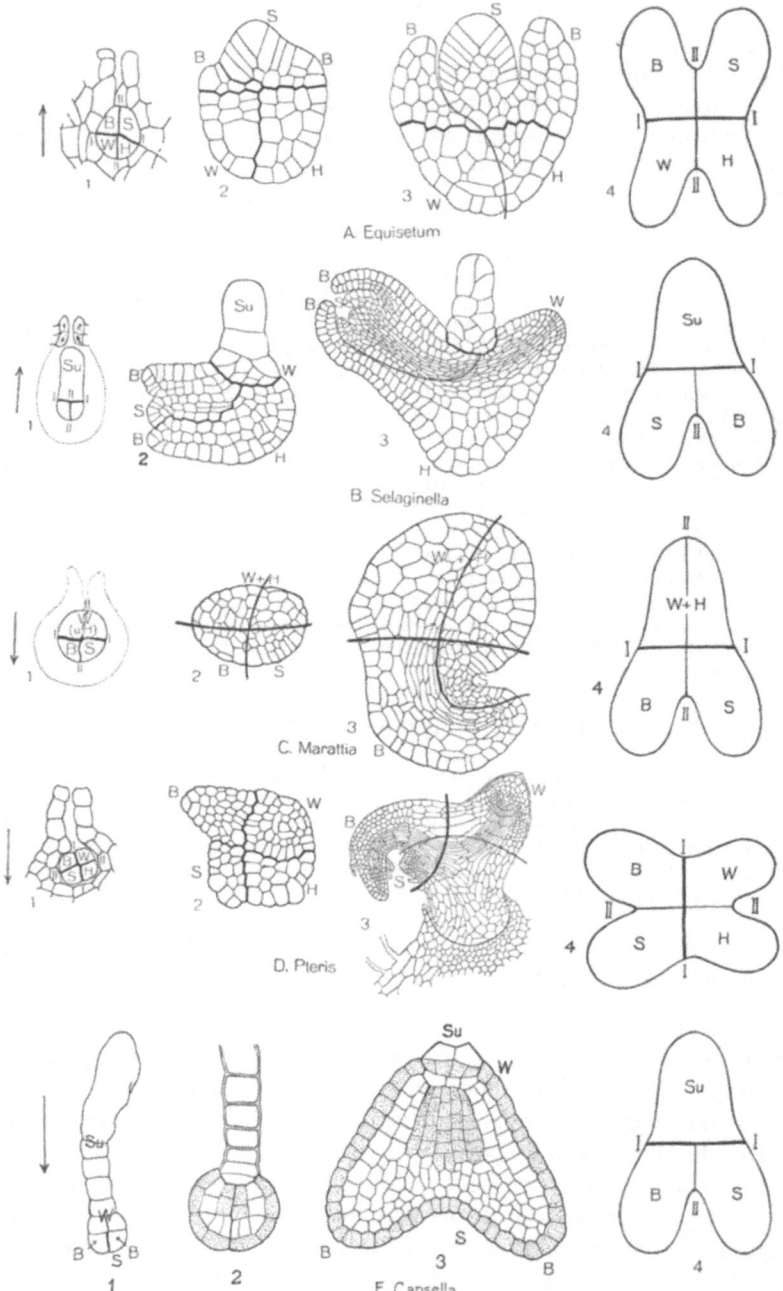

FIG. 6. Vergleichende Embryonalentwicklung der Pteridophyten und der Angiospermen. — 1–3 = jeweils 3 Entwicklungsstadien im Längsschnitt, 4 = Schemata der Organbezirke eines Embryos nach der 2. Zellteilung. (nach ZIMMERMANN 1930, Abb. 32). B = Blatt, S = Spross, W = Wurzel, Su = Suspensor, H = Haustorium.

(Zur leichteren Parallelisierung sind im Schema der Angiospermen die verschiedenen Querwände des Suspensors nicht berücksichtigt).

Diese doppelte „Gabelung" des Embryos kehrt in ihren Grundzügen so regelmässig bei allen Kormophytengruppen (und übrigens auch bei Bryophyten wie *Anthoceros*) wieder, sie stimmt so weitgehend mit der gabeligen Grundstruktur der. Ur-Archegoniaten und Ur-Pteridophyten (vgl. Fig. 3) überein, dass man wohl im d o p p e l t - g e g a b e l t e n E m b r y o e i n e g e m e i n s a m e A h n e n e i g e n t ü m l i c h k e i t d e r K o r m o p h y t e n sehen darf. Ein derartiger Embryonalaufbau ist für uns besonders interessant, weil er eine ontogenetische Gleichwertigkeit von Spross und Gegenpol (in vielen Fällen von Spross und Wurzel) sowie vom ersten Blatt und dem übrigen Spross (einschliesslich des 2. Blattes) ausdrückt.

Die prinzipielle Gleichwertigkeit von Sprosspol und Gegenpol erlaubt sogar noch bei einem heutigen Pteridophyten (*Marsilia*) einen experimentellen Austausch beider Pole. Je nach der Lage im Raume wird immer der obere Pol zum Sprosspol, der untere zum „Gegenpol" bzw. hier Wurzelpol. Auch wandeln sich Kriechtriebe (besonders deutlich bei den *Psilotales*) vielfach unmittelbar zu Sprossen.

Die Gleichwertigkeit von Blatt und Spross am Embryo hat LEITGEB für *Marsilia* dahin gekennzeichnet, dass beide Organe beim Erscheinen über der Oberfläche des Embryo als Höcker von einander nicht zu unterscheiden sind und ganz den Eindruck von Dichotomie machen.

Man darf also wohl annehmen, dass bei den Ur-Pteridophyten, der (wie bei *Equisetum* — Fig. 6A —, *Tmesipteris* und *Isoetes* „exoskop" orientierte) Sprosspol unmittelbar zu den einheitlich als Gabeltrieben gestalteten Organen wurde. D.h. der Sprosspol bildete zunächst die aufstrebenden Phylloide, und der „Gegenpol" die ebenfalls gabelig verzweigten und den Phylloiden äusserst ähnlichen Kriechtriebe, wie wir sie etwa bei *Rhynia* (Fig. 3) finden. Später, als die Telome sich differenzierten und teilweise übergipfelten, wurde die eine Gabelhälfte des Sprosspols zum 1. Blatt (Kotyledo) und die andere Hälfte zum eigentlichen Primär-Spross (der natürlich wie bei den Dikotylen) ebenfalls mit einem weiteren Kotyledo beginnen kann. Ob es unter den fossilen Formen Fälle gibt, in denen normalerweise auch nach der phylogenetisch erworbenen Übergipfelung beide Gabelhälften des Sprosspols einen selbständigen Spross geliefert haben, muss dahingestellt bleiben. *Tmesipteris* zeigt als Regelform noch einen Anklang.

Die Organe des „Gegenpols" behielten offenbar noch länger den gabeligen Grundaufbau bei, auch als sich specifischere unterirdische Organe, wie die (von Anbeginn sich gabelnden) Stigmarien und die Wurzeln aus ihnen entwickelten. HIRMER (1934) hat so m.E. mit gutem Grund noch für die Lepidophyten angenommen, dass der Embryo am Gegenpol unmittelbar gegabelt war und diese Gabeläste zu den Stigmarien wurden. [1]

---

[1] Ich stimme HIRMER (1934) auch darin bei, dass an sich auch eine 2. (von TROLL in idealistisch-morphologischer Form entwickelte) Ansicht phylogenetisch denkbar wäre, dass nämlich von vornherein bei den Pteridophyten der „Gegenpol"

Ausserdem hat die Phylogenie jedoch den „Gegenpol" des Embryos be-fähigt, auch noch verschiedenartige Embryonalorgane zu liefern, die die Ernährung des Embryos aus dem Gametophyten (auf dem der Pterido-phyten-Sporophyt bekanntlich zunächst „parasitiert".) ermöglichen. Es sind das das „Haustorium", ein Organ zur unmittelbaren Nahrungs-zufuhr, sowie der „Suspensor", der den jungen Embryo in eine geeignete Ernährungslage bringt. (Auf die wurzelhaubenähnliche Verlängerung des Suspensors der Angiospermen kann nur anlässlich Fig. 6 hingewiesen werden.)

So ist bei den ja allgemein von den übrigen Pteridophyten am stärksten abweichenden heutigen Lycopsiden (vgl. Fig. 6, *Selaginella*) im Laufe der Phylogenie die Embryogenese am stärksten abgewandelt worden. Hier gabelt sich vor allem der Gegenpol nicht mehr in 2 Organe, sondern er wird als ganzes zum Suspensor.

Im übrigen haben die Organ-determinierenden Reize offensichtlich im Laufe der Phylogenie mehrfach ihre Lagebeziehung zu den Achsen der Eier, bzw. der Archegonien gewechselt. Die beiden Pole können völlig vertauscht werden. Das geht wohl aus den in Fig. 6 wiedergegebenen Em-bryonalstadien (exoskope Lagerung Fig. 6A, endoskope Lagerung Fig. 6B, C u. E) zur Genüge hervor. Die determinierenden Reize werden „verlagert". Solche phylogenetische „Verlagerungen" determinierender Reize müssen wir ja vielfach annehmen (vgl. z.B. ZIMMERMANN 1935). Sie entsprechen den modernen entwicklungsphysiologischen „Homoplasie"-Vorstellungen.

## B. Phylogenie des Pteridophyten-Sprosses

Die Phylogenie ist verschieden für den makrophyll belaubten „Ptero-spross", (Fig. 7*F*), wie wir den Spross-Typ der pteridophytischen Pteropsi-den einschliesslich der Pteridospermen nennen wollen, und für den mikro-phyll belaubten „Lycospross" bzw. „Sphenospross" der Lycopsiden bzw Articulaten (Sphenopsiden) Die aussen-morphologische Umbildung (etwa die Bildung der Sprossachse durch Übergipfelung) hängt zwar jeweils aufs innigste mit einer anatomischen Umgestaltung zusammen, und beide Wandlungen machen sich gegenseitig erst verständlich. Aus technischen Gründen nehme ich aber auch hier die Darstellung der äusseren Gestalt vorweg.

---

Haustorium-artige Bildungen erzeugt hätte. Dass solche Haustorien sehr alte Bildungen sind, ist kaum zu bezweifeln. Wenn man aber entsprechend der „homo-morphen" Entstehung des Generationswechsels annimmt, dass der Gametophyt und Sporophyt ursprünglich selbständige Bildungen waren, kann das Haustorium nur eine sekundäre Bildung sein. Die phylogenetische Meinungsdifferenz besteht also nur in der Frage, *wann* sich das Haustorium herausgebildet hat, ob es also etwa entsprechend dieser Theorie schon sofort beim Parasitisch-werden des Sporophyten entstanden ist. Wie betont, ist mir — in Übereinstimmung mit HIRMER — diese Auffassung unwahrscheinlich.

a) *Pterospross (äussere Gestalt)* (Fig. 7F)

Für die Entstehung des Pterosprosses (Fig 8) liegen so viel Daten vor, dass heute in fast übereinstimmender Weise die Beteiligung folgender 4

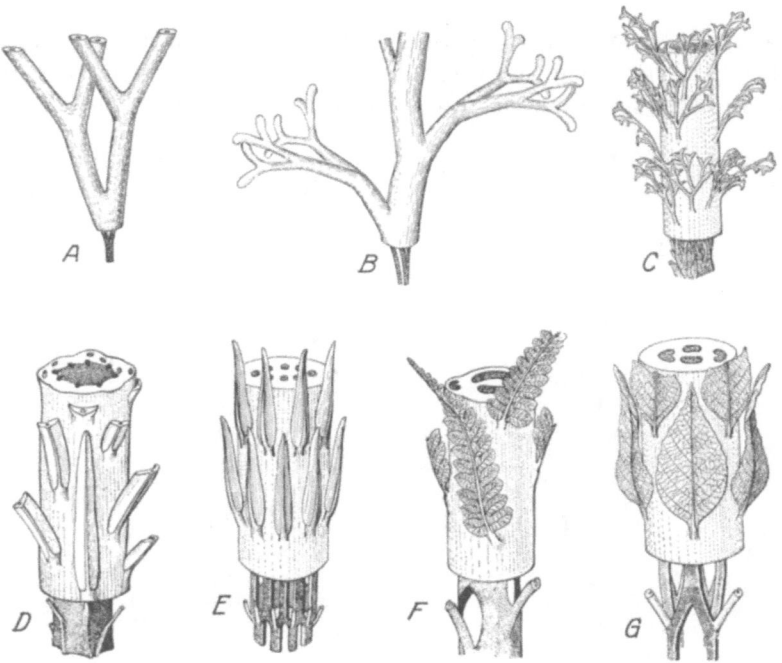

FIG. 7. — Die wichtigsten Spross-Typen: *A*. Ausgangstyp (Gabeltriebe), Holzkörper = Protostele; *B*. Übergangsform zum Gabelblatt-Spross; *C*. Gabelblatt-Spross, Holzkörper = Plektostele; *D*. Lycospross. Holzkörper = Aktinostele; *E*. Sphenospross. Holzkörper = Eustele; *F*. Pterospross. Holzkörper = Siphonostele; *G*. Pterospross (Anklang an die Angiospermen-Ausgestaltung: „Angiospross") Holzkörper = Eustele. — Das Protoxylem ist jeweils durch einen dunklen Punkt angedeutet.

Vorgänge, die wohl „Elementarprozessen" (vgl. ZIMMERMANN 1935, S. 244) entsprechen, angenommen werden:

   1) der A g g r e g a t i o n,

   2) der Ü b e r g i p f e l u n g und

   3) der seitlichen V e r w a c h s u n g („S y n g e n i e") der Telome und Mesome,

   4. der R e d u k t i o n von Organen und Organteilen.

Gerade hier ist es nützlich, dass wir uns die einleitend (S 558) erwähnte elementare Erkenntnis, worauf die phylogenetische Wandlung beruht, klar vor Augen stellen: So wie bei der Vererbung nicht „Eigenschaften" sondern „Erbfaktoren" übertragen werden, so wandeln sich auch in der Phylogenie nicht die fertigen Organgestalten bzw. ihre Eigenschaften ineinander, sondern es wandeln sich die dazugehörigen Erbfaktoren. Wenn wir also etwa sagen, die einzelnen Telome (vgl. die

Gestalt des Ur-Pteridophyten Fig. 3) seien in der Phylogenie „aggregiert", „übergipfelt" oder „verwachsen", so ist dies nur eine abgekürzte Ausdrucksweise. Tatsächlich haben sich in der Phylogenie die Erbfaktoren gewandelt, so dass bei den
Nachkommen dieser Ur-Pteridophyten im Verlauf der jeweiligen Ontogenie anstelle der freien Gabeltriebe „aggregierte", „übergipfelte" oder „verwachsene"
Teile aufgetreten sind.

Die **Aggregation** führt zur Anhäufung von Telomen oder Telom-
ständen dadurch, dass die sie tragenden Mesome sich verkürzen oder aus-
fallen. Die Mesome erfahren also hinsichtlich ihrer Streckungsfähigkeit

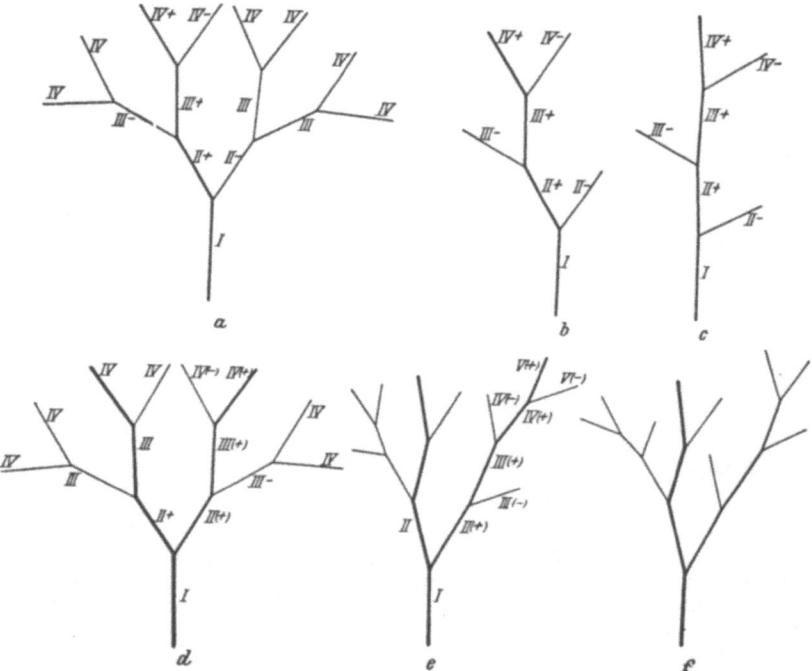

FIG. 8. Entstehung eines Fiederblattes oder Sprosses durch Übergipfelung
(Schema). Obere Reihe einfache Fiederung. Untere Reihe zusammengesetzte
Fiederung. Aus ZIMMERMANN 1930, S. 61, Abb. 20.

eine Arbeitsteilung. Einige bleiben streckungsfähig wie alle Mesome der
Ur-Pteridophyten, andere bleiben kurz.

Schon allein eine lokalisierte Aggregation kann zu einer Art Stamm-
sowie Zweig- und Blattbildung führen, wie das *Pseudosporochnus
Krejčii* [1]), eine Übergangsform zwischen Rhyniaceen und Pteropsiden, an-

---

[1]) KRÄUSEL und WEYLAND (1933 S. 17) meinen, dass meine (u.a. auf „*Pseudosporochnus Krejčii*) sich stützenden merkmalsphyletischen Ansichten über die
Stammbildung, z.B. über die zunehmende Stammhöhe, „soweit das Devon in
Frage kommt, auf recht schwacher Grundlage stehen". Sie begründen das damit,
dass manche z.B. von POTONIÉ als *Pseudosporochnus Krejčii* angesprochene

schaulich macht. Meist verbindet sich die Aggregation aber mit anderen Umbildungsvorgängen, wie bei primitiven Pteridophytenblättern (z.B. *Spenophyllum*, Fig. 13 u.a.) deutlich ist. Ein besonders deutliches Beispiel

FIG. 9. Entstehung eines Fiederblattes mit wechselständig gestellten Fiedern (obere Reihe) und mit gegenständig gestellten Fiedern (untere Reihe).

Fossilien vielleicht gar nicht zu dieser Art gehört hätten, und dass die betr. Stämme vielleicht kein Sekundärholz besessen hätten.

Bis zum Beweis des Gegenteils wird man KRÄUSEL u. WEYLAND in den beiden letztgenannten Annahmen beipflichten und es als eine zweifellos sehr verdienstvolle Aufgabe bezeichnen dürfen, solche Fragen zu klären. Ich sehe aber nicht recht, welchen Einfluss eine solche veränderte Erkenntnis auf meine merkmalsphyletischen Schlüsse haben könnte. Wer glauben sollte, dass eine palaeobotanischsystematische Umstellung eines Fossils aus einer Kormophytengruppe in eine andere einen derartigen merkmalsphylogenetischen Schluss beeinflussen könnte, der hat den Kernpunkt der merkmalsphyletischen Schlussweise nicht erfasst. Denn diese ist unabhängig von solchen Sonderfragen der Systematik und der Nomenklatur (z.B., wie hoch ein Stamm sein muss, ob er Sekundärholz besitzen muss, damit man von einem „Bäumchen" reden darf, vgl. dazu die verschiedene Auffassung von MOLL und BREMEKAMP).

der Blattbildung durch Aggregation (und Verwachsung) sind übrigens die Blätter von *Cordaites*. Bei dieser palaeozoischen Samenpflanze sind die Gabelungen derart an der Blattbasis gehäuft, dass die Blattfläche parallelnervig erscheint.

Auf rhythmisch wiederholter Aggregation bzw. „Ausfall von Internodien" beruht auch ein sekundärer Umbildungsprozess im Gefolge der Übergipfelung, d.h. die gegenständige bzw. wirtelige Seitenorganstellung an ursprünglich wechselständig gebauten Fiederblättern und Sprossen (Fig. 9 D-F und ZIMMERMANN 1935).

Die **Übergipfelung** (Fig. 8 u. 9) ist ein für Pterosprosse besonders typischer, von POTONIÉ erkannter Umbildungsprozess. Die beiden Schwestertriebe (Mesome oder Telome) einer Gabel, die bei den Ur-Pteridophyten noch durchweg gleichwertig waren, werden ungleichwertig. Der eine Schwestertrieb (in Fig. 8 mit + bezeichnet) wird kräftiger und stellt sich als Achsenglied vertikal, „er übergipfelt". Der andere Schwestertrieb (in Fig. 8 mit — bezeichnet) wird als Seitenorgan beiseite gedrängt, „er wird übergipfelt".

In einer Sprossachse bilden so die übergipfelnden Mesome die Internodien, und die übergipfelten Mesome bilden die Basis, bzw. Stiele der ansitzenden Laubblätter. Innerhalb des Blattes bilden die übergipfelnden Mesome die Rhachis, die übergipfelten Mesome bzw. Telome dagegen die Fiedern. Diese sind übrigens bei durchgeführter Übergipfelung genau so fiedrig gebaut wie das ganze Blatt (Fig. d-f).

Angedeutet ist die Übergipfelung schon bei manchen zwischen den Psilophyten und Pteropsiden vermittelnden Formen, wie den mitteldevonischen Gattungen *Aphyllopteris* und *Protopteridium*. Vollendet sehen wir sie in einem Spross, der spiralgestellte Fiederblätter trägt.

Die gemeinsame Herkunft von Blatt und Sprossachse macht es verständlich, dass namentlich bei den ältesten Pteridophyten vielfach zwischen beiden nicht unterschieden werden kann, und dass auch noch bei heutigen Farnwedeln oft Achsenmerkmale anklingen. Ja sogar bei Angiospermen charakterisieren gemeinsame Eigentümlichkeiten den Spross- und Blattaufbau, wie z.B. UITTIEN und ZIMMERMANN 1934*b*, S. 199 ff. ausgeführt haben.

Bemerkenswert ist die Umbildung der Auszweigungsebene. Bei der für Spross und Blatt gemeinsamen Ausgangsform war noch keine Auszweigungsebene bevorzugt. Wie bei einem Busch standen die Seitentriebe nach allen Richtungen ab, die Verkettung war radiär. Die „typischen" Sprosse und Blätter, welche sich aus dieser Urgestalt [1] herausentwickelt haben,

---

[1] Diese phylogenetische Anschauung von einer radiären Grundstruktur der Ur-Gestalt sowohl für den Spross wie für das Blatt steht natürlich sowohl methodisch wie sachlich in ausgesprochenem Gegensatz zur BRAUNschen, und z.B. neuerdings von TROLL vertretenen Richtung der idealistischen Morphologie, die den radiär gestalteten Stengel und das *a limine* flächige Blatt als „wesentlich verschiedene Teile" ansieht.

unterscheiden sich bekanntlich hinsichtlich ihrer Auszweigungsebenen, die sie bei der Übergipfelung erhalten haben. Die S p r o s s e sind entsprechend ihrem ursprünglich aufrechten Wuchs (von sekundären Umbildungen abgesehen) r a d i ä r. Ihre Auszweigungen stehen, wie die Gabeln der Ur-pteridophyten, allseitig ab. Beim lichtfangenden B l a t t dagegen werden die übergipfelten Auszweigungen in eine e i n z i g e E b e n e gelagert.

Die heute vorherrschenden „Typen" schliessen selbstverständlich die Existenz von abweichenden Formen namentlich in früheren Zeiten nicht aus. Im Gegenteil, wir kennen gerade als Belege der phylogenetischen Umbildung eine Reihe von Abweichungen. Z.B. ordnen sich bei *Protopteridium hostimense* KREJČI und bei manchen „flabellaten" Lycopsiden-Sprossen auch noch nicht übergipfelte Triebe in einer einzigen Ebene. Das zeigt, dass Übergipfelung und diagrammatische Verschiebung selbständige Elementarprozesse sind, die sich zeitlich unabhängig voneinander wandeln können. Ferner werden wir unten eine Anzahl von „Wedeltypen" zu nennen haben, bei denen die flächige Ausbildung noch nicht fertig durchgeführt ist. Selbst für die basalen Fiederchen der devonischen *Archaeopteris*-Arten, die auch in ihrer Gestalt als *Cyclopteris*-ähnliche „Aphlebien" abweichen, vermutet NATHORST (1904 S. 17) eine radiäre Stellung. Und bei der Verkettung von Sporangien und Phylloiden ist eine „cruciate" Verzweigung sogar recht häufig (vgl. *Ophioglossaceae* und Lycopsiden). Schliesslich sei daran erinnert, dass selbst heute noch Blätter wie die von *Tmesipteris* und *Iris* „reiten" bzw. sich in einer zur „typischen" senkrechten Ebene ausbreiten.

Sowohl für den Spross wie für das Blatt herrscht dabei die Regel, dass die aufeinanderfolgenden Auszweigungen möglichst fern von einander entspringen, dass sie also möglichst nicht übereinander stehen. Das wird im Raume (bzw. beim Spross) durch die Limitdivergenz ($137\frac{1}{2}$ Grad), in der Ebene (bzw. beim Blatt) durch eine $\frac{1}{2}$-Stellung (180 Grad, alternierende Fiederung) erreicht.

Besonders die zahllos überlieferten Pteridophylle zeigen auch viele, hier zu übergehende Einzelheiten dieses Übergipfelungsprozesses, wie die im Palaeozoikum fast ausnahmslos vorherrschende und aus der Übergipfelung sich ergebende Katadromie (Fig. 8*e*), während bei den heutigen Pteridophyllen sich vorzugsweise ein anadromer Aufbau (Fig. 8*f*) vorfindet. (Näheres vgl. ZIMMERMANN 1930, S. 61 ff.).

Die **Syngenie** (seitliche Verwachsung) ist am anschaulichsten innerhalb der Blätter zu verfolgen. Der einfachste Fall besteht in einer rein parenchymatischen Verwachsung, wie sie etwa durch die Fiederchen von *Pecopteris* (Fig. 10 und 11*o*) ausreichend gekennzeichnet ist. Es können aber auch die Leitbündel miteinander „verschmelzen", d.h. das Parenchym wird zwischen ihnen nicht mehr ausgebildet. Lokale Verschmelzungen führen durch Maschenbildung zur Netzaderung, die zwar für den „Angiospross" (vgl. Fig. 7*G*) besonders typisch ist, aber auch schon spärlich seit dem O. Karbon innerhalb der Pteridophyllen (vgl. Fig. 11*r*, *u*, *x*) auftritt. Über die Verschmelzung in der Sprossachse vgl. unten S. 591 ff. —

Die phylogenetische **Reduktion** von Organen und Organteilen besteht darin, dass diese in der Ontogenie nicht mehr ausgegliedert werden. Teilweise und völlige Reduktion sind sehr verbreitete phylogenetische Veränderungen. Aggregation und Syngenie sind Spezialformen der Reduktion.

**Pteridophylle.** Die Haupttypen, der schon lange von den Palaeo-
botanikern unterschiedenen palaeozoischen *Pteridophylle* sind besonders
anschauliche Beispiele für die verschiedenartige Kombination der soeben
genannten Umbildungsvorgänge. Allerdings klingt gerade bei diesen palaeozoischen Pteridophyllen die Organisation der Ur-Pteridophyten, namentlich die gabelige Verkettung der Triebe noch vielfach nach. So sind bei
altertümlichen Wedeln (aber auch bei heute „lebenden Fossilien", wie
*Matonia* und *Dipteris*) die Wedelachsen 1. oder folgender Ordnung (vgl.
Fig. 10*B*) gegabelt. Gegabelt sind ferner meist die letzten Auszweigungen,

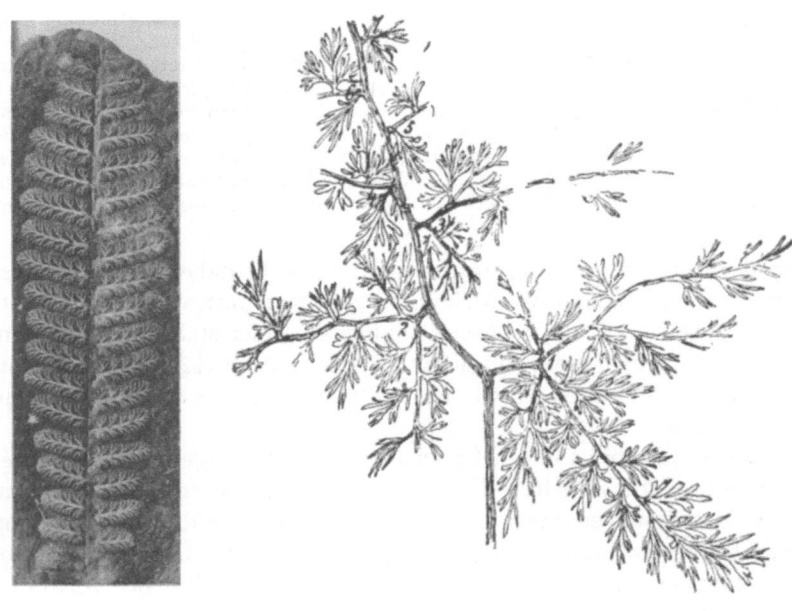

A                              B

FIG. 10. Pteridophylle. — *A. Pecopteris vestita* LESQ. = *P. lamurensis* var. *all-
gheny:* Kittanning Group Mazon Creek, Illinois, Westphal D. Nach einer freundlichst von Herrn DARRAH überlassenen Photographie. — *B. Palmatopteris furcata*
f. *typica,* als Beispiel der *Diplotmema*-Verzweigung. O. Karbon. Aus POTONIÉ 1904
Lief. II, 21, S. 3. — Typisch für *Palmatopteris* ist die kräftige Entwicklung der
katadromen Abschnitte bei starker Verkürzung der katadromen „Internodien".

bzw. wir erkennen hier eine Gabelnervatur (Fig. 10*A*). Auch die an
palaeozoischen Wedeln so vielfach festzustellende „flexuose" Rhachis (z.B.
Fig. 10*B* links) ist ein Anklang an die Achsengabelung.

Wir wollen die Verknüpfung der verschiedenen Elementarprozesse erläutern,
indem wir an Hand der Fig. 11 die Fiederchen der wichtigsten palaeozoischen Pteridophyllen kurz charakterisieren.

I. A r c h a e o p t e r i s-G r u p p e. Ohne ausgeprägte Übergipfelung innerhalb der Fiederchen.

1) *Archaeopteris* (*a*) DAWSON. Charakteristisch für Devon. Gabeltyp. Phylloide

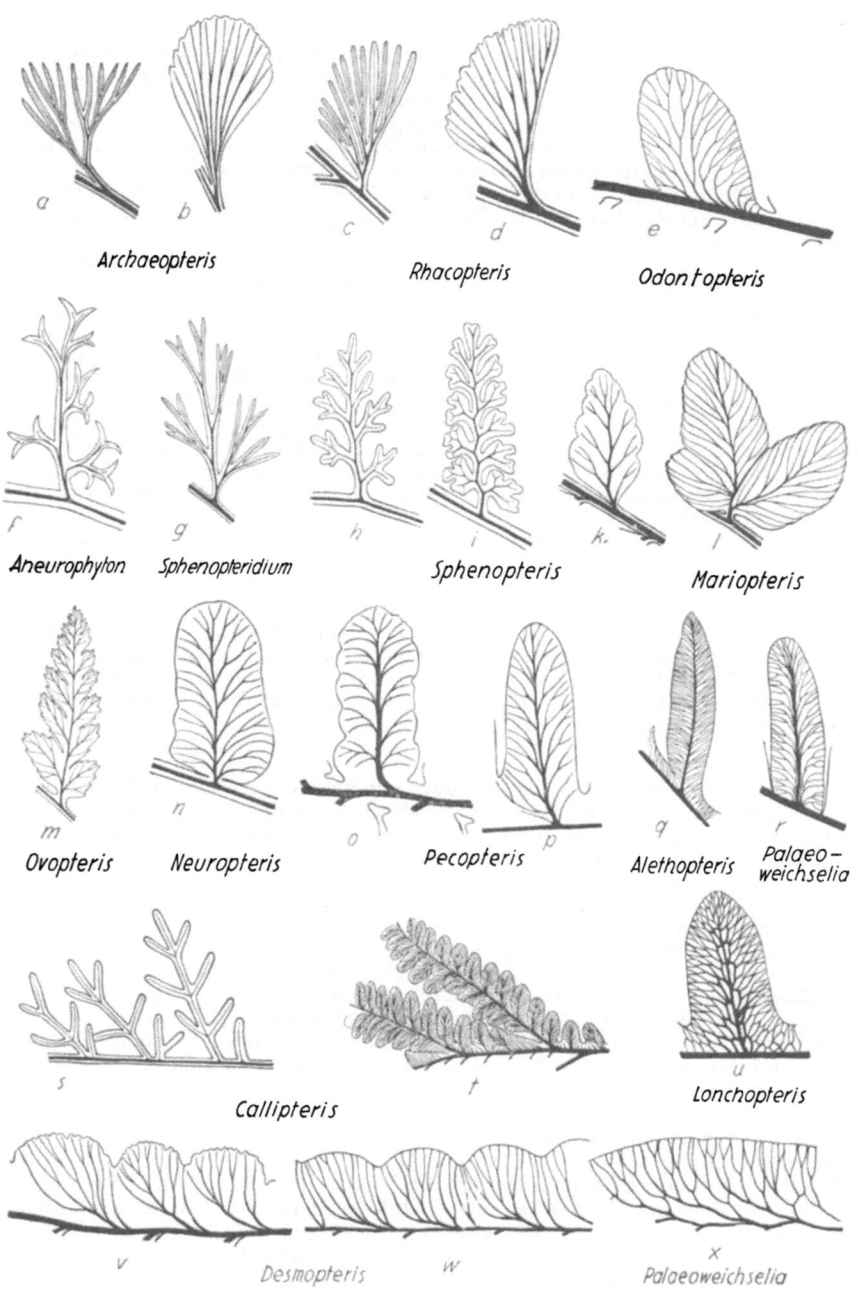

FIG. 11. Fiederchen palaeozischer Pteridophylle.

1. Reihe Archaeopteris-Gruppe: a) *Archaeopteris fissilis* SCHMALH., O. Devon, aus NATH. 1904, Taf. II, 9, Vergr. 2 ×. — b) *Palaeopteridium (Archaeopteris) Reussi* ETTINGH., O. Karbon, Aus KIDST. 1923, Taf. 45, Fig. 1a,

Vergr. 3,2 ×. — *c*) *Rhacopteris petiolata* GOEPP., U. Karb., aus KIDSTON 1923, S. 213, Textfig. 10 u. Taf. 53, Fig. 3*a*, nat. Gr. — *d*) *Rhacopteris lindsaeformis* BUNB., U. Karb., Aus KIDSTON 1923, Taf. 49, Fig. 4, Vergr. 1,8 ×. — *e*) *Odontopteris alpina* (STERNB.) GEIN., O. Karb., aus POTONIÉ 1904, Lief. II, 22, Vergr. 2,3 ×.

2. R e i h e   S p h e n o p t e r i s - G r u p p e: *f*) *Aneurophyton germanicum* KRÄUSEL u. WEYLAND, U. Devon, Kirberg. Original nach von Herrn Prof. WEYLAND freundlichst überlassenem Material unter Benutzung der Zeichnungen und Angaben in KRÄUSEL und WEYLAND 1923 u. 1926, Vergr. 1,4 ×. — *g*) *Sphenopteridium* (*Rhodea*) *Smithi* (KIDST.), U. Karb., aus KIDSTON 1923, Taf. 57, Fig. 2*a*, Vergr. 1,6 ×. — *h*) *Sphenopteris* (*Diplotmema, Palmatopteris*) *furcata* var. *compacta* (GOTH.) BRGT., Duttweiler, Original Geol. Inst. Tüb.Vergr. 1,6 ×. — *i, k*) *Sphenopteris Hoeninghausi* BRGT. var. *larischiformis* POT., O. Karb., aus POTONIÉ 1890, Taf. 9, Fig. 1*a*, Vergr. 2,3 ×. — *Sphenopteris Norinii* HALLE, Perm, aus HALLE 1927 Taf. 12, Fig. 5, Vergr. 2,3 ×.— *l*) *Mariopteris muricata* ZEILLER, O. Karb., Westphal. Saarbr. Sch. Gersweiler bei Saarbrücken., Orig. Geol. Inst. Tüb. Vergr. 0,4 ×.

3. R e i h e   P e c o p t e r i s - G r u p p e (ausser *q, u. r*): *m*) *Ovopteris pecoptevoides* LANDESCR., O. Karb., Rotl., aus POTONIÉ 1906, Lief. IV, 61. Vergr. 1,4 ×. — *n*) *Neuropteris*. — *o*) *Pecopteris vestita* (vgl. Fig. 10), Vergr. 1,8 × — *p*) *Pecopteris Armasi*, ZEILLER, aus KIDSTON 1925, Taf. 135 ,Fig. 2*a*, Vergr. 4,3 ×. — *q*) *Alethopteris lonchitica* UNG., O. Karb., aus POTONIÉ 1913, Lief. IX, 161, Vergr. 1,4 ×. — *r*) *Palaeoweichselia Defrancei* POT. u. GOTH., O. Karb., nach POTONIÉ 1909, Lief. VI, 116, Vergr. 3×.

4. R e i h e   C a l l i p t e r i s   u.   L o n c h o p t e r i s: *s*) *Callipteris strigosa* ZEILLER, Rotliegendes, aus ZEILLER 1898, Taf. 4, Fig. 5, Vergr. 1,4 ×. — *t*) *Callipteris conferta* BRGT., Rotliegendes, aus POTONIÉ 1898, Vergr. 0,5 ×.— *u*) *Lonchopteris Bricei* BRGT., O. Karb., aus GOTHAN (POTONIÉ 1909) Lief. VI, 118,Vergr. 1,6×.

5. R e i h e   S e i t l i c h e   F i e d e r v e r s c h m e l z u n g (T a e n i o p t e r i s   G r u p p e   u.a.): *v*) *Desmopteris serrata* GOTH., O. Karbon, aus POTONIÉ 1906, IV, 65, Vergr. 1,8 ×. — *w*) *Desmopteris integra* GOTH., O. Karb., aus POTONIÉ 1906, IV, 65, Vergr. 1,8 ×.— *x*) *Palaeoweichselia yuani* SZE, (Fiederchenausschnitt) O. Karb., aus SZE 1933, Taf. VI, Fig. 6, Vergr. 5,5 ×.

oft deutlich aggregiert und verschieden stark miteinander verwachsen. Formen mit starker Verwachsung und kräftiger Entwicklung der in der Fiederchen-Achse liegenden Partien werden bei rhombischem Umriss als *Palaeopteris* [1]) oder als *Adiantites* GOEPP. und bei mehr rundlichem Umriss als *Cardiopteris* SCHIMPER unterschieden (vgl. unter 3 auch die ähnliche *Neurodontopteris*).

2) *Rhacopteris* SCHIMP. (*c* u. *d*) Gabeltyp durch eine verstärkte Ausbildung der anadromen Abschnitte eines Fiederchens modifiziert. („einseitige Symmetrie" nach GOEBEL 1928, S. 250 ff.) Phylloide verschieden stark miteinander verwachsen. Aehnlich ist schon der „*Spiropteris*"-Typ bei *Protopteridium* (M. Devon).

3) *Odontopteris* BRGT. (*e*) In der Aderung noch Gabeltyp, Phylloide wie bei *Adiantites* und *Cardiopteris* stark verwachsen und Mittelpartie des Fiederchens kräftig entwickelt. Fiederchenbasis breit verwachsen, wodurch der Umriss pecopteridisch wird. Bei der ähnlichen *Neurodontopteris* ist — als Parallelform zu *Neuropteris* — die Fiederchenbasis eingeschnürt.

II.   S p h e n o p t e r i s - G r u p p e. Deutliche Übergipfelung. Aber basale Phylloidgruppen immer noch kräftig entwickelt, wodurch sich ein keilförmiger Gesamtumriss ergibt.

4) *Sphenopteridium* SCHIMPER (= *Rhodea* PRESL) (*g*) SCHIMP. Phylloide völlig frei und schmal, so dass Fiederchen letzter Ordnung einnervig sind. Aehnlich, aber wohl nicht ganz in einer Ebene ausgebildet ist *Aneurophyton* (*f*).

5) *Sphenopteris* BRGT. Phylloide etwas flächig verbreitet oder mehr oder minder miteinander verwachsen (*h–k*).

---

[1]) Die Definition für diese einseitige Asymmetrie GOEBELs, dass sich nur ein Gabelast, etwa der linke, jeweils weiter gabele, ist allerdings nicht ganz zutreffend. Die Asymmetrie kommt bei solchen Pteridophyllen durch eine ungleich grosse Streckung und ungleich starke Gabelung zustande.

6) *Mariopteris* ZEILLER (*l*) Phylloide meist ziemlich stark verwachsen. Im übrigen aber insofern primitiv, als namentlich in basalen Teilen der Fiedern die Übergipfelung noch nicht völlig durchgeführt ist. So ist z.B. an den basalen Fiederchen der basale katadrome Abschnitt meist sehr stark entwickelt, d.h. fast so kräftig wie der ganze übrige Teil des Fiederchens, wodurch das Fiederchen noch annähernd gegabelt aussieht. Ähnlich zeigt auch der Gesamtwedelaufbau „*Diplotmema Struktur*" (vgl. Fig. 10). D.h. je 2 Gabelpaare von Fiedern sind aggregiert, bzw. die katadromen Fiedern 2. Ordnung, die an die Gabelung anschliessen, sind besonders kräftig entwickelt.

7) *Diplotmema* STUR. (und ähnlich auch *Palmatopteris* POT.) mit dem eben genannten „diplotmematischen" Wedelaufbau, aber typischen *Sphenopteris*-Fiederchen, d.h. mit deutlicher Übergipfelung und meist schwächerer Verwachsung der Phylloide als bei *Mariopteris*.

III. P e c o p t e r i s - G r u p p e. Phylloide eines Fiederchens stark bis vollständig verwachsen. Die basalen Phylloidgruppen so weit reduziert, dass sie nicht mehr kräftiger entwickelt sind als die von den mittleren Abschnitten der Fiederchenachse ausgehenden. Dadurch lädt das Fiederchen an der Basis nicht mehr so breit aus wie bei *Sphenopteris*.

8) *Neuropteris* BRGT. (*n*) Fiederchen an der Basis eingeschnürt, also nicht mit der tragenden Achse verwachsen. Sonst ähnlich wie *Pecopteris*. Die *Cyclopteris* BRGT. genannten Fiederchen haben sich als Teil von *Neuropteris*-Wedel u.a. Pteridophyllen erwiesen. *Ovopteris* (*m*) vermittelt durch eiförmige Fiederchen zwischen *Sphenopteris* und *Neuropteris*.

9) *Pecopteris* BRGT. (*o, p*) Seitenkonturen des Fiederchens relativ parallel, Parenchym der Basis mit der tragenden Achse breit verwachsen.

IV. A l e t h o p t e r i s - C a l l i p t e r i s - G r u p p e. Merkmale der *Pecopteris*-gruppe. Die basalen Phylloidgruppen der Fiederchen sind aber auch in ihrer Aderung mit der tragenden Achse verwachsen, so dass diese Fiederchen bzw. ihre Adern als „Nebenadern" oder „Zwischenfiedern" [1]) aus den Achsen entspringen.

10) *Alethopteris* STERNBG. (*q*). Nur Nebenadern.

11) *Callipteridium* WEISS Umrisse der Fiedern pecopteridisch. Nebenadern u. Zwischenfiedern.

12) *Callipteris* BRGT. (*s, t*) Phylloide teilweise nicht (*s*), teilweise stark (*t*) verwachsen. Meist Nebenadern, immer Zwischenfiedern.

*Eremopteris* SCHIMPER umfasst wenige charakteristische Formen, die hierhergerechnet werden können, da sie Zwischenfiedern besitzen, aber durch Gabelnervatur zu *Archaeopteris* und durch Fiedergestalt zu *Sphenopteris* vermitteln.

V. T a e n i o p t e r i s - G r u p p e. Sehr weitgehende Verwachsung der Phylloide zu einem schwach oder gar nicht unterteilten Wedel.

13) *Desmopteris* (*v, w*) POT. bzw. *Validopteris* P. BERTR., zwischen *Pecopteris* bzw. *Alethopteris* und *Taeniopteris* vermittelnde Formen.

14) *Taeniopteris* BRGT. und *Lesleya* LESQ. Charakter der Gruppe. Die meisten (namentlich mesozoischen) Formen zu den *Cycadophyten, Benetittales* usw. gehörig.

VI. N e t z n e r v a t u r g r u p p e. Der Uebersichtlichkeit halber fasse ich hier einige Parallelformen zu den vorangehenden Gruppen zusammen, die sich durch eine anastomosierende Verwachsung der Leitbündel in den Blattflächen auszeichnen.

15) *Linopteris* PRESL Parallelform zu *Neuropteris*.

16) *Palaeoweichselia* (*r*) POT. u. GOTH. Parallelform zu *Pecopteris*.

17) *Lonchopteris* BRGT. (*u*) Parallelform zu *Alethopteris*.

18) *Pecopteridium* P. BERTR. Parallelform zu *Callipteris*.

19) *Gangamopteris* Mc COY, ähnlich wie *Glossopteris*, aber noch ohne deutliche Mittelader.

20) *Glossopteris* BRGT. Parallelform zu *Taeniopteris* mit völlig zu einem einzigen Wedel verwachsenen Phylloiden.

Anhangsweise sei noch auf die Aphlebien-artigen Bildungen [2]) verwiesen, Basalorgane der Fieder, die oft noch typisch primitive Organisation auf-

---

[1]) Zwischenfiedern hatte schon *Archaeopteris fissilis* (Fig. 11a).

[2]) Vgl. HIRMER 1927, S. 573 u. 690 sowie BROWNE 1935, S. 428*f*.

weisen, wie wenig verwachsene oder übergipfelte Phylloide. Ein charakteristischer Fall sind z.B. die katadromen Basis-Fiederchen von *Ovopteris Weissi*, die nach POTONIÉ (1903, Nr. 8) „aphleboid", gestaltet sind, d.h. einerseits aus stark zerschlitzten Teilen zusammengesetzt und andererseits wie bei *Mariopteris* gegen die katadrome Seite massiger entwickelt sind.

Eine prinzipiell ähnliche Bildung ist die Entstehung von Deckblättern als „Tragblätter" achselständiger Seitensprosse. Wie SCOTT (1912, S. 60) für *Ankyropteris* (*Zygopteris*) ausführte, wird das 1. Blatt des übergipfelten Seitensprosses zum Deckblatt. D.h. infolge Aggregation wird das unterste Internodium so weitgehend unterdrückt, dass das 1. Blatt an der übergipfelten Hauptachse zu sitzen kommt (vgl. Fig. 19*a*).

### b) *Lycospross und Sphenospross* (Fig. 7*D* und *E*).

Die Phylogenie dieser mikrophyll beblätterten Sprosse ist mehr umstritten als die Phylogenie des Pterosprosses. Doch lässt sich auch hier diese Frage wenigstens klären, wenn man sich, wie immer bei naturwissenschaftlich umstrittenen Fragen, zunächst vor Augen hält, welche Erklärungsm ö g l i c h k e i t e n überhaupt vorliegen und sich dann zur wahrscheinlichsten Erklärung bekennt. —

Drei (nur drei!) Möglichkeiten gibt es, wie ein Lycospross oder Sphenospross in der Phylogenie entstanden sein könnte:

1) entsprechend der Syngenie-Hypothese (Verwachsungs-Hypothese, Fig. 12*A*).

2) entsprechend der Reduktions-Hypothese (Fig. 12*B*).

3) entsprechend der „Epigenie"-Hypothese (Fig. 12*C*).

Wenn wir eine dieser 3 Hypothesen ablehnen, sagen wir damit, dass eine der beiden anderen Hypothesen wahrscheinlicher sei. Wir beschränken uns bei der Besprechung zunächst auf die besonders umkämpfte Phylogenie der Lycosprosse. —

Nach der S y n g e n i e - H y p o t h e s e (Verwachsungshypothese), die ich für die wahrscheinlichste halte, waren Ausgangsformen entweder noch Pteridophyten vom *Rhynia*-Typ, oder sie besassen schon mit Gabelblättern belaubte Sprosse wie Fig. 7*C* (und 12*A*). Die Verwachsung (Syngenie) bestand dann darin, dass die durch Übergipfelung seitlich gestellten gabeligen Seitenorgane so mit der übergipfelnden Sprossachse „verwuchsen", dass nur die äussersten Blattspitzen, die Telome, als einnervige Nadel- oder Schuppenblätter freiblieben (Fig. 12*A*).

Nehmen wir rein gabelige Ausgangsformen wie die Ur-Pteridophyten an, so bedeutet das, dass sich die Syngenie gleichzeitig mit der Übergipfelung abgespielt hat [1]. Für die ältesten Vertreter des Lycospross-Types, wie z.B. für *Asteroxylon* kann man an eine derartige Entstehung denken. Für die

---

[1] Vgl. z.B. *Aphlebiopteris* GOTH. u. F.Z.

typischen Lycopsiden ist mir allerdings eine andere Modifikation der Syngenie-Hypothese wahrscheinlicher: dass sich zunächst durch Übergipfelung

ein gabelblättrig belaubter Spross wie Fig. 7C gebildet hat, und dass erst später die Syngenie eingetreten ist. Denn bei den ältesten Lycopsiden finden sich ja typische Sprosse mit Gabelblättern.

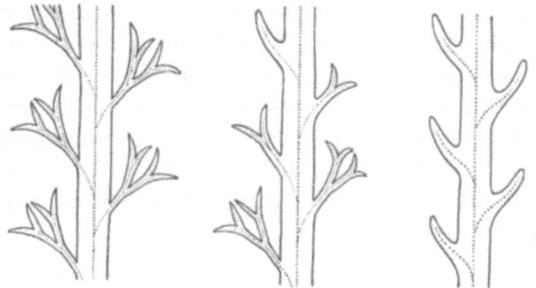

*A.* Syngenie-Hypothese

Wahrscheinlich hat jeweils bei der Umbildung durch Syngenie eine gewisse Reduktion eingesetzt d.h., die Zahl der Gabelungen an den Seitenorganen ging entsprechend der anschliessend geschilderten Reduktionshypothese zurück. Überhaupt stehen sich Syngenie- und Reduktionshypothese durch die gemeinsam angenommene „makrophylle" Ausgangsform (bzw. ihren gabelig-thallophytischen Vorläufer) gegenseitig nahe, sie stehen aber im Gegensatz zur Epigenie-Hypothese. Um den Anteil der Syngenie- und Reduktionshypothese zu sondern, müsste man die phylogenetische Umbildung mehr in den Einzelheiten verfolgen, als das zur Zeit möglich ist.

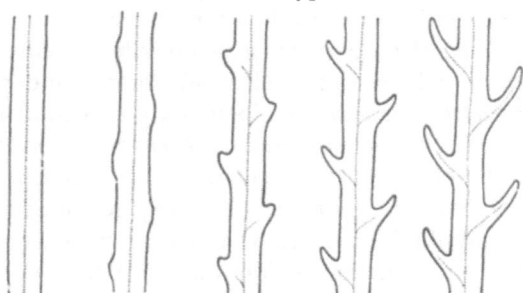

*B.* Reduktions-Hypothese.

*C.* Epigenie-Hypothese.

FIG. 12. Phylogenie des „mikrophyllen" Lycosprosses nach den drei einzigen möglichen Hypothesen.— Dargestellt sind jeweils die Ahnen- und Endform sowie einige Zwischenstufen. Bei der Reduktionshypothese sind die Zwischenstufen an der gleichen Sprossachse wiedergegeben. Die punktierte Linie gibt den Stelenverlauf, bzw. den Verlauf des Protoxylems wieder.

Um Missverständnisse, wie sie zum Beispiel VAN TIEGHEMS Verwachsungstheorie oder POTONIÉ's Perikaulomtheorie ausgelöst haben, zu vermeiden, muss nochmals unterstrichen werden, dass es sich bei phylogenetischen „Verwachsungen" zu denen auch z.B. die Sympetalie, die unterständigen Fruchtknoten u.a. gehören, um kein unmittelbares ontogenetisches Zusammenwachsen selbstständiger Teile handelt. Sondern sowohl bei der phylogenetischen Ausgangs- wie bei der Endform entstehen natürlich derartige Sprosse ontogenetisch aus einem ungegliederten Vegetationskegel. Es unterbleibt aber am abgeleiteten Lycospross die äussere Ausgliederung von Teilen, die beim Ahnentyp noch frei und selbständig waren. Nur z.B. im Leitbündelverlauf kann sich (wie beim ungeteilten Flächenblatt) noch die ehemalige Zusammensetzung des Gebildes verraten, weil Elementarprozesse, die beim Ahnentyp die äussere Gliederung gestaltet haben, nun die Anatomie der Sprossachse gestalten.

Die Reduktions-Hypothese (Fig. 12B) schliesst eng an die Syngenie-Hypothese an. Beide Hypothesen nehmen die gleiche Ahnenform an: den Gabelspross. Zum Unterschied ist aber bei der Reduktions-Hypothese jedes einzelne einnervige Nadelblatt ein Homologon eines makrophyllen Blattes, also hier eines ganzen Gabelblattes.

Die Epigenie-Hypothese (vgl. auch Fig. 19c) ergibt sich, wenn wir keine makrophyll beblätterte Urformen annehmen wollen. Sie rechnet damit, dass die Blätter gleich wie Emergenzen („enations") zu einer zunächst ungegliederten Achse hinzukommen und erst sekundär durch Ausbildung je eines Blatt-Nerven Anschluss an das Leitbündelsystem der Achse gewinnen. Ausgesprochenermassen haben z.B. LIGNIER und DOMIN (u.a. 1929 und 1931) diese Hypothese, übrigens für sämtliche Pteridophytenblätter, vertreten. Ähnlich hat sich auch BOWER 1935 für den Lycospross ausgesprochen.

Zugunsten einer makrophyllen Ausgangsform und der Syngenie-Hypothese sprechen vor allem

1) das Vorherschen makrophyller Formen bei den ältesten (devonischen und unterkarbonischen) Vertretern der Lycopsiden (z.B. *Protolepidodendron, Barrandeina, Sterzelia* u.a., vgl. ZIMMERMANN 1930, S. 130) und der Articulaten (*Hyeniales, Sphenophyllales, Asterocalamitaceae*), während diese Formen später nicht mehr vorkommen oder doch (wie z.B. *Sphenophyllum*) auf eigenartige Seitenzweige am Stammbaum der Pteridophyten beschränkt bleiben. Ja, es scheint mir bedeutungsvoll, dass innerhalb der gleichen „Gattungen" die älteren Vertreter gelegentlich deutlicher verzweigte Blätter besitzen als die jüngeren, wie das selbst ein Gegner der Syngenie-Hypothese, KRÄUSEL (1936, S. 310) für die Gattung *Protolepidodendron* ausführt.

2) Die Eigenart des Stelärbaus, gerade bei den ältesten mikrophyllen Formen wie *Asteroxylon*, ZIMMERMANN 1930a, S. (45) und *Drepanophycus*, HALLE 1936, S. 9. Die Aktinostele führt hier die „Verwachsung" des Blatt-Leitbündelsystems mit der aus aus Gabelarmen sich zusammensetzenden Achsen-Stele bzw. den gegabelten Blattspursträngen (ähnlich Fig. 12A rechts) unmittelbar vor Augen. Auch HALLE betont, wie schwer dieser Befund mit der früher von ihm selbst vertretenen Epigenie-Hypothese zu vereinbaren ist.

3) Die Verknüpfung von vegetativen und fertilen Teilen in den den

Blättern homotopen [1]) Sporophyllen, wie sie z.B. für die Lycopsiden und viele Sphenopsiden charkteristisch ist. Diese Verkettung von Phylloiden und Sporangien lässt für beide Telome eine übereinstimmende Geschichte erwarten. Das aber würde bei Annahme der Epigenie-Hypothese die ganz unbegründete Annahme veranlassen, dass auch die Sporangien Emergenz-Charakter hätten. Dagegen ist die Ableitung eines derartigen Sporophylls aus den Ur-Sporangienständen der Kormophyten sehr leicht (Fig. 18).

4) Bei manchen Pflanzengruppen (vor allem bei Angiospermen) ist unverkennbar, dass reduzierte nadel- und schuppenförmige Blattbildungen auf diesem Wege der Flächenverkleinerung von einem makrophyllen Typ aus zustande gekommen sind.

Ich kenne dagegen kein einziges p h y l o g e n e t i s c h e s Argument zugunsten der Epigenie-Hypothese. Andeutungsweise wurde von ihren Anhängern schon angeführt, dass auch in der Ontogenie die mikrophyllen Blätter (wie alle Seitenorgane!) epigenetisch, als „enations", d.h. als Ausstülpungen an einem einheitlichen Vegetationspunkt entstehen. Wollte man mit diesem Argument ernst machen, so müsste man annehmen, dass alle Blätter (auch die makrophyllen) sowie die Seitenzweige in der Phylogenie epigenetisch entstanden seien. Gerade dieser Vergleich zeigt aber deutlich, dass die epigenetische Neubildung der Blätter am Vegetationspunkt eine kaenogenetische Abweichung vom „Biogenetischen Grundgesetz" ist. Weiter wird darauf hingewiesen, dass mikrophylle Vertreter, wie *Baragwanathia* und *Asteroxylon* schon sehr früh auftreten, schon unter den ältesten Landpflanzen im Silur und Altdevon. Dieses Faktum stimmt an sich zweifellos. Aber ein so frühes Auftreten erschwert natürlich *jede* Erklärung der Mikrophyllie. Denn wir müssen die Entstehung derartiger mikrophyller Sprosse in eine fossil nicht belegte Vergangenheit zurückverlegen. In ähnlicher Weise ist auch die Tatsache, dass mikrophyll beblätterte ausgebildete Sprosse von *Asteroxylon*, *Psilotum* und Verwandten in unbeblätterte Achsen, wie Kriechtriebe und Sporangienträger übergehen, mit j e d e r Hypothese zu vereinigen. Ja die Art und Weise dieses Übergangs spricht sogar zugunsten der Syngenie-Hypothese. Der Stelärbau ist genau derjenige, den man entsprechend der Syngenie-Hypothese im Bereich einer ausklingenden Übergipfelung erwarten darf. D.h. wir finden in dieser Übergangsregion Aktinostelen mit Verminderung der Zahl der Arme bzw. zuletzt eine Protostele. (ZIMMERMANN unpubl.)

Jedenfalls scheinen mir diese zugunsten der Epigenie-Hypothese heranzuziehenden Argumente viel schwächer als die zugunsten der Syngenie — bzw. Reduktionshypothese sprechenden, nämlich

1) Das Vorherrschen der makrophyll beblätterten Formen in den später mikrophyll beblätterten Gruppen,

2) Die Aktinostele alter mikrophyller Typen wie *Asteroxylon*,

3) Der Platz von Sporangien an Stellen wo Mikroblätter stehen,

4) Die gleichartige Umbildung in sicher erkannten Gruppen.

---

[1]) Da sich bei einer phylogenetischen Betrachtungsweise die herrschenden Homologiebegriffe infolge selbständiger Abwandlungen der Elementarprozesse sowohl der Organstellung wie der Organgestaltung nicht mehr verwenden lassen, (vgl. ZIMMERMANN 1935), so zerlege ich den Homologiebegriff in seine beiden Grundkomponenten:

H o m o t o p i e für nicht abgewandelte Stellung der Organe (Lagebeziehung) aber etwa abgewandelte Gestalt, und

H o m o m o r p h i e für nicht abgewandelte Gestalt, aber etwa abgewandelte Zytologie, Stellung u. dgl.

Wenn Kräusel noch neuerdings (1936, S. 326), ohne neue Argumente versichert, dass die von ihm verfochtene Epigenie-Hypothese „nicht erschüttert" sei, so glaube ich, dass ihre Position nie sehr fest war.

Der Sphenospross unterscheidet sich in seiner äusseren Morphologie im wesentlichen durch die wirtelige Stellung der Phylloide. Die primäre Bildung des Mikroblattes auf dem Wege der Reduktion (bzw. Syngenie) scheint mir hier noch deutlicher als beim Lycospross. Hoeg (1931) hat eine entsprechende Ansicht vertreten, und Kräusel (1936) lehnt sie nicht völlig ab. Die sekundäre Umbildung zur Wirtelstellung dürfte dann auf dem Wege der Aggregation, d.h. durch Internodienverkürzung von einem wechselständig beblätterten Spross aus vor sich gegangen sein. Bezeichnenderweise sind z.B. die ältesten Vertreter, die *Hyeniales* z.T. noch teilweise wechselständig beblättert. Auch der 3-gliedrige Wirtel von *Sphenophyllum* ist offensichtlich durch eine solche rhythmische Aggregation entstanden. Ausgehend von solchen Wirtelsprossen mit niederer Wirtelzahl mögen dann durch syngenetischen Einschluss von Blättern bzw. Blattbasen Wirtel mit höherer Wirtelzahl entstanden sein, wie das P. Bertrand (1924) für die *Sphenophyllum*arten mit 6 strahliger Stele nachgewiesen hat, und wie das auch für Calamiten wahrscheinlich ist.

In altertümlichen Calamiten-Stämmen (*Asterocalamites* aus dem U. Karbon) laufen die Leitbündel noch von Internodium zu Internodium durch. Damit liegen noch die Abgangsstellen der (gegabelten) Blätter übereinander. Bei den späteren typischen Calamiten sind die Knoten komplizierter gestaltet; die Leitbündel anastomosieren hier, so dass sie in zwei aufeinanderfolgenden Internodien alternieren. Die Achsen der Sphenopsidenblüten zeigen nach Untersuchungen von Eames, Browne, Barratt und Santschi einen ähnlichen Aufbau. Allerdings ist die anatomische Struktur noch weitgehend primitiv. Wie bei *Asterocalamites* laufen selbst bei heutigen Equiseten die Leitbündel (bzw. ihr Protoxylem) noch durch, alternieren bzw. anastomosieren also nicht. Die einzelnen zu den „Sporophyllen" führenden Leitbündel sind Protostelen mit mesarch gelegenem Protoxylem, also wie die Blattspurstränge bei *Asteroxylon* u.a.

Interessant ist die Weiterbildung des Sphenosprosses innerhalb der Gattung *Equisetum*, z.B. die Arbeitsteilung in fertile und sterile Sprosse (*E. arvense* u.a.). Einzelheiten hat Schaffner zusammengestellt.

Die Zahl der nicht leicht unterzubringenden Spross-Typen ist übrigens namentlich in älterer Zeit grösser, als das die Einteilung in Lycosprosse und Sphenosprosse vermuten liesse. So sind aus dem Oberdevon, bzw. Unterkarbon Pflanzen mit dicht gedrängten Wirteln von alternierenden Mikroblättern bekannt (*Zimmermannia eleutherophylloides* Goth. u. F.Z., bzw. *Eleutherophyllum mirabile* Stur). Auch muss wegen anderer Einzelfragen, z.B. über solche Sondertypen oder wegen der spezifischen Ausgestaltung der Lepidophyten auf die Darstellungen bei der Anatomie bzw. Morphologie u. Systematik oder auf die angegebene Spezialliteratur verwiesen werden. Nur sei eine Selbstverständlichkeit unterstrichen. Vielleicht sind mikrophylle Sprosse auf verschiedenen Wegen entstanden; und wenn auch m.E. die bestbekannten mikrophyllen Sprosse wie die typischen Lycopsiden und Sphenopsiden sowie *Asteroxylon* entsprechend der Syngenie-Hypothese entstanden sind, so mag doch für manche ähnliche Bildungen wie die offenbar vielfach Drüsencharakter besitzenden „Dornen" altdevonischer *Psilophyton*arten (vgl. z.B. LANG 1931) u.a. die Epigenie-Hypothese zutreffen.

S p h e n o - B l a t t. Relativ mannigfaltig ist die Umbildungsweise der Phylloide zu Blättern bei den Articulaten („Spheno"-Blätter). Hier treten als Elementarprozesse fast ausschliesslich die Syngenie und die Reduktion auf. Für die Syngenie ist die Unwandlung der *Sphenophyllum*-Blätter ein auch statistisch sicher zu verfolgendes Beispiel (Fig. 13). Die ältesten Blätter, wie das oberdevonische *Sphenophyllum subtenerrimum* NATH., hatten noch freie Phylloide, während später die Phylloide zu einem meist gezähnten Flächenblatt verwachsen sind, wie bei *Sphenophyllum Thonii* MARR. aus dem Rotliegenden.

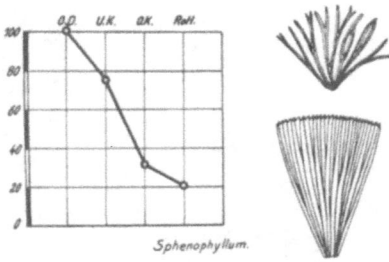

FIG. 13. Das S p h e n o p h y l l u m - B l a t t als Beispiel der Syngenie. — Die Kurve zeigt die Abnahme der Artenzahlen mit freien Phylloiden wie bei *Sphenophyllum subtenerrimum* NATH. (oben) aus dem Ob. Devon und entsprechende Zunahme der Artenzahl mit verwachsenen Phylloiden wie bei *Sphenophyllum Thonii* MARR. (unten) aus dem Rotliegenden.
Abszisse: Geologische Zeiten vom Ob. Devon bis zur Jetztzeit.
Ordinate: Prozentsatz der bekannten Arten mit freien Phylloiden.
Aus ZIMMERMANN 1930*a*, S. (41) Abb. 3.

Verwachsungen und Reduktion der einzelnen Phylloide kennzeichnen fernerhin die Blattbildung innerhalb der *Equisetales*. Bei *Asterophyllites* sind die Phylloide ganz frei, bei *Annularia* höchstens an der Basis ein wenig verwachsen. Eine Becherbildung aus den Blattbasen setzt dann bei der in der Gondwanaflora charakteristischen Gattung *Phyllotheca* ein. Sie ist bei heutigen Equiseten bekanntlich durch die fast vollständig zu einer Scheide verwachsenen und gleichzeitig stark reduzierten Phylloide vollendet. *Schizoneura* repräsentiert insofern eine sehr charakteristische Sonderbildung, als hier die Phylloide in zwei Gruppen als „gegenständige" Blätter vollständig verwachsen sind.

### c) *Achsenanatomie*

Wir besprechen vor allem die Stele, genauer den Holzkörper, der sich in charakteristischer Weise Hand in Hand mit der äusseren Umbildung gewandelt hat. Auch die anderen Gewebeformen würden an sich eine ähnliche Besprechung verdienen. Doch ist ihre Wandlung schon wegen der meist schlechteren fossilen Erhaltung weniger gut bekannt. Auf einzelne Fragen gehen wir im Abschnitt *d*) ein.

U m b i l d u n g   d e r   P r i m ä r g e w e b e. So wie die Gabeltriebe (Telome) der Ur-Pteridophyten (Rhyniaceen) die Ausgangsform der äusseren Gestalt sind, so ist auch die zugehörige Achsenstruktur, die Protostele der Rhyniaceen der Ausgangspunkt für die Umbildung der Stele [1]. „Pro-

---

[1] Vgl. dazu JEFFREY 1897, TANSLEY (1908), ZIMMERMANN 1930 S. 72 ff. und BOWER 1935, S. 565.

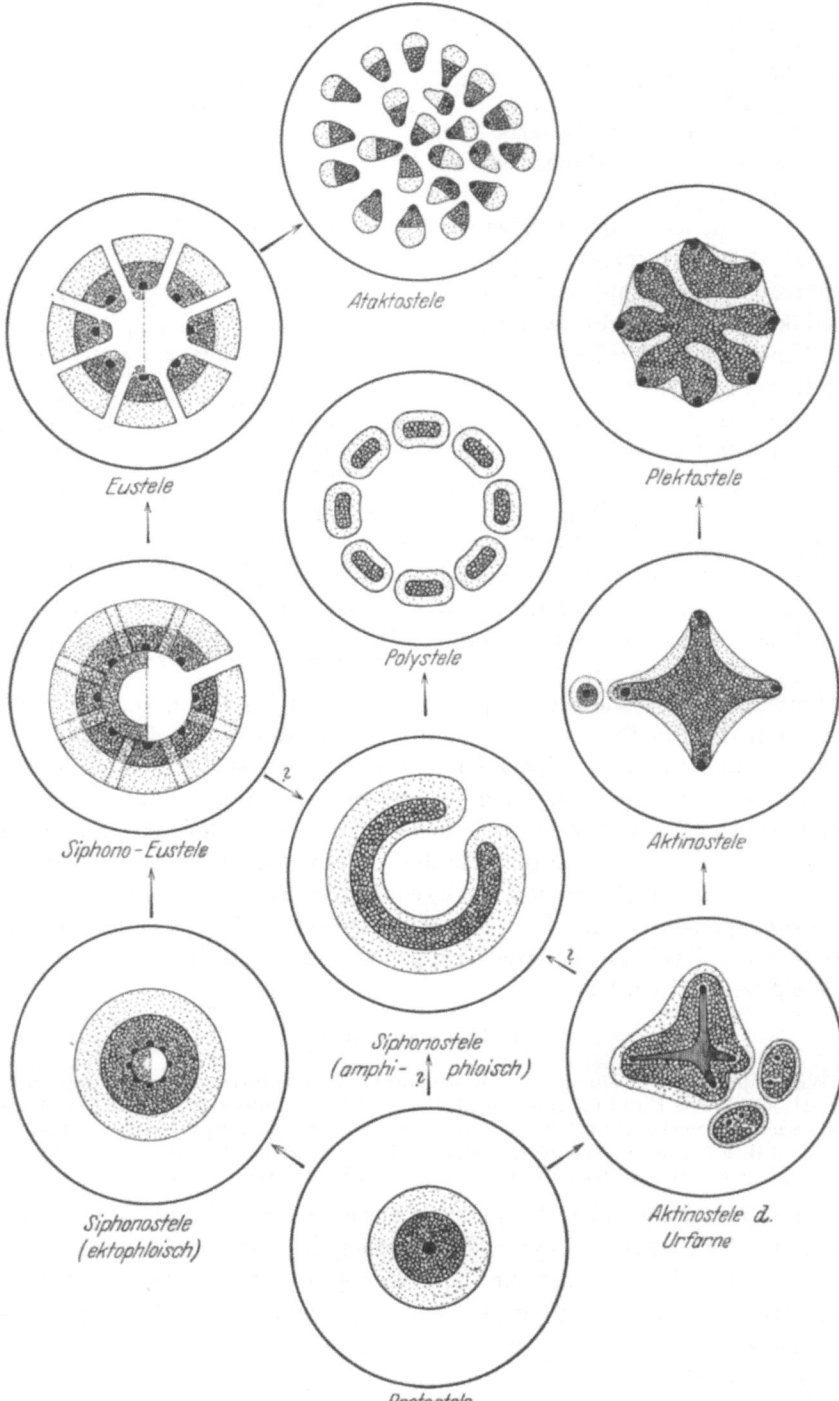

FIG. 14. Phylogenie der Stele. — Aus ZIMMERMANN 1930, S. 75.

Fig. 15. Ableitung der verschiedenen Holztypen in Zusammenhang mit der Blattstellung. — Unten die primitiven Formen, oben die abgeleiteten. Links für die Dreier-Reihe (z.B. *Articulaten*). — Mitte für die Vierer-Reihe (untere Hälfte z.B. *Coenopteridales*, obere Hälfte Weiterbildung z.B. *Articulaten*. — Rechts für Fünfer-Reihe. (viele *Pteropsiden*, z.B. *Osmundales*). Vom H o l z ist nur das Metaxylem (schwarz) und das Protoxylem (weisser Punkt) wiedergegeben. Die B l ä t t e r sind im allgemeinen als primitive Gabelblätter dargestellt. Oben auch die Übergänge zum Nadel- und Fiederblatt.

tostele" nennt man eine einfache zentral gelegene Tracheiden-Säule, umgeben von einem Mantel wenig differenzierter, längsgestreckter Zellen, die als prospektives Phloem bezeichnet werden kann.

Die Umbildung zum Stamm setzt 2 anatomische Ziele bzw. Konstruktionsaufgaben:

Vermehrung der wasserleitenden und der festigenden Elemente (letztere namentlich in der Peripherie: Biegungsfestigkeit!).

Vermehrung des Speichergewebes für Assimilate, das als Neubildung, also ausser der bei den Rhyniaceen schon vorhandenen Rinde, nun auch im Stammzentrum, innerhalb der Stele auftritt.

Diese Ziele werden durch Umbildung der Stele nach zwei verschiedenen Konstruktionsprinzipien, auf zwei verschiedenen Wegen erreicht (Fig. 14 u. 15). Der eine Weg führt über die Aktinostele zur Auflösung der einheitlichen Stele in eine Plektostele (und Polystele), der andere unter Markbildung (zusamt Markstrahlbildung) über die Siphonostele zur Eustele (und Polystele?). Im übrigen wirkt sich die Umbildung der Stele zu zergliederten Formen des Holzteiles vor allem bei grossen Pteridophyten mit dicken Achsen aus, worauf Bower (z.B. 1930) hingewiesen hat.

Die Aktinostelen-Umbildung trifft man bis zu einem gewissen Grade bei fast allen Pteridophytengruppen, wenn sie auch für mikrophylle Sprosse besonders charakteristisch ist. Denn die Aktinostele (Fig. 14 unten) ist eigentlich die notwendige Folge der Syngenie, welche, wie wir sahen, mikrophylle Sprosse entstehen liess. D.h. infolge der Syngenie verschmelzen die durch Verzweigung verketteten Stelen, bzw. es lagern sich die Stelen von Seitenorganen der Hauptstele gewissermassen als Verstärkungsleisten an (vgl. Fig. 7D, sowie z.B. auch ZIMMERMANN 1930, Abb. 28 und 1930a). Damit wird natürlich der Stelenquerschnitt zu einem mehrarmigen Stern, 3-, 4-, 5- usw. -armig (vgl. Fig. 15) je nach der Zahl der untereinander oder mit der Hauptachse verwachsenden Blätter bzw. Phylloidgruppen.

Man kann diesen Vorgang auch anders ausdrücken: die Abzweigungen der Stelen, die zunächst nur wenig tiefer lagen, als die Abzweigungen der betr. Organe, werden in den betr. Hauptachsen basalwärts verlagert, ja gewissermassen in die sich oft zu Seitenstelen völlig auflösende Hauptstele hinein [1]).

Die phylogenetischen Zusammenhänge zwischen den Stelen-Typen: Protostele-Aktinostele-Plektostele werden besonders deutlich, wenn wir das Protoxylem (die ontogenetisch zuerst angelegten Holzpartien) verfol-

---

[1]) Diese Auffassung vereinigt sowohl Vorstellungen der POTONIÉschen „Perikaulomtheorie" sowie Anschauungen von CAMPBELL, SCHOUTE u.a. über die Fortpflanzung eines Reizes von den Blättern nach dem Stamm. Eine einwandfreie Diskussion dieser sehr komplizierten Fragen wäre auch hier nur auf der Basis des hologenetischen Zusammenhanges möglich.

gen. Jeder Stelenarm (bzw. genauer die dazugehörige Seitenrippe) läuft natürlich nach oben als Leitbündel zu einem Blatt bzw. Phylloid aus. Bei primitiven Pteridophyten wie *Asteroxylon* bewahren solche „Blattspurstränge" wie auch die zugehörigen Arme der Aktinostele noch ganz den für ein primitives Phylloid charakteristischen Protostelenbau, z.B. durch gabelige Verkettung und durch die zentrale Lagerung des Protoxylems. (Vgl. unten S. 596).

Wenn die Zerklüftung des Holzkörpers durch diese mit der Seitenorganbildung zusammenhängenden Reize noch weiter geht, so dass der gesamte Holzkörper sich in einzelne Stränge auflöst, spricht man von einer Plektostele. Sie findet sich in schwacher Ausprägung schon bei *Stauropteris*, deutlicher aber namentlich bei manchen Lycopodien.

Die E u s t e l e n - U m b i l d u n g ist typisch für die makrophyll beblätterten Sprosse. Sie ist in einer geradezu klassischen Reihe bei den *Osmundales* zu verfolgen, wie KIDSTON und GWYNNE-VAUGHAN sowie SAHNI herausgearbeitet haben. Die Eustele ergibt sich durch Umwandlung der zentralen Tracheiden einer Stele (z.B. einer Protostele usw.) in Mark, sowie durch Umwandlung radialer Zellzüge zwischen Mark und Rinde (die „Markstrahlen") gleichfalls in parenchymatisches Gewebe. Sofern nur die Hauptachsenstele ein Mark zur Aufnahme der Assimilate aus den grossflächigen Wedeln erhält, ergibt sich als Übergangsglied zwischen Protostele und Eustele die typische, allerdings wohl seltene Siphonostele (Fig. 14). Wenn auch die Seitenstelen namentlich auf ihrer Oberseite Mark führen, wird der Holzzylinder durch die oberhalb der Seitenstelen abgehenden Markstrahlen in die einzelnen Leitbündel der Eustele zerklüftet.

Als Ausgangstyp der *Osmundales*-Reihe mit noch ziemlich unveränderter Protostele in Hauptachse und Seitenorganen (Wedeln bzw. Blattursträngen) ist neuerdings von SAHNI (1932) *Grammatopteris Baldaufi* (BECK) HIRMER bestätigt worden. *Thamnopteris, Zalesskya*, sowie verschiedene *Osmundites*-Arten aus dem Perm und Mesozoikum zeigen dann nach den Untersuchungen von KIDSTON u. GWYNNE-VAUGHAN die genannten Stufen der Siphonostele und der Eustele (Einzelheiten s. die zitierten Originalarbeiten, sowie POSTHUMUS 1924, HIRMER 1927, sowie ZIMMERMANN 1930, S. 211 ff, und 1930a).

Allerdings sind auch hier Komplikationen beobachtet, so entsteht innerhalb der Stele nicht nur Parenchym sondern manchmal (z.B. bei *Osmunda cinnamomea* und vielen anderen Farnen) Siebteil und eine Endodermis. Ferner brauchen nicht immer alle Teile eines Stelenquerschnittes ontogenetisch gleichzeitig ausgebildet zu werden. Auf eine derartige starke Abwandlung der Stelen-Ontogenie hat beispielsweise CAMPBELL (1931) bei *Ophioglossum moluccanum* aufmerksam gemacht. Hier wird die Hauptmasse der Achsenstele (Siphonostele) erst verspätet, im Gefolge einer adventiv verzögerten Ausbildung des Sprossvegetationspunktes, herausdifferenziert. Infolge davon führt hier zunächst nur ein einziges kollaterales Leitbündel als Ausschnitt aus der künftigen Siphonostele einerseits in den Kotyledo und andererseits in die monarche Stele der Primärwurzel hinein.

Als eine andere Modifikation dieser Markbildung kann man den „gemischten" Holzkörper bezeichnen, wie er z.B. beim rezenten Farn *Platyzoma* (BOWER 1923–28) erreicht ist, und auch im „gemischten" Mark mancher *Lepidodendron* Arten angebahnt wird. Hier werden nicht sämtliche zentral gelegenen Tracheiden einheitlich zu Markzellen, sondern mehr oder minder unregelmässig zerstreute Zellen behalten den tracheidalen Charakter.

Sehr komplizierte Umbildungen sind ferner die Stelen der Psaronien, wie besonders RUDOLPH (1906) gezeigt hat (vgl. oben S. 482) und viele andere Farnstämme,
worauf hier nicht näher eingegangen zu werden braucht, da es sich im Prinzip um
die zuvor geschilderten Prozesse handelt. (Vgl. z.B. ZIMMERMANN, 1930, S. 205ff).

In den älteren „vergleichenden", z.T. als phylogenetisch bezeichneten Stelärbe-
trachtungen finden sich, ausgehend von den nicht-phylogenetischen Anschauungen
VAN TIEGHEMS und STRASBURGERS, auch sehr detaillierte Ansichten über die Art
und Weise solcher „Umbildungen". Es wird hier diskutiert, ob die Zerklüftung und
Verwachsung der Stelen, die „schizostelische" oder „gamostelische" Struktur wirklich auf „Spaltung" bzw. „Verwachsung", auf Einwandern („intrusion", „sinking
in") oder auf einer „Umwandlung" („degradation", „transformation") o. dgl. von
Gewebeteilen beruhe. Solche Diskussionen sind m.E. nur möglich bei einer noch un-
geklärten genetischen Grundanschauung, die übersieht, dass bei den phylogenetischen Umbildungen überhaupt nicht die Gewebepartien „wandern" oder sich
„wandeln", sondern dass die entsprechenden Erbfaktoren sich verändern.

Die Meinungsdifferenzen in diesen Sonderfragen haben es aber verschuldet, dass
die längst, vor allem von JEFFREY in den Grundzügen klar dargestellte *phylogenetische* Stelärtheorie noch nicht die ihrer Beweislast zukommende Anerkennung
gefunden hat.

Kombination der anatomischen Elementarprozesse.
Bei etwas höher differenzierten Pteridophyten kombinieren sich sehr oft
diese Umbildungsvorgänge, insbesondere die Aktinostelen- und die Eustelen-Umbildung. Es empfiehlt sich hier, einige Beispiele herauszugreifen.

So kann man bei den baumförmigen Lycopsiden, den
Lepidodendren und Sigillarien verfolgen, wie innerhalb einer meist deutlichen Aktinostele Mark, ja bei letzteren sogar Markstrahlen auftreten.
(Übrigens hat auch die Stele der ältesten *Osmundales* und anderer Pteridophyten einen gewissen Aktinostelen-Charakter, wenn auch die „Arme"
wegen einer relativ schwachen Syngenie nicht sehr ausgeprägt sind). Man
kann daher die Stelärentwicklung der Lycopsiden als Beispiel der Umbildung sowohl zu Eustelen, wie zu Aktinostelen bezeichnen.

Die Weiterbildung der Stele, namentlich die Kombination einer Aktino-
stelen- und Eustelen-Umbildung hat eine Reduktion des Metaxylems zur
Folge. Bis auf wenige Ausnahmen (vgl. SINNOTT 1909) ist bei den späteren
Lycopsiden-Stelen das „zentrifugale" (in den Stelenarmen vor dem Protoxylem gelegene) Metaxylem verschwunden. Das Protoxylem dieser Aktino-
stelenarme liegt so nicht mehr „mesarch" sondern „exarch". Dagegen ist
für die Lycopsiden (und für alle echten Wurzeln) typisch, dass „zentripe-
tales" Metaxylem erhalten bleibt, selbst wenn hier Mark ausgebildet wird.
Umgekehrt ist innerhalb der Pteropsidenstelen auch das zentripetale Metaxylem meist reduziert und kann nur als Ausnahme (z.B. in den Blattstielen
der Cycadeen und in vielen Pteridospermen-Stelen) festgestellt werden. Bei
völlig fehlendem zentripetalem Metaxylem liegt das Protoxylem „endarch"
wie in den typischen Dikotylensprossachsen.

Articulaten. Ähnliche Kombinationen und Umbildungsvorgänge
charakterisieren wohl auch die Anatomie des Stammes der Articulaten oder des „Spheno"-Stammes. Ich möchte übereinstimmend mit
der von JEFFREY, POTONIÉ 1897 S. 205 und GWYNNE-VAUGHAN herausge-
arbeiteten Theorie annehmen, dass zunächst auch für den Sphenostamm

eine durch Übergipfelung und schwache Syngenie zustandgekommene, wenigarmige Aktinostele ein Ausgangspunkt war. *Sphenophyllum* mag ein charakterisches Beispiel sein. P. BERTRAND (1924) hat mit Recht darauf hingewiesen, wie von dieser Form aus durch weitere Syngenie von Gabelblättern die Zahl der Arme vermehrt werden konnte. Diese Vorgänge entsprechen ungefähr der Figur 14, linke Reihe, untere Hälfte. Weitere Umbildungen, repräsentiert im Calamiten-Stamm, ergaben sich dann durch das Auftreten von Mark und Markstrahlen nach dem Eustelen-Typ, sowie durch die „Karinalhöhlen" (anstelle des Protoxylems und anschliessender Metaxylem-Partien). Als Ersatz für das, praktisch genommen, nicht mehr ausgebildete Primärholz trat bei den fossilen Stämmen das Sekundärholz ein, während die heutigen krautigen Equiseten überhaupt nur sehr wenige Tracheiden ausbilden.

P r i m o f i l i c e s. Besonders instruktiv für die Verknüpfung all der genannten Umbildungsprozesse sowohl in ihrer extern-morphologischen wie in ihrer anatomischen Auswirkung sind ferner die oberirdischen Organe der in die Ordnungen „*Cladoxylales* und *Coenopteridales*" bzw. als „*Primofilices*" vereinigten [1]) altpalaeozoischen Pteropsiden.

Phylogenetisch besehen heben sich hier 3 Formenkreise heraus:

Der eine ist charakterisiert durch *Cladoxylon scoparium* KRÄUSEL u. WEYLAND aus dem Mitteldevon. In vegetativer Hinsicht ist er ausgezeichnet durch einen deutlichen Gegensatz von Sprossachse und Blättern, d.h. durch eine weitgehende Übergipfelung und Verwachsung der zu kleinen Gabelblättern gewordenen Seitenorgane. Dementsprechend zeigt die Hauptachsenstele, ähnlich wie bei *Asteroxylon*, eine aus regelmässig gabelig verketteten Armen bestehenden vielstrahligen Aktinostelen-Querschnitt.

Ein zweiter Formenkreis umfasst die kulmischen, von P. BERTRAND neuerdings eingehender bearbeiteten *Cladoxylon*-Arten, soweit sie sich als Stämme mit reich gegliederten wedelähnlichen Seitenorganen (früher als *Hierogramma*, *Syncardia* und *Clepsydropsis* beschrieben) ausgewiesen haben. Als Vorläufer gehört hierher wohl *Asteropsis noveboracensis* DAWSON aus dem Oberdevon. Dieser Formenkreis erscheint einesteils insofern weniger umgewandelt, als der Gegensatz zwischen Hauptachsen und Seitenorganen weniger ausgeprägt ist als im ersten Formenkreis. D.h. die genannten Formen bilden eine gleitende Reihe von der Hauptachse (*Cladoxylon taeniatum*, *mirabile* u.a.) zu den durch die unregelmässige Plektostele hauptachsenähnlichen Seitenorganen 1. Ordnung, (*Hierogramma*), weiter zu den gleichfalls noch radiär gebauten, aber nur mit 4 strahligen Plektostelen versehenen Seitenorgane 2. Ordnung (*Syncardia*), die schliesslich in die 2 strahligen bzw. in den letzten Auszweigungen protostelischen *Clepsydropsis*-Phylloide auslaufen. Die ganze Pflanze machte also wohl durch die weitgehend radiären Seitenorgane noch einen ziemlich

---

[1]) ZIMMERMANN 1930. Es dürfte sich empfehlen, diese Klasse durch eine Ordnung *Protopteridales* (vgl. Fig. 20) zu erweitern.

buschigen Eindruck. Entsprechend der reicheren Gliederung der Seitenorgane hat sich in den Armen der Stelen jeweils eine Art Mark entwickelt.

Ein dritter Formenkreis ist durch *Stauropteris* (S. 481) charakterisiert, Hier ist die Syngenie ganz erheblich schwächer ausgeprägt. In Verbindung mit der dementsprechend geringeren Zahl der Stelen-Arme (nur vier) hat Stauropteris eine grosse Regelmässigkeit des Aufbaus erworben, ähnlich wie er sich bei den letzten Auszweigungen der Cladoxylen vom zweiten Formenkreis auftritt. D.h. bei *Stauropteris* (vgl. Fig. 17 A—C) sind in ganz regelmässiger Weise die beiden elementaren phylogenetischen Umbildungsprozesse, die Übergipfelung und die Verwachsung miteinander gekoppelt. Einerseits zeigt der Stauropteris-Spross die typische Übergipfelung, die zu seitlich gestellten Gabeltrieben führt. Andrerseits sind aber an diesen Gabeltrieben die basalen Mesome jeweils bis zu ihrer ersten Gabelung mit der Hauptachse verwachsen, sodass an derselben Stelle der Hauptachse 2 Seitenäste entspringen (Fig. 17). Da die aufeinanderfolgenden Seitenäste alternieren, ergeben sich insgesamt 4 Reihen von Seitenästen. Die Stele ist übereinstimmend mit dieser äusseren Umgestaltung umgebildet zu einer vierstrahligen Aktinostele (ähnlich wie Fig. 14 Mitte rechts). Denn infolge der Syngenie verschmelzen die zwei Stelen der Gabeltriebe jeweils so stark mit der Hauptachsenstele, dass sie entweder gar nicht mehr unterscheidbar sind, oder als paarige Flügelleisten die Hauptachsenstele verstärken.

Also die Seitenäste selbst sind im Prinzip wie die Hauptachse gebaut und verzweigt, bis auf die letzten Auszweigungen, die einfach gegabelte Phylloide vom *Rhynia*-Typ sind. So ist dieser ganze gleichzeitig Sprossachsen Charakter tragende ,,Wedel'' wie ein Busch radiär gebaut.

Aehnliche, zwischen den 3 Formenkreisen vermittelnde Baupläne charakterisieren auch die übrigen *Primofilices.* Manchmal sind hier auch mehr typisch entwickelte Sprossachsen bekannt, und umgekehrt gewinnen die Wedel mehr den typischen Blattcharakter, indem sich die Seitenorgane, vor allem höherer Ordnung wie echte Farnwedel nur noch in einer einzigen Ebene ausbreiten. Die noch bei *Stauropteris* wie bei den Ur-Pteridophyten drehrunden Phylloide werden

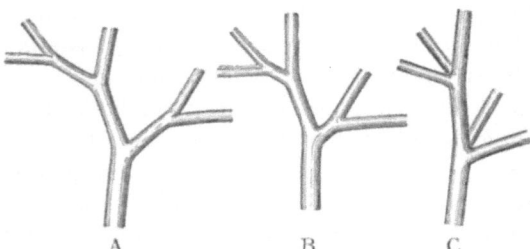

FIG. 16. Phylogenie des Sprossaufbaues bei *Stauropteris* durch Übergipfelung und gleichzeitige Aggregation des Basal-Mesoms des gabeligen Seitenorgans. — *A.* Gabeltriebe; *B.* Übergangsform; *C. Stauropteris.*

flächenförmig verbreitert. Auch bilden sich ähnlich wie bei *Cladoxylon* in den Stelen charakteristisch gelagerte Markpartien aus.

All diese *Primofilices* zusammen lassen aber eine Ausgangsform vermuten, die ähnlich wie *Cladoxylon* durch nicht ganz regelmässige (und schwächer als bei den bekannten *Cladoxylon*-Arten vollzogene) Syngenie und Uebergipfelung abgeleitet war. D.h. diese Zwischenform zwischen den typischen Psilophyten und den *Primofilices* hatte wohl auch eine Aktinostele mit unregelmässig gabeligen Armen und ähnlich gebaute (aber natürlich abnehmend schwächer differenzierte) Seitenorgane. Vielleicht gehört diese Ausgangsform in den anatomisch noch ungenügend bekannten mitteldevonischen Formenkreis von *Protopteridium* und *Aphyllopteris*, die man als *Protopteridales* (vgl. Fig. 20) zusammenfassen könnte.

S e k u n d ä r e   U m b i l d u n g e n im Stamm. Die bisher geschilderten Umbildungsvorgänge genügen aber nicht, um einen kräftigen Stamm zu gestalten, wie er bei den im Palaeozoikum vielfach baumförmigen Pteridophyten vorhanden war. Dieser Stamm musste entsprechend der allmählichen Grössenzunahme des Baumes sowohl zur vermehrten Wasserzufuhr, wie aus mechanischen Gründen verstärkt werden. Heute wird mindestens bei den echten „Holzbäumen" der Coniferen und der Angiospermen, die sekundäre Verstärkung vor allem durch das Dickenwachstum des Holzes erzielt. Auch die palaeozoischen Pteridophyten-Bäume haben zum Teil (z.B. Lepidodendren, Sigillarien, Sphenophyllen und Calamiten) ebenfalls die Fähigkeit zu sekundärem Dickenwachstum erworben, indem sich anstelle der peripheren Tracheiden des Primärholzes ein embryonal bleibendes Gewebe, das Kambium, ausbildete, das dann in der bekannten Weise in radialen Reihen Sekundärholz und sekundäres Phloem bildet.

Aber daneben treten doch auch ganz andere Konstruktionstypen auf; z.B. ging die hauptsächlichste Verdickung bei den Lepidodendren in der Rinde vor sich. Sie waren „Rindenbäume", da die ebenfalls durch ein Kambium sich verdickende Rinde mehr als 90% der Stamm-Masse ausmachte und auch die Rindenskulptur samt Ligula ein eigentümliches Aufnahmesystem für Regenwasser ausbildete (vgl. ZIMMERMANN 1930 S. 134 f).

Mit einer ganz anderen Form der Stammbildung hat uns SAHNI bei den *Coenopteridales*, z.B. bei *Austroclepsis australis* bekannt gemacht. Hier verflechten sich eine Anzahl dünner, wiederholt gegabelter Stämmchen, Blattstiele und Wurzeln zu einem Scheinstamm. Und schliesslich sei, als ein bis heute erhaltener Typ, der Stamm der Baumfarne erwähnt, bei dem das dünne eigentliche Stämmchen durch einen Mantel aus den Blattbasen und dazwischen abwärts strebenden Wurzeln verstärkt ist.

Umgekehrt hat GOEBEL (1930) gezeigt, dass die Stele fast bis zum Verschwinden reduziert werden kann, z.B. in den vegetativen Organen von *Trichomanes Motleyi* und *Tr. minutissimum*.

### d) *Sonstige Gewebe und Organe der Sprosse.*

S t o m a t a. Wir müssen uns auf einigeHinweise beschränken.Beim H a u t-
g e w e b e ist vor allem die Herausdifferenzierung der Spaltöffnungen bemerkenswert. Sie hat sich offenbar sehr früh, schon bei der Bildung der ältesten Landpflanzen vollzogen. Alle bekannten Pteridophyten (einschl. der
anatomisch bekannten Psilophyten) besitzen schon wohl differenzierte
Spaltöffnungen an ihren Sprossen. Leider fehlt noch für die Weiterent-
wickelung der Spaltöffnungen innerhalb der Pteridophyten eine so glänzende Darstellung, wie wir sie für die Gymnospermen FLORIN danken. So
kann hier nur festgehalten werden, dass der Anordnung nach die Spaltöffnungen der palaeozoischen Pteridophyten alle dem „haplocheilen" Typ
FLORINS angehören. Die verschiedensten Gruppen (z.B. *Equisetales* am
Stamm) haben die Sporangien in Reihen angeordnet, die Lepidophyten
stellen sie in Rillen der Blätter usf.

Die Schliesszellen selbst sind bei den Rhyniaceen noch wenig verändert,
d.h. es sind etwas verlängerte parenchymatische Zellen, deren gegen die
Spalte gelegener Kutikularrand das „Hörnchen" bildet (Archetypus). Von
dieser Grundform aus haben sich in den verschiedenen Pteridophyten-
Gruppen sehr charakteristische Gestalten der Schliesszellen herausgebildet,
z.B. die so auffälligen Formen der durch Leisten verstärkten *Equisetum*-
Schliess-Zellen, oder der schon bei den Lepidophyten auftretende *Lycopodium*-Typ (vgl. z.B. PORSCH, KRAUS, RIEBNER und ZIMMERMANN 1926).

E m e r g e n z e n. Eine andere Umgestaltung der Oberfläche sind die
schon relativ früh (z.B. bei *Coenopteridales*) auftretenden Haare (Ausstülpungen von Epidermiszellen) sowie die viel später namentlich bei jüngeren
Farnen vorgefundenen Spreuschuppen und andere typische „Emergenzen".
(Vgl. z.B. BOWER 1934 S. 1093).

In der R i n d e haben sich schon bei *Asteroxylon* beachtenswerte Inter-
zellularensysteme (Trabekulargewebe) herausdifferenziert, die auch später
wiederholt in ähnlicher Weise uns entgegengetreten. Eine ganz besonders
charakteristische Ausbildung der äusseren Rinde finden wir bei den Lepidophyten, sowohl in ihren „Blattpolstern" wie in einer borkenähnlich sich
differenzierenden äusseren Stammumkleidung. SEWARD (1932) hat auf die
Herausbildung dieser Rindenorgane in der Reihe *Haplostigma-Cyclostigma*-
typische Lepidophyten hingewiesen.

Durch eine sehr eigenartige Differenzierung der verschiedenen Rindenzellen in ihrer Ausgestaltung und namentlich in ihrem Streckungsvermögen
kommt die z.B. für *Lyginodendron* charakteristische „*Dictyoxylon*"-Rinde
zustande.

S e k u n d ä r e  U m b i l d u n g. Auch als Ganzes besehen, sind die
vegetativen Organe der Pteridophyten, nachdem sie sich in „typischer"
Weise herausdifferenziert haben, noch weitgehend umgemodelt worden.
Ich erinnere nur an die Umbildung von Blättern und Blatteilen zu Haken

bei *Gigantopteris* (HALLE 1929), an die Arbeitsteilung der Blätter z.B. beim Articulaten-Seitenzweige im Gefolge der Dorsiventralität (HALLE 1928), an die „Heterophyllie" mancher Sphenophyllen und vieles ähnliche mehr. Gerade diese weitere Umbildung („Metamorphose") der Organe, die schon vom „typischen" Spross ausgeht, ist ja das klassische Feld der Morphologie Die Resultate sind hier auch meist so eindeutig, dass auf diesem Gebiete wenig Meinungsdifferenzen bestehen. Die Schwierigkeiten liegen bei den früheren Umwandlungen, wo eine phylogenetische Morphologie mit der an rezenten Organismen gebildeten „Typus"-Vorstellung wenig anfangen kann.

### e) *Phylogenie der Wurzeln und ähnlicher Organe.*

Leider fehlt noch eine eingehendere Untersuchung, so dass wir uns hier auf einige Hinweise beschränken müssen.

Ausgangsform sind, wie wir sahen, wohl sicher die mit Rhizoiden versehenen gegabelten Kriechtriebe der Urpteridophyten. Es scheint, als ob zunächst keine vertikal abwärts steigenden Organe vorhanden gewesen sind. Daraus ergäbe sich die — auch reizphysiologisch sehr interessante — Folgerung (ZIMMERMANN 1927, S. 230), dass phylogenetisch der positive Geotropismus sich aus dem Plagiogeotropismus entwickelt hat.

Von diesen Kriechtrieben der Ur-Pteridophyten aus, haben sich verschiedene unterirdische Organe entwickelt, von denen wir hier drei Gruppen besprechen wollen.

Als 1. Abwandlung können wir schon innerhalb der Rhyniaceen, z.B. bei *Hornea*, die Ausbildung einer knolligen Basis feststellen. Ähnliche Bildungen beobachten wir z.B. beim „Protokorm" heutiger Lycopsiden.

Eine 2. Abwandlung zeigen die Lycopsiden z.B. in den „Stigmarien", den unterirdischen Organen der baumförmigen Lepidophyten samt „Appendices". Wir wiesen schon oben (S. 575) darauf hin, dass sie ontogenetisch wohl unmittelbar aus dem „Gegenpol" des Embryos entstanden sind. Ihre Anatomie veranlasst zur Annahme, dass sie sich phylogenetisch durch Übergipfelung und Verwachsung der Seitenorgane („Appendices") in prinzipiell der gleichen Weise herausgebildet haben, wie die Lycosprosse. Auch die Stigmarien sind nämlich mit einer deutlichen Aktinostele versehen, wobei die Arme der Aktinostele unmittelbar in die monarchen Stelen der Appendices auslaufen. Zumal die Appendices exogen an den Stigmarien entstehen, also die Stele (einschl. Sekundärholz) und das Rindengewebe sich unmittelbar aus den Stigmarien in die Appendices fortsetzen, ist das Bild der ungleich gewordenen Partner einer Gabelung noch unverkennbar. Innerhalb der Stigmarien-Gruppe hat sich dann meist in einer ähnlichen Reihe, wie wir sie oben für die Lepidophytenstämme schilderten, unter Rückbildung des zentripetalen Primärholzes Mark ausgebildet (WEISS 1932).

Aehnlich gebaut und auch noch relativ wenig abgewandelt gegenüber den ursprünglichen gabeligen Kriechtrieben sind die Wurzeln von *Isoetes*, die oft schon — wie mir scheint mit Recht — mit den Stigmarien verglichen worden sind. Doch auch die Wurzelträger der Selaginellen, die als Schwestertriebe eines Sprosses durch Gabelung hervorgehen, haben noch weitgehend den Charakter dieser alten Kriechtriebe erhalten. Das ist bei der ursprünglichen Gleichwertigkeit von Spross und Wurzel nicht überraschend.

Der oft unternommene Versuch, diese Stigmarien und Appendices in die alternativen Begriffsschemata: Wurzel/Spross (bzw. Wurzelstock) einzugliedern (vgl. z.B. Weiss 1932, Troll 1934, Hirmer 1934) ist von vornherein aussichtslos, wenn man mit einer solchen Einordnung einen naturwissenschaftlichen, d.h. phylogenetischen Sinn verbinden will. Denn diese im Karbon wohl ausgebildeten Stigmarien leiten sich bestimmt nicht von Organen her, welche an Pflanzen mit einem „typisch' entwickelten Gegensatz von echten Wurzeln und Sprossen sassen.

Auch waren die Stigmarien wohl kaum die Ahnenstadien e c h t e r W u r z e l n, die wir als

3) Abwandlungsgruppe der Kriechtriebe aufführen können. Diese Wurzeln haben sich aber in prinzipiell ähnlicher Weise vor allem bei den Pteropsiden aus den Kriechtrieben heraus entwickelt. D.h. die Aktinostele ist m.E. auch hier als ein Zeichen von Übergipfelung und Verwachsung von Seitenorganen aufzufassen, wobei die Seitenorgane (Seitenwurzeln) auch hier an die Arme der Hauptachsen-Stele mit ihren eigenen Stelen anschliessen. Abweichend von der Verkettung solcher Organe an den Urtrieben erfolgt die Auszweigung in Wurzeln „endogen''. Doch handelt es sich hier zweifellos um eine sekundäre Umbildung, die auf einem z.B. von Troll (1934) aufgezeigten Wege sich auch als phylogenetische Wandlung abgespielt hat.

Während bei den Lepidophyten die Stigmarien, bei den Angiospermen meist die Hauptwurzeln vom Gegenpol ausgehen, sind bei den Pteridophyten im allgemeinen Sekundärwurzeln, die unterhalb der Blätter entspringen, ausgebildet.

## § 7. Phylogenie des Sporophyten (Fortpflanzungsorgane).

### A. Sporangien.

Die Phylogenie der Sporangien ist als Abwandlung des Generationswechsels in ihren 3 Stufen: Isosporie, Heterosporie und Samen bzw. Pollen schon behandelt (Oben S. 566 f). Die Phylogenie des Sporangienstieles ergibt sich aus der Umbildung der Sporangienstellung (unten unter *B*). So bleibt für diesen Abschnitt nur die Phylogenie der Sporangienwandung und einiger Spezialorgane (Columella, Elateren u. dgl.).

Die ältesten überlieferten Pteridophyten-Sporangien haben alle einen so einheitlichen, mit den Rhyniaceen weitgehend übereinstimmenden Bau, dass dieser zweifellos der Ausgangsform für die Pteridophyten-Phylogenie (vgl. oben S. 566) nahestand. Diese „Ur-Sporangien" der Pteridophyten (Fig. 5) sind typische Telome; d.h. die Enden der Triebe laufen im fertilen Zustand in einen etwas keulig angeschwollenen Behälter für die Sporen aus. Die äusseren (ca 3—6) Zellagen des Sporangiums bilden die Wandung. Das innere Gewebe zerfällt in die Sporen bzw. zunächst in die Sporen-mutterzellen. Die „Ur-Sporangien" sind also wegen der mehrschichtigen Wandung „eusporangiat". Die einzelnen Wandschichten waren wohl bei den allerältesten Typen nicht besonders differenziert. Nur die Epidermis schloss wohl wie immer bei den Kormophyten etwas fester zusammen als die anderen Zellagen; ihre Zellwände mögen auch vielfach etwas kräftiger gewesen sein als die inneren Rindenlagen. Doch fehlte eine ausgesprochene Annulus-artige Bildung.

Eine D i f f e r e n z i e r u n g des S p o r a n g i e n i n h a l t e s setzte schon relativ frühzeitig ein. So wird bei den devonischen Psilophyten-Sporangien von *Sporogonites exuberans* [1]) und *Hornea Lignieri* nicht mehr der ganze Sporangien-Inhalt in Sporenmasse verwandelt, sondern die Stele des tragenden Stieles setzt sich hier in einem offensichtlich der Sporenernährung dienenden Zentralstrang, in einer Columella fort (Fig. 5). Eine solche Bildung fehlt bei heutigen Pteridophyten-Sporangien, findet sich aber bei Moosen (die *Hornea*-Form insbesondere bei *Sphagnum*).

Ich glaube aber nicht, dass man aus dieser gemeinsamen Columella schliessen dürfte, dass *Hornea* mit den Moosen enger als mit den übrigen Psilophyten verwandt sei. Und der Gedanke einer „sippenphylogenetischen [2]) Zwischenform" zwischen gleichaltrigen Gattungen ist ja von vornherein unphylogenetisch. Er entstammt dem nicht-phylogenetisch eingestellten, rein systematischen Vorstellungskreis, der die gleichzeitig lebenden Organismen in eine lineare Stufenfolge einordnet. Viel-mehr betrachte ich diese Columella als eine durch den gemeinsamen Grundaufbau der Archegoniaten-Sporophyten (Gewebegliederung der Telome) nahe gelegte Pa-rallelbildung. Eine schwach ausgebildete Columella besassen übrigens auch die Sporangien von *Lepidostrobus*.

Eine andersartige Arbeitsteilung des Sporangieninhaltes finden wir bei den Elateren von *Equisetum*. Sie sind offensichtlich eine sehr junge Erwer-bung. Mindestens in den wohl erhaltenen Calamiten-Sporangien fehlen sie. Wie weit sie in späteren Artikulaten-Sporangien etwa nur wegen mangel-hafter Fossilisation nicht mehr erkennbar sind (vgl. HALLE 1908 und HAR-RIS 1931 über Equisetites), ist im Einzelfall schwer zu entscheiden.

Auch die Sporangienwandung hat sich differenziert und zwar in 2 Haupt-Typen:

E u s p o r a n g i a t e r  T y p. Er ist im Grunde schon bei *Rhynia* aus dem Mitteldevon angedeutet. Der Unterschied gegenüber dem für alle

---

[1]) HALLE, 1936 rechnet mit der Möglichkeit, dass die Columella bei *Sporogonites* durch ungleiche Fossilisation vorgetäuscht wird.

[2]) In bezug auf einzelne Merkmale und Organe (also „merkmalsphyletisch") kann selbstverständlich ein Organismus auch zwischen gleichaltrigen Organismen-gruppen vermitteln.

bekannten Sporangien anzunehmenden Ur-Typ besteht im wesentlichen in einer stärkeren Differenzierung der Wandungs-Zellagen, die auch bei der Reife alle noch vorhanden sind. Ontogenetisch ist dieser Typ weiterhin dadurch charakterisiert, dass im Augenblick der Emporwölbung aus dem tragenden Organ eine grössere Anzahl von Zellen beteiligt ist. Das entspricht noch durchaus dem primitiven Telomcharakter, aber auch der Bildungsweise von kormophytischen Seitenorganen (Blättern, Seitensprossen usw.).

Als weitere Differenzierung finden wir bei solchen eusporangiaten Sporangien meist eine Verstärkung der Wandung (mit Ausnahme der Aussenwand der Epidermiszellen). Bei *Rhynia* sind noch alle Epidermis-Zellwände einigermassen gleichmässig verstärkt, während bei *Asteroxylon* und bei manchen *Coenopteridales*, wie *Stauropteris* der Scheitel von dieser Verdickung ausgenommen ist, so dass wir hier deutlich eine (bei *Zosterophyllum myretonianum* nach LANG spaltförmig verbreiterte) Oeffnungsstelle präformiert finden. Die verdickten Zellen beginnen damit deutlich als „Annulus'' lokalisiert zu werden. Entsprechend der ursprünglich allgemein verdickten Epidermis sind jedoch die „Annulus'' Zellen bei den altertümlichen Formen meist in einer grösseren Zahl vorhanden als später. Sie können recht verschieden gelagert sein, wodurch dann auch die Oeffnung z.B. von einem apikalen Porus zu einem in der Regel vom Apex ausstrahlenden Längsriss umgewandelt sein kann. Für die Einzelheiten dieser Endstadien der phylogenetische Sporangienumbildung kann auf den speziellen Teil verwiesen werden.

Eine andere Spezifizierung der Epidermiszellen besteht darin, dass nicht mehr die gesamte Zellwandfläche verstärkt wurde, sondern nur ein radial gestelltes Rippensystem. Diese — offensichtlich für den Kohäsionsmechanismus der Sporangienöffnung besonders wirkungsvolle — Umbildung der Epidermis als „Faserschicht'' ist sehr ausgeprägt bei den Calamiten-Sporangien. Bei Pteridophyten-Sporangien finden sich solche verdickte Zellen nur als „Exothecien'', in der Epidermis. „Endothecien'' wie bei Angiospermen, *Ginkgo* und *Pseudolarix* (vgl. ZIMMERMANN 1930, S. 235) fehlen.

Die innerste, unmittelbar an die Sporenmutterzellen (an das „sporogene Gewebe'') grenzende Zellschicht wird zu einem besonderen Ernährungsbzw. Drüsengewebe, dem „Tapetum'', das bei *Rhynia* übrigens besonders resistent war.

L e p t o s p o r a n g i a t e r T y p. Er ist eine jüngere Umbildung der Sporangien, die sich vor allem bei den nach ihm benannten Farnen findet. Er stellt sich im Gefolge einer Bergung der Sporangien ein. Sein Wesenszug ist das Schwächerwerden der Sporangienwandung, mindestens im Reifezustand. Hand in Hand damit geht eine besonders spezialisierte Ausbildung des Annulus, sowie eine Verlagerung der Öffnungsstelle meist zu einem Querriss. Dementsprechend lassen sich unter den heutigen leptosporangiaten Farnen 2 Gruppen nach GOEBEL unterscheiden:

1) Sporangiis longicidis (Die Sporangien öffnen sich mit einem Längsspalt): *Osmundaceae, Schizaeaceae, Gleicheniaceae.*

2) Sporangiis brevicidis (Die Sporangien öffnen sich mit einem schief oder transversal zur Längachse gestellten Querspalt): *Cyatheaceae, Hymenophyllaceae, Polypodiaceae.*

Die Wege, auf denen diese Umbildungen vor sich gegangen sind, insbesondere die Frage einer etwaigen Achsenverschiebung, sind hier leider noch sehr wenig untersucht, so dass ich mich auf einen Hinweis der speziellen Endstadien beschränken muss. (Vgl. auch GOEBEL 1930).

Deutlich ist dagegen eine Abwandlungslinie der Entstehungs- und Reifezeit der Sporangien eines Sorus, wie sie BOWER dargelegt hat. Bei den eusporangiaten Farnen und auch bei altertümlichen Leptosporangiaten entstehen und reifen die Sporangien ungefähr zur gleichen Zeit („Simultaneous Sorus" bei *Schizaeaceae, Gleicheniaceae* und *Matoniaceae*). Die Mehrzahl der heutigen leptosporangiaten Farnen besitzt jedoch den „Mixed Sorus". Die Sporangien entstehen und reifen zu sehr verschiedenen Zeiten und ohne eine durch ihre Stellung festgelegte Reihenfolge. Der „Gradate Sorus" (z.B. bei den *Dicksoniaceae* und *Loxsomaceae*) mit strengerer Gesetzmässigkeit der Entstehungszeit vermittelt wohl zwischen den beiden Gruppen. Der phylogenetische Fortschritt der ungleichen Reifungszeit der Sporangien liegt offensichtlich in der längeren Dauer einer Sporenproduktion.

## B. Sporangienstellung. (Fig. 18).

Die Sporangien können entweder einzeln für sich stehen oder vereinigt sein zu Sporangienständen, die (je nach ihrer Grösse und nach ihrem Ausbildungsgrad) Synangien, Sori, Sporophylle, Blüten u.dgl. genannt werden. Die phylogenetische Umbildung der Sporangienstellung verlief nach denselben Grundprinzipien wie die der vegetativen Organe. Das ist schon deshalb nicht überraschend, weil bei den ältesten Pteridophyten die einzelnen Sporangien ganz offensichtlich den vegetativen Trieben, den Phylloiden „homotop" sind. Daher bereitet es auch keine Schwierigkeit, diese altertümlichen Sporangien und Phylloide im gemeinsamen Begriff „Telom" zusammenzufassen.

Ausgangsform für die Sporangienstände ist wiederum die thallophytische Organisationsform gabelig verketteter Telome bei den Ur-Pteridophyten (ähnlich den Rhyniaceen, vgl. Fig. 3). Solche gabelige Sporangienstände finden sich übrigens noch heute bei den Samenständen von *Ginkgo* und bei den Mikrosporophyllen von *Ricinus.*

GOEBEL und HIRMER vertreten allgemein das „peltate Sporophyll" als Ausgangsform für die kormophytischen Sporangienstände. Das typische „peltate Sporophyll' so wie es auch in allen Zeichnungen festgehalten ist, ist durch die Anatropie der Sporangienstiele (vgl. Fig. 18) ausgezeichnet. Nach einer Diskussionsbemerkung HIRMERS auf dem Amsterdamer Botanikerkongress 1935 ist jedoch für das HIR-

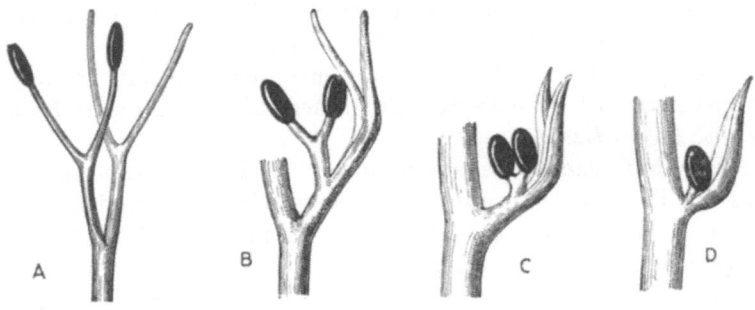

Lycopsida.

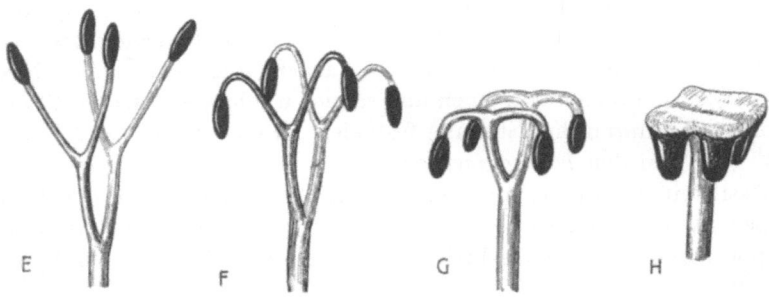

Articulata.

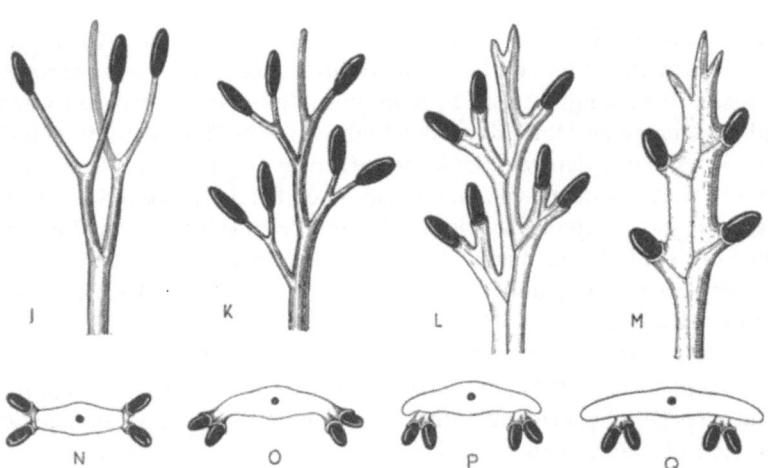

Pteropsida.

FIG. 17. Phylogenie der wichtigsten Sporophylltypen.
A–M = Gesamtansichten; N–Q = Querschnitte.

MERsche „peltate Sporophyll" diese Anatropie und die damit im Zusammenhang stehende plattenförmige Verbreiterung der Sporangienträger nicht wesentlich. Die restlichen Merkmale eines solchen „peltaten Sporangiums", insbesondere die radiäre Anordnung der Sporangien stimmen mit meiner Urform der Sporangien-stände (ZIMMERMANN 1930 und 1934) überein. Ich freue mich über diese sachliche Übereinstimmung mit HIRMER, halte aber die Bezeichnung „peltates Sporophyll" für eine Ausgangsform ohne Anatropie der Sporangien für unzweckmässig. Denn mit dem Begriff „peltates Sporophyll" war bisher die Vorstellung der Anatropie all-gemein verbunden. Daher bleibe ich bei meiner 1930 gewählten Ausdrucksweise: „gabeliger Telomstand" bzw. gabeliger Sporangienstand.

Die Umbildung zu den Sporophyllen und anderen abgeleiteten Sporangien-ständen vollzog sich in prinzipiell der gleichen Weise wie die Umbildung der vegetativen Sprosse bzw. Blätter. Daraus ergibt sich die schon oft be-tonte „Homologie" von vegetativen und fertilen Kormophytenorganen, etwa der Laubblätter und Sporophylle, der vegetativen Sprosse und Blüten u.dgl. Wir finden also wiederum bei der Umwandlung fertiler Pteridophy-tenorgane folgende „Elementarprozesse":

1) die Aggregation (typisch z.B. im „Sorus"),

2) die Übergipfelung (fiederiges Sporophyll),

3) die Verwachsung (Syngenie) entweder der Sporangien selbst zu Sy-nangien oder der tragenden Mesome zu Blattflächen, Plazenten, schild-förmigen Bildungen usw.

4) Die Reduktion.

Dazu kommen noch spezifisch fertile Umbildungsvorgänge, die auf eine Bergung der Fortpflanzungsorgane hinauslaufen:

5) Die Anatropie der Sporangien:

6) „Verschiebungen" der Sporangien, z.B. auf die Blattunterseite, durch stärkeres Wachstum des deckenden Organs. (Vgl. z.B. Fig. 18 N–Q).

Angedeutet sind solche Umbildungen schon bei den Psilophyten, ohne dass sich allerdings hier charakteristische Sporophylle u. dgl. herausdiffe-renziert hätten. So sind bei *Hornea* gelegentlich zwei Sporangien durch „Aggregation" stark genähert, oder durch „Syngenie" gewissermassen als Synangien seitlich verwachsen. Auch die anderen Elementarprozesse, ins-besondere die Übergipfelung (*Hostimella pinnata* LANG) und die Anatropie (*Psilophyton*-Arten) sind vertreten.

Für die typischen S p o r o p h y l l e, die die einzelnen Pteridophyten-Abteilungen charakterisieren, ist jedoch ein bestimmtes Zusammenspiel der „Elementarprozesse" charakterisch. Wir können da unterscheiden:

Das „L y c o - S p o r o p h y l l", in typischer Ausprägung ein einner-viges Blatt mit achselständigem Sporangium (Fig. 18D). Wie die Bezeich-nung andeutet, hat es sich insbesondere bei den Lycopsiden herausgebildet, ohne darum den anderen Pteridophyten-Abteilungen (vgl. *Sphenophyllum*- und ♀ Coniferen-Sporophylle) ganz zu fehlen.

Die Ausgangsform solcher Lyco-Sporophylle ist offensichtlich (wie z.B. SYKES ausgeführt hat) ein Telomstand, der einerseits einen vegetativen und andererseits einen fertilen Abschnitt besitzt. (Fig. 18A). Da sich mindestens bei den typischen „Lycosporophyllen" der vegetative Abschnitt nach aussen und der fertile nach innen kehrt, kann man (ohne damit eine phylogenetische „Spaltung" zu behaupten) hier von einer „serialen Spaltung", besser von einem serialen „Gespaltensein" reden.

Als makrophylle Parallelbildung finden wir eine solche Verknüpfung z.B. bei den *Ophioglossales*. Im allgemeinen ist aber die Zahl der in einem Lycosporophyll vereinigten Telome gering, entsprechend einer mikrophyllen Organisation der betr. Pflanze. Die weitere Umbildung verläuft gleichfalls offensichtlich unter Reduktionserscheinungen, sowohl der Sporangienzahl wie auch der tragenden Mesome.

Durch Aggregation (Fig. 18C) werden in der Regel die Sporangien untereinander und mit dem vegetativen Abschnitt inniger verbunden. Insbesondere verschwindet der für den Sporangienteil gemeinsame Stiel, so dass die Sporangien gewissermassen auf der Fläche des verschiedengestalteten vegetativen Sporophyllabschnittes zu stehen kommen. Diese Zwischenstufe eines vegetativen, blattähnlich ausgebildeten Sporophyllteiles und ihm aufsitzenden Sporangien ist charakterisch für manche Sporophylle der *Noeggerathiales* (einschl. *Tingia*), von *Sphenophyllum*[1]) (z.B. *Sph. majus*), oder auch von *Protolepidodendron* und den von BASSLER als „*Cantheliophorides*" zusammengefassten palaeozoischen Lepidophyten. Für alle diese altertümlichen Übergangsformen zu typischen Lycosporophyllen ist charakterisch, dass der fertile und der vegetative Teil noch einen gemeinsamen „Stiel", einen „Sporangiophor", (vgl. BOWER 1908) besitzen. Wie wohl nicht weiter ausgeführt zu werden braucht, bedeutet dann eine weitere Verkürzung dieses Sporangiophors, bzw. des Sporophyllstiels, die Blattachselstellung der Sporangien oder des einen, infolge einer Reduktion übriggebliebenen Sporangiums. Ja, es kann sogar zu einer noch weiter gehenden Verwachsung des Sporangiums mit der Sprossachse kommen, so dass die Sporangien etwas oberhalb des Sporophylls (bzw. seines vegetativen Teils) an der Sprossachse ansitzen. Für das typische Lycosporophyll ist dann schliesslich die Reduktion auf 1 Phylloid und 1 Sporangium charakteristisch. (Fig. 18 D).

Natürlich ist die Reihenfolge, in der diese einzelnen Prozesse der phylogenetischen Umbildung („Elementarprozesse") sich abgespielt haben, in den verschiedenen Stammreihen der Pteridophyten sehr verschieden gewesen. Verschieden ist auch der erreichte Umbildungsgrad.

Z.B. *Psilotum* und *Tmesipteris* sind heute noch ausgezeichnet durch 3 und 2 synangial verwachsene Sporangien. Ja gelegentlich finden wir auch bei rezenten Formen noch typische „Sporangiophore" (vgl. z.B. SYKES 1908 und LANG 1908).

Die vorgetragene Entstehungsweise eines Lycosporophylls entspricht der Reduktions- und Syngeniehypothese, (die sich nicht unterscheiden, wenn man die

---

[1]) Nach SCOTT 1929, S. 83 sind diese Sporangienstände übrigens entgegen HIRMER vorzugsweise gabelig gebaut.

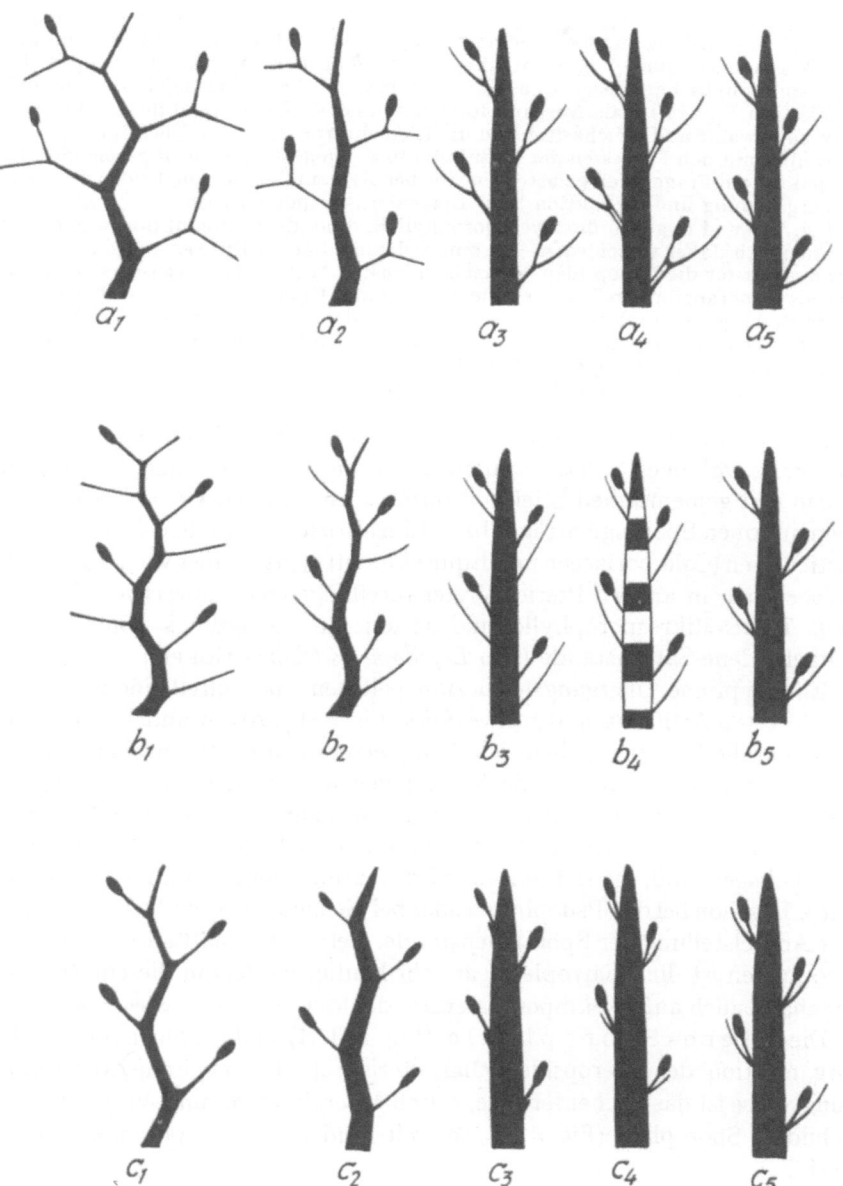

Fig. 18. Die 3 Möglichkeiten einer Phylogenie eines Lyco-Sporophylls bzw. eines Zapfens mit achselständigen Sporangien: *a*) aus einem Telomstand, der serial verkettete Sporangien und Blätter enthält, durch Wegfall des „Sporangiophors" (entspricht der Reduktionshypothese bzw. den „Sporangiophorhypothesen"und den Hypothesen der „serialen Teilung"). — *b*) aus einem Telomstand, der übereinander an der Achse selbständige Sporangien und Blätter enthält, durch Zusammenrücken je eines Sporangiums und Blattes. — *c*) Aus einem reinen Sporangienstand durch allmähliches Hinzuwachsen der Blätter („Epigeniehypothese"). — Die jeweils ausfallenden Mesome sind in $a_4$ und $b_4$ hell gezeichnet.

anatomische Umbildung ausser Acht lässt). Sie entspricht also unserer obigen (S. 586) Ausführung für die vegetativen Lycoblätter. Eine kritische Behandlung der Frage muss auch hier sämtliche denkmöglichen sonstigen Ableitungen des Lycosporophylls berücksichtigen (vgl. Fig. 19) und die wahrscheinlichste herauswählen. In Fig. 19 ist als Ausgangsform der gabelige Telomstand der Ur-Pteridophyten gewählt und gleichzeitig sind die Elementarprozesse der Übergipfelung unmittelbar mit den Prozessen der Reduktion bzw. Epigenie gekoppelt gezeichnet. Es sei jedoch die Frage offen gelassen, in welcher Reihenfolge sich die beiden Prozesse (Übergipfelung und Reduktion bzw. Epigenie) gewandelt haben.

Fig. 19a und b, geben die Lycosporophyll-Bildung durch Reduktion wieder. Es können sich dabei verschiedene Mesome reduziert haben. Entweder wie in a (d.h. bei der mir für die Lycopsiden wahrscheinlichsten Wandlungsweise) verschwinden die als „Sporangiophore" ausgebildeten Mesome. Oder — wie das vielleicht für wirtelständige Sporophylle möglich ist — es werden entsprechend Fig. 19b Mesome der Zapfenachse reduziert. Fig. 19c macht wohl nähere Ausführungen über die epigenetische Entstehung eines Tragblattes zu den Sporangien entbehrlich. Sie ist mir, wie oben, S. 589, ausgeführt, unwahrscheinlich.

Das peltate Sporophyll ist — bei der üblichen Begriffsabgrenzung, vgl. oben S. 605 — durch Anatropie aller einzelnen Sporangien gegen den gemeinsamen Stiel und durch schildförmige Verbreiterung der gemeinsamen Sporangienträger bzw. Einzelstiele entstanden. Es ist für die Articulaten (Calamariaceen und Equisetaceen) typisch, aber wiederum auch gelegentlich in andern Pteridophytenabteilungen nachzuweisen: z.B. heutige *Taxus*-Mikrosporophylle und rhaetische, neuerdings von HARRIS aufgefundene Samenstände (von *Lepidopteris Ottonis* GOEPP.).

Eine typische Übergangsform zum peltaten Sporophyll finden wir bei den ältesten Articulaten, den *Hyeniales* (Fig. 18F). Abgewandelte Formen, bzw. Parallelformen ergeben sich dann, wenn nicht jeder einzelne Sporangienstiel gegen die gemeinsame Achse gebogen ist, sondern alle Sporangien nach einer Seite anatrop sich krümmen. Dies kann auf zweierlei Weise bewirkt werden. Sei es, dass die Sporangien einheitlich vom gemeinsamen Stiel weg gebogen sind, sei es dass sich jedes einzelne Sporangium für sich biegt, wie z.B. schon bei den Psilophyten oder bei vielen Sphenophyllen im Gefolge der Achselstellung der Sporangienstände. Gerade bei blattachselständigen Sporangien ist die Anatropie sogar sehr häufig. Es darf in diesem Zusammenhang auch auf die ♀ Sporangienstände der Coniferen erwiesen werden.

Die Ptero-Sporophylle (Fig. 18 I-M) sind für die makrophylle Organisation der Pteropsiden charakteristisch. Die typische Ausgestaltungsweise ist das flächenförmige, durch Übergipfelung und Verwachsung gebildete Sporophyll (Fig. 18 L, M) mit randständigen Sporangien, bzw. Sori.

Gerade hier ist es besonders deutlich, dass sich die einzelnen Elementarprozesse der phylogenetischen Abwandlung: die Übergipfelung, Verwachsung usw. ziemlich selbständig abwandeln können. Wir wollen die Selbständigkeit der Elementarprozesse in ihrer phylogenetischen Abwandlung nur an der Ausbildung von Sori verfolgen.

Es gibt Pteropsiden wie *Cladoxylon scoparium* KRÄUSEL u. WEYLAND (vgl. oben S. 480), bei denen die Sporangienstände sich ohne Sorusbildung unmittelbar in flächenförmige Sporophylle (bei *Cladoxylon* noch vom Gabeltyp) zusammenziehen. Häufiger wird aber bei den Pteropsiden die radiäre Verkettung in den Endauszeigungen der Sporangienstände beibehalten. Die Sporangien werden nur durch Aggregation in eine dicht gedrängte Gruppe ohne deutlich unterscheidbare Mesome, also in einen S o r u s zusammen gezogen. Es ist nützlich, hier die 3 Differenzierungsformen der Sporangienstände hinsichtlich ihrer Verkettungsebenen gegenüberzustellen:

Bei dem
S o r u s, den Endauszweigungen, bleibt die radiäre Verkettung erhalten.

Bei dem
S p o r o p h y l l, also bei mittel-umfangreichen Sporangienständen wenigstens der „typischen" Pteropsiden verketten sich die Mesome (und Telome, soweit diese nicht zu Sori vereinigt sind) in einer Ebene. Bei der B l ü t e bleibt die radiäre Verkettung erhalten. Diese Verkettungsweise hat uns bei Besprechung der neuen Blütentheorie von THOMAS zu beschäftigen.

P. BERTRAND hat neuerdings (1934) dargelegt, dass selbst bei einer relativ einheitlichen Phyllomgestaltung wie beim palaeozoischen *Pecopteris*-Blatt eine ganz verschiedene Sporangienstellung vorkommen kann. Es lassen sich hier unterscheiden:

„ssp." *Senftenbergia* mit „acrostichoider" Sporangienanordnung, d.h. Sporangien isoliert und nicht in Sori zusammengezogen.

„ssp." „*Asterotheca*, typische Sori mit relativ selbständigen Sporangien auf der Blattunterseite,

„ssp." *Zeilleria*, Sori aus synangial verwachsenen Sporangien marginal gestellt;

„ssp." *Acitheca*, Sori aus synangial verwachsenen Sporangien auf der Blattunterseite.

Denken wir ferner an die Möglichkeit, dass die Sporangien echte Sporangien oder auch Samen sein können (*Pecopteris Pluckeneti*), oder dass umgekehrt innerhalb der „Familie" der *Schizaeaceae* eine übereinstimmende Sporangienform und Sporangienstellung sich mit einer sehr grossen Zahl von Phyllomgestaltungen verbinden kann (vgl. z.B. BOWER 1926 und 1935). Dann wird klar, dass die Aggregation zu Sori sich selbständig in verschiedenen Reihen herausdifferenziert hat, dass aber auch die übrigen Elementarprozesse der Phyllomgestaltung ebenso selbständig verlaufen können.

S o r i u n d S y n a n g i e n. Doch hat THOMAS (1935) mit Recht darauf aufmerksam gemacht, dass innerhalb der Pteropsiden die Zusammenziehung der Sporangien zu Sori relativ früh einsetzt. Er hat darauf seine Anschauungen über eine Phylogenie der Pteropsidenblüte gegründet. (Vgl. unten S. 613). Im übrigen haben sich die Sori namentlich bei den leptosporangiaten Farnen in sehr mannigfaltiger Weise durch Entstehung von Spezialorganen, wie den die Nahrung aufnehmenden Plazenten und von Hüll-organen (Indusien [1]) u. dgl.) weiter entwickelt. Ein Teil dieser indusien-artigen Bildungen ist bei dem (unten geschilderten) „Verschiebungspro-zess" der Sori auf die Blattunterseite entstanden. Andere Indusien haben wohl Emergenz-Charakter; doch muss wegen Einzelheiten auf die Schil-derungen der speziellen Abschnitte verwiesen werden. Bei nicht so stark geborgenen Sporangiengruppen, namentlich des eusporangiaten Types, wird offenbar vielfach die Widerstandsfähigkeit des einzelnen Sporangiums durch die synangiale Verwachsung gefördert.

Besonders eigentümlich sind die neuerdings von HALLE (1933) eingehend untersuchten Synangien aus den Pteridospermengattungen *Witleseya* u.a., die so stark verwachsen sind, dass samenähnliche Gebilde entstehen.

Sporophyllgestaltung. Im allgemeinen werden die Einzel-sporangien und die Sori bzw. Synangien der Pteropsiden zu typischen Spo-rophyllen vereinigt. Diese können mit der Assimilation dienenden Phyl-loiden kombinierte, „gemischte" Sporophylle sein, wie wir sie schon bei palaeozoischen Farnen (z.B. bei *Archaeopteris*), aber auch bei manchen Samenpflanzen (z.B. vielen Pteridospermen) finden. Je nach dem Grad dieser Vereinigung mit Phylloiden können wir 3 verschiedene Typen der Sporophylle herausstellen:

1) die Sporangien bzw. Sori bilden r e i n   f e r t i l e   S p o r o p h y l l e ohne Kombination mit Phylloiden („Holosporophylle" nach GOEBEL). Die strenge Arbeitsteilung in rein vegetative „Trophophylle" und rein fer-tile Sporophylle scheint nicht ursprünglich zu sein. Jedenfalls sind die an-deren Typen fossil früher belegt. Heute ist das „Holosporophyll" z.B. bei *Matteuccia Struthiopteris, Aneimia millefolia* u.a. vertreten.

2) Die Sporangien bzw. Sori bilden f e r t i l e   A b s c h n i t t e, die mit (nur aus Phylloiden zusammengesetzten) sterilen Abschnitten kombiniert sind. Je nach der gegenseitigen Lagebeziehung werden hier „akrotone" Sporophylle mit apikalem fertilen Abschnitt, „mesotone und basitone" Sporophylle mit mittlerem und basalem fertilem Abschnitt unterschieden. Solche Sporophylle sind schon sehr alt, z.B. bei *Archaeopteris*-Arten nachgewiesen. Eine Spezialform dieses Typs ist das Sporophyll der *Ophio-glossaceae*, bei dem zwei gleichwertige Abschnitte, ein fertiler und ein steriler, miteinander vereinigt sind.

3) Die M e s o m e (b z w.   d e r e n   p h y l o g e n e t i s c h e   D e r i-v a t e)   s i n d   a l s   P h y l l o i d e ausgebildet. Hieraus ergeben sich die typischen wie Laubblätter gestalteten „Farnwedel" und ähnliche Bildun-gen, die in irgendeiner Weise die Sporangien „tragen". Auf ihre Entstehung und weitere Umbildung gehen wir unten ein.

W e i t e r b i l d u n g   d e s   P t e r o - S p o r o p h y l l s. Sehr häufig

---

[1]) Schon der vermutlich silurische Psilophyt *Yarravia* hatte nach LANG und COOK-son eine becherförmige Umhüllung seiner Sporangien.

sind die Sporophylle jedoch auch weiter entwickelt als nur bis zum Pteropsiden-Typ mit durch Übergipfelung fiederig (bzw. durch Verwachsung marginal) gestellten Sporangien (Fig. 18 N-Q). Bei den heutigen und vielen fossilen Farnen werden die Sporangien vielfach auf die Fläche (meist auf die Unterseite) verlagert. Es ist das typische Bild der meisten heutigen leptosporangiaten Farnwedel [1]) oder auch der Gymnospermen-Mikrosporophylle. Oder die Sporophyllfläche krümmt sich einwärts, so dass die Sporangien in eine angiosperme Höhlung gelagert werden. Wenn die Sporangien dabei — wie beim Typus dieser Umbildung, beim Fruchtknoten mit marginaler Plazenta — randständig sitzen, kommen sie an die Verwachsungsnaht. Aehnlich ist wohl auch die Umbildung bei den Sporokarpen der Marsiliaceen. Selbst in den Fällen, wo die Sporangien schon auf die Sporophyllfläche verlagert waren, kann sich ein solches Fruchtknoten-ähnliches Gebilde entwickeln, wie THOMAS (1935) für die Entwicklungsreihe der mesozoischen Samenpflanzen von Pteridospermen über *Unkomasia* zu den *Caytoniales* gezeigt hat.


## C. Blütenbildung.

Die Sporophylle sind wohl bei primitiven Pteridophyten im allgemeinen mit vegetativen Phyllomen untermischt gestanden. Das ist bei der Homotopie und bei der parallelen Abwandlung von vegetativen und fertilen Telomen durchaus verständlich. Finden wir doch sogar innerhalb der Phyllome häufig vegetative und fertile Abschnitte!

In der Regel sind jedoch schon an altertümlichen Pteridophytensprossen die Endauszweigungen durch eine Anhäufung von Sporophyllen bzw. von sonstigen Sporangienständen ausgezeichnet. Von solchen Urformen ausgehend, haben sich in mindestens [2]) 4 selbständigen Stammreihen der typischen Pteridophyten: bei den Lycopsiden, bei den *Sphenophyllales*, bei den *Noeggerathieae* und bei den *Equisetales* echte Blüten (Zapfen) entwickelt. Das Prinzip dieser Zapfenbildung beruht auf zwei Elementarprozessen. Einerseits sind die im Zapfen vereinigten Phyllome nun rein fertil oder als Hilfsorgane für die Fortpflanzung umgebildet. Andererseits verkürzt sich die Sprossachse, bis die Phyllome dicht gedrängt übereinanderstehen. Allerdings haben z.B. nicht alle Lycopsiden solche Blüten. Wir finden innerhalb der heutigen Gattung *Lycopodium* (u.a.) sogar noch den „*Selago*-Typ"; d.h. die von den vegetativen Phyllomen wenig unterschiedenen Sporophylle stehen relativ locker und bilden eine Zone, die von rein vegetativen Blättern abgelöst wird.

---

[1]) Als Ausnahme z.B. auch bei *Osmunda regalis*.

[2]) Wahrscheinlich mehr; denn z.B. bei den Lycopsiden ist die Blütenbildung vermutlich mehrfach erworben.

Innerhalb der makrophyll belaubten Pteropsiden-Abteilung hat sich auf pteri-
dophytischer Basis vermutlich keine Blüte herausgebildet. Wir brauchen daher
dies Problem hier nur zu streifen. Die alte ČELAKOVSKÝsche „Strobilus"-theorie, die
im Grunde der oben ausgeführten Bildungsweise von Zapfen-blüten entspricht,
ist mit gewissen Abwandlungen von den meisten Phylogenetikern bis in die letzte
Zeit auch für die Entstehung der Pteropsidenblüten angenommen worden.

Erst kürzlich hat THOMAS (1935) eine grundsätzlich neue Auffassung in die
Diskussion geworfen. Er leitet die Blüten nicht mehr von einem Spross ab, dessen
Phyllome Sporophylle sind, sondern von einem Sorus. Es ist einerseits sicher sehr
verdienstlich, zu prüfen, ob die hergebrachten Ansichten nicht auf völlig falschen
Voraussetzungen beruhen, besonders verdienstlich, wenn die Kritik sich auf so
bedeutsame Neufunde stützen kann, wie die Samenpflanzenfunde von THOMAS.
Andererseits krankt schon die klassische Theorie der Blütenentstehung an der
Schwierigkeit, dass die Ausgangformen für die Pteropsidenblüte sich vielleicht gar
nicht in die Schemata: „Spross", „Sorus" u. dgl. einzuordnen lassen. Man muss
m.E. auch hier, wie so oft in phylogenetischen Dingen, ganz scharf die Frage des
natürlichen Entwicklungsvorganges und die Frage der Technik der Begriffsein-
ordnung auseinanderhalten.

Auch für die Pteropsiden ist die Umbildung zur Blüte offensichtlich ein Prozess,
der ausgeht von den wenig differenzierten radiären Sporangienständen der Ur-
Pteridophyten, und der beispielsweise zu den heutigen Phanerogamen-Blüten geführt hat. Dabei ist u.a. auch bei den Pteropsiden ein Teil der Mesome als „Sporophylle" in eine Ebene eingerückt. Die entscheidende Frage scheint mir nur die zu
sein: W a n n ist diese Sporophyllbildung, d.h. die Umwandlung der radiären Verkettung innerhalb der Sporophylle zu einem flächigen Gebilde erfolgt? Erst relativ
spät, als die Aggregation die gesamten für die künftige Blüte verwandten Sporangien schon Sorus-artig zusammengezogen hatte? Oder früher?

Kurz die Diskussion läuft — wenn wir uns frei halten von einem scholastischen
Streit um Begriffe — auf eine Erörterung hinaus, in welcher R e i h e n f o l g e
die Elementarprozesse der Sporangienstellung sich bei der Blütenbildung gewandelt haben. Und hier scheint mir ein sehr berechtigter Kern der THOMAS'schen Anschauungen der zu sein, dass tatsächlich für Pteropsiden, die eine Blüte erworben
haben, die Sorusbildung sehr frühzeitig einsetzte. Aber andererseits möchte ich annehmen, dass bevor sich die Längsachse sämtlicher zu einer Blüte vereinigten Sporangienstände als Blütenachse verkürzt hat, schon entsprechend der Strobilus-
theorie typische Sporophylle wie bei den Pteridospermen entstanden waren.

Jedenfalls für die Pteridophyten-Zapfen der Lycopsiden, *Noeggerathieae*
und Articulaten, auf die sich die THOMAS'sche Theorie ja gar nicht bezieht,
scheint mir die „Strobilus"theorie durchaus begründet. Den entscheiden-
den Beweis sehe ich in den Fossilfunden, namentlich im ursprünglichen
Vorherrschen der locker gebauten und von vegetativen Blättern durchsetz-
ten Sporophyllsprosse.

Über die Detailfragen der Articulaten-Blütenphylogenie, die eine ähnliche Sonderung von Begriffen und phylogenetischen Tatsachen erfordern,
kann ich wohl auf meine früheren Ausführungen (ZIMMERMANN 1930 S. 173
ff) verweisen.

### § 8. Sippenphylogenie der Pteridophyten. — Der Stammbaum Fig.
20 gibt in Verbindung mit der zuvor geschilderten Umbildung der einzelnen
Merkmale die von mir angenommenen sippenphylogenetischen Zusammenhänge so weitgehend wieder, dass eine Erläuterung sich auf einige kritische
Punkte beschränken kann.

Solch ein Stammbaum, oder überhaupt jede sippenphylogenetische Aussage drückt eigentlich in erster Linie die Überzeugung aus, in welcher Reihenfolge sich die einzelnen Merkmale geändert, bzw. die charakteristischen Eigenschaften herausdifferenziert haben. Wenn wir beispielsweise die Abteilungen der Lycopsiden und der Articulaten unterscheiden, und ihnen (in engerem Anschluss an die Articulaten) die Pteropsiden anschliessen, so drücken wir damit aus, dass sich nach unserer Auffassung die den einzelnen 3 Abteilungen gemeinsamen Eigenschaften relativ früh herausdifferenziert haben. Wir nehmen also an, dass die Tendenz zur Mikrophyllie und zur Blattachsenstellung der Sporangien sich sehr frühzeitig, schon bei den ge-

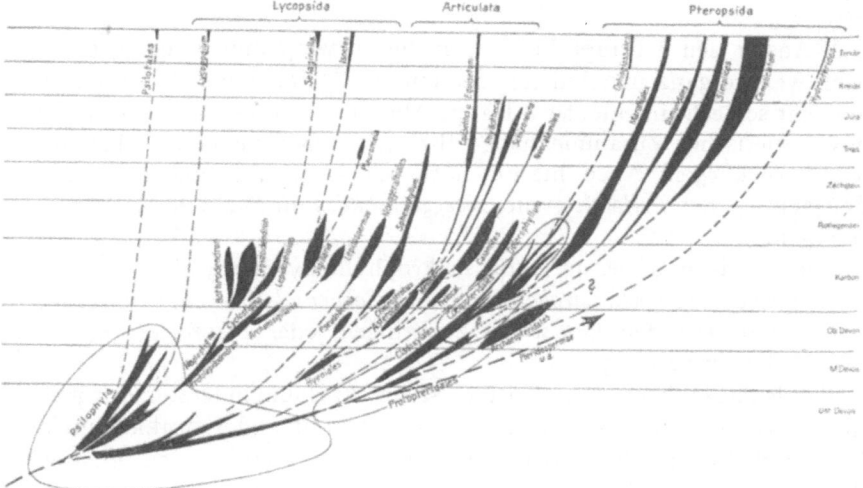

FIG. 19. Stammbaum der Pteridophyten.

meinsamen Ahnen der Lycopsiden angebahnt hatte, wenn sie auch, wie wir ausgeführt haben, bei diesen ältesten „Lycopsiden" keineswegs schon zur „typischen" Lycopsiden-Gestaltung geführt hatte. Und ähnliches gilt für die wirtelige Blattstellung sowie für die Neigung der anatropen Bergung der Sporangien bei den Articulaten, oder für die vorherrschende Neigung zur Makrophyllie, zur Wedelbildung bei den Pteropsiden.

Jedenfalls schliesse ich mich der heute ziemlich allgemein durchgedrungenen Auffassung von JEFFREY an, dass die Hauptmasse der Pteropsiden zu 3 bzw. 4 Hauptgruppen, die zweckmässigerweise mindestens als Abteilungen zu bezeichnen sind, zusammmengefasst werden können:

Die Lycopsiden, die Articulaten und die Pteropsiden, wozu dann eine in neuerer Zeit erst genauer bekannt gewordene gemeinsame Ur-Abteilung, die Psilophyten kommt. Allerdings knüpfen sich für jeden Phylogenetiker an eine solche Gruppierung von vornherein zwei fast selbstverständliche Annahmen:

1) Jede dieser Gruppen kann aus sehr weitgehend getrennten Stammreihen bestehen, die sich vielleicht schon im Psilophyten-Stadium voneinander getrennt haben. Ich habe diese Möglichkeit z.B. bei den Lycopsiden
dadurch angedeutet, dass die eligulaten Vertreter wie *Lycopodium* sehr
frühseitig von den ligulaten Lycopsiden abspalten. Sicherlich müsste ein
noch mehr ins Detail gehender Stammbaum — wie er für eine Übersichtsdarstellung der Pteridophyten technisch unmöglich ist — noch viel mehr
solche Parallel-Stammreihen aufzeichnen. Z.B. sind die Hydropterides bestimmt keine ganz einheitliche Gruppe. In wieviel Gruppen sie allerdings
phylogenetisch zu gliedern sind, und wo sie im einzelnen anschliessen, ist
noch ungeklärt (vgl. dazu ZIMMERMANN 1930 und die dort zitierte Literatur).

2) Ausser den 3 Hauptabteilungen dürfen wir damit rechnen, dass es
noch weitere ganz unabhängige Stammreihen gibt, deren Vertreter sich
nur sehr schlecht in eine der 3 Haupt-Abteilungen pressen lassen, weil die
phylogenetischen Zusammenhänge allzuweit, schon bei den Ur-Psilophy-
ten, zurückliegen. Auch hierfür habe ich — gewissermassen nur als
Beispiele — die *Psilotales*, die *Noeggerathiales* und *Dolerophyllum* ein-
getragen.

Vielleicht ist es ferner unrichtig, makrophylle Typen wie die *Sphenophyllales* (einschl. der *Cheirostrobales* und *Pseudoborniales*) in eine engere Verwandtschaft mit den *Equisetales* zu bringen, oder *Pleuromeia* mit den
Lycopsiden zu vereinigen.

Gerade das letztgenannte Beispiel ist ein gutes und in letzter Zeit mehrfach diskutiertes Beispiel (vgl. z.B. MÄGDEFRAU und HIRMER), um die
Schwierigkeiten solcher sippenphylogenetischer Aussagen zu erläutern.
Der auch von mir, in Übereinstimmung mit MÄGDEFRAU und HIRMER ge-
zeichnete Anschluss (Fig. 20) von *Pleuromeia* an die Lycopsiden wird ge-
stützt durch die grosse Übereinstimmung im vegetativen Gesamtaufbau.
Wir nehmen also bei dieser Stammbaumkonstruktion an, dass die vegetativen Grundeigentümlichkeiten wie die Stammbildung mit mikrophyller Beblätterung, die Stigmarien usw. sich zuerst in einer für die Lepidophyten
charakteristischen Weise herausdifferenziert hätten, so dass wir einen karbonischen Ahn von *Pleuromeia*, den uns ein günstiger Zufall in die Hände
gespielt hätte, zu den Lepidophyten stellen würden.

Die Schwierigkeit liegt in der Sporophyllform: als einzige Ausnahme unter den bekannten Lycopsiden hat *Pleuromeia* die Sporangien höchst-
wahrscheinlich auf der Blattunterseite!

Eine wissenschaftliche Diskussion ist nur unter Berücksichtigung des
hologenetischen Rahmens möglich. Derartige Sporophylle — einerlei ob
die Sporangien oben oder unten sitzen — gehen am Vegetationspunkt aus
undifferenzierten (wenn auch meist schon determinierten) Höckern, ähnlich der Fig. 21 hervor. Beim Lycopsidentyp wird der obere und beim Pleu-
romeiatyp der untere Abschnitt fertil. Wenn, unserer Annahme entspre-

chend, all die verschiedenen Sporophylltypen sich auf einen einheitlichen, wenig differenzierten Telomstand zurückführen lassen, muss auch bei der Entstehung des *Pleuromeia*typs eine Umdeterminierung in irgend einer Form stattgefunden haben: Das embryonale Gewebe, das — entsprechend der Fig. 21 — mindestens in einer hologenetischen „Seitenkette" aus der Oberseite Sporangien liefert, liefert in der zu *Pleuromeia* führenden hologenetischen Kette die Sporangien aus der Unterseite [1]).

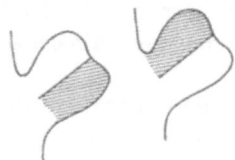

FIG. 20. Embryonalzustände von Sporophyllen. Links Sporangienteil (schraffiert) unten, rechts oben.

Die verschieden beantwortbare Frage ist nur die: W a n n hat diese Umdeterminierung stattgefunden? Da kommen zwei Zeitpunkte in Frage:

Entweder, wie das der oben geschilderten MÄGDEFRAU-HIRMERschen Ansicht entspricht: N a c h der Herausbildung des Lycopsiden-Typs als eines Ahns von *Pleuromeia*. Ein derartiger phylogenetischer Ablauf wäre durchaus möglich. Nur fehlen leider für eine solche Auffassung alle fossilen Belege. Es fehlen vor allem entsprechende Übergangsformen innerhalb der Lycopsiden. Denn wenn eine solche späte Umdeterminierung stattgehabt hätte, dann dürfen wir — in Übereinstimmung mit gut bekannten ähnlichen Umbildungsvorgängen — annehmen, dass die Umdeterminierung sich nicht mit einem Schlage irreversibel vollzogen hat. Vielmehr ist wahrscheinlicher, dass es zunächst auch Übergangsformen gegeben hat, die etwa im selben Zapfen ober- und unterständige Sporangien vereinigt hatten. Ob nun solche Übergangsformen nur wegen der bekannten Mangelhaftigkeit der pflanzlichen fossilen Überlieferung aus dem oberen Perm und der unteren Trias unbekannt sind, oder ob sie überhaupt nicht existiert haben, müssen wir dahin gestellt sein lassen.

Oder aber die verschiedenartige Determinierung der Sporangienstellung von *Pleuromeia* und den anderen Lycopsiden hatte auf einem so frühen Zeitpunkt stattgefunden, dass die Entscheidung, ob Ober-oder Unterseite zu Sporangien determiniert werden, noch nicht streng festgelegt war. Kurz diese zweite Auffassung führt die verschiedenartige Sporangienform von *Pleuromeia* und den Lycopsiden bis auf einen psilophyten-ähnlichen Urtyp zurück. Damit aber nehmen wir an, dass sich zuerst die Sporophyllgestalt herausdifferenziert hätte, und dass sich erst nachträglich die übereinstimmende vegetative Grundgestalt von *Pleuromeia* und den Lepidophyten gewandelt hätte. Auch für diese 2. Auffassung fehlen fossile Belege, so dass mir eine sichere Entscheidung zwischen den zwei Auffassungen heute nicht möglich scheint.

Kurz, das sippenphylogenetische *Pleuromeia*-Problem gehört zu jenen

---

[1]) THOMSON, 1909 hat in entsprechender Weise bei der Conifere *Saxegothaea* ♀ Sporangien auf der Sporophyllunterseite beobachtet.

sippenphylogenetischen Einzelproblemen, die deshalb so umstritten sind, weil der entscheidende Teil der Merkmalsphylogenie, hier der Sporophyll-phylogenie, hinsichtlich des Wandlungszeitpunktes noch ungeklärt ist.

Überhaupt muss auch hier eines unterstrichen werden: Im Gegensatz zu den auch in vielen Einzelheiten wohl begründeten merkmalsphyletischen Anschauungen, sind die sippenphyletischen Anschauungen gerade in ihren Einzelheiten noch sehr wenig geklärt. Auch mein Stammbaum Fig. 20 soll nur ein Anhaltspunkt sein für künftige phyletische Forschung.

# INDEX OF PLANT NAMES

# INDEX OF PERSONAL NAMES